ROUTLEDGE HANDBOOK OF ECOLOGICAL AND ENVIRONMENTAL RESTORATION

Ecological restoration is a rapidly evolving discipline that is engaged with developing both methodologies and strategies for repairing damaged and polluted ecosystems and environments. During the last decade the rapid pace of climate change coupled with continuing habitat destruction and the spread of non-native species to new habitats has forced restoration ecologists to re-evaluate their goals and the methods they use. This comprehensive handbook brings together an internationally respected group of established and rising experts in the field.

The book begins with a description of current practices and the state of knowledge in particular areas of restoration, and then identifies new directions that will help the field achieve increasing levels of future success. Part I provides basic background about ecological and environmental restoration. Part II systematically reviews restoration in key ecosystem types located throughout the world. In Part III, management and policy issues are examined in detail, offering the first comprehensive treatment of policy relevance in the field, while Part IV looks to the future. Ultimately, good ecological restoration depends upon a combination of good science, policy, planning and outreach – all issues that are addressed in this unrivalled volume.

Stuart K. Allison is the Watson Bartlett Professor of Biology and Conservation, and Director of the Green Oaks Field Study Center at Knox College, Galesburg, Illinois, USA. He is the author of *Ecological Restoration and Environmental Change* (Routledge, 2012).

Stephen D. Murphy is Professor and Director of the School of Environment, Resources and Sustainability at the University of Waterloo, Ontario, Canada. He is the editor-in-chief of *Restoration Ecology*.

ROUTLEDGE HANDBOOK OF ECOLOGICAL AND ENVIRONMENTAL RESTORATION

Edited by
Stuart K. Allison and Stephen D. Murphy

LONDON AND NEW YORK

from Routledge

First published 2017
by Routledge

2 Park Square, Milton Park, Abingdon, Oxfordshire OX14 4RN
52 Vanderbilt Avenue, New York, NY 10017

Routledge is an imprint of the Taylor & Francis Group, an informa business

First issued in paperback 2019

British Library Cataloguing-in-Publication Data
A catalogue record for this book is available from the British Library

Library of Congress Cataloging in Publication Data
Names: Allison, Stuart K., editor. | Murphy, Stephen D., editor.
Title: Routledge handbook of ecological and environmental restoration /
edited by Stuart K. Allison and Stephen D. Murphy.
Other titles: Handbook of ecological and environmental restoration
Description: London ; New York : Routledge, 2017. | Includes bibliographical
references and index.
Identifiers: LCCN 2016047488| ISBN 978-1-138-92212-9 (hbk) |
ISBN 978-1-315-68597-7 (ebk)
Subjects: LCSH: Restoration ecology.
Classification: LCC QH541.15.R45 R68 2017 | DDC 333.73/153—dc23
LC record available at https://lccn.loc.gov/2016047488

ISBN: 978-1-138-92212-9 (hbk)
ISBN: 978-0-367-35240-0 (pbk)

Typeset in Bembo
by FiSH Books Ltd, Enfield

CONTENTS

Contents

CONTRIBUTORS

Scott R. Abella, Assistant Professor, School of Life Sciences, University of Nevada Las Vegas, Las Vegas, Nevada, USA.

Susan C. Adamowicz, Land Management Research and Demonstration Biologist, United States Fish and Wildlife Service, Rachel Carson National Wildlife Refuge, Wells, Maine, USA.

Stuart K. Allison, Professor, Department of Biology, Knox College, Galesburg, Illinois, USA.

Juan J. Armesto, Professor, Department of Ecology, Pontifical Catholic University of Chile, Santiago, Chile.

Susan Baker, Professor, Cardiff School of the Social Sciences and Sustainable Places Research Institute, Cardiff University, Cardiff, Wales, UK.

Jonathan D. Bakker, Associate Professor, School of Environmental and Forest Sciences, College of the Environment, University of Washington, Seattle, Washington, USA.

Alex Baumber, Scholarly Teaching Fellow, Faculty of Transdisciplinary Innovation, University of Technology Sydney, Australia.

Sean M. Bellairs, Senior Lecturer, Research Institute for the Environment and Livelihoods, Charles Darwin University, Darwin, Northern Territory, Australia.

Brock Blevins, GIS Analyst, NASA Applied Remote Sensing Training Program (ARSET), Joint Center for Earth Systems Technology (JCET), University of Maryland, Baltimore County, Baltimore, Maryland, USA.

James Blignaut, Professor, Department of Economics, University of Pretoria, Pretoria, South Africa.

Keith Bowers, Landscape Architect, Restoration Ecologist, President and Founder, Biohabitats, Inc., Baltimore, Maryland, USA.

Ben Brown, Founder, Blue Forests, PhD Candidate, Research Institute for the Environment and Livelihoods, Charles Darwin University, Darwin, Northern Territory, Australia.

David M. Burdick, Associate Research Professor, Department of Natural Resources and the Environment, University of New Hampshire, Durham, New Hampshire, USA.

Marcela A. Bustamante-Sánchez, Professor, Department of Forestry Science, University of Concepción, Concepción, Chile.

Alton C. Byers, Senior Research Associate, Institute for Arctic and Alpine Research (INSTAAR), University of Colorado, Boulder, USA.

Robert Cabin, Associate Professor, Department of Environmental Studies, Brevard College, Brevard, North Carolina, USA.

Michael A. Chadwick, Lecturer, Department of Geography, King's College London, London, UK.

Robin L. Chazdon, Professor, Department of Ecology and Evolutionary Biology, University of Connecticut, Storrs, Connecticut, USA.

Young D. Choi, Professor, Department of Biological Sciences, Purdue University Northwest, Hammond, Indiana, USA.

An Cliquet, Associate Professor, Department of European, Public and International Law, Ghent University, Ghent, Belgium.

Loren D. Coen, Research Professor, Department of Biological Sciences and Harbor Branch Oceanographic Institute, Florida Atlantic University, Fort Pierce, Florida, USA.

Eric Conklin, Director of Marine Science, The Nature Conservancy, Honolulu, Hawaii, USA.

Heather A. Cray, Graduate Student, School of Environment, Resources and Sustainability, University of Waterloo, Waterloo, Canada.

Mark Dobrowolski, Principal Rehabilitation Officer, Iluka Resources Ltd, Perth, Western Australia, Australia and Adjunct Lecturer, School of Biological Sciences, The University of Western Australia, Perth, Australia.

Joan Dudney, Graduate Student, Department of Environmental Science, Policy and Management, University of California, Berkeley, California, USA.

Beatriz Duguy Pedra, Professor, Department of Evolutionary Biology, Ecology and Environmental Sciences, University of Barcelona, Barcelona, Spain.

Peter W. Dunwiddie, Affiliate Professor, School of Environmental and Forest Sciences, University of Washington, Seattle, Washington, USA.

Stephen R. Edwards, Advisor to the Chair, Resilience and Social Learning, IUCN Commission on Ecosystem Management, Baker City, Oregon, USA.

Mirijam Gaertner, Research Coordinator, Center for Invasion Biology, Department of Botany and Zoology, Stellenbosch University, Stellenbosch, South Africa.

Lauren M. Hallett, Postdoctoral Research Scholar, Department of Ecology and Evolutionary Biology, University of Colorado, Boulder, Colorado, USA.

Brice B. Hanberry, Research Ecologist, Grassland, Shrubland, and Deserts, Rocky Mountain Research Station, Rapid City, South Dakota, USA.

Boze Hancock, Senior Scientist-Marine Habitat Restoration, The Nature Conservancy, c/o University of Rhode Island, Graduate School of Oceanography, 215 South Ferry Road, Narragansett, Rhode Island, USA.

Tibor Hartel, Associate Professor, Environmental Science Department, Sapientia Hungarian University of Transylvania, Cluj-Napoca, Romania.

Liam Heneghan, Chair and Professor of Environmental Science and Studies, Institute for Nature and Culture, DePaul University, Chicago, Illinois, USA.

Oisín Heneghan, Research Assistant, Department of Environmental Science and Studies, DePaul University, Chicago, Illinois, USA.

Eric S. Higgs, Professor, School of Environmental Studies, University of Victoria, Victoria, British Columbia, Canada.

Richard Hobbs, Professor, IAS Distinguished Fellow, School of Biological Sciences, The University of Western Australia, Perth, Western Australia, Australia.

Patricia Holmes, Ecologist, Environmental Management Department, City of Cape Town, Cape Town, South Africa.

Darwin Horning, Assistant Professor, School of Environmental Planning, University of Northern British Columbia, Canada.

Austin T. Humphries, Assistant Professor, Department of Fisheries, Animal and Veterinary Sciences, University of Rhode Island, Kingston, Rhode Island, USA.

Lindsey B. Hutley, Professor of Environmental Science, Research Institute for the Environment and Livelihoods, Charles Darwin University, Darwin, Northern Territory, Australia.

Stephen T. Jackson, Director, Department of the Interior Southwest Climate Science Center, U.S. Geological Survey, Tucson, Arizona, USA.

Theresa B. Jain, Research Forester, US Forest Service, Rocky Mountain Research Center, Moscow, Idaho, USA.

Erik Jeppesen, Professor, Department of Bioscience, Aarhus University, Silkeborg, Denmark.

John M. Kabrick, Research Forester, US Forest Service, Northern Research Station, University of Missouri, Columbia, Missouri, USA.

Paul A. Keddy, Independent Scholar, Lanark County, Ontario, Canada.

Todd Keeler-Wolf, Senior Vegetation Ecologist, California Natural Diversity Database, California Department of Fish and Game, Sacramento, California, USA.

Benjamin O. Knapp, Assistant Professor, Department of Forestry, University of Missouri, Columbia, Missouri, USA.

Timo Kuuluvainen, Principal Investigator, Department of Forest Sciences, University of Helsinki, Helsinki, Finland.

David Lamb, Honorary Professor, School of Agriculture and Food Science, Center for Mined Land Rehabilitation, University of Queensland, Brisbane, Queensland, Australia.

Kemit-Amon Lewis, Coral Conservation Manager, The Nature Conservancy, US Virgin Islands, USA.

Zhengwen Liu, Professor, Nanjing Institute for Geography and Limnology, Chinese Academy of Sciences, Nanjing, China.

Stephanie Mansourian, Environmental Consultant, Mansourian.org, Gingins, Switzerland.

Michael J. McTavish, Graduate Student, School of Environment, Resources and Sustainability, University of Waterloo, Waterloo, Canada.

Alexander L. Metcalf, Research Assistant Professor, College of Forestry and Conservation, University of Montana, Missoula, Montana, USA.

Elizabeth Covelli Metcalf, Assistant Professor, Department of Society and Conservation, University of Montana, Missoula, Montana, USA.

Jakki J. Mohr, Regents Professor of Marketing and Gallagher Distinguished Faculty Fellow, School of Business Administration, Department of Management and Marketing, University of Montana, Missoula, Montana, USA.

Ladislav Mucina, Professor Iluka Chair in Vegetation Science and Biogeography, School of Biological Sciences, The University of Western Australia, Perth, Australia and Department of Geography and Environmental Sciences, Stellenbosch University, Stellenbosch, South Africa.

Sandra Cristina Müller, Adjunct Professor, Department of Ecology, Universidade Federal do Rio Grande do Sul, Porto Alegre, Brazil.

Stephen D. Murphy, Professor and Director of the School of Environment, Resources and Sustainability, University of Waterloo, Waterloo, Canada.

Jessica Hardesty Norris, Technical Writer, Biohabitats Inc., Baltimore, Maryland, USA.

Gerhard Ernst Overbeck, Professor, Department of Botany, Universidade Federal do Rio Grande do Sul, Porto Alegre, Brazil.

Stephen Packard, Ecological Restoration Pioneer and Visionary, Northbrook, Illinois, USA.

Michael P. Perring, Postdoctoral Researcher, Forest & Nature Lab, Department of Forest and Water Management, Ghent University, Belgium and Adjunct Postdoctoral Research Associate, School of Biological Sciences, The University of Western Australia, Australia.

Karel Prach, Professor, Department of Botany, Faculty of Science USB, České Budějovice, and Institute of Botany, Czech Academy of Science, Trebon, Czech Republic.

Charles Price, Adjunct Lecturer, School of Biological Sciences, The University of Western Australia, Perth, Western Australia, Australia.

José M. Rey Benayas, Professor, Departamento de Ciencias de la Vida, Universidad de Alcalá, Alcalá de Henares, Spain.

Jilliane Segura, Graduate Student, Research Institute for the Environment and Livelihoods, Charles Darwin University, Darwin, Northern Territory, Australia.

Benjamin Smith, Graduate Student, Earth and Environmental Dynamics Research Group, Department of Geography, King's College London, London, UK.

Cecilia Smith-Ramírez, Professor, Institute of Conservation, Biodiversity and Territory, University of Austral Chile, Valdivia, Chile and Institute of Ecology and Biodiversity, Santiago, Chile.

Martin Søndergaard, Senior Researcher, Department of Bioscience, Aarhus University, Silkeborg, Denmark.

Andrew Spaeth, Forest Program Director, Sustainable Northwest, Portland, Oregon, USA.

Erica N. Spotswood, Postdoctoral Research Scholar, Department of Environmental Science, Policy and Management, University of California, Berkeley, California, USA.

Rachel Standish, Senior Lecturer in Ecology, School of Veterinary and Life Sciences, Murdoch University, Perth, Western Australia, Australia.

John A. Stanturf, Senior Scientist, Center for Forest Disturbance Science, US Forest Service Southern Research Station, Athens, Georgia, USA.

Katharine Suding, Professor, Department of Ecology and Evolutionary Biology, University of Colorado, Boulder, Colorado, USA.

Péter Török, Associate Professor, Department of Ecology, University of Debrecen, Debrecen, Hungary.

Elizabeth Trevenen, Graduate Student, School of Biological Sciences, The University of Western Australia, Crawley, Western Australia, Australia.

Alberto Vilagrosa, Fundación CEAM, Department of Ecology, University of Alicante, Alicante, Spain.

Carol L. Williams, Research Scientist, Center for Agroforestry, University of Missouri, Columbia, Missouri, USA.

ACKNOWLEDGEMENTS

An edited volume like this one is very much a group effort. We are tempted to say a team effort, but the word team implies a group that is close-knit and has worked together for a long time towards a common goal. While the authors of the many chapters in this book share the common goal of understanding and advancing the practice of ecological and environmental restoration, we are certainly not a close-knit group. Many of the authors are frequent colleagues and friends of the editors, and via this handbook we have gotten to know many others who previously we knew only through publications and reputation.

First and foremost we must thank all of the authors of the chapters in this volume for their willingness to contribute a chapter despite no promise of any reward beyond the satisfaction of producing a good piece of work. We especially appreciate the kindness of strangers who worked with us despite not knowing us well or in person. All of the authors have been extremely patient throughout the process of putting the book together and have quickly answered the many queries we had for them as we reviewed chapters and put everything together.

We extend a huge thank you to our editors at Routledge – Tim Hardwick and Ashley Wright. They have been encouraging, supportive, and have provided many excellent suggestions that helped improve the book. They have also been patient as we worked to get everything ready for publication. This book would never have been completed without their comfort and confidence in our ability to succeed with the project. We also thank Hamish Ironside for copy-editing the entire book. Special thanks to Karl Harrington and everyone at Fish Books, who did the typesetting of the handbook.

Finally, many, many thanks to our colleagues and families who have supported and encouraged us at every step of the way. We put this book together in the hope that it will inspire a new generation of restorationists so that our children and students will live in a world of beautiful, functional landscapes and ecosystems that benefit the entire planet, we humans and all of our fellow beings on this wonderful Spaceship Earth.

1

INTRODUCTION

What next for restoration ecology?

Stephen D. Murphy and Stuart K. Allison

There have been previous edited volumes which provided a broad overview of the field of ecological restoration and which identified contemporary theory, practice and potential future directions for the field (Perrow and Davy 2002; van Andel and Aronson 2006). But the practice of ecological restoration and the science of restoration ecology are both rapidly evolving and much has changed in the past 10 to 15 years. In particular, we have become increasingly aware of the quickening pace of environmental change, a pace that threatens to continue to increase and which may indeed outpace our ability to restore some ecosystems. Thus this book was put together with the aim of both surveying current practice and identifying future opportunities and problems that will arise in our rapidly changing world.

The many authors in this book represent the state of the art of ecological restoration and the state of the science of restoration ecology. The most commonly used definition of ecological restoration comes from the Society for Ecological Restoration's *Primer on Ecological Restoration*:

> Ecological restoration is the process of assisting the recovery of an ecosystem that has been degraded, damaged, or destroyed.
>
> *(SER Science and Policy Working Group 2004)*

This definition is further developed in the *Primer* by an accompanying statement that expands on the goals of restoration:

> Ecological restoration is an intentional activity that initiates or accelerates the recovery of an ecosystem with respect to its health, integrity, and sustainability. Frequently, the ecosystem that requires restoration has been degraded, damaged, transformed or entirely destroyed as the direct or indirect result of human activities … Restoration attempts to return an ecosystem to its historic trajectory.
>
> *(SER Science and Policy Working Group 2004)*

The historical development of the practice of ecological restoration is difficult to trace (and thus somewhat contested) but certainly the practice began hundreds of years before the definition and also long before the well-documented early prairie restorations initiated at the

University of Wisconsin in the 1930s (Hall 2005; Allison 2012). Early restorations were carried out for a variety of reasons including practical concerns such as ensuring a continued supply of lumber and erosion control, aesthetic considerations such as the maintenance of a beautiful landscape, the desire to preserve lost or declining habitat, the need for humans to reconnect with nature, and a moral duty to repair what was damaged via human activity (Jordan 2003).

The notion of 'reconnecting with nature' may sound too idealistic – especially if one focuses on the technical aspects of restoration ecology – but the reason why the field of restoration ecology began was from a sense of ethics. Philosophers and pundits of science from Karl Popper to Peter Medawar have consistently argued that a science (like restoration ecology) does not emerge wholly formed and isolated from its social context. Most restoration ecologists likely entered the profession because they wished to right wrongs. This may smack of noblesse oblige and some may argue it is naïve, imperialistic, full of hubris, or fraught with a thousand other sins. While as restoration ecologists we should heed the call to examine our own motives, we should not lose sight that what drives us is a sense of ethics and empathy for the diversity of organisms and ecosystem functions – perhaps ecosystems are valuable for their services but let us not narrow or impoverish our world view to only such concerns. The opportunity to test theories that surround restoration ecology has just begun as the discipline has matured from 'stamp collecting' to that of a predictive science.

In the following chapters, readers will find a rich picture of the technical aspects of restoration ecology commingled with a strong sense of ethical underpinnings. The traditional case-based and scale-based approaches are still quite valid and also offer opportunities to test theories of population, community, and landscape restoration – to name a few. But despite the sometimes self-fulfilling term 'restoration' as a means of returning to the past, readers will find much about the emergent approaches that push disciplinary boundaries. Work on restoration ecology as a business or restoration ecology as an economic influence is something many have considered but few have explored – our fellow contributors will change that and perhaps change our ways of thinking. Trying to set goals and thinking about reasonable endpoints for a restoration project is becoming increasingly challenging as we see predictions that local climates will undergo significant changes in the next 50 to 100 years, while we know that some ecosystems like forests may take hundreds of years to return to pre-disturbance conditions even with the accelerated succession possible via restoration. How can we adjust our goals and maintain stakeholder interest in restoration and their confidence in our ability to restore ecosystems given the rapidly changing conditions? Will we accept the idea of restoration as a process of continual change? Thus it becomes even more important for scientists to learn to express themselves clearly in a manner that engages all stakeholders and is truly inclusive and respectful to all (Olson 2009). Our contributors will encourage us to expand our audience and the repertoire of tools we use to reach out to others.

Tony Bradshaw – as one of the founders of the discipline of restoration ecology – said to an audience of undergraduates in 1986, 'Your generation can learn from mine, but you are the future of this notion we call restoration or rehabilitation.'[1] Some of those in attendance are now leaders and a new generation beyond them is ascending – and some of those will be found in these pages. *Semper procedendum sine timore.*

Note

1   Recorded by Stephen D. Murphy, who was among the audience.

References

Allison, S. K. (2012) *Ecological Restoration and Environmental Change: Renewing Damaged Ecosystems*, Routledge, Abingdon, UK.

Hall, M. (2005) *Earth Repair: A Transatlantic History of Environmental Restoration*, University of Virginia Press, Charlottesville, VA.

Jordan, W. R. III (2003) *The Sunflower Forest: Ecological Restoration and the New Communion with Nature*, University of California Press, Berkeley, CA.

Olson, R. (2009) *Don't Be Such A Scientist: Talking Substance in an Age of Style*, Island Press, Washington, DC.

Perrow, M. R. and A. J. Davy (eds) (2002) *Handbook of Ecological Restoration*, Cambridge University Press, Cambridge, UK.

SER Science and Policy Working Group (2004) *The SER Primer on Ecological Restoration*, Society for Ecological Restoration, Washington, DC.

Van Andel, J. and J. Aronson (eds) (2006) *Restoration Ecology*, Blackwell Publishing, Malden, MA.

PART I

The basis for ecological restoration in the twenty-first century

2

CONSIDERING THE FUTURE

Anticipating the need for ecological restoration

Young D. Choi

Many of the Earth's natural characters have been altered or lost due to human development during the Anthropocene. To meet the demands of resource consumption for an ever-increasing human population and welfare, more than 60 per cent of the Earth's lands have already been converted or modified for human use (Hurtt *et al.* 2006), oceans have been subjected to exploitation of resources and pollution (Lotze *et al.* 2006), and the composition of atmospheric gases has been altered greatly with no or very little sign for reversing these changes. Human population growth, although slowing in recent decades, is still expected to grow at least for most of this century. Our continued expansion of our ecological footprint will only exacerbate the depletion of the Earth's natural capital. Moreover, the alterations in biogeochemical cycles of carbon, nitrogen and other elements have led to drastic changes in the environment for air, land and water quality (MA 2005; Clewell and Aronson 2007; Finzi *et al.* 2011). With these changes, it is not certain whether the Earth can keep evolving, stocking natural capital, and providing ecosystem services as it did before the appearance of industrial age *Homo sapiens*.

The idea of ecological restoration has been conceived and pioneered by early scientists and practitioners. For example, the reestablishment of tallgrass prairie by a group of Civilian Conservation Corps workers under a vision from Aldo Leopold has been regarded as the first-ever known attempt of ecological restoration in North America (Jordan *et al.* 1987a). Other examples of ecological restoration across the world in the twentieth century may include but are not limited to reclamation and revegetation of mined lands, afforestation and reforestation, conversion of old fields to grasslands, and mitigation of lost or altered wetlands. With the century-long (or much longer) tradition of ecological restoration (Palmer *et al.* 2006; Court 2012), 'restoration ecology' has emerged as a new discipline of applied ecology in the later part of twentieth century (Jordan *et al.* 1987a), and its emergence has been welcomed as a new way to meet numerous needs for ecological research and natural resource conservation (Bradshaw 1983; Jordan *et al.* 1987b; Dobson *et al.* 1997; Choi 2004; Choi *et al.* 2008). This chapter addresses such needs in five areas: conservation of biodiversity and evolutionary heritage, recovery of natural capital, enhancement of ecosystem services, a laboratory for testing ecological theories, and reconnection of human culture and nature.

Conserving biological diversity and evolutionary heritage

Conservation of biological diversity is among the top reasons for ecological restoration (Bradshaw 1983; Jordan *et al.* 1987b; Dobson *et al.* 1997). The current rate of species extinction is estimated to be 1,000 to 10,000 times greater than the normal rate, and habitat loss appears to be the leading cause of the extinctions in modern times. Conservation of biological diversity is essential not only to sustain the Earth's evolutionary heritage but also to shape the ecosystems of the future, because new biotas of the future emerge from the evolution of current species. Therefore, restoration of lost habitats is more than a way of species conservation (Wilson 1988).

Habitat restoration becomes more important for potential pole-ward migration of species in the wake of global climate change. IPCC (2014) predicts that the mean global surface temperature may increase 0.3–4.8°C by 2100. The pole-ward movements of species have already been documented (La Sorte and Thompson 2007; Somero 2010), and these kinds of movement would likely continue, particularly in the northern hemisphere. However, many of the species are subjected to major impediments in their migration attempts. Migration rates of certain species, especially sessile plants, are very slow. For example, Davis (1981) noted that many tree species in eastern North America have moved less than 400 metres per year to the north since the retreat of Wisconsinian glaciers. For example, balsam fir (*Abies balsamea*) and the nearly extinct American chestnut (*Castanea dentata*) moved less than 200 metres a year. Such slow-moving species would likely have no or very little chance to migrate north under the rapidly rising surface temperature.

Moreover, the impediments against species migration are often aggravated by highly fragmented habitat patches due to agricultural and urbanized landscapes (Lindenmayer and Fischer 2013). Habitat restoration on north-south migration routes is now urgent to allow the Earth's biotas to respond to the global climate change. For these reasons, ecological restoration is not just a way to conserve biological diversity, it is a proactive strategy to guard the processes of natural evolution so they may continue to proceed in the future.

Restocking natural capital

Natural capital is Earth's stock of natural resources that provide a wide array of goods (e.g. energy, food, fibres, timber and water) to human societies and economies. Like financial capital, its interest may accumulate or drop as the amount of stock increases or decreases, respectively (Costanza and Daly 1992), and thus the stocks of natural capital should be maintained at or above the level that does not deplete the resource (Clewell and Aronson 2007). The stocks of natural capital have been reduced to meet the demand for resource consumption from ever-increasing human population across the world. In many cases, depletion of natural resources has reached the level below which the Earth can no longer replenish them via natural processes (MA 2005).

For instance, marine fishery stock has declined drastically during the past decades due to overfishing and there is no or little sign of recovery (Branch *et al.* 2011). Global grain production has increased more than three times since 1960 (Nierenberg and Spoden 2012). However, this increase was mainly driven by energy input from combustion of fossil fuels, crop cultivation with petrochemical fertilizers and pesticides at the expense of natural capital in grasslands, forests, and wetlands (Tilman *et al.* 2002; Mulvaney *et al.* 2009). Tropical rainforests once covered 14 per cent of the Earth's land surface with more than 80 per cent of all living species but their cover was reduced to 6 per cent along with a large loss of biological diversity. IUCN

(2012) determined that more than 60 per cent of the 63,837 rainforest species assessed were critically endangered, endangered, threatened, or vulnerable to extinction. Nearly all of the grasslands and virgin forests and more than 50 per cent of the wetlands in the continental United States were converted for other uses such as agriculture, industrialization and urbanization, leaving very little room for them to recover by themselves (Mitsch and Gosselink 2007; Tilman *et al.* 2011).

Costanza *et al.* (1997) reported that the goods and services provided by the world's natural capital in 16 major biomes are worth US$16–54 trillion. However, many of them have been depleted and degraded – according to the Millennium Assessment report, more than 60 per cent of the goods and services have been lost (MA 2005). Aronson *et al.* (2007) urged that such degradation be halted and depleted capital needs to be restocked. This is a compelling justification for restoring natural capital. In this sense, ecological restoration is a necessity for restocking natural capital to sustain human civilization.

Enhancing ecosystem services

Along with the Earth's natural capital, ecosystem services are one way to characterize the rationale for restoration (Perring *et al.* 2011). Ecosystem services refer to the benefits that humans receive from nature. The benefits may include provision of goods from natural capital, regulation of ecosystem processes such as climate control, air and water purification and waste disposal, and enhancement of cultural values such as spiritual refreshment and discovery of new scientific knowledge. Like natural capital, degradation of ecosystem services has coincided with the expansion of human dominance of the Earth (MA 2005; Clewell and Aronson 2007). Massive combustion of fossil fuel has led to a major alteration in the global carbon cycle and climate. The changes of global climate cause a variety of ecosystem responses, such as desertification that may lead to reduction in primary productivity and pole-ward movements of species that may bring drastic socioeconomic ramifications in agriculture, forestry, fisheries, and other land–water uses. Particularly, the destruction of boreal and tropical forests would not only reduce the capacity of the Earth's vegetation to absorb atmospheric carbon through photosynthesis but also would convert them from carbon sinks to sources as the soils of deforested lands release carbon dioxides, methane and nitrous oxide to the atmosphere, further exacerbating the degradation of global carbon cycle (IPCC 2014). Restored grasslands, wetlands, forests and others may not only restock natural capital but also sequester atmospheric carbon, slow the process of global climate change, and mitigate the biogeochemical and hydrologic cycles that have been impaired.

For example, the Mississippi River watershed encompasses nearly 3 million hectares, approximately 40 per cent of the continental United States (Mitsch and Gosselink 2007). A vast majority of the watershed lands were converted to farmlands for agriculture upon the arrival of European settlers a few centuries ago. Such conversions have eliminated a vast majority of the grasslands, wetlands, woodlands and forests that existed prior to European settlement. Agricultural practices, along with loss of riparian wetlands, have led to significant alterations in the river's nitrogen dynamics. Nitrate and ammonia are brought from the farmlands to the river channel by eroded soils and surface runoff, causing eutrophication (Donner 2003). The polluting nutrients are further transported by the river, which already has lost its capacity for 'self-cleansing' of pollutants due to destruction of riparian wetlands, to the Gulf of Mexico. Consequently a gigantic 'dead zone' of hypoxia, covering more than 20,000 km^2 off the coast of Louisiana, allows no or very few aerobic organisms to survive (Turner *et al.* 2007; David *et al.* 2011). For this reason, restoration of wetlands has been advocated to reduce the nutrient loads in the Mississippi (Hey and Philippi 2002; Zedler 2003; Mitsch *et al.* 2005).

Coupled with global climate change, withdrawal of surface- and groundwater for irrigation to agricultural lands has altered the hydrologic cycle of the Mississippi River and other watersheds (Hey and Philippi 2002; Raymond *et al.* 2008). In particular, destruction of wetlands has reduced the capacity of land surfaces to hold water and of aquifers to be recharged after withdrawal of groundwater for irrigation of agricultural land (Steward *et al.* 2013). As the problems of eutrophication and surface- and groundwater depletion prevail all over the domesticated lands of world (Postel 1998; Wada *et al.* 2010), the need for ecological restoration is greater than ever to replenish the Earth's freshwater capital.

Testing ecological theories

Until the later part of the last century, the science of ecology was long dominated by 'descriptive' studies, and the description was often a compilation of 'telephone directory' lists of taxonomic species and correlations of what was observed (Harper 1982). Harper (*ibid.*), borrowing the words of Nobel laureate physicist Ernest Rutherford, argued that descriptive study alone is not sufficient to allow ecology to develop into a mature theory-generating science with predictive capacity. Manipulative experiments, which investigate cause–effect relationships, develop, test, validate and establish theories of ecological science, are essential. In this respect, ecological restoration appears to be a natural laboratory to test ecological theories because it is an experiment that manipulates numerous factors (e.g. preparing site conditions, determining assemblage of species to be reintroduced). For this reason, Bradshaw (1983) noted that ecological restoration is 'an acid test' of ecological science.

Indeed, the practice of ecological restoration has provided numerous theories and models of ecology testing in both field and laboratory settings since the emergence of the discipline of restoration ecology. For example, Palmer *et al.* (1997) stated, '(community) ecological theory may play an important role in the development of a science of ecology', bringing mutual benefits for both restoration and basic ecological research. Classical concepts of succession, such as monoclimax (Clements 1916), individualistic (Gleason 1926), and continuum (Bray and Curtis 1957) models, have been scrutinized by numerous observations of restoration trajectories (Walker and Del Moral 2003; Choi 2004). Assembly rules (Diamond 1975) have resurfaced for testing their applicability to ecological restoration (Temperton *et al.* 2007). In that context, there have been tests of – and support for – alternative successional models, such as the self-design model (Mitsch *et al.* 2012), and centrifugal model (Wisheu and Keddy 1992).

Biodiversity–ecosystem functioning (BEF) is another hypothesis that is being tested in restoration ecology. Originally conceived by MacArthur (1955) and Elton (1958), it hypothesizes that enhanced biological diversity can promote and stabilize ecosystem functions (Schulz and Mooney 1993; Naeem and Li 1997; Tilman *et al.* 1997; Hooper *et al.* 2005; Schindler *et al.* 2015). Most restoration projects aim to enhance biological diversity of target communities. In such restored communities with enhanced biological diversity, the BEF can be tested by measuring certain ecological function(s), such as stabilized primary production in restored prairies, flood prevention by improved water retention in mitigated wetlands, and reinforced sequestration of atmospheric carbon with reforestation. So far, based on the results from a few field tests (e.g. Temperton *et al.* 2007; Marquard *et al.* 2009; Doherty *et al.* 2011), the validity of the BEF hypothesis is still controversial. Such controversy is indeed a justification to test the utility of restored biological diversity for promoting ecosystem services (Choi *et al.* 2008; Perring *et al.* 2011).

In the midst of such scrutiny, restoration ecology itself has emerged as a nursery for development of its own concepts and theories. The concept of ecological restoration was solidified (SER 2004), after the seminal explication of the terms restoration, rehabilitation and replacement by Bradshaw (1983). Whisenant (1999) suggested two levels of threshold in response to degradation of ecosystem function: biotic and abiotic. If the degradation is biotic (e.g. loss of native species), biotic manipulations (e.g. reintroduction of the lost species) should be the key restoration practice. Otherwise, restoration efforts need to focus on removing the degrading factors and repairing the abiotic environmental conditions (Hobbs 2002). This concept of degradation thresholds became a basis of the 'biotic, abiotic and socioeconomic filter model' (Hobbs and Harris 2001; Hobbs and Norton 2004) and 'dynamic environmental filter model' (Fattorini and Halle 2004) for ecosystem reassembly.

In the wake of global changes, the use of traditional successional models for re-establishing historical ecosystems has been questioned because of ecological regime changes occurring at an unprecedented combination of extent and pace. This is why Hobbs and Norton (2004) suggested 'alternative state models,' as opposed to the historical successional models, as a guide for restoration. The alternative (stable) state model was further elaborated by Suding *et al.* (2004) as they constructed models that explicated how ecosystem changes 'flip' suddenly and irreversibly and that the restoration path back to some semblance of the previous state would likely be along a different trajectory.

Choi (2004, 2007) and Choi *et al.* (2008) repeatedly have called for a shift in the paradigm of ecological restoration from 'historic' to 'futuristic' ('anticipatory ecology' *sensu* Murphy 2005) because the ecosystems that are restored based on the past environment would not be sustainable in the future environment. Consistent with this forward-focused approach, the concept of 'novel ecosystem' has emerged. Hobbs *et al.* (2009) argued that the drastic alterations in the biotic and abiotic conditions made restoration of historic reference systems extremely difficult or impossible. At least some of the restored ecosystems may well be novel and most probably will be a hybrid between the novel and historical reference systems under the altered environment. Although different from the original assemblage of species and environmental conditions, 'novel ecosystem' may restore some ecological functions that once occurred in historic reference systems, otherwise subjected to degraded states (Doley *et al.* 2012). This concept has drawn some criticism from a fear that it can undermine the need, rationale and legitimacy of on-going restoration practices (Murcia *et al.* 2014), though Standish *et al.* (2013) had anticipated those arguments as well.

Reconnecting humanity with nature

> Conservation is a state of harmony between men and nature.
>
> *(Aldo Leopold 1949)*

> Ecological restoration is to rebuild a harmonious relationship between human society and nature.
>
> *(John Cairns 1994)*

Humans have evolved from nature and thus are a member of the Earth's community (Leopold 1949). Jordan writes:

> Human beings are social species. For such a species, relationship with nature is not a personal matter but is necessarily mediated by the community. The solitary individual, King Lear's 'unaccommodated man,' is an ecological and spiritual nonentity, as helpless

and as ecologically irrelevant as a solitary honey bee, cut off not only from the human community, but from the larger community of other animals and plants as well.

(Jordan and Lubick 2012)

In this sense, human civilization originated from nature. However, ironically, the process of civilization has resulted in the domination of the environment by humanity and the separation of human culture from nature (Boyden 2004). Despite such dissociation, our desire to remain as a part of nature has never ended as evidenced by numerous pro-nature activities such as backpacking, nature hikes, mountain climbing, wildlife watching, nature-mimic landscaping, and reading and watching books, photos, movies, and TV programs on nature subjects (Cairns 2002). However, according to Jordan (1986), none of these activities provides 'complete immersion in nature' as restoration activities do. Jordan and Lubick (2012) argue that ecological restoration is repayment of our debt to nature and our obligation to participate in the economy of nature for 'exchange of gifts' between humans and nature. Their point is this: so far, we have taken 'gifts (as natural resources and services)' from nature, and the consequences of 'taking gifts' are degradation of nature. For this reason, it is our moral responsibility to give 'gifts (ecological restoration)' back to nature.

Ecological restoration is the key to the development of our relationship with nature (Jordan 2003). Activities of ecological restoration are an opportunity for us to participate in the process of nature recovery and to experience 'personal transcendence' and 'spiritual renewal' of minds (Clewell and Aronson 2007). At the same time, these activities are a return of 'gifts' from us to nature. This concept of 'returning gifts' sets a new paradigm of nature conservation from 'defensive' to 'offensive'. Ecological restoration calls for 'proactive creation' to let ecosystems evolve under a harmonious combination of nature and human culture in the future, rather than 'passive protection' of what is left after human dominance as in the past. In addition, these restoration activities often occur in the places where human dominance is prevalent. This is an opportunity for us to engage, experience and learn about nature in close proximity to our own neighbourhood, not necessarily in remote wilderness. Should the restored nature in our neighbourhood attract people to experiences with nature, many remote areas that are highly valued for conservation, such as Yosemite and Yellowstone National Parks, would be subjected to fewer visits and be more protected from anthropogenic disturbances (Jordan 2003 cited by Woodworth 2013). Ecological restoration appears to be a win–win case for both humans and nature.

References

Aronson, J., S. J. Milton and J. N. Blignaut. 2007. Restoring natural capital: definition and rationale. Pages 1–8 in J. Aronson, S. J. Milton and J. N. Blignaut (eds), *Restoring natural capital: science, business, and practice*. Island Press, Washington, DC.

Boyden, S. V. 2004. *The biology of civilization: understanding human culture as a force in nature*. New South Publishing, Atlanta, GA.

Bradshaw, A. D. 1983. The reconstruction of ecosystems. *Journal of Applied Ecology* 10: 1–17.

Branch, T. A., O. F. Jensen, D. Ricard, Y. Ye and R. Hilborn. 2011. Contrasting global trends in marine fishery status obtained from catches and from stock assessments. *Conservation Biology* 25: 777–786.

Bray, J. R. and J. T. Curtis. 1957. An ordination of the upland forest communities of southern Wisconsin. *Ecological Monographs* 27: 325–349.

Cairns, J. Jr. 1994. Ecological restoration: re-examining human society's relationship with natural systems. The Abel Wolman Distinguished Lecture. National Research Council, Washington, DC.

Cairns, J. Jr. 2002. Rationale for restoration. Pages 1–23 in M. R. Perrow and A. J. Davy (eds), *Handbook of ecological restoration, volume 1: principles of restoration*. Cambridge University Press, Cambridge.

Choi, Y. D. 2004. Theories for ecological restoration in changing environment: toward 'futuristic' restoration. *Ecological Research* 19: 75–81.

Choi, Y. D. 2007. Restoration ecology to the future: a fall for new paradigm. *Restoration Ecology* 15: 351–353.

Choi, Y. D., V. M. Temperton, E. B. Allen, A. P. Grootjan, M. Halassy, R. J. Hobbs, M. A. Naeth and K. Torok. 2008. Ecological restoration for future sustainability in a changing environment. *Ecoscience* 15: 53–64.

Clements, F. E. 1916. *Plant succession: an analysis of the development of vegetation.* Publication 242. Carnegie Institution of Washington, Washington, DC.

Clewell, A. F. and J. Aronson. 2007. *Ecological restoration: principles, values, and structure of an emerging profession.* Island Press, Washington, DC.

Costanza, R. and H. E. Daly. 1992. Natural capital and sustainable development. *Conservation Biology* 6: 37–46.

Costanza, R., R. D'Arge, R. De Groot, S. Farber, M. Grasso, B. Hannon, K. Limburg, S. Naeem, R. V. O'Neill, J. Paruelo, R. G. Raskin, P. Sutton and M. Van Den Belt. 1997. Value of the world's ecosystem services and natural capital. *Nature* 387: 253–260.

Court, F. E. 2012. *Pioneers of ecological restoration: the people and legacy of the University of Wisconsin Arboretum.* University of Wisconsin Press, Madison, WI.

David, M. B., L. E. Drinkwater and G. F. McIssac. 2011. Sources of nitrate yields in the Mississippi River Basin. *Journal of Environmental Quality* 39: 1657–1667.

Davis, M. B. 1981. Quaternary history and the stability of forest communities. Pages 132–153 in D. C. West, H. H. Shugart and D. B. Botkin (eds), *Forest Succession Concepts and Applications.* Springer-Verlag, New York.

Diamond, J. M. 1975. Assembly of species communities. Pages 342–344 in M. L. Cody and J. M. Diamond (eds), *Ecology and evolution of communities.* Harvard University Press, Cambridge, MA.

Dobson, A., A. D. Bradshaw, and A. J. M. Baker. 1997. Hope for the future: restoration ecology and conservation biology. *Nature* 227: 515–522.

Doherty, J. M., J. C. Callaway and J. B. Zedler. 2011. Diversity–function relationships changed in a long-term restoration experiment. *Ecological Applications* 21: 2143–2155.

Doley, D., P. Audet and D. Mulligan. 2012. Examining Australian context for post-mined land rehabilitation: reconciling a paradigm for the development of natural and novel systems among post-disturbance landscapes. *Agriculture, Ecosystems and Environment* 163: 85–93.

Donner, S. 2003. The impact of cropland cover on river nutrient levels in the Mississippi River Basin. *Global Ecology and Biogeography* 12: 341–355.

Elton, C. S. 1958. *The ecology of invasions by animals and plants.* Chapman & Hill, New York, New York.

Fattorini, M. and S. Halle. 2004. The dynamic environmental filter model: how do filtering effects change in assembling communities after disturbance? Pages 96– 14 in V. M. Temperton, R. J. Hobbs, T. Nuttle and S. Halle (eds), *Assembly rules and restoration ecology.* Island Press, Washington, DC.

Finzi, A. C., A. T. Austin, E. E. Cleland, S. D. Frey, B. Z. Houlton, and M. D. Wallenstein. 2011. Responses and feedbacks of coupled biogeochemical cycles to climate change; examples from terrestrial systems. *Frontiers in Ecology and the Environment* 9: 61–67.

Gleason, H. A. 1926. The individualistic concept of the plant association. *Bulletin of Torrey Botanical Club* 53: 1–20.

Harper, J. L. 1982. *After description.* British Ecological Society, London.

Hey, D. L. and N. S. Philippi. 2002. Flood reduction through wetland restoration: the Upper Mississippi River Basin as a case history. *Restoration Ecology* 3: 4–17.

Hobbs, R. J. 2002. The ecological context: a landscape perspective. Pages 24–45 in M. R. Perrow, and A. J. Davy (eds), *Handbook of ecological restoration, volume 1: principles of restoration.* Cambridge University Press, Cambridge.

Hobbs, R. J. and J. A. Harris. 2001. Restoration ecology: repairing the Earth's ecosystem in the new millennium. *Restoration Ecology* 9: 239–246.

Hobbs, R. J. and D. A. Norton. 2004. Ecological filters, thresholds, and gradients in resistance to ecosystem assembly. Pages 72–95 in V. M. Temperton, R. J. Hobbs, T. Nuttle and S. Halle (eds), *Assembly rules and restoration ecology.* Island Press, Washington, DC.

Hobbs, R. J., E. Higgs, and J. A. Harris. 2009. Novel ecosystems: implications for conservation and restoration. *Trends in Ecology and Evolution* 24: 599–605.

Hooper, D. U., F. S. Chapin III, J. J. Ewel, A. Hector, P. Inchausti, S. Lavorel, J. H. Lawton, D. M. Lodge, M.

Loreau, S. Naeem, B. Schmid, H. Setälä, A. J. Symstad, J. Vandermeer, and D. A. Wardle. 2005. Effects of biodiversity on ecosystem functioning: a consensus of current knowledge. *Ecological Monographs* 75: 3–35.

Hurtt, G. C., S. Frolking, M. G. Fearon, B. Moore, E. Shevliakova, S. Malyshev, S. W. Pacala, and R. A. Houghton. 2006. The underpinnings of land-use history: three centuries of global gridded land-use transitions, wood harvest activity, and resulting secondary lands. *Global Change Biology* 12: 1208–1229.

IPCC. 2014. *Climate change 2014: synthesis report*. IPCC Technical Support Unit, Bilthovan, The Netherlands.

IUCN. 2012. IUCN Red List of Threatened Species: 2012 release. Retrieved from www.iucnredlist.org.

Jordan, W. R. III. 1986. Restoration and the reentry of nature. *Restoration and Management Notes* 4: 2.

Jordan, W. R. III. 2003. *The sunflower forest*. University of California Press, Oakland, CA.

Jordan, W. R. III and G. M. Lubick. 2012. *Making nature whole: a history of ecological restoration*. Island Press, Washington, DC.

Jordan, W. R. III, M. E. Gilpin, and J. D. Aber. 1987a. *Restoration ecology: a synthetic approach to ecological research*. Cambridge University Press, Cambridge.

Jordan, W. R. III, R. L. Peters, and E. B. Allen. 1987b. Ecological restoration as a strategy for conserving biological diversity. *Environmental Management* 12: 55–72.

La Sorte, F. A. and F. R. Thomson. 2007. Poleward shifts in winter ranges of North American birds. *Ecology* 88: 1803–1812.

Leopold, A. 1949. *A Sand County Almanac*. Oxford University Press, Oxford.

Lindenmayer, D. B. and J. Fischer. 2013. *Habitat fragmentation and landscape change*. Island Press, Washington, DC.

Lotze, H. K., H. S. Lenihan, B. J. Bourque, R. H. Bradbury, R. C. Cooke, M. C. Kay, S. M. Kidwell, M. X. Kirby, C. H. Peterson, and J. B. C. Jackson. 2006. Depletion, degradation, and recovery potential of estuaries and coastal seas. *Science* 312: 1806–1809.

MA. 2005. *Ecosystems and human well-being; general synthesis*. Island Press, Washington, DC.

MacArthur, R. H. 1955. Fluctuation of animal populations, and a measure of community stability. *Ecology* 36: 533–536.

Marquard, E., A. Weigelt, V. M. Temperton, C. Roscher, J. Schumacher, N. Buchmann, M. Fischer, W. W. Weisser and B. Schmid. 2008. Plant species richness and functional composition drive overyield in a six-year grassland experiment. *Ecology* 90: 3302.

Mitsch, W. J. and J. G. Gosselink. 2007. *Wetlands* (4th edition). John Wiley & Sons, Hoboken, NJ.

Mitsch, W. J., J. W. Day, L. Zhang, and R. R. Lane. 2005. Nitrate-nitrogen retention in wetlands in the Mississippi River Basin. *Ecological Engineering* 24: 267–278.

Mitsch, W. L., L. Zhang, K. C. Stefanik, A. M. Nahlik, C. J. Anderson, B. Bernal, M. Hernandez, and K. Song. 2012. Creating wetlands: primary succession, water quality changes, and self-design over 15 years. *Bioscience* 62: 237–250.

Mulvaney, R. L., S. A. Khan, and T. R. Ellsworth. 2009. Synthetic nitrogen fertilizers deplete soil nitrogen: a global dilemma for sustainable cereal production. *Journal of Environmental Quality* 38: 2295–2314.

Murcia, C., J. Aronson, G. H. Kattan, D. Moreno-Mateos, K. Dixon, and D. Simberloff. 2014. A critique of the 'novel ecosystem' concept. *Trends in Ecology and Evolution* 29: 548–553.

Murphy, S. D. 2005. Concurrent management of an exotic species and initial restoration efforts in forests. *Restoration Ecology* 13: 584–593.

Naeem, S. and S. Li. 1997. Biodiversity enhances ecosystem reliability. *Nature* 390: 507–509.

Nierenberg, D. and K. Spoden. 2012. Global grain production at record high despite extreme climatic events. Retrieved from www.worldwatch.org/global-grain-production-record-high-despite-extreme-climatic-events-0.

Palmer, M. A., R. F. Ambrose, and N. L. Poff. 1997. Ecological theory and community restoration ecology. *Restoration Ecology* 5: 291–300.

Palmer, M. A., D. A. Falk, and J. B. Zedler. 2006. Ecological theory and restoration ecology. Pages 1–10 in D. A. Falk, M. A. Palmer and J. B. Zedler (eds), *Foundations of restoration ecology*. Island Press, Washington, DC.

Perring, C., S. Naeem, F. S. Ahrestani, D. E. Bunker, P. Burkill, G. Canziani, T. Elmqvist, J. Fuhrman, F. M. Jaksic, Z. Kawabata, A. Kinzig, G. Mace, H. Mooney, A. Prieur-Richard, J. Tschirhart and W. Weisser. 2011. Ecosystem services, targets, and indicators for the conservation and sustainable use of biodiversity. *Frontier in Ecology and the Environment* 9: 512–520.

Postel, S. 1998. Water for food production: will there be enough in 2025? *Bioscience* 48: 629–663.

14

Raymond, P. A., N. H. Oh, R. E. Turner, and W. Broussard. 2008. Anthropogenically enhanced fluxes of water and carbon from the Mississippi River. *Nature* 451: 449–452.

Schindler, D. E., J. B. Armstrong, and T. E. Reed. 2015. The portfolio concept in ecology and evolution. *Frontiers in Ecology and the Environment* 13: 257–263.

Schulz, E. D. and H. A. Mooney. 1993. *Biodiversity and ecosystem function*. Springer-Verlag, Berlin, Germany.

SER. 2004. The SER international primer on ecological restoration. Society for Ecological Restoration Science & Policy Working Group. Retrieved from www.ser.org/content/ecological_restoration_primer.asp (accessed 21 April 2015).

Somero, G. N. 2010. The physiology of climate change: how potentials for acclimating and genetic adaptation will determine 'winners' and 'losers.' *Journal of Experimental Biology* 213: 912–920.

Standish R. J., A. Thompson, E. S. Higgs, and S. D. Murphy. 2013. Concerns about novel ecosystems. Pages 296–309 in R. Hobbs, E. Higgs and C. Hall (eds), *Novel ecosystems: intervening in the new ecological world order*. John Wiley & Sons, Hoboken, NJ.

Steward, D. R., P. J. Bruss, X. Yang, S. A. Staggenberg, S. M. Welch and M. D. Apley. 2013. Tapping unsustainable groundwater stores for agricultural production in the High Plains Aquifer of Kansas, projections to 2110. *Proceedings of the National Academy of Science*. 110: E3477–E3486.

Suding, K, N., K. L. Grass and G. R. Houseman. 2004. Alternative states and positive feedbacks in restoration ecology. *Trends in Ecology and Evolution* 19: 46–53.

Temperton, V. M., P. N. Mwangi, M. Scherer-Lorenzen, B. Schmid and N. Buchmann. 2007. Positive interactions between nitrogen-fixing legumes and four different neighbouring species in a biodiversity experiment. *Oecologia* 151: 190–205.

Tilman, D., D. Wedin and J. Knops. 1997. Productivity and sustainability influenced by biodiversity in grassland ecosystems. *Nature* 379: 718–720.

Tilman, D., K. G. Cassman, P. A. Matson, R. Naylor and S. Polasky. 2002. Agricultural sustainability and intensive production practices. *Nature* 418: 671–677.

Tilman, D., C. Balzer, J. Hill and B. L. Befort. 2011. Global food demand and the sustainable intensification of agriculture. *Proceedings of National Academy of Science* 108: 20260–20264.

Turner, R. E., N. N. Rabalais, R. B. Alexander, G. McIssac and R. W. Howarth. 2007. Characterization of nutrient, organic carbon, and sediment loads and concentration from the Mississippi River into the northern Gulf of Mexico. *Estuaries and Coasts* 30: 773–790.

Wada, Y. L., L. P. H. Van Beek, C. M. Van Kempen, J. W. T. M. Reckman, S. Vasak and M. F. P. Bierkens. 2010. Global depletion of groundwater resources. *Geophysical Research Letters* 37(20): article 044571.

Walker, L. R. and R. Del Moral. 2003. *Primary succession and ecosystem rehabilitation*. Cambridge University Press, Cambridge.

Whisenant, S. G. 1999. *Repairing damaged wildlands: a process-oriented, landscape-scale approach*. Cambridge University Press, Cambridge.

Wilson, E. O. 1988. *Biodiversity*. National Academy of Science, Washington, DC.

Wisheu, I. C. and P. A. Keddy. 1992. Competition and centrifugal organization of plant communities: theories and tests. *Journal of Vegetation Science* 3: 147–156.

Woodworth, P. 2013. *Our once and future planet*. University of Chicago Press, Chicago, IL.

Zedler, J. B. 2003. Wetlands at your service: reducing impacts of agriculture at the watershed scale. *Frontiers in Ecology and the Environment* 1: 65–72.

3

THE PRINCIPLES OF RESTORATION ECOLOGY AT POPULATION SCALES

Stephen D. Murphy, Michael J. McTavish and Heather A. Cray

Restoration at population scales cannot be done in isolation

While restoration ecology is probably best considered as a cross-scalar effort, its origins and practice are often firmly in the camps of the more disciplinary levels of domains like population ecology. When we use the term 'cross-scalar', we refer to the notion that ecosystem functions (processes like nutrient and water cycling or interactions between organisms and their environment) and structures (the genetic and species diversity of organisms or the size and physiognomy of habitats) are not definable or restricted to molecular, population, community, landscape, or ecological regime domains. One recent paper that captures this nicely is Rose *et al.* (2015). They examined cross-scalar ecological restoration impacts on fish populations and communities in the context of ecological modelling (a topic of much discussion in this chapter). Their main message was that successful restoration ecology starts with an understanding and communication of the major steps involved at different scales – population, community, and landscape – and they fulfilled a much more ambitious objective of discussing all of these in terms of best practices for management within restoration ecology. We will restrict ourselves here to population scales, but that context by Rose *et al.* (*ibid.*) is what ultimately drives these discussions.

We can conceptualize that restoration ecology is really about the changes in evolutionary ecology – how drivers like natural selection, genetic drift, and phylogenetic constraints are changed by humans and how humans may then try and manipulate them further to repair ecosystem damage. However, the traditional oeuvre of restoration ecology is still entrenched in population scales – rescuing endangered species – because legal instruments tend to focus solely on this scale. There is focus at ecological community scales because the pioneers of restoration ecology were mainly from that school of thinkers – Aldo Leopold, John Curtis, Norman Fassett, and others were often focused even further on prairies and the community drivers like fire. This book as a whole will not restrict itself to population scales but it is important that a chapter be devoted to reviewing and discussing these given their prominence. Our cue is the classic paper by Montalvo *et al.* (1997) – the paper is nearing its twentieth anniversary but it is so well written that the ideas it reviews and the notions it inspired are relevant today; readers will see that this chapter builds on their excellent strategic paper.

The fundamentals of understanding population dynamics: Genetics and evolution

We start with some reminders of basic terms because we have found that not all restorationists will have a strong background in ecology. The basis of ecology is evolution and its theories; we cannot do justice to the complexity of the contemporary theoretical framework.

Evolution is the change in the frequencies of heritable genes over time. This will be influenced by factors including: mutations incurred during mitosis, random gene assortments during meiosis (producing gametes) and recombination during fusion of gametes, the relative benefit or detriment created by genes interacting with and within the whole environment during each generation or cohort, the response of genes to drivers that favour some over others (selection – natural and human directed), the response of genes to random influences like some organisms dying because of an accident while some less robust ones happen to survive (genetic drift), the interactions of genes with each other and the varied influences on the expression of genes (many mechanisms like epistasis and pleiotropism) and the constraints on gene inheritance and structure imposed by the evolutionary history and developmental processes (roughly, evolutionary developmental biology – 'evo-devo').

We can speak of genes in terms of their encapsulation in genotypes (all the genes in an individual), their expression in individuals (phenotypes), and the entire genetic complement of a species (a genome). While there is tendency to limit evolution to the concept of the non-random factor of natural selection favouring well-adapted genes/genotypes/phenotypes, this is not correct because the preponderance of neutral mutations and the influences of genetic drift, interactions, and evo-devo are quite important. Despite breathless reporting to the contrary, most organisms' genetic complement and their expression is not a history of optimization or excising useless or even detrimental genes or gene products. Organisms are filled with junk DNA that does no harm and thus is not selected out, genes that are co-opted but suboptimal for functionality, and functions and structures that are reflections of evo-devo (like the human eye – the octopus eye is much more efficient and reflects an evolutionary history less constrained than our own).

But what is a population – how do we define or delineate it?

Traditionally, populations are considered to exist when there are a group of phenotypes that contain genotypes that are similar enough to allow for successful sexual reproduction and survivorship of offspring. As readers will discover, this is problematic for many organisms because they do not require sexual reproduction to survive. Further, populations are normally considered to be constrained and defined by some form of sympatry – they live near enough to one another to interact with some regularity and likelihood of breeding. This compounds the problem because this still relies on the notion of sexual reproduction and now it refers to some vague notion of being close enough to likely breed.

If we now think of populations as being genetically similar enough to breed, we probably assume they are from the same species – defined again in terms of being able to produce viable offspring. But species are not immutable (evolution eventually or even suddenly leads to new species arising from ancestral ones), there are some that are classified as different species yet produce viable cross-species hybrids, and some species are rarely – perhaps never – sexual. Species were often defined more by morphological characteristics that belie the complexity of breeding systems and the molecular basis of life.

Still, one can argue that many populations (and the species they are part of) are reasonably well-defined in the sense that many species do reproduce sexually – often or not, that relative to the vast diversity of life on Earth, many species are well defined enough genetically and phenotypically that they do not mate or produce viable hybrids, and that populations often are definable by studying the gene identities and frequencies, and the barriers to interaction. Practically, restoration ecologists often do not consider the nuances of what a population actually represents and it may not matter to success in many cases; however, we would be remiss in not alerting readers – even beginners – to the issues that arise because nature is not as easily compartmentalized as humans would like it to be.

What is population ecology?

Population ecology bleeds into other scales of restoration ecology because it is based on genetic assortment, differentiation and diversity; ultimately these are the bases for how we define species and hence how we track how species interact to form ecological communities. Because populations are affected by spatial factors as well as time, we can examine populations at landscape scales (meta-populations – populations that are separated spatially and their interactions are defined by their ability to overcome spatial constraints or take advantage of spatial facilitations like physical corridors connecting habitats). In restoration ecology, populations are not treated any differently than in general ecology. We can start with the basics of population dynamics – the main demographic variables of birth, death, migration rates.

How do we measure demographic variables when studying restoration of populations?

While eponymous and therefore self-explanatory, the actual study of birth, death, migration rates in restoration ecology reveals some nuances. An important concept is that unlike humans, 'birth' in the many organisms that a restoration ecologist studies has multiple meanings. It can mean what humans expect – two individuals mate; their genes were randomly assorted during meiosis and recombination, providing increased genetic diversity as long as the two who mated are relatively unrelated. But many organisms have more complex mating behaviours. Some plants self-fertilize while others cannot. Many organisms reproduce asexually: fission, budding, fragmentation, sporogenesis, agamogenesis (no male gamete needed), and a large range of vegetative reproduction in plants.

The range of mating systems found in organisms can make birth rates hard to discern since some of these processes happen many times in a short period (short generation times) and others take much longer – it is not a case where one calendar year or even one generation truly form a unique cohort of individuals. Even death can be hard to measure; it can be difficult to detect when cryptic organisms die (it is not easy to measure bacterial death rates for example) and even with organisms like plants, algae, or fungi, do we measure death rates based on when the genetically unique individual ('genet') finally dies or when a given asexually reproduced 'ramet' dies? And can we easily tell the difference between death and dormancy – this is not easy with organisms that undergo sporogenesis or ones with some type of dormancy, especially if the dormant structure is hidden, like a seed or spore, or a tuber, corm, or rhizome that is underground. Migration can be fuzzy too – pollen and asexual forms can travel on wind, animals, or human conveyances long distances and it can be hard to track them at all, much less their success at fertilization (pollen) or survival.

For a restoration ecologist, the basic information needed to gauge the need for restoration and the success of restoration can be more elusive than the layperson realizes – it is challenging, though there are useful approaches and we can measure population dynamics. Restoration ecologists can use standard tools like molecular markers to track the origins and dispersal of genes within genotypes of populations. Still, even with modern techniques for markers, it can still be very expensive and requires gaining an adequate sample of source and destination populations. Indeed, the basic goal that was perhaps implicit in the origins of restoration ecology is the same today, except more explicit – we want heterogeneity and variation at the genetic and phenotypic level of source and destination populations.

Falk *et al.* (2006) provided a detailed review of the measurements used by restoration ecologists studying population dynamics in order to meet the goal of genetic and phenotypic diversity.

Intriguingly, there is an operational caveat to a goal of population-level diversity – if a site is extremely degraded and therefore in dire need for ecological restoration, it may be useful to introduce populations that are less diverse and more amenable to being able to establish under extreme conditions. Populations of organisms that are able to sequester compounds like organo-metals, polyaromatic hydrocarbons, or concentrated acids often have low genetic diversity because only a few will survive under such extreme selection pressures. Such conditions tend to favour homozygosity for alleles on genes that confer an ability to sequester toxins. This creates an apparent 'stress paradox' because such homozygosity reduces potential adaptation response so stress-tolerant genotypes should go extinct quickly. Ironically, once stress-tolerant genotypes and phenotypes have reduced toxicity to levels other organisms can withstand they create a new successional pathway that actually dooms the stress-tolerant populations.

However, the low genetic and phenotypic diversity is not as low as some might assume. This is because during the time they are under stress from toxicity, they survive because mutation rates will increases under stress – some will be able to adapt to successively less toxic conditions, sexual recombination increases under stress – there will be new genetic combinations also able to adapt successively to less toxic conditions, and many have transposons that allow for rapid mobile response to new environmental conditions. Genetic linkage, epistasis, pleiotropism, and phenotypic plasticity can also allow for some increases in genetic or phenotypic diversity even while the overall genetic diversity is still low under stressful conditions. The paradox is that the same selection pressures can favour low genetic diversity because it augments survival during stress and yet favours increases in diversity – and that latter outcome then helps some part of each organism's genotype remain in the population once conditions are less stressful.

The larger principle the stress paradox portends is the practical question of how one copes with inbreeding and outbreeding depression. Inbreeding depression occurs when organisms that have very similar genetic compositions – they are close relatives – mate and their offspring survive and mating between close relatives (and their genotypes) is rampant. While some plants are extreme inbreeders – self-compatible and mate with themselves – many organisms have biochemical and behavioural barriers that discourage or prevent inbreeding.

Bear in mind that the need for ecological restoration is often created because populations of a species have become very low – and inbreeding then is a means of last resort, even with attendant problems. The main problems arise because the genetic diversity of a population is so low that:

- It is vulnerable to extinction because if the environment changes, the entire population may be disfavoured by natural selection.
- Genetic drift can have disproportionate impacts in that some desirable genes may be lost because of random factors – this is a small risk if the genome has many genes and alleles but is a large risk if there are few to begin with.

- Deleterious mutations can accumulate quickly in low diversity populations. This creates a genetic bottleneck – the low genetic diversity hampers the survival of populations and perhaps the entire species if it is a widespread occurrence.

The response of restoration ecologists to this situation is usually to either translocate new genotypes from nearby populations or to begin a captive breeding or nursery programme using new genotypes from nearby populations. The latter is used if the situation is so dire that there is a need to ensure that successful mating of unrelated organisms occurs. However, if the numbers of organisms of a species is already so low that genetic diversity is practically nil, then the efforts will fail.

There is some promise that if DNA can be extracted from samples of preserved (dead) specimens from museum collections, then it can be reintegrated into a modern genome of species or at least populations. This is still in early stages but one can read about efforts to bring about 'de-extinction' of species such as the thylacine. For now, the best one can do if populations are too low worldwide is to promote hybridization between closely related species (not individuals) if their chromosomes will align properly during fertilization and produce viable offspring. Both methods can be controversial even under desperate circumstances and some argue that they are not ethical under any circumstances; it is not true restoration because the original species will still be extinct, it is not true restoration if the hybrids would not exist outside of a breeding programme (species are not sympatric), or it delays the inevitable extinction while risking source populations or introducing a new type of species to environments where it may disrupt existing community-level interactions. And this assumes hybrids are viable. In cases like *Panthera*, most male hybrids are sterile but a few are fertile – like the males produced from female lions and male leopards.

The hybrid question underscores a problem often neglected by restoration ecologists – outbreeding depression; this occurs when two organisms are from populations that should or could be able to produce offspring, but (a) they cannot do so at all because their chromosomes are not able to align during fertilization, (b) they produce sterile offspring for similar reasons, (c) they produce weak offspring because the chromosomes align poorly, causing genetic damage, or (d) they produce offspring poorly suited to local conditions.

This is why restoration ecologists must focus on source populations – and here the question of the provenance, manipulating source populations, and the genetic differentiation of those populations is of great concern for any organism – plant, animal, fungi, or otherwise (Hufford and Mazer 2003; Rice and Emery 2003; McKay *et al.* 2005; Armstrong and Seddon 2007; Weeks *et al.* 2011). They often should be geographically close on the assumption that most dispersal is relatively slow and local so that even if several hundred years have passed since populations migrated, there has still been some gene flow between them and they are not so isolated as to be nearing the point where their local genomes are too divergent or even nearing speciation thresholds. This may not apply to long distance migrants and even apparently sedentary organisms like plants can have some long distance dispersal via pollen, seeds, or vegetative structures like pieces of rhizomes being transported by wind, water, animals or human conveyances.

Restoration ecologists also have to take care in how they measure the state and function of populations. If one uses proper sampling techniques – and what is proper depends on the context of the research or desired outcome of restoration – it is feasible to census most populations. For herbs and forbs, we probably will use a stratified random sample using transects and quadrats and strive to minimize sampling bias, including autocorrelation. For many animals, we will do some form of mark and recapture or mark and monitoring via radio-collars, barcodes, drones, airplanes, or satellites; again, we will strive to minimize bias but must be aware that our initial capture can make an animal trap shy or trap happy should we want to repeat their capture for measurements. If we're doing monitoring from aircraft, we will have to be careful – and be able – to determine which animals we've already counted in a given period of time so we do not repeat counts and over-estimate population sizes.

But a census only tells us how many. It does not tell us if the population is viable. For that, we need to determine the effective population size – how many are fertile now and currently able to breed successfully, how many actually do breed successfully, and how many future organisms should be able to breed successfully. Depending on breeding system, we may need to know how many female or female-expressed organisms exist and then the same for males/male-expressed organisms. It may appear odd to see the word 'expressed' but we remind readers that many organisms are not dioecious – they have mixed expressions of what humans would call genders and even humans and other dioecious species have some range of expression of sexual organs (and behaviour in animals).

Thus, we can census (determine $n_{sampled}$ and $N_{estimated}$ – the sampled and estimated total population[1]). We can sample more thoroughly and determine N_e – the effective population size that, usually, represents the number of organisms that do mate and produce viable and fertile offspring – though it can represent potential numbers that are known to be able to mate. This would be further enhanced in populations more reliant on sexual reproduction if we also knew the numbers of female and male individuals or the relative expression of functional female and male reproductive capacity – an extended N_e.

Population models in restoration ecology

We want to use our samples of populations in restoration ecology – and conservation ecology, for that matter – to help us determine if our restoration efforts are likely to bear success, if they are bearing success, or if they did bear success. For that, we usually use several approaches but we often will model our populations – we create population models that either represent what has happened to population dynamics already or we predict what might happen to population dynamics in the future. This could mean that we represent populations mathematically and stop there. It could mean that we use that mathematical expression further – we try to create scenarios or perhaps even more concrete predictions about the likely future of the size or composition of populations. These still will be tied to the mathematical functions but they will normally become more complex mathematically and more realistic ecologically.

These population models can be expressed in different ways. One approach with a long history is to use matrix algebra – a means of expressing and calculating repeated algorithms. This was quite useful in the eras before personal computers were economical, powerful, and ubiquitous and even after that, the structure of matrices is very similar to how even modern analysis programs input data. At the risk of your editors seeming even older than we are, the prehistoric era before the advent of small, powerful, personal computers lasted until the mid-1990s in many places – and still exists in some regions today. This is another reason the matrix algebra approach is still used today – it allows for consistency in data expression and analysis across the decades of data collection and recording where matrices were used for most of that time period. The ability to use the same basic approach is important for reasons clear to anyone who experiences the frustration of new devices that are not backwards-compatible. This is why the literature is replete with references to population models that are based on such arcane terms as 'the Leslie matrix' or 'the Leftkovich matrix'.

The core of population models is not so much their mathematical expression as their assumptions. 'The Leslie matrix' and 'the Leftkovich matrix' differ on that basis. We usually start with the simplest population model – a linear relationship that adds organisms born or immigrating and subtracts organisms dying or emigrating during a time interval expressed as $(t, t + 1)$

$$N_{t+1} = N_t + B_{(t, t+1)} - D_{(t, t-1)} + I_{(t, t+1)} - E_{(t, t+1)}$$

Again, it is very difficult to sample even these variables accurately. We might try to write a model that focuses on the main outcomes of population dynamics of one gender – as if all species were dioecious; this often is focused on females because there usually are fewer female gametes in populations as they are more expensive, energetically, to produce:

$$N_{t+1} = N_{t(\text{reproductive females})} \times S_{t, t+1(\text{reproductive females})} + N_{t(\text{pre-reproductive females})} \times S_{t, t+1(\text{pre-reproductive females})}$$
$$\times [S_{t, t+1(\text{pre-reproductive females})} / S_{t, t+1(\text{reproductive females})}]$$

In this model, N is the number of females and S is the survival rate of females. The measurements are based on current measurements (now = $t + 1$) and prior measurements, generally expressed as an interval between now $(t + 1)$ and the earlier time (t) (that interval is often assumed to be annual – one year – or one reproductive/breeding cycle).

This can be simplified further if the measurements exclude any possible immigrants or emigrants and also assume that resources are not limited. In fact, those assumptions – combined with the exclusive focus on one gender expression (female) – are the basis for the often cited Leslie population model (also called the Leslie matrix model if matrix algebra is used). This is what leads to an exponential population growth model – which is not realistic but just like learning to count, this is what allowed population ecologists to build more sophisticated and realistic population models.

The first step in that history was to focus more on the stages rather than ages. Instead of assuming that all organisms' life history was tethered to human calendars or even a seasonal cycle, the Leslie model was modified to account for the basic difficulty in properly calculating the age of many organisms, the fact that many organisms reach reproductive maturity based on their size-stage (usually this occurs once they have enough resources to reach a certain size) rather than age, the ability of many organisms to effectively 'age backwards' in the sense that they might reproduce vegetatively and the daughter organisms are clones but smaller than the parent or they could become dormant. This more advanced approach is called the Leftkovitch model of population dynamics.

While both of the above models were – and still are – popular because of their simplicity, that is their very drawback. Again, they usually focus on one expressed gender, do not consider any resource limitation, and do not consider the existence of immigration or emigration. Both therefore assume that any age or stage are subject to the same fecundity, mortality, and growth rates – all of those are also not true in most cases.

More traditionally, we express population changes in mathematical notation that focuses on the key variables of N (symbolizing actual population size in this case), the constant, intrinsic growth rate (r) of a population, and the carrying capacity, K. While this still simplifies the ecological world by assuming that r and K are constants – they never change regardless of genetics or environment – this leads to useful approaches in population modelling. While r is not really a constant, we can conceptualize r as representing the maximum growth rate of a population – which can happen in the real ecological world, if only for a short time; it can be expressed as the exponential growth rate:

$$N_{t+1} = N_t + rN_t$$

Any population growing near the maximum value of r is likely to be one that is, not surprisingly, termed 'r-selected'. This means that there are periods of time when it is evolutionary advantageous to produce massive numbers of offspring quickly – short generation times exist. Bacteria are an obvious example; so too are fungi. Plants that have annual lifecycles are slower but consistent with this model. In no case is the growth rate maintained at maximum r – competition within or between species for resources, diseases, predation, herbivory will all contribute to a slowing of the population growth.

Some of these causes may be a function of density – how many organisms exist in a given space; if so, they are density-dependent variables and this means the probability of the variable affecting population dynamics increases with density. Diseases would be one example. Of course a variable may be density-independent and there will be more variation in the probability it will affect a population. This is often the case with abiotic limits to population growth – a drought's impact is not dependent on the population density if the drought is wide-spread and the occurrence of a drought is not completely deterministic and is therefore not predictable either. This latter notion alludes to yet another broader issue – whether populations are more affected by deterministic (non-random or at least constrained) variables or stochastic ('random') variables.

A restoration ecologist can exploit this knowledge. If the organisms are beneficial, then it may be inexpensive and fast to establish key components of ecosystems in restoration ecology. And this is usually true. A useful strategy in restoration ecology is to introduce beneficial bacteria, fungi, and annual plants – among other r-selected organisms – to a degraded start to speed the whole process. We still need to be careful about source material and maximize genetic diversity within species' populations and we'd need to spread the material around but this is a major first step in ecosystem restoration. We need to quickly increase populations of desirable organisms and ones – like bacteria and fungi – that will be needed to re-initiate and maintain processes like nutrient cycling. This is what our research group does – if we compare success of restoration at sites where we 'inoculate' soil with beneficial bacteria and fungi versus sites where we do not and simply hope these re-colonize from nearby source populations, the inoculated sites are restored much quicker. Here we measure the pace of ecological restoration as a functional response of NO_3 concentrations in what was a situation where it was an eroded, depleted soil; we planted 12 herbaceous and forb species at a site where 6 replicates where inoculated and 6 were left un-inoculated as a control. Figure 3.1 has been simplified (no error bars) for presentation as an example but the variation was such that by the time 2012 arrived, the inoculated sites had significantly higher concentrations of the limiting resource of NO_3.

To be more realistic, we should acknowledge that populations will in fact be limited by some factors – resources, diseases, random events. Considering this, there is a fundamental equation – the Verhulst equation – that expresses population dynamics as the population change based upon the interaction of the maximum population growth rate and resource limitation, as represented by the carrying capacity. One version of the equation can be symbolized as follows:

$$N_{t+1} = rN_t(1 - N_t/K)$$

This creates a curve that is also known as the logistic equation – it is a sigmoidal shape that shows a rapid growth phase that is truncated at an asymptote. The example shown in Figure 3.2 is a more realistic one than is often shown in textbooks where there is a perfect logistic curve – that really never happens with real data. The example is still a simple one where fungal

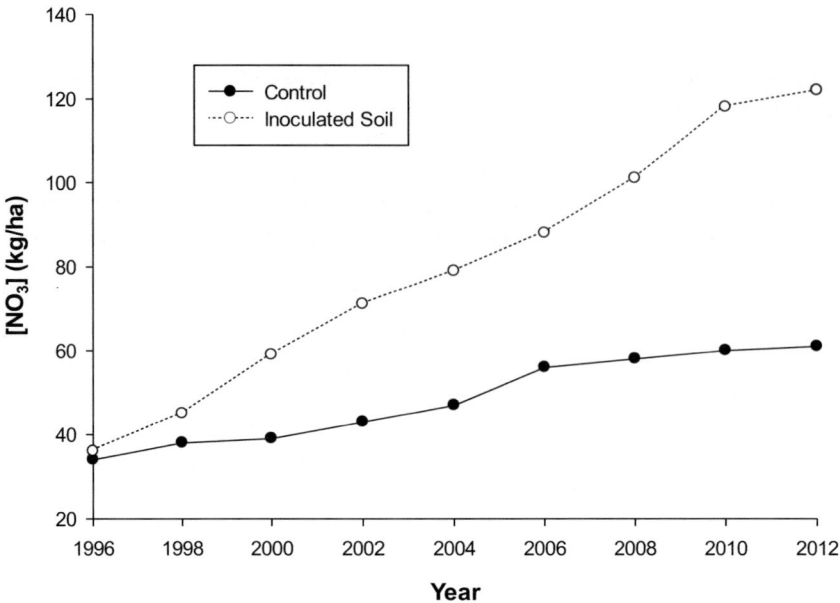

Figure 3.1   The contrast between the amount of NO$_3$ accumulated in soil that was/was not inoculated with symbiotic and transformational bacteria and fungi

Source: Murphy (unpublished data)

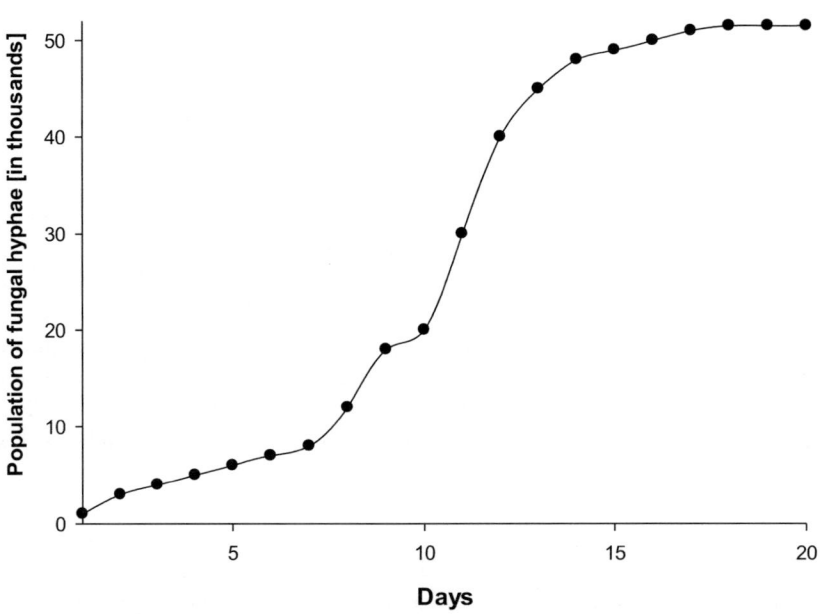

Figure 3.2   Logistical growth in a population of fungal hyphae grown in petri dishes

24

hyphae were grown in petri dishes from an initial population of 1 hypha (and spores that will produce hyphae since they are given nutrients in the petri dishes); not all species of fungi will show logistical growth in their hyphae but this species – *Ischnoderma resinosum* – does (under laboratory conditions, at least).

The asymptote (here at about 51,000 hyphae between 15–20 days) occurs when the carrying capacity is reached; again, reaching carrying capacity is caused by the real-world limitations like resource scarcity or disease – in this example, it was because the nutrients provided were nearly used up and so was the space available.

The theoretical distribution does not allow the population to exceed the asymptote at *K*; in reality, populations can overshoot this theoretical cap and they may even 'crash' to a much lower population if they die en masse in a short period of time.

For a restoration ecologist, this means we need to consider how many restored populations – and their size – an ecosystem can support. If we have too much of a good thing – introduce too many organisms – we might exceed *K* and create a serious problem if the population crashes because the excess over-consumed resources at a level that precludes recovery of the population and/or affects populations of other species. It also means that some of the populations we restore will be ones that will need to be able to compete with the faster growing *r*-selected populations; often, we wait until the *r*-selected populations have re-established ecosystem conditions that support the *K*-selected populations to avoid that competition and possible thwarting of restoration in the typical ecosystem where there are more *K*-selected populations as an ecosystem matures through time.

Even in harsh conditions where one might expect there to be selection pressures for *r*-selected species, the strategy of restoring with *r*-selected populations first and waiting several years to restore the *K*-selected species works. Our research group has done this in recently abandoned farmlands on sandy soil where water and nutrients quickly are lost once farming stops (Murphy *et al.*, unpublished data). The best approach is to inoculate with fast growing populations of micro-invertebrates, bacteria, fungi, and other protists to re-establish the nutrient and water cycles before the sandy soil erodes or crusts. Once accomplished, *r*-selected grasses and herbs are seeded and transplanted to provide a fast-growing population of plants to anchor the soil, and begin returning carbohydrates to the protists and micro-invertebrates via symbiosis or decomposition. The plants feed the soil organisms and then the soil organisms recycle the nutrients so new populations can grow.

Some minor paedogenesis may occur but the basic outcome is that nutrient and water cycles are restored and the soil becomes near capacity for resources. This will allow more *K*-selected plant species to colonize if they are near enough or to be seeded or translocated. During this time, macroinvertebrates and perhaps vertebrates will begin to colonize or be able to be restored by human intervention. This will normally be an accelerated process assuming ecological restoration is implemented – rather than just waiting for recolonization from nearby source populations.

If we had simply restored all types of populations at once, failure would likely have been the outcome because the *K*-selected species would not have sufficient resources to survive long-term but probably would survive long enough to reduce the ability of *r*-selected species to acquire sufficient resources to survive long-term. Another possibility is that simultaneously saturating a site like this with *r*- and *K*-selected populations will result in some surviving *r*-selected populations but these could be undesirable exotic species able to withstand harsher conditions of resource limitation and initial competition with *K*-selected species. We have not experienced this type of failure ourselves but have monitored sites where this has happened – in contrast to successful ecological restoration (Figure 3.3).

Figure 3.3 On the left is a successful restoration after 10 years of staggered introduction of *r*- and *K*-species' populations. On the right is a failed restoration ecology design that tried to introduce all species simultaneously – the result was dominance by exotic species like *Alliaria petiolata* after 10 years

Aside from the fact that both models maintain all variables as constants and don't yet account for the complexity of birth, death, immigration rates, and emigration rates (temporal variation), they are also oversimplified because they don't account for spatial variation. Population dynamics are influenced by the size of their habitat and the connectivity between habitats, where generally larger and better connected habitats will increase the potential population size. This is why restoration ecologists usually focus on what are called 'spatially explicit population models'. These can be quite simple in that they can represent a geographical area as a simply polygon like a rectangle that is composed of smaller rectangles ('cells') as shown in Figure 3.4.

The one shaded area represents either an individual organism or it could represent a whole population. Modelling population dynamics then depends on the factors already discussed now that spatial dispersion is added into the model. This could be still quite simple – any individual or population can move anywhere in two-dimensional space. It could be a bit more complex and realistic – any individual or population can only move certain distances, certain distances within a certain time, certain directions, or under certain conditions (like it can only move to unoccupied cells). We can then add more conditions – movement is only allowed to

Figure 3.4 A simple 'cellular automata' population surface – two dimensional, simple polygons, and the 'automata' term means that we could program the computer to allow the shaded area to move next to any of the unoccupied cells at any time

cells where the habitat is suitable for survival of a given population or individual. And we can go further still – the space is a two-dimensional complex polygon or it is an actual map of a geographical space (usually mapped with GIS), or it is a three-dimensional map of actual geographical space because there will be abilities to move underground, up into trees, below the water and so on. Our population or individual moves through the world that is familiar and realistic, as in Figure 3.5.

The individual-based models are best because each individual will encounter a range of selection pressures that will shape the population as a whole but even with modern technology of genetic markers, barcoding tags and drones, it can be difficult to do this if populations are large and dense and if their life cycles are complex (e.g. not r-selected or at least not annual, where all adults die after only one year at most).

We also have to account for the reality alluded to earlier in discussing the Leslie and Leftkovich models – that there will be differential birth, death, immigration, and emigration rates between each cohort of populations. Yes, there will be constraints on how much variation exists – because there is a fundamental value of r – but unless one is working in a

Figure 3.5 An actual aerial photograph of a landscape matrix. The 'cells' are now a mix of regular and irregular polygons (the farms are more regular – the yards of the housing development are surprisingly irregular because these are estate-type developments), there are physical barriers like roads to movement of some organisms, and the ecological conditions of each part of the matrix will be variable between locales and across time. The ability to restore a population (or several populations) of a species if a farm should be abandoned and restoration desired will be challenged by the complexity of migrations or other interactions of movement of individuals between populations or habitats

laboratory under carefully controlled environmental conditions, each individual in each cohort (age- or stage-based generations within populations) will experience differential selection pressures, causing whole-cohort differences in population dynamics.

This is why population models normally use some form of sensitivity analysis. We test to determine which variables – like birth, death, immigration, or emigration rate – change our population models the most; we determine which variables the population model is most sensitive too. In simpler models, we simply can change these variables across a large range to test sensitivity. We usually constrain that range to ecologically realistic conditions (e.g. if we have never recorded greater than 20 per cent of the population dying in a given time period, that may be considered our worst-case scenario).

However, in restoration ecology, we are usually not dealing with historical or typical conditions existing at the time of restoration so we usually include the extreme values of population variables as best- and worst-case scenarios. Somewhere in between best and worst, we would like to determine the most probable set of circumstances under conditions of restoration versus no restoration. That can be difficult unless we have experience with a given type of ecological restoration already.

Still, a population model will allow us to understand how fast we might expect a restored population to grow, to set goals to maintain a certain size of population over time, and to act if we see populations getting too big or too small as we monitor an ongoing restoration project. The last one is challenging because how do we know when to intervene during restoration – populations might be able to rebound if there is a crash? Once again, our ability depends on experience and early experimentation; this is why we do smaller scale experiments to determine fundamentals of the population under restoration and non-restoration conditions if we have no experience or comparative, reliable, accessible, and appropriate case examples.

In using population models in restoration ecology, we also have to be cognizant of how the variables are constructed. By that, we mean what do the variables represent and what is their mathematical expression. If we are measuring survivorship and fecundity as indicators of success or failure in restoring populations, we have to be careful about including these blithely in a single population model. Survivorship ranges from 0.0 to 1.0 (0% to 100%). Fecundity – especially in *r*-selected populations – might be in the millions numerically. This will distort the relative importance of the two variables of fecundity and survivorship because their mathematical expressions are different – a percentage versus a raw number that grows exponentially. Survivorship is actually more important in the sense that it tells us the outcome of a cohort after restoration and indicates the potential intrinsic growth of a population in the next cohort – granted that emigration by natural means of dispersal or via more restoration intervention will affect the actual outcome. Thus, we test the reality of our population models by measuring elasticity – what is the proportional effect of a variable caused by its mathematical expression and what should be the proportionate effect based on demographic and perhaps ecological importance. Here too the problem of life cycles arises because elasticity will not be constant cohort to cohort.

The formulae for sensitivity and elasticity analysis often are expressed in the mathematical notation of matrix algebra, and can look quite intimidating or cryptic. Expressing these in a wordier but perhaps more tenable fashion, we can calculate sensitivity as:

- $s = d_{\text{population growth rate}} \, / \, d_{\text{(population variable)}}$

The 'd' symbolizes that this involves calculus – we calculate the partial derivative of a population's growth rate as it changes with a given population variable. Examples:

- $d_{\text{population growth rate}}$ / $d_{\text{birth rate}}$
- $d_{\text{population growth rate}}$ / $d_{\text{survivorship}}$
- $d_{\text{population growth rate}}$ / $d_{\text{emigration}}$

We can test how s (sensitivity) varies with each variable above.

We need to determine sensitivity because we then calculate elasticity with that value. The simplified formula for elasticity (e) is:

- $e = s \times$ (population variable measured/population growth rate)

This means that elasticity measures the proportional sensitivity of a given population variable.

Remember that we would compare sensitivity and elasticity for all population variables that we measured to understand how the proportional importance/effect of each variable is exaggerated by our analysis, which ones are under-estimated, and which ones are basically accurate.

Examples of population models used in attempts to restore populations

The population models can get more elaborate but they can become so complex as to be rendered intractable even with modern computing knowledge and technology. However, there are a series of useful elaborations of the basic population models, starting with the exponential and Verhulst logistical model.

One good example is from Cromsigt *et al.* (2001) wherein they compared the utility of several still reasonably simple population models in determining impacts of using source populations of black rhinos that were reasonably stable to restore populations in other areas. They compared how many black rhinos would be needed for success and what translocation of the various numbers of rhinos would do to the source population.

Specifically, it had been determined that in the mid-1990s, the total worldwide black rhino population was 2500 individuals. About 1050 of these were in South Africa where poaching was less common; the rest of Africa had scattered and smaller sub-populations. To restore and thus conserve the species by restoring other African sub-populations, translocating rhinos from South Africa was proposed but this would only work if the source population was really as large as 1050 individuals. There was concern that it may not have been accurate because of risks of double counting when aerial surveys are done and the very issue that there are low population numbers meant expenses in finding them, resulting in high yearly variance in estimates of N – much less anything like N_e.

This study used an approach that almost any reasonably educated student in restoration ecology could do by second or third year – they used Microsoft Excel to determine and minimize the errors between census data and their population model data – they basically minimized the sums of squares to generate best estimates of initial size N that will then be used to predict all $N(t_i)$ values. This was to account – as best they could – for any census errors. They then compared several population models:

- exponential model: $N_{t+1} = N_t + rN_t$
- Verhulst logistic model: $N_{t+1} = rN_t(1 - N_t/K)$
- Fowler's model: $N_{t+1} = rN_t(1 - [N_t/K]^n)$
- Verhulst logistic model with translocation of 'h' numbers of rhinos: $N_{t+1} = rN_t(1 - N_t/K) - h$
- Fowler's model with translocation of 'h' numbers of rhinos: $N_{t+1} = rN_t(1 - [N_t/K]^n) - h$

'*h*' represents the number of rhinos that would be translocated – it can vary from 0 to all of the rhinos available but the realistic numbers would be at least in the double digits so the new population would survive.

'*n*' is used to model the real-world situation where as a population of black rhinos gets closer to its carrying capacity, *K*, the population is increasingly affected by density of rhinos; it is a way to express that there is a density-dependence that alters population dynamics but that dependence is not usually constant (if it was, then $n = 1$ and this reduces back to the simpler Verhulst logistic equation because $x^1 = x$).

What they found was that the basic Fowler model was the most applicable and accurate but that adding in the even more (potentially) accurate variable of number of translocated individuals only worked for one of the two game reserves they studied. What this told them was that when the translocation variable was not important, the population census was over-estimated because it should have made a difference and improved accuracy of the population model. They also found that the exponential model was almost as good as the Fowler model in the same game reserve where adding the translocation variable did not improve accuracy. This may mean that the population there is still growing as opposed to being near *K*.

Generally, they found that each model gave quite different values for the maximum population size – the logistical model indicated that up to 50 per cent of the rhinos could be removed (and translocated) whereas the Fowler model indicated that only 10–15 per cent of the rhinos could be removed (and translocated). Given that translocation can harm the source population if it is too large and insufficient translocation can result in failure in restoration for the target/sink population, this is a very important issue – make a mistake and both source and sink populations of rhinos might drop and you might just become the cure that was worse than the problem. Implicitly, Cromsigt *et al.* (2001) did consider the gender and age variables as well but the basic question of numbers to be translocated was the main issue.

The problem of sufficient translocation to a new target or sink population and the corollary of ensuring sufficient individuals remain in the source population relates to the problem of estimating the minimum viable population (MVP) size – how many organisms are needed to maintain a population that can successfully produce new generations of fertile offspring that continue to survive. Once again, this number should be tempered by our knowledge of the breeding system – if there is (as usual) a minimum and maximum age or stage threshold for reproduction and if sexual reproduction (aside from self-fertilization) is important, then once again we would want to know the extended N_e. For MVP calculations, one needs to know how much genetic variation exists, how much of this is expressed, whether the genetic and phenotypic expressions are relatively equally distributed and how this is influenced by natural selection, genetic drift, patch size and proximity effects and the probability of stochastic or deterministic factor.

Interestingly, we can provide a 'ballpark' estimate for entire groups of organisms, based on empirical work experience. For example, in using plants to help restore habitats, we tend to harvest and translocate 200+ seeds from a random sample of all dispersal agents (like seeds) except for vertebrate dispersal agents where translocation of 200–500 seeds is a typical range because animals tend to collect them from the same plant (Falk *et al.* 2006). This is not as theoretically sound as we would like and certainly caution is urged until replicate studies that formally study MVP are completed for a given species–environment combination because demographic and environmental variation will affect the number to be translocated.

Finally, we also would like to know whether a population is likely to go extinct sooner rather than later or what the long-term probability of extinction of a population or entire species might be. For this, we turn to population viability analysis (PVA). This relies upon most of the variables and processes/analyses we already have discussed so an example should help illustrate how this works. Ferreras (2001) and Ferreras *et al.* (2001) studied the Iberian lynx. While we are simplifying the study here, he basically showed that lynx habitat was critically low and that the small N and N_e of isolated populations of lynx meant that translocation was needed. Lynx can be too sensitive to allow for successful human transport, hence translocation would be encouraged by some type of habitat restoration within the agricultural landscape and then reconnecting restored or existing habitats via corridors (this steers us into some landscape ecology elements but that is inevitable). However, this begged the question of whether some other management was needed (i.e. reduce mortality in lynx).

Ferreras *et al.* (2001) asked just that broader question – their question focused on whether the goal should be to restore lynx habitat or reduce lynx mortality. They used PVA to model the risk of extinction of this species and then to determine which management options – restoration of habitat, reduced poaching, reduced road kills – were most effective. The standard approach – reduce human-caused mortality – did not sufficiently reduce the risk of mortality of lynx, hence habitat restoration was likely needed. Their study showed the nuances of management needed. Using restoration ecology techniques to improve K in source populations was effective at reducing extinction risk to lynx. Oddly, these same techniques were not very effective at improving K and reducing extinction risk if applied to sink populations unless those populations also experienced the total removal of all human-caused mortality – and that is not likely. The most effective method was still to increase connectivity between isolated populations, hence in that sense the outcome of effectively 'restoring' habitat by increasing connections between smaller habitats works because it increases one, again effective, larger habitat.

Summary

Overall, restoration ecologists have many tools at their disposal to examine the outcome and effectiveness of restoration as measured at population scales. There are challenges in terms of gathering sufficient and reliable samples and building population models that capture the range of variation in the expression and meaning of fundamental demographic variables of birth, death, immigration, and emigration. One must be careful in weighting variables (examine elasticity) and determining which ones are more important (examine sensitivity). As molecular scale tools improve, it may become easier to identify and classify the range of genetic and phenotypic variation within populations of different species and this can give restoration ecologists more confidence that their efforts at restoring populations will succeed in reconstituting the diversity needed to support a self-sustaining population that can cope with rapid or slow environmental changes.

Note

1   Often just '*n*' and '*N*' are used, but we prefer terms to be explicit.

References

Armstrong DP, Seddon PJ. 2007. Directions in reintroduction biology. *Trends in Ecology and Evolution* 23: 20–25.

Cromsigt JPGM, Hearne J, Heitkönig IMA, Prins HHT. 2001. Using models in the management of black rhino populations. *Ecological Modelling* 149: 203–211.

Falk DA, Richards CM, Montalvo AM, Knapp EE. 2006. Population and ecological genetics in restoration ecology. Pages 14–41 in Falk DA, Palmer MA, Zedler JB (eds), *Foundations of restoration ecology*. Island Press, Washington, DC.

Ferreras P. 2001. Landscape structure and asymmetrical inter-patch connectivity in a metapopulation of the endangered Iberian lynx. *Biological Conservation* 100: 125–136.

Ferreras P, Gaona P, Palomares F, Delibes M. 2001. Restore habitat or reduce mortality? Implications from a population viability analysis of the Iberian lynx. *Animal Conservation* 4: 265–274.

Hufford KM, Mazer SJ. 2003. Plant ecotypes: genetic differentiation in the age of ecological restoration. *Trends in Ecology and Evolution* 18: 147–155.

McKay JK, Christian CE, Harrison, S, Rice KJ. 2005. 'How local is local? A review of practical and conceptual issues in the genetics of restoration. *Restoration Ecology* 13: 432–440.

Montalvo AM, Williams SL, Rice KJ, Buchmann SL, Cory C, Handel SN, Nabhan GP, Primack R, Robichaux RH. 1997. Restoration biology: a population biology perspective. *Restoration Ecology* 5: 277–290.

Rice KJ, Emery NC. 2003. Managing microevolution: restoration in the face of global change. *Frontiers of Ecology and Environment* 1: 469–478.

Rose KA, Sable S, DeAngelis DL, Yurek S, Trexler JC, Graf W, Reed DJ. 2015. Proposed best modeling practices for assessing the effects of ecosystem restoration on fish. *Ecological Modelling* 300: 12–29.

Weeks AR, Sgro CM, Young AG, Frankham R, Mitchell NJ, Miller KA, Byrne M, Coates DJ, Eldridge MDB, Sunnucks P, Breed MF, James EA, Hoffmann AA. 2011. Assessing the benefits and risks of translocations in changing environments: a genetic perspective. *Evolutionary Applications* 4: 709–725.

4

LANDSCAPE-SCALE RESTORATION ECOLOGY

Michael P. Perring

Introduction

For restoration ecologists to successfully achieve their goals, they need to be mindful of the landscape scale (i.e. the processes and patterns in the matrix beyond a target patch, and the abiotic and biotic responses to these properties). In this chapter, I outline fundamental ecological reasons why this is the case. In addition I explain how land degradation, environmental change and biodiversity loss have led to targets that imply broad-scale restoration, and elucidate practical developments, in the planning stages and 'on-the-ground', that are aiding restoration at the landscape scale. Throughout my elaboration, it will be apparent that restoration at that scale poses challenges: challenges to prioritizing and cost-effectively implementing schemes given competing demands for space and limited resources; challenges for those attempting to restore systems in the field; and challenges to communities, governments and other stakeholders as to what landscape-scale restoration can aim for in the changing environments of the Anthropocene. At the same time, landscape-scale restoration offers opportunities to bolster natural, social and economic capital. In my opinion, these opportunities will be most effectively realized when restoration is carried out with a future-oriented view that is mindful of history, and that works with, rather than against, the context of changing environments and the needs of a growing global human population. I don't believe it will be sufficient for restoration to act in a piecemeal way (i.e. restoring patch by patch in a manner that ignores surrounding and potentially competing land uses), even when using forward-looking principles. To fully grasp opportunities, restoration needs to work with a multi-functional landscape perspective which will involve collaboration and co-operation with multiple, potentially conflicting, stakeholders. This chapter aims to inform the reader as to the why and wherefore of landscape-scale restoration.

The ecological imperative for a landscape perspective to restoration

One of the early proponents of restoration ecology as a recognized discipline stated that the acid test of ecological understanding would be our ability to restore functioning ecosystems (Bradshaw 1983). At that time, restoration was a site-based, predominantly plant-focused discipline (Young 2000) with little appreciation of any need to consider directionally changing environmental conditions or global biodiversity loss (Perring *et al.* 2015). Since that time, there has been increased awareness of the scale of land degradation and land clearance for agriculture and urban development, greater concentration on the magnitude and rate of environmental changes such as nitrogen deposition and climate variation, and confirmation of large losses of unique elements of biodiversity from across the globe (Vitousek *et al.* 1997). The concern that these changes could lead to impaired human well-being through degradation of ecosystem services (Millennium Ecosystem Assessment 2005), as well as adverse effects on biota more broadly, has led to ambitious global restoration targets, including Aichi Target 15 (restoring more than 15% of degraded ecosystems by 2020). The practical means of achieving these commitments through, for example, the Bonn Challenge (www.bonnchallenge.org), has led to the development of initiatives such as the Global Partnership on Forest Landscape Restoration, and calls to commit to the values of ecological restoration (Suding *et al.* 2015). Although these factors give proximate reasons for restoration at a broad scale, targets could arguably be achieved through cumulative site-by-site restoration (i.e. patch-based approaches). Patch-based approaches, although sometimes successful (Holl and Crone 2004), ignore fundamental ecological reasons for restoration at the landscape scale – in other words, the ecological imperative for a landscape perspective to restoration.

There are at least four inter-related fundamental ecological reasons for considering the landscape in restoration. First, the landscape context of any restored area for the functioning of an ecosystem matters because an ecosystem is not a closed entity. Second, the overall habitat cover in any given landscape likely affects the ability of any restored area to support viable populations of fauna and flora. In combination, these two reasons suggest that those undertaking restoration need to think about broader landscape properties in their efforts to maintain evolutionary potential, including whether habitats of differing age are required for organism survival. Third, certain ecologically important processes (e.g. dispersal and disturbance) act at the landscape scale and their successful reinstatement necessitate a landscape perspective. These three elements would need to be considered in a temporally unvarying environment (i.e. without directional change); the reality that ecosystems are being affected by multiple global environmental changes adds a fourth reason to consider the landscape in restoration: movement of biota in response to these environmental drivers will be impeded or aided depending on the environmental matrix. In a 'hostile' matrix for any given species, restoration would need implementing to allow movement. The characteristics of this restoration would need to keep the changing environment and the landscape context in mind.

Landscape context and restoration

In examining the landscape scale, which by definition incorporates processes across spatially defined mosaics and the abiotic and biotic responses to these processes (Bell *et al.* 1997), interactions between process and pattern need considering. Not only can process lead to patterns, but patterns themselves can influence process. In other words, the spatial patterning of ecosystems across landscapes can have ecological effects (Turner 1989). Effects of the landscape on an ecosystem occur because a functioning ecosystem is not a closed entity: energy and matter flow in, around, and out of systems, as organisms pursue their livelihoods, and climatological and geomorphological processes play out. This openness means that the broader landscape context will affect how a restored ecosystem functions and its trajectory through time (Ehrenfeld and Toth 1997). For restoration, this means that the target of restoration, and the expectations of what constitutes a 'properly' functioning system, will likely differ depending on whether it is surrounded by remnant vegetation, or it is in an agricultural mosaic with plentiful remnant fragments and hedgerows, or clearly isolated by being situated in an intensively used and cleared agricultural landscape or surrounded by a built urban environment (Bell *et al.* 1997).

The ecological importance of spatial landscape context has been highlighted in meta-population research (Hanski 1998), debates about single large or several small reserves (McCarthy *et al.* 2011), and the potential of habitat corridors to influence connectivity and thus boost species richness (and by implication maintain ecosystem function) (Damschen *et al.* 2006). Landscape corridors have received a lot of attention, from a theoretical (Earn *et al.* 2000) and practical perspective (e.g. Damschen *et al.* 2006) and for their potentially negative ecological effects (Haddad *et al.* 2014). These debates have often been couched in the context of conservation and land fragmentation. Restoration ecology can learn lessons from them, since restoration is an additional tool in a conservationist's toolbox given reserves will likely not be sufficient to avert the loss of global biodiversity (e.g. Laurance *et al.* 2012). Indeed, ideas in restoration have recently shifted from consideration of corridors to the importance of the landscape matrix and how stakeholders manage the whole landscape (Hobbs *et al.* 2014; Lindenmayer *et al.* 2008). There will be context specificity as to what arrangement of habitats will work best for particular species and/or particular restoration goals, and how the matrix will influence outcomes. However, Driscoll *et al.* (2013) suggested there are three core matrix effects that can generally be considered in influencing ecological outcomes: movement and dispersal, resource availability, and the abiotic environment. They argued that these effects can be modified by five dimensions – the spatial and temporal variation in matrix quality, its spatial scale, the longevity and demographic rates of species relative to the temporal scale of matrix variation, and adaptation (*ibid.*). These matrix properties will influence the success of restoration initiatives and demand that restorationists consider the landscape scale.

Despite the predominant and understandable focus on spatial relations across a landscape the temporal dimension needs to be clearly acknowledged in its own right. In the extreme, patches may have a similar current landscape context, but their history can be markedly different which in turn will lead to different contemporary composition and dynamics (Ramalho and Hobbs 2012) and thus potentially alternative appropriate restoration interventions. Ecological memory and land use history can have a marked effect on contemporary ecosystem properties (Foster *et al.* 2003; Ogle *et al.* 2015), with directional environmental change potentially interacting with these legacies of the past to influence future ecosystem dynamics (Perring *et al.* 2016; Ryan *et al.* 2015). These facets suggest that restoration ecologists need to be mindful of the ecological history of a landscape when determining appropriate goals and in attempting to understand ecosystem function, as well as valuing history for reasons highlighted by Higgs *et al.* (2014).

Landscape thresholds and restoration

One particular landscape context variable that is worthy of further elucidation is whether thresholds exist that mark a boundary between the ability of a landscape to support species or ecosystems, or not. At ecosystem and regional scales there is clear evidence for threshold responses in populations, organisms and ecosystem properties to land cover changes or environmental factors e.g. lakes suddenly turning turbid in response to increasing nutrient loads (Carpenter 2005). For land cover, most research has concentrated on the likelihood of habitat fragmentation driving threshold changes in ecosystem properties such as species richness or population sizes (see examples in Brook *et al.* 2013). In aggregate, this research suggests that at low levels of habitat loss, abundance will decline proportionally with the amount of suitable habitat in the landscape, providing connectivity is retained. However, non-linear changes (i.e.

threshold responses) emerge as patches become more isolated and patch size becomes too small to sustain a local population and rapid losses of abundance occur and thus eventually extinction. In general, whether or not linear change occurs across a range of variation, or threshold responses, depends on spatial homogeneity of the landscape, the interconnectivity of ecosystem responses and the spatial distribution of drivers (*ibid.*). Modelling and empirical evidence suggests that fragmentation thresholds for biotic variables, including phylogenetic integrity, can exist between 10 and 30 per cent of original habitat cover (Andrén 1994; Banks-Leite *et al.* 2014; Swift and Hannon 2010).

This (depressing) framing of thresholds in biotic response to habitat loss, leading to eventual extinction, can be reversed by considering (the positive message of) ecological restoration. Findings from the landscape threshold literature can be used to ask: Can habitat be restored to certain coverage levels in the landscape to maintain viable populations? Will this reinstate the potential for more resilient trajectories of ecosystem development and organismal evolution, albeit with the context of changed global environments? The answer to these questions may depend on whether hysteresis in response to previous fragmentation exists. If hysteresis is present, much greater levels of habitat coverage are required to enable future persistence of populations than the 10–30 per cent cover previously suggested to maintain viability. This may occur because current patterns following habitat destruction may reflect an unpaid extinction debt (Tilman *et al.* 1994); that is, the 10–30 per cent coverage value has been incorrectly specified. On the other hand, restored areas could take time to accrue populations due to immigration credit (Jackson and Sax 2010), such that given sufficient time, restoring habitat patches in the landscape to a lower aggregate level than at first appears necessary could allow the re-establishment of evolutionary viable populations. The existence of an extinction debt, and whether restoration may prevent it being paid, could depend on the taxa under consideration and its associated abundance distribution, as well as the connectivity of the landscape (Kitzes and Harte 2015).

What is certain is that any landscape threshold figure will be context dependent from a landscape and species perspective. The tipping point, in either direction and providing one exists for any given organism or community, will depend on landscape characteristics such as the quality of the matrix, the suitability of patch habitat and patch turnover rates (Swift and Hannon 2010). Species themselves will respond differently to the creation of new habitat through restoration, depending on their ability to utilize and enter any matrix, dispersal and reproductive rates, and response to the restored habitat as it ages. Indeed, for certain species to persist, habitat of different ages may be required (e.g. old trees to provide nesting hollows for vertebrates with adjacent young stands providing cover and energy resources), and time lags associated with revegetation need elucidating (Vesk *et al.* 2008). For multi-species restoration, evidence is mounting that heterogeneous landscapes will support more viable populations of animals than homogeneous ones – for instance, the maintenance of different faunal species in dry forests in North America requires areas that have burned with different fire intensities or equivalent management surrogates (Fontaine and Kennedy 2012). The danger of simplifying the idea of thresholds to a simple binary yes/no answer to the question of whether there is enough habitat of a certain type in the landscape, has led some to recently argue that threshold modelling should predominantly be used for comparative exercises of area-sensitivity or the identification of environmental threats (van der Hoek *et al.* 2015). These authors argue that nominal amounts should only be used if thresholds are detailed, species-specific and translated to conservation targets particular to the study area (*ibid.*).

Landscape processes and restoration

The preceding two sub-sections have concentrated on how *patterns* in the landscape can influence ecological process, the persistence of species and the properties of ecosystems, and thus influence restoration decisions. There are also fundamental reasons to consider landscape-scale *processes* in dictating ecological dynamics in restored areas (Menz *et al.* 2013). Such processes include dispersal of biota (which is an oft-cited reason for implementing corridors), broad-scale disturbance, or the movement of water. This large-scale framing of processes has perhaps been particularly evident in the restoration of aquatic systems. Here, processes are considered from the scale of individual river reaches, to regional wetland systems (e.g. Culotta 1995; Gunderson *et al.* 1995). For instance, tackling aquatic dispersal to enable restoration requires more than considering problems created by dams. In the Great Lakes Basin of Canada and the United States, barriers created by road crossings, which were far more numerous than dams, had a large impact on the potential to restore fish communities. Only tackling the dams, in the absence of the broader landscape context of road crossings, would fail to provide sufficient restoration of connectivity to restore the target communities (Januchowski-Hartley *et al.* 2013).

Disturbance processes can also act at landscape scales. Implementing disturbance at such scales is arguably critical for the successful management of landscapes, as shown by fire mosaics in areas such as the Northern Territory of Australia (Bradstock *et al.* 2005) and dry forests of the United States (Fontaine and Kennedy 2012). Other landscapes that may benefit from the reintroduction or modification of disturbance regimes, or novel interventions that may be a surrogate for previous disturbances (Shackelford *et al.* 2013), include pine barren ecosystems (Radeloff *et al.* 2000) and rangeland ecosystems (Cingolani *et al.* 2013). Implementation of appropriate disturbances may also ultimately improve the resilience of systems to environmental changes and prevent ossification (Harris *et al.* 2006). Indeed, changed disturbance regimes and altered groundwater can be considered as environmental alterations in their own right, and such changes, together with those of other environmental drivers, provide the fourth fundamental ecological reason to consider the landscape in ecological restoration.

Environmental change and landscape-scale restoration

The current rate and extent of environmental change also demands that the landscape scale be considered in restoration for fundamental ecological reasons. Although the connectivity created by landscape-scale restoration is desirable in the absence of a directionally changing abiotic environment, rising temperatures, heightened nitrogen deposition and increasing atmospheric carbon dioxide concentrations (to name a few) demand that organisms have the capability to move, or adapt or potentially face extinction, in response to environmental change. Although movement may be achieved through managed translocations to new habitat in areas that are presumed to be suitable into the future, and allow persistence at the time of translocation (Vitt *et al.* 2009), increasing connectivity through the landscape potentially allows organisms to independently respond to changes. This rationale at least partially underlies restoration schemes such as Peniup in Gondwana Link in south-west Western Australia (Jonson 2010) (Figure 4.1a), and conservation initiatives such as Yellowstone to Yukon (Figure 4.1b). In the absence of such connectivity, and without more drastic interventions, it is unlikely that species will be able to migrate in some heavily cleared landscapes (Renton *et al.* 2012).

Figure 4.1 Examples of ambitious landscape-scale restoration endeavours: (a) Gondwana Link in south west Western Australia and (b, opposite) Yellowstone to Yukon (Y2Y) in western North America. Such initiatives are becoming more common as the ecological imperative for landscape-scale restoration, explained in the main text, is becoming more widely acknowledged

Notes: Gondwana Link has strategically purchased properties through different non-governmental organizations (e.g. Bush Heritage, Greening Australia) while also encouraging conservation covenants on privately owned land, to enable ongoing restoration and protection of a swathe of Australian bushland, from Margaret River to Kalgoorlie. Y2Y aims to protect and connect the system of wild lands and waters stretching from Yellowstone to Yukon, dividing the area into core areas and linkage zones, while meeting the needs of people and nature (see http://y2y.net). Initiatives include road bridges (inset) and tunnels to improve landscape connectivity.

Sources: (a) Rowan Woods, Australian National University; (b) Parks Canada Agency

Implementing restoration at the landscape scale

Lessons from fundamental ecology and the reality of multiple environmental changes dictate that the landscape scale be considered when implementing restoration. As these lessons have been taken on board by restorationists (whether academic or practical, and sometimes one and the same), and restoration targets have become global in extent, more research has focused on implementation strategies for restoration at landscape scales. I go on to explore conceptual and technological developments in the design, prioritization and cost-effective implementation of restoration across the landscape. I also discuss the incorporation of multiple stakeholder perspectives in restoration, given that restoring land necessarily involves land-use change decisions potentially involving multiple actors.

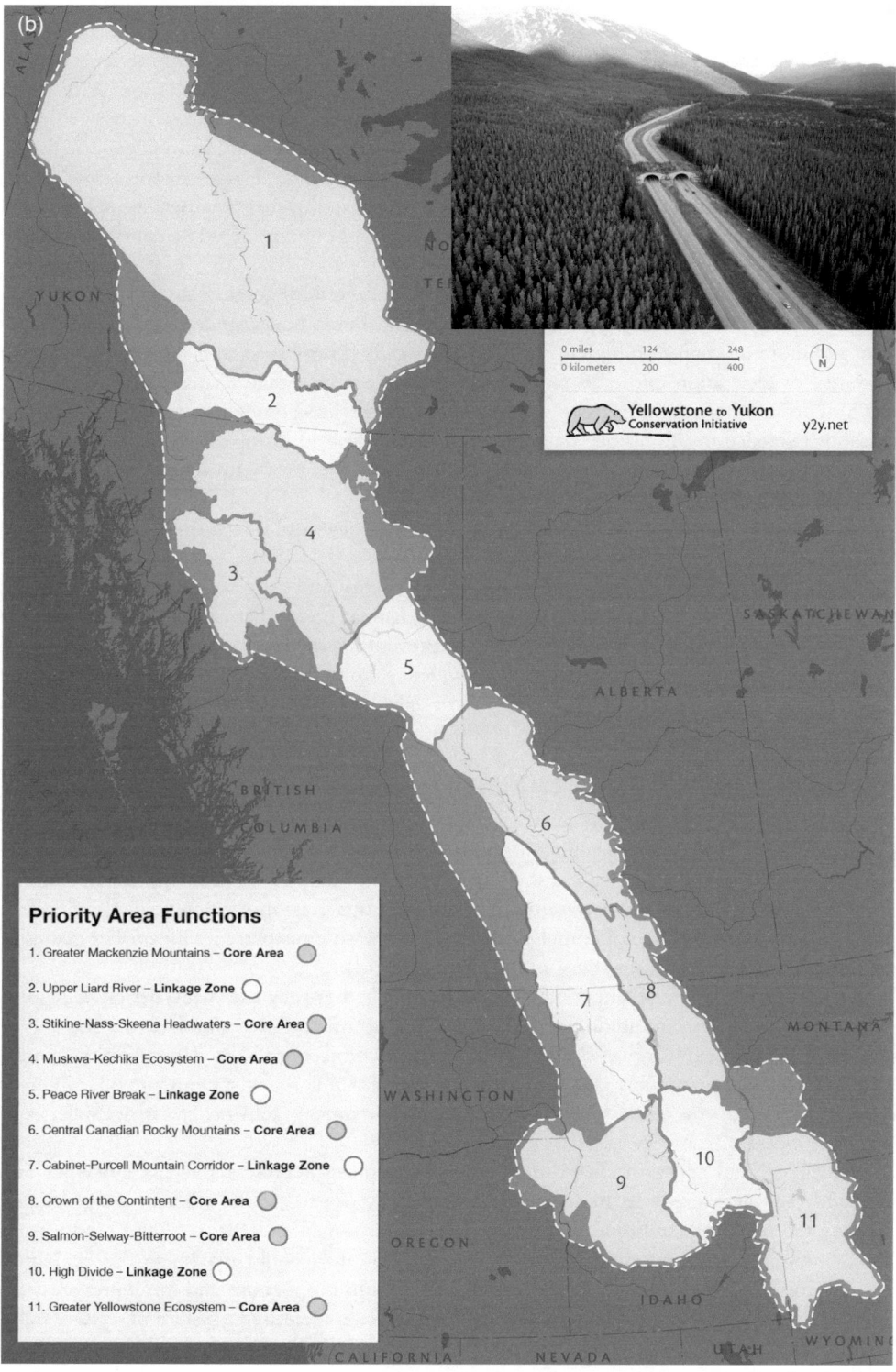

Priority Area Functions

1. Greater Mackenzie Mountains – **Core Area**
2. Upper Liard River – **Linkage Zone**
3. Stikine-Nass-Skeena Headwaters – **Core Area**
4. Muskwa-Kechika Ecosystem – **Core Area**
5. Peace River Break – **Linkage Zone**
6. Central Canadian Rocky Mountains – **Core Area**
7. Cabinet-Purcell Mountain Corridor – **Linkage Zone**
8. Crown of the Contintent – **Core Area**
9. Salmon-Selway-Bitterroot – **Core Area**
10. High Divide – **Linkage Zone**
11. Greater Yellowstone Ecosystem – **Core Area**

Figure 4.1   (b)

Design and prioritization of restoration across the landscape

As debate in conservation and restoration has moved on from single large or several small reserves, to habitat corridors and the condition of the landscape matrix, questions have arisen as to how to design and prioritize restoration across the landscape. Clearly, answers to these questions will depend on the specific goals of any given project, but at least one generic design principle includes consideration of likely future environmental changes in species selection. Species selection will potentially be aided by trait-based approaches to target species that will respond well to current and potential future environmental conditions, while delivering desired ecosystem functions (Laughlin 2014).

Once species are selected, propagules need to be collected for active restoration. It has been suggested that in the design of projects, seed sourcing should concentrate less on local collection and more on capturing high quality and genetically diverse seed to maximize the adaptive potential of restoration efforts to current and future environmental change (Broadhurst *et al.* 2008). Further, these authors argue that the use of generalized guidelines for seed movement without reference to life history traits, spatial distribution, and historical factors will continue to restrict restoration success through poor management decisions with respect to seed collection and deployment.

Once species are chosen and propagules collected, they could be introduced using island design principles, rather than actively restoring across whole landscapes (Rey Benayas *et al.* 2008). Such an approach is grounded in nucleation theory, and suggests biodiversity outcomes at the landscape scale can be enhanced by creating habitat nodes that can then spread into the surroundings, provided that thresholds that may prevent natural recruitment are overcome (e.g. Standish *et al.* 2007). Recent experimental evidence from tropical forests suggests that such 'applied nucleation' is a promising strategy to accelerate forest recovery to a similar degree as plantation-style restoration (Zahawi *et al.* 2013), while good results have also been demonstrated in Mediterranean agricultural systems (Rey Benayas *et al.* 2008). The idea of using islands in restoration design has also permeated grassland projects in Australia. However, in this situation, high nutrient levels in surface soils that are preventing desired species recovery are removed through progressive scalping of soil layers. Scalped soil is formed into islands with the deepest layers now at the surface and sown with complex mixes of broadleaf species. In contrast, the scalped matrix areas can be sown with complex native grass mixes potentially leading to a restored grassland landscape of complex native grassy sward interspersed with smaller islands of broadleaf complexity (Gibson-Roy and McDonald 2014).

The implementation of all projects, regardless of whether they use island design principles, needs to consider current abiotic and biotic constraints to restoration both at the site and in the wider landscape (Hulvey *et al.* 2013). The design of projects will also depend on the condition of surrounding systems and the uses that they are put to, as systems provide different benefits, as well as sometimes delivering unwanted outcomes (Hobbs *et al.* 2014; Zedler *et al.* 2012).

Principles from conservation planning, such as adequacy, representativeness, efficiency and flexibility are increasingly harnessed with computing power to aid the efficient design of, and identify the need for, landscape restoration projects. Crossman and Bryan (2006) used integer programming to demonstrate a proof-of-concept that they could maximize the ecological benefit from landscape restoration by satisfying minimum proportions and minimum areas for each environment type. Interestingly, their optimal solution identified a system of scattered sites with an inadequate landscape structure. To overcome this, they suggested 'impedance surfaces' of existing native vegetation, riparian areas and road corridors be used to preclude the model from selecting low priority sites (*ibid.*). In later work, they integrated many spatial datasets and models describing elements of natural capital, the degradation of stocks, and economic value, and identified cost-effective hotspots to restore natural capital and enhance landscape multi-functionality (Crossman and Bryan 2009). Restoration needs across a forest landscape were also identified with the aid of satellite mapping, using a combination of the natural range of variability (NRV) due to past disturbance, and current forest structure in different biophysical settings across the Pacific Northwest of the United States (Haugo *et al.* 2015). Although there are limitations associated with the NRV concept given changing environmental conditions and the need to manage federal forest for more than just ecological values, this approach can potentially facilitate the ability of local land managers to incorporate a regional-scale, multi-ownership context into forest management and restoration while future ranges of variability can be analysed in more forward-looking projects (*ibid.*).

The resources for restoration are not limitless, so projects require prioritization. Prioritization will at least partly depend on land being made available for restoration, the vagaries of human nature and, even in countries with strong governance, political whim as to the demand for restorative action. However, techniques exist to inform restoration priorities, or at least clarify decision-making processes. As well as those already discussed (e.g. Haugo *et al.* 2015), hierarchical approaches that consider geomorphic context and rarity of ecosystems can be used in conjunction with assessments of the viability of restoring disturbed systems (Palik *et al.* 2000), while fixed budgets can be incorporated into prioritization decisions based on landscape conditions and likely restoration effectiveness (Wilson *et al.* 2011). Recent work has shown that these ecological approaches need to be combined with land parcel cost estimates to provide a return-on-investment framework to landscape-scale restoration (Torrubia *et al.* 2014), since prioritizing actions without taking costs into account can lead to inefficient outcomes (Ando *et al.* 1998). In Torrubia *et al.*'s (2014) analysis, restoration priority was determined iteratively for Washington ground squirrels by identifying sites with the highest improvement score for enabling connectivity and increasing matrix permeability. In a second analysis, they selected barriers with the largest improvement score but only per dollar of combined purchase and restoration cost. In comparing these two scenarios, they showed that spending a fixed amount on restoration that is prioritized with land and restoration costs in mind would yield 36% more connectivity benefit, and, because of lowered parcel costs, twice as much area restored than scenarios that only considered the ecology (*ibid.*). These approaches could be applied more broadly for other target species or ecosystems and allow efficient prioritization of restoration activities at the landscape scale.

Cost-effective 'on-the-ground' implementation

Restoration at the landscape scale needs to be implemented cost-effectively given limited resources. This imperative lies behind the prioritization approaches elucidated previously. However, restoration implementation on the ground at the landscape scale also demands that costs are minimized as far as possible to maximize the return on investment. Although some projects can turn to large pools of human labour to implement restoration at scale (with varying degrees of success) – examples include China's Grain for Green Project (Cao *et al.* 2009) or South Africa's Working for Water program (McConnachie *et al.* 2012) – and volunteers are a crucial component of the restoration economy (Cunningham 2002; Egan *et al.* 2011), other projects turn to technology to aid cost-effective implementation. In particular, since many landscape-scale projects are reliant on quality seed input, much recent effort has been directed towards identifying and overcoming bottlenecks to recruitment by considering all components of a chain-of-seed-use (James *et al.* 2013). This includes the use of X-rays to determine seed viability while developing guidelines for effective seed storage (Martyn *et al.* 2009), pre-treatments to alleviate dormancy (Merritt *et al.* 2007) and the development of seed-enhancement technologies, such as embedding seeds in a soil matrix composed of compounds known to assist seedling establishment (Madsen *et al.* 2012; Figure 4.2). As an example of the latter approach, seeds can be embedded in a pod of activated carbon. This carbon allows native seedlings to emerge unscathed in landscapes that have been sprayed with herbicide to target undesirable species that would otherwise outcompete the sown native species (Madsen *et al.* 2014). Although technological approaches have appeal, lessons from ecology can also be borne in mind to help cost-effectively achieve restoration targets from the outset, e.g. the use of mycorrhizal inoculation to reinstate key plant–soil–microbe interactions and improve plant establishment and growth, taking account of other plant–soil feedback processes, considering nurse plant interactions, or giving explicit consideration to faunal components from initiation (Perring *et al.* 2015).

Figure 4.2   In recent times, landscape-scale restoration in the north of Western Australia has seen a marked increase in plant establishment of many key species through the implementation of improved seed management and topsoil application. (a) Before restoration; (b) after restoration. (c) Seed enablement technologies, such as extruded seed pellets, embed seeds in a 'pellet' or 'pillow' consisting of soil and other compounds to encourage germination and boost seedling survival (a video of their manufacture can be sourced from www.plants.uwa.edu.au/research/restoration-seedbank-initiative). (d) These technologies are being experimentally tested to ensure future large-scale restoration endeavours using seeds are repeatable and cost-efficient

Sources: (a, b) Brad Stokes, BHP Billiton Iron Ore; (c, d) Todd Erickson

Agricultural technology has also been increasingly harnessed to implement landscape-scale restoration, with modifications to seeding machines to allow the sowing of multispecies mixes or target taxa (Jonson 2010; St Jack *et al.* 2013). There are even discussions about the use of unmanned aerial vehicles being used to both collect and distribute seeds, as evidenced by the recent workshop on automated forest restoration (FORRU 2015), although legislative barriers may need to be overcome to fully realize such vehicles' potential (Allan *et al.* 2015). Other barriers to landscape-scale restoration may be cost-effectively overcome by solution scanning (Sutherland *et al.* 2014).

The foregoing highlights the multi-disciplinary nature of cost-effectively implementing restoration, for example knowledge of plant physiology, chemistry, engineering and agronomy are all utilized. Inter-disciplinarity also extends to studying climate to try to maximize organism survival and establishment; for instance, planting only in years when rains are predicted to be good given the current status of the El Niño Southern Oscillation (Holmgren and Scheffer 2001). Restoration enterprises need to be flexible such that projects are not instigated that would fail due to insufficient attention being paid to such factors. Despite the potential for technologies to aid cost-effective landscape-scale restoration initiatives, humans are a vital component of the restoration enterprise (Egan *et al.* 2011), and too great a reliance on technology could compromise this human aspect, further contributing to the insidious 'extinction of experience' (Miller 2005).

Multiple stakeholder perspectives

Implementing restoration at the landscape scale requires consideration of the social sphere and recognition of multiple stakeholder perspectives. There are cogent arguments that successful restoration will be more likely if capacity in, and overlap among, ecological, technological, and social spheres is enhanced (Jacobs *et al.* 2013). For instance, although technological advances in blight resistance allow the potential reintroduction of *Castanea dentata* to eastern North America, restoration will likely only be successful if people's concerns with genetic modification, as well as the altered ecological reality of contemporary forests (McEwan *et al.* 2011), are taken into account (Jacobs *et al.* 2013). More generally, landscape-scale restoration will need to acknowledge and deal respectfully with multiple perspectives of restoration, nature and people's role in both, while also taking account of power relations (Egan *et al.* 2011). Participation in the restoration process is empowering, and potentially leads to better quality decisions of greater durability although there is limited empirical research in a restoration context (Reed 2008; van Marwijk *et al.* 2012). Costa Rica shows evidence of successful landscape restoration following an inclusive and heterarchical approach (Reyes 2011). The examples of Gondwana Link, Yellowstone to Yukon, and, at a smaller scale, the Chicago Wilderness Vision (Conservation Fund undated) all point to the need for collaboration and dialogue.

More pragmatically, implementing restoration at the landscape scale needs to take account of land tenure arrangements. Restoration will likely be easier when the land is in one management unit such as a large pastoral property. In instances where ownership, and presumably management oversight, occurs in smaller units than the landscape scale, co-operation will be required among multiple landholders and managers to successfully implement landscape-scale restoration. These multiple managers will likely have conflicting targets and suggested approaches (Gobster 2001). Any conflicts could potentially be resolved through dialogue and a clear demonstration of the requirement for a landscape-scale perspective in ecological restoration. In particular, landscape-scale restoration requires an elucidation of benefits and costs to each of the landholders, and to the landscape as a whole. It is only recently that non-ecological benefits are being taken account of in measures of restoration success (Aronson *et al.* 2010; Wortley *et al.* 2013).

Stakeholders may be more likely to 'buy-in' to restoration if they appreciate how they can contribute to a multi-functional landscape. This integrated landscape perspective (e.g. Chazdon *et al.* 2009) is an opportunity for restoration since an overall goal of landscape multi-functionality could allow the provision of multiple benefits that would not occur at the patch scale (Jarchow and Liebman 2011). These benefits may provide a means to fund restoration at the landscape scale via payments for ecosystem services schemes (van Noordwijk *et al.* 2012). Such schemes, providing they give benefit to individual landowners, may make it more likely for landscape-scale restoration to be adopted.

Concluding remarks: challenges and opportunities for landscape-scale restoration

The development of ecology as a discipline has shown the necessity to include landscape processes to more completely understand process and pattern in ecosystems. Furthermore, landscape patterns themselves have been shown to influence ecological processes. This developing ecological knowledge is increasingly, and necessarily, being applied to restoration projects as the imperative to restore, due to land degradation, environmental change and policy initiatives, has led to restorationists tackling bigger and bolder projects. Such projects bring with them challenges, such as how to prioritize scarce resources, what to target in an era of environmental change and given competing demands for land, and how to practically implement restoration at the landscape scale. However, while researchers and practitioners continue to tackle these challenges, landscape-scale endeavours provide opportunities for social and economic, as well as ecological, gains. All of these gains, some of which have only recently been elucidated, will need emphasizing for landscape-scale restoration to garner support beyond its traditional base, and truly live up to its promise.

Acknowledgements

Thanks to David Freudenberger (Australian National University), Fraser Los (Yellowstone to Yukon Conservation Initiative), Todd Erickson (The University of Western Australia) and Brad Stokes (BHP Billiton Iron Ore) for supplying figure legend information and images, and Amanda Keesing (Greening Australia) for assisting with the preparation of Figure 4.1a. Thanks to the Parks Canada Agency for permission to use the inset image on Figure 4.1b and BHP Billiton Iron Ore for permission to use Figures 4.2a and 4.2b. Finally, my thanks to Stuart Allison and Pat Kennedy for valuable comments on a previous draft.

References

Allan, B. M., Ierodiaconou, D., Nimmo, D. G., Herbert, M. and Ritchie, E. G. (2015) Free as a drone: ecologists can add UAVs to their toolbox. *Frontiers in Ecology and Environment* 13, 354–355.

Ando, A., Camm, J., Polasky, S. and Solow, A. (1998) Species distributions, land values, and efficient conservation. *Science* 279, 2126–2128.

Andrén, H. (1994) Effects of habitat fragmentation on birds and mammals in landscapes with different proportions of suitable habitat: a review. *Oikos* 71, 355–366.

Aronson, J., Blignaut, J. N., Milton, S. J. *et al.* (2010) Are socioeconomic benefits of restoration adequately quantified? A meta-analysis of recent papers (2000–2008) in *Restoration Ecology* and 12 other scientific journals. *Restoration Ecology* 18, 143–154.

Banks-Leite, C., Pardini, R., Tambosi, L.R. *et al.* (2014) Using ecological thresholds to evaluate the costs and benefits of set-asides in a biodiversity hotspot. *Science* 345, 1041–1045.

Bell, S. S., Fonseca, M. S. and Motten, L. B. (1997) Linking restoration and landscape ecology. *Restoration Ecology* 5, 318–323.

Bradshaw, A. D. (1983) The reconstruction of ecosystems: Presidential address to the British Ecological Society, December 1982. *Journal of Applied Ecology* 20, 1–17.

Bradstock, R. A., Bedward, M., Gill, A. M. and Cohn, J. S. (2005) Which mosaic? A landscape ecological approach for evaluating interactions between fire regimes, habitat and animals. *Wildlife Research* 32, 409–423.

Broadhurst, L. M., Lowe, A., Coates, D. J., Cunningham, S. A., McDonald, M., Vesk, P. A. and Yates, C. (2008) Seed supply for broadscale restoration: maximizing evolutionary potential. *Evolutionary Applications* 1, 587–597.

Brook, B. W., Ellis, E. C., Perring, M. P., Mackay, A. W. and Blomqvist, L. (2013) Does the terrestrial biosphere have planetary tipping points? *Trends in Ecology and Evolution* 28, 396–401.

Cao, S., Chen, L. and Yu, X. (2009) Impact of China's Grain for Green Project on the landscape of vulnerable arid and semi-arid agricultural regions: A case study in northern Shaanxi Province. *Journal of Applied Ecology* 46, 536–543.

Carpenter, S. R. (2005) Eutrophication of aquatic ecosystems: Bistability and soil phosphorus. *PNAS* 102, 10002–10005.

Chazdon, R. L., Harvey, C. A., Komar, O. *et al.* (2009) Beyond reserves: A research agenda for conserving biodiversity in human-modified tropical landscapes. *Biotropica* 41, 142–153.

Cingolani, A. M., Vaieretti, M. V., Giorgis, M. A., La Torre, N., Whitworth-Hulse, J. I. and Renison, D. (2013) Can livestock and fires convert the sub-tropical mountain rangelands of central Argentina into a rocky desert? *The Rangeland Journal* 35, 285–297.

Conservation Fund (undated) Chicago Wilderness Vision. Retrieved from www.conservationfund.org/projects/chicago-wilderness-region (accessed 13 December 2016).

Crossman, N. D. and Bryan, B. A. (2006) Systematic landscape restoration using integer programming. *Biological Conservation* 128, 369–383.

Crossman, N. D. and Bryan, B. A. (2009) Identifying cost-effective hotspots for restoring natural capital and enhancing landscape multifunctionality. *Ecological Economics* 68, 654–668.

Culotta, E. (1995) Bringing back the Everglades. *Science* 268, 1688–1690.

Cunningham, S. (2002) *The restoration economy*. Berrett-Koehler, San Francisco, CA.

Damschen, E. I., Haddad, N. M., Orrock, J. L., Tewksbury, J. J. and Levey, D. J. (2006) Corridors increase plant species richness at large scales. *Science* 313, 1284–1286.

Driscoll, D. A., Banks, S. C., Barton, P. S., Lindenmayer, D. B. and Smith, A. L. (2013) Conceptual domain of the matrix in fragmented landscapes. *Trends in Ecology and Evolution* 28, 605–613.

Earn, D. J. D., Levin, S. A. and Rohani, P. (2000) Coherence and conservation. *Science* 290, 1360–1364.

Egan, D., Hjerpe, E. E. and Abrams, J. (eds) (2011) *Human dimensions of ecological restoration. Integrating science, nature and culture*. Island Press, Washington, DC.

Ehrenfeld, J. G. and Toth, L. A. (1997) Restoration ecology and the ecosystem perspective. *Restoration Ecology* 5, 307–317.

Fontaine, J. B. and Kennedy, P. L. (2012) Meta-analysis of avian and small-mammal response to fire severity and fire surrogate treatments in US fire-prone forests. *Ecological Applications* 22, 1547–1561.

FORRU (2015) Automated forest restoration (AFR) workshop. Retrieved from www.forru.org/en/content.php?mid=4855 (accessed 10 January 2016).

Foster, D., Swanson, F., Aber, J., Burke, I., Brokaw, N., Tilman, D. and Knapp, A. (2003) The importance of land-use legacies to ecology and conservation. *BioScience* 53, 77–88.

Gibson-Roy, P. and McDonald, T. (2014) Reconstructing grassy understories in south-eastern Australia: Interview with Paul Gibson-Roy. *Ecological Management and Restoration* 15, 1–12.

Gobster, P. H. (2001) Visions of nature: conflict and compatability in urban park restoration. *Landscape and Urban Planning* 56, 35–51.

Gunderson, L. H., Light, S. S. and Holling, C. S. (1995) Lessons from the Everglades. *BioScience* Science and Biodiversity Policy Supplement, S66–S73.

Haddad, N. M., Brudvig, L. A., Damschen, E. I. *et al.* (2014) Potential negative ecological effects of corridors. *Conservation Biology* 28, 1178–1187.

Hanski, I. (1998) Metapopulation dynamics. *Nature* 396, 41–49.

Harris, J. A., Hobbs, R. J., Higgs, E. and Aronson, J. (2006) Ecological restoration and global climate change. *Restoration Ecology* 14, 170–176.

Haugo, R., Zanger, C., DeMeo, T. *et al.* (2015) A new approach to evaluate forest structure restoration needs across Oregon and Washington, USA. *Forest Ecology and Management* 335, 37–50.

Higgs, E., Falk, D. A., Guerrini, A. *et al.* (2014) The changing role of history in restoration ecology. *Frontiers in Ecology and Environment* 12, 499–506.

Hobbs, R. J., Higgs, E., Hall, C. M. *et al.* (2014) Managing the whole landscape: historical, hybrid and novel ecosystems. *Frontiers in Ecology and Environment* 12, 557–564.

Holl, K. D. and Crone, E. E. (2004) Applicability of landscape and island biogeography theory to restoration of riparian understorey plants. *Journal of Applied Ecology* 41, 922–933.

Holmgren, M. and Scheffer, M. (2001) El Niño as a window of opportunity for the restoration of degraded arid ecosystems. *Ecosystems* 4, 151–159.

Hulvey, K. B., Standish, R. J., Hallett, L. M. *et al.* (2013) Incorporating novel ecosystems into management frameworks. In *Novel ecosystems: Intervening in the new ecological world order* (R.J. Hobbs *et al.*, eds), pp. 157–171. John Wiley and Sons, Chichester.

Jackson, S. T. and Sax, D. F. (2010) Balancing biodiversity in a changing environment: extinction debt, immigration credit and species turnover. *Trends in Ecology and Evolution* 25, 153–160.

Jacobs, D. F., Dalgleish, H. J. and Nelson, C. D. (2013) A conceptual framework for restoration of threatened plants: the effective model of American chestnut (*Castanea dentata*) reintroduction. *New Phytologist* 197, 378–393.

James, J. J., Sheley, R. L., Erickson, T., Rollins, K. S., Taylor, M. H. and Dixon, K. W. (2013) A systems approach to restoring degraded drylands. *Journal of Applied Ecology* 50, 730–739.

Januchowski-Hartley, S. R., McIntyre, P. B., Diebel, M., Doran, P. J., Infante, D. M., Joseph, C. and Allan, J. D. (2013) Restoring aquatic ecosystem connectivity requires expanding inventories of both dams and road crossings. *Frontiers in Ecology and Environment* 11, 211–217.

Jarchow, M. and Liebman, M. (2011) Maintaining multifunctionality as landscapes provide ecosystem services. *Frontiers in Ecology and Environment* 9, 262.

Jonson, J. (2010) Ecological restoration of cleared agricultural land in Gondwana Link: lifting the bar at 'Peniup'. *Ecological Management and Restoration* 11, 16–26.

Kitzes, J. and Harte, J. (2015) Predicting extinction debt from community patterns. *Ecology* 96, 2127–2136.

Laughlin, D. C. (2014) Applying trait-based models to achieve functional targets for theory-driven ecological restoration. *Ecology Letters* 17, 771–784.

Laurance, W. F., Carolina Useche, D., Rendeiro, J. *et al.* (2012) Averting biodiversity collapse in tropical forest protected areas. *Nature* 489, 290–294.

Lindenmayer, D., Hobbs, R. J., Montague-Drake, R. *et al.* (2008) A checklist for ecological management of landscapes for conservation. *Ecology Letters* 11, 78–91.

Madsen, M. D., Davies, K. W., Williams, C. J. and Svejcar, T. J. (2012) Agglomerating seeds to enhance native seedling emergence and growth. *Journal of Applied Ecology* 49, 431–438.

Madsen, M. D., Davies, K. W., Mummey, D. L. and Svejcar, T. J. (2014) Improving restoration of exotic annual grass-invaded rangelands through activated carbon seed enhancement technologies. *Rangeland Ecology and Management* 67, 61–67.

Martyn, A., Merritt, D. and Turner, S. (2009) Seed banking. In *Plant germplasm conservation in Australia: Strategies and guidelines for developing, managing and utilising ex situ collections* (C. A. Offord and P. F. Meagher, eds), pp. 63–86. Australian Network for Plant Conservation, Canberra.

McCarthy, M. A., Thompson, C. J., Moore, A. L. and Possingham, H. P. (2011) Designing nature reserves in the face of uncertainty. *Ecology Letters* 14, 470–475.

McConnachie, M. M., Cowling, R. M., van Wilgen, B. W. and McConnachie, D. A. (2012) Evaluating the cost-effectiveness of invasive alien plant clearing: A case study from South Africa. *Biological Conservation* 155, 128–135.

McEwan, R. W., Dyer, J. M. and Pederson, N. (2011) Multiple interacting ecosystem drivers: toward an encompassing hypothesis of oak forest dynamics across eastern North America. *Ecography* 34, 244–256.

Menz, M. H., Dixon, K. W. and Hobbs, R. J. (2013) Hurdles and opportunities for landscape-scale restoration. *Science* 339, 526–527.

Merritt, D. J., Turner, S. R., Clarke, S. and Dixon, K. W. (2007) Seed dormancy and germination stimulation syndromes for Australian temperate species. *Australian Journal of Botany* 55, 336–344.

Millennium Ecosystem Assessment (2005) *Ecosystems and human well-being: Biodiversity synthesis.* World Resources Institute, Washington, DC.

Miller, J. R. (2005) Biodiversity conservation and the extinction of experience. *Trends in Ecology and Evolution* 20, 430–434.

Ogle, K., Barber, J. J., Barron-Gafford, G. A. *et al.* (2015) Quantifying ecological memory in plant and ecosystem processes. *Ecology Letters* 18, 221–235.

Palik, B. J., Goebel, P. C., Kirkman, L. K. and West, L. (2000) Using landscape hierarchies to guide restoration of disturbed ecosystems. *Ecological Applications* 10, 189–202.

Perring, M. P., Standish, R. J., Price, J. N. *et al.* (2015) Advances in restoration ecology: rising to the challenges of the coming decades. *Ecosphere* 6, article 131.

Perring, M. P., De Frenne, P., Baeten, L. *et al.* (2016) Global environmental change effects on ecosystems: the importance of land-use legacies. *Global Change Biology* 22(4), 1361–1371.

Radeloff, V. C., Mladenoff, D. J. and Boyce, M. S. (2000) A historical perspective and future outlook on landscape scale restoration in the Northwest Wisconsin Pine Barrens. *Restoration Ecology* 8, 119–126.

Ramalho, C. E. and Hobbs, R. J. (2012) Time for a change: dynamic urban ecology. *Trends in Ecology and Evolution* 27, 179–188.

Reed, M. S. (2008) Stakeholder participation for environmental management: A literature review. *Biological Conservation* 141, 2417–2431.

Renton, M., Shackelford, N. and Standish, R. J. (2012) Habitat restoration will help some functional plant types persist under climate change in fragmented landscapes. *Global Change Biology* 18, 2057–2070.

Rey Benayas, J. M., Bullock, J. M. and Newton, A. C. (2008) Creating woodland islets to reconcile ecological restoration, conservation, and agricultural land use. *Frontiers in Ecology and Environment* 6, 329–336.

Reyes, J. E. (2011) Public participation and socioecological resilience. In *Human dimensions of ecological restoration. Integrating science, nature and culture* (D. Egan *et al.*, eds), pp. 79–92. Island Press, Washington, DC.

Ryan, E. M., Ogle, K., Zelikova, T. J., Lecain, D. R., Williams, D. G., Morgan, J. A. and Pendall, E. (2015) Antecedent moisture and temperature conditions modulate the response of ecosystem respiration to elevated CO_2 and warming. *Global Change Biology* 21, 2588–2602.

St Jack, D., Hesterman, D.C. and Guzzomi, A. L. (2013) Precision metering of *Santalum spicatum* (Australian Sandalwood) seeds. *Biosystems Engineering* 115, 171–183.

Shackelford, N., Renton, M., Perring, M. P. and Hobbs, R. J. (2013) Modeling disturbance-based native invasive species control and its implications for management. *Ecological Applications* 23, 1331–1344.

Standish, R. J., Cramer, V. A., Wild, S. L. and Hobbs, R. J. 2007 Seed dispersal and recruitment limitation are barriers to native recolonization of old-fields in western Australia. *Journal of Applied Ecology* 44, 435–445.

Suding, K., Higgs, E., Palmer, M. *et al.* (2015) Committing to ecological restoration. *Science* 348, 638–640.

Sutherland, W. J., Gardner, T., Bogich, T. L. *et al.* (2014) Solution scanning as a key policy tool: identifying management interventions to help maintain and enhance regulating ecosystem services. *Ecology and Society* 19, 3.

Swift, T. L. and Hannon, S. J. (2010) Critical thresholds associated with habitat loss: a review of the concepts, evidence, and applications. *Biological Reviews* 85, 35–53.

Tilman, D., May, R. M., Lehman, C. L. and Nowak, M. A. (1994) Habitat destruction and the extinction debt. *Nature* 371, 65–66.

Torrubia, S., McRae, B. H., Lawler, J. J., Hall, S. A., Halabisky, M., Langdon, J. and Case, M. (2014) Getting the most connectivity per conservation dollar. *Frontiers in Ecology and Environment* 12, 491–497.

Turner, M. G. (1989) Landscape ecology: the effect of pattern on process. *Annual Review of Ecology and Systematics* 20, 171–197.

van der Hoek, Y., Zuckerberg, B. and Manne, L. L. (2015) Application of habitat thresholds in conservation: considerations, limitations, and future directions. *Global Ecology and Conservation* 3, 736–743.

van Marwijk, R. B. M., Elands, B. H. M., Kampen, J.K., Terlouw, S., Pitt, D. G. and Opdam, P. (2012) Public perceptions of the attractiveness of restored nature. *Restoration Ecology* 20, 773–780.

van Noordwijk, M., Leimona, B., Jindal, R. *et al.* (2012) Payments for environmental services: Evolution toward efficient and fair incentives for multifunctional landscapes. *Annual Review of Environment and Resources* 37, 389–420.

Vesk, P. A., Nolan, R., Thomson, J. R., Dorrough, J. W. and MacNally, R. (2008) Time lags in provision of habitat resources through revegetation. *Biological Conservation* 141, 174–186.

Vitousek, P. M., Mooney, H. A., Lubchenco, J. and Melillo, J. M. (1997) Human domination of Earth's ecosystems. *Science* 277, 494–499.

Vitt, P., Havens, K. and Hoegh-Guldberg, O. (2009) Assisted migration: part of an integrated conservation strategy. *Trends in Ecology and Evolution* 24, 473–474.

Wilson, K. A., Lulow, M., Burger, J. *et al.* (2011) Optimal restoration: accounting for space, time and uncertainty. *Journal of Applied Ecology* 48, 715–725.

Wortley, L., Hero, J.-M. and Howes, M. (2013) Evaluating ecological restoration success: A review of the literature. *Restoration Ecology* 21, 537–543.

Young, T. P. (2000) Restoration ecology and conservation biology. *Biological Conservation* 92, 73–83.

Zahawi, R. A., Holl, K. D., Cole, R. J. and Reid, J. L. (2013) Testing applied nucleation as a strategy to facilitate tropical forest recovery. *Journal of Applied Ecology* 50, 88–96.

Zedler, J. B., Doherty, J. M. and Miller, N. A. (2012) Shifting restoration policy to address landscape change, novel ecosystems, and monitoring. *Ecology and Society* 17, 36.

5

UNDERSTANDING SOCIAL PROCESSES IN PLANNING ECOLOGICAL RESTORATIONS

Stephen R. Edwards, Brock Blevins, Darwin Horning and Andrew Spaeth

Introduction

Ecology focuses on understanding the interrelationships between and among the biotic and abiotic components of an ecosystem. Metaphorically, ecology is about the pieces of a puzzle and how they fit together. The ecosystem is the picture – or the sum of the parts – that is revealed when the puzzle pieces are put together in their proper positions. Carrying the metaphor forward, our picture is constantly changing because the complex of structures is constantly evolving over time in response to independent influences such as climate, stochastic events, and human interventions. Consequently the ecosystem is more like a very slow-motion picture than a photo, and what we see is a snapshot when we look at an ecosystem at a particular moment in time.

Humans are an integral part of every ecosystem on the planet. Personal interest and commitment often transcends legal 'ownership' of ecosystems today. These interests – and associated commitments – are highly stratified and may involve local, provincial/state, national, and global stakeholders, with each tier wielding varying degrees of influence over decision-making and ultimately the actions taken. The involvement of people who, by their decisions and actions, influence ecological processes is now formalized through the term 'social-ecological' which acknowledges the duality of the human–nature relationship (Holling 2001).

> If we save the living world, we will also automatically save the physical world, because in order to achieve the first we must also achieve the second. But if we only save the physical world, which appears our present inclination, we will ultimately lose both.
>
> *(Wilson 2012: 294)*

One consequence of this ever-widening circle of stakeholders, often with widely varying and conflicting views of how an ecosystem should be managed or restored, makes management – and hence restoration – much more complicated. These complex social-ecological systems fall into a class of problems referred to as wicked problems. The terminology and initial exploration of wicked problems was first used in fields like engineering and the planning activities associated with large-scale processes. Characteristics of wicked problems are provided in Table 5.1 (based on Conklin 2005).

Table 5.1 Wicked problems

Characteristics	Explanation
1. You don't understand the problem until you have developed a solution.	As potential solutions are applied, they in turn provide insights into new dimensions of the problem – a process that repeats itself with each cycle.
2. Wicked problems have no stopping rule.	As there is no definitive solution to the problem it is not possible to reach an endpoint where the problem can be declared 'solved'.
3. Solutions to wicked problems are not right or wrong.	As there is no definitive solution, actions must be gauged in subjective terms – better, worse, seems to work – that are assessed in a social context.
4. Every wicked problem is essentially unique and novel.	A solution that works in one situation most likely will not work in a similar situation; differences in diversity and variety of the factors influencing a proposed solution will yield different results with each problem.
5. Every solution to a wicked problem is a 'one-shot operation'.	This is the embedded conundrum: you cannot 'learn about a problem without trying solutions, but every solution you try … [has costs] … and has lasting unintended consequences which are likely to spawn new wicked problems'.
6. Wicked problems have no given alternative solutions.	Because each situation is unique, and because there is no single solution, dealing with a wicked problem requires creativity; however, the effectiveness of the solution will be judged in qualitative terms in a social context.

Source: Conklin (2005)

In the context of an ecological restoration wicked problems are implicit in the dynamics of both social and ecological components of an ecosystem. While the processes of both are inextricably linked, our knowledge and understanding of the overlapping and interwoven relationships across these systems at different scales is incomplete, resulting in the need for innovative approaches to planning, management, and decision-making. Simply put, to undertake a restoration project today is as much about how the social component is addressed as it is about the ecology and science relevant to the restoration goals.

This chapter is about how to employ an understanding of social dynamics to help develop more successful plans for restoration actions. Our approach to restoration is not prescriptive. We recognize that concepts emanating from the conservation and restoration communities that are guided by local knowledge, socio-economic and cultural goals, and desired outcomes, are as important as the science of ecology that underlies restoration practice.

Stakeholders at various levels often maintain influence, to varying degrees, that inform ecosystem management decisions. In cases where both the social and ecological aspects of an ecosystem are addressed there is a much higher chance of achieving durable solutions. This type of balanced approach ultimately relies on an understanding of the inter-relationship between social systems and the array of ecological services that are provided by an ecosystem. The latter of which provides an important lens to explore approaches to restoration that is linked to both socially and ecologically desired outcomes.

Conservation framework for restoration

The conservation community is about advocating for restoration of ecosystem functions and processes to preserve critical and sensitive species and to generally protect biodiversity. Two conceptual frameworks that have broad acceptance in the international conservation community can provide a framework that is useful for goal setting, planning and managing restorations that balance social and ecological needs:

- the Ecosystem Approach (UN Convention on Biological Diversity 2000); and
- the categories of ecosystem services (Millennium Ecosystem Assessment 2005).

The Ecosystem Approach provides a set of principles designed to promote more successful management of ecosystems (see Box 5.1). The Ecosystem Approach was adopted by the Parties of the Convention on Biological Diversity at their 5th Conference of the Parties (2000; Nairobi, Kenya) and subsequently many nations have embraced the principles in national policies and guidelines.

The twelve principles of the Ecosystem Approach convey the optimal conditions in which ecosystem management, including restoration, should be undertaken. Nevertheless, it is important to note that not all principles will apply in all situations, in part, due to the context-dependent nature of management and restoration work.

Box 5.1 Principles of the Ecosystem Approach

1 The objectives of management of land, water, and living resources are a matter of societal choice.

2 Management should be decentralized to the lowest appropriate level.

3 Ecosystem managers should consider the effects (actual or potential) of their activities on adjacent and other ecosystems.

4 Recognizing potential gains from management, there is usually a need to understand and manage the ecosystem in an economic context. Any such ecosystem-management programme should:
   (a) reduce those market distortions that adversely affect biological diversity;
   (b) align incentives to promote biodiversity conservation and sustainable use; and
   (c) internalize costs and benefits in the given ecosystem to the extent feasible.

5 Conservation of ecosystem structure and functioning, in order to maintain ecosystem services, should be a priority target of the Ecosystem Approach.

6 Ecosystems must be managed within the limits of their functioning.

7 The Ecosystem Approach should be undertaken at the appropriate spatial and temporal scales.

8 Recognizing the varying temporal scales and lag-effects that characterize ecosystem processes, objectives for ecosystem management should be set for the long term.

9 Management must recognize that change is inevitable.

10 The Ecosystem Approach should seek the appropriate balance between, and integration of, conservation and use of biological diversity.

11 The Ecosystem Approach should consider all forms of relevant information, including scientific and indigenous and local knowledge, innovations and practices.

12 The Ecosystem Approach should involve all relevant sectors of society and scientific disciplines.

The second framework provides a typology for the different classes of services that ecosystems provide (Millennium Ecosystem Assessment 2005). These classes and specific examples of the services are listed in Figure 5.1. This assessment was undertaken between 2001 and 2005 and involved 16 international institutions and more than 1,360 specialists, which reflects broad acceptance of this terminology across the scientific community. The framework is included here to provide a straightforward approach to:

1   identify the various categories of services that ecosystems deliver;
2   provide a standard language to facilitate communications; and
3   ensure that *all* of an ecosystem's services are considered, including the supporting and regulating services – without which the provisioning and cultural services would not be possible.

Nevertheless, this approach is not without contention as some people fundamentally disagree with the premise that the outcomes of ecosystems processes be considered 'services' as it implies that nature exists to serve humans. For example, one perspective is that the products of nature are to be cherished and not monetized; that they are 'gifts' at a spiritual level. At the other end of the spectrum, some people use 'service' as a basis for quantifying a value for the benefits to humans, wildlife and natural systems that are delivered. Those seeking to establish values for the various services ecosystems provide argue that using the language of economists will have greater influence in national and regional planning and policies governing ecosystem conservation, management, and restoration.

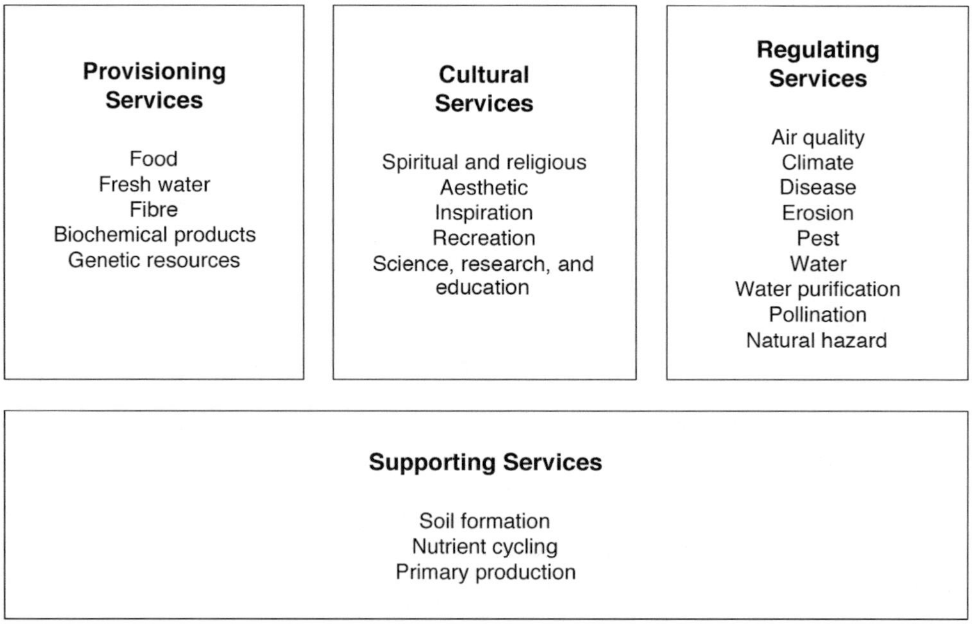

Provisioning Services	**Cultural Services**	**Regulating Services**
Food	Spiritual and religious	Air quality
Fresh water	Aesthetic	Climate
Fibre	Inspiration	Disease
Biochemical products	Recreation	Erosion
Genetic resources	Science, research, and education	Pest
		Water
		Water purification
		Pollination
		Natural hazard

Supporting Services
Soil formation
Nutrient cycling
Primary production

Figure 5.1   Categories of ecosystem services

Understanding the different classes of services that a particular ecosystem is capable of providing informs discussions about the goals of a restoration and represents an important step in laying out an approach to develop a restoration plan that is sensitive to social values and interests while also respecting the scientific and technical requirements of a restoration project.

Assessing ecosystem risk and resilience

Restorations span a range of approaches; no two will ever be identical in execution because conditions and the interests of the people involved vary substantially from place to place and over time. Nevertheless, full restoration depends to a greater or lesser extent on what is entailed in setting the stage for the biotic and abiotic components to engage in the ecological processes that deliver the services associated with the desired outcome. Restoration efforts are dependent upon current ecological conditions, past disturbance and management regimes, and a capacity to plan and execute management objectives. Further, particularly in cases that occur in ecosystems of high social and ecological value, establishing project goals and objectives is likely to happen through a stakeholder engagement process – from purely collaborative with shared decision-making to advisory and consultative-based approaches.

Viewed through a social lens, planning a restoration project is best begun by (i) developing a mutual understanding of the purpose and goals of the project, (ii) conducting a critical evaluation of the relationships between current and desired ecological conditions, and (iii) determining the potential overlapping and competing social and cultural values and interests of affected parties. An independent facilitator can be helpful in guiding the process, which may take considerable time. During these facilitated discussions the focus should be on contextualizing the scope and scale of the restoration and guiding the stakeholders' discussions toward mutually agreed goals and objectives of the restoration. Beginning with overarching goals and objectives may naturally lead to more deliberate discussions of the actions to be taken to achieve the objectives.

In some cases restoration of ecosystem capacities may occur simply by removal of the factors that are negatively affecting the abiotic–biotic interactions. This could be as straightforward as removing ungulate browsers – usually stock animals like sheep and goats – from an area and allowing the native vegetation to recover sufficiently on its own to sustain the array of services characteristic of the natural system. This approach has borne substantial benefits in developing countries as diverse as Pakistan, Kenya, and Zimbabwe, where the primary goal was to provide natural resources that would contribute to rural development and where funds are very limited. A good example is the Shinyanga project in Kenya where 300,000 hectares was restored when authority for the region was delegated to rural villagers in the region who applied traditional usage patterns in the degraded area (Barrow 2014; Smith 2016). This is not to say that simply changing some 'bad behaviour' may always result in the anticipated outcome or require further action. For example, in cases where the degradation of ecological capacity is also being affected by the presence of undesirable invasive species, achievement of restoration goals would also require extirpation of the invasives.

Ecosystem restoration at scale

Where the scope of the problem far exceeds the capacities of the institutions to address the problem – often referred to as 'scale mismatches' – a process is required to either increase capacities or limit objectives and priorities to the area that can be addressed within the human and financial resources that are available (Cumming *et al.* 2006). In such cases the stakeholders involved will need to adopt criteria to help frame the more limited objectives and priorities.

The scale and context of the restoration within the larger landscape should be the focus with a critical eye towards scale mismatches – differences in ecosystem challenges relative to physical and geopolitical boundaries that will have to be crossed in order to achieve the desired solution. Determination of the resources available and the limitations of the human resources who will be engaged in the restoration – locally and at the appropriate broader scale – will also be an important consideration.

In most cases, restoration activities occur at a more limited scale, like a short reach of river, a stretch of coastal shoreline, or a drainage basin within a larger ecosystem context. Pragmatically, restorations are limited by the capacity and scale of the institutions that govern and manage the ecosystem as well as the human and financial resources that are available.

Tools to assess the state and capacity of ecosystems

To help decide how to proceed in such situations we look at two tool sets that provide insight into the state and capacity of an ecosystem. The first gives an objective assessment of the risk that the ecosystem is likely to collapse if present conditions are not addressed. The second provides an assessment of the resilience of the ecosystem – the capacity of the ecosystem to sustain delivery of its services while undergoing change.

The metrics provided in Box 5.2 are contained within the Red List of Ecosystems project being undertaken by the International Union for the Conservation of Nature (IUCN) (Rodriguez *et al.* 2015). Knowing the ecosystem services provided by the subject ecosystem, and how those services compare with historical records is crucial for stakeholders to engage in a restoration planning process. This ensures that the less obvious services are considered with those services that often garner most of the interest of stakeholders – such as recreational opportunities or harvest of high value products like timber (Keith *et al.* 2015). These metrics allow an ecosystem manager to assess the probability that a particular ecosystem could collapse or the features that define the system could be lost.

Box 5.2 IUCN criteria for assessing the status of an ecosystem

To determine the risk, two metrics are considered:

- the amount of reduction in the distribution of an ecosystem based on persistent negative impacts; and
- the degree to which an ecosystem distribution is becoming more restricted spatially, including fragmentation, as well as in relation to its capacity.

Two other metrics which provide additional insights are:

- the amount of degradation of the abiotic environment (e.g. ocean acidification or soil loss); and
- the amount of disruption of biotic processes and interactions (e.g. pollination, nutrient cycling).

These metrics are based on the state of the expanse (declining, expanding) of the ecosystem and key ecological processes (decreasing or maintaining) that sustain the overall capacity of the ecosystem. In those cases where there is high probability that reduction in the expanse of the ecosystem could happen due to the increased presence of specific disturbance agents (e.g. beetle attacks on coniferous forests), land managers may conclude that there is a high risk that the ecosystem is poised for collapse. Ideally corrective actions are taken to reduce that risk and increase the likelihood that the system will persist. If on the other hand, no corrective action is possible or taken and the ecosystem collapsed, the ability of managers to design efficacious restoration treatments is severely diminished.

The second tool – resilience assessment – provides insight into the capacity of an ecosystem to sustain delivery of needed services following disturbance, whether fire, flood, or some other natural or human-induced event. For the purposes of this paper, resilience is defined as 'The capacity of a system to absorb disturbances and reorganize while undergoing change so as to retain essentially the same function, structure, identity, and feedbacks' (Resilience Alliance 2010).

Holling (2000) thought of resilience as 'the capacity to create, test and maintain adaptive capability', which implies the need to manage an ecosystem in the context of changing environmental conditions – a fundamental requirement if an ecosystem is to be managed to ensure continued delivery of the basic services on which people depend.

Disturbance is a regular part of ecosystem processes and can be viewed as having both positive and negative effects on the natural world. For example, a flood will deposit rich nutrients in surrounding areas, wildfires release important nutrients, and a windstorm may create snags for wildlife. Alternatively, 'extreme' disturbance events cause uncharacteristic harm and may result in shifts in the state of an ecosystem and its function and capacity. For example, a wildfire occurring at very high intensity and severity in some forest ecosystems can lead to loss of topsoil, sedimentation of streams, increased risk of invasive species spread, and soil hydrophobicity (Ice *et al.* 2004).

Undertaking an assessment of the resilience of an ecosystem also provides a means to understand the social considerations of restoration by addressing such questions as:

- What are the social values associated with the area to be restored?
- Are they culturally significant?
- If the social values are culturally significant, to whom and why?

Exploring resilience among other related questions can help uncover key pieces of information that will inform the most appropriate approach to restoration, particularly if the solutions are to be lasting.

Keep in mind that while you may determine the resilience of an ecosystem at a particular point in time, you cannot predict the resilience of an ecosystem in the future nor can you accurately report the resilience of an ecosystem in previous times. All ecosystems are constantly changing and evolving under the influences of any number of factors, both endogenous and exogenous, which at the same time are also often changing. Given the complexity of these systems – the variety and number of ecological processes, the diversity and linkages among the biotic and abiotic, and the dynamic nature expressed as a reflection of the sum of the parts, there is an inherent capacity to persist under existing environmental conditions. Substantial changes occur when those conditions shift dramatically.

The relevance of resilience in relation to ecosystem management, including restoration, is highlighted in Table 5.2 that lists several attributes of resilience that are relevant to ecosystem restoration. Based on these and other attributes of resilience, procedures have been developed

Table 5.2 Selected attributes of resilience

Attribute	Explanation	Metric
1. Diversity	The greater the diversity in social-ecological systems the greater the resilience of the system.	The greater the diversity in natural conditions, (e.g. species, habitats, trophic levels, abiotic elements), social contexts (e.g. stakeholder interests, institutional affiliations, motivations), and uses (e.g. recreational, spiritual, commercial) the greater the resilience of the system.
2. Ecological variability	Variation in different ecological zones within a management area contributes to resilience.	A mosaic of different ecosystems such as grassland, pine-dominated forest, fir-dominated forest, wetlands, and water flows across a landscape will foster greater resilience in the landscape.
3. Modularity	Disconnectedness within ecological, social, and physical components of the system promotes resilience.	The greater the modularity the greater the resilience. If different elements of an ecosystem are overly connected, there is greater risk of 'shock' affecting all elements. For example, a system of firebreaks or roads provides modularity; a mosaic of forest types, possibly with different stand configurations would contribute to greater modularity and resilience.
4. Slow variables	Slow (sometimes called 'controlling') variables are fundamental to socio-ecological processes. Examples of slow variables that affect ecological processes are soil composition, nutrient cycles, and carbon cycle. Examples of social-based slow variables would be cultural values, laws, and policies.	If the slow variables are acknowledged in policy and management actions, greater resilience will be promoted.
5. Relatively fast feedbacks	Having the means to detect change or increased risk quickly in the state of a social-ecological system fosters resilience.	A monitoring scheme that enables ecosystem managers and other stakeholders to detect deviations from desired outcomes promotes resilience. Such feedback could be used to avoid undesirable transformations in the state of the social-ecological system – while providing a basis for adapting management actions.
6. Social capital	Promotion and engagement of social institutions in developing and implementing management actions and facilitating communications with diverse stakeholders contributes to resilience.	Institutional settings that foster diverse stakeholder involvement in the management of an ecosystem provide means to develop broader social capital, which in turn fosters greater resilience.

Table 5.2 continued

Attribute	Explanation	Metric
7. Innovation	Allowing innovative approaches in different areas would foster learning and resilience.	Policies and management that promote innovative approaches to problem solving can contribute to more resilient socio-ecological systems.
8. Overlap	Overlap in governance or 'nested enterprises' promotes resilience.	Where different property rights apply (e.g. private, state, and federal) across a landscape or ecosystem resilience is enhanced.
9. System reserves	Setting aside 'no-harvest/use' zones or drawing on new knowledge/ scholarship will foster resilience.	No harvest zones promote resilience by preserving genetic diversity and seed/gene resources that may be important in re-establishing communities. Human knowledge that can be mobilized in response to a particular problem may foster recovery and improve resilience.

Note: Many of the explanations and examples provided for each attribute are based on *Basic Resilience Assessment: A Practitioners' Guideline For Learning About Resilience While Doing A Resilience Assessment* (Jones 2013). This is a working draft that has been circulated for application, discussion and development beginning in late March 2013.

to assess resilience in treatment areas or to monitor the effect of actions on the resilience of the social-ecological system (Resilience Alliance 2010).

Clearly one characteristic that frames all ecosystem restoration is the *uncertainty* of knowing the ecological processes supporting the ecosystem as well as the *unpredictability* of some aspects of the social component of planning and executing a restoration, especially given the complex and multi-scalar interactions across time and space that are associated with these systems. Further, changing climatic regimes will contribute additional uncertainty for the foreseeable future as practitioners and interested parties work to restore ecosystems.

Because restoration affects complex ecological and social systems, neither of which is fully understood, there are further risks including:

- conflict among stakeholders;
- unintended consequences due to the interconnectedness and complex nature of biological systems;
- pursuit of goals and/or objectives dominated by economic wants rather than ecological needs;
- external forces such as climate change having unanticipated effects on the system; and/or
- restoration actions result in only partial solutions creating further and unintended consequences on the ecosystems intended for repair or restoration.

In this chapter we have provided a set of tools that can help a manager to effectively address the social dynamic associated with planning and executing a restoration.

- *Wicked problems* (Table 5.1) – by understanding that the process of restoration will be plagued with wicked problems managers can be prepared to handle the unpredictability that such problems bring; know that traditional 'fix' solutions will not apply; that the process to be followed will be incremental and iterative, with each cycle providing new insights and challenges, but greater understanding of the whole of the problem.

- *Ecosystem Approach* (Box 5.1) – provides a set of principles that can serve as the basis for engaging with stakeholder groups. They establish basic standards within which the restoration approach would be developed and ensure that the stakeholders have a common understanding of the 'rules' under which the restoration will be planned and executed.
- *Ecosystem services* (Figure 5.1) – provides a common language for discussing the full array of services that ecosystems deliver; ensures that the supporting and regulating services that are crucial to the sustained delivery of provisioning and cultural services are factored into decisions in the design and execution of the restoration plan.
- *Risk of collapse* (Box 5.2) – The IUCN Red List of Ecosystem criteria to assess the risk of an ecosystem collapsing provide a relatively simple way to document the state of an ecosystem and may prove valuable in substantiating the need for restoration.
- *Resilience assessment* (Table 5.2) – Understanding the relative resilience of an ecosystem provides an insight into an ecosystem's capacity to withstand change and still deliver the services on which the ecosystem and people depend. Of equal importance is the fact that the resilience assessment process ensures that the full spectrum of stakeholders are identified and engaged in the process, which provides a basis for setting goals and objectives, and sharing views while recognizing the integral relationships stakeholders have with ecosystems.

These five tools provide a robust toolbox for an ecosystem manager to design, plan, and execute an ecosystem restoration that ensures that the social perspective is integrated into a restoration plan. To provide a focus for applying these tools we recommend that you consider some simple yet broad principles within which to plan a restoration. For example, Suding *et al.* (2015) articulate four principles that are based on scientifically valid approaches to ecosystem restoration. They argue that by adopting these principles selective bias can be avoided and the prospect of achieving successful restoration of ecological processes will be increased. In the context of our emphasis on the importance of restoration sustaining the core ecosystem services on which the ecosystem, including humankind, depends we put forward four comparable principles phrased in terms of desired outcomes of the restoration:

1   The restoration will 'increase the ecological integrity' of the ecosystem.
2   The ecosystem will be capable of maintaining the diverse array of ecological processes and services necessary to sustain the ecosystem in the long term.
3   Knowledge of the historical processes and services associated with the ecosystem have informed and guided the restoration process.
4   Stakeholders – at all levels – recognize and support restoration actions that ensure sustained delivery of all ecosystem services on which they depend.

By recognizing and adopting principles like these, and employing the tools most relevant and applicable to your situation in the planning and implementation steps of restoration, there will be a far greater chance of meeting both ecological and societal needs of a restoration.

Technology – another set of tools

As the underlying science of ecology has grown so have advances in relevant technologies, ranging from the ease of data sharing and the ability to rapidly communicate successes and failures, to supercomputing, remote sensing via satellite and drone, digital mapping and spatial analysis. The use of such technologies has shown great benefit in priority setting, ecosystem assessments, providing insights into historical baseline conditions, observing indicators of biodiversity, and monitoring restoration effects (Cloern *et al.* 2011; DiGennaro *et al.* 2012; Cabello *et al.* 2012; He *et al.* 2007).

Planning and monitoring of restoration projects generally occur in a geographic context. Tools such as remotely sensed data (i.e. earth observations from satellite and airborne platforms) and Geographic Information Systems (GIS) are therefore fundamental in determining the scale and scope of a restoration (Oetter *et al.* 2014; Hestir *et al.* 2008; Macmillan *et al.* 1998). Measurements of abiotic and biotic and socio-economic parameters associated with a landscape, such as human population densities, protected areas, industrial land use, and private land ownership can also be considered. All provide information that would be helpful in designing and implementing a restoration project. With today's concerns about changing climatic conditions, such tools also can be used to identify shifts in vegetative cover across landscapes that will be invaluable in designing a restoration. Climate models, combined with the increased accuracy of spatial data and computer storage and processing capabilities, make it possible to factor in the variability in species compositions that will be associated with changing climates.

While technologies have the ability to greatly aid restoration activities there are also limitations that should be considered. For example, in relation to remote sensing, strengths include the ability to acquire spatially continuous data over large, landscape-scale areas, which provide a context for planning a restoration at a much finer scale. While access to remote-sensing data was a handicap, that is no longer true. With the adoption of an Executive Order (Obama 2013) an 'open data policy' was adopted that calls on US government agencies to make available certain classes of information, including satellite imagery, and many nations have begun to follow suit. Limitations today relate to availability of time-sequenced images, non-contiguous area coverage and the relatively coarse spatial resolution of remote-sensed imagery. Nevertheless, with each new platform and sensor developed these limitations should be mitigated.

Lowered cost of data storage, cloud computing, and finer resolution of GIS data will mean that information access and use will be done more efficiently; information can be shared more easily with colleagues anywhere on the planet; proposed actions, monitoring data, and conclusions of assessments can be shared more easily with regulators, stakeholders, and the public. It is expected that communications technologies will continue to increase efficiencies and access and more countries will adopt open data policies; more web-based tools, interactive online mapping, increased data sharing, and more open source software will be available to access and integrate remotely sensed data in models that will be capable of forecasting future ecosystem extent and function under different management and climatic scenarios. There is no doubt that the restoration of ecosystems and the services they provide will require people working on the ground; but it is equally true that those involved in restoration will have greater access to data and information that will better inform decisions and actions in the field.

Applying technologies in restoration of ecosystems

For the sake of illustrating how these tools can be effective, consider two approaches to restoration that are common today: First, a traditional ecosystem restoration approach, which seeks to achieve a hypothesized historic range of variability. This is the basic premise of most US Forest Service restoration planning. The assumption in such cases is that what existed in the past is the most appropriate goal and more so than anything that can be achieved through a path-dependent approach (Higgs 2003).

An alternative approach is path-dependent where the restoration is designed to achieve a new equilibrium based on discrete goals and objectives – such as reduction of fuels in a forest or stabilizing soil conditions to prevent erosion, mud slides, and reservoir siltation. The risk associated with the path-dependent approach, given that stakeholder engagement is a given, is that the goals and objectives can be influenced in ways that may have little to do with ecological processes that provide the capacity of an ecosystem to sustain delivery of the basic supporting and regulating services that are crucial to maintaining the ecosystem itself. For example, a restoration project with a specific goal of reforestation that replaces a multi-species forest with a monoculture plantation reduces diversity (decreasing resilience hence increasing risk) and can result in the establishment of what are called 'novel' ecosystems. Ecological processes will continue, but at a much diminished capacity in relation to the broader spectrum of services that diverse and resilient ecosystems deliver.

Either approach may be facilitated with the use of technologies. However, given diverse stakeholder involvement in the process, it is crucial that their use be understood, their value/role in the process be acknowledged, and that they be used in accordance with the guiding principles the stakeholders have adopted. Examples of how technologies can be helpful are:

- mapping land vegetation patterns and threats to distribution and degradation of function;
- measuring distributions of ecosystem components;
- evaluating ecosystem function;
- assessing indicators of resilience and the variability associated with changing climates; and
- monitoring the effects of the restoration process and concluded restoration.

A good starting point to use technology would be in establishing the scope and scale of the restoration. While restorations are most often planned around a particular site of limited scale, it is important to understand how the conditions at the prospective restoration site relate to the larger landscape. It would not be surprising, when looking at the broader landscape, that there would be many areas that would benefit from restoration and having a 'big picture' understanding of the landscape along with the interconnectedness of the ecosystems across the landscape will be invaluable in developing the project goals and objectives. Maps of the landscape constructed through cost-effective remote-sensing technologies retrieved either by satellite or airborne platforms (aircraft or UAV) provide an excellent means to identify the scope of local stakeholders to engage with; facilitate dialogue with the stakeholders to develop a common understanding of the state of the landscape; identify specific threats to ecosystem components in the landscape; and the relevance of restoration to address these identified issues. By having an overview of the situation, field assessments can be planned for specific sites.

Keep in mind that the process of developing common understanding will be iterative and that management of the planning process must be adaptive at each step in the process to optimize input and manage the 'wickedness' (Table 5.1) of such processes. In addition, this initial phase of planning provides an opportunity to introduce basic principles of ecosystem management (Box 5.1) and discussions of the array of ecosystem services (Figure 5.1) that are being provided – and through discussions with the stakeholders to use these tools to facilitate adoption of planning principles and initial ideas about goals and objectives of the restoration that are contextualized in relation to the services provided by the ecosystems.

Whether following a traditional approach or a path directed approach, restoring ecological processes (those interactions between and among biotic and abiotic elements) that are necessary to deliver the full spectrum of ecosystem services is an essential goal of ecosystem restoration. The decline in ecological functionality, as compared to a reference state appropriate to the system, can be measured through a number of remote-sensing methods. Remotely sensed data products can be used to evaluate regulatory and supporting services such as carbon sequestration, net primary production, and nutrient cycling. Parameters such as net primary production, or rates of photosynthesis and Leaf Area Index (LAI), can be used to estimate aboveground biomass and carbon sinks. These metrics can be used to monitor the loss of crucial supporting and regulatory services provided by an ecosystem.

The interaction between the atmosphere and terrestrial vegetation begins on the leaf surface of plants and remote sensors like ASTER (Advanced Spaceborne Thermal Emission and Reflection Radiometer) or MODIS (Moderate Resolution Imaging Spectroradiometer) aboard the Aqua and Terra Satellites and other optical sensors can be used to measure spectral reflectance from flora (Heiskanen 2006; Yamaguchi *et al.* 1998; Knyazikhin *et al.* 1998).

Ecosystem supporting, regulating, and provisional services can be assessed by using LiDAR data (Light Detection and Ranging) to measure biomass accumulation and carbon cycling by taking into account the height of vegetation. Estimates of net primary production for ecosystems can be derived from MODIS vegetation indices or LiDAR data by feeding its products into established ecosystem process models (Maselli *et al.* 2011). LiDAR's ability to measure vertical vegetation structure, in combination with spectral satellite imagery, will be highly useful for evaluating stand volume, age determination, vegetation vigour and chlorophyll, variables that can predict net primary production.

Global nitrogen parameters can be detected in the atmosphere and tracked globally through an atmospheric sensor such as the Ozone Monitoring Instrument (OMI) aboard the Aura Satellite (a NASA mission aimed at measuring atmospheric pollution). Indicators of oceanic nitrogen and carbon can be detected and monitored by linking increased photosynthesis from runoff nutrients with sea surface temperature measurements.

MODIS and NASA Landsat mission imagery is very useful for mapping and monitoring the state of coral reef and mangrove ecosystems, which provide coastal protection from natural hazards such as hurricanes and typhoons and support rich assemblages of biodiversity. Monitoring the extant of these ecosystems over time allows one to assess relative risk at particular sites (see Box 5.2 criteria: rapid decline in area and increase in fragmentation).

Other measures of ecosystem function that can be evaluated through remote sensing are trophic complexity, species richness, and community structure – all of which relate to degree of biodiversity and hence resilience of the system. Hyperspectral and multispectral imagery by satellites such as IKONOS – a commercial satellite platform and – NASA EO-1 offer spatial resolutions that are being evaluated in their ability to detect species assemblages and possibly even individual vegetation species (Turner *et al.* 2003) using ever increasing spectral libraries. However, it should be noted that satellite imagery alone, due to spatial and temporal coverage gaps, remains a limiting factor. The addition of drones and UAVs can greatly increase coverage for the hyperspectral imagery gathered by these sensors.

MODIS-derived greenness measures such as NDVI (Normalized Differential Vegetation Index) can serve as a proxy for healthy, growing vegetation. Vegetation phenology, indicated by NDVI data products, refers to the seasonal cycles of greening up influenced by climate and can have the ability to show the effects of invasive plant or insect species upon a landscape. Optical sensors capture light from the spectrum of wavelengths that are naturally reflected from the earth's surface, referred to as bands. Commonly used band combinations can highlight differences in vegetation, and from that land-cover, or ecosystem types can be distinguished and evaluated for changes in extent over time. The type and patterns in changes in seasonal NDVI can diagnose the presence of invasive species among native vegetation. By examining spectral bands, bare ground for example (indicated primarily by the short-wave infrared band) and healthy vegetation (bands corresponding to the near infrared wavelength) can be identified. By identifying bare ground following a disturbance, whether manmade or natural, used in combination with digital elevation maps, practitioners can identify regions susceptible to erosion where mitigation-oriented restoration efforts can be aimed. These would be cost-effective tools to monitor potential for erosion or mudslide events after fires covering large areas in mountainous forests.

Loss of key species, spread of invasive species, or persistent pollution associated with an ecosystem can be indirectly measured using remote-sensing instruments. Vegetation indices have been used as a surrogate for assessing fecundity for top predators by density of herbivore prey if the relationship between prey species and particular classes of vegetation is known (Carroll *et al.* 2006). These methods can also be employed to assess the extent and possible fragmentation of habitat for species.

In the context of assessing the state of an ecosystem, its near-term historic range in variability can be established by comparing remote-sensing imagery across a time line beginning in the early 1970s that is available from the NASA Landsat mission library. Discussions of the results of the analysis, along with maps and databases derived from photo interpretation and satellite imagery along with field assessments can assist in developing common understanding of restoration needs and help establish an initial assessment of scale and scope of a restoration and planning of projects aimed at restoring the extent of an ecosystem. Characteristic vegetation that comprises the ecosystem can be determined by land-cover maps from local to landscape scales. Time sequenced land-cover assessments provide a means to identify shifts in forest species composition; insect predation patterns; and other influences that affect the status of an ecosystem. Resilience within ecosystems can also be evaluated using these technologies by assessing diversity and ecological variability within a region (species and habitat) through land-cover mapping. Fast feedback such as the near-real-time satellite imagery can be accessed through online portals as the data processing methods are continually reducing data latency times to supply to end users.

Early in the planning phase it would be advantageous for the stakeholders to consider alternative restoration scenarios, based on agreed alternative outcomes. Remotely sensed data within a GIS (Geographic Information System) allows for earth observations to be collected along large scales in a consistent manner in standardized geographic data. Such data can be stored and accessed through online or local databases. Spatial analyses using various models can provide stakeholders different insights into possible effects of restoration activities. For example, maps of soil types and elevation constructed from open, online databases can be used to determine the suitability for alternative approaches to reforestation activities.

Ultimately, the indicators used to plan and monitor on-the-ground restoration activities can be aided greatly by spatial analysis provided by GIS and the remote-sensing data collected through the various satellite and airborne platforms. Spatial data ranging from soil conditions, vegetative condition, endangered species ranges, rainfall, terrain, to land use change and habitat suitability can be measured directly, modelled or inferred. The technology to communicate, study, plan, monitor, and execute restoration activities should only continue to expand.

Conclusions

This chapter has focused on the social dimensions of planning and executing ecosystem restoration. Where personal interest and commitment transcends legal 'ownership', ecosystem services provide an important lens for exploring various approaches to environmental restoration that is inevitably linked, both socially and ecologically, to desired outcomes. Conducting a restoration project begins with a handful of critical steps, which are not required to be carried out in some prescribed order but, in totality, represent the necessary actions required to move from an assortment of puzzle pieces to seeing the ecosystem as a whole. These actions include:

- identify ecosystem services provided by a landscape with a focus on the supporting and regulating services;
- define the scale and context of the restoration – how many hectares, what are the boundaries, and who are the actors; and
- assess available resources – human, financial, and natural capital among others – that can be used to advance restoration work.

To assist those responsible for planning a restoration we have introduced five tools to address the complex social factors that often have greater influence over a restoration than the science-oriented specialists. We begin with an introduction to 'wicked problems' – those problems that defy simple solutions and when solutions are applied the conditions change and inevitably new problems arise, nevertheless, taking that initial action often incrementally increases understanding. The checklist of characteristics and explanations provided in Table 5.1 will help managers identify these complex problems and thus provide insights into how to address them in the planning and design process.

The Ecosystem Approach provides broad principles for guiding ecosystem management actions, including restorations. Their principal value though lies in providing managers with general concepts to help establish common ground and guidelines among diverse stakeholders that can be used when initiating dialogues. They generally frame the social context in which ecosystem management is most effectively pursued.

Above all, ecosystem services provide an important lens through which managers can frame ecological restoration goals and objectives and inform current and future practitioners of restoration. Knowing the ecosystem services provided by the subject ecosystem, and how those services compare with historical records is crucial for stakeholders to engage in a restoration planning process. This ensures that the less obvious services are considered with those services that often garner most of the interest of stakeholders – such as recreational opportunities or harvest of high-value products like timber.

Ecosystem restoration that results in durable solutions and ecological benefits is guided by local knowledge, socio-economic and cultural goals, and desired outcomes. To help inform management decisions it is important to recognize and understand the four types of services: provisioning, regulating, cultural, and supporting, that constitute the basis for healthy and resilient socio-ecological systems. Globally accepted guiding principles – namely the Ecosystem Approach advanced by the Convention on Biological Diversity and the Millennium Ecosystem Assessment – provide critical considerations in planning and executing ecosystem restoration.

The IUCN Red List criteria provide an unbiased means to assess the state of an ecosystem, which in turn can reinforce arguments in support of the restoration and clarify the objectives of the intervention.

And finally, the assessment of the resilience of an ecosystem provides a way to consider the capacity of both the social and ecological components of the system to withstand change while sustaining delivery of needed ecosystem services. The assessment process provided by the Resilience Alliance (see References) is a great tool to engage stakeholders in a guided analysis of the resilience (both ecological and social) associated with a particular ecosystem. Further it provides a means for setting goals in terms of achieving greater resilience in the ecosystem, which would help frame restoration objectives.

Ecological restoration today is an interdisciplinary challenge, in large part, because an understanding of the natural sciences alone will not lead to the restoration of ecosystems. Human systems are inextricably connected to ecological systems and restoration decision-making occurs within a social and cultural context. Recognizing and understanding this fundamental relationship will facilitate the restoration of resilient ecosystems capable of maintaining the diverse array of ecological processes and services necessary to sustain life on the planet.

References

Barrow, E. 2014. *Shinyanga Forest: Retrofitting Resilience to the Shinyanga Forest Landscape Restoration Case Study*. IUCN, Gland, Switzerland.

Cabello, J., Fernández, N., Alcaraz-Segura, D., Oyonarte, C., Piñeiro, G., Altesor, A., Delibes, M. and Paruelo, J. M. 2012. The ecosystem functioning dimension in conservation: insights from remote sensing. *Biodiversity Conservation* 21: 3287–3305.

Carroll, C., Phillips, M. K., Lopez-Gonzalez, C. A. and Schumaker, N. H. 2006. Defining recovery goals and strategies for endangered species: the wolf as a case study. *BioScience* 56(1): 25–37.

Cloern, J. E., Knowles, N., Brown, L. R., Cayan, D., Dettinger, M. D., Morgan, T. L., Schoellhamer, D. H., Stacey, M. T., van der Wegen, M., Wagner, R. W. and Jassby, A. D. 2011. Projected evolution of California's San Francisco Bay-Delta-River System in a century of climate change. *PLoS ONE* 6(9): e24465.

Conklin, J. 2005. *Dialogue Mapping: Building Shared Understanding of Wicked Problems*, pp. 1–20. John Wiley, Hoboken, NJ. Retrieved from http://cognexusgroup.com/wp-content/uploads/2013/03/wicked-problems.pdf.

Cumming, G. S., Cumming, D. H. M. and Redman, C. L. 2006. Scale mismatches in social-ecological systems: causes, consequences, and solutions. *Ecology and Society* 11(1): 14. Retrieved from www.ecologyandsociety.org/vol11/iss1/art14.

DiGennaro, B., Reed, D., Swanson, C., Hastings, L., Hymanson, Z., Healey, M., Siegal, S., Cantrell, S. and Herbold, B., 2012. Using conceptual models in ecosystem restoration decision making: an example from the Sacramento-San Joaquin River Delta, California. *San Francisco Estuary and Watershed Science* 10(3): 1–15.

He, H. S., Dey, D. C., Fan, X., Hooten, M. B., Kabrick, J. M., Wikle, C. K. and Fan, Z. 2007. Mapping pre-European settlement vegetation at fine resolutions using a hierarchical Bayesian model and GIS. *Plant Ecology* 191(1): 85–94.

Heiskanen, J. 2006. Estimating aboveground tree biomass and leaf area index in a mountain birch forest using ASTER satellite data. *International Journal of Remote Sensing* 27(6): 1135–1158.

Hestir, E. L., Khanna, S., Andrew, M. E., Santos, M. J., Viers, J. H., Greenberg, J. A., Rajapakse, S. S. and Ustin, S. L. 2008. Identification of invasive vegetation using hyperspectral remote sensing in the California Delta ecosystem. *Remote Sensing of Environment* 12(11): 4034–4047.

Higgs, E. 2003. *Nature by Design: People, Natural Process, and Ecological Restoration*. MIT Press, Cambridge, MA.

Holling, C. S. 2000. Theories for sustainable futures. *Conservation Ecology* 4(2): 7. Retrieved from http://www.ecologyandsociety.org/vol4/iss2/art7.

Holling, C. S. 2001. Understanding the complexity of economic, ecological, and social systems. *Ecosystems* 4: 390–405.

Ice, G.G., Neary, D. G. and Adams, P.W. 2004. Effects of wildfire on soils and watershed processes. *Journal of Forestry* September: 16–20.

Keith D. A., Rodríguez, J. P., Brooks, T. M., Burgman, M. A., Barrow, E. G., Bland, L., Comer, P. J., Franklin, J., Link, J., McCarthy, M. A., Miller, R. M., Murray, N. J., Nel, J., Nicholson, E., Oliveira-Miranda, M. A., Regan, T. J., Rodríguez-Clark, K. M., Rouget, M. and Spalding, M. D. 2015.The IUCN Red List of Ecosystems: motivations, challenges and applications. *Conservation Letters* 8(3). Retrieved from http://onlinelibrary.wiley.com/doi/10.1111/conl.12167/epdf.

Knyazikhin, Y., Martonchik, J. V., Myneni, R. B., Diner, D. J. and Running, S. W. 1998. Synergistic algorithm for estimating vegetation canopy leaf area index and fraction of absorbed photosynthetically active radiation from MODIS and MISR data. *Journal of Geophysical Research* 103(24): 32,257–32,276.

Macmillan, D.C., Harley, D. and Morrison, R. 1998. Cost-effectiveness of woodland ecosystem restoration. *Ecological Economics* 27(3): 313–324.

Maselli, F., Chiesi, M., Fibbi, L. and Moriondo, M. 2011. Use of ground and LiDAR data to model the NPP of a Mediterranean pine forest. *Remote Sensing Letters* 2(4): 309–316.

Millennium Ecosystem Assessment. 2005. *Ecosystems and Human Well-being: Current State and Trends: Findings of the Condition and Trends Working Group*, vol. 1 (ed. R. Hassan, R. Scholes and N. Ash). Island Press, Chicago, IL.

Obama, B. 2013. Executive Order: making open and machine readable the new default for government information. 9 May. Retrieved from https://obamawhitehouse.archives.gov/the-press-office/2013/05/09/executive-order-making-open-and-machine-readable-new-default-government-.

Oetter, D. R., Ashkenas, L. R., Gregory, S. V. and Minear, P. J. 2004. GIS methodology for characterizing historical conditions of the Willamette River flood plain, Oregon. *Transactions in GIS* 8: 367–383.

Resilience Alliance. 2010. *Assessing Resilience in Social-Ecological Systems: Workbook for Practitioners, Version 2.0.* Resilience Alliance. Retrieved from www.resalliance.org/resilience-assessment.

Rodríguez, J. P., Keith, D. A., Rodríguez-Clark, K. M., Murray, N. J., Nicholson, E., Regan, T. J., Miller, R. M., Barrow, E. G., Bland, L. M., Boe, K., Brooks, T. M., Oliveira-Miranda, M. A., Spalding, M. and Wit, P. 2015. A practical guide to the application of the IUCN Red List of Ecosystems criteria. *Philosophical Transactions of the Royal Society B* 370: 20140003. http://dx.doi.org/10.1098/rstb.2014.0003

Smith, M. 2016. *Collaboration among How Business, Government and NGOs Could Be the Key to Living with Turbulence and Change in the 21st Century.* IUCN, Gland, Switzerland.

Suding, K., Higgs, E., Palmer, M., Callicott, B., Anderson, C. B., Baker, M., Gutrich, J. J., Hondula, K. L., LaFevor, M. C., Larson, B. M. H., Randal, A., Ruhl, J. B. and Schwartz, K. 2015. Committing to ecological restoration: Efforts around the globe need legal and policy clarification. *Science* 348(May): 638–640.

Turner, W., Spector, S., Gardiner, N., Fladeland, M., Sterling, E. and Steininger, M. 2003. Remote sensing for biodiversity science and conservation. *Trends in Ecology and Evolution* 18(6): 306–314.

UN Convention on Biological Diversity. 2000. Decision V/6, of the 5th Meeting of the Conference of the Parties to the Convention on Biological Diversity, Nairobi, Kenya, 15–26 May 2000.

Wilson, E. O. 2012. *The Social Conquest of Earth.* New York: Liveright.

Yamaguchi, Y., Kahle, A. B., Tsu, H., Kawakami, T. and Pniel, M. 1998. Overview of advanced spaceborne thermal emission and reflection radiometer (ASTER). *IEEE Transactions on Geoscience and Remote Sensing* 36(4): 1062–1071.

6

THE ROLE OF HISTORY IN RESTORATION ECOLOGY

Eric S. Higgs and Stephen T. Jackson

Introduction

Restoration ecology is fundamentally a historical pursuit. The very essence of the word *restoration* implies a return to past conditions. To restore, according to one of the several meanings recorded in the *Oxford English Dictionary*, is 'the action of returning something to a former owner, place, or condition'. A variant and now obscure spelling, 'restauration', suggests a complicated lineage that describes the recovery of health or proper condition (Higgs 2003). There is also *instauration*, used by Francis Bacon in the early seventeenth century to create something new from old by adding enlightenment and truth. Thus, what we now think of as restoration generally, and ecological restoration specifically, has a significant cultural and linguistic lineage.

At its root, restoration of anything – ecosystems, works of art, civil order – implies a return of some desirable feature that has been lost or damaged. To the uninitiated, the restoration of ecosystems resembles the restoration of damaged works of art or old buildings: the goal is to take it back to the way it once was. The problem is that ecosystems are constantly changing in response to internal processes and external drivers. If ecosystems were paintings, imagine what it would be like to restore a degraded painting that is constantly shifting in design and form and in ways that are not fully comprehensible.

Marcus Hall opens his comparative history of ecological restoration in Europe and North America with an account of the restoration of the Sistine Chapel (Hall 2005). Painstaking work over several years resulted in the removal of centuries of grime and also various additions to the original ceiling paintings by Michelangelo. The results were shocking: the vibrant colours and uncloaked cherubs defied public perception about what it is *supposed* to look like. Despite the relatively unchanging conditions of human artefacts, there are changes not only in the artefact itself but how the artefact is appreciated. This is the essence of the complication faced by those who undertake restoration of any kind: there is the fact of what was and how it changed over time, but at least as important is how the conditions are *perceived*. Understanding history in restoration ecology requires awareness of culturally laden ideas that make up an understanding of an ecosystem. Ecosystems are restored by reference to historical conditions, but the choice of which conditions to honour are rooted in cultural ideas about what is appropriate. A decision, for example, to control rapid ingrowth of Douglas-fir (*Pseudotsuga menziezii*) trees within a grove of Garry oak (*Quercus garryana*) trees on southern Vancouver Island, Canada, is shaped by understanding of historical data and also the aesthetic and policy premium placed on restoring threatened oak habitat (MacDougall *et al.* 2004). Ecological and cultural understanding shapes what we value about these ecosystems. Such understanding of ecosystems historically is nearly always fragmentary and imperfect, and even systematic, expert inferences from data are shaped by cultural notions about what is important to study and how those data are interpreted.

In this chapter, we examine the ways in which historical knowledge is used in restoration, mindful that what we gather as facts about the past are interpreted in the present with an eye to guiding the future. Our first task is explaining the concept of historical fidelity and its role in restoration, and how rapid environmental and ecological changes are forcing a reconsideration of how historical knowledge is used in shaping restoration. We argue that in spite of rapid change, historical knowledge in its many forms will continue to be valuable in restoration science and practice, and that its value will even increase. Nevertheless, the ways in which historical knowledge will be used in restoration in the future will change, and we offer some speculations about how this may play out.

Historical fidelity

Historical fidelity generally is the degree of correspondence between our understanding of the past and the reality of that past. This concept, however, has both epistemological and practical elements. In the epistemic domain, historical fidelity describes our best understanding of past historical conditions of an ecosystem based on evidence pulled from incomplete, imperfect, and fragmentary sources including palaeoecology, archival records, historical photographs, early descriptions, old-growth remnants, and legacy data (Egan and Howell 2001; Higgs 2003; Cole *et al.* 2010; Jackson 2012b). Perfect fidelity of inference – that is, perfect correspondence between the actual past state(s) of a system and our inferential understanding of the past state(s) – is impossible: we cannot know everything about an ecosystem's past, and indeed sometimes evidence is scarce or non-existent. An implicit goal in ecological restoration continues to be a responsible search for reasonable historical evidence to establish appropriate goals for restoration. The practical element of historical fidelity is concerned with the correspondence between outcomes of restoration practice with inferred understanding of past conditions. In this sense, restoration practice aspires to achieve specific goals based on understanding of past conditions, and fidelity of practice is a measure of how well outcomes meet these goals.

Why place such a premium on historical evidence? The most obvious answer is that historical conditions can define a pre-disturbance condition in which the ecosystem presumably demonstrated greater health or integrity. The *condition* of an ecosystem includes more than structural aspects (e.g., composition of species), but the processes and dynamics that characterize the distinctive qualities of an ecosystem. Past conditions then provide a model for how that ecosystem *ought* to look and function (Stephenson 1999; Falk 2006). This points to a subtler explanation of the role of history in ecological restoration: as an arbiter of appropriate goals. Imagine that the choice of goals is left to human intentions, even if those intentions are ecologically benevolent. The results might be goals that more directly benefit human interests rather than reflecting the conditions of the ecosystem irrespective of human interests. Thus, by resorting to historical evidence, there is an anchor point, or set of potential goals mostly outside of human considerations (well, almost: the choice of historical conditions may reflect human perceptions of what is good or right).

History as a model makes restoration seem deceptively simple. It is not, and largely because of the difficulty in obtaining reliable, comprehensive, and detailed historical data. Documentary evidence is often ambiguous, incomplete, or absent, and analysis is shaped by the hypotheses used to guide the study. For instance, work on repeat photography in Jasper National Park in the late 1990s that led to one of the world's most extensive repeat photography initiatives, the Mountain Legacy Project (explore.mountainlegacy.ca; Trant *et al.* 2015; Higgs 2003), was spurred by a lack of historical documentation and fragmentary records. Similarly, oral histories of individuals who might have known something about the pre-disturbance conditions of an ecosystem are critical resources, but too seldom collected and archived for easy access. Palaeoecological evidence may be absent for any number of reasons (e.g., no sedimentary archives, no studies attempted, existing studies leave critical questions unanswered). All records require interpretation, and pitfalls and ambiguities can arise in inferring both fundamental descriptive properties of past ecosystems (Jackson 2012b) and causal mechanisms responsible for those properties (Jackson and Blois 2015). For example, the drivers underlying a rapid, rangewide decline in hemlock populations in eastern North America during the mid-Holocene, an intensively studied palaeoecological phenomenon, remain enigmatic (Booth *et al.* 2012), in part because of the likelihood of multiple mechanisms operating at different spatial and temporal scales (Jackson and Blois 2015). In urban areas, the rapid pace of development, and often many layers of sequential development, make the discovery and reconstruction of precise historical conditions a monumental task. The Welikia project used painstaking historical research alongside spatial modelling to create a similar representation of New York City's boroughs prior to European colonization (Sanderson 2013). Finally, historical understanding is often location-specific, and application of that understanding to other sites or across broader scales often bears a hidden assumption of uniformity or spatial stationarity of past patterns. For example, scattered, isolated old-growth forest remnants in eastern North America may not represent the diversity and heterogeneity of pre-clearance forests across the landscape.

A realistic approach to using historical information is to think of it in terms of *reference conditions*. This idea has circulated in ecological restoration for several decades. Much like the choice of references to support arguments in a scholarly article, historical references are used in support of particular goals for an ecosystem. References can be partial or inconclusive, but they do at least provide a starting point. White and Walker (1997) proposed a simple schematic for reference information based on spatial and temporal information. Reference data may be available for the specific site to be restored, or they may be missing and need to pulled in from another location (Figure 6.1). Historical data may be missing, and sometimes the easiest and most reliable reference information is available from undisturbed sites that may or may not be proximate. And, as is often the case, several lines of evidence are brought to bear (e.g., fragmentary historical data from the site itself, as well as the conditions of a nearby intact ecosystem). Definition of reference conditions is inevitably judgmental, requiring weighing of multiple sources of evidence and perspectives.

Restoration ecologists and practitioners deploy references in a variety of ways, ranging from those that are more-or-less precise as regards historical data to those that use references as a starting point for goals and objectives. While rare, precise historical references are central to a limited range of projects, such as the Antietam National Historic Battlefield that recreated and now manages the ecological conditions of the battlefield to mimic as closely as possible those experienced in the 1862 battle (see www.nps.gov/anti/index.htm). This kind of project is an exception that proves the rule that historical references are more generally used in the way citations are used as references in a research article or book: the evidence is used variably and flexibly to interpret appropriate goals for a restoration project. The Redstreak restoration

	Same place	**Different place**
Same time	*Contemporary status; assessment of disturbance*	*Determination of extent of disturbance and potential for restoration*
Different time	*Site-specific reference*	*Analogous reference information*

Figure 6.1 Reference information about a site to be restored can be gathered from one or several sources to create a composite view of a pre-disturbance or intact ecosystem. Hiers *et al.* (2012) extend this to a dynamic reference to address 'a rapidly changing no-analogue future'

Source: adapted from White and Walker (1997); Higgs (2003)

project undertaken by Parks Canada in Kootenay National Park focused on resetting the structural forest conditions of the nineteenth century to reflect the longer term historical range of variability. The open, mature Douglas-fir-dominated forest conditions had given way to dense stands of even-aged forests that created significant fire risk for the nearby campground and community (Radium Hotsprings), and open-mature forests provide high quality forage for Rocky Mountain Sheep away from the verge of the adjacent highway. Thus, historical conditions derived from fire history studies (Kubian 2013), historical and repeat photographic surveys (see www.mountainlegacy.ca), and local knowledge resulted in forest thinning and subsequent low intensity regular fires (see www.pc.gc.ca/eng/pn-np/bc/kootenay/natcul/redstreak.aspx). Increasingly, restoration projects require adaptive and wide-ranging approaches to meet the challenges of rapidly shifting environmental and ecological conditions (Keenleyside *et al.* 2012). Emerging approaches such as a 'dynamic reference' (Hiers *et al.* 2012), or using an 'extended historical range of variability' (eHROV) (Jackson 2012a), are attempts at using historical references in less literal ways. Rapid change might point to a diminishing role for historical data and knowledge, but this is not likely the case (see 'Varieties of historical knowledge' later in this chapter).

It is not only technical reasons that propel restoration scientists and practitioners toward historical knowledge. While most restoration projects are committed to retaining species and ensuring ecological integrity and resilience, restoration scientists and practitioners also know that cultural reasons play important roles in shaping the value proposition for historical knowledge. There are three cultural reasons that stand out, but there are likely others that hold importance. First, *nostalgia*, or the 'bittersweet longing for the past', shapes much of how people think about past ecological (and cultural conditions; Higgs 2003). For example, nostalgia grips people when they reflect on the loss of a favourite childhood haunt or playspace. What we decide to restore is shaped to an extent by how people remember their experiences of particular ecosystems and landscapes. Nostalgia is often dismissed as an unrealistic longing for something that cannot be, but ecological restoration opens up hope that some of what is lost may be regained. Nostalgia provides a partial antidote for shifting baseline syndrome, which occurs when present (altered) conditions form the baseline for shaping future decisions in restoration and conservation. Nostalgic remembrances recall earlier baselines, and propel a search for deeper meaning of the past.

Second, people are also shaped by attachment to *place*. The connections people form to ecosystems are described as situated by their attachment to place, which is a complicated term describing a host of cultural relationships people established with localities: seasonality, topography, viewscapes, vegetation cover, charismatic species, smells, cultural activities and celebration, and so on. For some, a home place is what defines deep relationships to a place, and informs how and why restoration takes place. While not precisely correlated, the greater a sense of place the greater likelihood ecological restoration would be undertaken to address degradation. Third, *time depth* describes intergenerational connections to place. Having a long connection to a place will generally inspire greater concern for restoration and conservation (Turner 2005). It is also the case that older ecosystems often convey greater value for ecosystems; 'old growth' forests, for example, enhance a commitment to restoration because they show what can be lost without careful conservation and how much effort is involved in restoring them. The cultural resonance of nostalgia, of place, and of history runs deeply and packs emotional power. Heritage and related notions (homeland, heartland, motherland, fatherland) are typically a mix of natural and cultural elements.

Challenges to historical knowledge

A model of 'classical' restoration, or 'restoration 1.0', is being replaced by a more flexible version of restoration ('restoration 2.0') that proposes a shift from historical information as a template to a guide for setting restoration goals and objectives (Higgs *et al.* 2014). Restoration practitioners have long been aware that site constraints and information availability shapes the practical outcome of projects. For much of the last three decades, historical references were viewed as providing a determinate guide for restoration. There are four reasons why this conventional view is being challenged.

First, the appropriate temporal conditions are seldom obvious. It was common, and remains so in many settings, to speak of determinate times in the past to represent in restoration projects (e.g., pre-colonial intervals in North America; an interval prior to a specific disturbance; a period preceding disappearance of megaherbivores or top carnivores). This poses difficult questions: Which point in the past is the appropriate one to use as the reference point? Who gets to decide? Is the past really a state – or is it a process? Historical and palaeoecological records indicate that ecological systems evolve and can be laden with historical contingencies (Jackson *et al.* 2009; Jackson 2012a), and contemporary old-growth ecosystems bear legacies of past events (Hiers *et al.* 2012). Finally – with apologies to William Faulkner – is the past even past? Ecological processes unfold across many temporal scales (Jackson and Blois 2015), and any point in time, including the present, may be somewhat arbitrary. For example, northward migrations of several tree species in the central Rockies has been underway for several millennia, and haven't reached an obvious endpoint (see Norris *et al.* 2016 for a recent example). A helpful distinction exists between the past, which simply happened and is not subject to particular interpretations, and history, which 'indicate[s] the interpretation of both human and ecological pasts, recognizing that such interpretations are constantly changing in response to new knowledge' (Higgs *et al.* 2014). To understand properly the role of history in restoration is to acknowledge that any deployment of historically contingent goals in restoration is a consequence of distinctly human beliefs and priorities. Thus, a decision to return prescribed fire to a fire-dependent forested ecosystem may be motivated by public safety, a nostalgia for a landscape before industrial activity, recovery of a mosaic that honours indigenous land management, or a combination of these goals (Higgs and Hobbs 2010).

Second, during this same period (since the 1980s) there developed an unequivocal understanding of directional environmental and ecological drivers of ecosystem change. Climate change is the most commonly recognized form of directional change, but there are multiple drivers – biogeochemical alterations, land conversion, biological invasions – that contribute to steady change and also fluctuations in ecosystems and landscapes. Of particular consequence for restoration scientists and practitioners are invasive species, which result in transformations of ecosystems and often in combination with other drivers (Meyerson and Mooney 2007; Walther *et al.* 2009). The potential for unexpected thresholds in ecosystem response to drivers (Bestelmeyer *et al.* 2011) creates significant uncertainties in how best to plan for restoration. Are historical references salient in an era of marked change? Can the future be modelled in such a way as to make restoration prescriptions match goals in coming decades? These uncertainties pose a significant challenge to restoration practice, and imply that historical knowledge may be less valuable for restoration in the future. Paradoxically, though, history may actually increase in relevance with high-magnitude or abrupt environmental and ecological changes. Although specific changes underway now or in the future may go beyond historical bounds of specific existing systems, history can provide insights into kinds and consequences of high-magnitude or rapid changes, applicable in more general terms. In this sense, history expands our range of experience, providing examples and models of changes unlike those of the relatively recent past.

Third, and closely related to the second point, is that the historical range of variability (HROV), a concept that has provided a useful approach to determining a flexible range of interpreting ecological history for restoration goals, may be undermined by rapid directional change (Cole *et al.* 2008; Keane *et al.* 2009). In fact, historical studies indicate that natural climatic and ecological variations in the past have been non-stationary, with directional change at various timescales, sometimes with slow or rapid switches in direction (Jackson *et al.* 2009; Milly *et al.* 2008). Human activities, including climate change, have added new magnitudes of directional change that are likely to exceed historical experience (Steffen *et al.* 2015). The dynamic reference concept (Hiers *et al.* 2012; Kirkman *et al.* 2013) provides a framework for incorporating evolving environmental conditions and dynamic responses, incorporating historical knowledge, monitoring, and emergent ecological novelty.

Fourth, the solution to the challenge of fixing historical points in a changing world can at least be partly addressed by allowing multiple potential trajectories for an ecosystem (Balaguer *et al.* 2014). However, the idea of addressing degradative pressures on an ecosystem and allowing it to find a suitable path diminishes the significance of historically contingent trajectories. This may be a problem in cases where rare species, ecosystem or landscape elements are in play. It opens concerns about how different trajectories would be arbitrated in relation to shifting human beliefs about appropriate conditions. However, this is no different from other decisions routinely made in restoration, wherein competing and often conflicting values are adjudicated. Furthermore, palaeoecological studies indicate that multiple trajectories can occur naturally, contingent on particular climatic, disturbance, and recruitment events (Jackson *et al.* 2009; Jackson and Blois 2015). In such cases, selecting a particular historical benchmark for restoration may be arbitrary (Jackson and Hobbs 2009).

In an era of rapid change, it seems the deck is stacked against models of restoration that embed strong historical ideas. Or, is it? Certainly, rapid change is likely to shift the composition of ecosystems, and in this respect strict historical adherence will prove less directly helpful. Understanding the historical processes that give rise to ecosystem functions, including those recognized as providing valuable ecosystem services, remains a critical role for history. In the next section, we explore an emerging way of thinking about historical knowledge that may increase the relevance of historical knowledge in future restoration activities.

Varieties of historical knowledge

Historical knowledge is diverse, comprising many strands of data and approaches to interpretation. This insight led an interdisciplinary team of historians, ecologists, and philosophers to consider historical knowledge as a *genus* comprising various species of historical information (Higgs *et al.* 2014). Early attempts at taxonomy yielded more than twenty distinct types of historical information drawn from a wide variety of sources. Refinement led to nine types that fit into three broader categories: history as information and reference; history as enriching cultural connections; and, history as revealing the future (Table 6.1). What proved surprising even for those well versed in the deployment of historical knowledge in restoration was the extent to which history supports efforts at projecting future conditions (as scenario, experiment, and virtue of the future) for ecosystems and the importance of cultural forms of historical knowledge (as place, a form of redress for past harms, and as a governor, or limit, on excessive human ambition). For each of the nine types, and based on professional judgment the nine co-authors assessed whether their role in restoration was likely to decrease, increase, or remain the same. The counterintuitive result was that dividing historical knowledge into component types of knowledge showed that five of the nine types were 'intensifying' (legacy, governor, scenario, experiment, and virtue), two were holding steady (reference, redress), only one was diminishing (range of variability), and one was unclear (place). This taxonomy and

Table 6.1 Categories and types of historical knowledge interpreted in terms of contemporary and likely future roles

Categories	*Historical knowledge* Types	*Role in contemporary restoration (Restoration v. 1.0)*	*Likely role in the future (Restoration v. 2.0)*
History as information and reference	(1) History as reference	Strong	Steady
	(2) History as range of variability	Strong	Diminishing
	(3) History as legacy	Moderate	Intensifying
History as enriching cultural connections	(4) History as place	Strong	Unclear
	(5) History as redress	Strong	Steady
	(6) History as governor	Weak	Intensifying
History as revealing the future	(7) History as scenario	Moderate	Intensifying
	(8) History as experiment	Moderate	Intensifying
	(9) History as virtue	Weak	Intensifying

Note: These roles were ascertained by the authors, based on experience and informed speculation, and are advanced as a starting point for further debate and refinement.

Source: adapted from Higgs *et al.* (2014)

assessment of the future role of the various types is a starting point for debate about the role of history in restoration ecology. There seems little question, however, that historical knowledge will contribute significantly to ongoing restoration efforts in the future, and this result seems almost counterintuitive.

In Table 6.1, nine types of historical knowledge are placed in three main categories. It is striking that just three of nine types are what is commonly associated with restoration (history as information and reference category). The remaining types are mostly self-explanatory. 'History as redress' refers to ecological and environmental damage that sets up obligations on restoration scientists and practitioners to address past harm. 'History as governor' is the quality of historical knowledge and research that tends to slow down (literally and figuratively) restoration goal setting as a counter to overly ambitious and exuberant projects that fail to understand past conditions and trajectories. 'History as virtue' refers to a potential character virtue of the future, whereby restorations scientists, and practitioners will embed historical knowledge in their practices despite, or perhaps because of, rapid change (see also Higgs 2012).

The changing role of history in restoration ecology

It is a turbulent time for restoration and allied practices such as conservation and reclamation. Rapidly changing conditions, the risks of tipping points in ecosystems, the emergence of novel environments and ecosystems, are challenging conventional approaches. Calls for a 'restoration 2.0', (Higgs *et al.* 2014) or a new and broader approach to intervention ecology (Hobbs *et al.* 2011) suggest that the role of history in restoration will change as the conditions influencing ecosystems outpace traditional adaptation. Of all the types of historical knowledge, one stands out as playing a critical role in a changed world: history as governor. Historical knowledge of the past serves to anchor human ambitions in an understanding of the past, which has the advantage of compelling thought before action. Indeed, restoration has operated with the advantage of treating ecosystems with respect to their history, however interpreted, and this provides some measure of ethical grounding. Negating or ignoring historical knowledge opens ecosystem intervention to less constrained human ambitions: enhancing ecosystem services, functional ecosystems, and green infrastructure in the place of natural systems. History teaches important lessons about ecosystems. By maintaining a focus on the myriad forms of historical knowledge that support restoration, it is more likely that our practices will be better measured, more thoughtful, and ultimately more responsive to the dynamics of rapid change.

Acknowledgements

We are grateful to Kevin Hiers and Tom Swetnam for their reviews of an earlier version of this chapter.

References

Balaguer L, Escudero A, Martín-Duque JF, Mola I and Aronson J, 2014. The historical reference in restoration ecology: re-defining a cornerstone concept. *Biological Conservation* 176: 12–20.

Bestelmeyer BT, Ellison AM, Fraser WR, Gorman KB, Holbrook SJ, Laney CM, Ohman MD, Peters DPC, Pillsbury FC, Rassweiler A, Schmitt RJ and Sharma S, 2011. Analysis of abrupt transitions in ecological systems. *Ecosphere* 2(12): p.art129.

Booth RK, Brewer S, Blaauw M, Minckley TA and Jackson ST, 2012. Decomposing the mid-Holocene *Tsuga* decline in eastern North America. *Ecology* 93: 1841–1852.

Cole DN, Yung L, Zavaleta ES, Aplet GH, Stuart Chapin III F, Graber DM, Higgs ES, Hobbs RJ, Landres PB, Millar CI, Parsons DJ, Randall JM, Stephenson NL, Tonnessen KA, White PS, and Woodley S, 2008. Naturalness and beyond: protected area stewardship in an era of global environmental change. *The George Wright Forum* 25(1): 36–56.

Cole D, Higgs E and White P, 2010. Historical fidelity: maintaining legacy and connection to heritage. In D Cole and L Yung (eds), *Beyond naturalness: rethinking park and wilderness stewardship in an era of rapid change*. Washington DC: Island Press, pp. 125–141.

Egan D and Howell E, 2001. *The historical ecology handbook: a restorationist's guide to reference ecosystems.* Washington, DC: Island Press.

Falk, DA, 2006. Process-centred restoration in a fire-adapted ponderosa pine forest. *Journal for Nature Conservation* 14(3–4): 140–151.

Hall M, 2005. *Earth repair: a transatlantic history of environmental restoration.* Charlottesville, VA: University of Virginia Press.

Hiers JK, Mitchell RJ, Barnett A, Walters JR, Mack M, Williams B and Sutter R, 2012. The dynamic reference concept: measuring restoration success in a rapidly changing no-analogue future. *Ecological Restoration* 30(1): 27–36.

Higgs E, 2003. *Nature by design: people, natural process, and ecological restoration.* Cambridge, MA: MIT Press.

Higgs E, 2012. History, novelty, and virtue in ecological restoration. In A Thompson and J Bendik-Keymer (eds), *Ethical Adaptation to Climate Change.* Cambridge, MA: MIT Press.

Higgs ES and Hobbs RJ, 2010. Wild design: principles to guide interventions in protected areas. In DN Cole and L Yung (eds), *Beyond naturalness: rethinking park and wilderness stewardship in an era of rapid change.* Washington, DC: Island Press, pp. 234–251.

Higgs E, Falk DA, Guerrini A, Hall M, Harris J, Hobbs RJ, Jackson SJ, Rhemtulla JM and Throop W, 2014. The changing role of history in restoration ecology. *Frontiers in Ecology and the Environment* 12(9): 499–506.

Hobbs R, Hallett L, Ehrlich P and Mooney H, 2011. Intervention ecology: Applying ecological science in the twenty-first century. *BioScience* 61(6): 442–450.

Jackson ST, 2012a. Conservation and resource management in a changing world: extending historical range of variation beyond the baseline. In JA Wiens *et al.* (eds), *Historical environmental variation in conservation and natural resource management.* New York: John Wiley, pp. 92–109.

Jackson ST, 2012b. Representation of flora and vegetation in Quaternary fossil assemblages: known and unknown knowns and unknowns. *Quaternary Science Reviews* 49: 1–15.

Jackson ST and Blois JL, 2015. Community ecology in a changing environment: perspectives from the Quaternary. *Proceedings of the National Academy of Sciences* 112: 4915–4921.

Jackson ST and Hobbs RJ, 2009. Ecological restoration in the light of ecological history. *Science* 325: 567–569.

Jackson ST, Betancourt JL, Booth RK and Gray ST, 2009. Ecology and the ratchet of events: climate variability, niche dimensions, and species distributions. *Proceedings of the National Academy of Sciences* 106: 19685–19692.

Keane R, Hessburg P, Landres P and Swanson FJ, 2009. The use of historical range and variability (HRV) in landscape management. *Forest Ecology and Management* 258(7): 1025–1037.

Keenleyside K, Dudley N, Cairns S, Hall C and Stolton S, 2012. *Ecological restoration for protected areas.* Gland, Switzerland: IUCN. Retrieved from http://data.iucn.org/dbtw-wpd/edocs/PAG-018.pdf.

Kirkman LK, Barnett A, Williams BW, Hiers JK, Pokswinski SM and Mitchell RJ, 2013. A dynamic reference model: a framework for assessing biodiversity restoration goals in a fire-dependent ecosystem. *Ecological Applications* 23: 1574–1587.

Kubian R, 2013. Characterizing the mixed-severity fire regime of the Kootenay Valley, Kootenay National Park. MSc thesis, University of Victoria.

MacDougall AS, Beckwith BR and Maslovat CY, 2004. Defining conservation strategies with historical perspectives: a case study from a degraded oak grassland ecosystem. *Conservation Biology* 18(2): 455–465.

Meyerson L and Mooney H, 2007. Invasive alien species in an era of globalization. *Frontiers in Ecology and the Environment* 5(4): 199–208.

Milly PCD, Betancourt J, Falkenmark M, Hirsch RM, Kundzewicz ZW, Lettenmaier DP and Stouffer RJ, 2008. Stationarity is dead: whither water management? *Science* 319: 573–574.

Norris J, Betancourt JL and Jackson ST, 2016. Late Holocene expansion of Ponderosa pine in the central Rockies. *Journal of Biogeography* 43: 778–790.

Sanderson EW, 2013. *Mannahatta.* New York: Harry N. Abrams.

Steffen W, Richardson K, Rockström J, Cornell SE, Fetzer I, Bennett EM, Biggs R, Carpenter SR, de Vries W, de Wit CA, Folke C, Gerten D, Heinke J, Mace GM, Persson LM, Ramanathan V, Reyers M and Sörlin S, 2015. Planetary boundaries: guiding human development on a changing planet. *Science* 347(6223): DOI: 10.1126/science.1259855.

Stephenson NL, 1999. Reference conditions for giant sequoia forest restoration: structure, process, and precision. *Ecological Applications* 9(4): 1253–1265.

Trant AJ, Starzomski BM and Higgs E, 2015. A publically available database for studying ecological change in mountain ecosystems. *Frontiers in Ecology and the Environment* 13(4): 187.

Turner NJ, 2005. *The Earth's blanket*. Vancouver: Douglas & McIntyre.

Walther G-R, Roques A, Hulme PE, Sykes MT, Pyšek P, Kuhn I, Zobel M, Bacher S, Botta-Dukat Z, Bugmann H, Czucs B, Dauber J, Hickler T, Jarisik V, Kenis M, Klotz S, Minchin D, Moore, M, Netwig W, Ott J, Panov V, Reineking B, Robinet C, Semenchenko V, Solarz W, Thuiller W, Vila M, Vohland K and Settele J, 2009. Alien species in a warmer world: risks and opportunities. *Trends in Ecology and Evolution* 24(12): 686–693.

White P and Walker J. 1997. Approximating nature's variation: selecting and using reference information in restoration ecology. *Restoration Ecology* 5(4): 338–349.

7

SOCIAL ENGAGEMENT IN ECOLOGICAL RESTORATION

Susan Baker

Introduction

Public participation in environmental policy is supported by national and international agencies and organizations, including the United Nations. An extensive literature exists on the theme, particularly in developing world settings. Much of this is connected with the field of common pool resources, focusing on designing incentives and institutions to combine sustainable livelihood activities and conservation practices (Ostrom 1990). This literature has been influential in establishing best practices and developing innovative incentive mechanisms to promote community participation and overcome related problems of environmental degradation and poverty.

Like conservation management, ecological restoration has proved a fertile ground for grassroots engagement and is strongly supported by an array of institutions and voluntary actors and groups, operating across multi levels of governance. This paper explores participation in restoration, viewing goals and assessing outcomes using a political science lens, taking a governance approach and a sub-disciplinary, policy analysis perspective. It views restoration as both a tool of public policy, aimed at multiple benefits such as climate change mitigation and adaptation, and as a practice, as exercised in large-scale restoration initiatives and local-level restoration projects.

The focus is on civil society engagement in the making and delivery of public policy. We refer to this as participation, distinguishing it from societal engagement more generally, the latter that may not necessarily be directly engaged in arenas of public policy making. Participation, understood in this sense, occurs through institutional settings that bring together various actors at some stage in the policy-making process (van de Hove 2000). Participatory practices can vary considerably and are best seen as being located along a continuum, ranging at their extremes from allowing only a minor, consultative role for non-state actors to more deliberative processes in which actors have a major say in shaping policy goals through dialogue and social learning (van Zeijl-Rozema *et al.* 2008).

Participation contributes to the governance of public affairs. Governance focuses on the administrative and process-oriented elements of governing. As a form of social steering, it is undertaken in three 'classical' ways, referred to as hierarchies, markets and networks governance styles. Public policy making typically involves some partnership combination of state agents, market actors and civil society groupings. Hierarchies see the state, that is, governments and bureaucracies including subnational governments such as municipalities and local authorities, utilize laws and regulations, so-called 'command and control', to bring about desired ends. State steering also sees the utilization of planning, budgetary allocation and other coordinating and distribution actions. This top-down approach is typically combined with market tools, such as taxation and subsidies. Network governance sees the engagement of economic interests and societal groups, a form that has taken on a new impetus in the face of rising societal challenges, including those stemming from global environmental change. It includes the involvement of economic actors through, for example, voluntary agreements and covenants as well as public/private partnerships for service delivery. The participation of civil society groups in restoration policy and practice forms the main subject matter of this chapter. It explores the significance of this participation for ecological restoration, focusing on how it contributes to environmental awareness as well as societal benefits. The paper is also mindful of how participation combines with other governance styles and thus of the need to consider the implications this has for societal participation in restoration.

We begin by examining the prospects of participation before turning to the barriers to participatory practices. Here attention is given to how institutional factors, including policy styles and political cultures, shape participatory opportunities. Participation is then examined in the context of the full range of governance styles.

Participation as beneficial governance

The study of participation constitutes a vast arena of research, spanning several disciplines and issue areas (van Deth 2014). Scholars from the discipline of philosophy and environmental ethics have provided keen insight into how engagement in restoration can heighten environmental consciousness, fostering community connection with the land (Jordan 2000). Societal engagement also holds the potential for a positive, local human–nature relationship (Light and Higgs 1996: 236). Thus, rather than consuming nature, through high-impact, expensive 'leisure' activities, engagement in ecological restoration projects can provide an alternative space for low-impact, non-consumptive engagements with nature (Jordan 2000: 33).

These views can be complemented by a political science perspective, which sees participation as both driving and being driven by values associated with democracy, in particular contemporary participatory principles. These principles give a heightened role to active citizenship, as opposed to more traditional forms of representative democracy giving a more passive role for the citizen within the political system. Participation is also seen to support the discursive quality of the democratic space, helping to free it from cultural domination, power relations and non-rational attitudes, that is, as supporting deliberative forms of democracy (Dryzek 2005). These arguments draw upon the claim that deliberation, cooperation and learning are stimulated and facilitated by participatory practice. The development of more open, inclusive and participatory approaches that promote dialogue and deliberation helps develop active environmental citizenship, enabling people to take more responsible actions, including in their daily lives (Dobson 2003). This spills over into society more generally, helping to foster sound environmental values and behaviour, a stance that resonates with that provided by scholars from within environmental ethics.

Participation is also linked to goal objectives. Ever since the publication of the Brundtland Report (WCED 1987), the participation of both economic and social actors has come to be seen as a necessary quality of sustainable development governance (Baker 2015). Participation is also a means of enhancing political stability by encouraging more accountable, less corrupt forms of governing. In this context, it is indicative of what is referred to as 'good governance'. Good governance is the practice of managing public affairs in ways that are participatory, consensus oriented, accountable, transparent, responsive, effective and efficient, equitable and inclusive and that follow the rule of law (*ibid.*). Meeting such criteria is increasingly used by major donors and international financial institutions as a conditionality clause in the granting of aid and loans, including those that relate to conservation and environmental management in developing countries.

Taking a sub-disciplinary perspective from within political science, that of policy analysis, adds further insights. This highlights an array of instrumental arguments associated with participatory practices. Participation is seen to improve the knowledge base available to policy makers, increasing the range of policy solutions and thus creating more effective implementation outcomes. It is also seen to produce greater stakeholder 'buy in', reducing the risk of policy failure. Participation is also purported to give a voice to marginalized social groups that might otherwise be excluded from influencing policy-making (Gouldson and Bebbington 2007: 6). There are also claims that participation results in policy responses that better take account of local circumstances, knowledge and capacities (Gunningham 2009: 146). Thus, restoration efforts that draw upon participatory practices are seen as more legitimate, thereby helping to reduce the risk of conflict over priorities, decisions and outcomes. This is important because 'even if ecologists can provide theoretical and technical input to answer questions about what goes where and how to accomplish it, the ultimate success of restoration efforts rely on cultural acceptance' (Gobster and Barro 2000: 186). Such acceptance is not only needed to start a project, but also to ensure on-going commitment to, rather than subsequent neglect of, restoration efforts. It is also argued that participation is necessary because ecological restoration raises issues of a value-laden character (Higgs 2003). Given the myriad nature of these value differences, agreement on the objectives of ecological restoration policy is more likely when these objectives are reached through participatory practices.

Acceptance of restoration can often be more forthcoming if projects build upon strong emotional or cultural attachments that people and communities have to a place. Place can be understood as a series of locales or settings wherein everyday life activities occur. But it is also a unique community, landscape, and moral order. A strong sense of 'belonging' to a place is evidenced by everyday behaviour, such as participating in place-related affairs. While some sense of place is a prerequisite for social solidarity and collective action, place attachment can itself be multifaceted (Agnew 2011). Place attachment can be expressed in diverse and even contradictory ways and is often related to how people, or social groups, use or experience the place, be it for leisure, commercial purposes or for the purposes of ritual and healing. This all affects social attitudes towards restoration, and the willingness of different groups to take part in a project (Baker *et al.* 2014). Different users will have different expectations about what is to be restored and how that can underpin their continued use of the place. For example, they might want picturesque planting, or paths, or want a particular type of landscape to be restored. This makes it essential to understand and respect the 'place' to which people have attachments before undertaking ecological restoration (Ryan 2000: 224). Participation can play a role here in allowing those different voices and understandings to be aired and listened to, helping to ensure successful restoration outcomes.

While these instrumental advantages are important in their own right, and documented in other literature, a political perspective will see how these combine to move society beyond an adversarial approach to environmental problems by shaping new forms of relationship between the state and the polity (Paehlke 1996). This is no less important for restoration as for other environmental policies, given that they are often mired in controversy, particularly at the project level. By facilitating partners to understand better the interests of others, participation helps develop an integrated vision and thus a shared agenda for the future (Gunningham 2009: 161). It also provides a platform for policy makers to spread awareness of the benefits of restoration to society.

This latter claim is closely linked to arguments that participation helps support a new, reflexive relationship between expert and lay understanding of issues, one that promotes learning about different expertise, perspectives and values. This resonates with recognition of the strong role that practitioners have always played in ecological restoration. Here participation has been shown to help in problem co-construction as well as identification of, and agreement upon, actions needed. Participation provides the opportunity for both instrumental learning, through the acquisition of new skills, information and knowledge, alongside communicative learning about how to cooperate with others in solving collective problems. This can include developing a sense of group solidarity. A study of civil society participation in river restoration in the UK found that that these two forms of learning intertwined to support local-level river restoration projects (Petts 2007). Such learning is often more forthcoming when restoration targets local community problems, such as flooding.

This brings attention to claims that participation not only helps policy making directly but is also able to bring wider societal benefits, especially to disadvantaged social groups. Participation helps to give rise to societies that are built on a sense of common purpose, collective engagement and mutual support. Some have described this as a potential that engagement in restoration activities has for 'emancipatory egalitarianism' (Light and Higgs 1996). Another way to approach this issue is to see community participation in restoration initiatives as building social capital, through enabling the development of networks within communities that foster the progress and exchange of skills and knowledge, in turn contributing to local-level empowerment. Social capital is here understood in its classic sense to be features of social organization, such as trust, norms and networks, which improve the efficiency of society by facilitating coordinated actions (Putnam 1993). This supports efforts to ensure more equitable outcomes from restoration initiatives, including through ensuring access to and distribution of natural resources arising from these initiatives. This can open up restoration as a tool for poverty alleviation though practices that promote equity. Social capital can also promote communication and learning across communities, facilitating the scaling up of restoration initiatives to a landscape level. Scaling up is not a simple summation of actions, but needs to pay attention to the broader context, culture and heterogeneity of communities (Bebbington 1999). The literature has often neglected the importance of community culture and normative values in shaping social acceptance of restoration practices (Waylen *et al.* 2009).

Not only is participation seen as a means of building understanding and commitment for collective policy making, but it can also be seen as an end in itself, that is, as an aspect of the necessary and richer alternatives to lives centred on material consumption (Kemp *et al.* 2005: 16). Restoration has a distinctive advantage in this respect. It may be easier to get local people involved in ecological restoration projects – as opposed to getting the public to commit to more abstract principles such as the promotion of sustainable development. This is because ecological restoration projects are content specific, local in focus, pragmatic and immediate. Viewed from the perspective of participatory democracy, the advantage is that projects can be organized around community activities to address communal concerns.

Overcoming barriers to participation

Regardless of the strong arguments mounted in its favour, participation is not without its problems. Even where opportunities for participation are available, not all local communities can be organized and not all interests have, or can obtain, a voice. The problem with participation is that those who fail to participate may not be properly represented. A growing literature addresses this from the perspective of conservation more generally, exploring inter-community connections and links between communities and higher-level organizations. This has high-lighted the role that intermediary organizations play in overcoming barriers to local-level community participation in conservation initiatives (Bosselmann and Lund 2013). Intermediaries fall into a number of different categories, including international organizations such as the Society for Ecological Restoration, regionally based organizations such as the European Centre for River Restoration, a European network of national centres, organizations, institutions and individuals supporting river restoration throughout greater Europe; national organizations, such as the British Land Reclamation Society, a registered charity concerned with land reclamation, rehabilitation and restoration; and an array of independent consultants and professional service providers and looser networks for information exchange and events. Research suggests a necessary role for such intermediary organizations in supporting commu-nity groups (see Van Lente *et al.* 2003). They play a key role in facilitating capacity enhancement, often a necessary prerequisite for participation. They can also help adapt conser-vation initiatives to the cultural values and social practices of communities (Petheram and Campbell 2010). They also support vertical links between communities and government agen-cies (Garcia-Lopez 2013), including by providing a focal point for access by new entrants and act as a conduit through which outside actors can engage over the longer term (Bird and Barnes 2014).

This turns attention to the role of intermediary organizations in developing further, shared projects, and spreading the benefits of community approaches more widely. They do this by facilitating horizontal links between communities (Taylor 2010), helping to up-scale the build-ing of social capital. This enables participation to be conceived not only in terms of its usefulness for local projects, but also in terms of how it can support the linking of restoration initiatives across spatial scale.

The role of meso-level intermediaries in facilitating participation and shaping its outcomes at a broader landscape level can be conceptualized using ideas associated with networks and social capital. Hargreaves *et al.* (2013), drawing on Geels and Deuten (2006), suggest four key roles for intermediaries:

1   aggregation and 'decontextualizing' of knowledge from local projects into more abstract and mobile forms;
2   creation of institutional infrastructure, such as forums and newsletters etc., which serve as a repository and circulation of abstract knowledge;
3   coordination and framing, in which collective knowledge starts guiding local projects; and
4   brokering and managing partnerships with external parties (see Bird and Barnes 2014 for a fuller elaboration).

These actions build and thicken networks that can help integrate individual local restoration initiatives into wider coordinated strategies that deal with environmental problems at wider spatial scale. This can enable local restoration initiatives to go beyond localized objectives to take a strategic perspective on outcomes, including the need to use restoration to promote higher-scale ecosystem connectivity.

Empirical work supports these insights, including research on two initiatives in the Lower Kinabatangan floodplain in Sabah, where communities are involved in forest restoration with the aim of establishing habitat corridors in an area that has experienced severe habitat degradation and fragmentation (Hamza and Mohamad 2012). Community participation in forest restoration in Sabah has followed an integrated conservation and development model, led primarily by NGOs. A number of projects developed in isolation, leading in turn to more recent moves to link these projects within wider, landscape-level conservation strategies. Research has demonstrated the role of NGOs in facilitating this wider programme, while also revealing both the cooperation and conflicts between NGO, private-sector, community and governmental actors (Bloor 2014). The role of women in a community watershed restoration project in the north of Sabah is also being recognized for its capacity for lesson learning with respect to community-led forest restoration, leading to the transfer of best practice to higher, regional scales (Asian Philanthropy Forum 2014). These findings provide useful insights into the role of meso-level intermediary organizations from empirically driven perspectives.

While intermediaries can act as facilitators of, and channels for, marginalized groups to engage in participatory processes, their potential political biases and the complex and often competing interactions between different intermediary organizations needs to be taken into account. The extent to which such actions succeed in adapting restoration practices in ways that foster long-term stewardship of the environment and embed such practices into the core of community livelihood strategies remains uncertain.

To understand more fully the factors that shape whether participatory processes can engage or have significance beyond the project level, it is important to pay attention to the broader governance context within which community restoration projects are situated. The literature on participation has often ignored this dimension, although we can draw upon a well-developed body of literature within political science that cautions against the danger of treating the local level as a 'black box' disconnected from the global, international and national governance contexts with which localities are framed and actions play out (Bulkeley and Betsill 2005). Seeing participation in this wider institutional context reveals the political and social interactions through which ecological restoration goals, and environmental goals more generally, are determined and revised, collective decisions are enforced and resources are authoritatively allocated (Meadowcroft 2009).

Accounting for the institutional context

The capacity for participatory processes to provide a meaningful forum for deliberation is heavily dependent upon several constraining factors operating within the system of public administration and its related governance processes. One way to conceptualize this is to take a policy analysis perspective, investigating the type of formal access to policy making that is given across different stages in the policy-making process and how this shapes participatory outcomes (Baker and Eckerberg 2013).

This lens reveals how the state or its agents can limit the role of 'outsiders' in restoration policy and in particular projects. Allowing participation at early stages in the policy-making process can ensure that societal groups play a role not only in defining the nature of the problem that restoration is designed to address, but also in shaping the policy response. This enables collective choices about what is to be done, by whom, when and using what policy tools, such as subsidies and grants. Participation at later stages in the process can often see societal groups restricted in the scope of their actions, often limited to acting as implementation agents for decisions made higher up the administrative system. This can limit their capacity to shape restoration policy and practice (Baker and Eckerberg 2013).

Irrespective of what stage of the policy-making process participation takes place, all ecological restoration policy decisions and outcomes are best seen as the result of negotiations between different interest groups. Like all political interactions, this involves the negotiation of trade-offs among potentially competing objectives (Meadowcroft 2009). The outcomes of such negotiations can become skewed by the unequal distribution of power, including financial resources, between participants. They can become the arena for the expression of narrow, vested interests. For example, local interests can capture participatory processes, intent on the spatial and temporal displacement of environmental problems to other regions or to future generations. Similarly, economic interests can also capture process. There is thus no *a priori* reason to believe that participation leads to better environmental decisions, especially with respect to long-term, strategic planning.

Combined, such stresses can weaken the ability of participatory processes to reflect wider, collective interests. This can undermine their credibility and reduce the democratic *legitimacy* of participation for ecological restoration, including at the project level. This is all the more so given that participatory practices have long since been criticized because they are weak in terms of traditional political accountability and representation (Gouldson 2009). Critics ask: who do these participating groups represent? Who voted for them? While they act with authority, how can they be held accountable? Who is in charge here? Arguments thus abound for the state to carry the mantle for ensuring the legitimacy of participatory arrangements. This includes overseeing the inclusion of broad societal interests in the policy-making process and dealing with coordination problems that arise, especially at higher scale, from such engagements (Black 2008).

Account has also to be taken of the institutional constraints that are placed upon participatory processes (Hallstrom 2004). One way is to look at the 'opportunity structures' that exist within the political system, and more particularly the policy process, to influence policy making. Although scholars have different visions of political opportunity components (see Koopmans 1996), the 'openness' of government is seen as a key factor in determining opportunity (Eisinger 1973). Openness relates to whether people are incentivized to engage with state bureaucracy and administrative structures, including whether they believe that they have the potential to shape policy outcomes. Approaches that deal with political opportunity structure and political openness do not necessarily limit attention to institutions, but take into account the whole political context, and even some economic variables, which open windows of opportunity for certain participatory practices to occur. In the following text, we see the significance of these wider considerations in our treatment of 'managed realignment' restoration policy.

It is important not to posit a simple, positive relationship between openings in the political structure, including in relation to the system of public administrations, and the capacity of participation to influence policy outcomes. It is also important to distinguish the opportunities for participation to shape local outcomes from the opportunities to influence policy change at higher levels. A wide range of influences operating across different governance levels, including obligations incurred under global and regional governance regimes, past decision and prior spending commitments, legislation and regulation, budgetary allocation, interest group mediation, alongside a myriad of other factors shape public policy. As such, the political environment provides both consistent and variable influences across outcomes (Meyer and Minkoff 2004). These differences can be seen when we look at restoration from a comparative perspective, that is, where restoration has been used as a tool for the delivery of policy commitments across several countries operating in collaboration. It is important to take this comparative perspective, not least because restoration is now a tool in the hands of several collaborative governance regimes, such as the European Union (EU) and the CBD, where it is increasingly put to use as a mechanism for addressing collectively negotiated global environmental goals.

Restoration in the North Sea coast represents an example of countries that share a geographically connected area which faces common restoration problems, but have been dealt with in contrasting ways according to different national contexts. The countries on the North Sea coast of the EU face a number of both common and country-specific coastal management challenges. The coastline of this region is primarily low-lying and areas inland are susceptible to flooding. This problem is exacerbated by the threat of sea level rises as a result of climate change. The region also faces problems of coastal habitat loss (Esteves 2014). Salt marshes in particular are subject to 'coastal squeeze', where rising water levels degrade the seaward side, but sea walls prevent compensating colonization of new salt marsh on the landward side (French 2006). In addition, these countries fall under the common administrative framework of the EU, and are required to deal with these problems under the auspices of an array of regulatory requirements embedded in the EU Birds, Habitats, Water and Floods directives (CEC 2011). Managed realignment has been offered as a solution, repositioning existing 'hard' sea defence to a more landward location, allowing space for the creation of intertidal habitat, with the resultant increase in the intertidal zone allowing increased flood water storage and wave attenuation (Möller *et al.* 1999). This restoration practice has gained a great deal of policy traction in the region, allowing managed realignment to bring about ecosystem restoration for coastal flood risk reduction, while also helping habitat creation for the purposes of meeting various EU directives. Here restoration combines a range of different policy objectives into a single policy approach.

The way in which country-specific policy styles, including the openness of its political structures, shapes this managed realignment can be best illustrated by contrasting initiatives in the UK and the Netherlands, which have the largest coastlines in this region. In common with much of the literature on restoration, the majority of publications focus on the restoration of ecological structure and function, and how these can be maximized through improved techniques and their application (see for instance Wolters *et al.* 2005 for the whole region, French 2006 for the UK and Bakker *et al.* 2002 for the Netherlands). The limited literature that deals with the political and institutional implications of managed realignment shows that the UK and the Netherlands differ substantially.

The UK is considered to have a relatively well-coordinated system of government across its different tiers, including at regional and local levels, and as being a pioneer in introducing environmental policy integration, that is, the integration of environmental considerations into other policy fields (Russel and Jordan 2010). In this context, managed realignment represents a potential means of combining the multiple objectives of environmental policy with social and economic ones, an approach that has received considerable government support since the publication in 2004 of the Department of Environment, Food and Rural Affairs report, *Making Room for Water* (DEFRA 2004). Since then, the UK Environment Agency has stated as a target that 10 per cent of the UK coastline should be realigned by 2030 (CCC 2013). NGOs such as the National Trust, the Royal Society for the Protection of Birds, and local wildlife trusts have been at the forefront of promoting managed realignment in the region (RSPB 2002; Essex Wildlife Trust 2005; National Trust 2014). Nevertheless, in spite of these favourable preconditions, several barriers impede the integration of environmental and flood defence policy into a coordinated managed realignment strategy. First, there is a lack of resources and an inconsistency in compensatory mechanisms for paying landowners on whose land managed realignment takes place. Second, inflexibility within traditional institutions, such as the planning system and local authorities, have shown them ill-equipped to deal with the unfamiliarity of managed realignment and consequently have been unable to coordinate it at a wider spatial level. This has meant that to date implementation has been largely piecemeal. Third is resistance from the public and interest groups, such as farmers' unions, due to unfamiliarity with the potential benefits and lack of public consultation. Because of these barriers, a significant gap exists between policy targets and rate of implementation progress. This has led to recommendations for more coordinated spatial planning between localities, more substantial and transparent compensation mechanisms backed up by more consistent valuation techniques, and more effective means of raising public awareness and support through participatory processes (Ledoux *et al.* 2005; CCC 2013).

In the Netherlands, a different picture emerges, where distinctive institutional barriers to managed realignment arise from the long history of reclaiming land from the sea. This has created an entrenched cultural perception of the need to 'hold the line' against the sea that often equates managed realignment with perceptions of retreat and therefore defeat (Wiering and Arts 2006). Traditionally Dutch water management has been characterized by a closed technocratic and top-down governance style that is dominated by semi-autonomous water broads. This has limited coordination between coast defence and environmental policy (van der Brugge *et al.* 2005). It has also restricted the opportunities for bottom-up civil society participation. Since the 1990s, because of the acute threat to the Netherlands of sea level rises associated with climate change and the experience of severe floods in 1995, the Dutch policy position on coastal management has begun to change (Kabat *et al.* 2009). The transition towards a policy approach less dominated by 'hard' engineered coastal defences is characterized by new discourses under the terms 'living with water', 'building with nature' and 'eco-engineering' (Esteves 2014). This brings greater emphasis on spatially integrated coastal management, more attention to the wider benefits of ecosystem services, longer term planning and wider involvement of scientists, NGOs and the public (van Konigsveld and Mulder 2004). As a result, coastal management has become less narrowly focused and more open to the engagement of different interests and policy approaches.

In both examples, institutional factors shape the participation of communities in coastal management restoration policy. In contrast to the UK where certain 'insider' environmental groups have historically played a key role in shaping environmental conservation policy, including in relation to land use, the Dutch case shows the centrality of technocratic institutions restricting participatory opportunities. However, environmental threat, or what we might refer to as external drivers, in both cases has shifted policy approaches not only to softer engineering methods but also towards greater public engagement. Policy failure, in the UK case in relation to lack of social acceptability and in the Dutch case in relation to the failure to deliver on flood defence, saw greater willingness to engage in new participatory forms of learning.

Commingling governance

This chapter opened by discussing participation as one means of governance, typically combined with other governance styles in the making and shaping of public policy. Although there are strong arguments mounted in favour of participation, there is still a gap, however, between support and advocacy for participation and actual practice. Empirical research reveals that the extent of genuine devolution of power to civil society actors is limited (Baker and Eckerberg 2008) and the state has been shown to retain 'a very high degree of discretionary intervention and direct control' over participatory processes (Gunningham 2009: 160). It acts as a gatekeeper, determining who can participate, defines the terms of participation and is often the final voice that is heard in decision-making outcomes. Indeed, this state coordination and oversight is a prerequisite of democratic practice, even if that practice is evolving into new forms. There is also evidence that precisely because opening up restoration projects to a wider range of social actors makes the oversight and management of restoration initiatives harder not easier, participation itself needs additional administrative capacity within state agencies to cope with the processes (Jordan *et al.* 2003: 222). Thus, participation confirms the need for, but also enhances, hierarchical steering. It is best seen to *commingle* with traditional, hierarchical governance. Here, we are reminded that laws and regulations shape the framework within which restoration takes place (Baker and Eckerberg 2008).

While some attention has been given to the role of hierarchical governance, this chapter has paid less attention to the way in which economic instruments used for restoration purposes facilitate partnerships between organizations from the governmental, private and non-governmental sectors. This leaves us uncertain as to how commingling across the three classic governance styles, including markets, takes place and with what consequences for participatory practices. It is to this we now turn.

Market instruments are applied to restoration initiatives through the use of economic incentives that attempt to incorporate restoration into economic decision making, such as for example in relation to land use practices. Take mitigation banking as an example, a form of economic instrument built on the principle that land development in areas of natural habitat should result in 'no net loss' of biodiversity and ecosystem function. Wetland mitigation in the USA represents a mature programme that sees state steering alongside private sector involvement (Hallwood 2006; Robertson 2004; Ruhl *et al.* 2009). Emerging primarily as a regulatory approach, the USA has evolved well-established market mechanisms leading to an extensive private-sector mitigation banking industry. Mitigation banking is less well developed in Europe, although the EU has recently introduced a consultation process on the introduction of a 'no net loss' initiative that promotes the use of market instruments. In Germany, revisions to the regulations in 2002 and 2009 gave more room for the creation of mitigation banks and private-sector involvement. This aspect remains less well developed than in the USA and government agencies have retained greater control in Germany (Rundcrantz and Skärbäck 2003). Nevertheless, situations where mitigation banking forms part of wider landscape-level conservation strategy that involves the participation of NGOs and local communities as well as a combination of different types of ecosystem service credits do exist. Two such cases have been identified in the USA, in the Willamette Watershed in Oregon and Chesapeake Bay on the Atlantic Seaboard (Madsen *et al.* 2011). The latter in particular presents an example of combining larger scale approaches to mitigation banking with the use of economic instruments alongside wider community and civil society participation to tackle ecological problems in Chesapeake Bay, which in addition has involved coordinated action across five state governments. The question nonetheless remains as to how to marry this marketization to other drivers of participation, including those that are motivated by the need to foster a long-term stewardship approach to the use of natural resources that go beyond short-term financial calculations. More generally, different motivations for restoration bring with them different political and social potentials, including participatory opportunities. Ecological restoration undertaken to comply with planning permits for development projects, as in the USA and Germany, are more often associated with privatized and profitable forms of economic activity and less with the practice of participation for societal benefits and democratic enhancement (Baker and Eckerberg 2013).

A sizable body of literature, mainly originating from the fields of environmental and ecological economics, has emerged on the role of payments for ecosystem services (PES) for the purposes of restoration. The use of this market tool has been subject to considerable research in the case of Costa Rica, given that this country demonstrates a highly developed application. Research has revealed that poorer landowners generally face barriers to taking advantage of PES schemes, including institutional barriers such as transaction costs, lack of secure tenure, start-up capital, capacity, knowledge and awareness, and leadership, alongside cultural suspicion of outsiders (Zbinden and Lee 2005). More recent programmes have evolved to deal with these criticisms, including shifting payments to target areas of highest biodiversity need and poverty, while relaxing restrictions on formal land title. In addition, a new agro-forestry scheme allows poor landowners to engage in forest restoration and small-scale agriculture on the same plots of land. These have led to some improvement in the participation of poorer landowners, although lack of data means that it is difficult to evaluate properly relative success (Porras 2010). A limited number of publications have also highlighted the role of NGOs and agricultural associations in facilitating the participation of poorer land owners in Costa Rica's PES programme. These have highlighted the value of intermediary meso-level organizations in providing financial and technical support, as well as raising aware-ness of the programme in more isolated rural areas (Russo and Candela 2004; Bosselmann and Lund 2013). Irrespective of this advance, we can see how market governance does not act alone – both the use of such mechanisms and the capacity of different groups to participate in them are shaped by institutional barriers. These barriers require state steering to address their unintended and unanticipated consequences.

Conclusion

Successful ecological restoration, including in relation to societal acceptance, requires engage-ment in collective decision making among civil society, economic interests and governmental authorities. Operating across three governance styles is a necessary part of the on-going process of formulating policy objectives and ensuring their relevance to the socio-economic, political and cultural contexts in which ecological restoration takes place. However, the potential for democratic engagement in ecological restoration can be restricted, particularly given institu-tional constraints and associated power imbalances. Furthermore, the fact that restoration projects may stem from different underlying rationales, including planning mitigation, can favour one style over another, with negative consequences – unless appropriate, top-down checks and balances are put in place to address these negative externalities.

We conclude with a question: can participation facilitate better restoration outcomes? Our answer needs to consider whether participation helps in policy formation and delivery outcomes that take account of the interests of a wide societal base. Our answer must also consider the perspective of spatial scale, and the extent to which participation enables restora-tion to be linked horizontally between different projects and the jurisdictions that govern them, as well as vertically in ways that connect local, national and international authorities so that they can promote spatially coordinated policy actions. Our answer must also be given from the perspective of temporal scale, in that we need to consider whether participation is geared towards long-term intergenerational goals and values, as opposed to being driven by the desire to meet local and immediate concerns. An institutional perspective brings our attention to whether participation supports restoration as a tool for the integration and mainstreaming of environmental considerations, rather than allowing restoration to be an instrument that remains subsidiary to economic objectives. The answer to each part of that question is neither prom-ised nor denied in participatory practices for restoration. In each case, the answer is context dependent, shaped by institutional systems and their associated policy processes. From a polit-ical science perspective, participation in ecological restoration is neither good nor bad, but is instead a governance tool shaped by political practice.

Acknowledgement

I am grateful to Dr Richard Bloor, working with me as a research assistant during the summer of 2014, for his research on some of the empirical material presented here, including in relation to managed realignment in the North Sea. This research was part funded by the Swedish Research Council for Environment, Agricultural Sciences and Spatial Planning, Formas, Umeå University and the Swedish University of Agricultural Science, for the project 'Ecosystem restoration in policy and practice: restore, develop, adapt' (RESTORE), Grant number 2009–450, www.restore-project.org.

References

Agnew, J. (2011) Space and place, in J. Agnew and D. Livingstone (eds), *Handbook of Geographical Knowledge*, Sage, London, pp. 316–330.

Asian Philanthropy Forum (2014) Women lead a new green economy in Sabah, Malaysia, retrieved from www.asianphilanthropyforum.org/women-lead-new-green-economy-sabah-malaysia.

Baker, S. (2015) *Sustainable Development*, Routledge, Abingdon.

Baker, S. and Eckerberg, K. (2008) Introduction: in pursuit of sustainable development at the sub-national level: the 'new' governance agenda, in S. Baker and K. Eckerberg (eds), *In Pursuit of Sustainable Development: New Governance Practices at the Sub-national Level in Europe*, Routledge, London, pp. 1–26.

Baker, S. and Eckerberg, K. (2013) A policy analysis perspective on ecological restoration, *Ecology and Society*, vol 18, no 2, art 17.

Baker, S., Eckerberg, K. and Zachrisson, A. (2014) Political science and ecological restoration, *Environmental Politics*, vol 23, no 3, pp. 509–524.

Bakker, J. P., Esselink, P., Dijkema, K. S., van Duin, W. E. and De Jong, D. J. (2002) Restoration of salt marshes in the Netherlands, *Hydrobiologia*, vol 478, pp. 29–51.

Bebbington, A. (1999) Capitals and capabilities: a framework for analyzing peasant viability, rural livelihoods and poverty, *World Development*, vol 27, no 12, pp. 2021–2044.

Bird, C. and Barnes, J. (2014) Scaling up community activism: the role of intermediaries in collective approaches to community energy, *People, Place and Policy*, vol 8, no 3, pp. 208–221.

Black, J. (2008) Constructing and contesting legitimacy and accountability in polycentric regulatory regimes, *Regulation and Governance*, vol 2, pp. 137–164.

Bloor, R. (2014) Forest governance and forest conservation in Sabah, Malaysian Borneo, PhD thesis, Cardiff University, Wales, UK.

Bosselmann, A. and Lund, J. (2013) Do intermediary institutions promote inclusiveness in PES programs? The case of Costa Rica, *Geoforum*, vol 49, pp. 50–60.

Bulkeley, H. and Betsill, M. M. (2005) Rethinking sustainable cities: multilevel governance and the 'urban' politics of climate change, *Environmental Politics*, vol 14, no 1, pp. 42–63.

CCC (2013) Regulating services – coastal habitats, from report managing the land in a changing climate, in *Managing the Land in a Changing Climate: Adaptation, Sub-Committee Progress Report 2013*. HMSO, Committee on Climate Change, London.

CEC (2011) *Towards Better Environmental Options for Flood Risk Management*, CEC, DG Env. D.1 (2011) 236452, Commission of the European Communities, Brussels.

DEFRA (2004) *Making Space for Water*, DEFRA, London, retrieved from http://archive.defra.gov.uk/environment/flooding/documents/policy/strategy/strategy-update.pdf.

Dobson, A. (2003) *Citizenship and the Environment*, Oxford University Press, Oxford.

Dryzek, J. (2005) *The Politics of the Earth*, Oxford University Press, Oxford.

Eisinger, P. (1973) The conditions of protest behavior in American cities, *American Political Science Review*, vol 81, pp. 11–28.

Essex Wildlife Trust (2005) *Abbotts Hall Farm: Lessons Learned from Realignment*, Essex Wildlife Trust, Colchester, UK.

Esteves, L. (2014) *Managed Realignment: A Viable Long Term Coastal Management Strategy?* Springer, New York.

French, P. (2006) Managed realignment: the developing story of a comparatively new approach to soft engineering, *Estuarine, Coastal and Shelf Science*, vol 67, pp. 409–423.

Garcia-Lopez, G. (2013) Scaling up from the grassroots and the top down: the impacts of multi-level governance on community forestry in Durango, Mexico, *International Journal of the Commons*, vol 7, no 2, pp. 406–43.

Geels, F. and Deuten, J. J. (2006) Local and global dynamics in technological development: a socio-cognitive perspective on knowledge flows and lessons from reinforced concrete, *Science and Public Policy*, vol 33, no 4, pp. 265–275.

Gobster, P. H. and Barro, S. C. (2000) Negotiating nature: making restoration happen in an urban park context, in P. H. Gobster and B. Hull (eds), *Restoring Nature: Perspectives from the Social Sciences and Humanities*, Island Press, Covelo, CA, pp. 185–207.

Gouldson, A. (2009) Advances in environmental policy and governance, *Environmental Policy and Governance*, vol 19, pp.1–2.

Gouldson, A. and Bebbington, J. (2007) Corporations and the governance of environmental risk, *Environment and Planning C: Government and Policy*, vol 25, pp. 4–20.

Gunningham, N. (2009) Environment law, regulation and governance: shifting architectures, *Journal of Environmental Law*, vol 21, no 2, pp. 179–212.

Hallstrom, L. K. (2004) Eurocratising enlargement? EU elites and NGO participation in European environmental policy, *Environmental Politics*, vol 13, no 1, pp.175–193.

Hallwood, P. (2006) Contractual difficulties in environmental management: the case of wetland mitigation banking, *Ecological Economics*, vol 63, pp. 446–451.

Hamzah, A. and Mohamad, H. (2012) Critical success factors of community based ecotourism: case study of Miso Walai homestay, Kinabatangan, Sabah, *The Malaysian Forester*, vol 75, no 1, pp. 29–42.

Hargreaves, T., Hielscher, S., Seyfang, G. and Smith, A. (2013) Grassroots innovations in community energy: the role of intermediaries in niche development, *Global Environmental Change*, vol 23, pp. 868–880.

Higgs, E. (2003) *Nature by Design: People, Natural Processes and Ecological Restoration*, MIT Press, Cambridge, MA.

Jordan, W. R. (2000) Restoration, community, and wilderness, in P. H. Gobster and R. Bruce Hull (eds), *Restoring Nature: Perspectives from the Social Sciences and Humanities*, Island Press, Washington, DC, pp. 21–36.

Jordan, A., Wurzel, R. K. W. and Zito, A. R. (2003) 'New' instruments of environmental governance: patterns and pathways of change, *Environmental Politics*, vol 12, no 1, pp. 1–24.

Kabat, P., Fresco, L. O., Stive, M. J. F., Veerman, C. P., Van Alphen, J. S. L. J., Parmet, B. W. A. H., Hazeleger, W. and Katsman, C. A. (2009) Dutch coasts in transition, *Nature Geoscience*, vol 2, pp. 450–452.

Kemp, R., Parto, S. and Gibson, R. B. (2005) Governance for sustainable development: moving from theory to practice, *International Journal of Sustainable Development*, vol 8, nos 1–2, pp. 13–30.

Koopmans, R. (1996) New social movements and changes in political participation in western Europe, *West European Politics*, vol 19, pp. 28–50.

Ledoux, L., Cornell, S., O'Riordan, T., Harvey, R. and Banyard, L. (2005) Towards sustainable flood and coastal management: identifying drivers of, and obstacles to, managed realignment, *Land Use Policy*, vol 22, pp. 129–144.

Light, A. and Higgs, E. S. (1996) The politics of ecological restoration, *Environmental Ethics*, vol 18, pp. 227–247.

Madsen, B., Carroll, N., Kandy, D., and Bennett, G. (2011) *Update: State of Biodiversity Markets*, Forest Trends, Washington, DC, retrieved from www.ecosystemmarketplace.com/reports/ 2011_update_sbdm.

Meadowcroft, J. (2009) What about the politics? Sustainable development, transition management, and long term energy transitions, *Policy Sciences*, vol 42, no 4, pp. 323–340.

Meyer, D. S. and Minkoff, D. C. (2004) Conceptualizing political opportunity, *Social Forces*, vol 82, no 4, pp. 1457–1492.

Möller, I., Spencer, T., French, J. R., Leggett, D. J. and Dixon, M. (1999) Wave transformation over salt marshes: a field numerical modelling study from north Norfolk, England. *Estuarine, Coastal and Shelf Science*, vol 49, pp. 411–426.

National Trust (2014) *Shifting Shores: Adapting to Climate Change*, National Trust, Swindon, retrieved from www.nationaltrust.org.uk/shiftingshores.

Ostrom, E. (1990) *Governing the Commons: The Evolution of Institutions for Collective Action*, Cambridge University Press, Cambridge.

Paehlke, R. (1996) Environmental challenges to democratic practices, in W. M. Lafferty and J. Meadowcroft (eds), *Democracy and the Environment: Problems and Prospects*, Edward Elgar, Cheltenham, pp. 18–38.

Petheram, L. and Campbell, B. (2010) Listening to locals on payments for environmental services, *Journal of Environmental Management*, vol 91, pp. 1139–1149.

Petts, J. (2007) Learning about learning: lessons from public engagement and deliberation on urban river restoration, *The Geographical Journal*, vol 173, no 4, pp. 300–311.

Porras, I. (2010) *Fair and Green? The Social Impacts of Payments for Environmental Services in Costa Rica*, International Institute for Environment and Development, London.

Putnam, R. D. (1993) *Making Democracy Work: Civic Traditions in Modern Italy*, Princeton University Press, Princeton, NJ.

Robertson, M. (2004) The neoliberalisation of ecosystem services: wetland mitigation banking and problems in environmental governance, *Geoforum*, vol 35, pp. 361–373.

RSPB (2002) *Seas of Change*, Royal Society for the Protection of Birds, Sandy, UK.

Ruhl, J. B., Salzman, J. and Goodman, I. (2009) Implementing the new ecosystem services mandate of the section 404 compensatory mitigation program, a catalyst for advancing science and policy, *Stetson Law Review*, vol 38, pp. 251–272.

Rundcrantz, K. and Skärbäck, E. (2003) Environmental compensation in planning: a review of five different countries with major emphasis on the German system, *European Environment*, vol 13, pp. 204–226.

Russel, D. and Jordan, A. (2010) Environmental policy integration in the UK, in A. Goria, A. Sgobbi and I. von Homeyer (eds), *Governance for the Environment: A Comparative Analysis of Environmental Policy Integration*, Edward Elgar, Cheltenham, pp. 157–178.

Russo, R. and Candela, G. (2004) Payment of environmental services in Costa Rica: evaluating impact and possibilities, *Tierra Tropical*, vol 2, no 1, pp. 37–48.

Ryan, R. L. (2000) A people-centred approach to designing and managing restoration projects: insights from understanding attachment to urban natural areas, in P. H Gobster and R. Bruce Hull (eds), *Restoring Nature: Perspectives from the Social Sciences and Humanities*, Island Press, Washington, DC, pp. 208–228.

Taylor, P. L. (2010) Conservation, community, and culture? New organizational challenges of community forest concessions in the Maya Biosphere Reserve of Guatemala, *Journal of Rural Studies*, vol 26, pp. 173–184.

Van de Hove, S. (2000) Participatory approaches to environmental policy-making: the European Commission Climate Policy Process as a case study, *Ecological Economics*, vol 33, no 3, pp. 457–472.

Van der Brugge, R., Rotmans, J. and Loorbach, D. (2005) The transition in Dutch water management, *Regional Environmental Change*, vol 5, pp. 164–176.

Van Deth, J. W. (2014) A conceptual map of political participation, *Acta Politica*, vol 49, no 3, pp 349–367.

Van Koningsveld, M. and Mulder, J. (2004) Sustainable coastal policy developments in the Netherlands: a systematic approach revealed, *Journal of Coastal Research*, vol 20, no 2, pp. 375–385.

Van Lente, H., Hekkert, M., Smits, R. and van Waveren, B. (2003) Roles of systemic intermediaries in transition processes, *International Journal of Innovation Management*, vol 7, no 3, pp. 247–279.

Van Zeijl-Rozema, R., Cörvers, R., Kemp, R. and Martens, P. (2008) Governance for sustainable development: a framework, *Sustainable Development*, vol 16, pp. 410–421.

Waylen, K. A., Fischer, A., McGowan, P. J. K., Thirgood, S. J. and Miller-Gulland, E. J. (2009) Effect of local cultural context on the success of community-based conservation interventions, *Conservation Biology*, vol 24, no 4, pp. 1119–1129.

WCED (1987) *Our Common Future*, Oxford University Press, Oxford.

Wiering, M. and Arts, B. (2006) Discursive shifts in Dutch river management: 'deep' institutional change or adaptation strategy?, *Hydrobiologia*, vol 565, pp. 327–338.

Wolters, M., Garbutt, A. and Bakker, J. P. (2005) Salt-marsh restoration: evaluating the success of de-embankments in north-west Europe, *Biological Conservation*, vol 123, pp. 249–268.

Zbinden, S. and Lee, D. (2005) Paying for environmental services: an analysis of participation in Costa Rica's PSA Program, *World Development*, vol 33, no 2, pp 255–272.

PART II

Restoring key ecosystems

8

RESTORATION AND ECOSYSTEM MANAGEMENT IN THE BOREAL FOREST

From ecological principles to tactical solutions

Timo Kuuluvainen

Introduction

Natural ecosystems are inherently complex, resilient and adaptive (Levin 1998). Millions and thousands of years of evolution of life on earth have led to flourishing ecosystems of amazing diversity and complexity. This also applies to forest ecosystems (Filotas *et al.* 2014). Endurance and ability to adapt to changing environmental conditions, and to recover even from catastrophic events, are convincing evidence of the resilience of forest ecosystems. However, human activities are changing ecological systems at local, regional and global scales at unprecedented rates (Rockström *et al.* 2009; Steffen *et al.* 2011a, 2011b). For example, it has been estimated that on a global scale the rate of species extinction, mostly driven by habitat destruction, is from 100 to 1000 times faster than it would be naturally. Although habitat destruction and species extinction rates are fastest in the tropical forests, habitat destruction, decline of biodiversity and local species extinctions are prevalent phenomena also in the boreal forest (Hanski 2000; Bradshaw *et al.* 2009). These are mostly driven by direct human impact due to resource extraction, notably timber harvesting, but increasingly so because of the rapid warming of climate at high latitudes (Scheffer *et al.* 2012). This risks the health of boreal forest ecosystems and leads to a growing plea for forest restoration and ecosystem-based management (Halme *et al.* 2013; Moen *et al.* 2014).

As it is increasingly evident that we have entered the era of Anthropocene, where human activities have an impact on the global system (Steffen *et al.* 2011a), the challenges of sustainable use, restoration and conservation of forest ecosystems become unprecedented (Hobbs *et al.* 2011). We are forced to seek answers to the most fundamental questions: how to live and fulfil our needs and aspirations, and safeguard those of the coming generations, without risking the long-term integrity of the ecological systems we depend on. This view, also called the ecosystem approach, sets ecosystems, their health, biodiversity and conservation as the basis for the realization of all dimensions of sustainability (CBD 2004). The two crucial and interrelated questions for long-term sustainability of forest use are: (1) to what extent applied forest management systems provide suitable habitats for species populations and their dynamics, and (2) how well the habitat-shaping ecosystem processes, such as disturbances, successions and nutrient cycling, are maintained in the long run?

The boreal forest covers about 30 per cent of the global forest area and around two-thirds of this is considered managed forest (Brandt *et al.* 2013). The southern parts of the boreal zone are most heavily managed and the primeval forests have by and large disappeared. The remaining natural forests are mostly located in the northern and high-altitude low-productivity areas. Only in Canada, Russia and Alaska do some larger intact boreal forests still prevail (Potapov *et al.* 2008). This change has been caused by the pervasive historical impact of humans on forests and more recently by industrial-scale forestry (Burton *et al.* 2010). Large-scale timber harvesting for sawmills first relied on the access to high quality timber from pristine forests that first seemed endless (Linder and Östlund 1998; Keto-Tokoi and Kuuluvainen 2014). In areas where this resource was depleted, management of secondary growth forests has become an issue (Kuuluvainen 2009; Holm 2015).

In regions such as Scandinavia, forests have been subjected to intensive utilization and management especially after World War II. The roots of this 'Scandinavian model' of forest management, which is basically plantation forestry (Holm 2015), can be traced back to the early-nineteenth-century German scholars of forest science. The dominant management practice is to grow even-aged homogeneous stands which are periodically treated with thinning and harvested by clear cutting when annual growth culminated (Mielikäinen and Hynynen 2003; Holm 2015). At present this management system, which mainly aims at efficient logistics and timber production, dominates the structure of forest landscapes in countries like Sweden and Finland. This intensive model of management has proved successful in increasing timber yields compared with unmanaged forest, and it has been viewed favourably by foresters around the circumboreal zone (Kuuluvainen 2009; Holm 2015).

But there is another side of the coin. It has become evident that forest structures and processes on which biodiversity depends, can be seriously impoverished at the same time that the timber resources and growth are maintained or even increased (Kuuluvainen 2009; Bergeron and Fenton 2012). Although some measures to safeguard biodiversity and ecosystem services have been taken in intensively managed forests, there has been a declining trend of biodiversity and this trend is envisaged to continue (Bradshaw *et al.* 2009; Kuuluvainen 2009). Most vulnerable are those species that are strictly confined to old natural forest habitats and their characteristics, or need large patches of natural forest (Kuuluvainen and Siitonen 2013). This indicates that the development of timber resources, which is often used as a general indicator of sustainability, may not be suitable for this purpose when including ecological and social aspects. A healthy forest ecosystem is simply much more than fast-growing trees.

In this chapter forest restoration is treated within a wide perspective so that it covers aspects of ecosystem management. This kind of approach has also been termed as the *restorative imperative* by Burton and Macdonald (2011) and *intervention ecology* by Hobbs *et al.* (2011).

Can nature be used as a guide?

There is ample evidence that intensive forest utilization, following the Scandinavian model, has in many regions profoundly changed native forest structural characteristics, and led to loss of natural habitat and species diversity (Bergeron and Fenton 2012; Kuuluvainen and Siitonen 2013). As an alternative to the traditional monolithic even-aged management, silviculture and restoration inspired by structures and dynamics of the natural forest have been suggested as an approach to maintain biodiversity and ecosystem resiliency (Bauhus *et al.* 2009; Gauthier *et al.* 2009; Kuuluvainen and Grenfell 2012). The logic here is that by restoring and maintaining stand and landscape structures that are sufficiently similar to those habitats occurring in natural forests, it would be possible to reduce the adverse effects of timber harvesting on biodiversity and key ecological processes (Bergeron *et al.* 2002). Although the logic is appealing, emulating natural forests is not an easy task and faces several challenges (Kuuluvainen and Grenfell 2012).

First, defining natural ecosystem reference conditions and their variability is often difficult or even impossible due to strong historical human impact in many boreal regions. It has also been questioned whether past ecosystem conditions can be used as a reference for future ecosystem restoration and management in a rapidly changing climate (Keane *et al.* 2009). Second, it is important to keep in mind that the goals and methods of restoration and sustainable forest management cannot be directly derived from the 'natural forest', because they should ultimately be determined through societal debate and democratic decision-making processes (Klenk *et al.* 2009). What is 'natural' may not be socially or culturally acceptable or sustainable, and vice versa. To take an extreme example: strict emulation of natural boreal forest might mean that very little timber could be extracted from the forest to fulfill social needs of local people or raw-material needs for industry. Thus, forest restoration and sustainable management in most cases represents a compromise between different competing needs and aspirations of people and interest groups.

In spite of such important caveats, understanding how the forest functions as an ecological system in the absence of human interference forms an important basis and reference for understanding what kind of management options are possible in general (Keane *et al.* 2009). Under conditions of limited ecological understanding, the natural forest can offer an objective 'benchmark' for setting both short-term and possibly long-term goals for restoration and management of ecosystems (*ibid.*). The fact that natural ecosystems have proved to be extremely resilient gives a compelling example and further motivation to try to understand their functioning and to apply this knowledge in forest restoration and management.

It is important to recognize that there can be different levels of ambition in practicing restoration and ecosystem management (Figure 8.1). The traditional approach has been to pay attention to where the natural forest and species can currently be found and then protect the best, often rather small such areas. A good example from the Scandinavian boreal forests is the key biotope strategy where small habitats that are assumed to be valuable for biodiversity are set aside (Timonen *et al.* 2011). This approach has been criticized because species do not survive as individuals but as interacting dynamic populations or metapopulations (Hanski 2000). Then the question becomes whether the habitats remain in the long term and form a functional network where the species can move from one habitat to another and sustain viable (meta)population dynamics.

Another challenge arises from the fact that habitats are not static entities but evolve through time. Habitats are created and renewed by ecological processes such as disturbances and successions that often occur at the landscape or regional scale in the boreal forest (Halme *et al.* 2013). Then the challenge is to understand and maintain the critical spatial processes that replenish habitats and species populations. The highest level of ambition is perhaps that of maintaining resilient ecosystems, i.e. systems that are able to recover from and adapt to rapidly changing environmental conditions and potentially unprecedented disturbance events (Drever *et al.* 2006). Thus, when setting goals for restoration and ecosystem management the level of ambition needs to be carefully and realistically considered.

Forest health: biodiversity, resilience and adaptive capacity

A generally accepted goal of forest restoration and sustainable management is a healthy forest ecosystem. Forest health can be defined as maintenance of desirable function and processes, biodiversity, resistance to biotic and abiotic disturbances and ability for renewal after

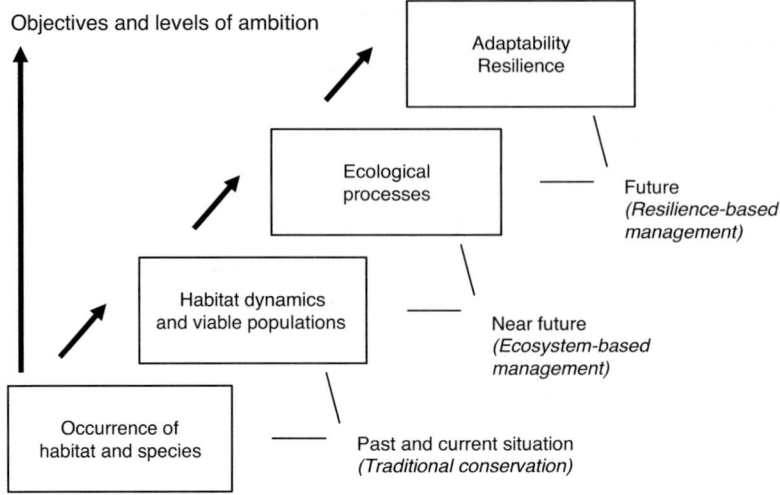

Figure 8.1 There can be different levels of ambition in ecosystem restoration and management. At the basic level the interest is in where the natural forest and species occur, and to try to safeguard them. Incorporating (meta)population dynamics requires the consideration of spatial and temporal dimensions. This is also true for accounting for all ecological processes but a more comprehensive approach is needed. Finally, facing the questions of ecosystem adaptability and resilience in the face of global change is even more challenging

disturbances (Venier *et al.* 2014). In short, forest health can be indicated by three interrelated ecosystem properties: *biodiversity*, *resilience* and *adaptive capacity*, which ensure the long-term availability of a wide variety of ecosystem goods and services (Drever *et al.* 2006). *Biodiversity* can be defined as the variation of ecosystem structures and processes, and species (including genetic variation) in space and time. *Resilience* denotes the ability of ecosystems to recover from disturbances (*ibid.*). *Adaptive capacity* refers to the long-term ability of ecosystems to evolve and adjust to changing environmental conditions, either through short-term changes in structure and species composition (acclimation) or through long-term changes in the genetic pool of the organisms (adaptation). The adaptive capacity aspect is increasingly important as human utilization is modifying both global environmental conditions and forest ecosystems at an unprecedented rate and at the same time we are likely to enter an era of rapid global change with possibly drastic consequences on environmental conditions around the globe (Steffen *et al.* 2011a).

Biodiversity (e.g. species and genetic diversity) is important because it enhances resilience of ecosystem functioning. When a perturbation occurs, a higher species number would allow higher functional redundancy. This means that if one species is removed from the ecosystem another one can take its part in the ecosystem and ecosystem functioning remains unchanged. There is empirical evidence to support this claim (Cardinale *et al.* 2006).

In general it can be expected that species populations respond to management disturbances and to decreases in original habitat area in a nonlinear manner, so that with environmental change the population size first declines slowly, but after a certain point a rapid decline occurs. According to this model there is a threshold value of habitat area (>0) for population existence. This can be applied to both animal and plant populations.

If the crashing population is that of a keystone species, the consequences for ecosystem structure and functioning can be rapid and devastating: ecosystem resilience may be lost and the ecosystem may end up in a qualitatively different steady state (Suding *et al.* 2004). A similar result can follow when the disturbance exceeds the resilience limits of the ecosystem. An example could be, for example, permanent deforestation of the boreal forests by fire cycles short enough to prevent trees from attaining sexual maturity.

In general, boreal forest ecosystems appear to be relatively resilient against such catastrophic regime shifts, at least when compared to forests in tropical and arid climates. This may be because the species have evolved during the past 2–3 million years under repeated glaciation and de-glaciation cycles. Thus most species have adapted to harsh environmental conditions and prevalent disturbances where the ability to colonize new barren lands and withstand environmental fluctuations has been a key to success. However, examples of catastrophic state shifts do exist also from boreal forest ecosystems. For example, some high-altitude forests have not regenerated after extensive clear cuttings or severe fires. An example of a shift in ecosystem state due to natural causes is provided by the deforestation of 1000 km^2 of mountain birch forest in the 1960s in northern Fennoscandia due to a massive outbreak of the autumnal moth (*Epirrita autumnata*) and subsequent reindeer browsing of the sprouting birches.

Such catastrophic shifts in ecosystems, if they are caused by management actions, provide an ultimate example of failure in sustainability of management. Ecosystem state shifts can also result from severe droughts in the warming climate (Scheffer *et al.* 2012). In both cases, the restoration of the original ecosystem may be extremely difficult and slow, or even impossible.

Ecosystem adaptive capacity

Biodiversity and resilience are dynamic ecosystem characteristics. They are formed and renewed through ecological processes that shape the habitats and environmental conditions for all forest-dwelling biota and they often take place at the landscape scale (Holling 2001). The two main driving processes that alternate in forest ecosystem dynamics are *disturbances* and *successions*. These two interlinked processes create and maintain spatial and temporal heterogeneity of habitats and environmental conditions, on which biodiversity ultimately depends (Kuuluvainen 2009).

Disturbances demolish structures that have been built and accumulated in successions over long periods of time, from tens to hundreds of years. Disturbances create new opportunities for a wide variety of organisms. On the other hand, long-lasting successions ensure the landscape-level continuity and relative stability of conditions that many species are adapted to. When a major disturbance occurs in a landscape the undisturbed areas in late successional stages can act as an 'ecological memory' facilitating dispersal of species to areas recovering from disturbances.

The disturbance–succession cycle not only affects the existing habitat mosaic and species dynamics, but provides the ecosystem a possibility to experiment, evolve and adapt in a constantly changing environment (Levin 2005; Filotas *et al.* 2014). Because of this the disturbance–succession cycle has been also called the *adaptive cycle* (Holling 2001), which has facilitated the maintenance and resilience of forest ecosystems in the changing environment through millennia. Thus the adaptive cycle is of pivotal importance in maintaining *adaptive capacity* of ecosystems (Puettmann 2014). Ecosystems with high adaptive capacity are capable of reconfiguring and evolving in rapidly changing environments without experiencing state shifts or losing their essential functions. On the other hand, ecosystems with low adaptive capacity are susceptible to losing their basic functions or at risk to experience catastrophic state shifts. Understanding of the ecological (resilience) and evolutionary (adaptive capacity) functions of the disturbance–succession cycle is a prerequisite for developing management that is sustainable in the long term (Levin 2005).

Ecosystem complexity

Traditionally the boreal forest has been considered as a simple system with a low number of tree species and slow predictable development. At first glance this view may seem justified, if just the number of tree species is considered and compared with tropical forests. The picture changes when considering the ecosystem as a whole (Levin 2005). A single stand of boreal forest in Fennoscandia is estimated to contain around 2500–5000 species with lots of inter-species interactions (Kuuluvainen and Siitonen 2013). Thus the boreal forest ecosystem exhibits a daunting complexity of interaction webs, which are in turn interacting with the dynamic physical systems, such as climate. Therefore boreal forests can justifiably be characterized as complex systems (Levin 1998; Drever *et al.* 2006; Filotas *et al.* 2014), meaning that forests are composed of spatially and temporally heterogeneous populations of species and individuals that interact locally and develop through evolution based on the outcome of those interactions (Levin 2005).

What are the practical implications of this ecological complexity for restoration and ecosystem forest management? The first implication is that a holistic approach is compulsory. For example, in forest landscapes it is necessary to consider the spatial heterogeneity composed of both managed and protected parts. Secondly, the applied conceptual models and landscape classification systems must be realistic enough to address ecologically important detail and variability in the ecosystem. Overly simplistic conceptual approaches, such as the conventional stand-level management paradigm, easily overlooks critical detail and interactions at spatial scales higher (landscape patterns) and lower (within-stand structures, microhabitats) than a tree stand. Thus, management must address important structures and processes at multiple scales, from decaying logs to landscape patterns of habitats (Puettmann 2014).

Thirdly, forest ecosystems are always in the process of change and so they must be managed taking their long-term changes into consideration. Biodiversity and resilience are ultimately based on the adaptive cycle of disturbances and successions. Indeed, forests are complex systems and the challenge is to assimilate this complexity as a basis for forest ecosystem restoration and management (Messier *et al.* 2013).

Lessons from the wild forests

Fire and other natural disturbances

The first step in practising forest ecosystem restoration and ecosystem-based management is to try to understand to what extent the current management system maintains naturally occurring ecosystem structures and processes (Gauthier *et al.* 2009). For this we need to understand the similarities and differences between forest structure and dynamics of unmanaged forests and those of managed forests (Kuuluvainen 2002).

The forest disturbance regime (i.e. the spatio-temporal occurrence pattern of varied disturbances in a given area) is a key factor shaping boreal forest ecosystem structures and processes, and availability of ecosystem services at the landscape scale. The disturbance agents in unmanaged forests are varied and include fires, storms, pathogenic fungi, insect outbreaks and effects of some mammals, such as moose and beaver (Kuuluvainen 2002).

It was traditionally considered that in boreal forests fires are generally stand-replacing, killing all trees. It was also assumed that such fires would naturally occur repeatedly with a roughly 100-year cycle (Payette 1992). This conception has also been used to legitimize clear cutting with a roughly 100-year cutting cycle as a nature-based harvesting method (e.g. Sirén 1955). However, such a static generalized view of the natural disturbance regime is not supported by accumulated scientific evidence (Kneeshaw *et al.* 2011). Originally, the conception of the dominant role of stand-replacing fires may have arisen among practising forest managers because large severe fires, although rare, attracted strong public attention. Another reason may have been that the early research results on fire ecology came from North America where stand-replacing fires indeed are common in continental areas (Payette 1992).

More recent research suggests that the natural fire regimes in the boreal forest exhibit significant geographic variation due to climatic variability and differences in tree species life-history characteristics (Kneeshaw *et al.* 2011; Rogers *et al.* 2015). Even within a landscape there can be high spatial and temporal variability in fire occurrence, severity and impact on the forest ecosystem (Kuuluvainen 2002). Important factors influencing fire regime and fire behaviour include climate (from maritime to continental), and fire resistance or serotiny of trees. In addition, in the boreal forest fire areas often exhibit fine-scale patchiness because of mosaics of wet paludified and drier upland sites which are intermixed in the landscape. Under these conditions significant areas of trees within a landscape often escape fire as unburned patches. In addition, in Eurasia fire severity is often low (surface fires) and a large part of the bigger trees survive forming an important legacy structure and seed source after fire (Wallenius *et al.* 2004; Rogers *et al.* 2015).

The intrinsically long fire cycles that are characteristic to landscapes also mean that disturbance agents other than fire play a major role in the dynamics of unmanaged forests (Kneeshaw *et al.* 2011; Kuuluvainen and Aakala 2011). In the absence of severe disturbances, trees die due to competition, fungi, insects, and snow loads which are frequently related to old age and senescence of trees. This type of tree mortality creates small-scale heterogeneity into forest structure.

Stand dynamics and landscape structures

The view that boreal forest would everywhere be naturally dominated by severe stand-replacing fires was challenged in the early 1990s, when the importance of small-scale or gap dynamics in some parts of the boreal forest was 'rediscovered' (Kuuluvainen 1994). First this led to another simplified conception of boreal forest dynamics, emphasizing the importance of either large-scale catastrophic versus small-scale gap-phase dynamics. These two types of disturbance dynamics were seen as nested so that 'small cycle' (gap) dynamics are specific to the late-successional phases of 'large cycle' forest dynamics (Kuuluvainen 2009).

However, this model has proven to be too simplistic to be able to account for the diversity of boreal forest dynamics in Fennoscandia (Kuuluvainen and Aakala 2011). Research suggests more diversified dynamics of the boreal forest with variable disturbance phenomena operating across multiple space and time scales, initiating a wide range of non-converging successional pathways, leading to high natural variability and heterogeneity of stand and landscape structures. While acknowledging this complexity, a relatively simple but more realistic classification of natural types of boreal forests dynamics has been suggested based on the prevailing disturbance regime, the main disturbance categories being stand-replacing and partial disturbances, and disturbances operating at fine scales (Angelstam and Kuuluvainen 2004). Based on this, a minimum of three main types of forest dynamics can be separated:

* even-aged dynamics, related to stand-replacing disturbances;

- cohort dynamics, related to partial disturbances; and
- gap dynamics, related to tree mortality at fine scale.

In boreal Fennoscandia, a growing body of evidence emphasizes the natural prevalence of varied uneven-aged forest structures and dynamics. In *Picea*-dominated stands occurring on moist sites this is due to inherently long fire rotations (Pitkänen *et al.* 2002, 2003; Wallenius *et al.* 2010). In dry *Pinus*-dominated sites, where fires are more prevalent but therefore also less intense, large *Pinus* trees often survive fires because of their thick heat-insulating bark. This, and the enhancement of regeneration after fire, result over time in forest structures consisting of several age cohorts of trees, surviving or regenerating after fires (Wallenius *et al.* 2004). Thus, cohort dynamics and the gap dynamics would likely dominate in most forest landscapes under natural dynamics (Pennanen 2002; Kuuluvainen and Aakala 2011).

For sustaining native biodiversity through ecosystem management, the landscape-level forest age structure is of paramount importance. Studies of unmanaged forest dynamics and landscape structures in boreal Fennoscandia emphasize the role of non-stand-replacing disturbances and the resulting complex stand structures and dynamics at the landscape level. The resulting landscape is dominated by a more or less continuous forest cover characterized by big old trees and versatile dead wood.

Restoration and management versus natural forest dynamics

Even-aged management based on timber harvesting by clear cutting is currently the dominant method all across the boreal zone. However, as becomes evident from the previous discussion, this type of management poorly matches with the natural forest structures and dynamics (Figure 8.2; Bergeron *et al.* 2002; Kuuluvainen and Aakala 2011). Clear cutting means that the two naturally common types of forest structure and dynamics, the uneven-aged forests

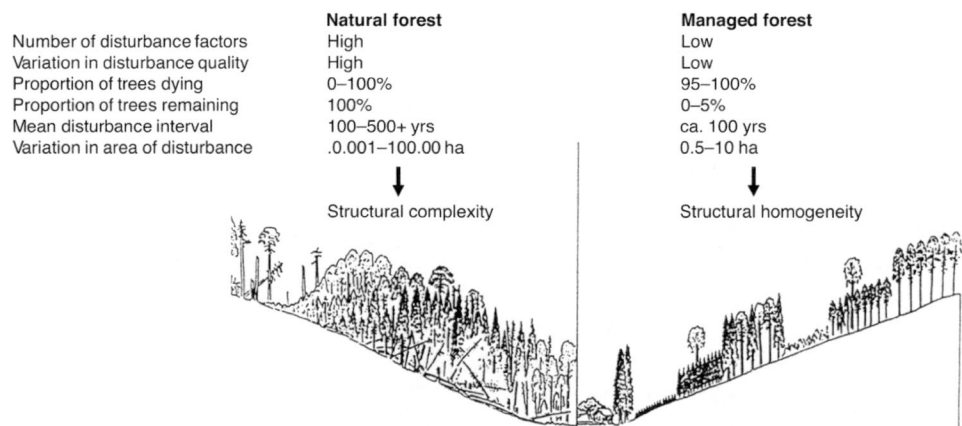

	Natural forest	Managed forest
Number of disturbance factors	High	Low
Variation in disturbance quality	High	Low
Proportion of trees dying	0–100%	95–100%
Proportion of trees remaining	100%	0–5%
Mean disturbance interval	100–500+ yrs	ca. 100 yrs
Variation in area of disturbance	.0.001–100.00 ha	0.5–10 ha
	Structural complexity	Structural homogeneity

Figure 8.2 A comparison of essential disturbance parameters between natural and managed boreal forest and illustration of their impact on forest structure. In natural forests the large variation in disturbance parameters and disturbance legacies lead to high structural complexity at multiple scales in the forest, while the restricted variation in disturbances in managed forest leads to structural homogeneity and potentially to decline in biodiversity and ecosystem resilience

characterized by small-scale gap dynamics and the multi-cohort stands driven by mixed-severity disturbances, have by and large been eradicated from managed forest landscapes. As a consequence, forest landscape structures have been simplified and homogenized and they have moved outside their natural range of variability (NRV) (Bergeron *et al.* 2010; Kuuluvainen 2009). This also means that management based on clear cutting is unable to create the *coarse filter* which is considered important for ecosystem management (Hunter 1991, 1993).

The most fundamental change in forest ecosystem conditions, brought about by management, is perhaps the decline in the amount of old forest and trees (Linder and Östlund 1998; Kuuluvainen 2009), which is also indicative of the fragmentation of the original habitat and dramatic increase in forest edge environments. A concomitant decrease in forest structural variability and number of big trees and dead wood has taken place. For example, in Finland the amount of dead wood in managed forests is only 1–10 per cent of the natural levels. This has posed a severe threat to biodiversity because in Fennoscandia dead wood harbours about 1/4 (approximately 6000 species) of forest-dwelling species (Siitonen 2001). Also in young stands the structural and compositional variability has decreased due to homogenization of forests through tree planting and thinning. In addition, in Scandinavia, natural disturbances such as fires have been largely eliminated; other natural disturbances are scarce as well. Thus effective restoration and ecosystem management should aim at emulating natural disturbances to maintain the basic characteristics and dynamics of forest habitats at various scales (i.e. the coarse filter; Hunter 1991, 1993; Kuuluvainen and Grenfell 2012).

Leading principles

Preserve or restore natural habitats or their characteristics at multiple scales

In most cases habitat loss is the main driver behind declining species populations. Therefore, maintaining the natural habitats and/or their critical properties (i.e. the coarse filter) should be the leading principle of restoration and ecosystem management. In some cases restoration may be planned for the conservation of just one or a few focal species. However, this should not be taken as a general approach, as the goal usually is to maintain viable populations of all native species and all ecosystem functions, not just some large 'sexy' species. Because there generally is a lack of knowledge of species' habitat requirements, the species-by-species planning approach is not generally suitable as an overarching strategy for restoration and ecosystem management. Instead, a more realistic and balanced approach may be to try to maintain the habitats as close to their natural state as possible when simultaneously fulfilling other management objectives (Kuuluvainen and Siitonen 2013).

Preservation or restoration of habitats should be considered at multiple scales. Often forest management operations are planned using a forest stand or compartment as a basic unit, without taking into account the spatial context. However, it is evident that measures to protect biodiversity should always incorporate the landscape level, because many important ecological processes occur at this scale. Such phenomena are, for example, movement and home ranges of many animals, dispersal of species and disturbance dynamics. However, although a larger scale such as that of a landscape is necessary, restoring habitat at various scales should be the goal. This is because biodiversity is tied to habitats occurring at different spatial scales in the forest ecosystem, from micro-fungi on dead tree trunks to large predators needing regional-scale habitat planning for conservation.

Because of the multiple scales involved, the idea of *hierarchical planning* should always be applied in restoration and ecosystem management. Hierarchical planning means thinking and acting simultaneously at different hierarchical levels of a forest ecosystem. Preferably at least four spatial scales should be considered: microhabitat, stand, landscape and regional scales. Microhabitats are small objects which are important for species, such as snags, logs, stumps, big and hollow trees and uprooting spots. Stand-level characteristics to consider are tree species composition, tree size distribution, spatial pattern of trees and amount and quality of dead wood. Important landscape-level attributes are distribution of stand ages or forest successional stages, amount of protected area and connectivity of habitats. Consideration of regional scale takes into account geographical variation in land use history, climate, geomorphology, disturbance regimes, and their effect on ecosystem functioning.

Special attention is needed for aquatic ecosystems embedded in the forested landscape since they and the surrounding riparian forested zones are usually hot spots of biodiversity in the landscapes (e.g. Nummi and Kuuluvainen 2014). Such aquatic ecosystems are brooks, rivers, beaver ponds, lakes, springs, vernal pools, paludified areas and bogs. Their riparian habitats are characterized by high plant species diversity, which hosts high overall organism diversity. Such moist forests usually escape fires and are characterized by small-scale gap disturbances and continuity of dead wood. As such they act as natural corridors in the landscape and sources of species colonizing post-disturbance sites. Whether these riparian habitats should be left outside management or not depends on the possibility of maintaining their important functions in the landscape.

Manage for processes rather than for structures

Forest ecosystems are constantly changing in structure and function in a cyclic manner due to repeated disturbances and the following successions (the adaptive cycle) (Kuuluvainen 2009). This emphasizes that for maintaining favourable ecosystem conditions, ecosystem resilience and adaptive capacity, managers should pay special attention to the spatial and temporal processes that create and maintain structures in the landscape in the long term. Designing restoration and sustainable management as inspired by natural disturbances provides an example of this approach (Bergeron *et al.* 2002; Kuuluvainen and Grenfell 2012).

Consider managed and protected areas in combination

The traditional approach to protect nature has been to establish specific set-aside protection areas, which can be small (e.g. key biotopes) or large and variably connected (Figure 8.1). The success of this approach depends on the area, quality and connectedness of the protection area network, and on the habitat quality of the managed forest matrix. If the protected areas cover a sufficiently large amount of all ecosystem types, they can ensure long-term ecological functioning, including connectedness and disturbance dynamics. However, in many regions the protected areas are small and/or geographically unevenly distributed (Potapov *et al.* 2008). They are often isolated and poorly represent the natural range of ecosystem types and species communities. For example in Finland and Sweden the protected areas are primarily located in the low productivity northern or high altitude forests with low levels of native biodiversity compared to more productive, low altitude or southern areas. Thus, the representativeness of the reserve network is typically far from adequate, particularly in the southern regions where natural biodiversity would naturally be highest (Angelstam *et al.* 2011).

The problem may also be that there are no guarantees that species and habitats will in the future be found where they are found today. There are several reasons for this. First, the habitats may be too small (e.g. key biotopes) and affected by harmful edge effects to maintain their ecological characteristics in the long run. They may also be the last isolated refuges left for species and the remaining small populations can be doomed to local extinction (the extinction debt). Second, habitats may be destroyed by unexpected severe disturbance events, especially if the protected areas are small. Thirdly, habitats may also be slowly transformed to less preferable states due to lack of natural processes or environmental change.

Therefore, for safeguarding biodiversity, it is necessary to consider landscapes consisting of protected areas and managed forests in concert. Even in situations where a relatively large proportion of unmanaged or protected forest exists, a majority of forest area and species will typically be under the impact of forest management. Because of this, the way the managed forest is treated is of fundamental importance for biodiversity conservation. Treatment of managed forest largely determines the amount, configuration and connectivity of available habitats at the landscape scale. The treatment of managed forest in the vicinity of protected areas should be carried out to reduce harmful edge effects and hence to improve the functioning of the protected area as habitat.

The coarse and fine filter, and historical range of variability

The coarse and fine filter concept distinguishes two complementary strategies for sustaining biodiversity. The coarse filter strategy is congruent with the ecosystem management approach in that it emphasizes the importance of maintaining natural ecosystem structures (Hunter 1991, 1993). This relates to considering the historical range of variability (HRV) of the ecosystem based on the knowledge of the past disturbance regimes (Landres *et al.* 1999; Keane *et al.* 2009). The assumption is that by restoring or maintaining landscape conditions within a range that the organisms have adapted to will most likely conserve biodiversity and maintain sustainable ecosystems. The coarse filter approach does not necessarily consider only reserves, but rather recognizes ecological processes and the dynamic distribution of habitats across the whole landscape over time. The coarse filter strategy has been suggested to improve planning efficiency as it avoids the pitfalls of species by species planning.

A complementary fine filter approach focuses on individual species or fine scale elements of diversity which are critical in conserving elements of diversity that are not sufficiently accounted for by the coarse filter approach. This approach tries to safeguard those species that 'fall through' the coarse filter. Thus the fine filter approach consists of developing specific conservation strategies for focal species, which are considered to be at particular risk under the coarse filter approach. Examples of fine filter strategies could be providing nesting boxes for cavity-nesting birds and mammals (e.g. the flying squirrel) when hollow trees are not available.

Natural disturbance emulation

Natural disturbance emulation (NDE) has become an important concept when aiming at implementing the coarse filter approach in forest ecosystem restoration and management (Figure 8.3; Angelstam 1998; Bergeron *et al.* 2002; Kuuluvainen 2002). According to this approach, management actions are planned to emulate natural disturbances and their outcomes (the coarse filter) in the forest landscape. This should be done at multiple scales, from dead wood microsites to landscape patterns, to provide natural habitat availability for different organisms. The potential of the natural disturbance approach becomes most pronounced in situations where the protection area network is clearly insufficient.

Applying the natural disturbance emulation in forest restoration and management requires adequate knowledge of the natural dynamics of forest structures at the landscape scale (i.e. the natural range of variability; Figure 8.3; Landres *et al.*1999; Kuuluvainen 2002). However, in

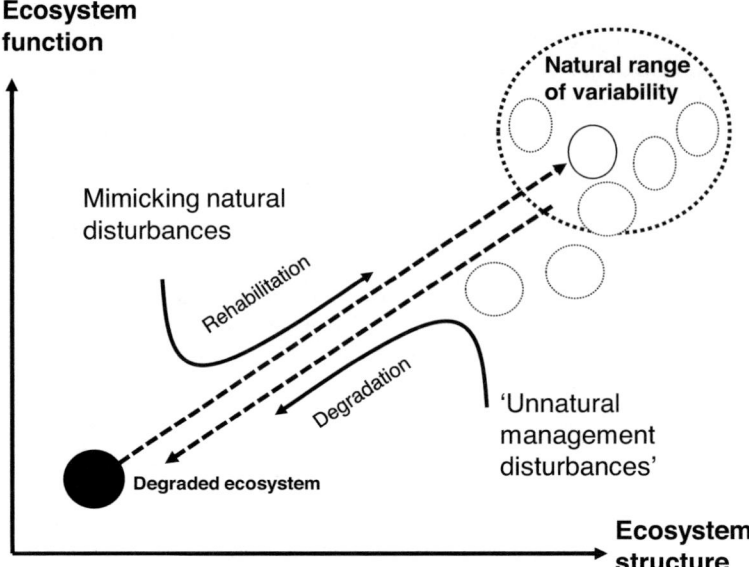

Figure 8.3 Restoration means rehabilitating ecosystem structure and function closer to their natural range of variability. According to the natural disturbance emulation approach this can be done by replacing previous 'unnatural disturbances' created by previous even-aged management with more diverse harvesting and management practices inspired by natural disturbances

many areas this is challenging because of long-term human impact and because the protected forest fragments that can be used as references are too small to harbour natural ecosystem dynamics. For example, it has been found that the stand structures and species composition in small protected areas in southern Finland differ considerably from stands in more naturally dynamic landscapes in Russian Karelia (Lilja and Kuuluvainen 2005; Nordén *et al.* 2013).

Even without human influence, forest landscape structure is affected by local geomorphic features. Moreover, a natural forest has no equilibrium but changes in structure and dynamics as a function of variation in climatic conditions and disturbance dynamics, thus exhibiting its natural range of variability (Landres *et al.* 1999). However, with these caveats in mind, it is possible to define the main features of natural forest for a specific area using information from multiple sources including (1) retrospective analyses based on historical materials, (2) biological archives using dendroecological and palaeoecological methods, (3) research on geographically close and ecologically similar but more natural forests, and (4) integrative modelling (Pennanen 2002).

Tactical models

Tactical models are needed to translate abstract theoretical and strategic concepts, such as the coarse filter and natural disturbance emulation (Figure 8.3), into practical forest management. So far, only a limited number of such tactical models have been proposed and used for the boreal forest. These models are tactical in the sense that they do not merely state the strategy to emulate natural forest dynamics, but go a step further and make more concrete, practical proposals on how to decide what, when and how to act. Examples of such tactical approaches are provided by (1) the *patch-corridor-matrix model*, (2) the *ASIO-model* and (3) the *multi-cohort model*.

The patch-corridor-matrix model

The patch-corridor-matrix model is based on the idea that landscapes can be conceptualized as mosaics of separable patches (the patch–mosaic paradigm). This paradigm views forested landscapes as composed of three basic elements: (a) valuable *habitat patches*, which are considered to differ clearly from their surroundings; in forested landscapes these are typically high-quality habitat or key-biotope patches; (b) *corridors or stepping-stones*, which should allow the movement of organisms between the high-quality habitat patches; (c) surrounding *matrix* which is often assumed to be unsuitable or at least low-quality habitat for the organisms of interest (Forman 1995). The background of this model is in the *theory of island biogeography* (MacArthur and Wilson 1967) and on the more recently developed *metapopulation theory* (Hanski 2000). The patch-corridor-matrix model has been widely applied as a basic planning approach in forest landscape management in Finland and Sweden on both public and private forest lands.

The application is straightforward: the valuable habitat patches (a), corridors and stepping stones (b) are either set aside or managed in a way that maintains their natural characteristics, while 'business-as-usual' management is practised in the remaining matrix area (c). If the set-aside patches represent habitats of low natural disturbance probability, such as moist sheltered depressions, this could be seen to represent some form of natural landscape emulation. However, the patch-corridor-matrix model's representation of the landscape is static and extremely simplified and does not therefore conform to the contemporary understanding of cross-scale complexity of structure and dynamics of forest ecosystems (Filotas *et al.* 2014). It thus remains unclear to what extent the course filter principle is realized.

The ASIO model

The ASIO model was developed in Sweden in the 1990s for the purpose of emulating natural disturbances in forest management (Rülcker *et al.* 1994). The model was inspired by research on the pivotal role of natural disturbances, especially that of fire, in the northern boreal forests (e.g. Angelstam 1998). The model is founded on research indicating that in boreal Fennoscandia, fire occurrence is related to site fertility and soil moisture (e.g. Wallenius *et al.* 2004), so that natural fire frequency ranges from low in moist herb-rich forests to high fire frequency in dry sandy heaths dominated by Scots pine (Figure 8.4). The acronym of the model was derived from four perceived parts on this site-type gradient qualitatively describing fire occurrence; namely,

- **a**bsent
- **s**eldom
- **i**nfrequent and
- **o**ften.

The ASIO model prescribes various management and cutting regimes for each category which are congruous with the assumed frequency and effects of fire (or absence of fire) on stand structure and succession (Figure 8.4). In practice, it was suggested that herb-rich sites with high

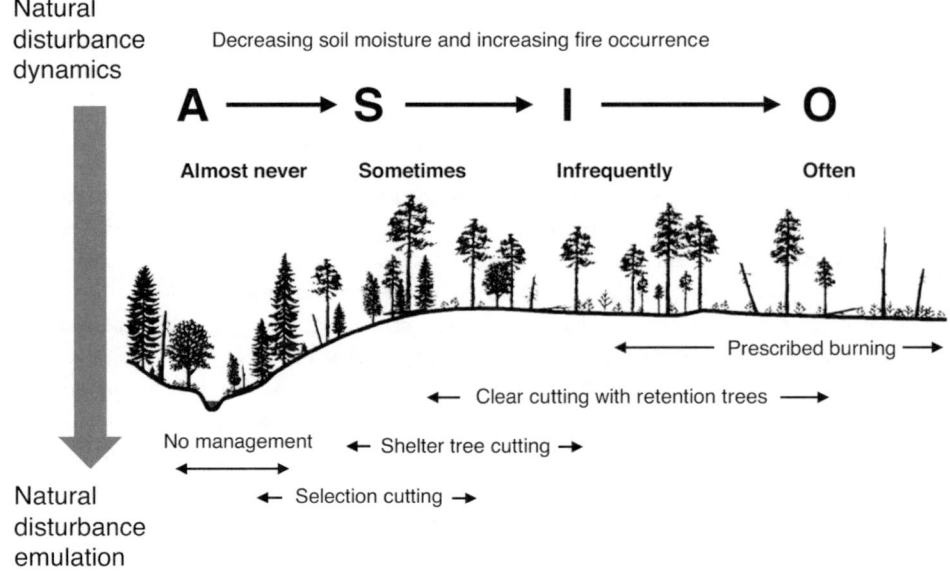

Figure 8.4   Illustration of the principle of the ASIO model in forest restoration and ecosystem management. The landscape is divided into four categories based on variation in site moisture and its assumed impact on fire occurrence (i.e. where fire is 'absent', occurs 'seldom', 'infrequently' or 'often'). The ASIO model can be implemented with various management and cutting regimes for each category, which are supposed to emulate the frequency and effect of fire (or absence of fire) on stand structure and succession

Source: drawing by J. Karsisto

biodiversity and conservation value are set-asides, while cuttings and management of variable types and intensities were suggested for the remaining landscape (Angelstam 1998).

In practical forest management, the ASIO model has often been used more as a 'theoretical reference', with limited ambition to use the model as a real restoration or ecosystem management tool. Instead, conventional low-retention clear cutting is used across most of the forest (S-, I- and O-classes) with the loose justification of 'emulating' high intensity fire dynamics. This is not, however, emulating natural fire effects on forests, which we know to be much more variable (cf. Kuuluvainen 2009).

The multi-cohort model

The multi-cohort model was suggested by Bergeron *et al.* (1999, 2002). The model is based on fundamental research on forest fire and successional dynamics carried out in eastern Canada (Bergeron *et al.* 2010; Figure 8.5). The approach explicitly considers the implications of multiple spatial scales from landscape to stand structure in natural disturbance emulation. The dominant disturbance agent to be considered is fire of variable periodicity and severity. The fundamental idea is that after a stand-replacing fire, forest succession moves through distinct phases, which can be termed 'structural cohorts', under the potential influence of further disturbance events (Figure 8.5). The natural proportions, or the range of variability, of different structural cohorts can be derived from a historical analysis of past spatio-temporal landscape

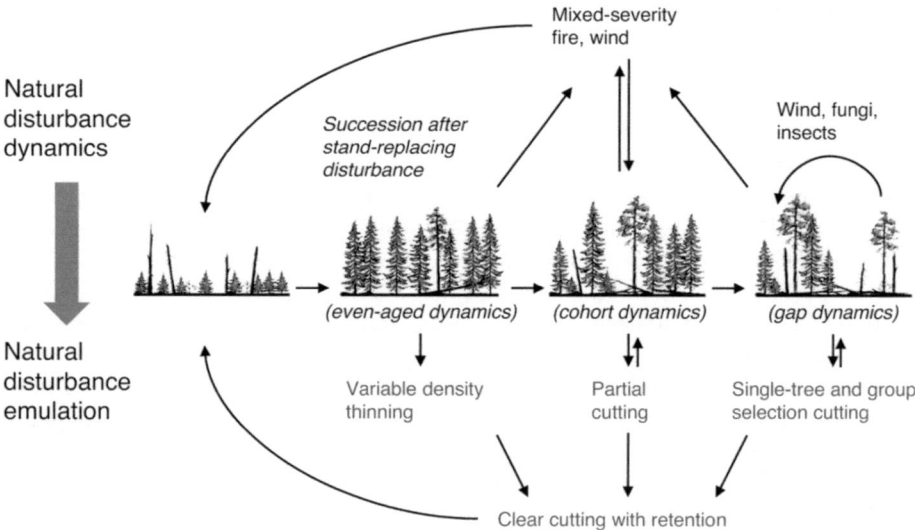

Figure 8.5   Illustration of the principle of the multi-cohort model in forest restoration and ecosystem management. The model aims to emulate natural disturbances and succession in order to maintain the structural features typical of the natural forest at stand and landscape scales. In this approach, the forest area is divided into 'structural cohorts' representing different post-fire stand successional stages. Different cutting methods and intensities, with variable final harvest rotations, are applied to maintain landscape-level forest structures and age distributions similar to those that would exist under a natural disturbance regime

Sources: Bergeron *et al.* (2002); drawing by J. Karsisto

disturbances (Bergeron *et al.* 2002). Under conditions where stand-replacing disturbances are infrequent, the landscape is dominated by old forest, whereas young forest prevails when severe fires are frequent.

Based on the analysis of natural forest landscape age structure and disturbance regime, different silvicultural approaches are applied at the stand scale in order to maintain the natural proportions and spatial pattern of different 'structural forest cohorts' in the landscape (Figure 8.5, Bergeron *et al.* 2002). Stands are subjected to different cutting treatments depending on which 'structural cohort' they should emulate. Younger stands can be treated with variable density thinnings and partial cuttings, to emulate the natural self-thinning process. Older stands, representing late-successional stages, can be subjected to partial, gap or selection cutting to emulate small-scale disturbances and associated stand structures. Finally, clear-felling with appropriate retention is applied to emulate stand-replacing disturbances and to regulate the proportion of young forest-age classes in the landscape.

It is suggested that by using this approach it is possible to maintain the forest landscape within its natural range of structural and compositional variability (NRV; Figure 8.3). As with the ASIO model, the importance of areas set aside for conservation is highlighted. These areas could be surrounded by buffer zones with a high level of tree retention. A distinction from the ASIO model is that in the original multi-cohort model fires are assumed to occur more or less randomly across the landscape (Bergeron *et al.* 2002), whereas in the ASIO model the fire frequency was assumed to be dependent on site type (see Figure 8.4). Although neither of these assumptions may be strictly realistic, they may reflect differences in the fire ecology of forests between northern North America and northern Eurasia (Kneeshaw *et al.* 2011; Rogers *et al.* 2015).

Cross-scale hierarchical implementation

The different strategic and tactical approaches to forest restoration and ecosystem management can also be seen as complementary components of a hierarchical regional or landscape management approach (Kuuluvainen and Grenfell 2012). The patch–corridor–matrix model defines the network of permanent protection areas, together with corridors and 'stepping stones', which are the skeleton components of a hierarchical planning approach. Natural disturbance emulation (NDE) approaches, represented by ASIO and multi-cohort models, provide tactical tools to create a desired variability and dynamics of forest structures across the landscape. This kind of landscape design would create synergy between the different approaches. For example, by using NDE principles in matrix forest management, the connectivity properties of protected areas and the landscape as a whole could be improved.

The core of the hierarchical approach is formed by the protected area network, which is needed to sustain sensitive species that are specialized to natural forest habitat. Restoration can be used in those parts of protected areas which in the past have been affected by human utilization. In the managed forest the natural disturbance emulation (NDE) can be applied to maintain landscapes within their natural range of variability (thus implementing the coarse filter strategy; Kuuluvainen and Grenfell 2012). In the vicinity of protected areas, it may be useful to create buffer zones by applying silviculture that has a strong restoration aspect. This kind of 'restoration silviculture' would aim at creating ecologically important natural-like early successional stages with higher dead wood levels (e.g. 30–60 m³/ha) than in ordinary managed forest, while still harvesting most of the timber and ensuring rapid regeneration (Vanha-Majamaa *et al.* 2007). This kind of 'hybrid approach' with restoration emphasis could enhance the functioning of protected areas and thus improve the efficiency of the overall strategy for sustainable forest management.

In the hierarchical approach, partial and local-scale disturbances are emulated by maintaining continuous cover forests and fine-scale structural heterogeneity. At the landscape scale, clear cutting with retention is used to imitate severe disturbances, to the extent they would occur naturally, while care is taken to leave sufficient amount of retention trees. Methods like variable density thinning are used to enhance heterogeneity in young stands to imitate the high variability of stand successional pathways characteristic of natural forest development. The management approach needs to be comprehensive. This means that in restoration and management, the forested landscape as a whole, must be taken into consideration simultaneously (Lindenmayer and Franklin 2002).

Conclusions

The health of the vast boreal forest biome is increasingly threatened by the combined effects of human activities and the anticipated drastic changes in climatic conditions at high latitudes. This risks the long-term availability of important ecosystem services at global scale and the livelihoods of forest-based communities and economies at local scales. This situation calls for international collaboration and action. There exist restoration and ecosystem management strategies and methods that can be used to protect forest health and prevent ecosystems from further degradation. However, current international agreements and regional market mechanisms do not provide sufficient incentives to implement effective measures at local and global scales (Moen *et al.* 2014).

Successful forest restoration and ecosystem management must be based on fundamental understanding of the functioning of forest ecosystems and the processes maintaining their diversity and resiliency, but also on knowledge of their socio-ecological system dynamics (Drever *et al.* 2006; Kuuluvainen 2009; Filotas *et al.* 2014). A useful conceptual framework for action is provided by adaptive systems perspective, which can incorporate both social and ecological factors (Messier *et al.* 2013). Understanding of the ecological and evolutionary functions of disturbance–succession cycles (i.e. the adaptive cycles) should form the basis of restoration and ecosystem management strategies. This is because ecosystem adaptive cycles are the key for maintaining ecosystem biodiversity, resiliency and adaptive capacity in the long term.

Although restoration and ecosystem management emphasize the importance of ecological understanding of the 'natural forest', a distinction between nature and culture in forest restoration should be avoided. Nature can be used as a useful guide and important reference, but it does not tell us what successful restoration and ecosystem management is. The goals of forest restoration and ecosystem management must be sought through societal dialogue and democratic processes. This means that conceptions such as 'emulation of natural disturbance' or 'historical range of variability' should be regarded as tools and guidelines that can be taken into account and respected when striving for successful restoration and ecosystem management.

References

Angelstam, P. K. (1998) Maintaining and restoring biodiversity in European boreal forests by developing natural disturbance regimes. *Journal of Vegetation Science*, vol. 9, no. 4, pp. 593–602.

Angelstam, P. and Kuuluvainen, T. (2004) Boreal forest disturbance regimes, successional dynamics and landscape structures – a European perspective. *Ecological Bulletins*, vol. 51, pp. 117–136.

Angelstam, P., Andersson, K., Axelsson, R., Elbakidze, M., Jonsson, B. G. and Roberge, J.-M. (2011) Protecting forest areas for biodiversity in Sweden 1991–2010: the policy implementation process and outcomes on the ground. *Silva Fennica*, vol. 45, no. 5, pp. 1111–1133.

Bauhus, J., Puettmann, C. and Messier, C. (2009) Silviculture for old-growth attributes. *Forest Ecology and Management*, vol. 258, pp. 525–537

Bergeron, Y. and Fenton, N. (2012) Boreal forests of eastern Canada revisited: old growth, nonfire disturbances, forest succession and biodiversity. *Botany*, vol. 90, pp. 509–523.

Bergeron, Y., Harvey, B., Leduc, A. and Gauthier, S. (1999) Forest management guidelines based on natural disturbance dynamics: stand- and forest-level considerations. *Forestry Chronicle*, vol. 75, no. 1, pp. 49–54.

Bergeron, Y., Leduc, A., Harvey, B. D. and Gauthier, S. (2002) Natural fire regime: a guide for sustainable management of the Canadian boreal forest. *Silva Fennica*, vol. 36, pp. 81–95.

Bergeron, Y., Cyr, D., Girardin, M. P. and Carcaillet, C. (2010) Will climate change drive 21st century burn rates in Canadian boreal forest outside its natural range of variability: collating global climate model experiments with sedimentary charcoal data. *International Journal of Wildland Fire*, vol. 19, pp. 1127–1139.

Bradshaw, C. J. A., Warkentin, I. G. and Sodhi, N. S. (2009) Urgent preservation of boreal carbon stocks and biodiversity. *Trends in Ecology and Evolution*, vol. 24, pp. 541–548.

Brandt, J. P., Flannigan, M. D., Maynard, D. G. and Thompson, I. D. (2013) An introduction to Canada's boreal zone: ecosystem processes, health, sustainability, and environmental issues. *Environmental Reviews*, vol. 21, pp. 207–226.

Burton, P. J. and Macdonald, S. E. (2011) The restorative imperative: Challenges, objectives and approaches in restoring naturalness in forests. *Silva Fennica*, vol. 45, no. 5, pp. 843–863.

Burton, P. J., Bergeron, Y., Bogdanski, B. E. C., Juday, G. P., Kuuluvainen, T., McAfee, B. J., Ogden, A., Teplyakov, V. K., Alfaro, R. I., Francis, D. A., Gauthier, S. and Hantula, J. (2010) Sustainability of boreal forests and forestry in a changing environment. In G. Mery, P. Katila, G. Galloway, R. I. Alfaro, M. Kanninen, M. Lobovikov and J. Varjo (eds), *Forests and Society – Responding to Global Drivers of Change*. International Union of Forest Research Organizations (IUFRO), Vienna, Austria, pp. 249–282.

Cardinale, B. J., Srivastava, D. S., Duffy, J. E., Wright, J. P., Downing, A. L. Sankaran, M. and Jousseau, C. (2006) Effects of biodiversity on the functioning of trophic groups and ecosystems. *Nature*, vol. 443, pp. 989–992.

CBD (2004) *The Ecosystem Approach, (CBD Guidelines).* Secretariat of the Convention on Biological Diversity, Montreal. Retrieved from www.cbd.int/doc/publications/ea-text-en.pdf.

Drever, C. R., Peterson, G., Messier, C., Bergeron, Y. and Flannigan, M. (2006) Can forest management based on natural disturbances maintain ecological resilience? *Canadian Journal of Forest Research*, vol. 36, pp. 2285–2299.

Filotas, E., Parrott, L., Burton, P. J., Chazdon, R. L., Coates, K. D., Coll, L., Haeussler, S., Martin, K., Nocentini, S., Puettmann, K. J., Putz, F. E., Simard, S.W. and Messier, C. (2014) Viewing forests through the lens of complex systems science. *Ecosphere*, vol. 5, no. 1, pp.1–23, article 1.

Forman, R. T. T. (1995) *Land Mosaics: The Ecology of Landscapes and Regions.* Cambridge University Press, Cambridge.

Gauthier, S., Vaillancourt, M.-A., Leduc, A., De Grandpré, L., Kneeshaw, D., Morin, H., Drapeau, P. and Bergeron, Y. (eds) (2009) *Ecosystem Management in the Boreal Forest.* Les Presses de l'Université du Québec, Québec.

Halme, P., Allen, K. A., Aunins, A., Bradshaw, R. H. W., Brumelis, G., Cada, V., Clear, J. L., Eriksson, A.-M., Hannon, G., Hyvärinen, E., Ikauniece, S., Iršénaité, R., Jonsson, B. G., Junninen, K., Kareksela, S., Komonen, A., Kotiaho, J. S., Kouki, J., Kuuluvainen, T., Mazziotta, A., Mönkkönen, M., Nyholm, K., Olden, A., Shorohova, E., Strange, N., Toivanen, T., Vanha-Majamaa, I., Wallenius, T., Ylisirniö, A.-L. and Zin, E. (2013) Challenges of ecological restoration: Lessons from forests in northern Europe. *Biological Conservation*, vol. 167, pp. 248–256.

Hanski, I. (2000) Extinction debt and species credit in boreal forests: modelling the consequences of different approaches to biodiversity conservation. *Annales Zoologici Fennici*, vol. 37, pp. 271–280.

Hobbs, R. J., Hallett, L. M., Ehrlich, P. R. and Mooney, H. A. (2011) Intervention ecology: applying ecological science in the twenty-first century. *Bioscience*, vol. 61, no. 6, pp. 442–450.

Holling, C. S. (2001) Understanding the complexity of economic, ecological, and social systems. *Ecosystems*, vol. 4, pp. 390–405.

Holm, S.-O. (2015) A management strategy for multiple ecosystem services in boreal forest. *Journal of Sustainable Forestry*, vol. 35, no. 4, pp. 358–379.

Hunter, M. L. (1991) Coping with ignorance: the coarse filter strategy for maintaining biodiversity. In K. A. Kohm (ed.), *Balancing on the Brink of Extinction.* Island Press, Washington, DC, pp. 266–281.

Hunter, M. L. Jr. (1993) Natural fire regimes as spatial models for managing boreal forests. *Biological Conservation*, vol. 65, pp. 115–120.

Keane, R. E., Hessburg, P. F., Landres, P. B. and Swanson, F. J. (2009) The use of historical range and variability (HRV) in landscape management. *Forest Ecology and Management*, vol. 258, pp. 1025–1037.

Keto-Tokoi, P. and Kuuluvainen, T. (2014) *Primeval Forests of Finland, Cultural History, Ecology and Conservation.* Maahenki, Helsinki.

Klenk, N. L., Bull, G. Q., and MacLellan, J. I. (2009) The 'emulation of natural disturbance' (END) management approach in Canadian forestry: a critical evaluation. *Forestry Chronicle*, vol. 85, pp. 440–445.

Kneeshaw, D., Bergeron, Y. and Kuuluvainen, T. (2011) Forest ecosystem structure and disturbance dynamics across the circumboreal forest. In A. C. Millington, M. B. Blumler and U. Schickhoff (eds), *The Sage Handbook of Biogeography.* Sage, Los Angeles, CA, pp. 263–280.

Kuuluvainen, T. (1994) Gap disturbance, ground microtopography, and the regeneration dynamics of boreal coniferous forests in Finland: a review. *Annales Zoologici Fennici*, pp. 31, 35–51.

Kuuluvainen, T. (2002) Natural variability of forests as a reference for restoring and managing biological diversity in boreal Fennoscandia. *Silva Fennica*, vol. 36, pp. 97–125.

Kuuluvainen, T. (2009) Forest management and biodiversity conservation based on natural ecosystem dynamics in northern Europe: the complexity challenge. *Ambio*, vol. 38, pp. 309–315.

Kuuluvainen, T. and Aakala, T. (2011) Natural forest dynamics in boreal Fennoscandia: a review and classification. *Silva Fennica*, vol. 45, no. 5, pp. 823–841.

Kuuluvainen, T. and Grenfell, R. (2012) Natural disturbance emulation in boreal forest ecosystem management: theories, strategies and a comparison with conventional even-aged management. *Canadian Journal of Forest Research*, vol. 42, pp. 1185–1203.

Kuuluvainen, T. and Siitonen, J. (2013) Fennoscandian boreal forests as complex adaptive systems. Properties, management challenges and opportunities. In C. Messier, K. J. Puettman and K. D. Coates (eds), *Managing Forests as Complex Adaptive Systems: Building Resilience to the Challenge of Global Change.*

Earthscan, Abingdon, pp. 244–268.

Landres, P. B., Morgan, P. and Swanson, F. J. (1999) Overview of the use of natural variability concepts in managing ecological systems. *Ecological Applications*, vol. 9, pp. 1179–1188.

Levin, S. A. (1998) Ecosystems and the biosphere as complex adaptive systems. *Ecosystems*, vol. 1, pp. 431–436.

Levin, S. A. (2005) Self-organization and the emergence of complexity in ecological systems. *BioScience*, vol. 55, pp. 1075–1079.

Lilja, S. and Kuuluvainen, T. (2005) Structure of old *Pinus sylvestris* dominated forest stands along a geographic and human impact gradient in mid-boreal Fennoscandia. *Silva Fennica*, vol. 39, pp. 407–428.

Lindenmayer, D. B. and Franklin, J. F. (2002) *Conserving Forest Biodiversity: A Comprehensive Multiscaled Approach*. Island Press, Washington, DC.

Linder, P. and Östlund, L. (1998) Structural changes in three mid-boreal Swedish forest landscapes, 1885–1996. *Biological Conservation*, vol. 85, pp. 9–19.

MacArthur, R. H. and Wilson, E. O. (1967) *The Theory of Island Biogeography*. Princeton University Press, Princeton, NJ.

Messier, C., Puettmann, K. J. and Coates, K. D. (eds) (2013) *Managing Forests as Complex Adaptive Systems: Building Resilience to the Challenge of Global Change*. Routledge, Abingdon.

Mielikäinen, K. and Hynynen, J. (2003) Silvicultural management in maintaining biodiversity and resistance of forests in European boreal zone: case Finland. *Journal of Environmental Management*, vol. 67, pp. 47–54.

Moen, J., Rist, L., Bishop, K., Chapin III, F. S., Ellison, D., Kuuluvainen, T., Petersson, H., Puettmann, K. J., Rayner, J., Warkentin, I. G. and Bradshaw, C. J. A. (2014) Eye on the taiga: removing global policy impediments to safeguard the boreal forest. *Conservation Letters*, vol. 7, no. 4, pp. 408–418.

Nordén, J., Penttilä, R., Siitonen, J., Tomppo, E. and Ovaskainen, O. (2013) Specialist species of wood-inhabiting fungi struggle while generalists thrive in fragmented boreal landscapes. *Journal of Ecology*, vol. 101, pp. 701–712.

Nummi, P. and Kuuluvainen, T. (2014) Forest disturbance by an ecosystem engineer: beaver in boreal forest landscapes. *Boreal Environment Research*, vol. 18 (suppl. A), pp. 13–24.

Payette, S. (1992) Fire as a controlling process in the North American boreal forest. In H. H. Shugart, R. Leemans and G. B. Bonan (eds), *A Systems Analysis of the Global Boreal Forest*. Cambridge University Press, New York, pp. 144–169.

Pennanen, J. (2002) Forest age distribution under mixed-severity fire regimes – a simulation-based analysis for middle boreal Fennoscandia. *Silva Fennica*, vol. 36, pp. 213–231.

Pitkänen, A., Huttunen, P., Jugner, K. and Tolonen, K. (2002) 10,000 year local forest fire history in a dry heath forest site in eastern Finland, reconstructed from charcoal layer records of a small mire. *Canadian Journal of Forest Research*, vol. 32, pp. 1875–1880.

Pitkänen, A., Huttunen, P., Tolonen, K. and Jugner, K. (2003) Long-term fire frequency in the spruce dominated forests of Ulvinsalo strict nature reserve, Finland. *Forest Ecology and Management*, vol. 176, pp. 305–319.

Potapov, P., Yaroshenko, A., Turubanova, S., Dubinin, M., Laestadius, L., Thies, C., Aksenov, D., Egorov, A., Yesipova, Y., Glushkov, I., Karpachevskiy, M., Kostikova, A., Manisha, A., Tsybikova, E. and Zhuravleva, I. (2008) Mapping the world's intact forest landscapes by remote sensing. *Ecology and Society*, vol. 13, no. 2, article 51.

Puettmann, K. J. (2014) Restoring the adaptive capacity of forest ecosystems. *Journal of Sustainable Forestry*, vol. 33, pp. 15–27.

Rockström, J., Steffen, W., Noone, K., Persson, Å., Chapin III, F. S., Lambin, E. F., Lenton, T. M., Scheffer, M., Folke, C., Schellnhuber, H. J., Nykqvist, B., Wit, C. A., Hughes, T., van der Leuuw, S., Rodhe, H., Sörlin, S., Snyder, P. K., Constanza, R., Svedin, U., Falkenmark, M., Karlberg L., Corell, R. W., Fabry, W. J., Hansen, J., Walker, B., Liverman, D., Richardson, C., Crutzen, P. and Foley, J. A. (2009) A safe operating space for humanity. *Nature*, vol. 461, pp. 472–475.

Rogers, B. M., Soja, A. J., Goulden, M. L. and Randerson, J. T. (2015) Influence of tree species on continental differences in boreal fires and climate feedbacks. *Nature Geoscience*, vol. 8, pp. 228–234.

Rülcker, C., Angelstam, P. and Rosenberg, P. (1994) Naturlig branddynamik kan styra naturvård och skogskötsel in boreal skog. *SkogForsk Resultat*, vol. 8, pp. 1–4.

Scheffer, M., Hirota, M., Holmgren, M., Van Nes, E. H. and Chapin, F. S. (2012) Thresholds for boreal biome transitions. *Proceedings of the National Academy of Sciences USA*, vol. 109, pp. 21384–21389.

Siitonen, J. (2001) Forest management, coarse woody debris and saproxylic organisms: Fennoscandian boreal forests as an example. *Ecological Bulletins*, vol. 49, pp. 11–41.

Sirén, G. (1955) The development of spruce forest on raw humus sites and its ecology. *Acta. Forestalia Fennica*, vol. 62, p. 363.

Steffen, W., Grinevald, J., Crutzen, P. and McNeill, J. (2011a) The Anthropocene: conceptual and historical perspectives. *Philosophical Transactions of the Royal Society of the Academy*, vol. 369, pp. 842–867.

Steffen, W., Persson, Å., Deutsch, L., Zalasiewicz, J., Williams, M., Richardson, K., Crumley, C., Crutzen, P., Folke, C., Gordon, L., Molina, M., Ramanathan, V., Rockström, J., Scheffer, M., Schellnhuber, H. J. and Svedin, U. (2011b) The Anthropocene: from global to planetary stewardship. *Ambio*, vol. 40, p. 739.

Suding, K. N., Gross, K. L. and Houseman, G. R. (2004) Alternative states and positive feedbacks in restoration ecology. *Trends in Ecology and Evolution*, vol. 19, pp. 46–53.

Timonen, J., Gustafsson, L., Kotiaho, J. S. and Mönkkönen, M. (2011) Hotspots in a cold climate: Conservation value of woodland key habitats in boreal forests. *Biological Conservation*, vol. 144, pp. 2061–2067.

Vanha-Majamaa, I., Lilja, S., Ryömä, R., Kotiaho, J. S., Laaka-Lindberg, S., Limdberg, H., Puttonen, P., Tamminen, P., Toivanen, T., and Kuuluvainen, T. (2007) Rehabiliting boreal forest structure and species composition through logging, dead wood creation and fire: the EVO experiment. *Forest Ecology and Management*, vol. 250, pp. 77–88.

Venier, L. A., Thompson, L. D., Fleming, R., Aubin, I., Trofymov, J. A., Langor, R., Sturrock, R., Patry, C., Outerbridge, R. O., Holmes, S. B., Haussler, S., De Grandpre, L., Chen, H. Y. H., Bayne, L., Arsenault, A. and Brandt, J. P. (2014) Effect of natural resource development on the terrestrial biodiversity of Canadian boreal forests. *Environmental Reviews*, vol. 22, pp. 457–490.

Wallenius, T., Kuuluvainen, T. and Vanha-Majamaa, I. (2004) Fire history in relation to site type and vegetation in eastern Fennoscandia, Russia. *Canadian Journal of Forest Research*, vol. 34, pp. 1400–1409.

Wallenius, T. H., Kauhanen, H., Herva, H. and Pennanen, J. (2010) Long fire cycle in northern boreal Pinus forests in Finnish Lapland. *Canadian Journal of Forest Research*, vol. 40, pp. 2027–2035.

9

RESTORATION OF TEMPERATE BROADLEAF FORESTS

John A. Stanturf

Introduction

The temperate forest lies in the zone between the polar and sub-tropical climate zones, between approximately 25° and 50° latitude in the northern and southern hemispheres. The temperate forest can be divided into cold and warm types; the cold temperate or boreal forests are dominated by evergreen conifers and the warm temperate forests by broadleaf species (Box and Fujiwara 2015). Mixed broadleaf forests may contain lesser amounts of coniferous trees species. Regionally, the warm temperate forests are divided into many types based on composition and relative dominance of ubiquitous species. The climate of the temperate broadleaf and mixed forests is decidedly seasonal with warm and cool seasons and sometimes a distinct dry season. Temperate broadleaf forests dominate the natural vegetation of New Zealand and are found in the southern cone of South America, southeastern Australia and Tasmania, southern China, Korea, Japan, mid-elevations of the Himalayas, eastern North America and northern Europe. Temperate deciduous forests are a subgroup of temperate broadleaf forests distinguished by their shedding of leaves in the autumn. Temperate deciduous forests occur in the Northern Hemisphere, primarily eastern North America, eastern Asia and northern Europe, but including the Hyrcanian forests of Iran and Azerbaijan that occur along the coast of the Caspian Sea and the northern slopes of the Alborz mountain range. The ancient fruit and nut forests of central Asia contain the wild ancestors of many domesticated species (Cantarello *et al.* 2014).

Temperate deciduous broadleaf forest species have broad leaves, as opposed to the needle-like leaves of conifers. Evergreen broadleaf trees have narrower laurophyllous leaves, blending into the sclerophyllous leaves of drier forest types found in Mediterranean climatic regions. Characteristic dominant genera include the oaks (*Quercus* spp.), beeches (*Fagus* spp.), maples (*Acer* spp.) and birches (*Betula* spp.) in the northern hemisphere with many taxa common to eastern North America and eastern Asia (Wen 1999); *Nothofagus* and *Eucalyptus* species are endemic in the Southern hemisphere. The closed forests are dominated by tall trees (30 to 60 m), although some *Eucalyptus* spp. are very tall (exceeding 80 m). Lesser trees and shrubs occur in the understorey and the ground flora can be quite diverse. Glaciation, megafaunal collapse, predator extirpation, conflicts, European colonization, and global trade have influenced the composition and structure of the temperate broadleaf forest (Figure 9.1).

Figure 9.1 Allegheny hardwood forests. Deciduous broadleaf forests on the Allegheny Plateau in the eastern United States, dominated by shade-intolerant species including *Prunus serotina*, resulted from widespread clearing of old forests and locally extirpating a native herbivore (white-tailed deer, *Odocoileus virginianus*)

Source: S. Stout

Temperate broadleaf forests have been exploited for timber and charcoal, and their inherent high productivity targeted them for clearing for agriculture and pasture. The remaining forests are mostly fragmented, secondary forests; even those that have never been cleared often were grazed by domestic livestock. Wildfire and fire suppression have shaped the temperate broadleaf forest of many regions. In central Europe particularly, extensive areas of broadleaf forests were converted to conifer plantations. The potential for restoration of temperate broadleaf forests lies mainly in mosaic landscapes where human pressure is moderate (between 10 and 100 people per km²) and forest stands are intermingled with other land uses (Stanturf 2015). The following sections describe general strategies for restoring temperate broadleaf forests and provide examples from specific biomes.

What are the main restoration needs?

The approach taken to restoration is situational, certainly depending upon the desired ending point but more importantly, the starting point (Stanturf *et al.* 2014a, 2014b). Restoration is usually undertaken with a goal of increasing sustainability by enlarging the area of specific ecosystems, enhancing biodiversity or repairing ecosystem functions (Stanturf *et al.* 2014a). Specific objectives commonly include timber, wildlife habitat for game species, aesthetics, biological diversity, nongame mammals and birds, endangered animals and plants, protection of water quality and aquatic resources, recreation, carbon sequestration, or some combination (Ciccarese *et al.* 2012). The starting point reflects the cause and extent of degradation, whether forest cover is lacking and the degree of soil legacy remaining (Baer *et al.* 2013; Stanturf 2016). Severe abiotic disturbances such as mining or accelerated erosion from agriculture combine the effects of removal of forest cover and greatly diminished soil legacy. Thus the starting point for restoration must first overcome the abiotic threshold (i.e. lack of native soil) as well as the biotic threshold of no forest species present. A less degraded starting point may be conversion from abandoned agriculture or pasture; degradation is less because soil legacy is greater (Baer *et al.* 2013).

Other abiotic disturbances, including severe wildfire, fire suppression or altered inundation regime may not be accompanied by changes in land use although the forest cover may be reduced or lost. Biotic disturbances from invasive plants, pathogens or herbivores (often exotic and inadvertently or intentionally introduced) similarly alter forest cover by changes in structure, composition or both. Forest management objectives in the past many times included homogenization of structure, composition or both by conversions to plantations or by management that reduced landscape diversity and militated against older age classes.

Diverse goals, ecological and social contexts, and resources available result in a myriad of restoration approaches at the stand and landscape scales. Although cessation of the degradation drivers is critical, simply halting degradation does not guarantee restoration success. Because of the magnitude of the restoration need (Minnemayer *et al.* 2011; Stanturf 2015), low-cost approaches such as reliance on natural regeneration are attractive but it must be understood that passive approaches are not free of cost and may not deliver the outcomes desired within an acceptable timeframe (Zahawi *et al.* 2014). Measures of success may be defined by ecosystem properties (Keddy and Drummond 1996; Ruiz-Jaén and Aide 2005) and techniques may be compared by cost-effectiveness (Kimball *et al.* 2015). Ecological restoration selects target reference sites and the properties of these targets have a major impact on the cost-effectiveness of techniques that attempt to re-create these properties, as does environmental variation (*ibid.*).

Reclamation after severe abiotic disturbance

Reclamation of surface mined landscapes date back to the nineteenth century, although these early efforts focused simply on revegetation (Macdonald *et al.* 2015; Stanturf *et al.* 2014b). Modern methods emphasize returning the land surface to something approaching pre-mining contours and handling overburden and topsoil so as to create a suitable rooting medium. Native species are favored, although site conditions may dictate use of non-natives to catalyze native restoration (Parrotta *et al.* 1997). Planting seedlings is generally preferred because other methods (native recolonization or direct seeding) are slower and variable, exposing the surface to erosion or capture by weedy species. Nevertheless, sowing mixtures of native tree seed and using soil amendments has proven successful in reclaiming jarrah (*Eucalyptus marginata*) forests on bauxite mined sites in Australia (Grant and Koch 2007). Natural recolonization can be aided by avoiding surface soil compaction (that usually accompanies surface grading) or deep ripping, by creating heterogeneous microtopography to increase surface roughness, and by adding woody debris to create microsites (Macdonald *et al.* 2015; Prach and Hobbs 2008).

The experience of two countries, Ireland and South Korea, illustrate the changing attitudes toward reclamation of severely degraded sites. Industrial peatlands in Ireland, mined for energy production, become uneconomical after outcrops of the underlying mineral soil are reached (Renou-Wilson and Byrne 2016). In the last century, reclamation of these cutaway peatlands for commercial forest production overcame challenging site conditions and demonstrated that commercial forest crops could be established. In this century, the goals of forest management in Ireland have changed toward greater emphasis on sustainability and ecosystem services and increasingly peatland areas are given over to open wetland systems (*ibid.*).

Overharvesting and illegal cutting of native Korean forests during the Japanese Occupation (1910–1945) and the Korean War (1950–1953) left large areas of the country denuded and subject to severe erosion (Lee *et al.* 2016). Reforestation on a massive scale began in 1959, mostly by planting fast-growing non-native species to meet local fuelwood needs. The planted area is now estimated to cover 70 per cent of the forest land in South Korea. The emphasis shifted in the 1960s and 1970s toward planting commercial timber species (*Larix laempferi* and *Pinus koraiensis*). The resulting simple stand structures with non-native species are now recognized to result in low biodiversity and stability and declining productivity; emphasis is shifting toward use of native species and natural regeneration (*ibid.*). Many of the early plantings failed due to lack of early tending or off-site planting and the native *Quercus mongolica* and *Pinus densiflora* are recolonizing the failed plantations.

Reconstruction after agriculture, grazing

Clearing temperate broadleaf forests for agriculture and pasture have occurred throughout the biome, at different rates and times (Stanturf 2016). Changing market conditions, demographic shifts and agricultural policies after World War II have resulted in abandonment or programs to retire marginal lands. Usually infertile, often sandy soils were abandoned first (e.g. Madsen *et al.* 2016), but other causes such as backwater flooding (Gardiner and Oliver 2005) or restitution of collectivized land to private owners in post-Soviet nations (Jõgiste *et al.* 2016) created the need for restoration. Recolonization using natural regeneration has occurred spontaneously as well as intentionally. Afforestation is an active approach using single or multiple species plantings in a variety of spatial arrangements and sometimes using a faster growing nurse species to facilitate development of a slower growing species (Stanturf *et al.* 2014a).

Recolonization using natural regeneration

Forests develop spontaneously after agricultural abandonment as seeds disperse from forest fragments, soil seed banks, pasture trees, windbreaks and ditch banks. The former fields or pastures may go through prolonged development of herbaceous or shrub stages, or open woodland stages before a closed canopy forest develops. In this way, many secondary forests were formed in developed countries such as France (Balandier and Prévosto 2016) and the US (Flinn *et al.* 2005). Species composition, stocking and stem quality may be deficient for meeting some restoration goals. On the one hand, the length of time before forest conditions are achieved means considerable delay before some ecosystem services are provided. On the other hand, early successional habitat is lacking in many managed landscapes so that a lengthy period of 'old field succession' (as it is called in North America) may be a conservation goal (Swanson *et al.* 2010).

Using natural regeneration to intentionally reconstruct open land has gained recent prominence in light of the 2.2 billion ha of land in need of forest landscape restoration (Minnemayer *et al.* 2011). Decreased cost relative to traditional planting, increased biodiversity and the greater likelihood that locally adapted genotypes will establish are some benefits of relying on recolonization, but the necessary conditions must be present. First, sources of the desired species must be available. Deforestation, loss of dispersal agents (especially for heavy-seeded species that rely on birds and mammals for dispersal), insects or diseases that prevent seed production and high-grading of source forest stands are all factors that work against recolonization. Second, multiple factors can prevent establishment of desirable species even when seed sources are available. Sites may be captured by invasive plants that prevent establishment, uncontrolled herbivores may preferentially browse seedlings, and species sensitive to fire can be eliminated by altered fire regimes. Degraded site conditions can create inhospitable habitat for desired species including soil salinization, flooding or salt water intrusion and desertification. Some inhibiting conditions can be overcome by competition control, soil amendments or fencing; lack of seed sources requires planting or direct seeding (Fischer *et al.* 2016).

Restoration of pasture land and grazed woodlands has been accomplished using natural regeneration where a few mother trees provide seed and fencing excludes grazers. So-called minimum interference management in New Zealand relies on natural regeneration of indigenous species and requires exclusion of cattle and sheep and control of mammalian pests (Norton 2009). There are over 100 exotic naturalized deciduous trees and shrubs growing in combination with native species on abandoned agricultural land and they may be managed as novel ecosystems (Allen *et al.* 2013). Restoration of native *Eucalyptus* woodlands by recolonization in Australia was highest in ungrazed sites yet occurred under intermittent grazing (Dorrough and Moxham 2005). Remnant paddock trees are seed sources but are themselves in decline. Yates and Hobbs (1997) found that observed response to reduced grazing pressure was variable and advocated fencing off farmland adjacent to remnant woodlands to overcome any seed supply and dispersal limitations. They noted that this approach would at best be slow and at worst would result in woodlands dominated by a few natives and exotics (*ibid.*).

Afforestation

Afforestation, or planting to establish forests on land formerly in other land use, has long been used to establish commercial plantations, often with exotic species. In the restoration context, afforestation is a method used to meet restoration goals and mostly uses native species. Some of the tools and techniques developed for establishing commercial plantations are useful for restoration plantings, possibly with alterations (Stanturf *et al.* 2014a). Critical decisions include choice of species, site preparation, planting designs, stocking levels, appropriate stock types and competition control (*ibid.*). Landscape-scale afforestation can meet multiple restoration goals and provide ecosystem services including enlarging and connecting forest remnants, thereby creating connectivity; creating diversity at the stand and landscape levels by varying planting designs; adding buffers along watercourses that protect water quality; and possibly altering wildfire behavior by adding 'green' firebreaks of relatively inflammable vegetation (*ibid.*; Stanturf 2015).

Restoration of riverine broadleaf forests of the Lower Mississippi Alluvial Valley (LMAV) in the southern US illustrates the complexity of restoration in a mixed-ownership landscape

(Gardiner and Oliver 2005). Deforestation began in the 1800s and draining of wetland areas intensified in the 1900s, resulting in a loss of critical wildlife and fish habitat, decreased water quality, reduced floodwater retention and increased sediment loads. By the 1980s less than 20 per cent of the original forest was left. Although data on restoration are scant and scattered, best estimates are that between 1993 and 2008, an estimated 405,000 ha were afforested and almost 162,000 ha of existing forests protected through easements. Restoration by federal and state programs, initially on publicly owned land, was driven by the goal of enlarging and enhancing wildlife habitat, maintaining navigability of rivers and flood control. Beginning in the 1990s, restoration on private lands for the same objectives was funded by federal and state programs. Increasingly carbon sequestration objectives have added incentives to restoration.

The initial design was widely spaced plantings or sowing of heavy-seeded *Quercus* species, assuming wind-dispersed species would increase stocking. Many failures occurred in the early days due to improper species selection for site conditions, poor seedling quality and inadequate planting techniques. Improvements over time included expanding the choice of species from a few *Quercus* spp. to multiple species plantings, the use of nurse crops, better matching species to inundation regimes and higher quality seedlings. A nurse crop technique was developed as a more intensive alternative to the standard planting bareroot seedlings or direct-seeding acorns. Interplanting the fast-growing, native species eastern cottonwood (*Populus deltoides*) to nurse Nuttall oak (*Quercus texana*) was successful in rapidly developing forested conditions and vertical structure, demonstrating that environmental benefits can be obtained quickly by more intensive efforts. Experience in the LMAV suggests that complete restoration of species-rich forests with complex structures will require multiple interventions over time. Interventions sequenced to take advantage of native recolonization and stand development processes can be successful if limitations to dispersal distance (in this case, 100 m) are clearly recognized. Substantial functionality can be obtained in a short time using innovative techniques such as interplanting.

Large-scale restoration using afforestation has occurred at different times in Europe (Weber 2005), China (Wenhua 2004) and Korea (Lee *et al.* 2016). The severity of degradation on many sites meant that restoration began by planting conifers, often non-natives, able to tolerate harsh conditions. In Denmark, for example, the original forests were cleared for agriculture and livestock grazing. Centuries of shifting cultivation and grazing prevented natural regeneration and encouraged the emergence of heather (*Calluna vulgaris*) and other ericaceous plants. Periodic burning of the heathland to enhance grazing further prevented seedlings from establishing. Degradation became so severe by 1800 that acute shortages of wood and food developed. Fires and overgrazing had exposed the underlying sandy soils and drifting sand created dunes that occasionally covered villages and farms. Restoration efforts began by suppressing heathland burning and fencing livestock to encourage recolonization from remnant stands and planting non-native conifer species able to tolerate poor soils and early season frosts. Mountain pine (*Pinus mugo*) was planted on the poorest sites. Today, the Danish government favors establishing multifunctional forests with greater emphasis on broadleaf and mixed plantings (Madsen *et al.* 2016).

The collapse of the Soviet Union and re-independence of the Baltic States of Estonia, Latvia and Lithuania significantly changed land use as collective farmland was first abandoned then restituted to private owners (Jõgiste *et al.* 2016). This upheaval resulted in large areas of farmland becoming available for forest through recolonization and afforestation. Pioneer species such as birch (*Betula* spp.), alder (*Alnus* spp.), willow (*Salix* spp.) and aspen (*Populus tremula*) have colonized many areas. Afforestation has long been practiced in the Baltics but until recently, it was mostly spruce (*Picea* spp.) and pine (*Pinus* spp.) for commercial timber production. New afforestation objectives of enhancing environmental quality and biodiversity have shifted species selection toward broadleaf species on public land. Timber production remains paramount on managed private land with conifers such as Norway spruce (*Picea abies*) although silver birch (*Betula pendula*) for veneer and hybrid aspen (*Populus* × *wettsteinii*) on short-rotation for pulpwood are attractive (Jõgiste *et al.* 2016). While this is not ecological restoration, species diversity and landscape diversity have increased.

Rehabilitating existing forests

Increased interest globally in more sustainable and environmentally friendly forest management has manifested as ecological forestry, close-to-nature silviculture, conservation forestry and continuous cover forestry; all intending to create more 'natural' forest conditions (Stanturf *et al.* 2014b). This translates into restoration goals of greater compositional and structural complexity and landscape heterogeneity, emphasizing native species and emulating natural disturbance regimes. Existing forests lacking desired species composition or complex structures may require wholesale replacement (i.e. conversion to another forest type), augmentation by planting desired or removing undesirable species, or partial harvests to transform structure. Rehabilitation includes intervening after catastrophic disturbances such as severe wildfire or blowdown. The starting point may be a commercial plantation, a previous planting of indigenous species with persistent non-native species, or a stand degraded by exploitive harvesting, wildfire, insect or disease outbreaks (Stanturf *et al.* 2014a). In Europe, efforts to meet the targets of the Convention on Biological Diversity are driving restoration toward complex landscapes that may be more resilient in the future (Halme *et al.* 2013). Methods include species reintroductions, controlled burning, creation of deadwood and variable density thinning (*ibid.*; Stanturf *et al.* 2014a).

Rehabilitating species composition

Altering the species composition of an existing forest usually involves two types of activity: adding or removing species and manipulating the overstorey to create appropriate resource conditions (e.g. light, moisture or nutrients). Adding species may be through planting, direct-seeding or colonization from outside sources. Removing exotic or ruderal species may be required so that desired species can occupy the vacant niche. Overstorey manipulations require complete or partial removals (Stanturf *et al.* 2014a).

Conversions from conifer plantations to broadleaf forests in Europe have been extensive, driven in part to create more wind-stable forests (Hahn *et al.* 2005; Spiecker *et al.* 2004; Zerbe 2002). For ecological and economic reasons the conversion of coniferous monocultures, consisting mainly of pine (*Pinus* spp.) and spruce (*Picea* spp.), into natural broadleaf forest stands has included the integration of natural regeneration (Fischer *et al.* 2016). The conversion process is carefully planned and requires consideration of questions including which sites to convert, what are the target forest types, and which silvicultural methods to employ. In Central and Northern Europe, severe winter storms have added urgency to the conversion process as large areas of conifer plantations have been destroyed or damaged and need immediate attention (Spiecker *et al.* 2004).

Conifer plantations were established in many countries in response to real or perceived timber shortages. In the UK, for example, most of the native forests were long ago cleared for agriculture and the few remnant forests had been managed in some way. Large areas afforested during the twentieth century were primarily with non-native conifers. Towards the end of the last century, interest grew in restoring these Plantations on Ancient Woodland Sites (PAWS) to semi-natural forests of mostly broadleaves (Harmer and Morgan 2009). Natural regeneration is the preferred method and used mostly on upland sites in Scotland. Where planting is required, mixtures of species typical of the target woodland type are used, frequently requiring protection from browsing animals. Because most restoration is for ecological reasons, weeds are controlled by manipulating light levels by canopy control rather than herbicides (Harmer *et al.* 2005). Increasingly, adaptation to changing climate is considered in restoring PAWS and creating new woodlands (Harmer *et al.* 2016).

Dispersed plantings in small patches (nucleation or cluster planting) have been suggested to overcome the high financial cost of planting large areas (Stanturf *et al.* 2014a). Such methods that rely on natural regeneration to fill in by dispersal have produced mixed results in fragmented landscapes (Kasel *et al.* 2015; Bustamante Sánchez and Armesto 2012) but may be more reliable in landscapes with lower pressure from ruderal species (e.g. Albornoz *et al.* 2013). Australian conifer plantations, predominantly *Pinus radiata*, resulted from afforestation (Kasel *et al.* 2015) and some have been converted to native *Eucalyptus* spp. primarily by planting (Kasel 2008). Even with a developing overstorey of native species, restoration of native understorey and shrub species are limited by lack of seed supply and recruitment failure, with persistent non-natives because of seed rain from surrounding areas (Kasel *et al.* 2015).

Many secondary broadleaf forests lack desired species composition because of one or more conditions. Some were high-graded in the past, whereby highly valued species were selectively removed and can no longer regenerate. Other species have been lost because of heavy browsing pressure from ungulates or domestic livestock. Along with other disturbances such as fire, insects, diseases or competing understorey vegetation, singly or in combination, there is often a need to add species or create conditions more favorable for seedlings to establish and advance regeneration to develop. Enrichment or underplantings, sometimes requiring overstorey or midstorey reduction or removal, are ways to enrich species composition (Dey *et al.* 2012). Heavy seeded species such as *Quercus*, *Fagus* and *Nothofagus* species especially may be difficult to regenerate without planting. For example, *Nothofagus* forests in the Chilean Andes Mountains have been high-graded and adding to the difficulty of restoration is an understorey of bamboo (*Chusquea culeou*) that prevents regeneration of the light demanding *N. dombeyi* and *N. alpina* (Donoso *et al.* 2015). Underplanting in created gaps produced good initial growth in these two species. Light availability had an increasing effect on growth, indicating the need to intervene again to reduce the canopy and increase light to the understorey (*ibid.*).

Rehabilitating structure

Managing forest stands to restore structural heterogeneity is an important goal for ecological management (Franklin *et al.* 2007). Creating complex stand structure, as well as age class diversity in the landscape, may be necessary to overcome a legacy of even-aged silviculture. Many commercial temperate broadleaf timber species are intolerant of shade and are managed best using even-aged methods. In regions of North America, for example, many forests were harvested at about the same time (i.e. within one or two decades) and the naturally regenerated second-growth forests are of the same age and may continue to be managed under even-aged systems. In Europe, many semi-natural *Quercus* and *Fagus* forests were planted. Transforming these forests to multiple age classes and complex structures may be accompanied by attempts to increase the composition of shade-tolerant species.

The simplest approach to transform stands with simple even-aged structure to more complex multi-cohort structure is to retain shade- or mid-tolerant trees at regeneration harvest.

These trees will remain through the next rotation in a managed stand. Other approaches of partial harvests based on gap- and patch-dynamics are more challenging and may shift composition to more shade-tolerant tree species (Nyland 2003). A long time may be required (many decades to centuries) to make the transition. Another approach called variable density thinning or VDT can increase spatial heterogeneity of structure. One popular method of VDT is known as 'skips and gaps' thinning. With this approach, unthinned areas (referred to as 'skips') and heavily thinned patches ('gaps') are created, along with intermediate levels of thinning and residual density in the rest of the stand (Stanturf *et al.* 2014a). Structural diversity can also be achieved through deliberate creation of decadence and retention of deadwood (*ibid.*).

Changing economic and social contexts may render traditional stand management systems obsolete and present opportunities for restoring different structure. Coppice management of broadleaf species such as *Quercus* for charcoal and firewood in Japan was abandoned in the 1960s when industry switched to fossil fuels (Nagaike *et al.* 2005). Conversion of abandoned coppice stands to high forests required changes in structure and composition. Artificial canopy gaps were created; the size of gap depended upon the proportion of the shade-tolerant *Fagus crenata*. A difficulty in stands with high proportions of *F. crenata* was to maintain shade-intolerant *Quercus* spp., thus requiring larger gaps. Soil preparation and planting or direct-seeding was used if seed sources of desirable species were lacking (*ibid.*).

Rehabilitating processes

Ecological processes are physical, chemical, and biological actions or events linking organisms to their environment and involve transfers of material and energy through the landscape. Direct efforts to restore disrupted ecological processes have focused mostly on fire or inundation regimes. Indirect attempts to restore disturbance processes have manipulated vegetation to affect structure and composition in ways that are similar to the effects of disturbance.

Natural fire regimes have been altered in many fire-adapted forest types and restoring fire is motivated by ecological and safety reasons. Restoring natural fire regimes at the landscape scale in mosaic landscapes is difficult because of roads, habitation and past suppression. Current fuel conditions often are too hazardous simply to re-introduce prescribed burning. Restoration treatments may require some combination of reducing dense overstorey or midstorey stems by mechanical or chemical means, conducting multiple low-intensity prescribed burns over several seasons to reduce fine fuel accumulation, planting ecologically appropriate herbaceous and graminoid species or converting the overstorey to more fire-adapted species. Many *Quercus* spp. are adapted to frequent disturbance such as periodic low-intensity fires; young seedlings allocate much biomass to developing root systems and basal buds are located below the soil surface. Older trees develop thick bark that insulates cambium from heat damage (Phillips *et al.* 2012). In oak-pine and oak-hickory forest types in eastern North America, for example, fire suppression has altered stand structures (denser) and species composition (increased fire-sensitive, decreased fire-adapted understorey species). Overstorey and midstorey reductions followed by repeated fires are used to develop needed advance *Quercus* regeneration. Different systems have been developed to address the regeneration problem, commonly pairing shelterwood treatments with periodic burning. Timing (dormant or growing season) and frequency of burning must be optimized not only to secure advance oak regeneration but also to favour fire-adapted herbaceous and graminoid understorey (*ibid.*).

Wet forests in low-lying areas have significant social and ecological values; they often represent unique habitats locally and globally. In many countries, these forests have been heavily impacted by humans. Restoring functioning in wet forests may require hydrologic modifications and may be part of a larger effort to restore natural river flows (e.g. Hughes *et al.* 2012). Whatever the scope of restoration, the first step is an objective examination of the extent to which hydroperiod can be truly restored. Fully restoring hydroperiod may be impractical because of cost, incompatibility with current land uses or conflict with navigation. Even though fully restoring a natural inundation regime may be infeasible, attempts to create site diversity have focused on modifying microtopography and inundation. Reversing drainage by plugging drains or removing flood control structures may provide partial hydrologic restoration. Successful planting must match species tolerances to inundation regime (Figure 9.2).

Looking to the future

Global climate change will increase the need for restoration of temperate broadleaf forests (Stanturf 2015). Climate change may degrade plants directly through increased weather variability and extreme events. Species with limited range, particularly those that are rare or endangered under the present climate, will be at greatest risk. Secondary effects from altered fire regimes and introduction of invasive exotic organisms could impact those species unaffected by altered climate. Temperature increases that alter insect phenology may cause novel outbreaks of native insects. The daunting challenge for restorationists is how to restore under great uncertainty and rapid change. Restoration focusing on returning to the historic condi-

Figure 9.2 Planting on permanently flooded sites. Restoring some wet forests requires innovative techniques such as planting in flooded, unstable substrate. Pruning the roots of bareroot seedlings of *Fraxinus pennsylvanica* and *Taxodium distichum* facilitates forcing them into the soil

Source: W. Inabinette

tions of the past, or even restoring to current conditions, may be maladapted to the rapidly changing present and anticipated future. Current definitions of native plants and natural disturbance regimes may require reassessment. As extreme climate events increase in frequency and intensity, assisted migration will gain credence as a way to counter extirpation of local populations (Dumroese *et al.* 2015; Stanturf 2015).

Several approaches have been proposed to adapt restoration of forested landscapes to climate change (Stanturf 2015). An incremental or no-regrets approach entails pursuing endpoints based on the best available understanding of contemporary conditions, focusing on using native species of local sources and reducing current stressors. Restoring diversity of composition, structure, and function to simplified, production-oriented forests provides the flexibility to intervene in the future and shift development toward better adapted conditions. Taking these actions would be beneficial regardless of whether the future environment is drastically changed. A more anticipatory approach to climate change would emphasize reducing vulnerability to future stressors by accepting or intentionally using non-native species that are functional equivalents to natives (Davis *et al.* 2011), extending native ranges or translocating species outside their native range (Stanturf *et al.* 2014a; Dumroese *et al.* 2015) and accepting and learning how to manage novel (emergent or no-analog) ecosystems that arise. A more transformative approach, intervention ecology, has been advocated to replace ecological restoration (Hobbs *et al.* 2011). Transformation would be achieved by intentionally creating novel ecosystems, using biotechnology to create transgenic species to replace extinct keystone species or genotypes better adapted to future climate (e.g. *Castenea dentata*; Jacobs *et al.* 2013) or synthetic biology to create designer organisms with heretofore unknown capabilities (Stanturf 2015). Which adaptation strategy to adopt and how soon to shift to new restoration goals are difficult decisions rife with uncertainty but the objective of restoring robust, resilient forests remains the same.

References

Albornoz, F. E., Gaxiola, A., Seaman, B. J., Pugnaire, F. I. and Armesto, J. J. 2013. Nucleation-driven regeneration promotes post-fire recovery in a Chilean temperate forest. *Plant Ecology*, 214:5, 765–776.

Allen, R. B., Bellingham, P. J., Holdaway, R. J. and Wiser, S. K. 2013. New Zealand's indigenous forests and shrublands. In J. Dymond (ed.), *Ecosystem Services in New Zealand: Condition and Trends*, 34–48. Lincoln, NZ: Manaaki Whenua Press.

Baer, S. G., Heneghan, L. and Eviner, V. T. 2013. Applying soil ecological knowledge to restore ecosystem services. In D. Wall *et al.* (eds), *Soil Ecology and Ecosystem Services*, 377–393. Oxford, UK: Oxford University Press.

Balandier, P. and Prévosto, B. 2016. Forest restoration in the French Massif Central Mountains. In J. A. Stanturf (ed.), *Restoration of Boreal and Temperate Forests*, 2nd edition, 337–354. Boca Raton, FL: CRC Press.

Box, E. O. and Fujiwara, K. 2015. Warm-temperate deciduous forests: concept and global overview. In E. O. Box and K. Fujiwara (eds), *Warm-Temperate Deciduous Forests around the Northern Hemisphere*, 7–26. New York: Springer.

Bustamante Sánchez, M. A. and Armesto, J. J. 2012. Seed limitation during early forest succession in a rural landscape on Chiloé Island, Chile: implications for temperate forest restoration. *Journal of Applied Ecology*, 49:5, 1103–1112.

Cantarello, E., Lovegrove, A., Orozumbekov, A., Birch, J., Brouwers, N. and Newton, A. C. 2014. Human impacts on forest biodiversity in protected walnut-fruit forests in Kyrgyzstan. *Journal of Sustainable Forestry*, 33:5, 454–481.

Ciccarese, L., Mattsson, A. and Pettenella, D. 2012. Ecosystem services from forest restoration: thinking ahead. *New Forests*, 43:5–6, 543–560.

Davis, M. A., Chew, M. K., Hobbs, R. J., Lugo, A. E., Ewel, J. J., Vermeij, G. J., Brown, J. H., Rosenzweig, M. L., Gardener, M. R. and Carroll, S. P. 2011. Don't judge species on their origins. *Nature*, 474:7350, 153–54.

Dey, D. C., Gardiner, E. S., Schweitzer, C. J., Kabrick, J. M. and Jacobs, D. F. 2012. Underplanting to sustain future stocking of oak (*Quercus*) in temperate deciduous forests. *New Forests*, 43:5–6, 955–978.

Donoso, P. J., Soto, D. P. and Fuentes, C. 2015. Differential growth rates through the seedling and sapling stages of two Nothofagus species underplanted at low-light environments in an Andean high-graded forest. *New Forests*, 46:5, 885–895.

Dorrough, J. and Moxham, C. 2005. Eucalypt establishment in agricultural landscapes and implications for landscape-scale restoration. *Biological Conservation*, 123:1, 55–66.

Dumroese, R. K., Williams, M. I., Stanturf, J. A. and St Clair, J. B. 2015. Considerations for restoring temperate forests of tomorrow: Forest restoration, assisted migration, and bioengineering. *New Forests*, 46:5–6, 947–964.

Fischer, H., Huth, F., Hagemann, U. and Wagner, S. 2016. Developing restoration strategies for temperate forests using natural regeneration processes. In J. A. Stanturf (ed.), *Restoration of Boreal and Temperate Forests*, 2nd edition, 103–64. Boca Raton, FL: CRC Press.

Flinn, K. M., Vellend, M. and Marks, P. 2005. Environmental causes and consequences of forest clearance and agricultural abandonment in central New York, USA. *Journal of Biogeography*, 32:3, 439–452.

Franklin, J. F., Mitchell, R. J. and Palik, B. 2007. *Natural Disturbance and Stand Development Principles for Ecological Forestry*. General Technical Report NRS–19. Newtown Square, PA: US Forest Service Northern Research Station.

Gardiner, E. S. and Oliver, J. M. 2005. Restoration of bottomland hardwood forests of the Lower Mississippi Alluvial Valley, USA. In J. A. Stanturf and P. Madsen (eds), *Restoration of Boreal and Temperate Forests*, 1st edition, 235–251. Boca Raton, FL: CRC Press.

Grant, C. and Koch, J. 2007. Decommissioning Western Australia's first bauxite mine: Co evolving vegetation restoration techniques and targets. *Ecological Management and Restoration*, 8:2, 92–105.

Hahn, K., Emborg, J., Larsen, J. and Madsen, P. 2005. Forest rehabilitation in Denmark using nature-based forestry. In J. A. Stanturf and P. Madsen (eds), *Restoration of Boreal and Temperate Forests*, 1st edition, 299–317. Boca Raton, FL: CRC Press.

Halme, P., Allen, K. A., Auniņš, A., Bradshaw, R. H., Brūmelis, G., Čada, V., Clear, J. L., Eriksson, A.-M., Hannon, G. and Hyvärinen, E. 2013. Challenges of ecological restoration: lessons from forests in northern Europe. *Biological Conservation*, 167, 248–256.

Harmer, R. and Morgan, G. 2009. Storm damage and the conversion of conifer plantations to native broadleaved woodland. *Forest Ecology and Management*, 258:5, 879–886.

Harmer, R., Thompson, R. and Humphrey, J. 2005. Great Britain – conifers to broadleaves. In J. A. Stanturf and P. Madsen (eds), *Restoration of Boreal and Temperate Forests*, 1st edition, 319–338. Boca Raton, FL: CRC Press.

Harmer, R., Watts, K. and Ray, D. 2016. A hundred years of woodland restoration in Great Britain: changes in the drivers that influenced the increase in woodland cover. In J. A. Stanturf (ed.), *Restoration of Boreal and Temperate Forests*, 2nd edition, 299–320. Boca Raton, FL: CRC Press.

Hobbs, R. J., Hallett, L. M., Ehrlich, P. R. and Mooney, H. A. 2011. Intervention ecology: Applying ecological science in the twenty-first century. *BioScience*, 61:6, 442–450.

Hughes, F. M., del Tánago, M. G. and Mountford, J. O. 2012. Restoring floodplain forests in Europe. In J. Stanturf, P. Madsen and D. Lamb (eds), *A Goal-Oriented Approach to Forest Landscape Restoration*, 393–422. Dordrecht: Springer.

Jacobs, D. F., Dalgleish, H. J. and Nelson, C. D. 2013. A conceptual framework for restoration of threatened plants: the effective model of American chestnut (Castanea dentata) reintroduction. *New Phytologist*, 197:2, 378–393.

Jõgiste, K., Metslaid, M. and Uri, V. 2016. Afforestation and land use dynamics in the Baltic States. In J. A. Stanturf (ed.), *Restoration of Boreal and Temperate Forests*, 2nd edition, 187–199. Boca Raton, FL: CRC Press.

Kasel, S. 2008. Eucalypt establishment on former pine plantations in north-east Victoria: An evaluation of revegetation techniques. *Ecological Management and Restoration*, 9:2, 150–153.

Kasel, S., Bell, T. L., Enright, N. J. and Meers, T. L. 2015. Restoration potential of native forests after removal of conifer plantation: A perspective from Australia. *Forest Ecology and Management*, 338, 148–162.

Keddy, P. A. and Drummond, C. G. 1996. Ecological properties for the evaluation, management, and restoration of temperate deciduous forest ecosystems. *Ecological Applications*, 6:3, 748–762.

Kimball, S., Lulow, M., Sorenson, Q., Balazs, K., Fang, Y.-C., Davis, S. J., O'Connell, M. and Huxman, T. E. 2015. Cost-effective ecological restoration. *Restoration Ecology*, 23:6, 800–810.

Lee, D. K., Park, P. S. and Park, Y. D. 2016. Forest restoration and rehabilitation in the Republic of Korea. In J. A. Stanturf (ed.), *Restoration of Boreal and Temperate Forests*, 2nd edition, 217–231. Boca Raton, FL: CRC Press.

Macdonald, S. E., Landhäusser, S. M., Skousen, J., Franklin, J., Frouz, J., Hall, S., Jacobs, D. F. and Quideau, S. 2015. Forest restoration following surface mining disturbance: challenges and solutions. *New Forests*, 46:5–6, 703–732.

Madsen, P., Jensen, F. A. and Fodgaard, S. 2016. Afforestation in Denmark. In J. A. Stanturf (ed.), *Restoration of Boreal and Temperate Forests*, 2nd edition, 201–216. Boca Raton, FL: CRC Press.

Minnemayer, S., Laestadius, L. and Sizer, N. 2011. *A World of Opportunity*. Washington, DC: World Resource Institute.

Nagaike, T., Yoshida, T., Miguchi, H., Nakashizuka, T. and Kamitani, T. 2005. Rehabilitation for species enrichment in abandoned coppice forests in Japan. In J. Stanturf and P. Madsen (eds), *Restoration of Boreal and Temperate Forests*, 1st edition, 371–381. Boca Raton, FL: CRC Press.

Norton, D. A. 2009. Ecological restoration in New Zealand: current trends and future challenges. *Ecological Management and Restoration*, 10:2, 76–77.

Nyland, R. D. 2003. Even-to uneven-aged: the challenges of conversion. *Forest Ecology and Management*, 172:2, 291–300.

Parrotta, J. A., Turnbull, J. W. and Jones, N. 1997. Catalyzing native forest regeneration on degraded tropical lands. *Forest Ecology and Management*, 99:1–2, 1–7.

Phillips, R. J., Waldrop, T. A., Brose, P. H. and Wang, G. G. 2012. Restoring fire-adapted forests in eastern North America for biodiversity conservation and hazardous fuels reduction. In J. Stanturf, P. Madsen and D. Lamb (eds), *A Goal-Oriented Approach to Forest Landscape Restoration*, 187–219. Dordrecht: Springer.

Prach, K. and Hobbs, R. J. 2008. Spontaneous succession versus technical reclamation in the restoration of disturbed sites. *Restoration Ecology*, 16:3, 363–366.

Renou-Wilson, F. and Byrne, K. A. 2016. Irish peatland forests: lessons from the past and pathways to a sustainable future. In J. A. Stanturf (ed.), *Restoration of Boreal and Temperate Forests*, 2nd edition, 321–335. Boca Raton, FL: CRC Press.

Ruiz-Jaén, M. C. and Aide, T. M. 2005. Vegetation structure, species diversity, and ecosystem processes as measures of restoration success. *Forest Ecology and Management*, 218:1, 159–173.

Spiecker, H., Hansen, J., Klimo, E., Skovsgaard, J. P., Sterba, H. and von Teuffel, K. 2004. *Norway Spruce Conversion-Options and Consequences*. Leiden, The Netherlands: Brill.

Stanturf, J. A. 2015. Future landscapes: opportunities and challenges. *New Forests*, 46:5–6, 615–644.

Stanturf, J. A. 2016. What is forest restoration? In J. A. Stanturf (ed.), *Restoration of Boreal and Temperate Forests*, 2nd edition, Boca Raton, FL: CRC Press.

Stanturf, J., Palik, B. and Dumroese, R. K. 2014a. Contemporary forest restoration: A review emphasizing function. *Forest Ecology and Management*, 331, 292–323.

Stanturf, J. A., Palik, B. J., Williams, M. I., Dumroese, R. K. and Madsen, P. 2014b. Forest restoration paradigms. *Journal of Sustainable Forestry*, 33 Sup1, S161–S94.

Swanson, M. E., Franklin, J. F., Beschta, R. L., Crisafulli, C. M., DellaSala, D. A., Hutto, R. L., Lindenmayer, D. B. and Swanson, F. J. 2010. The forgotten stage of forest succession: early-successional ecosystems on forest sites. *Frontiers in Ecology and the Environment*, 9:2, 117–125.

Weber, N. 2005. Afforestation in Europe: lessons learned, challenges remaining. In J. A. Stanturf and P. Madsen (eds), *Restoration of Boreal and Temperate Forests*, 1st edition, 121–135. Boca Raton, FL: CRC Press.

Wen, J. 1999. Evolution of eastern Asian and eastern North American disjunct distributions in flowering plants. *Annual Review of Ecology and Systematics*, 30, 421–455.

Wenhua, L. 2004. Degradation and restoration of forest ecosystems in China. *Forest Ecology and Management*, 201:1, 33–41.

Yates, C. J. and Hobbs, R. J. 1997. Temperate eucalypt woodlands: a review of their status, processes threatening their persistence and techniques for restoration. *Australian Journal of Botany*, 45:6, 949–973.

Zahawi, R. A., Reid, J. L. and Holl, K. D. 2014. Hidden costs of passive restoration. *Restoration Ecology*, 22:3, 284–287.

Zerbe, S. 2002. Restoration of natural broad-leaved woodland in Central Europe on sites with coniferous forest plantations. *Forest Ecology and Management*, 167:1, 27–42.

10

TEMPERATE GRASSLANDS

Karel Prach, Péter Török and Jonathan D. Bakker

Temperate grasslands are generally dominated by graminoid vegetation and have less than 10 per cent cover of trees and shrubs. These ecosystems occur around the world, provide important ecological functions, and often have high biodiversity. We review the types of temperate grasslands, consider why they have been lost or degraded and why they are being restored, summarize common restoration methods, and end with several examples of such restoration efforts.

Types, origin and present distribution of temperate grasslands

We distinguish primary and secondary temperate grasslands on the basis of the factors that conditioned their existence and maintain them. Among the secondary grasslands we distinguish pastures and hay meadows.

Primary grasslands

Primary grasslands occur where the establishment of woody plants is restricted by natural processes. These restrictions may be climatic (e.g. duration of the dry season), edaphic (low water holding capacity and/or high salt content of substrates), or disturbance-related (fire, avalanches, grazing) or combinations thereof. Extensive primary grasslands whose existence is conditioned by the macroclimate represent zonal biomes. For detailed descriptions see Archibold (1995) or Gibson (2009). Below we highlight a few key types:

- *Euro-Asia*: Steppes stretch from central Europe to central Asia and vary from forest-steppes to very short-grass steppes with some transitions to semi-desert formations. They often exhibit high species diversity especially at scales up to several square meters. Key genera are *Stipa* and *Festuca*; *Artemisia* are also typical in short-grass steppes. Many steppes have been ploughed, particularly where they occurred on fertile chernozem soils; some spontaneously recovered after ploughing ceased but rarely reached their original species composition (Dengler *et al.* 2014). Nowadays, undisturbed steppe vegetation mostly occurs on extreme sites such as south-facing hillsides, in transitional areas towards semi-deserts, or at man-made ancient cultural and burial monuments such as kurgans.

- *North America*: Prairies occur primarily in the centre of North America. Several types of prairie are distinguished, ranging from tallgrass prairie in the south and east to shortgrass prairie in the north and west. Warm-season grasses increasingly dominate as humidity and temperature increase (Gibson 2009). Key genera include *Andropogon*, *Hesperostipa* and *Bouteloua*. The historical influence of Native Americans is difficult to ascertain but it was likely significant. The arrival of European immigrants in the 1800s resulted in the conversion of many prairies to intensive agriculture, especially in the tall-grass prairie region, and introduced many alien plants. Nowadays, there are many programs and projects to restore prairies. The world's first intentional restoration project was established in 1936 to restore a tall-grass prairie near Madison, Wisconsin.
- *South America*: The Southern Cone of South America contains large expanses of temperate grasslands, including pampas, campos and steppe (Zuleta *et al*. 2015). Compared to North American prairies, there is little historical evidence of disturbances such as fire or grazing that would have maintained them. However, livestock were introduced several centuries ago, and have been a dominant aspect of the ecology of these systems since then: Zuleta *et al*. (*ibid*.) estimate three-quarters of the grasslands in Argentina are subject to livestock ranching. Intensive grazing is practiced in some areas, and other areas are being lost to afforestation (Six *et al*. 2014).

Secondary grasslands

Secondary grasslands are maintained and/or created by cultural practices such as mowing, burning, hay harvesting, and grazing by domestic livestock.

European secondary grasslands are described in Veen *et al*. (2009) and briefly also in Dengler *et al*. (2014). They started to develop during the Neolithic period, beginning there approximately 7500 years before present (BP). They had precursors in earlier primary grasslands and in treeless openings within forests that were probably maintained predominantly by free grazing of wild animals (Vera 2000). Hay meadows appeared in temperate Europe during the Bronze Age, some 4000 years BP. Over the millennia since then, changes in management have altered the composition of these pastures and meadows, though these changes pale compared to the degradation they experienced in the second half of the twentieth century (see below).

In North America, the relative importance of natural factors and the cultural practices of Native Americans before European settlement are difficult to ascertain, though there is evidence that some grasslands were strongly shaped by cultural practices. In the Pacific Northwest, for example, edaphic conditions are highly suitable for forests yet there were estimated to be ~100,000 ha of prairies, with unique flora and fauna, when European settlers arrived around 1850. These areas were maintained by Native American practices such as burning and the harvesting of foods such as *Camassia* bulbs (Dunwiddie and Bakker 2011). In some cases, the ecotone between grassland and forest has been stable enough that it can be detected by examining soil properties (Hegarty *et al*. 2011). Some of the temperate grasslands in the Midwestern and eastern United States were created by anthropogenic clearing and grazing that began several hundred years ago (Gibson 2009).

Reasons why temperate grasslands have been lost or degraded

Loss of grasslands

Besides direct destruction by building and other construction activities, there are three widespread reasons grasslands have been lost:

- *Ploughing.* With increasing crop production, especially of cereal grains, many grasslands were converted to arable land. Large-scale ploughing of primary steppes in the former Soviet Union happened during the communist era. Secondary grasslands were ploughed in some parts of Europe between 1950 and 1980, reducing their original area by up to a third. Compared to unploughed grasslands, arable land exhibits lower water retention capacity, lower water filtering effects and is more susceptible to erosion. Moreover, many grasslands were ploughed and then re-seeded by species-poor commercial seed mixtures containing a few productive grasses and legumes.
- *Spontaneous forest succession.* Forests may spontaneously develop when grasslands no longer experience the disturbances that hinder tree seedling establishment. Wind or animal dispersed woody plants (e.g. *Betula, Populus, Pinus, Picea*) are often the first to colonize these areas.
- *Technical afforestation.* In some regions, grasslands have been intentionally planted with trees, usually fast-growing species that provide pulp or timber.

Grassland degradation

We usually define degradation when there is undesirable changes in species composition, especially decrease of diversity, increase of unwanted competitors and/or weeds, and deterioration of ecosystem functioning. Grasslands can be degraded in many ways:

- *Cessation of management.* If the cultural practices, such as cutting, grazing or burning, that maintain a secondary grassland cease, the area can rapidly undergo succession towards woodland. Changes can occur in a few decades. In some cases, competitive herb or grass dominants may expand and preclude establishment of woody species for a long time, but diversity of grassland species also decreases.
- *Fragmentation.* Fragmentation reduces the size of remnant patches and increases their isolation, thus limiting landscape-scale dispersal. As fragmentation increases, the areas around the patches can increasingly alter the patches by, for example, serving as sources of weeds.
- *Altered water regime.* Many wet grasslands were drained to increase the amount and quality of fodder production. Draining alters the species composition of the grassland and can decrease ecosystem functions such as water filtration.
- *Eutrophication.* Nutrients may be added to grasslands directly by mineral or organic fertilizers or indirectly by aerial deposition, mostly of nitrogen, or fertilizer run-off from cropland areas. Species diversity usually declines due to the expansion of competitive, nutrient-demanding species.
- *Altered frequency and intensity of management.* Grasslands that are overgrazed or cut too frequently have reduced biodiversity as they support only a few resistant species. Conversely, however, diversity also decreases if management is not intensive enough. Management should be heterogeneous in kind, space, and time; cutting or grazing large areas at once may drastically reduce insect populations and cause large-scale homogenization of the vegetation.

Why restore grasslands? Possibilities and limitations

Grasslands provide many ecosystem services, so their restoration can be profitable for humans.[1] However, grassland restoration is also a moral challenge or obligation. Reasons to restore temperate grasslands include:

- *To increase local biodiversity.* Biodiversity is a multi-faceted term, encompassing variation within species, the wide range of living organisms (plants, birds, microbes, etc.), and communities. Grasslands with higher biodiversity usually exhibit better ecosystem functioning. Moreover, many people agree that there are ethical reasons to preserve and restore biodiversity (Lanzerath and Friele 2014).
- *To enhance connectivity.* When conducted on sites between extant grasslands, restoration can create corridors and increase landscape-scale connectivity. This connectivity is necessary to facilitate the migration of species for which the matrix vegetation is a barrier.
- *To increase productivity.* In areas where grasslands have been lost to desertification, for example, grassland restoration can increase the availability of fodder for livestock.
- *To decrease erosion.* Grasslands are generally dominated by perennial species whose roots hold the soil in place and reduce erosion compared to arable lands.
- *To increase water quality and quantity.* Water that drains into grassland soils rather than running off via surface flow can recharge aquifers and decrease the flood risk in lower parts of a watershed. Grassland soils can also immobilize pollutants and nutrients held in solution in the water.
- *To sequester carbon and counteract climate change.* Chernozem and other soils, which develop beneath grasslands, accumulate high amounts of carbon, hence act as carbon sinks that can 'lock up' what otherwise would be mass releases of greenhouse gases.
- *To restore the aesthetic and cultural values of landscapes.* Grasslands have a subtle beauty, and also have important cultural connections. This is most notable for secondary grasslands that have developed from long-term human activities.

Efforts to restore grasslands face various obstacles and limitations. We group these broadly into natural and societal obstacles.

Natural obstacles include:

- *Spontaneous succession* has proceeded to such an advanced stage, such as a former grassland that is now forested, that restoration back to a grassland would be prohibitively expensive.
- *Seed bank depletion.* Seeds of desirable grassland species are not preserved in the soil seed bank. Many grassland species do not form persistent seed banks.
- *Limited species pool.* Target grassland species are not present in the surrounding landscape. Instead, weedy species that disperse from the matrix vegetation are a threat to the grassland community.
- *Species are unable to colonize the site.* Dispersal opportunities may be limited or blocked. Examples include altered floodplains where flooding does not occur to transport seeds, systems in which seeds are animal-dispersed but those animals are extirpated, or areas where the landscape includes barriers to movement such as woodlands that some grassland insect species cannot move through.
- *Establishment is limited.* Seeds or other propagules can reach a site, but species are unable to establish because of adverse abiotic or biotic site conditions.
- *Abiotic site conditions are so deeply changed* (e.g. the site is heavily eutrophied or the water regime is deeply altered or there has been extensive soil erosion) that restoration to a desirable stage is not possible without significant effort.

Societal obstacles include:

- Financial or physical limitations such as insufficient personnel or equipment.
- Unwillingness or inability of stakeholders to agree on restoration approaches and targets.
- Legislative obstacles.
- Insufficient knowledge about how to do restoration.

Methods of grassland restoration

We distinguish between restoration of extant degraded grasslands and restoration of grasslands on sites where they do not currently exist. Effective restoration efforts often integrate several of the listed methods. The efforts should be based on knowledge of the autecology of the species potentially forming the community, as they may also create conditions where undesirable species can establish.

Restoration of degraded grasslands

Restoration of degraded grasslands may include the following:

- *Removal of undesirable woody species.* Woody species are often undesirable in grasslands as they alter the physical structure and microenvironment. On the other hand, scattered woody species in grasslands usually increase heterogeneity and thus increase the diversity of grasslands. Their removal can be appropriate, especially where conservation strategies require enlarged spaces for grassland species or to increase connectivity between grassland patches. Removal can be accomplished by many different methods, including cutting, pulling, prescribed fires, and herbicide application. Appropriate techniques vary among systems and with the life history characteristics of the species, such as whether they resprout. Once woody species are removed, a regular grassland management regime should be initiated to prevent their re-establishment (see later text).
- *Control of invasive non-native species.* Especially in North American grasslands, invasive non-native species are a common challenge in restoration. Restoration often occurs at scales that are too large for effective manual control of these species, so managers rely on herbicides (Tu *et al.* 2001). Careful consideration of ancillary effects of herbicides is important. Herbicides that target particular types of plants can minimize off-target effects. For example, invasive grasses can be controlled in forb-dominated grasslands using grass-specific herbicides. When possible, the timing or spatial pattern of application is adjusted to capitalize on life history differences among undesirable invasive species and desirable native species. For example, the application of broad-spectrum herbicides soon after prescribed fires can be very effective at controlling invasive species that rapidly resprout without hindering native species which resprout more slowly (Stanley *et al.* 2011).
- *Adjustment or re-introduction of former/traditional management.* The re-introduction of former management regimes such as grazing or mowing is important for restoring many grasslands. Management regimes should not be uniform, but should be allowed to vary spatially and temporally to allow the survival and dispersal of plants and insects with varied life cycles. Abandoned grasslands are usually dominated by competitive plant species which suppress diversity. When appropriate management is re-established, these species usually decrease in cover, enabling the establishment of other species.
- *Manipulation of the water regime.* Former wet grasslands that have been drained may benefit

from an increase of water table by blocking ditches or removing drainage tubes. On the other hand, we also know examples where the diversity of species-rich managed grasslands decreased when drainage ditches were not regularly cleaned.

- *Manipulation of nutrient levels.* Many grasslands are overloaded by nutrients; reducing their nutrient levels can increase species diversity. Regular cutting and removal of biomass can gradually reduce nutrient levels, especially for nitrogen and available phosphorus. Grazing tends to redistribute rather than remove nutrients. Adding carbon can increase soil C:N ratios and reduce N uptake by plants (Török *et al.* 2011). Frequently used carbon sources are wood mulch, hay, and even sucrose. However, the effect of carbon addition on nutrient availability may be temporary as high microbial turn-over in the soil enables rapid mineralization.

- *Topsoil removal* can be very effective and immediately reduce nutrient levels in heavily eutrophied grasslands, but requires heavy machinery. In addition, this approach leads to problems about disposal of the removed material.

- *Creation of artificial gaps.* Some plants establish best in bare soil, which can be exposed by raking or harrowing.

- *Sowing seeds of desirable plants.* Many typical grassland species are seed limited (Seabloom *et al.* 2003). Seeds of desirable species may be obtained from wild populations or from seed production beds. A variety of seeding techniques are possible, including broadcasting, drilling, or hydroseeding. Species vary greatly in establishment rate and in the degree to which their establishment is affected by site preparation, though this information is often not known or at least not published. Where suitable reference sites exist, particularly those that do not contain undesirable non-native species, hay can be collected there and spread on the restoration site, thereby dispersing seeds of diverse species simultaneously. Decisions about which species to include in the seed mix can have long-term ramifications for how the grassland develops.

- *Planting desirable plants.* Species richness in grasslands can be rapidly increased by planting individual plants or perennial belowground structures (e.g. rhizomes, bulbs). By avoiding the germination and establishment phases of a plant's life, planting can greatly accelerate grassland recovery, particularly for species with good clonal growth. Planting is a cost- and manpower-intensive method, and therefore is most effective at small scales. Plants can be grown as plugs from seed in the nursery. In special cases, such as when a site is slated for destruction, it is possible to salvage established plants and move them to restoration sites.

- *Transfer of topsoil or turf.* When grasslands are slated for destruction another alternative is to salvage topsoil and spread it onto restoration sites. In specific cases we can also transfer turf or whole compact blocks, i.e. intact topsoil together with above- and belowground biomass, from a donor to a restored site. In addition to plant propagules, the turf or blocks may contain microbes and soil biota that are thus transferred to the restoration site. These benefits may ensure a much quicker recovery process. The expectation is that target species will spread from these blocks (used as stepping stones) across the restored site. However, it must be stressed that these methods are only appropriate in very specific circumstances as they strongly damage or destroy the donor site. In addition, they are costly and require considerable manpower and machinery. Even with careful planning and care during implementation, transplantation often heavily damages the transferred biota and often results in high mortality.

- *Transfer of desirable animals.* Insects are important for pollination and other processes in grasslands. Desirable insects or other invertebrates can be collected when at appropriate life stages and transferred to restoration sites, though this too can be thwarted by limited species and genetic pools and possible mismatches of genotypes and phenotypes to a restored environment.

Establishment of new grasslands

New grasslands are often created on arable lands but also in disturbed sites such as mining sites, road banks, ski runs, and brownfields. These areas may or may not have been grassland historically. Many of the restoration methods discussed previously are also appropriate here, but other methods may also be required.

- *Spontaneous succession.* There is potential for grasslands to recover spontaneously, without subsequent management, in sites with environmental conditions corresponding to those of primary grasslands (Hölzel *et al.* 2002). However, the respective species must occur in the surroundings and be able to colonize the sites.
- *Spontaneous succession and subsequent management.* In sites where woodland is the potential vegetation, it is more common that spontaneous succession must be accompanied by management such as regular cutting or grazing to prevent the establishment and spread of woody species and support establishment of grassland species. Generally, successful spontaneous grassland recovery can be expected in sites where (i) agricultural production (i.e. crop production) only lasted for a short time, (ii) adjacent grasslands can act as effective seed sources of target species, and (iii) the risk of infestation by weeds is low.
- *Site preparation.* The conditions at the time restoration begins will inform the type of site preparation required. Arable fields may require control of existing vegetation and the seed bank, but more impoverished sites such as mining sites and landfills may also require the installation of allochthonous topsoil. Methods of controlling the existing vegetation include prescribed fires, herbicides, tilling, and topsoil removal; these methods differ in their effects on the seedbank. It is important to take the time to ensure that good site preparation has been conducted before seeding occurs, as it is much more difficult to deal with weeds or to modify environmental conditions once desired species are established.
- *Seeding.* Seeding is the most commonly restoration method when establishing new grasslands. As noted previously, species selection is a very important restoration decision. Sowing a few productive grasses and legumes, as is common in some commercial seed mixes, is unlikely to be an effective restoration, though desired grassland species may eventually colonize the site (Török *et al.* 2011; Prach *et al.* 2015). The best option is to use a regional seed mix containing as many species as possible (Jongepierová 2008; Kiehl *et al.* 2014). Local propagules are likely to be better adapted to the local environmental conditions and may increase restoration success. Seeds can be collected in nearby reference sites or the constituent species can be cultivated. Seeds can be harvested by hand or by special harvesters (Kiehl *et al.* 2014).
- *Plant material transfer.* This includes harvested raw plant material or hay, raked litter, threshed material or hay-chaff containing the seeds of target species. This may be cheaper than seeding, though effectiveness can vary. A common issue is finding sufficiently large donor sites. Two factors to consider are the areas of the targeted and the donor sites, and when to collect and apply the plant material. The ratio between target and donor sites ranges from 1:2 to 1:10, depending on the species and seed richness of the vegetation in the donor site at the time of harvest (Kiehl *et al.* 2014). These methods can also be used to improve degraded grasslands (see previous section).
- *Topsoil removal.* Most former croplands are characterized by high residual nutrient levels in

the upper soil layers arising from the use of fertilizers. High nutrient loads favor weedy species after cultivation ceases, and can reduce the establishment of less-competitive grassland species adapted to nutrient-limited and stressed conditions. Under these conditions, it can be helpful to remove the upper 10-50 cm of soil. This also removes many of the weed seeds (Klimkowska *et al.* 2007). The method is very costly because it requires heavy machinery, though sometimes the removed topsoil can be sold for other agricultural purposes, thus offsetting the costs.

Examples of restoration of main grassland types

Eurasian steppes

Nearly all original zonal steppes in the Ukraine, Kazakhstan and Russia have been ploughed. In the 1990s, after communist rule fell, large portions of arable land in these countries were abandoned and underwent spontaneous succession. In the past decade, however, the intensity of agriculture has increased again (Hölzel *et al.* 2002). Kamp *et al.* (2011) reported from Kazakhstan that spontaneously developed and then extensively grazed habitats on ex-arable land resemble natural steppe and are convenient for biome-restricted bird species. An assessment of spontaneous recovery there is in progress (N. Hölzel, pers. comm.). In the Ukraine, there have been attempts to restore steppe by seeding abandoned land with native species (Charles 2010), though the future of these efforts is unclear given current Russian military attacks.

North American prairies

The general restoration strategy for North American prairies is to control invasive species, add seed of desired species, and identify the disturbance regime that will enable the community to persist. There is no single restoration treatment that can achieve all of these elements: treatment combinations are required, and often have to be applied repeatedly (Stanley *et al.* 2011). A case study illustrates these ideas.

The Pacific Northwest contained significant prairies at the time of European settlement in around 1850. Many of these areas were ploughed by European settlers to grow agricultural crops. Areas that were not cultivated often were subject to fire suppression; many of these are now forested. Around 2000, it was estimated that only 2–3 per cent of these grasslands were still dominated by native species (Dunwiddie and Bakker 2011). Prairie remnants are owned by numerous land managers, thus necessitating collaborative working relationships. Restoration and management of this ecosystem has been stimulated in large part by the listing of rare species by the federal government and associated support for the recovery of these species. Listed species include a plant (*Castilleja levisecta*), butterfly (*Euphydryas editha taylori*), bird (*Eremophila alpestris strigata*) and mammal (*Thomomys mazama pugetensis*).

Restoration efforts began in the 1990s and focused initially on the control of non-native species. During the first decade or two, a primary task was mowing to control *Cytisus scoparius*, a non-native leguminous shrub. That species is now largely under control, though some mowing and manual removal continue. Current control efforts include the use of grass-specific herbicides to control non-native invasive grasses such as *Arrhenatherum elatius* and of broad spectrum herbicides to control non-native invasive forbs such as *Potentilla recta* and *Senecio jacobaea* (N. Johnson, pers. comm.).

A key restoration goal in this system is to increase the quantity and diversity of native species. Initially, seed was collected by hand from wild populations and, because it was so laborious to collect, sown in the nursery to produce plugs that were then outplanted into degraded prairies. In recent years, wild-collected seed was used to establish seed increase beds that produce large quantities of seed. In 2014, for example, one conservation nursery produced about 1000 kg of seed and 344,000 plants of more than 100 prairie species (S. Smith, pers. comm.). The production of large quantities of seed enables restoration across larger areas and with more diverse species. Research in recent years demonstrated that abandoned agricultural lands can be effectively restored to prairie and examined the sensitivity of restoration success to variability among sites and seeding years (E. Delvin, pers. comm.; Figure 10.1). This work has also led to successful establishment of the rare *Castilleja levisecta*; the global population of this species increased by almost an order of magnitude from 20,000 plants in 2004–2010 to 186,000 in 2014 (J. Arnett, pers. comm.). Another focus in recent years has been the enhancement of butterfly habitat (M. Linders, pers. comm.).

A frequent fire regime has been re-instated on many prairies: in 2014, 90 burns totaling about 1,000 ha were conducted (M. McKinley, pers. comm.). These prairies are generally burnt on a 2–4 year rotation to reduce thatch accumulation and create microsites of bare ground in which species can establish. Frequent fires may also create important habitat for native annual species, which appear to be much less common now than they were historically (Dunwiddie *et al.* 2014). Burning is often followed soon thereafter by a broad-spectrum herbicide to control rapidly resprouting non-native species (Stanley *et al.* 2011).

Figure 10.1 Experimental plots used to test methods of restoring grassland on former agricultural land in the Pacific Northwest. Each plot in the foreground of this image is 40 m². Beyond them is a larger area that was restored a few years later and, in the far distance, are *Cytisus scoparius* and *Pseudotsuga menziesii* on the edge of the prairie. The plots were seeded in autumn 2008 and burned in autumn 2014; the photo was taken in spring 2015

Source: Photo by J. D. Bakker

Looking to the future, efforts in this ecosystem are likely to focus on enhancing plant species diversity in degraded prairies, habitat enhancement for rare species, understanding interactions between species and disturbance regimes (e.g. how heterogeneous should prescribed fire effects be to balance general management needs without harming insect larvae), and improving connectivity. Another opportunity is to work with private landowners and explore how live-stock grazing could be integrated into grassland management.

European secondary grasslands

In Europe, the first grassland restorations occurred in the countries with the most altered land-scapes due to intensive exploitation (i.e. the United Kingdom and the Netherlands). Degraded existing grasslands were the first restoration focus; later efforts examined the restoration of various types of grasslands on ex-arable land or other environments such as spoil heaps.

Early restorations focused on grasslands from the ends of the moisture gradient. Dry grass-lands, usually on calcareous soils, support a high diversity of plants and insects and thus were of high conservation interest (Dengler *et al.* 2014). Wet grasslands, often alluvial, were of interest because of efforts to restore degraded river floodplains and decrease flood risk (Joyce and Wade 1998). A similar pattern existed along the nutrient gradient: restoration interest focused primarily on oligotrophic grasslands or highly eutrophied grasslands. The former were of interest because of their fast disappearance due to eutrophication, abandonment and afforestation, while the latter were of interest due to their striking degradation and indication of excessive pollution.

Degraded existing grasslands

A classic and long-term restoration study was conducted in the Netherlands, where abandonment of species-rich chalk grasslands led to an expansion of the competitive grass *Brachypodium pinnatum* and subsequent deep decrease of diversity (Bobbink and Willems 1993). Willems and Bik (1998) conducted an experiment in 1970 and then 20 years later. Experimental sites were more degraded in 1990 than 1970, but restoration of high species richness was significantly faster in 1990 not only due to the lower start but also due to the increase size of the community species pool at the site as a result of appropriate management in the surrounding landscape during intervening decades (Figure 10.2). Their study demonstrates restoration success at the site and landscape scales.

Restoration of degraded wet grasslands in Western Europe was reviewed by Klimkowska *et al.* (2007). The most effective restoration required a combination of techniques such as rewetting, topsoil removal, and seed transfer. In the UK, restoration of grasslands is often connected to restoration of heathlands (Lowday and Marrs 1992). One project restored Estonian alvar grasslands overgrown by woody species (Pärtel *et al.* 1998). The authors emphasized the importance of proximity effects, concluding that species-rich grasslands could recover if the woody species were cut and regular extensive grazing occurred, as long as the local species pool had been maintained. The role of seed sources has also been emphasized in restored dry calcareous grasslands overgrown by shrubs in the French Prealps (Barbaro *et al.* 2001), species-rich grass-lands in northern France (Muller *et al.* 1998), and formerly abandoned grasslands restored by grazing in Sweden (Lindborg and Eriksson 2004). It seems that the character of the surrounding landscape is a key factor determining restoration success.

Figure 10.2 Species number as a function of the number of years since restoration began for plots where restoration began in either 1970 or 1990. All plots were the same size (2.25 m²). Speed of the restoration process as represented by the slope of the linear regression line was significantly different ($p < 0.05$) in 1970–1977 compared with 1990–1997

Source: after Willems and Bik (1998), with permission

Restoration of grasslands on ex-arable land

Effective spontaneous restoration of dry grasslands on ex-arable land was reported from drier sites in central and southern Europe (Ruprecht 2006). Assisted restoration, mostly by seeding or hay transfer, is a common restoration activity in recent years around Europe (Fagan *et al.* 2008; Kiehl *et al.* 2010, 2014). Studies at different spatial scales have been conducted in the UK, the Netherlands, France, Germany, Czech Republic and Hungary. In terms of practical restoration advice, studies are most important if they are conducted at large spatial scales, include multiple sites, and compare different restoration methods (Fagan *et al.* 2008; Török *et al.* 2011; Prach *et al.* 2013, 2015). We present here, as examples, two studies that we have been involved with.

Large-scale restoration of dry grasslands in the Carpathian Mountains

This study was conducted in the White Carpathian Mountains Protected Landscape Area and Biosphere Reserve, eastern Czech Republic (Jongepierová 2008; Prach *et al.* 2013, 2015). Thousands of hectares of dry grasslands in this region have been managed as hay meadows for several centuries (Hájková *et al.* 2011; Jongepierová 2008). However, many of these grasslands were ploughed, overfertilized or abandoned between 1950 and 1990. About 4,000 ha of semi-natural hay meadows remain; these are now protected under national legislation and are internationally recognized as Natura 2000 habitats (Jongepierová 2008). These grasslands are among the most diverse communities in the world at scales < 100 m² (Wilson *et al.* 2012) and were used as reference sites in this study.

Restoration goals especially included improved connectivity among the remnant grasslands. Plant species composition was compared between the reference sites and 82 dry grassland stands restored on ex-arable land. Restored sites were sown either with a species–rich regional seed mixture (44 species, >500 ha) or a species-poor commercial clover-grass seed mixture, or were left to experience spontaneous succession. The ordination results (Figure 10.3) demonstrate the convergence of grasslands restored by different methods towards reference grasslands. Soil characteristics, especially P content, had the strongest effects, followed by restoration method, proximity, and age.

Overall, regional seed mixtures were the best method to re-establish dry grasslands on ex-arable land, though spontaneous succession, and even regrassing with commercial seed mixes provided reasonable results at sites in close vicinity of reference sites. However, these two methods supported restoration trajectories towards rather mesic grasslands instead of targeted dry grasslands. Soil characteristics and landscape context need to be considered during restoration projects, along with the selection of proper restoration methods.

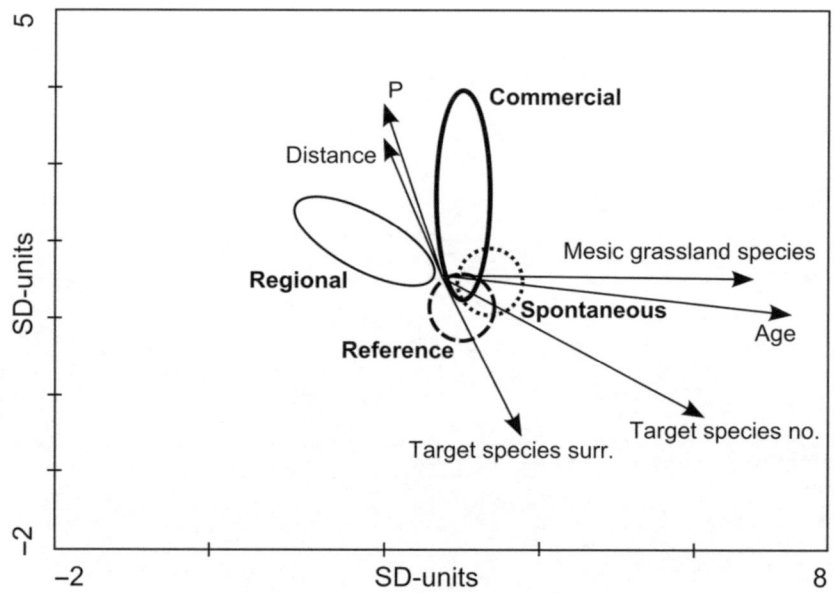

Figure 10.3 Unconstrained ordination (detrended correspondence analysis) of vegetation samples, community characteristics, and most significant environmental factors from grasslands restored by three different methods: using regional seed mixtures (Regional), commercial seed mixtures (Commercial) and spontaneous succession (Spontaneous). Vegetation samples from reference dry grasslands were passively projected (Reference), as well as the community characteristics and environmental factors: the number of target species typical for dry grasslands (Target species no.) and mesic grassland species both occurring in the restored grasslands; the number of target species occurring in the surroundings (Target species surr.), distance to the nearest reference grassland (Distance), total soil phosphorus (P) and time since restoration started (Age)

Landscape-scale restoration in Hortobágy National Park

In the western part of Hortobágy National Park (east Hungary), 760 ha of former croplands, alkali grasslands, and loess grasslands were recovered using low-diversity regional grass seed mixtures. The aim of the restoration was to eliminate croplands at high elevated places and to increase the landscape-scale connectivity of natural grassland habitats by using secondary restored grasslands to create green corridors. The seed mixtures contained the seeds of characteristic grasses of alkali and loess grasslands in the region (*Festuca pseudovina, F. rupicola, Poa angustifolia, Bromus inermis*) and were sown at a rate of 25 kg/ha between 2004 and 2008. The development of grassland vegetation was followed in permanent plots. Restoration success was influenced both by the seed mixture used and by site history (Török *et al.* 2012). Vegetation development progressed towards reference grasslands: within three years, the former croplands were dominated by the sown grasses and most weeds were effectively suppressed.

However, the seed bank remains dominated by weed seeds. In addition, regular management by mowing and/or grazing is necessary to sustain the desired vegetation composition – because of the high biomass produced in these grasslands, large-scale degradation of the vegetation may occur within a few years if management ceases. The area and fragmentation level of grasslands in the landscape strongly influenced long-term restoration success. The spontaneous colonization capacity of alkali grassland species is generally promising in the region, but loess grasslands are highly fragmented and degraded; thus, many loess specialists cannot reach the restored grasslands (but see Figure 10.4).

Figure 10.4 Restored grassland with high cover of spontaneously immigrated loess specialist *Dianthus pontederae*

Source: Photo by Orsolya Valkó

For loess grassland restoration, the application of hay transfer or the use of a high diversity seed mixture would be the most promising options. However, the use of a high diversity regional seed mixture will only be feasible once local grassland species begin to be propagated, as seeds of these species cannot be obtained from commercial sources (Török *et al.* 2011).

Conclusions

In terms of vascular plants, temperate grasslands are the most species-rich ecosystems in the world at scales of 1–100 m² (Wilson *et al.* 2012). Many other organisms are related to this diversity, especially insects. Moreover, these diverse grasslands provide important ecosystem services. Primary grasslands are a natural heritage, and secondary grasslands are also a cultural heritage. Thus, conservation and restoration of these grasslands is a challenge for ecologists. Based on the literature cited and our experiences, we conclude:

- Restoration seems to be easier on moderately nutrient-rich than on nutrient-poor or heavily eutrophied sites.
- Restoration is difficult if water and/or nutrient regimes have been deeply altered.
- Restoration is easier if target species still exist in the site itself or in its immediate surroundings.
- Some restoration measures can be profitable for one group of organisms and detrimental for some others, thus consultancy among experts is needed prior to restoration starting.
- Continuous management must be ensured in the case of secondary grasslands.
- Management activities should be spatially and temporally variable.
- Long-lasting monitoring should be ensured in any grassland restoration project.

Finally, we consider how climate change may affect grassland restoration. Areas that become warmer and drier, and thus have a prolonged dry season, may become more suitable for xerophilous and thermophilous herbaceous species and less suitable for woody species. This could ease the restoration of dry temperate grasslands. Experimental reductions in precipitation sometimes reduce the biomass of dominant competitors, usually graminoids, and consequently increase the diversity of forbs (Holub *et al.* 2013). However, restoration of mesic or wet grasslands, or to achieve other restoration targets such as increased fodder production, could be negatively impacted by climate change. Some areas that potentially become wetter may experience increased woody plant encroachment unless the management intensity is sufficient to prevent this. However, the cumulative effects of climate change on temperate grasslands may also depend on its interactive effects on disturbance regimes, such as fire microbial activity, decomposition rates, and nutrient uptake by plants, and are difficult to predict. It will be important to consider site-specific predictions of climate change and to adaptively manage sites based on those predictions and the site context. However, we also need more multisite observational and experimental studies to make firmer conclusions (Wu *et al.* 2011).

Acknowledgements

K. Prach was partly supported by GACR 17-09979S, RVO67985939, and the Fulbright Program. P. Török was supported by NKFIH K 119225 and a CAMPUS Hungary mobility grant during manuscript preparation. Orsolya Valkó kindly provided Figure 10.4. We also thank Ivana Jongepierová, Klára Řehounková, Karel Fajmon and Lenka Šebelíková for consultancy or technical help.

Note

1   During proof reading the following important publication was received: Blakesley D. and Buckley P. (2016). *Grassland Restoration and Management*. Pelagic Publishing, Exeter, UK.

References

Archibold O. W. (1995) *Ecology of World Vegetation*. Chapman & Hall, London.

Barbaro L., Dutoit T. and Cozic P. (2001) A six-year experimental restoration of biodiversity by shrub-clearing and grazing in calcareous grasslands of the French Prealps. *Biodiversity and Conservation* 10: 119–135.

Bobbink R. and Willems J. H. (1993) Restoration management of abandoned chalk grasslands in The Netherlands. *Biodiversity and Conservation* 2: 616–626.

Charles D. (2010) Renewing the post-Soviet steppe. *Science* 328: 1225.

Dengler J., Janišová M., Török P. and Wellstein C. (2014) Biodiversity of Palaearctic grasslands: a synthesis. *Agriculture, Ecosystems and Environment* 182: 1–14.

Dunwiddie P. W. and Bakker J. D. (2011) The future of restoration and management of prairie-oak ecosystems in the Pacific Northwest. *Northwest Science* 85: 83–92.

Dunwiddie P. W., Alverson E. R., Adam R. A., and Gilbert R. (2014) Annual species in native prairies of South Puget Sound, Washington. *Northwest Science* 88: 94–105.

Fagan K. C., Pywell R. F., Bullock J. M. and Marrs R. H. (2008) Do restored calcareous grasslands on former arable fields resemble ancient targets? The effect of time, methods and environment on outcomes. *Journal of Applied Ecology* 45: 1293–1303.

Gibson D. J. (2009) *Grasses and Grassland Ecology*. Oxford University Press, Oxford.

Hájková P., Roleček J., Hájek M., Horsák M., Fajmon K., Polák M. and Jamrichová E. (2011) Prehistoric origin of extremely species-rich semi-dry grasslands in the Bílé Karpaty Mts. (Czech Republic and Slovakia). *Preslia* 83: 185–204.

Hegarty J., Zabowski D. and Bakker J. D. (2011) Use of soil properties to determine the historical extent of two western Washington prairies. *Northwest Science* 85: 247–254.

Holub P., Fabšičová M., Tůma I., Záhora J. and Fiala K. (2013) Effects of artificially varying amounts of rainfall on two semi-natural grassland types. *Journal of Vegetation Science* 24: 518–529.

Hölzel N., Haub C., Ingelfinger M. P., Otte A. and Pilipenko V. N. (2002) The return of the steppe: large-scale restoration of degraded land in southern Russia during the post-Soviet era. *Journal for Nature Conservation* 10: 75–85.üøøüUË

Jongepierová I. (ed.) (2008) *Grasslands of the White Carpathian Mountains*. ZO ČSOP Bílé Karpaty, Veselí nad Moravou, Czech Republic.

Joyce C. B. and Wade P. M. (eds) (1998) *European Wet Grasslands. Biodiversity, Management and Restoration*. Wiley, Chichester.

Kamp J., Urazaliev R., Donald P. F. and Hölzel N. (2011) Post-Soviet agricultural changes predicts future declines after recent recovery in Eurasian steppe bird populations. *Biological Conservation* 144: 2607–2614.

Kiehl K., Kirmer A., Donath T. W, Rasran I. and Hölzel N. (2010) Species introduction in restoration projects – Evaluation of different techniques for the establishment of semi-natural grasslands in Central and Northwestern Europe. *Basic and Applied Ecology* 11: 285–299.

Kiehl K., Kirmer A., Shaw N. and Tischew S. (eds) (2014) *Guidelines for Native Seed Production and Grassland Restoration*. Cambridge Scholars Publishing, Newcastle upon Tyne.

Klimkowska A., van Diggelen R., Bakker J. P. and Grootjans A. P. (2007) Wet meadow restoration in Western Europe: A quantitative assessment of the effectiveness of several techniques. *Biological Conservation* 140: 318–328.

Lanzerath D. and Friele M. (eds) (2014) *Concepts and Values in Biodiversity*. Routledge, London.

Lindborg R. and Eriksson O. (2004) Effects of restoration on plant species richness and composition in Scandinavian semi-natural grasslands. *Restoration Ecology* 12: 318–326.

Lowday J. E. and Marrs R. H. (1992) Control of bracken and the restoration of heathland. 1. Control of bracken. *Journal of Applied Ecology* 29: 195–203.

Muller S., Dutoit T., Alard D. and Grévilliot F. (1998) Restoration and rehabilitation of species-rich grassland ecosystems in France: a review. *Restoration Ecology* 6: 94–101.

Pärtel M., Kalamees R., Zobel M. and Rosen E. (1998) Restoration of species-rich limestone grassland communities from overgrown land: the importance of propagule availability. *Ecological Engineering* 10: 275–286.

Prach K., Jongepierová I. and Řehounková K. (2013) Large-scale restoration of dry grasslands on ex-arable land using a regional seed mixture: establishment of target species. *Restoration Ecology* 21: 33–39.

Prach K., Fajmon K., Jongepierová I. and Řehounková K. (2015) Landscape context in colonization of restored dry grasslands by target species. *Applied Vegetation Science* 18: 181–189.

Ruprecht E. (2006) Successfully recovered grassland: a promising example from Romanian old-fields. *Restoration Ecology* 14: 473–480.

Seabloom E. W., Borer E. T., Boucher V. L., Burton R. S., Cottingham K. L., Goldwasser L., Gram W. K., Kendall B. E. and Micheli F. (2003) Competition, seed limitation, disturbance, and reestablishment of California native annual forbs. *Ecological Applications* 13: 575–592.

Six L. J., Bakker J. D. and Bilby R. E. (2014) Vegetation dynamics in a novel ecosystem: agroforestry effects on grassland vegetation in Uruguay. *Ecosphere* 5: art 74.

Stanley A. S., Dunwiddie P. W. and Kaye T. N. (2011) Restoring invaded Pacific Northwest prairies: management recommendations from a region-wide experiment. *Northwest Science* 85: 233–246.

Török P., Vida E., Deák B., Lengyel S. and Tóthmérész B. (2011) Grassland restoration on former croplands in Europe: an assessment of applicability of techniques and costs. *Biodiversity and Conservation* 11: 2311–2332.

Török P., Miglécz T., Valkó O., Kelemen A., Deák B., Lengyel S. and Tóthmérész B. (2012) Recovery of native grass biodiversity by sowing on former croplands: Is weed suppression a feasible goal for grassland restoration? *Journal for Nature Conservation* 20: 41–48.

Tu M., Hurd C. and Randall J. M. (2001) *Weed Control Methods Handbook: Tools and Techniques for Use in Natural Areas.* Wildland Invasive Species Team, The Nature Conservancy, Arlington, VA. Retrieved from www.invasive.org/gist/handbook.html.

Veen P., Jefferson R., de Smidt J. and van der Straaten J. (eds) (2009) *Grasslands in Europe of High Nature Value.* KNNV Publishing, Zeist.

Vera F. W. M. (2000) *Grazing Ecology and Forest History.* CABI Publishers, Wallingford.

Willems J. H. and Bik L. P. M. (1998) Restoration of high species density in calcareous grassland: the role of seed rain and soil seed bank. *Applied Vegetation Science* 1: 91–100.

Wilson J. B., Peet R. K., Dengler J. and Pärtel M. (2012) Plant species richness: the world records. *Journal of Vegetation Science* 23: 796–802.

Wu Z., Dijkstra P., Koch G. W., Peñuelas J. and Hungate B. A. (2011) Responses of terrestrial ecosystems to temperature and precipitation change: a meta-analysis of experimental manipulation. *Global Change Biology* 17: 927–942.

Zuleta G., Rovere A. E., Pérez D., Campanello P. I., Johnson B. G., Escartín C., Dalmasso A., Renison D., Ciano N. and Aronson J. (2015) Establishing the ecological restoration network in Argentina: from Rio1992 to SIACRE2015. *Restoration Ecology* 23: 95–103.

11

RESTORATION OF TEMPERATE SAVANNAS AND WOODLANDS

Brice B. Hanberry, John M. Kabrick, Peter W. Dunwiddie,
Tibor Hartel, Theresa B. Jain and Benjamin O. Knapp

Introduction

Savannas and woodlands are open forest phases that occur along a gradient between grasslands and closed canopy forests. These ecosystems are characterized by open to nearly closed canopies of overstorey trees, relatively sparse midstorey and understorey woody vegetation, and dense, species-rich ground flora. In contrast to closed forests, the dominant and codominant trees in the canopy of open forests often have large, spreading crowns. Relatively open structure allows light to the floor, which is critical for grasses and forbs; the greatest plant species diversity and turnover rates occur in the herbaceous layer of open forest ecosystems in temperate zones (Gilliam 2007). Open structure, old trees, and variation of light and microclimate are important for lichens (Paltto *et al.* 2011), insects (e.g. carabids; Taboada *et al.* 2011), birds (Hunter *et al.* 2001), mammals and other taxa.

Savannas and woodlands result from low-severity disturbances or conditions that restrict development of dense forests with closed canopies. Type and severity of disturbance vary according to region and temporally. In North America, low-severity fire, resulting from lightning or deliberate ignitions, historically prevented closed forest development, although forests also may be limited by grazing and browsing, land use practices, shallow soils, or drought. Large diameter fire-tolerant tree species survive surface fire, while herbaceous vegetation is favoured by fire compared to shrubs, small-diameter trees, and other woody vegetation that lose a considerable proportion of their energy reserves if their aboveground tissue is consumed (Brose *et al.* 2013). Tree species and other vegetation adapted to surface fire regimes often produce fine fuels to promote fire spread through litter that dries rapidly and decomposes slowly. Due to fire exclusion that began during the mid-1800s to mid-1900s, in combination with extensive harvest and land use conversion, most open forest ecosystems in North America have transitioned to closed forests that support a different suite of species than open forests. A combination of large herbivores, fire, and climate may have been important in maintaining the open woodland structure of Europe, while silvopastoralism and transhumance shepherding (e.g. seasonal movement) that often entail practices such as coppicing, pollarding, and grazing currently maintain wood-pastures (Vera 2000; Sandom *et al.* 2014; Plieninger *et al.* 2015).

In this chapter, we will describe state-of-the-art restoration techniques for open forest ecosystems that once occurred across large areas in North America (Figure 11.1) and Europe,

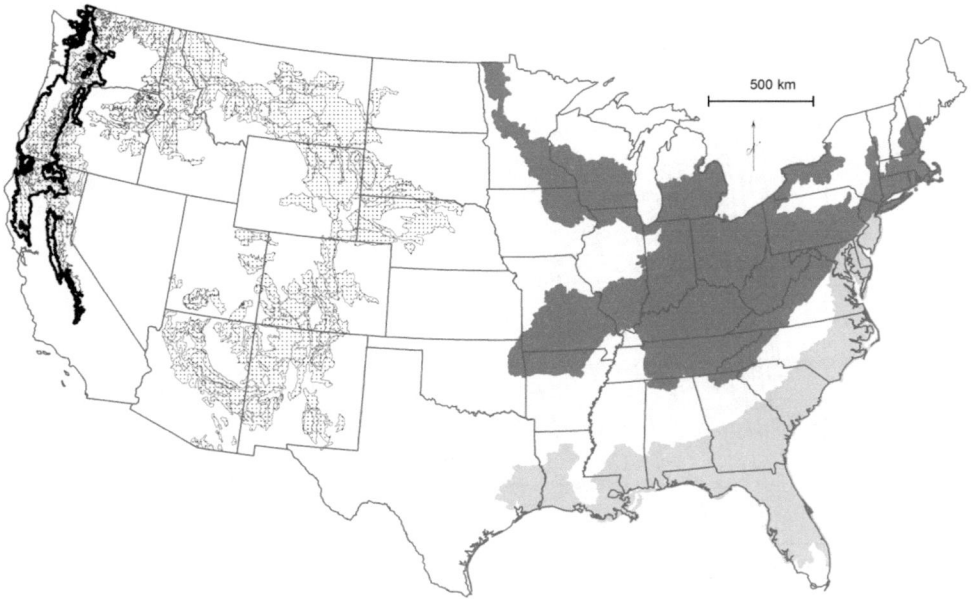

Figure 11.1   Eastern oak savannas and woodlands predominantly occurred in the central eastern United States (shaded dark grey), longleaf pine in the southern Coastal Plain (shaded light grey), ponderosa pine throughout the western United States (stippled), and Garry oak in the Pacific Northwest and into California (dark outline)

using a broad, systematic approach to address the complex issue of open woodlands for two continents. Savannas and woodlands are described in North America, Europe and portions of Australia, where synonyms include barrens and parklands in North America and wood-pastures in Europe. Savannas and woodlands are poorly described in temperate Asia, South America and Africa but appear to occur in localized areas within grassland ecosystems of steppes, pampas and veld. For North America, the structural threshold that separates temperate savannas from woodlands may vary regionally, but we recently suggested a maximum of 30 per cent stocking (or growing space occupied; Figure 11.2) and 100 trees ha^{-1} ≥12.7 cm diameter, with about 40–50 per cent canopy closure; the threshold between woodland and forest is approximately 75 per cent stocking and 250 trees ha^{-1} and more importantly, an open midstorey (Hanberry *et al.* 2014). For Europe, the crucial defining feature for wood-pasture systems is presence of grazing, which produces tree densities ranging from 10–100 ha^{-1} for oak wood-pastures to 200–500 trees ha^{-1} in some sub-alpine wood-pastures (Hartel *et al.* 2013; Garbarino and Bergmeier 2014). Despite variation, we suggest that the primary concern is restoration of the open forested state; in general, management of a greater area of open woodlands rather than a smaller area of savannas surrounded by closed forests will provide a greater return in terms of support of declining species and rare ecosystems.

Representative North American savannas and woodlands

Oaks were the dominant species of savannas and woodlands that historically covered large landscapes in the central eastern United States, particularly along the western side of eastern forests

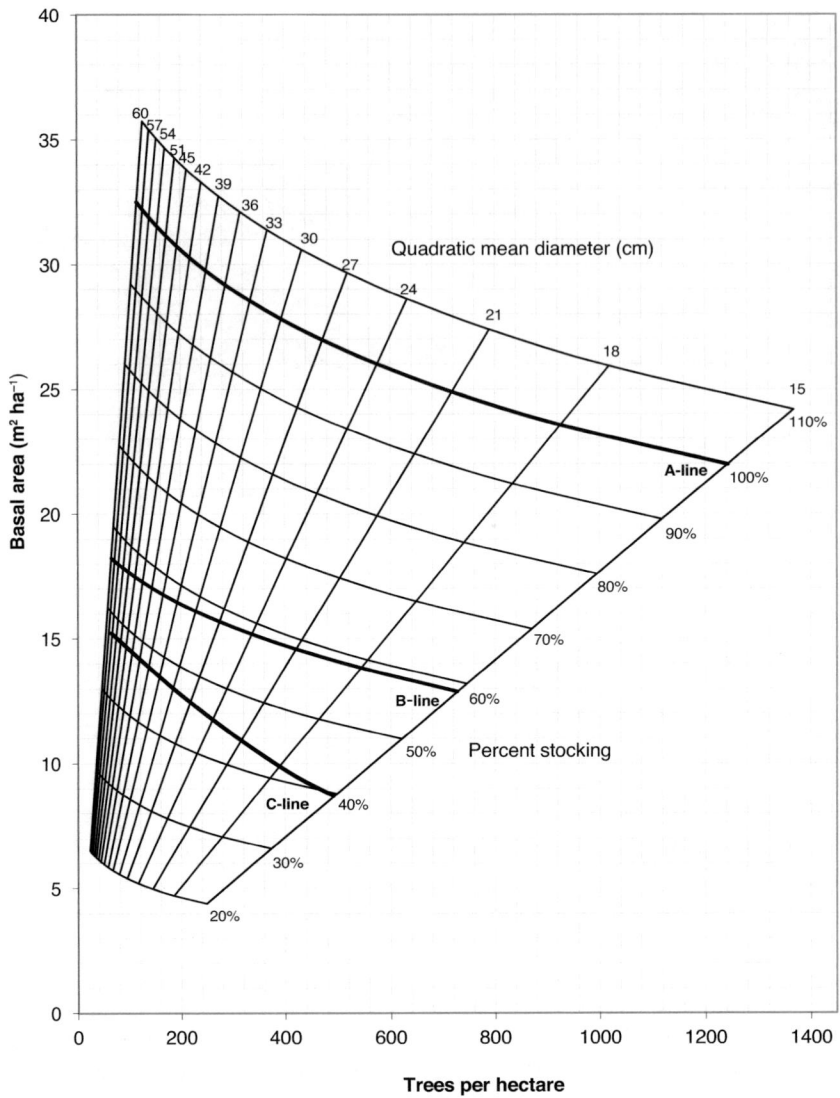

Figure 11.2 Stocking chart (Gingrich 1967) for oaks and hickories where the quadratic mean diameter at breast height is >18 cm. Maintenance operations in woodlands keep the stocking between 30 and 75 per cent with 'closed-canopy' woodlands maintained near the B line and 'open-canopy' woodlands maintained below the B line. At stocking levels less than 30 to 35 per cent the structure begins to resemble that of a savanna. For regenerating woodlands, stocking is reduced below the C line. The stocking chart can be used to convert stocking levels to a basal-area basis for managements units of a given average diameter

and on dry and nutrient-deficient sites that supported lower plant densities and slower growth (Kabrick *et al.* 2008; Figure 11.1). Generally, white oak (*Quercus alba*) was common along with black (*Q. velutina*), post (*Q. stellata*), chestnut (*Q. montana*), northern red (*Q. rubra*), blackjack (*Q. marilandica*), and bur (*Q. macrocarpa*) oaks, varying by location. Frequent surface fire maintained fire-tolerant oak species. Cotyledons of oak seedlings are belowground; if top-killed by fire, the cotyledons remain protected and provide some of the nourishment needed to resprout and remain in the stand. Oak seedlings also establish a large root system at the expense of early shoot growth. This larger root system enables oak seedlings to resprout readily after being top-killed. Numerous ground flora species are considered woodland indicators, particularly graminoids, sedges, and forb species in the genera *Lespedeza*, *Silphium*, *Solidago* and *Symphyotrichum*. Most woodland indicator species are herbaceous plants that produce flowers and seeds during the summer months and are adapted to ecosystems where light penetration is relatively high. Open oak forest ecosystems have transitioned to closed broadleaf forests composed of a variety of tree species, without frequent low severity surface fires that favoured oak species and non-woody plants.

Much of the southeastern United States historically was dominated by longleaf pine (*Pinus palustris*) savannas and woodlands, occupying an estimated 37 million hectares within the Coastal Plain, the fall-line sandhills, and mountains of central Georgia and Alabama. Longleaf pine trees can regenerate and recruit under fire return intervals of 2–3 years due to adaptations to frequent, low-severity fire. Longleaf pine seedlings develop a large root system while delaying height growth in their unique grass stage, and seedlings begin height growth when the root collar diameter is approximately 2.5 cm, at which point the cambium is protected by exceptionally thick bark. Frequent fire promotes the development of herbaceous ground flora, including bunchgrasses such as wiregrass (*Aristida* spp.) or bluestems (*Andropogon* spp., *Schizachyrium* spp.), which creates a positive feedback by contributing fine fuels for subsequent fires (Mitchell *et al.* 2009).

In the Pacific Northwest, drier areas in the lowlands of western Oregon, Washington, and coastal British Columbia historically supported extensive savannas and woodlands, largely defined by Garry oaks (*Quercus garryana*; Oregon white oak). Regular burning by Native Americans limited regeneration and growth of Douglas-fir (*Pseudotsuga menziesii*) and other rapidly-growing, taller conifers. Today, oak woodlands have been reduced to a few per cent of their historical extent. Douglas-firs have overtopped the oak canopy in many areas and understories have become dense thickets of native and exotic shrubs, particularly Himalayan blackberry (*Rubus armeniacus*) and Scotch broom (*Cytisus scoparius*), often resulting in the loss of ground flora.

Ponderosa pine savannas and woodlands cover approximately 1.6 million hectares that are dispersed across the western United States (McPherson 1997). These ecosystems experience periods of high temperatures with little or no precipitation causing plant stress (*ibid.*). Fire, by lightning and deliberate ignition, was the primary disturbance that sustained pine savannas. Since European settlement, overgrazing, fire exclusion, species introductions, and some timber harvesting changed ponderosa pine savannas (Jain *et al.* 2012). Between the mid-1800s through early 1900s, overgrazing by cattle, sheep, and horses promoted the spread of invasive plants that were introduced during settlement and compacted soil, resulting in reduced water infiltration and increased erosion, and ultimately changed plant community species composition and diversity (Belsky and Blumenthal 1997).

European wood-pastures

Wood-pastures in the European Union cover approximately 203,000 km², out of which approximately 109,000 km² are pastures with sparse trees (i.e. the typical 'savanna' type of landscape), 85,000 km² are pastures in open woodlands, and 9000 km² are pastures with cultivated trees (e.g. olive groves and fruit trees; Plieninger *et al.* 2015). Spain, Romania, and Portugal have the greatest coverage of wood-pastures and retain the most traditional land-use practices. Wood-pastures in the temperate and submeridional broadleaved woodlands (following Bergmeier *et al.* 2010) region contain a diversity of dominant tree species, including oaks (e.g. *Q. robur*, *Q. petraea*, *Q. pyrenaica*, *Q. pubescens*, *Q. fainetto*, *Q. cerris*, *Q. suber*) but also species such as *Carpinus betulus*, *Fagus sylvatica*, *Ulmus* spp., and a number of coniferous trees in the mountainous areas (e.g. *Pinus sylvestris*, *Abies alba*, *A. pinsapo*). Grazing livestock include cattle, buffalo, pig, horses, sheep, and goats (Bergmeier *et al.* 2010). Southern European wood-pastures are dominated by more xerophylous trees and shrubs. Sheep use is common in Spanish transhumance, and grazing may prevent fire in these ecosystems. Traditional grazing systems have changed in the majority of countries to become more intensive and without seasonal movement. Additionally, with the exception of cork production, wood and other tree products are rarely extracted from pastures; coppicing (cutting the stems at the basal level while allowing multiple stems to regenerate from the stump for harvest again in roughly 7 to 30 year cycles) and pollarding (applying the same intervention above the browsing height of the livestock) have been largely abandoned (Hartel *et al.* 2015) for modern silvicultural practices that produce timber and other forest products.

General restoration guidelines

First and foremost, clear and specific objectives are necessary for a restoration prescription that includes information about current condition, desired future condition, and social and economic context for the area planned for restoration. With clear objectives defined, identification of treatment and potential treatment combinations, timing of application, and proposed vegetative target thresholds or indicators (i.e. establishment of preferred species) is most efficient (Figures 11.3 and 11.4). Additionally, monitoring to determine management effectiveness will guide future decisions (i.e. adaptive management) and identify when a maintenance treatment is needed to sustain the ecosystem.

An inventory of tree composition and density, ground flora, and soil is essential to assess restoration and management potential (Table 11.1). Sites with the greatest restoration potential already contain tree species of interest in the overstorey. Presence of woodland indicator ground flora, which are a critical component to restoration and also provide fuel for frequent surface fires, is a valuable additional element that often is absent or highly depauperate. Low abundance or richness of woodland ground flora may indicate that the site has been eroded, grazed severely, ploughed, or has remained for many decades in a closed canopy state, although viable seeds and bulbs of understorey taxa may be present in the soil seedbank. Information about native vegetation may be included in a soil survey or plant associations (Peet 2006), and management units can be grouped using similar soils and stand conditions. Soils, terrain, and precipitation must support desired changes and match the desired plant species. In some regions, selection of certain topography and soils will make restoration more successful by reducing competition from other species. Indicators of historical wood-pastures in Europe are presence of (Kirby and Perry 2014):

* large trees (e.g. veteran pedunculate oaks);
* large herbivores;
* open woodland communities;
* archaeological features typical of wood-pastures;
* historical records and maps; and
* oral history.

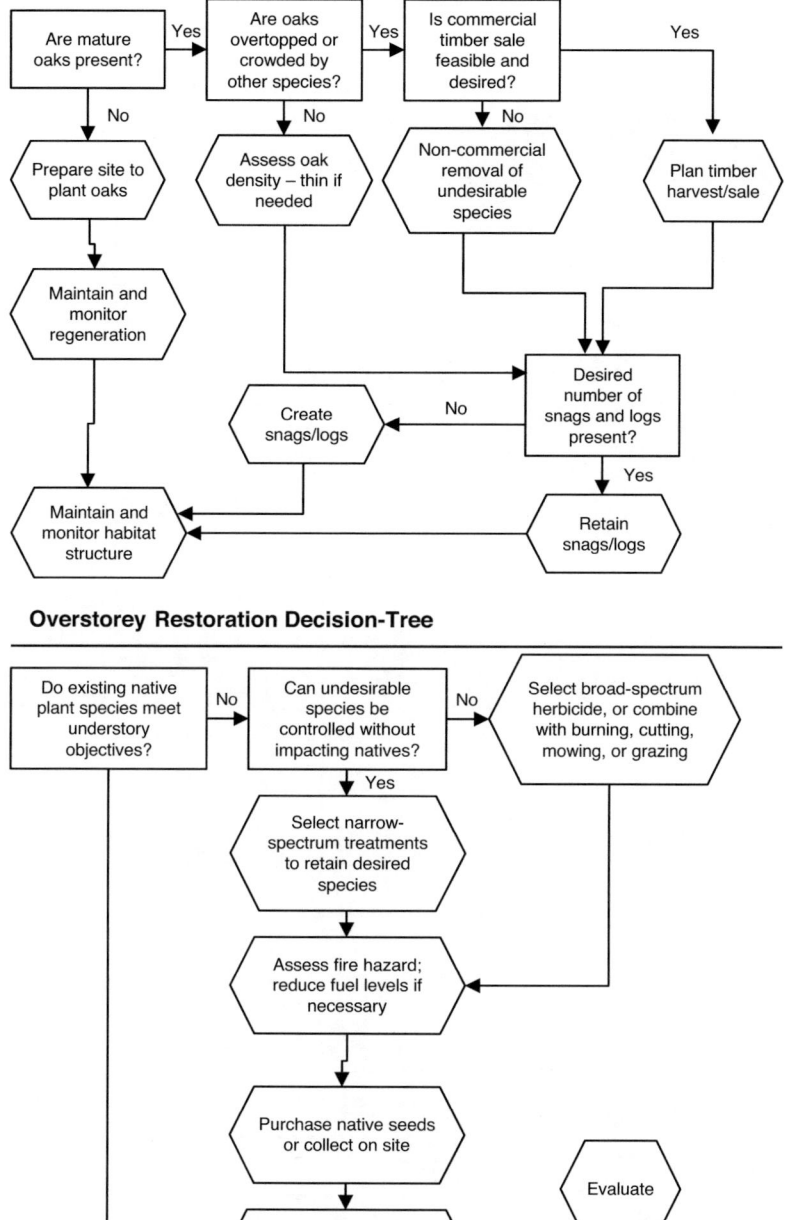

Overstorey Restoration Decision-Tree

Understorey Restoration Decision-Tree

Figure 11.3 Decision trees for overstorey and understorey restoration in oak savannas and woodlands of the Pacific Northwest

Source: adapted from Vesely and Tucker (2004)

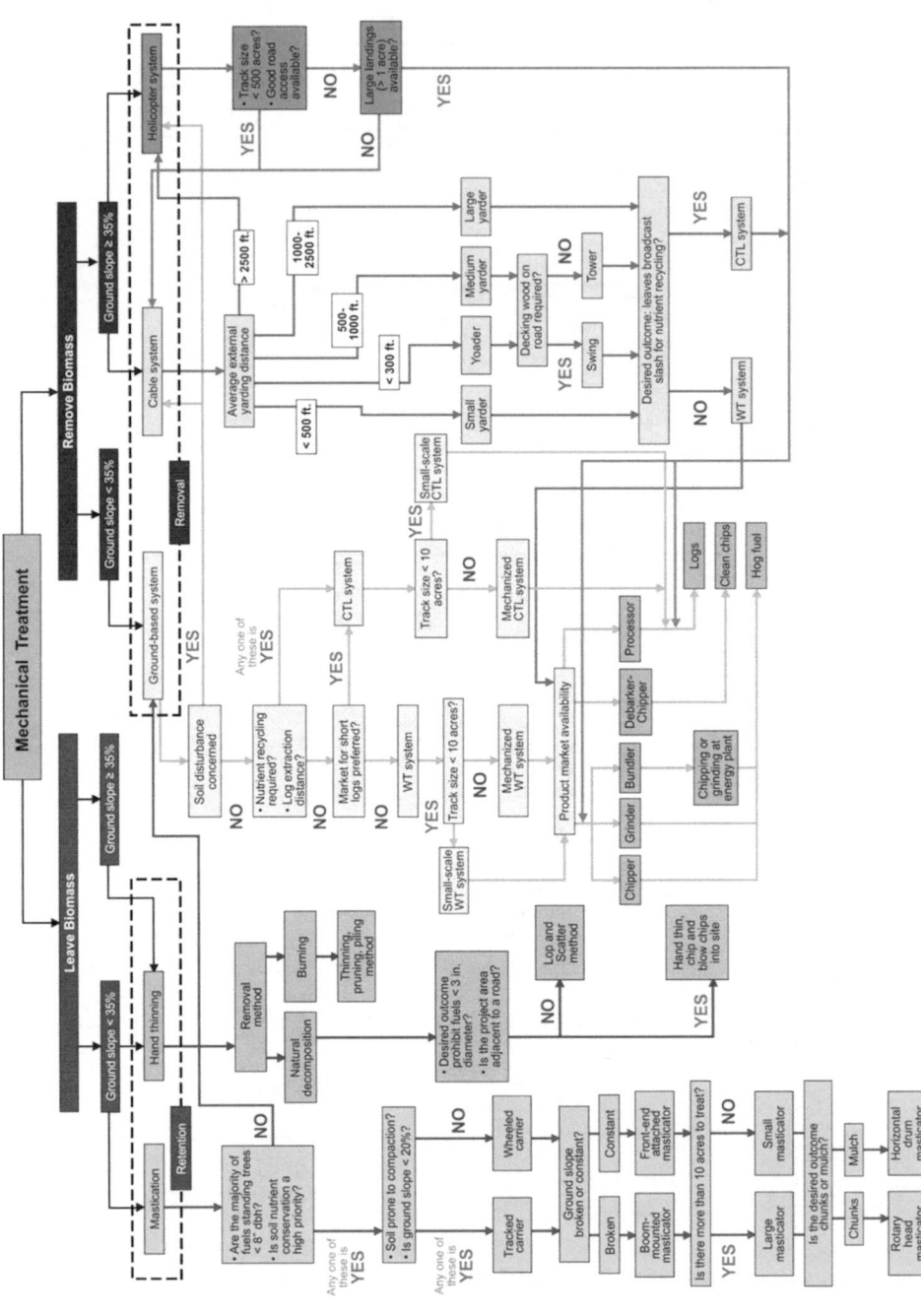

Figure 11.4 Decision tree for equipment selection during restoration of ponderosa pine in the western United States. Equipment selection should fulfil treatment objectives, work conditions under which treatment operations are implemented, and efficient use of limited budget

Source: adapted from Jain *et al.* (2012)

Table 11.1 Example inventory for managing woodlands and savannas in the eastern United States

Stand or mgt. unit	Soil series	Slope position	AWC (cm)	pH	Ecological site description	Tree density (ha⁻¹)	Basal area (m² ha⁻¹)	Stocking (Gingrich 1967)	Dominant tree species	Woodland indicators present
1	Nixa-Clarksville Complex, 1–3% slopes	Summits	17.8	5.0	*Quercus stellata–Quercus velutina/Amelanchier arborea–Vaccinium pallidum/Helianthus hirsutus–Schizachyrium scoparium*	951	18.9	87	*Quercus alba, Quercus velutina, Quercus stellata*	*Helianthus hirsutus, Parthenium integrifolium*
2	Coulstone-Clarksville complex, 3 to 8% slopes	Summits and shoulders	19.6	5.0	*Quercus velutina–Pinus echinata/Rhus aromatica–Vaccinium pallidum/Carex–Schizachyrium scoparium*	1119	22.3	99	*Quercus alba, Quercus velutina, Quercus coccinea*	*Lespedeza hirta, Tephrosia virginiana*
3	Clarksville very gravelly silt loam, 35 to 50% slopes, very stony	East-facing, backslopes	18.3	5.0	*Quercus stellata–Pinus echinata/Vaccinium arboreum Schizachyrium scoparium–Desmodium*	894	18.6	83	*Quercus alba, Quercus velutina, Quercus stellata*	*Vaccinium arboretum, Desmodium rotundifolium*
4	Leon fine sand, very deep, poorly drained, 0 to 2% slopes	Atlantic Coastal Plain flatwoods	10.2	5.1	*Pinus palustris–Pinus serotina/Gaylussacia dumosa/Vaccinium crassifolium–Aristida stricta* Woodland	222	23.0		*Pinus taeda*	*Aristida stricta, Gaylussacia dumosa*
5	Nankin sandy loam, very deep, well drained, 5 to 12% slopes	Eastern Gulf Coastal Plain	20.1	5.0	*Pinus palustris/Schizachyrium scoparium/Verbesina aristata* Loam-hill Woodland	371	27.6		*Pinus taeda Quercus marilandica Quercus falcata*	*Schizachyrium scoparium Sericocarpus asteroides Solidago odora*

Restoration and management of savannas and woodlands can be conceptualized as occurring in three different steps: restoration, maintenance, and recruitment. Stand conditions assessed in the inventory determine the initial step (Table 11.2). The restoration step applies when initiating management in woodlands where tree and shrub density is high or loss of species of interest has occurred. The maintenance step occurs where structure and composition of the tree canopy have been restored and treatments occur as needed to remove woody reproduction and enhance cover and diversity of ground flora. The recruitment step occurs when regeneration and recruitment of new trees is necessary or desirable. Lack of tree regeneration in wood-pastures is a commonly recognized threat in Europe (Bergmeier *et al.* 2010).

Vegetation can be removed through burning, mechanical, chemical, and grazing (or browsing) treatments. However, fire may be contraindicated without an ecological or cultural history, and fire and chemical use may be regulated. Significant change may require several years and multiple applications of one or more methods. Factors such as minimizing soil compaction, logistics, and topography may affect choices. Removing biomass may offset the cost of implementation if the biomass is a sellable product and infrastructure exists. Methods for vegetation removal can prepare sites for species reintroduction, but removal of one species may favour establishment of unwanted species; thus the surrounding area, potential seed sources, and time a site is susceptible to invasion also are important considerations.

Where current overstorey tree recruitment levels are insufficient, seedlings can be established by direct seeding or by planting bareroot or container stock seedlings. In many cases, overplanting will be necessary to allow for browsing mortality. Survival often can be enhanced by using tree shelters, cages, root guards, mulch, weed barrier cloth, and thorny shrubs to reduce browse and conserve water. Savanna and woodland restoration objectives for understorey communities are similar to grassland restoration. Fire and/or removal of dry biomass may be critical to stimulating germination of seeds in the seed bank. If planting or direct seeding is used, plants must be adapted to edaphic and climatic site conditions. Multi-species seed

Table 11.2 Management triggers and tools for oaks in the eastern United States

Management step	Assessment tools
When to initiate the restoration step	
Suitable site conditions	Soil survey, on-site investigation
Presence of mature overstorey trees	Stand examination
Overstorey stocking > 75%	Inventory and stocking chart
Woodland indicators present	Woodland indicator species list
Presence of deep litter	Stand examination
When to initiate the maintenance step	
Overstorey stocking < 75%	Inventory and stocking chart
Understorey sparse or absent	Stand examination
Woodland indicators abundant	Woodland indicator species list
When > three prescribed burns have been conducted	Management plan
When to initiate the recruitment step	
Stand exceeds designated rotation age	Management plan
Overstorey mortality is excessive	Stand examination
Desirable advance reproduction present	Reproduction survey

mixtures should be planted because they maintain diverse plant communities that are resilient to multiple disturbances, limit invasive plants, and enhance wildlife habitat. Select the seed mixture carefully for purity, viability, and similar competitive ability. Planting time is critical to ensure there is sufficient precipitation for establishment.

Restoration in North America

Eastern and western oaks

Restoration goals for canopy tree abundance in savannas and woodlands can be based on measures such as tree density, canopy cover, and stocking, which will differ by region and community type. In mature eastern woodlands with large diameter (>30 cm) trees, there will be about 75 to 300 canopy dominant or codominant trees ha^{-1}, which corresponds to a desirable stocking level for closed woodlands that ranges from 55 to 75 per cent and for open woodlands that ranges from 30 to 55 per cent. In the Pacific Northwest, Campbell (2004) has suggested a goal of 5–30 per cent canopy cover for savannas and 30–60 per cent for woodlands; Dunwiddie *et al.* (2011) reported a historical density of about 80 canopy dominant trees ha^{-1} in a Garry oak/Douglas-fir woodland in Washington.

In stands with many small trees, higher stocking levels (55 to 70 per cent) should be maintained until woody sprouts have been controlled by prescribed fire. Higher stocking levels will allow some light penetration to the forest floor but reduce the growth rate of woody understorey species. If release of small trees is not a concern or if the overstorey is capturing most of the light, or after control of small tree growth, then considerable thinning or girdling of canopy trees may be necessary (Harrington and Devine 2006). Mechanical or chemical thinning from below to remove the midstorey trees first, followed by canopy trees if necessary, will reduce stand density to an overall density or stocking appropriate to woodland or savanna restoration goals (see stocking chart; Figure 11.2). Unlike mechanical thinning, chemical thinning prevents trees from resprouting.

Prescribed fire and thinning treatments may be applied to reduce understorey and midstorey density and leaf litter accumulation, increasing sunlight to the ground. Initially, prescribed fire may not be possible before other treatments, due to loss of fine fuels or presence of shrubs that will increase fire intensity. Prescribed fire may be applied annually during this step but is more typically applied every other year or every three years to protect mineral soil and retain soil organic matter, which is needed for maintaining water infiltration and plant growth. Low-intensity prescribed fires are sufficient for removing some leaf litter and killing aboveground growth of small trees but will have little effect on overstorey and midstorey trees. Higher intensity prescribed burns will remove more leaf litter and top kill (i.e. kill aboveground growth) shrubs and trees up to 25 cm in diameter, thereby reducing overall stand density by thinning from below. Growing season burns are more lethal to woody vegetation than dormant season burns although high humidity and green fuels can reduce the rate of spread or leave large areas unburned and are more difficult to conduct. Herbicide control of invasive species as well as extensive seeding of native species may be necessary.

After stand structure and ground flora objectives have been achieved through restoration treatments, management shifts in emphasis to monitoring and maintenance of structure and composition using prescribed fire and periodic thinning as needed. Prescribed fire can enhance ground flora and reduce woody plant density in the understorey and alter behaviour of subsequent prescribed fire. The shift from a fuel type dominated by leaf litter and seedlings to a fuel type dominated by forbs and grasses causes prescribed fire to burn with a greater flame length and intensity and with a greater rate of spread. Prescribed fire intensity in this step needs to be based on consideration that younger and smaller trees are more susceptible to losses in commercial value when fire-scarred years prior to commercial harvests because of the longer period for decay to occur to damaged cambium. If higher levels of stocking are acceptable, then only prescribed fire needs to be applied to maintain an open midstorey.

The stocking level can be reduced through commercial harvesting if there is sufficient merchantable material to warrant a timber sale and if impacts from logging activities do not damage the understorey. Otherwise, non-commercial thinning from below can meet desired stocking levels. As a general rule, stocking will increase by about 1.3 per cent per year when reduced to below the B level (Figure 11.2). On good sites this may be as much as 3 per cent and on poor sites as low as 1 per cent (Dale and Hilt 1989). Woodlands thinned to 30 per cent stocking can be expected to reach canopy closure, or B-level stocking, in about 20 years. Maintaining variation in stocking throughout the woodland is desirable and stocking can be adjusted to different levels depending on local soil conditions and slope position.

During the recruitment step, reduction of overall stocking will release oak seedlings that have accumulated during the maintenance step. Where economically viable, commercial timber harvests should be conducted. Distribution of retained trees can be varied so that there are locations within the stand with large openings and other locations where trees remain and stocking is greater. Thinning from below is recommended and retained trees should be large-crowned pines and oaks that provide habitat for wildlife and are considered 'character' trees for the woodland. Additionally, partial shade provided by retained trees will reduce woody regrowth surrounding residual trees. Prescribed fire should be excluded until a portion of the reproduction cohort is sufficiently large to escape being top-killed by fire.

Southeastern pines

The restoration step in longleaf pine savannas and woodlands often focuses on establishing longleaf pine as a future canopy tree, reducing midstorey densities, improving ground flora composition and abundance, and perpetuating desirable characteristics of forest fuels. Because longleaf pine occurs in <3 per cent of its historic range, artificial regeneration commonly is required, and development of container-grown nursery seedlings has greatly improved survival of planted longleaf pine seedlings (South *et al.* 2005). Conventionally, artificial regeneration is preceded by harvesting existing canopy trees, but underplanting is possible in existing pine forests, after reducing canopy density to 5–11 m^2 ha^{-1} basal area (Mitchell *et al.* 2006; Knapp *et al.* 2013).

Mechanical treatments, for example using rotary mowers or chainsaws, reduce midstorey densities and increase light to the forest floor but also result in resprouting. Herbicides such as hexazinone, imazapyr, or triclopyr may also be required to reduce woody stem density. Regardless of herbicides or mechanical treatments, prescribed fire is critical for restoring herbaceous ground flora and ecosystem function (Brockway *et al.* 2009; Martin and Kirkman 2009). Agricultural legacies present additional restoration challenges by changing ground flora composition and physical and chemical soil characteristics. Herbicides such as hexazinone and sulfometuron methyl may be needed to control non-desirable, 'old-field weeds' to improve longleaf pine seedling establishment (Ramsey *et al.* 2003). Many desirable, native ground flora species are long-lived perennials that are slow to recolonize following soil disturbance. If the seedbank is depleted, seeds of native species may be broadcast or nursery-grown plugs of native species may be planted throughout the restoration site.

In longleaf pine ecosystems, the maintenance and recruitment steps can occur simultaneously due to fire tolerance of longleaf pine seedlings and saplings. A frequent fire regime of 2–5 years generally is appropriate for limiting development of broadleaf trees, enhancing herbaceous ground flora, and allowing for longleaf pine regeneration and recruitment. Single-tree selection methods that also include removal of small groups of trees create heterogeneous forest structure that allows for spatially variable patterns of tree regeneration (Mitchell *et al.* 2006). Canopy openings of 0.1–0.25 ha are large enough to release longleaf pine seedlings into larger size classes. Woodland objectives are often compatible with extended rotation lengths, which provide habitat for wildlife specialists (e.g. the red-cockaded woodpecker) and can increase market opportunities for timber (e.g. telephone poles).

Western pines

Restoration of ponderosa pine savannas and woodlands requires removing unwanted woody vegetation and invasive species followed by restoring plant communities and function, including reintroducing disturbance (McPherson 1997). If prescribed fire is used to remove excess vegetation and enhance preferred species expansion, then experience, intuition, and uncertainty are part of the prescription because unique conditions will influence the outcome (Sackett *et al.* 1993, 1996). Nonetheless, success is dependent upon target tree size and species and fire intensity, which is a function of fuel load, relative humidity, daytime temperature, fuel moisture, soil heating, and rate of spread (Jain *et al.* 2012). The season affects plant life cycles and phenology at the time of prescribed fire, influencing post-fire community trajectory (Knapp *et al.* 2009; Kerns *et al.* 2006). If woody vegetation has increased surface litter and humus, harvest plus multiple prescribed fires may be necessary to return the forest floor to desired conditions. There are periods when wildlife cannot escape a fire (such as during birth and rearing), which can influence when prescribed fire is applied, but specific details about seasonality and impact are rare and often anecdotal (Pilliod *et al.* 2006; Knapp *et al.* 2009). For maintenance, varying time between prescribed fires reduces selection for certain species and increases potential for diversifying species composition (Knapp *et al.* 2009).

Chemical control generally is an effective way to remove invasive species and specific herbicides target a unique species or group of species. Desired vegetation must respond to release and the herbicide must match the species targeted for removal. Multiple applications may be needed to control sprouting species. Soil texture can decrease herbicide effectiveness; sandy soils may drain quickly while clayey or loamy soils can quickly immobilize a soil-active herbicide. Herbicides become impractical with large and tall vegetation. Vegetation does not die immediately and once dead, the stems potentially may create a fuel hazard. Costs of chemical methods will depend on acreage being treated, mode of application, and type and amount of herbicide.

In contrast to overgrazing, targeted grazing is a controlled approach to change species composition through herbivory. Success is dependent on matching the appropriate type of livestock to the targeted vegetation. Goats and other woody browsers are not useful for grass or invasive forb control. Use of multiple species over sequential periods and matching grazing period to when the target plant is most vulnerable will result in greatest benefit. For example, sheep consume flower heads of leafy spurge early in the year and cattle consume the plant during the grazing season (Burritt and Frost 2006).

Restoration of European wood-pastures

Large herbivores were important in management of wood-pastures in many parts of lowland Europe (Figure 11.5). Crucial aspects of use of large grazers (cattle, horse, and water buffalo)

Figure 11.5 Oak wood-pasture grazed traditionally by water buffalo, horse, and cattle in Transylvania. Small sized ponds such as the one in the front, left side of the picture are crucial for water buffalo, which contribute to maintenance of these ponds. Moreover, these ponds are critical breeding habitats for the yellow-bellied toad, an endangered European amphibian

include herbivore selectivity for certain habitat types and plants, density and temporal variation of herbivores or grazing pressure (animal units, AU ha^{-1} year^{-1}), trampling and seed dispersion by herbivores, various strategies of plants for coping with herbivores, existence of grazing refugia (where the grazing pressure decreases, at least temporarily), light requirements of plants, and nutrient content of the soil (Olff *et al.* 1999). A grazing pressure of approximately 0.20–0.50 AU ha^{-1} generally allows tree regeneration in wood-pastures, but in some contexts, grazing pressure of 0.1–0.2 AU ha^{-1} or even lower is preferable, particularly for lands enrolled in payment packages for environmentally friendly farming. At herbivore densities ≥ 0.5 ha^{-1}, woody vegetation will not regenerate outside of protective structures; grazing pressure in Oostvaardersplassen (5,600 ha, the largest and oldest rewilding area of Europe, situated in the Netherlands) is extremely high (currently approx. 2.6 AU ha^{-1}) and is formed by Heck cattle, Konik horses, and red deer that are regulated by their own density; tree regeneration only occurs under protective structures in such conditions. A grazing rotation scheme where parts of the grazed woodlands are delineated and grazing temporarily is prohibited allows tree regeneration, similar to non-continuous grazing practices such as transhumance shepherding in Spain that allows oak regeneration in dehesas.

Coppicing and pollarding also contributed to the maintenance of open woodlands. Some trees are more responsive to coppicing and pollarding (e.g. *Quercus* spp., *Carpinus betulus*, *Fagus sylvatica*, *Fraxinus* spp., *Salix* spp., *Betula* spp.) while other trees are less responsive (e.g. coniferous species lack basal buds). Coppicing now is considered a highly inefficient forestry practice for timber production. The economic output of coppices can be enhanced by lengthening the coppice rotation, selective crown thinning around the best formed individual stems, and clearing undergrowth and thinning to select best stems, creating an open silvopastoral woodland structure suitable for grazing by domestic animals (Buckley and Mills 2015).

Removal of (often shade-tolerant) trees to re-create open light conditions around large trees is called 'haloing'. Sudden release of overgrown trees potentially may affect trees and epiphytes by wind damage and water stress. Clearance in 2–3 operations over 10 years may be beneficial for oaks of at least 250 years if the oak is densely overgrown and immediate clearing (in one operation) when the oak is moderately overgrown (Johannesson and Elk 2005).

Challenges

Savannas and woodlands have a long history of low-intensity disturbance and management. Cessation or intensification of disturbance and management will transform the ecosystems into either treeless grasslands or closed forests. Once changed, restoration of savannas and woodlands may take decades, even if canopy trees are present, and great financial investment. Restoration of savannas and woodlands is challenging because restoration typically requires removing unwanted vegetation, returning disturbance patterns (type, severity, frequency), and establishing desired tree and understorey species. Even with intervention, restoration is not guaranteed; therefore, it may be a better use of limited resources to first maintain existing savannas and woodlands. Each place, management period, and social-economic context is unique, thus no specific guidelines exist. However, clear and specific objectives (e.g. desired future condition within an ecological, social, and economic context) will guide restoration prescriptions, including tool selection, implementation period, and specific treatment parameters.

Immediate challenges include formal recognition of savannas and woodlands as multifunctional ecosystems where both trees and the herbaceous layer are critical for biodiversity. Low-severity historical disturbance and management regimes produced open forest ecosystems rather than the current landscape pattern of closed forest and cleared land. Open forest ecosystems and clearcuts provide comparable habitat for many early-successional species and we suggest directing resources to maintenance and restoration of open forests, rather than creation of transient clearcuts that fragment forests. There is a need for closer collaboration between various institutional sectors (such as forestry and agriculture) to find economic incentives for multifunctional low-intensity use of savannas and woodlands. While economic returns from grazing are great enough to maintain (or often intensify) grazing, maintenance of open forests by mechanical harvesting often is unprofitable.

We have hope in finding resources to restore a frequent, low-intensity disturbance regime in some areas even as we recognize that disturbances have changed over the last 100 years and will continue to change into the future. We know that people who recognize historical ecosystems have, in some cases, written them off as having no place under current and developing stressors. However, current and coming environmental changes may provide reasons to restore rather than challenges to the need for restoration. For example, increasing aridity due to climate change in some areas are likely to generate conditions that favour savannas and woodlands over closed forests. Restoration, if done well, increases biodiversity and a variety of ecosystem services; therefore, forest management and restoration share principles of maintaining biodiversity, structure, and function that prepare forests for an uncertain future (Hanberry *et al.* 2015). It is easier to devalue restoration of ecosystems that historically were widespread than to commit to restoration of open forest ecosystems and diverse species associated with open forests.

References

Belsky, J. A. and Blumenthal, D. M. (1997). Effects of livestock grazing on stand dynamics and soils in upland forests of the interior west. *Conservation Biology* 11: 315–327.

Bergmeier, E., Petermann, J. and Schroder, E. (2010). Geobotanical survey of woodpasture habitats in Europe: diversity, threats and conservation. *Biodiversity and Conservation* 19: 2995–3014.

Brockway, D. G., Outcalt, K. W., Estes, B. L. and Rummer, R. B. (2009). Vegetation response to midstorey mulching and prescribed burning for wildfire hazard reduction and longleaf pine (*Pinus palustris* Mill.) ecosystem restoration. *Forestry* 82: 299–314.

Brose, P. H., Dey, D. C., Phillips, R. J. and Waldrop, T. A. (2013). A meta-analysis of the fire-oak hypothesis: Does prescribed burning promote oak reproduction in eastern North America? *Forest Science* 59: 322–334.

Buckley, P. and Mills, J. (2015). Coppice silviculture: from the Mesolithic to the 21st century. In K. Kirby and C. Watkins (eds), *Europe's Changing Woods and Forests: From Wildwood to Cultural Landscapes*, pp. 77–92. Wallingford: CABI.

Burritt, E. and Frost, R. (2006). Chapter 2: Animal principles and practices. In K. Launchbaugh (tech. ed.). *Targeted Grazing: A Natural Approach to Vegetation Management and Landscape Enhancement*, pp. 11–21. Centennial, CO: American Sheep Industry Association, A. Peischel and D. D. Henry Jr.

Campbell, B. H. (2004). *Restoring Rare Native Habitats in the Willamette Valley: A Landowner's Guide to Restoring Oak Woodlands, Wetlands, Prairies, and Bottomland Hardwood and Riparian Forests*. West Linn, OR: Defenders of Wildlife.

Dale, M. W. and Hilt, D. E. (1989). Estimating oak growth and yield. In F. B. Clark (tech. ed.) and J. G. Hutchinson (ed.), *Central Hardwood Notes*, pp. 5.03-1–5.03-6. St Paul, MN: US Department of Agriculture, Forest Service, North Central Forest Experiment Station.

Dunwiddie, P. W., Bakker, J. D., Almaguer-Bay, M. and Sprenger, C. (2011). Environmental history of a Garry oak/Douglas-fir woodland on Waldron Island, Washington. *Northwest Science* 85: 130–140.

Garbarino, M. and Bergmeier, E. (2014). Plant and vegetation diversity in European wood-pastures. In T. Hartel and T. Plieninger (eds), *European Wood-Pastures in Transition: A Social-Ecological Approach*, pp. 113–131. Abingdon: Earthscan from Routledge.

Gilliam, F. S. (2007). The ecological significance of the herbaceous layer in temperate forest ecosystems. *BioScience* 4: 845–858.

Gingrich, S. F. (1967). Measuring and evaluating stocking and stand density in upland hardwood forests in the Central States. *Forest Science* 13: 38–53.

Hanberry, B. B., Jones-Farrand, D. T. and Kabrick, J. M. (2014). Historical open forest ecosystems in the Missouri Ozarks: Reconstruction and restoration targets. *Ecological Restoration* 32: 407–416.

Hanberry, B. B., Noss, R. F., Safford, H. D., Allison, S. K. and Dey, D. C. (2015). Restoration is preparation for the future. *Journal of Forestry* 113: 425–429.

Harrington, C. A. and Devine, W. D. (2006). *A Practical Guide to Oak Release*. General Technical Report PNW-GTR–666. Portland, OR: US Department of Agriculture, Forest Service, Pacific Northwest Research Station.

Hartel, T., Dorresteijn, I., Klein, C., Máthé, O., Moga, C. I., Öllerer, K., Roellig, M., von Wehrden, H. and Fischer, J. (2013). Wood-pastures in a traditional rural region of Eastern Europe: Characteristics, management and status. *Biological Conservation* 166: 267–275.

Hartel, T., Plieninger, T. and Varga, A. (2015). Wood-pastures in Europe. In K. Kirby and C. Watkins (eds), *Europe's Changing Woods and Forests: From Wildwood to Cultural Landscapes*, pp. 61–76. Wallingford: CABI Publishing.

Hunter, W. C., Buehler, D. A., Canterbury R. A., Confer, J. L. and Hamel, P. B. (2001). Conservation of disturbance-dependent birds in eastern North America. *Wildlife Society Bulletin* 29: 440–455.

Jain, T. B., Battaglia, M. A., Han, H., Graham, R. T., Keyes, C. R., Fried, J. S. and Sandquist, J. E. (2012). *A Comprehensive Guide to Fuel Management Practices for Dry Mixed Conifer Forests in the Northwestern United States*. Gen. Tech. Rep. RMRS-GTR–292. Fort Collins, CO: US Department of Agriculture, Forest Service, Rocky Mountain Research Station.

Johannesson, J. and Elk, T. (2005). Multi-purpose management of oak habitats. County administration of Östergötland, report. 104 p

Kabrick, J. M., Zenner, E. K., Dey, D. C., Gwaze, D. and Jensen, R. G. (2008). Using ecological land types to examine landscape-scale oak regeneration dynamics. *Forest Ecology and Management* 255: 3051–3062.

Kerns, B. K., Thies, W. G. and Niwa, C. G. (2006). Season and severity of prescribed burn in ponderosa pine forests: implications for understorey native and exotic plants. *Ecoscience* 13: 44–55.

Kirby, K. and Perry, S. (2014). Institutional arrangements of wood-pasture management. In T. Hartel and T. Plieninger (eds), *European Wood-Pastures in Transition: A Social-Ecological Approach*, pp. 254–270. Abingdon: Earthscan from Routledge.

Knapp, E. E., Estes, B. L. and Skinner, C. N. (2009). *Ecological effects of prescribed fire season: a literature review and synthesis for managers*. Gen. Tech. Rep. PSW-GTR–224. Albany, CA: US Department of Agriculture, Forest Service, Pacific Southwest Research Station.

Knapp, B. O., Wang, G. G. and Walker, J. L. (2013). Effects of canopy structure and cultural treatments on the survival and growth of *Pinus palustris* Mill. seedlings underplanted in *Pinus taeda* L. stands. *Ecological Engineering* 57: 46–56.

Martin, K. L. and Kirkman, K. L. (2009). Management of ecological thresholds to re-establish disturbance-maintained herbaceous wetlands of the south-eastern USA. *Journal of Applied Ecology* 46: 906–914.

McPherson, G. R. (1997) *Ecology and Management of North American Savannas*. Tucson, AZ: University of Arizona Press.

Mitchell, R. J., Hiers, J. K., O'Brien, J. J., Jack, S. B. and Engstrom, R. T. (2006). Silviculture that sustains: the nexus between silviculture, frequent prescribed fire, and conservation of biodiversity in longleaf pine forests of the southeastern United States. *Canadian Journal of Forest Research* 36: 2724–2736.

Mitchell, R. J., Hiers, J. K., O'Brien, J. and Starr, G. (2009). Ecological forestry in the southeast: understanding the ecology of fuels. *Journal of Forestry* 107: 391–397.

Olff, H., Vera, F. W. M, Bokdam, J., Bakker, E. S., Gleichman, J. M., de-Maeyer, K. and Smit, R. (1999). Shifting mosaics in grazed woodlands driven by the alternation of plant facilitation and competition. *Plant Biology* 1: 127–137.

Paltto, H., Nordberg, A., Norden, B., and Snall, T. (2011). Development of secondary woodland in oak wood pastures reduces the richness of rare epiphytic lichens. *PLoS ONE* 6: e24675.

Peet, R. K. (2006). Ecological classification of longleaf pine woodlands. In S. Jose, E. J. Jokela, D. L. Miller (eds), *The Longleaf Pine Ecosystem: Ecology, Silviculture, and Restoration*, pp. 51–93. New York: Springer.

Pilliod, D. S., Bull, E. L., Hayes, J. L. and Wales, B. C. (2006). *Wildlife and Invertebrate Response to Fuel Reduction Treatments in Dry Coniferous Forests of the Western United States: a Synthesis*. Gen. Tech. Rep. RMRS-GTR-173. Fort Collins, CO: US Department of Agriculture, Forest Service, Rocky Mountain Research Station. 34 p.

Plieninger, T., Hartel, T., Marin-Lopez, B., Beaufoy, G., Kirby, K., Montero, M. J., Moreno, G., Oteros-Rozas, E. and Van Uytvanck, J. (2015). Wood-pastures of Europe: Geographic coverage, social-ecological values, conservation management, and policy. *Biological Conservation* 190: 70–79.

Ramsey, C. L., Jose, S., Brecke, B. J. and Merritt, S. (2003). Growth response of longleaf pine (*Pinus palustris* Mill.) seedlings to fertilization and herbaceous weed control in an old field in southern USA. *Forest Ecology and Management* 172: 281–289.

Sackett, S., Haase, S. and Harrington, M. G. (1993). Restoration of southwestern ponderosa pine ecosystems with fire. In W. W. Covington and L. F. DeBano (eds), *Sustainable Ecological Systems: Implementing an Ecological Approach to Land Management*, pp. 115–121. Gen. Tech. Rep. RM-247. Fort Collins, CO: US Department of Agriculture, Forest Service, Rocky Mountain Forest and Range Experiment Station.

Sackett, S. S., Haase, S. M. and Harrington, M. G. (1996). Lessons learned from fire use for restoring southwestern ponderosa pine ecosystems. In W. Covington, P. K. Wagner (technical coordinators), *Conference on Adaptive Ecosystem Restoration and Management: Restoration of Cordilleran Conifer Landscapes of North America: Proceedings*, 6–8 June 1996, Flagstaff, AZ, Gen. Tech. Rep. RM-GTR–278, pp. 54–61. Fort Collins, CO: US Department of Agriculture, Forest Service, Rocky Mountain Forest and Range Experiment Station.

Sandom, C. J., Ejrnas, R., Hansen, M. D. D. and Svenning, J.-C. (2014). High herbivore density associated with vegetation diversity in interglacial ecosystems. *PNAS* 111: 4162–4167.

South, D. B., Harris, S. W., Barnett, J. P., Hainds, M. J. and Gjerstad, D. H. (2005). Effect of container type and seedling size on survival and early height growth of *Pinus palustris* seedlings in Alabama, U.S.A. *Forest Ecology and Management* 204: 385–398.

Taboada, A., Kotze, D. J., Salgado, J. and Tarrega, R. (2011). The value of semi-natural grasslands for the conservation of carabid beetles in long-term managed forested landscapes. *Journal for Insect Conservation* 5: 573–590.

Vera, F. W. M. (2000). *Grazing Ecology and Forest History*. Wallingford: CABI Publishing.

Vesely, D. and Tucker, G. 2004. *A Landowner's Guide for Restoring and Managing Oregon White Oak Habitats*. Salem, OR: US Department of the Interior.

12

RESTORING DESERT ECOSYSTEMS

Scott R. Abella

Introduction

Deserts include arid and semi-arid lands where evaporation plus transpiration (water lost from plants to the air) exceeds precipitation. The 2005 Millennium Ecosystem Assessment categorized dryland climates as hyper-arid, arid, semi-arid and sub-humid (Table 12.1). This classification was based on an aridity index, defined by how much greater evapotranspiration is than precipitation. Hyper-arid lands have at least 20 times more evapotranspiration than precipitation, indicating a moisture deficit. For comparison, humid climates often have less evapotranspiration than precipitation.

Hyper-arid, arid and semi-arid lands occupy one-third of Earth's 147,573,197 km² land area (Safriel *et al.* 2005). These drylands occupy most of Australia and Africa, much of Asia and western North America and dry portions of South America (Figure 12.1). Twenty per cent of Earth's human population lives in deserts, which increases to over one-third if drought-susceptible

Table 12.1 Classification of dryland climates and their characteristics based on the Millennium Ecosystem Assessment

Climate	Aridity index[a]	Major biome	Earth's land (%)	Human population (%)[c]
Hyper-arid	> 20:1	Desert	7	2
Arid	20:1 to 5:1	Desert	11	4
Semi-arid	< 5:1 to 2:1	Grassland	15	14
Sub-humid[b]	< 2:1 to 1.5:1	Woodland	9	15

Notes: [a] Ratio of annual evapotranspiration to precipitation. Evapotranspiration is the sum of evaporation and transpiration (water lost from plants to the air). All of these dry climates have ratios greater than 1, as evapotranspiration exceeds precipitation. This means there is a moisture deficit. Polar regions can also be dry but are not considered in this chapter as deserts.

[b] Sub-humid environments often support trees and savanna grasslands, but are sometimes grouped with deserts because they are susceptible to degradation during droughts.

[c] Percentage of the total human population living in each climate.

Source: Safriel *et al.* (2005)

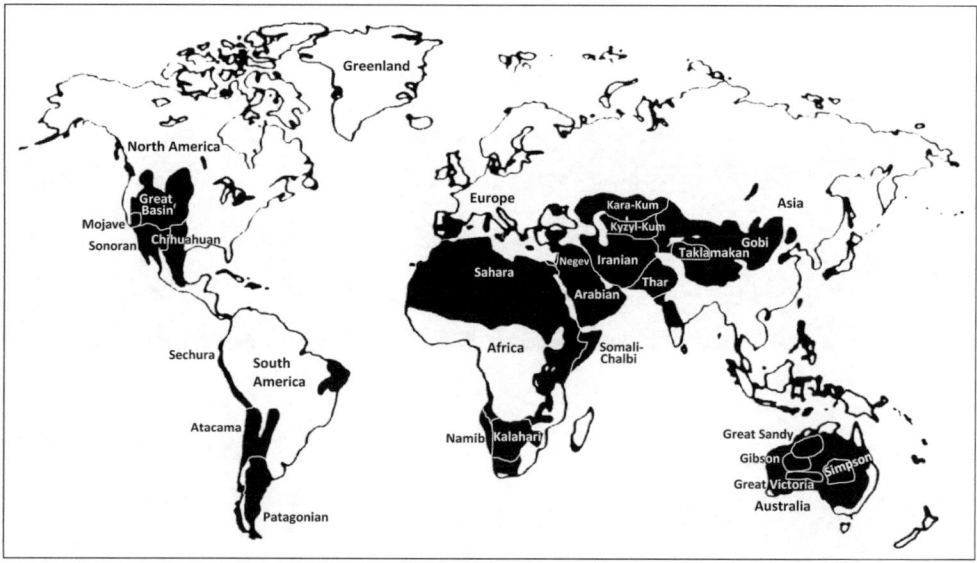

Figure 12.1 Location of hyper-arid, arid and semi-arid lands (all shaded in black), with examples of specific deserts highlighted

Source: map adapted from Ffolliott *et al.* (2001)

sub-humid lands are included. In addition to providing human habitat, 8 of 25 global biodiversity 'hotspot' regions are in drylands (*ibid.*). Despite their relatively low productivity, deserts (including sub-humid lands) store 27 per cent of Earth's soil organic carbon (*ibid.*). Remarkably, deserts further store 97 per cent of Earth's entire soil inorganic carbon. These examples illustrate that degradation of desert land, and its reversal through restoration, has both local and global implications.

While an idealized vision of a desert might be a hot, dry, sparsely vegetated land with sand dunes, this fits only a portion of the world's deserts. For example, the Great Basin Desert in western North America is cold most of the year, with freezing temperatures common, and receives snow (Figure 12.2). Climates within and among deserts are variable, with some characterized by greater precipitation or different seasonal patterns of rainfall than other deserts (Whitford 2002). A given desert can experience shifts in aridity through time with changes in climate or human land uses, such as increasing evaporation through alterations to the soil surface. A commonality is that ecological restoration is challenging in deserts, because deserts represent extremes of Earth's climates and precipitation is unreliable. It is not uncommon for some deserts to receive little or no rainfall for an entire year, or even multiple years.

Desert ecology principles paramount to restoration

Three principles of desert ecology germane to restoration include: (1) extreme spatial and temporal patterning of resource availability (e.g. water and nutrients), (2) unique nature and speed of vegetation change, and (3) prevalence of herbivory (eating of plant matter by animals) and granivory (eating of seeds by animals). An order of magnitude variation in the

Figure 12.2 Examples of arid and semi-arid ecosystems: (a) Great Basin Desert, a cold desert in the western United States, with the photo taken in Nevada with snow on surrounding hills; (b) Sonoran Desert of the southwestern United States and Mexico, showing the columnar cactus giant saguaro, with the photo taken in Arizona; (c) transitional semi-arid woodland in Western Australia; and (d) Gurbantunggut Desert northeast of Urumqi, Xinjiang Uygur Autonomous Region, northwestern China, with the dark area on the right side of the slope covered with biological soil crust

Source: photos by S. R. Abella

concentration of soil nutrients within a few meters of horizontal space is common in deserts (Padilla and Pugnaire 2006). This spatial patterning of nutrient availability often results from the distribution of perennial plants, below which soil nutrients accumulate. These 'fertile islands' also trap dust and seeds, provide shade, ameliorate extreme weather, and often support high biological activity compared to the interspaces between perennial plants. Distribution of large shrubs can partly regulate recruitment of annual and other perennial plants, because some plant species depend on fertile islands for germination and seedling establishment (Abella and Smith 2013).

Similar to their spatial patterning of biological activity, deserts are described as 'pulse systems' because brief periods of resource availability (e.g. following rainstorms) can influence an entire year's plant and animal activity (Whitford 2002). The extreme temporal variability of deserts can drive trajectories of desert restoration projects. A well-timed rain, for instance, could trigger plant establishment early in a restoration project to initiate a persistent trajectory that hinged upon presence of perennial plants. On the other hand, dry conditions could result in loss of restoration materials (e.g. seeds) and minimal restoration success.

Deserts may seem recalcitrant to change, but this masks appreciable short- and long-term change that can occur in desert vegetation. Vegetation change in deserts after disturbance does not necessarily fit an idealized 'succession' described for moister regions (Abella 2010). For example, a post-disturbance succession after removal of temperate forest may include initial colonization by annual plants, followed by perennial grasses or shrubs, and then trees to form a forest largely lacking annual plants. In deserts, annual plants are components of both early colonizing and mature ecosystems, though annuals may only be present in high-rainfall years. Species in some desert shrublands and grasslands re-sprout if their tops are killed. Thus, species of the mature community re-establish directly, without an intervening early colonizing community, not fitting traditional succession as a transition from one community to another. In other cases, short-lived perennial species, uncommon in mature vegetation, can initially colonize, eventually giving way to re-establishment of species of the mature community. While brief droughts can rapidly alter vegetation through death of certain perennial species, the pace of plant colonization after disturbance is generally slower in deserts than in moister environments.

Herbivory and granivory in certain deserts is extreme. While herbivory can have a major influence in temperate ecosystems, it can be so extreme in deserts as to remove plant cover completely, because there is little forage to begin with and plants grow slowly. One example is the Arabian Desert in the Middle East, where domestic camels almost remove plant cover entirely (Figure 12.3). Restoration will fail unless herbivory is accounted for or restoration sites are protected. Similarly, insects, mammals, and birds can consume large quantities of seeds, including those intended for restoration. The importance of granivory varies among deserts (Brown *et al.* 1979), and where it is prevalent, is a major consideration for restoration.

Figure 12.3 Context of restoration in the Arabian Desert of Kuwait: (a) fencing is essential to protect restoration sites from domestic camel grazing; (b) gazelle, one of several species of large animals that have been reduced or eliminated in the Arabian Desert; (c) 'oil lake' resulting from the Gulf War; (d) shrubland developing free from camel grazing inside a protected area, one of several possible reference communities for restoration

Source: photos by S. R. Abella

Why is restoration needed in deserts?

Deserts have long been used by humans for extractive purposes that reduce long-term productivity or generate negative off-site impacts, stimulating restoration as a land management tool. Many deserts are endowed with valuable rocks and minerals, making mining common in deserts. Restoration is mandated after mining in Australia, for example, to reduce negative off-site impacts (e.g. generation of dust), re-establish biodiversity, and maintain land productivity for other possible uses (Commander *et al.* 2013). Dryland agriculture has occurred in many deserts, with farms frequently abandoned or with declining productivity over time. Restoration may help stabilize eroding soil, or be performed in certain locations (such as near water sources) to help sustain productivity within a working agricultural landscape. Rangeland for grazing livestock is the most extensive land use of deserts and has been practised for thousands of years in some deserts such as the Arabian. Overgrazing is a leading cause of desertification, which is land degradation in arid lands that reduces land productivity and typically makes deserts even drier (e.g. by reducing capacity for rainwater to soak into soil). Humans also commonly withdraw groundwater or alter surface water flow, both of which can affect plant productivity and spur restoration. Invasion by non-native plants in certain deserts, such as the Mojave Desert in the United States, has changed desert fuels and corresponded with increased extent of wildfires. These fires destroy mature desert vegetation. Numerous other disturbances – such as roads no longer needed – are environments where restoration is conducted in deserts. Desired functional outcomes, like reducing dust or enhancing habitat for conservation-priority wildlife species, can also spur restoration. Degraded arid lands are missing key ecological functions and are liabilities locally and globally through their influence on the atmosphere.

Restoration goals and reference conditions

Considerations for whether to use restoration and guidelines for conducting restoration projects in deserts are similar to those for other ecosystems (Society for Ecological Restoration 2004). A first task is identifying if or how an area is degraded, such as based on whether key ecosystem components (e.g. surface soils, perennial plants) are missing. Another task is determining what type (or combination) of management is most appropriate to reverse the degradation. Restoration is just one of numerous possible management interventions along a continuum of management options. For example, constructing a fence to reduce livestock grazing and promote plant growth would usually be considered a management action but not active ecological restoration. Fencing plus actively performing treatments to assist soil recovery or planting native plants inside the exclosure, however, would often be considered restoration. Land management activities like habitat creation or land reclamation (e.g. using non-native plants for habitat or soil stabilization) can be useful for particular management objectives but should not be termed desert restoration.

As in temperate ecosystems, reference conditions underpin desert restoration. Reference conditions represent our understanding of the ecological conditions (e.g. species present, depth of the soil, and natural types of disturbance such as flooding) characterizing ecosystems relatively free from degradation. These conditions can be based on knowledge of an ecosystem before it was degraded, nearby less-degraded sites, or derived through modelling ecological processes like losses or gains of soil nutrients.

The extreme temporal variation of deserts complicates evaluations of reference conditions and care must be used to account for this. Long-term vegetation monitoring in Joshua Tree National Park in the Mojave Desert provides an example. Miriti *et al.* (2007) inventoried perennial plants in a 1 ha plot every five years from 1984 to 2004. The density of adult plants of the large shrub creosote bush (*Larrea tridentata*) changed little during the 20-year period, but populations of other shrubs and perennial forbs fluctuated (Figure 12.4). The shrub white bursage (*Ambrosia dumosa*) had a relatively constant number of individuals (1,555 to 1,714/ha) during the 1984 to 1999 inventories, but then plummeted to only 523 individuals in 2004. The forb desert globemallow (*Sphaeralcea ambigua*) had densities of 50 to 81 plants/ha between 1984 and 1999 before completely disappearing in 2004. Droughts occurred at the study site between 1988 and 1991 and 1999 and 2003. While the every-five-year plant census could not detect annual fluctuations, the 2004 inventory reflected mortality of perennial plants associated with the severe 1999–2003 drought (Miriti *et al.* 2007). If reference conditions at this site were assessed only in 2004, we would underestimate densities of most species relative to the previous 20 years and completely 'miss' desert globemallow. This example shows how recent weather could influence our perception of reference conditions and that stability of plant populations (including through droughts) varies among desert perennial species.

Techniques for restoring components of desert ecosystems

Particular components of desert ecosystems might be degraded or missing. Only one component might require restoration if the rest of the ecosystem is healthy, or several components could require restoration within a comprehensive ecological framework. The next sections discuss techniques for components most commonly requiring restoration.

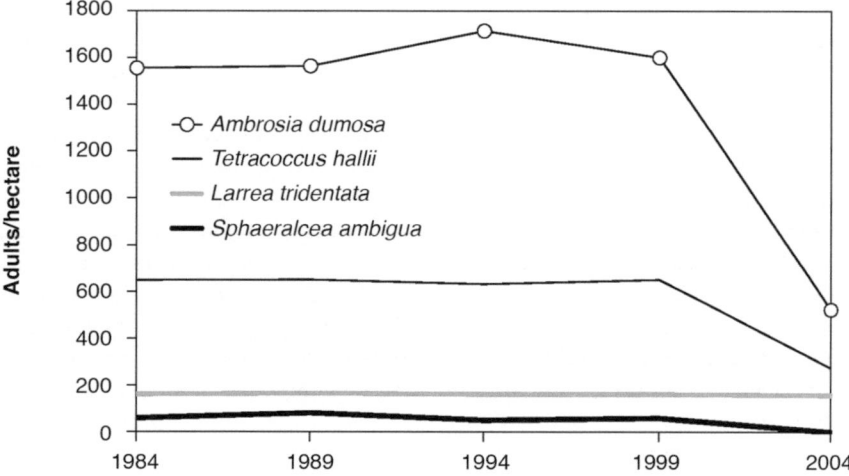

Figure 12.4   Fluctuating density and species composition of the perennial plant community of a site in Joshua Tree National Park, Mojave Desert, United States. Mortality of certain perennial species corresponded with extended dry periods, making consideration of variable weather patterns important in assessing reference conditions

Source: data from Miriti *et al.* (2007)

Soil

Soil formation is slower in deserts than in temperate ecosystems, and salvaging topsoil from a disturbed area for later reapplication can be among the most effective restoration strategies. The upper 5 cm of soil can contain almost all the viable seed bank of desert soils (Guo *et al.* 1998). Similarly, the upper 10 cm stores much of a desert soil's plant-available nutrients, organic matter, and microorganisms. Because of these concentrated resources in the topsoil, applying salvaged topsoil accelerated plant colonization better than seeding at decommissioned quarries in the Namib Desert (Burke 2008).

Several techniques can help increase efficiency and success of salvaging desert soils. Properties of the soils should be considered, including avoiding salvaging toxic soils and soils within or near infestations of non-native plants. Timing salvage to correspond with the end of maximum dispersal of native seeds within a year might help maximize retention of native seed banks. Salvaging only the upper 5 to 10 cm of soil can be critical in many deserts to avoid 'diluting' the organic-rich topsoil with subsoil (Scoles-Sciulla and DeFalco 2009). Ideally, topsoil would not be stored at all and immediately applied to a restoration site. This is rarely feasible, necessitating topsoil be stored. In the arid land of India, stored topsoil lost 27 per cent of its nitrogen during a year of storage and another 10 per cent during the second year (Ghose 2001). Keeping stockpiles short, perhaps less than 60 cm tall to avoid creation of anaerobic conditions, might help increase longevity of biotic components. Furthermore, treatments such as covering stockpiles in water-proof tarps or even planting vegetation on piles can extend longevity of soil biota if soil must be stored for a long time (Golos and Dixon 2014). Strategically re-applying topsoil where it is needed most and matching soil types with the donor and recipient sites likely makes the best use of limited amounts of salvaged soil.

Compacted and eroded soils are common legacies of arid land degradation. These soils are problematic because retention of seed and fine soil particles, holding of water, and accumulation of soil organic matter are compromised. Roughening compacted soil using machinery or hand tools can break up impermeable surfaces and create indentations to trap water and organic matter (Bainbridge 2007). In areas of severe wind erosion, constructing a series of parallel barriers (e.g. short fences) can slow wind speed and result in deposition and retention of wind-blown soil. Covering soil with mulches and organic matter also can help stabilize eroding soils and alleviate compaction. Materials used for cover can include mats, nets or networks of cylindrical structures (e.g. sticks) pushed into the soil, all of which can be of organic origin and biodegrade. Materials for mulching can include straw, wood chips, charcoal or gravel.

Trade-offs of different cover and mulch materials need to be considered. For example, wood chips can make soil nutrients unavailable to plants by providing food for soil microbes, which then uptake nutrients. This could be desirable for reducing nutrient-loving, non-native plants, but undesirable if growth of native plants is harmed. On the other hand, reduced nutrient availability might be compensated for by increased moisture retention under protective mulch. Whichever techniques are used, a key for restoration on compacted or eroding soil is promoting retention of natural soil structure and organic matter, facilitating establishment of plants and soil biota for long-term soil health.

Biological soil crusts

Located on or near the soil surface, biological soil crusts include bacteria, cyanobacteria, algae, mosses, liverworts, fungi and lichens (West 1990). Crusts in particular deserts or sites within deserts may only contain one or a few of these components. Moreover, the composition of crust can change during ecosystem development. Bacteria are often the initial colonizers after disturbance, and lichens the last. Not all desert soils have crusts – crusts are typically absent on naturally unstable soils (e.g. ephemeral stream channels, shifting sand dunes) and on soil perpetually disturbed by human activities, including through off-road vehicles or livestock grazing. Where crusts are found, they are major features of desert soils and contribute to soil stability, nutrient cycling, interactions with vascular plants (variously enhancing or reducing vascular plant establishment), soil temperature regimes and biodiversity. Loss of soil crusts can de-stabilize soil, accelerating soil erosion, and result in loss of carbon storage and nutrients.

Most research on biological soil crusts has focused on how disturbance affects them, but recent attention has turned toward restoring crusts (Bowker 2007). One example is in the Gobi Desert of Inner Mongolia, China. Wang *et al.* (2009) collected algae-dominated soil crust from relatively undisturbed sites, propagated the algae in a greenhouse, and created a slurry to increase the volume of inoculum. The slurry was then applied to eroding soil at field sites. Within one year, cyanobacteria covered 42 per cent of the soil surface (compared to 0 per cent on controls). Between the first and third years, different species of cyanobacteria colonized the soil surface, and mosses colonized by the second year. The resulting formation of biological soil crust changed soil functions. Soil organic carbon was < 1 g/kg in controls without inoculum, but was 9 g/kg by the third year on plots receiving inoculum. The cover of vascular plants also increased as the soil crusts formed. Effective restoration treatments are likely to vary with the type of crust, stability and properties of the underlying soil, and resources available.

Perennial plants

Perennial plants are a primary component targeted for restoration in deserts because of their importance to soil formation and other biota. Perennial plants can be restored through planting nursery-grown seedlings (outplanting), salvaging plants before disturbance for later replanting, seeding or facilitating natural recruitment (Abella and Newton 2009). Successful outplanting projects start with quality seed and seedlings grown under conditions appropriate for the species. At least 6 to 12 months of growth in a nursery are commonly required for seedlings to develop root systems providing the best chance of survival when planted at restoration sites. This makes advance planning, sometimes two years or more, necessary to collect seed and grow plants. Seedlings can be grown in a variety of containers, such as round plastic pots, deeper rectangular pots and biodegradable materials that can be planted without having to remove the soil and roots (Bainbridge 2007).

Understanding the ecology of planting sites and the plant species is vital to identify treatments required to enhance plant establishment. For example, herbivory is often intense, requiring that planting sites be fenced to exclude large herbivores or that individual plants be enclosed in shelters. Shelters, such as plastic cylinders, can provide protection from herbivory and ameliorate hot, dry weather. Other treatments, including irrigation, may be evaluated on a cost/benefit basis relative to simply planting more untreated plants. Irrigation is typically difficult to implement at remote restoration sites and can have unintended effects, such as promoting non-native plants and wetting biological soil crusts during time periods detrimental to their growth. With due consideration of these types of trade-offs, strategically delivering water or using slow-release irrigation gels (e.g. DriWater) has increased plant establishment at remote sites such as Saudi Arabian semi-arid woodlands (Aref *et al.* 2006). Not all species require treatments such as irrigation or shelter, so identifying species-specific needs helps increase efficiency of restoration.

Salvaging plants and facilitating natural recruitment both make use of existing plant material on site and can be cost-effective revegetation options. Unless salvaging at the donor and planting at the recipient site can be performed in one operation, the salvaged plants require care in nurseries similar to outplanting. Protecting naturally recruited seedlings to enhance their survival and growth is little studied but may be particularly suited for species difficult to propagate or where transporting plants is difficult.

Seeding perennial species is difficult, owing to infrequent conditions suitable for plant establishment, similar to why natural recruitment events are rare (Abella et al. 2012). Germination ecology is poorly understood in many deserts. Enhancing knowledge of germination ecology underpins identifying which species are amenable to seeding and what treatments are required for success. A general approach for enhancing seeding success includes: collecting quality seeds and storing them appropriately unless used immediately, conducting any treatments required to enhance germination, implementing any species-specific treatments for protecting seeds (e.g. coating them) or improving soil substrates, and timing seeding optimally within a year to coincide with favourable conditions. While still no guarantee of success, these procedures can provide the best chance for success if weather conditions are favourable for plant establishment. Additionally, pairing outplanting with seeding might maximize chances that some plants become established.

Annual plants

Annual plants can be restored through seeding and indirectly through establishing perennial plants or improving site conditions. Perennial plants in natural desert ecosystems typically facilitate recruitment of annual plants, and diversifying the perennial plant community can increase diversity of annual plants (Abella and Smith 2013). Improving knowledge of annual plant germination ecology and restoration is particularly important in changing desert climates, as the frequency and timing of weather conditions suitable for germination may shift.

Hydrology and springs

Humans have manipulated water flow in deserts for thousands of years to increase retention of water or concentrate it in certain areas for agriculture or domestic purposes. One example is the Negev Desert in Israel, where ancient rainwater harvesting systems of terraced hillsides, small canals, conduits (low rock structures to direct water) and cisterns remain visible on the landscape (Evenari et al. 1982). Some of these millennia-old water harvesting techniques are employed in contemporary restoration to retain water on site and enhance infiltration into soil (Bainbridge 2007).

Three of the main surface water hydrological patterns in deserts include: (1) water flow across the soil surface as sheet flow, (2) ephemeral stream channels flowing after rains, and (3) permanently flowing rivers and springs. Human disturbances that disrupt sheet flow include soil compaction that limits infiltration of some of the flowing water into the soil; tracks from off-road vehicles that form artificial flow paths; disturbances like fire or oil spills that create hydrophobic layers; and roads that divert flow or concentrate it into a few pathways. Ephemeral stream channels are disrupted by land-clearing disturbance and roads that cut across channels and sever connectivity (and without culverts to allow water to pass under the road). In addition to restoring plants and soils, several tactics can help re-establish natural drainage patterns. Decommissioning unwanted roads and removing roadside berms can stop roads from diverting surface flow and severing stream channels. Creating small water catchments on disturbed soil surfaces can retain water and provide locations for plant recruitment.

Desert springs and rivers are special cases in desert restoration because of their important functions and because availability of water is not necessarily as limiting at these sites as it is elsewhere. Major ways humans have altered springs is by piping water away from springs, excavating soil around springs in an attempt to store water, or pumping groundwater to lower the water table. In these circumstances, a main goal of restoration is reconfiguring how water is distributed on the site. For springs that have been piped or excavated, removing the pipes or filling in storage basins may be an initial restoration step. A lowered water table makes restoration of natural surface flow particularly difficult. Partial restoration may still be possible by re-contouring to lower the elevation of the spring to the new level of the water table, allowing outflow to the surface. At many springs, restricting access of livestock to project sites may be critical to allow re-establishment of plants, and this likely requires balancing access of native animals to the water source.

Restoring surface water in deserts can produce tremendous ecological benefits. For example, Patten *et al.* (2008) developed models for transitions from dry upland plant species composition to wetland plant composition based on changing the depth to the water table even slightly, at one-metre increments. Similarly, high biodiversity in aquatic invertebrates, fish and birds might depend on permanency of flowing springs and associated wetlands in central Australian arid lands (Box *et al.* 2008). Oases occupy only small portions of desert landscapes, but have a disproportionate restoration potential for enhancing biodiversity and ecosystem functions.

Animals

Several species of large animals are native to Middle Eastern deserts and provide examples of the challenges and opportunities for restoring large desert animals. Two examples are the gazelle (*Gazella* spp.) and the Arabian oryx (*Oryx leucoryx*) in the Arabian Desert, spanning several countries such as United Arab Emirates, Kuwait and Saudi Arabia. Gazelles are small antelopes that are herbivores and fast runners – some can run 100 km/hour in short bursts. The Arabian oryx is a medium-sized antelope, weighing about 70 kg, which roamed in herds searching for plants following rains and can go several weeks without water. Gazelles and oryx were hunted by humans for food for thousands of years, but catastrophic declines largely began in the early 1900s with the introduction of rifles, vehicles and intensified livestock grazing (Thouless *et al.* 1991). A major land use currently affecting gazelle and oryx, and their potential for restoration, is grazing by domestic camel. For example, the United Arab Emirates contained 250,000 mostly free-roaming camels in the mid-2000s, or three camels/km^2 (El-Keblawy *et al.* 2009). Camels today are kept primarily for racing, as camel racing competitions are a major cultural activity in the Arabian Desert.

Even where hunting and harassment by humans of gazelle and oryx have been curtailed, the animals face a situation of sparse to non-existent forage on camel rangelands. As a result, the main conservation approach has been to create fenced reserves and reintroduce gazelle and oryx inside. One example is the 225-km^2 fenced Dubai Desert Conservation Reserve, established by 2003 and occupying 5 per cent of Dubai (El-Keblawy *et al.* 2009). Camels still grazed in much of the reserve, but a 27-km^2 portion had only gazelles and oryx. As has been observed in some other fenced reserves containing only native animals, vegetation composition quickly differentiated between the areas open to camel grazing and those only open to native animals. The small shrubs ramram (*Heliotropium kotschyi*) and rattlebox (*Crotalaria aegyptiaca*) were only found with native animals, whereas the unpalatable large shrub rimth (*Haloxylon salicornicum*) and the sedge thenda (*Cyperus conglomeratus*) dominated camel rangelands.

Restoration of these animals must also fit the landscape context, as an example from central Saudi Arabia illustrated. The 2,200-km^2 Mahazat as-Sayd Protected Area was fenced in 1988 and oryx and gazelle were reintroduced in the 1980s and 1990s (Islam *et al.* 2010). Mortality of the animals was heavy in dry years, with 560 oryx and 2,815 gazelle dying between 1999 and 2008. Historically, populations likely moved hundreds of kilometres to locate forage after localized rains. Despite the relatively large size of the fenced reserve, the fence prevents natural roaming of the animals over a much larger area. To partly offset this, revised management plans included providing supplemental forage and water within strategic locations of the reserve (*ibid.*).

Case studies

Western Australia arid zone

Mining is a major land use in arid lands of Australia, and restoration is commonly performed after mining. One study recently compared outplanting and seeding for restoration on disturbed borrow pits at a mine on the Edel Peninsula, within the Shark Bay World Heritage Area (Commander *et al.* 2013). The area receives 22 cm/year of rainfall, and natural vegetation consists of desert shrubland. Based on a reference condition of nearby, undisturbed vegetation, seed was collected of three shrub species (*Acacia tetragonophylla*, *Atriplex bunburyana* and *Solanum orbiculatum*) in sufficient quantity for direct seeding and for propagating some plants in a nursery. To roughen the soil surface, the borrow pits were ripped with a grader. Some areas then received seed broadcast on the soil surface, while others received outplanting. Rainfall was only 68 per cent of average during the two-year study.

Researchers concluded that on these relatively small sites (~ 1 ha), outplanting more rapidly revegetated the soil than did seeding (*ibid.*). The percentage of seeds producing a seedling varied with timing of seeding and was highest where the soil was ripped. Survival of outplants varied among species (being highest in *Atriplex bunburyana* at 42 per cent) and was enhanced by ripping. Fertilizer, water-holding gel, and pruning minimally influenced survival. In dry periods, outplanting likely outperforms seeding, but if timing of seeding can be flexible, brief windows of moist conditions might enable seeding to be effective.

Mojave Desert, United States

Parks managed by the US National Park Service, such as Lake Mead National Recreation Area in the Mojave Desert, undergo periodic maintenance of highways traversing the parks. Restoration was conducted where an existing road was to be removed and re-routed nearby through intact desert (Abella *et al.* 2015a). The restoration approach included salvaging topsoil (upper 5 to 20 cm) and 2,105 individuals of 23 native perennial species including cacti, shrubs, forbs and grasses from segments to be destroyed by the new road route. After construction, topsoil was re-applied in 2010, salvaged plants were moved from temporary nurseries to planting sites, and survival of the salvaged plants (some of which received different irrigation treatments) was monitored over a 27-month period. The study period received close to the average of 16 cm/year of precipitation.

Half of the salvaged plants survived the process of salvage and one year of residence in a temporary nursery. About 27 per cent of individuals then survived at least 27 months at restoration sites. Species able to survive being salvaged also generally had among the best survival after planting back at field restoration sites. Cacti had nearly 100 per cent survival and did not require any supplemental irrigation. Several shrub species performed well – such as shadscale saltbush (*Atriplex confertifolia*) with 47 per cent survival and white bursage with 45 per cent survival. The forb desert globemallow also was successful, with 38 per cent survival. The type of irrigation (watering by hand, or DriWater as a slow-release gel) interacted with species differently. Both irrigation types similarly enhanced survival of white bursage by 1.4 times, whereas only hand watering increased survival of desert globemallow. The benefits of planting on salvaged topsoil were substantial: transplants exhibited 56 per cent survival on topsoil, compared to 25 per cent without topsoil. Topsoil alone (without irrigating plants) resulted in plant survival nearly equivalent to that produced by irrigating plants. The project met goals of visual restoration and rapidly revegetating severely disturbed soil, but understanding long-term survival of the planted species and their influence on natural recruitment is desirable (Figure 12.5).

Arabian Desert, Kuwait

On the Arabian Peninsula receiving an average of 12 cm of rainfall annually, Kuwait exemplifies progressive land degradation and the type of conditions desert restoration must ameliorate. Similar to elsewhere in the Arabian Desert, Kuwait experienced an increase in camel grazing,

Figure 12.5   Examples of restoration in the Mojave Desert, United States: (a) protecting outplanted perennial plants to revegeate disturbed slopes in Lake Mead National Recreation Area; (b) roadside revegetation in Joshua Tree National Park; (c) protecting outplanted creosote bush for post-wildfire restoration near Las Vegas, Nevada; and (d) establishing desert wetlands to provide the functions of watershed management and wildlife habitat

Source: photos by S. R. Abella

off-road vehicle use and resource extraction (mainly oil and mining) during the 1900s. This corresponded with lowered plant cover, shifts in plant composition toward unpalatable species, and soil degradation including increased wind erosion producing dust storms harmful to human health (Al-Hurban 2014). Invasion of Kuwait by the Iraqi military in 1990–1991 and the resulting Gulf War devastated already degraded desert land. In addition to damage from construction of military fortifications, movements of war vehicles across the desert, and explosions, 700 Kuwaiti oil wells and pipelines were destroyed by the retreating army. 'Oil lakes' formed across the desert and hardened (Figure 12.3).

Restoration on the current landscape must confront several challenges. First, additional surveys for land mines and unexploded ordinance left by the war are required before restoration can begin. Second, camel grazing is so intense that it must be ameliorated (e.g. by constructing and maintaining fences) before attempting restoration. Third, the legacy of human land use spans millennia, making evaluations of reference conditions difficult. For example, Brown and Al-Mazrooei (2003) found that within four years after fencing, the shrub arfaj (*Rhanterium epapposum*) re-established from underground stumps that had probably remained in the soil for decades. However, this shrubland may itself have been a product of decades to centuries of livestock grazing, and replaced an earlier *Acacia* woodland containing palatable grasses (Brown and Al-Mazrooei 2003). A current approach in Kuwait is to focus restoration within protected areas, survey potential restoration sites for unexploded ordinance, and evaluate a diverse mixture of species and community types for their restoration potential.

Functional outcomes and benefits of desert restoration

Expanding research on the functional outcomes of desert restoration is desirable to improve matching restoration treatments with specific goals and to explore the full potential of restoration. Some examples illustrate functional benefits that can be anticipated locally and globally if restoration in deserts expands.

In the Western Rajasthan region, India, planting the native shrubs rimth and phog (*Calligonum polygonoides*) on desertified land increased storage of organic carbon in the upper 20 cm of soil by 59 per cent (Rathore *et al.* 2015). By curtailing soil erosion and stimulating photosynthesis and accumulation of organic matter, desert restoration has high potential for sequestrating carbon and limiting release of carbon into the atmosphere.

In the western Mojave Desert of California, United States, abandoned eroding farmland created dust storms resulting in air quality violations and interfering with airplane and vehicle travel (Grantz *et al.* 1998). Revegetating these lands, through seeding and outplanting, reduced airborne dust by up to 99 per cent in the 1990s, significantly improving air quality.

Jilantai Salt Lake, in the Ulan Buh Desert, is one of the most economically important salt resources in China (Gao *et al.* 2002). However, salt production was compromised by encroachment of wind-blown sand accelerated through increased woodcutting and livestock grazing. Protection (fencing), combined with planting four native shrub and tree species, increased air humidity by 10 per cent (reducing effects of desertification) and reduced sand encroachment to the salt lake by 85 per cent (*ibid.*).

Restoration has high potential for enhancing habitat quality for desert animals, including threatened species. One example is the desert tortoise (*Gopherus agassizii*), a long-lived reptile ($\geq$ 50 years) of the western Sonoran and Mojave Desert in the United States. To improve forage quality and quantity, several fencing and seeding treatments were tested. Fencing and seeding pelletized seed (coated in a protective substance) increased by sixfold the native annual desert plantain (*Plantago ovata*), a food plant favoured by desert tortoises (Abella *et al.* 2015b).

With over one-fifth of the human population living in deserts and sharing habitat with an unknown number of invertebrate, animal and plant species, expanding restoration in degraded deserts is likely to produce numerous benefits. Future work to improve desert restoration could focus on refining understanding of reference conditions and restoration goals, further developing cost-effective treatments for meeting different goals across spatial and temporal scales, and monitoring functional benefits produced by restoration.

Acknowledgements

I thank Sharon Altman for adapting Figure 12.1, Lindsay Chiquoine for reviewing the section on biological soil crust, and Stuart Allison for helpful comments to improve the chapter.

References

Abella, S. R. 2010. Disturbance and plant succession in the Mojave and Sonoran Deserts of the American Southwest. *International Journal of Environmental Research and Public Health* 7:1248–1284.

Abella, S. R. and A. C. Newton. 2009. A systematic review of species performance and treatment effectiveness for revegetation in the Mojave Desert, USA. In A. Fernandez-Bernal and M. A. De La Rosa (eds), *Arid environments and wind erosion*, pp. 45–74. Nova Science Publishers, Hauppauge, NY.

Abella, S. R. and S. D. Smith. 2013. Annual-perennial plant relationships and species selection for desert restoration. *Journal of Arid Land* 5:298–309.

Abella, S. R., D. C. Craig and A. A. Suazo. 2012. Outplanting but not seeding establishes native desert perennials. *Native Plants Journal* 13:81–89.

Abella, S. R., L. P. Chiquoine, A. C. Newton and C. H. Vanier. 2015a. Restoring a desert ecosystem using soil salvage, revegetation, and irrigation. *Journal of Arid Environments* 115:44–52.

Abella, S. R., L. P. Chiquoine, E. C. Engel, K. E. Kleinick and F. S. Edwards. 2015b. Enhancing quality of desert tortoise habitat: augmenting native forage and cover plants. *Journal of Fish and Wildlife Management* 6(2):278–289.

Al-Hurban, A. E. 2014. Effects of recent anthropogenic activities on the surface deposits of Kuwait. *Arab Journal of Geoscience* 7:665–691.

Aref, I. M., L. I. El-Juhany and M. N. Shalby. 2006. Establishment of acacia plantation in the central part of Saudi Arabia with the aid of DRiWATER. In *2nd International Conference on Water Resources and Arid Environment*, 26–29 November 2006, pp. 110–118. King Saud University, Saudi Arabia.

Bainbridge, D. A. 2007. *A guide for desert and dryland restoration*. Island Press, Washington, DC.

Bowker, M. A. 2007. Biological soil crust rehabilitation in theory and practice: an underexploited opportunity. *Restoration Ecology* 15:13–23.

Box, J. B., A. Duguid, R. E. Read, R. G. Kimber, A. Knapton, J. Davis and A. E. Bowland. 2008. Central Australian waterbodies: the importance of permanence in a desert landscape. *Journal of Arid Environments* 72:1395–1413.

Brown, G. and S. Al-Mazrooei. 2003. Rapid vegetation regeneration in a seriously degraded *Rhanterium epapposum* community in northern Kuwait after 4 years of protection. *Journal of Environmental Management* 68:387–395.

Brown, J. H., O. J. Reichman and D. W. Davidson. 1979. Granivory in desert ecosystems. *Annual Review of Ecology and Systematics* 10:201–227.

Burke, A. 2008. The effect of topsoil treatment on the recovery of rocky plain and outcrop plant communities in Namibia. *Journal of Arid Environments* 72:1531–1536.

Commander, L. E., D. P. Rokich, M. Renton, K. W. Dixon and D. J. Merritt. 2013. Optimising seed broadcasting and greenstock planting for restoration in the Australian arid zone. *Journal of Arid Environments* 88:226–235.

El-Keblawy, A., T. Ksiksi and H. El Alqamy. 2009. Camel grazing affects species diversity and community structure in the deserts of the UAE. *Journal of Arid Environments* 73:347–354.

Evenari, M., L. Shanan and N. Tadmor. 1982. *The Negev: the challenge of a desert*. Harvard University Press, Cambridge, MA.

Ffolliott, P. F., J. O. Dawson, J. T. Fisher, I. Moshe, D. W. DeBoers, T. E. Fulbright, J. Tracy, A. Al Musa, C.

Johnson and J. P. M. Chamie. 2001. *Arid and semiarid land stewardship: a 10-year review of accomplishments and contributions of the International Arid Lands Consortium*. General Technical Report RMRS-GTR-89. US Department of Agriculture, Forest Service, Rocky Mountain Research Station, Fort Collins, CO.

Gao, Y., G. Yu Qiu, H. Shimizu, K. Tobe, B. Sun and J. Wang. 2002. A 10-year study on techniques for vegetation restoration in a desertified salt lake area. *Journal of Arid Environments* 52:483–497.

Ghose, M. K. 2001. Management of topsoil for geo-environmental reclamation of coal mining areas. *Environmental Geology* 40:1405–1410.

Golos, P. J. and K. W. Dixon. 2014. Waterproofing topsoil stockpiles minimizes viability decline in the soil seed bank in an arid environment. *Restoration Ecology* 22:495–501.

Grantz, D. A., D. L. Vaughn, R. Farber, B. Kim, M. Zeldin, T. Van Curen and R. Campbell. 1998. Seeding native plants to restore desert farmland and mitigate fugitive dust and PM10. *Journal of Environmental Quality* 27:1209–1218.

Guo, Q., P. W. Rundel and D. W. Goodall. 1998. Horizontal and vertical distribution of desert seed banks: patterns, causes, and implications. *Journal of Arid Environments* 38:465–478.

Islam, M. Z., K. Ismail and A. Boug. 2010. Catastrophic die-off of globally threatened Arabian oryx and sand gazelle in the fenced protected area of the arid central Saudi Arabia. *Journal of Threatened Taxa* 2:677–684.

Miriti, M. N., S. Rodríguez-Buriticá, S. J. Wright and H. F. Howe. 2007. Episodic death across species of desert shrubs. *Ecology* 88:32–36.

Padilla, F. M. and F. I. Pugnaire. 2006. The role of nurse plants in the restoration of degraded environments. *Frontiers in Ecology and the Environment* 4:196–202.

Patten, D. T., L. Rouse and J. C. Stromberg. 2008. Isolated spring wetlands in the Great Basin and Mojave Deserts, USA: potential response of vegetation to groundwater withdrawal. *Environmental Management* 41:398–413.

Rathore, V. S., J. P. Singh, S. Bhardwaj, N. S. Nathawat, M. Kumar and M. M. Roy. 2015. Potential of native shrubs *Haloxylon salicornicum* and *Calligonum polygonoides* for restoration of degraded lands in arid western Rajasthan, India. *Environmental Management* 55:205–216.

Safriel, U., Z. Adeel, D. Niemeijer, J. Puigdefabregas, R. White, R. Lal, M. Winslow, J. Ziedler, S. Prince, E. Archer, C. King, B. Shapiro, K. Wessels, T. Nielsen, B. Portnov, I. Reshef, J. Thonell, E. Lachman and D. McNab. 2005. Dryland systems. In R. Hassan, R. Scholes and N. Ash (eds), *Millennium Ecosystem Assessment: ecosystems and human well-being: current state and trends*, volume 1, pp. 623–662. Island Press, Washington, DC.

Scoles-Sciulla, S. J. and L. A. DeFalco. 2009. Seed reserves during surface soil reclamation in eastern Mojave Desert. *Arid Land Research and Management* 23:1–13.

Society for Ecological Restoration. 2004. *The SER international primer on ecological restoration*. Society for Ecological Restoration International, Tucson, AZ.

Thouless, C. R., J. G. Grainger, M. Shobrak and K. Habibi. 1991. Conservation status of gazelles in Saudi Arabia. *Biological Conservation* 58:85–98.

Wang, W., Y. Liu, D. Li, C. Hu and B. Rao. 2009. Feasibility of cyanobacterial inoculation for biological soil crusts formation in desert area. *Soil Biology and Biochemistry* 41:926–929.

West, N. E. 1990. Structure and function of microphytic soil crusts in wildland ecosystems of arid to semi-arid regions. *Advances in Ecological Research* 20:179–223.

Whitford, W. G. 2002. *Ecology of desert systems*. Academic Press, New York.

13

ECOLOGICAL RESTORATION IN MEDITERRANEAN-TYPE SHRUBLANDS AND WOODLANDS

Ladislav Mucina, Marcela A. Bustamante-Sánchez,
Beatriz Duguy Pedra, Patricia Holmes, Todd Keeler-Wolf,
Juan J. Armesto, Mark Dobrowolski, Mirijam Gaertner,
Cecilia Smith-Ramírez and Alberto Vilagrosa

Introduction

The Mediterranean-type ecosystems (further MTEs) are limited to five regions on Earth (Cowling *et al.* 1996): Mediterranean Basin, California, Central Chile, the Cape of South Africa and the Southwest (and partly South) Australia. These regions are characteristic of western ocean coastlines in warm-temperate latitudes characterized by descending water-deprived ethesial winds. They are invariably transitional between temperate forests and semi-deserts. Precipitation seasonality and prevalence of winter-rainfall/summer drought cycling are regular, although not exclusive to the MTEs (Blumler 2005; Rebelo *et al.* 2006). Fire has been part of the natural regeneration cycles and undoubtedly also evolutionary history of the scrublands and woodlands (perhaps except for the Chilean MTE) for millions of years. The Northern Hemisphere MTEs and the Central Chilean MTE are home to relatively young geologically and climatically dynamic landscapes. The MTEs of the African Cape and Australia are, on the other hand, geologically quiescent and climatically buffered – most of these regions qualify as Old Stable Landscapes (Hopper 2009; Mucina and Wardell-Johnson 2011).

MTEs are evolutionary hotbeds and musea: they are home to several global centres of biodiversity (Myers *et al.* 2000) and have about 20 per cent of total floristic diversity in an area covering just 5 per cent of the land surface. Vegetation of the MTEs is typically sclerophyllous shrublands, however (pine, oak, eucalyptus) woodlands are also important.

Besides the enormous biodiversity, the regions supporting MTEs have been under human pressure for a long time. Some (Mediterranean Basin and its eastern outposts in the Middle East) have been the cradle of agriculture and have seen the rise of many civilizations. Past and present human use put these ecosystems under pressure and where possible and feasible, restoration of these ecosystems emerged as one of the ways for their wise, future-oriented management. Each of the partial MTEs is exposed to multiple challenges of rehabilitation and a profound review of these is beyond the scope of this chapter. Therefore we have embarked on featuring the dominant rehabilitation focus in each MTE.

Mediterranean woodlands and shrublands

Challenges

Major degradation processes affecting the Mediterranean Basin (hereafter the Mediterranean) ecosystems are related to the long-term overuse of natural resources (overgrazing, woodland clearing, invasive alien species), increasing population pressure and associated political issues (such as urban sprawl and alteration of fire regimes) generating threats to sclerophyllous shrublands and woodlands in the entire Mediterranean. Paradoxically, land abandonment may play a negative role too. Steep moisture gradients spanning sub-humid and arid climate zones within the Basin, together with the natural environmental heterogeneity and the diversity of land-use histories underpin great ecosystem variability and hence, a wide range of degradation scenarios (Vallejo *et al.* 2012a). Desertification is affecting large areas in dry lands, reducing soil productivity (Vallejo *et al.* 2012c). In the past few decades, large intensive wildfires increased in frequency and intensity in the European Mediterranean landscapes, except in arid areas where fires are fuel-limited, imposing a serious threat both to natural ecosystems on one hand and to human life, property, and well-being on the other (Duguy *et al.* 2013; Moreno *et al.* 2013). Synergistic interactions between severe fires and ongoing degradation processes may be especially acute in vulnerable semi-arid ecosystems and trigger strong changes in ecosystem composition and structure (Moreno *et al.* 2013).

Climate change projections for the Mediterranean foresee increasing extreme temperatures and decreases in both rainfall and relative humidity (Kovats *et al.* 2014). These changes are predicted to foster longer fire seasons and more intense fires (Duguy *et al.* 2013). Under such conditions, post-fire regeneration will likely be impeded, hence diminishing resilience of plant communities (Delitti *et al.* 2005) and promoting opportunistic alien species (Lloret *et al.* 2003). Once the degradation thresholds have been crossed (e.g. by loss of keystone species), certain processes may reduce the ability of supporting spontaneous regeneration. Then, the vegetation, and ecosystems in general, may be restored only through human intervention in the form of restoration actions and manipulations (Vallejo *et al.* 2012a).

Some major challenges that ecological restoration in the Mediterranean is confronting include:

- definition of proper models of vegetation dynamics to use as a reference (e.g. natural, or 'historic', fire regimes), keeping in mind that in the light of global change local historical references may become of limited value while references from drier sites may become more appropriate (Fulé 2008);
- increasing need of long-term research and adaptive management leading to a better understanding of the abiotic and biotic factors acting as drivers of ecosystem functioning in order to identify processes hindering post-disturbance natural recovery and better selecting the areas and the ecosystem components or functions to be restored;
- promotion of better governance, based on broad social participation and diffusion networks in order to meet new demands on natural environment and spread best practices associated with land management (e.g. the promotion of prescribed fire, which entails a social, political and technical challenge, requires social acceptance and adequate regulation); and
- promotion of quality control and scientifically based evaluation of projects for optimizing restoration investments and delivering feedback from restoration experiences into the improvement of restoration processes.

Theoretical underpinning

Reforestation and afforestation have been the restoration actions traditionally implemented in the Mediterranean countries. Technically not considered ecological restoration projects as understood today, these actions addressed some of the broad aims of restoration, such as reduction of soil erosion and runoff, recovery of natural forests, and the like (Vallejo *et al.* 2012a). The 'tree-oriented' approach, based on Clementsian successional models and still applied by forest administrations, consists of (re)introducing one or several keystone species (generally pines), acting as 'ecosystem engineers' that are expected to modify the habitat and facilitate the establishment of late-successional species, thus fostering successional trajectories towards an ideal state, often identified as potential natural vegetation (Cortina *et al.* 2011).

This old paradigm recently shifted towards new approaches embracing successional state-and-transition models identifying likely trajectories and desired potential states on the basis of community composition and structure, ecosystem functioning and capacity to provide goods and services (Cortina *et al.* 2011). New projects generally consider a higher diversity of woody species, positive plant–plant interactions (i.e. facilitative effects), less aggressive plantation techniques, and smaller extent of the interventions.

Despite the high diversity of degradation and restoration needed in scenarios encountered across the Mediterranean, some objectives should be common to all restoration projects (Vallejo *et al.* 2012a), such as soil and water conservation, increase of ecosystem resilience to current and future disturbance regimes, promotion of native biodiversity while eradicating alien invasive species, and improvement of landscape quality and provision of ecosystem services.

Approaches to restoration

Under climate change projections, restoration strategies and techniques have to be adapted to increased drought stress and fire. In this sense, two types of interventions have probably received the most attention from ecological restoration research across the Mediterranean over the past few decades. These include *post-fire restoration* for mitigating or reversing negative fire impacts (often caused by novel combinations of fire regime and other disturbances in fire-dependent ecosystems), and *restoration of vegetation cover* in ecosystems affected by desertification and biodiversity loss in semi-arid areas.

Planning post-fire restoration requires an understanding of how the fire regime is affecting ecosystem fire resilience and the identification of the specific degradation processes triggered by fire (Vallejo and Alloza 2015). Restoration should address soil conservation in the short term (< 1 year) and the recovery of ecosystem integrity (function and structure, including biodiversity) together with ecosystem services in the longer term.

In ecosystems showing high erosion and runoff risk, with low short-term plant regeneration capacity, emergency rehabilitation actions are needed (Vallejo *et al.* 2012b). Two main (non-exclusive) soil protection techniques are used, such as *seeding* with fast-growing native species and, *mulching* with various kinds of organic materials.

In a second stage, vegetation recovery is the key factor for restoring soil productivity. Post-fire regeneration strategies of dominant species determine recovery rate (Keeley *et al.* 2012). Resprouting species allow a faster recovery of species composition and abundance than obligate seeders. The (re)introduction of native woody resprouters is thus recommended to increase fire resilience (Valdecantos *et al.* 2009).

The aspects to be considered for improving plantation success are plant species selection, nursery and planting techniques (Chirino *et al.* 2009; Duguy *et al.* 2013). Selection of species must be based on the natural flora and vegetation of the area and the specific biophysical characteristics of the site (Vallejo *et al.* 2012a). The number of species used in reforestation is increasing rapidly, moving from a reduced set of easy-to-grow species (mostly pines) to a large variety of native species (including shrubs).

Drought is the most critical factor hindering seedling survival across the Mediterranean (Vallejo *et al.* 2012c). Plantations have incorporated innovations for reducing seedlings' water stress (Chirino *et al.* 2009; Duguy *et al.* 2013; Vallejo *et al.* 2012c) by (1) increasing water-use efficiency (selection of drought-tolerant species and ecotypes, seedling preconditioning, improvement of below-ground performance and nutritional status), (2) increasing water supply (soil preparation and amendment for improving microsite conditions and resource availability), and (3) reducing water losses (tree shelters, mulching, microsite selection).

Some techniques may significantly enhance plant establishment in semi-arid ecosystems (Figure 13.1), particularly when combined in harsher sites (Kribeche *et al.* 2012; Valdecantos *et al.* 2014). Over the past decade, indeed, as advanced technologies were implemented, the relationship between seedling survival and drought length changed for experimental plantations in dry lands; survival rate doubled under a three-month drought (Vallejo *et al.* 2012a; Figure 13.2).

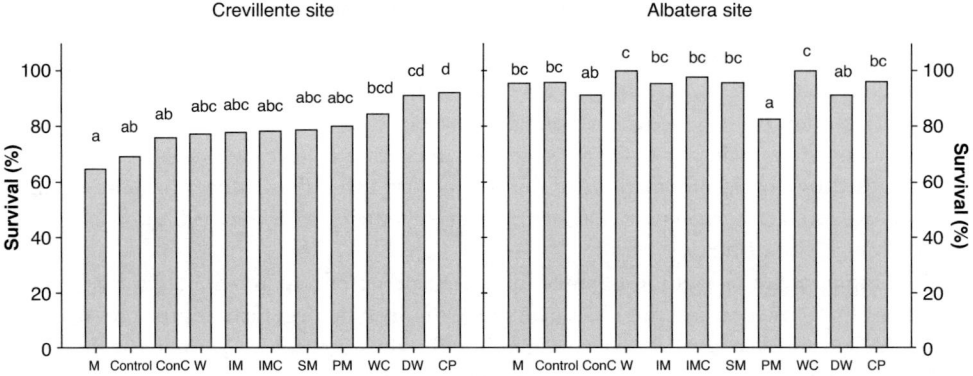

Figure 13.1 Survival of *Olea europaea* seedlings ten months after planting in relation to experimental treatments in two semi-arid stations (rainfall was 10 per cent higher in Albatera during the studied period). Control: traditional planting holes; M: Microcatchment; W: Deep water application (1.5L, twice); IM: M+waterproof surface upslope of holes; SM: M+stone mulch on soil surface; PM: M+plastic mulch on soil surface; DW: IM+stone mulch+2preferential water pathways; CP: 2.5L buried clay plot (filled twice); ControlC: Control+C; IMC: IM+C; WC: W+C, where +C is application of composted sewage sludge at an equivalent rate of 22.5Mg/ha. Different letters indicate significant differences by the log-linear analysis ($p < 0.05$)

Source: modified from Valdecantos *et al.* (2014)

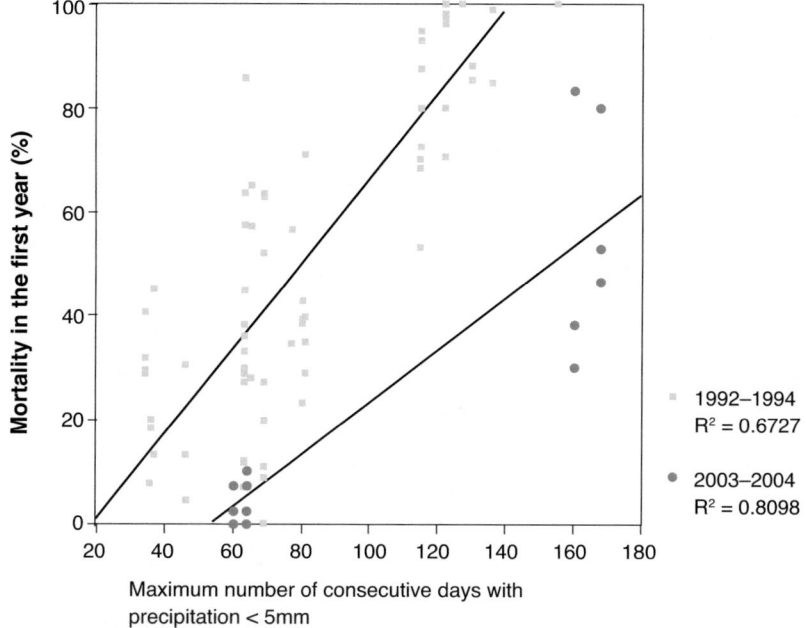

Figure 13.2 Seedling mortality in relation to the dry period length during the first post-plantation year, for several native species planted in eastern Spain. The first set of plantations (1992–1994) used conventional techniques at those times. The second set (2003–2009) used recent technical innovations

Source: modified from Vallejo *et al.* (2012a)

Californian coastal sage scrub and chaparral

Challenges to ecological restoration

Geographic set up and vegetation patterns

Mediterranean-type ecosystems occur in North America throughout the California Floristic Province (CFP) west of the Sierra Nevada and Cascade Mountains crest and the Transverse and Peninsular ranges from northwest Baja California (Mexico) to southwest Oregon.

California has the most extreme summer drought of all the Mediterranean-type climates (Vasey *et al.* 2014). Between May and September, there are almost no significant rain events. Its endemic shrublands have developed two divergent life history strategies which enable them to contend with this extreme summer drought: the shallow-rooted drought-deciduous approach, and the deep-rooted, evergreen sclerophyll strategy.

The major ecosystems of the CFP include California broadleaf and coniferous woodlands, California prairie (composed of annual and perennial grasses, graminoids and broad-leaved herbaceous species), and two shrubland types commonly known as coastal scrub and chaparral. The California *coastal scrub* is a drought deciduous shrubland formation dominated by short-lived shrubs with shallow, spreading root structures. Leaves tend to develop during the winter

---

Box 13.1 Restoration of drylands in Southeastern Spain: the combined role of site conditions and reforestation techniques

Southeastern Spain is one of the areas most affected by desertification in Europe. In the framework of the *Spanish Action Programme to Combat Desertification*, the Spanish Ministry of Environment and the Valencia Government Forest Service implemented in 2003 the Albatera restoration demonstration project, under scientific advice of CEAM, University of Alicante and CIDE-CSIC. Albatera site is a 25 ha catchment located in Alicante province. Land degradation was driven by the synergistic effect of past management and harsh environmental conditions, such as scarce (around 280 mm year⁻¹) and highly variable rainfall and erosion-prone soils.

The project aimed at putting into practice the best available restoration techniques for degraded semi-arid ecosystems, being an example of successful collaboration and technology transfer between the scientific community and stakeholders.

The main objectives of the project were (1) to repair ecosystem functioning by creating functional vegetation patches that contribute to the re-allocation of water, materials and nutrients, (2) to increase ecosystem diversity, stability and resilience, and (3) to prevent further surface and landscape degradation, soil erosion and off-site damage.

Based on functional characteristics, seven landscape units were identified and specific actions (species selection and restoration treatments) were designed for each. Eighteen native evergreen species (trees and shrubs) were selected. Plantation techniques aimed at maximizing water collection and conservation (micro-catchments, mulching), organic amendment (compost) and minimizing abiotic stress (tree-shelters). Although survival and growth rates of the introduced species were highly variable, six years after planting, survival was close to 50 per cent on south-facing slopes (where treatments were accumulated), against almost 10 per cent in the same area for past reforestations (Kribeche *et al.* 2012). Diversity and plant cover were higher (about 10 per cent) than in non-restored sites nearby. Some species had flowered and fruited, contributing to the recovery of the area. Soil loss decreased in all units over the monitoring period. Results show that suitable species selection and technological innovations improve reforestation outcomes, particularly in harsher sites. The cost-benefit analysis for best-technology actions yields a positive balance.

---

rainy season, and many are shed, or replaced, with smaller thicker leaves in the onset of summer drought (Keeley and Keeley 1984). The sclerophyllous evergreen shrubland known as California *chaparral* occurs on well-drained soils, often on steeper and rockier settings than coastal sage scrub. It consists of a variety of long-lived shrub species all of which tend to have deep roots and evergreen sclerophyllous leaves ranging from nanophyll to microphyll size. This vegetation is largely restricted to the CFP from SW Oregon, and south to northern Baja California, Mexico. California chaparral shrubs are divided into two main life-history strategies: (1) *seeders* tend to be relatively short-lived, storing seeds in a soil seed bank, and reproducing naturally through stand-replacing fire events, and (2) *resprouters* are long-lived (hundreds of years), and build up large carbohydrate stores in enlarged underground root structures (Keeley and Davis 2007).

Nature of disturbance sources

Natural fire frequencies in the shrublands of the CFP are not particularly high, since lightning is limited to the occasional summer monsoonal thunderstorm emanating from Southern

Mexico. Natural fire frequencies for many sage scrub and chaparral landscapes have been projected to be 20–50 years and fire occurred primarily in the late summer and fall. Recurring high-frequency fires can deplete the seed banks of many chaparral and sage scrub species, especially those that regenerate largely from seed.

Resprouting chaparral species (Keeley and Davis 2007) are able to survive periodic fires, but can persist for decades without them. A dense cover generally develops in the first decade after a fire, and these shrubs dominate within 30 to 40 years as they overtop shorter-lived or shorter stature species. Long intervals between fires and mesic conditions promote the development of mesic resprouting chaparral species.

The California shrublands are coincident with several of the largest urban centres in coastal California including the Los Angeles Basin, San Diego and the San Francisco Bay Area. An estimated 80–85 per cent of the pre-European extent of sage scrub has been eliminated from its southern California range (Reid and Murphy 1995), replaced with anthropogenic landscapes of ruderal or seral herbaceous vegetation, in some cases driven by increased atmospheric nitrogen deposition through automobile and industrial exhaust (Allen *et al.* 1998).

Motivation for ecological restoration

Many wildlands in California have been eliminated by rapid urban growth over the past 50–60 years. Increasing awareness of natural values of wildlands have motivated urban inhabitants to seek more sustainable solutions to urban growth compatible with maintaining natural populations of plants and animals, and natural processes in these landscapes. Certain subtypes of sage scrub and chaparral are recognized as rapidly decreasing habitat for endangered vertebrates.

Theoretical underpinning

Both vegetation formations tend to respond to different seed germination agents. For example, sage scrub species such as *Salvia* spp. respond to smoke-borne germination agents while chaparral seeds tend to respond to high heat (Keeley and Fotheringham 2001). Long intervals between fires are needed to develop mesic chaparral stands due to their short-lived shade tolerant seeds and resprouting response.

Sage scrub restoration is generally less complex than chaparral restoration. Many coastal sage scrub species are considered to be highly opportunistic in their germination requirements (*ibid.*). Several sage scrub species seeds have simple smoke-induced germination, but also produce some seedlings without fire. Coastal sage scrub species are also wind dispersed, and many are unpalatable to herbivores.

Compared to sage scrub, chaparral restoration efforts have more difficulty with establishment due to more specific relationships with mycorrhizal fungi (Horton *et al.* 1999), larger, fewer, seeds with more specific dispersal strategies, lower recruitment, and longer establishment times due to palatable higher nutrition foliage to herbivory and effects of drought.

However, like chaparral, coastal sage scrub is vulnerable to type-conversion. Abundant and ubiquitous introduced non-native annual grasses and herbs produce flashy fuels and when ignited create a continuous herbaceous fuel bed. Frequent fires are perpetuated by proximity to human ignition sources and atmospheric nitrogen deposition encouraging rapid re-growth and thatch production. The mycorrhizal root associations of coastal sage scrub can be disrupted by clearing the native shrub cover in conjunction with ploughing or 'deep ripping', also precipitating a type conversion to alien-dominated vegetation (Bozzolo and Lipson 2013).

Approaches to restoration

State-mandated regional conservation planning of the 1990s involved a significant portion of the natural range of coastal sage scrub. In southern California, many local government agencies have established objectives, standards and success criteria for coastal sage scrub restoration plans. These typically include the following principles:

1   Restore and raise the ecological condition of a disturbed site to a high-quality condition, equal to the pre-disturbance condition.
2   Use locally adapted plant material to establish a self-sustaining habitat with appropriate plant species richness, diversity and composition, based on the original vegetation, the physical characteristics of the site, the biological context of the site, and the nature and degree of disturbance to the original vegetation.
3   The restoration plan should clearly state the desired result of the habitat restoration for the particular site and set forth standards and success criteria. The condition of the restoration site has a strong bearing on the method of approach. Standard approaches have included conventional horticultural practices, such as irrigation, fertilizer, cages and soil amendments.

Applicability of restoration approaches to California scrub vary widely due to the climate, topography and existing site conditions of the project. Restoration settings include a wide range of possible situations from lightly disturbed (e.g. removal of original scrub and replacement by unwanted non-native cover), moderately disturbed (e.g. soil profile disturbed by cultivation but original topography intact), or heavily disturbed (e.g. all original vegetation removed, soil removed, topography modified). Depending upon the circumstances the following restoration practices are generally considered necessary to ensure a reasonable level of success.

Box 13.2 Leona Quarry restoration, Oakland, CA

A rock quarry since the early 1900s, this roughly 53 ha area has been reclaimed as partial mitigation for a housing project occupying the lower portion of the site in 2005. The upper two-thirds of the site was re-engineered to reduce geological impacts based on the location along the geologically active Hayward Fault, and to prepare the site for a chaparral restoration project. An approved county conservation measure established to reduce urban sprawl in the eastern portion of Alameda County sent developers scouting for available sites within the existing metropolitan zone. The project in the city of Oakland gained credit through a conservation easement for increasing habitat for the endangered Alameda Whipsnake, by reclaiming chaparral and sage scrub habitat lost during the expansion of the quarry. Following slope stabilization, slopes were blanketed with strips of coconut fibre and straw. The exposed face of the quarry was then covered with a layer of topsoil amended with compost. High-pressure hoses were used to spray a hydroseed mixture of native grass seeds, paper mulch and a glue agent. About 2000 holes were augered and filled with soil for the planting of coast live oak (*Quercus agrifolia*) and coastal sage scrub and chaparral species largely local to the adjacent natural area (Figure 13.3), including *Salvia mellifera, Eriogonum fasciculatum, Lotus scoparius,* and chaparral species such as *Adenostoma fasciculatum* and *Frangula californica*. Planting was arranged in a regular pattern that alternated chaparral and coastal scrub species, following hydroseeding of grassland species across the stabilized slope. The area was irrigated and maintained for

---

Figure 13.3   Landscape of the adjacent Chimes Creek (California) watershed used as the restoration model and local species source area with matrix of coastal scrub, chaparral and oak woodland. Newly prepared slope in upper right background May 2007, about one year post-planting. White-flowered shrubs in foreground are the regionally dominant chaparral shrub *Adenostoma fasciculatum*; small rounded trees are *Quercus agrifolia*

three years using a network of drip irrigation tubes. A five-year monitoring period was established to ensure native plant establishment and invasive species removal.

After planting species cover increased slowly for the first five years. During that time wire mesh protective cones were established around the slower-growing chaparral species (*A. fasciculatum, F. californica*). Following a major rainfall year in 2010 the cover increased appreciably (Figure 13.4). However, most of the cover came from increases in coastal scrub species and early seral species such as *Baccharis pilularis*, which seeded naturally from adjacent off-site areas.

Sage scrub species grew quickly from seed and comprised the majority of cover ten years after restoration. Invasive exotics have colonized the site particularly on the west side adjacent to a local off-site source of alien *Genista monspellieanus* and *Cortaderia jubata*.

Ten years post-site preparation, *Adenostoma fasciculatum*, the local chaparral dominant, established only in protected cages due to heavy preferential browse by native mule deer, while short-lived *Baccharis pilularis* (right) grew fast and did not require assistance establishing.

Vegetation was at a maximum in 2011 following a high rainfall season, but has actually declined since, due to four successive low rainfall years, and continued browsing by local mule deer. Based upon the first decade following restoration at this site, approximation of native chaparral structure dominated by *Adenostoma fasciculatum* will take many more years to achieve.

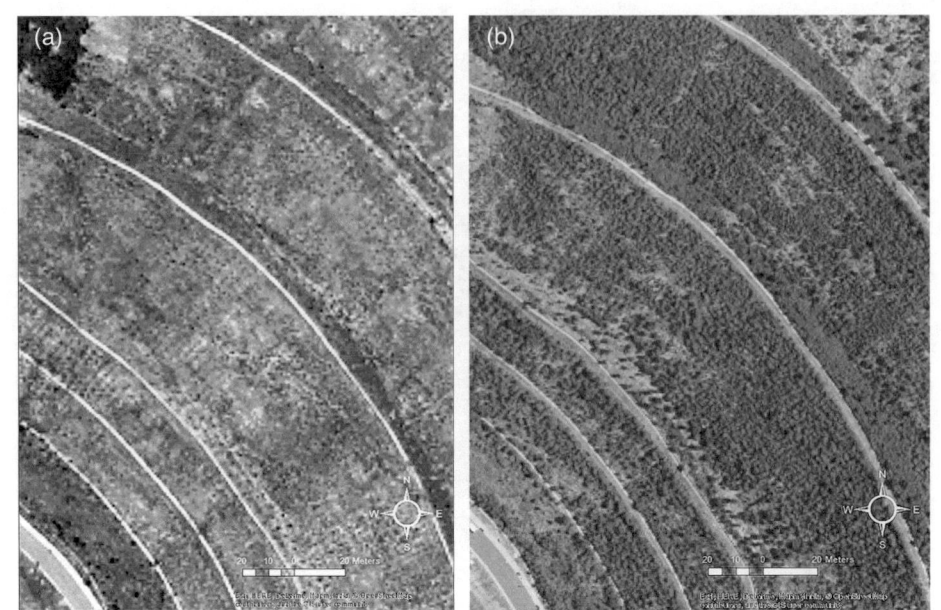

Figure 13.4   (a) August 2010 view of upper quarry five years following restoration, showing sparsely populated regular rows of shrub plantings, grass and herbaceous cover. Shrub cover is both native, naturally established *Baccharis pilularis* and potted and planted starts of *Adenostoma fasciculatum*. Thickly vegetated upper left corner is invasive exotic *Genista monspessulana*. (b) Shrub cover has increased as of this photo in July 2015. However, most is composed of naturally colonizing early seral species *Lotus scoparius* and *Baccharis pilularis*, not planted at the site. Of the planted sage scrub species, the highest cover was *Eriogonum fasciculatum*, followed by *Salvia mellifera* and *Artemisia californica*. The chaparral species comprised less than 1 per cent of total cover

Chilean sclerophyllous woodlands and matorral

Challenges

Anthropogenic disturbances have shaped the sclerophyllous forests and scrublands of Central Chile for almost five centuries. Such disturbances have been associated primarily with extensive use of firewood, cattle grazing, and more recently, agricultural and urban expansion, together with continuous degradation of vegetation by persistent burning. Fire, used by humans to clear vegetation for several centuries, greatly altered the historical disturbance regime in the Mediterranean-ecosystem region of Chile (Armesto *et al.* 2009).

The increased frequency of fire and logging in Central Chile has rejuvenated and changed the extent of remaining forests and scrubland, generating a predominance of areas dominated by shrubs such as *Acacia caven*, young forest stands (>50 years old) and a mosaic of small-size patches (<100 ha) (Schulz *et al.* 2010; Van de Wouw *et al.* 2011). Some native plant and animal species have declined to levels of quasi-extinction locally (e.g. *Beilschmedia berteroana*, *Beilschmedia miersii*, *Gomortega keule*; Hechenleitner *et al.* 2005). Efforts to restore community composition are faced with the problem that the scarcity and heterogeneity of native sclerophyllous forest remnants makes it difficult to define the reference systems to be used as targets for restoration. Knowledge of structure, composition, and environmental conditions of ancient remnant forest is typically unknown, but this knowledge is of great relevance for both restoration and rehabilitation planning.

A conceptual framework to guide restoration

Grounded in a conceptual framework based on knowledge of succession, disturbance, ecological filters and community assemblage rules (Temperton *et al.* 2004), we identify the factors that limit or facilitate spontaneous colonization in degraded areas of Mediterranean-type forest ecosystems in Chile and that may be applied to assist the regeneration of degraded areas. Based on the review of the existing scientific knowledge about forest ecosystems of Central Chile, we have put forward five general strategies to guide future restoration programmes in the region.

First, 'passive restoration' or assisted natural regeneration may be effective under certain circumstances. It should be favoured in degraded areas that are located near forest remnants that could act as propagule sources, or in south- and west-facing slopes, and other relatively humid sites in Central Chile (e.g. coastal areas, wet ravines). Such sites show lower seedling mortality and can recover from disturbance without active intervention. Depending on the specific site, sometimes it will be necessary to exclude herbivores (installing a fence). However, native vegetation could often recover even in the presence of herbivores (Fuentes-Castillo *et al.* 2012; Holmgren *et al.* 2000).

Second, the main limitations for successful establishment of native woody species in a degraded or open Mediterranean-type ecosystem derive from the combination of seasonal water stress, and the intensity of chronic disturbances such as fire and herbivory (Fuentes *et al.* 1983, 1984). Consequently, restoration strategies must be able to identify techniques to allow woody species to persist in the face of chronic stresses, which are part of the current environment. It will be advisable, therefore, to identify assemblages of species with complementary functional traits that may tolerate different kinds of stress and flexible characters that allow them to survive in open, degraded land (Laughlin 2014). For instance, if restoration is limited by high water stress, attributes such as hydraulic architecture, rooting depth or LAI could be related to differences in growth and survival. Traits that favour persistence should be selected and promoted within the species used in restoration.

Third, seed dispersal into open areas is often a limiting factor for succession and establishment. Because *dispersal vectors* are very important in the Mediterranean-type ecosystem of Central Chile (Armesto *et al.* 1987; Reid and Armesto 2011), succession can be initiated by planting species with fleshy fruits such as *Maytenus boaria*, *Lithraea caustica* or *Aristotelia chilensis*. These species will attract frugivorous birds to these areas to feed on their fruits, thus facilitating the arrival of fleshy propagules of other species and enhancing a diversity of colonizing species.

Fourth, *nurse effects* are an important plant–plant interaction facilitating plant establishment in stressful environments. It has been shown that the survival of woody seedlings is enhanced under the cover of pioneer shrubs (Becerra *et al.* 2011; Fuentes *et al.* 1984; Ovalle *et al.* 1999). In the case of sites without a shrubby cover, it may be advisable to actively promote the establishment of nurse shrubs.

Fifth, given the fact that *plant–bird interactions* often enhance the arrival of seeds of trees or shrubs into disturbed sites, and that nurse effects facilitate woody plant establishment, we propose that restoration programmes should follow a model of succession known as *nucleation* (Figure 13.5). In fact, this process spontaneously occurs in several places in the region (Fuentes

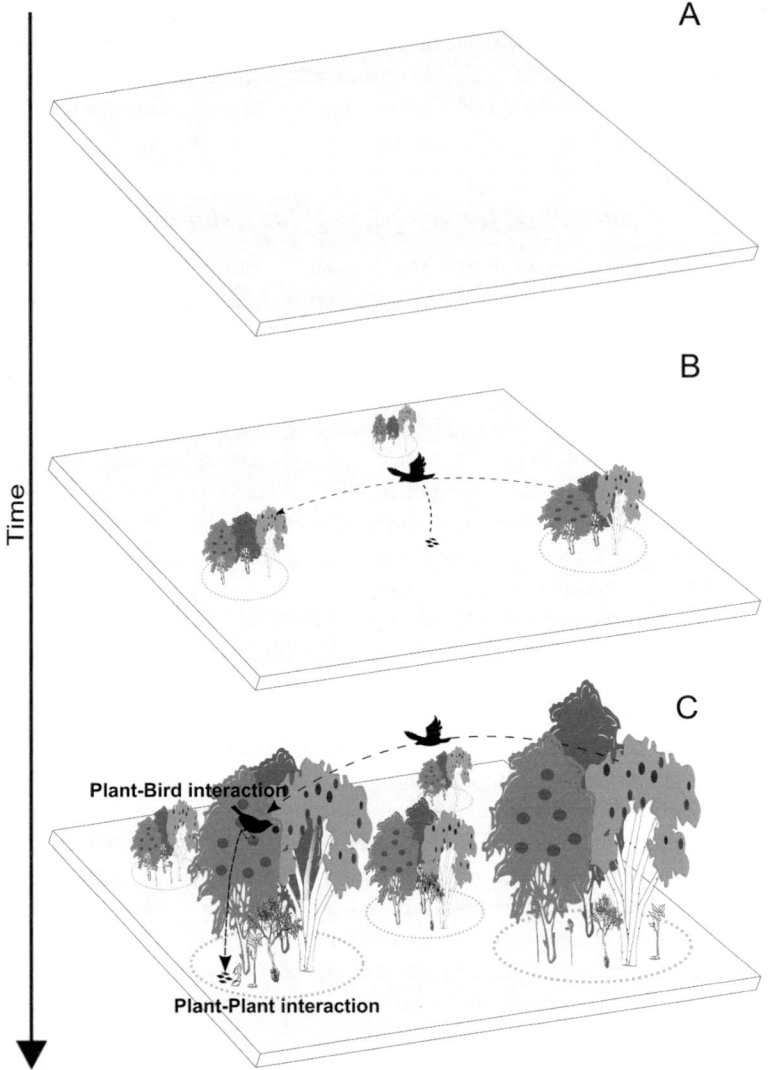

Figure 13.5 Graphic model of the nucleation process based on 'islands' or patches of planted trees. An area deprived from vegetation (A) is planted with blocks of multiple woody fleshy species (B). Once trees and shrubs have grown to sexual maturity (C), patches will expand and disperse seeds into neighbouring open areas

Source: modified from Benayas *et al.* (2008)

et al. 1984; Fuentes-Castillo *et al.* 2012). Under this view, the planting of woody pioneers in open areas should be done in clumps, or artificial 'islands' (Benayas *et al.* 2008), including species with fleshy fruits. In the long term, these initial nuclei could promote natural regeneration over large spatial scales.

Box 13.3 Partnerships between scientists and practitioners are effective means for success

Experiments of ecosystem restoration are commonly conducted by researchers working in the science of restoration ecology. Frequently, researchers conduct quantitative assessments of the initial and long-term success of their experiments, in contrast to similar experiments performed by practitioners, which are assessed mainly through qualitative descriptions.

We discuss two case studies in Central Chile, from the point of view of practitioners and scientists, to show that working in partnership and sharing data between these two groups is important for the success of future projects:

1   The ecological reserve Oasis de la Campana (www.reserva.cl/ proyectos_quienes.html), located in the Valparaiso Region, aims to create a space for the coexistence of humans and nature by conserving the natural vegetation and actively contributing to the recovery of human-degraded areas. To this end, they have worked on the recovery of 140 ha of degraded areas by hand-sowing *Acacia caven* seeds. Five years later, the open area has been filled by a dense cover of this shrub, regardless of the impact of herbivores (P. Moreno, unpublished report).
2   Experimental assays conducted by scientists in Valparaiso and the Metropolitan Region evaluated the effect of herbivore exclusion on woody regeneration. Fenced and unfenced plots were located in open pastures (covered by some individuals of *A. caven*). A higher proportion of woody seedlings became established under shrubs of *A. caven* in fenced plots than in unfenced plots (Miranda *et al.*, unpublished data).

From both examples, we conclude that it is clearly possible to increase the woody cover of areas devoid of vegetation by using low-cost techniques (hand sowing) to start up the recovery process. Once a shrub cover has been established, fencing may be necessary to enhance survival of established native trees. Therefore, combining results from both experiences has a greater value than each result taken separately.

Sclerophyllous shrublands of the Cape Region of South Africa

Introduction

The Cape Floristic Region (CFR) global biodiversity hotspot is located in the southwestern corner of South Africa and comprises only 4 per cent of the country. Mediterranean-type ecosystems are confined to western regions of the CFR while the eastern areas also receive a portion of summer rainfall. The major ecosystems of the CFR are fynbos and renosterveld sclerophyllous shrublands. Fynbos occurs on nutrient-poor substrate, primarily sandy soils derived from quartzite, sandstone, granite and rarely shale and limestone, whereas renosterveld shrublands occur on more nutrient-rich soils derived mainly from shale and granite parent materials under a drier rainfall regime (Rebelo *et al.* 2006). Strandveld shrublands that occur along the coast on alkaline soils, Cape Thicket and Afrotemperate forest confined to fire-protected habitats, are smaller vegetation types in the CFR and shall not be dealt with here.

In fynbos shrublands, fire and water availability are the major ecological drivers of recruitment and community structure, whereas in renosterveld shrublands mega-herbivore grazing also appears as a key driver (Radloff *et al.* 2013). Both fynbos and renosterveld shrublands require restoration strategies tailored to the nature of major drivers and their (re)assembly. Here we focus on alien-invaded fynbos shrublands as an example of restoration.

Ecological degradation and restoration

Nature of disturbance

Fynbos ecosystems are highly threatened, primarily as a result of direct habitat loss for agriculture, urban development and mining. The second largest threat is invasions by alien plants. Several alien tree species, including pines, hakeas and acacias, have invaded extensive areas and have out-competed the fynbos to form dense stands that alter ecosystem function, including changes to fire regimes (Wilson *et al.* 2014). Further ecosystem degradation is caused by inappropriate management such as too-frequent fires, over-harvesting and over-grazing (Holmes and Richardson 1999).

Motivation for ecological restoration

Ecological restoration in fynbos is motivated for two main reasons: first, to restore ecosystem function and thereby improve ecosystem services, such as water production from mountain catchments that are invaded by alien trees (van Wilgen *et al.* 2012); and second, to improve both function and community composition in support of biodiversity conservation in degraded areas of biodiversity networks (Rebelo *et al.* 2011). Owing to the high number of IUCN threatened plant species in the CFR (Raimondo *et al.* 2009), species restoration also is an important goal, particularly in the highly transformed lowlands.

Theoretical underpinning

Conceptual framework

A conceptual framework for restoration developed by Holmes and Richardson (1999) derived protocols from fynbos recruitment dynamics, community structure and ecosystem function. Predicted outcomes based on this framework were upheld in restoration studies in upland fynbos, following degradation by dense alien vegetation (Holmes *et al.* 2000) as well as topsoil disturbance that simulated mining (Holmes 2001).

In terms of recruitment dynamics, the majority of fynbos species have persistent soil-stored seeds that are stimulated to germinate by direct and indirect fire-related cues. Ephemeral geophytes persist as bulbs, corms and tubers. In upland fynbos, such species may survive two fire-cycles of dense alien invasion as dormant propagules, enabling autogenic restoration following alien clearance and fire. However, perennials that rely mainly on sprouting (e.g. shrubs with lignotubers) and shrubs with canopy-stored seeds become locally scarce or extinct, respectively, following dense alien invasion. In order to improve community structure, active restoration of such under-represented guilds is required.

Fynbos community re-assembly follows an auto-succession, whereby all components recruit after fire and are represented in the aboveground community for varying periods according to their respective lifespans (Bond and van Wilgen 1996). Therefore it is appropriate in active restoration to re-introduce all components together, in the immediate post-fire (or exposed soil) stage. Owing to high fynbos diversity, particularly of beta and gamma diversity (i.e. high turn-over along environmental and geographical gradients), practitioners should be careful to collect and re-introduce appropriately adapted, local taxa.

Many fynbos species are structurally and functionally similar. From an ecosystem function perspective, such analogues could be considered redundant, yet are thought to improve resilience to perturbations. From a restoration perspective, it is essential to re-instate a balance of the most important functional guilds, including species representing the main growth forms, regeneration and nutrient acquisition modes. Wherever possible, several species in each of these guilds should be introduced in order to improve resilience in the restored community.

Restoration thresholds

Research in lowland fynbos indicated that not all the protocols discussed previously were universally applicable, as regime shifts would occur more rapidly in low-altitude, particularly sandplain ecosystems. A restoration threshold model has been applied to conceptualize the different responses to degradation (Gaertner *et al.* 2012a). A threshold is recognized as the point at which the dominance of regulating feedbacks that maintain resilience switch to a dominance of positive feedbacks that lead to loss of resilience. Seed banks of perennial fynbos species are shorter-lived in the lowlands, possibly owing to intensive small mammal activity, limiting the potential for autogenic restoration (Holmes 2002). In lowland fynbos a biotic threshold may be passed following only one cycle of dense aliens, compared to two cycles in upland fynbos. The dominant invasive alien in the lowlands (Australian acacias) alters soil nutrient cycling processes (Gaertner *et al.* 2011), with nitrogen enrichment causing positive feedback and regime shift to a weedy, herbaceous community (Yelenik *et al.* 2004). In this case, a second, abiotic threshold has been passed and restoration interventions also must address the altered soil chemistry.

After more extreme degradation, such as in previously farmed land devoid of native propagules and with altered soil conditions, restoring community structure will be a more challenging goal. In some contexts it may be appropriate to modify the goal to one of restoring a particular ecosystem function.

Approaches to restoration

The theoretical frameworks outlined previously are useful in restoration planning (Figure 13.6). The first step is to identify an appropriate restoration goal. This will depend on several factors, such as extent of degradation, conservation importance, future land use and available budget. Restoring composition as well as ecological structure and function generally will be the most exacting goal requiring the highest resource inputs. Topsoil should be stripped ahead of mining operations and conserved, then replaced during restoration as it contains the soil-stored seed bank and microbial symbionts that greatly enhance restoration outcomes. Where topsoil has been lost or modified, the return of a small amount of topsoil, together with appropriate soil amelioration, can help to overcome the abiotic threshold to restoration (Holmes 2001).

Where soil-stored seed banks have been lost (e.g. ploughed fields) or depleted after long invasion by alien vegetation, a biotic threshold has been crossed and active restoration in the form of sowing and/or planting major fynbos structural components will be required to meet the goal of restoring ecological structure and function. This intervention should align with the natural post-fire recruitment regime in the autumn season, using fire-related stimuli to cue germination in sown seed. Where indigenous seed banks survive following dense alien

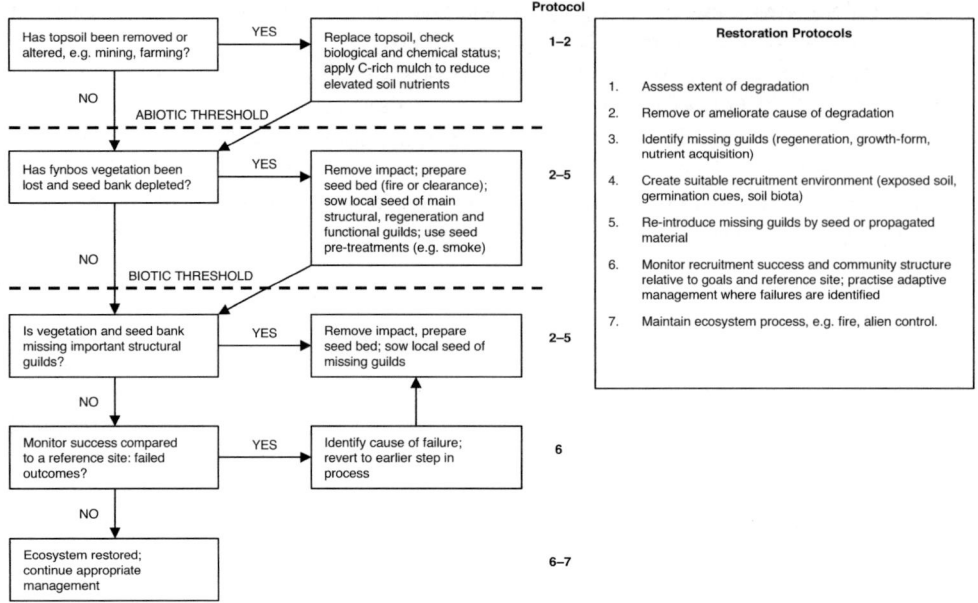

Figure 13.6 Flow diagram illustrating the degree of vegetation degradation and the minimum restoration actions required in order to restore fynbos community structure

invasion, the main intervention will be to re-introduce missing guilds, such as the overstorey proteoids with canopy-stored seeds. Ongoing follow-up control to ensure that invasive alien species do not re-establish will be essential. Restoration success will vary among sites and year of implementation, therefore it is important to set specific targets and timeframes that can be monitored towards meeting the restoration goal. This will assist in deciding when additional interventions may be required. A case study of restoration of Cape shrublands is outlined in Box 13.4.

Western Australian Mediterranean shrublands and woodlands

Ecological and evolutionary background

Within the structurally (and climatically) typical Mediterranean-type ecosystems of Western Australia (and smaller regions in South Australia, Western Victoria and New South Wales), the most iconic are the kwongan shrublands (incl. *Banksia* woodlands) on deep leached sandy regolith. Closed Jarrah and Wandoo woodlands on lateritic and shallow granite soils are yet another iconic vegetation type falling within the category of the MTE ecosystems. Technically, the latter ecosystem would qualify as dry forest, yet it has a number of features in common with the kwongan.

Besides the climatic characteristics common to all MTEs, especially the seasonal drought associated with high evapotranspiration and enhanced by frequent desiccating winds originating from the neighbouring semi-deserts and deserts, the major ecological challenge to restoration is posed by the soils supporting kwongan and woodlands; these are the oldest, and probably the nutrient-poorest soils on this planet (Lambers 2014). Fire-dependency of the

Box 13.4   Ecological restoration of old fields at Flower Valley, near Gansbaai, South Africa

Ecological and financial feasibility of active restoration was studied on three different invaded sandstone fynbos sites on Flower Valley Farm in the CFR with the aim to identify cost-effective ways of restoring functional native ecosystems following alien plant invasion (Gaertner *et al.* 2011; Figure 13.7). We tested mechanical clearing, burning, different soil restoration techniques and sowing of

Figure 13.7   Restoration study sites at Flower Valley Farm, Gansbaai, South Africa. (a) Eucalyptus plantation (*Eucalyptus conferruminata, E. cladocalyx* and *E. gomphocephala*). (b) Seedlings coming up two years after the restoration. (c) Acacia thicket (*Acacia cyclops, A. longifolia, A. mearnsii* and *A. saligna*). (d) Reference site with mountain fynbos on acidic Table Mountain sandstone. (e) *Pennisetum clandestinum* (kikuyu grass) field subject to experimental fire (f)

Source: photos by Brummer Olivier

native species. We also investigated the possibility of creating incentives for private landowners by introducing fynbos species for sustainable flower harvesting. Restoration was successful: diversity and evenness of native plant species increased significantly at all three sites, whereas cover of alien plants decreased. However, sowing of fynbos species had no significant effect on native cover, species richness, diversity or evenness in the recently invaded *Acacia* thicket (*Acacia cyclops, A. longifolia, A. mearnsii* and *A. saligna*) and a formerly ploughed kikuyu (*Pennisetum clandestinum*) grass field, implying that, after one cycle of invasion, the ecosystem was still sufficiently resilient to allow autogenic recovery. But introduction of native species improved ecosystem structure, particularly major fynbos growth-forms such as proteoids and ericoids shrubs that otherwise would have been under-represented.

Income from flower harvesting following active restoration consistently outweighed income following passive restoration, but the associated increase in income did not fully compensate the higher costs. In conclusion active restoration can be effective and financially more viable than passive restoration, depending on the invasion characteristics (Gaertner *et al.* 2011, 2012b).

Australian MTEs probably has the longest evolutionary history, and therefore changes to the fire regimes (such as protection of the rehabilitation sites from naturally occurring fires) might create a new problem for the rehabilitation.

The kwongan scrub and eucalyptus woodlands on nutrient-poor soils are extremely species rich (Mucina *et al.* 2014), home to old lineages, staggering number of endemic, rare and endangered plant species. They are a national biodiversity treasure deserving World Heritage status. Due to poor soils, agriculture has not impacted much on kwongan (except for some scrub types such as woodjil). Small areas of kwongan are under active mining targeting titanium-rich minerals (Eneabba, Cooljarloo) and some Jarrah forests have been targeted because of high-quality bauxite (e.g. Gardner 2001; Koch 2007).

Regional rehabilitation challenges

Post-mining rehabilitation in water-deprived Western Australia is naturally water constrained. Despite most of the flora of the kwongan shrublands and the eucalyptus woodlands being adapted to semi-arid conditions, the initial stages of the rehabilitation (spread of top-soil, seed broadcasting, planting) require considerable amounts of initial water input that might not be available especially in summer time. Low nutrients naturally do not pose a major constraint on the rehabilitation since most of the flora involved is well adapted (or well-exapted) to the low nutrient levels. These plants also sport a plethora of nutrient-acquisition strategies (Lambers *et al.* 2008). However, many of those involve special biotic interactions (mycorrhiza, rhizobia) and failure to restore the microbial life in soil and the microbial-plant interactions lead to failure in restoring populations of so called 'recalcitrant', often iconic, species groups (orchids, sedges and the like). There is not much knowledge on the role of fire in rehabilitations (but see for instance Roche *et al.* 1997), but scanty observations support an idea that controlled application of burning in progressive stages of the rehabilitation might be profitable to restore populations of some recalcitrant seeders.

Theoretical underpinning

Successful, scientifically based post-mining rehabilitation is rooted in understanding of (1) the nature of vegetation dynamical pathways of the impacted vegetation, (2) the processes of plant community assembly, (3) the interaction of the restored biotic community with its environment, (4) the technological tools used in the rehabilitation, and finally (5) skill in application of the scientific knowledge and technological power to formulate and execute the rehabilitation plan. The first three items invoke scientific knowledge as an important source of rehabilitation planning and execution; they are the theoretical underpinning of the entire process.

It appears that the world of post-mining rehabilitation (at least in Australia) has an obvious biodiversity-conservation focus motivated by recovering the 'lost' biodiversity patterns. On the other hand, the ultimate goal of post-mine rehabilitation from the point of view of mining-companies is the creation of a functioning, self-sustaining ecosystem (Mucina and Dobrowolski 2015). These are two different goals, underpinned by different scientific theories and sets of important trade-offs. We suggest that knowledge of the plant community assembly, especially the new approaches focusing on plant functional traits rather than taxonomy, should find serious consideration in setting new, realistic restoration targets, especially in recovery/ rehabilitation of species-rich 'recalcitrant' ecosystems.

Approaches to restoration on laterite and deep sands

Alcoa approach

Rehabilitation of Mediterranean-type eucalyptus jarrah woodlands (grading into dense forests) on extremely nutrient-poor soils is in the Australian context always cited a success story of rehabilitation (Banning *et al.* 2011; Grant and Koch 2007; Koch 2007). Perhaps this multi-layered and species diversity-focused approach is depicted best in a scheme presented in Figure 13.8. The approach takes into consideration a number of important ecological drivers and incorporates elements of landscaping as well as 'reconstructing of community structure', aimed at the major, politically important target – species diversity. How the reconstructed vegetation would behave in terms of ecosystem services remains to be seen since the pace of the vegetation-dynamic processes in slowly evolving systems is in discordance with the life span of a researcher or research-funding span. The rehabilitation of the jarrah woodlands and forests is surely an excellent example of cooperation between science and post-mining rehabilitation practice.

Iluka approach

Mining rehabilitation occurs in sequence with ore removal for mineral sands mining: topsoil removal and storage (or 'direct return' of topsoil where possible to other areas being rehabilitated); overburden removal and storage; removal and wet separation (physical separation by density) of the ore; return of sand/clay/mixed tailings to the pit; once tailings are drained, reshaped to an appropriate landform design; return of overburden then topsoil with its critical seed bank; broadcast of collected seed; surface stabilization against wind erosion with a cover crop, native mulch (now discontinued) or temporary chemical sticking agent; in-fill planting from nursery propagated material; and finally, monitoring of the establishing vegetation. Broad groupings of local floristic communities found on similar soils/landforms to those reconstructed

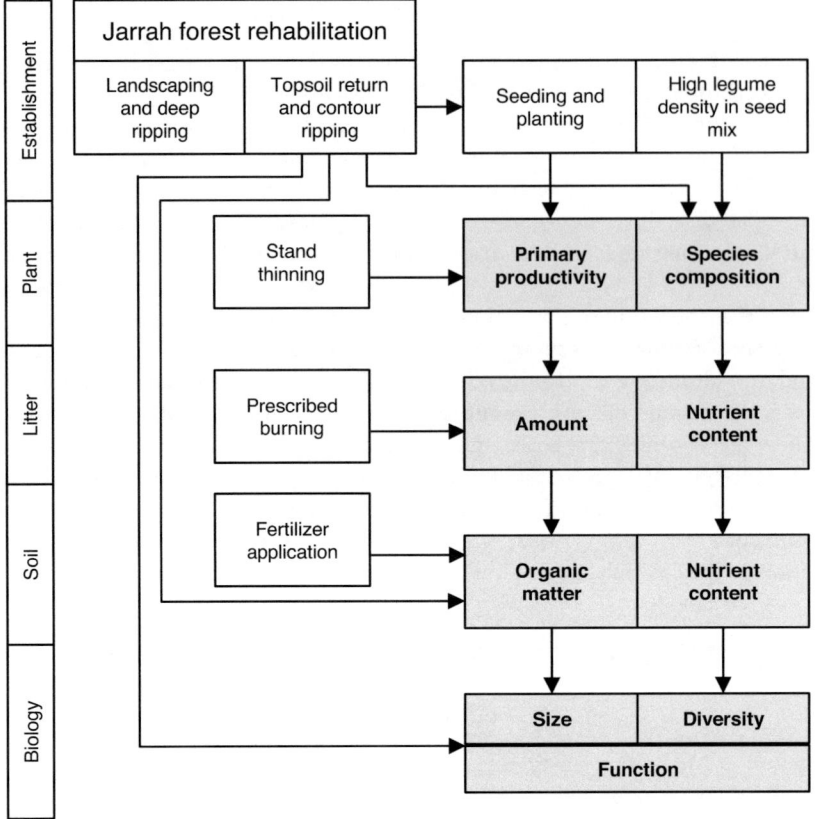

Figure 13.8   Schematic overview of the influence of management practices on ecosystem attributes in post-mining jarrah forest rehabilitation

Source: Banning *et al.* (2011)

in post-mining rehabilitation inform the choice of plant species from which seed is collected. This maximizes species diversity that can be selected in the knowledge that many species cannot be propagated from seed or in vegetative manner in nurseries.

New perspectives

Rehabilitation or restoration of species-rich systems showing a high level of functional complexity (rich spectrum of biotic interactions, specialized life histories of the constituent species, small-size populations, rarity etc.) is the ultimate frontier of the restoration science and practice.

Using the trait-focused approach to rehabilitation in predictive modelling of restoration and rehabilitation outcomes (Laughlin 2014) and its application in pot-mining rehabilitation should be broadened.

Acknowledgements

Laco Mucina thanks the Iluka Chair at the University of Western Australia for logistic support. This research was supported in part by an ARC Linkage grant LP150100339. Beatriz Duguy thanks A. Valdecantos and V. R. Vallejo for giving permission to reproduce some of their material, and in particular to V. R. Vallejo for his helpful comments and suggestions. Marcela A. Bustamante-Sánchez, Cecilia Smith-Ramírez and Juan J. Armesto thank project Fondef-Idea CA13I1027, project Native Forest Act 007, FONDECYT 11121452. Alberto Vilagrosa thanks projects SURVIVE (CGL-2011-30531-CO2-02), GRACCIE-NET (CTM2014-59111-REDC) and the PROMETEO program (DESESTRES 2014/038). CEAM is supported by the Generalitat Valenciana.

Credits

Laco Mucina co-authored (together with Mark Dobrowolski) the section on SW Australia and edited all sections and shaped the chapter. Beatriz Duguy Pedra and Alberto Vilagrosa co-authored the section on the Mediterranean Basin; Todd Keeler-Wolff wrote the section on Californian shrublands; Pat Holmes and Mirijam Gartner co-authored the part on the Cape (South Africa) shrublands, while Marcela A. Bustamante-Sánchez, Cecilia Smith-Ramírez and Juan J. Armesto co-authored the Chilean section of the chapter.

References

Allen, E. B., Padgett, P. E., Bytnerowicz, A. and Minnich, R. A. (1998) Nitrogen deposition effects on coastal sage vegetation of southern California, in *Proceedings of the International Symposium on Air Pollution and Climate Change Effects on Forest Ecosystems*, USDA Forest Service, Pacific Southwest Research Station, Riverside, CA.

Armesto, J. J., Rozzi, R., Miranda, P. and Sabag, C. (1987) Plant/frugivore interactions in South American temperate forests, *Revista Chilena de Historia Natural*, vol 60, pp. 321–336.

Armesto, J. J., Bustamante-Sánchez, M. A., Díaz, M. F., González, M. E., Holtz, A., Nuñez-Avila, M. and Smith-Ramírez, C. (2009) Fire disturbance regimes, ecosystem recovery and restoration strategies in Mediterranean and temperate regions of Chile, in A. Cerdà, P. Robichaud and R. Primlani (eds), *Fire Effects on Soil and Restoration Strategies*, Science Publishers, Enfield, NH.

Banning, N. C., Lalor, B. M., Grigg, A. H., Phillips, I. R., Colquhoun, I. J., Jones, D. L. and Murphy, D. V. (2011) Rehabilitated mine-site management, soil health and climate change, in B. P. Singh, A. L. Cowie and K. Yin Chan (eds), *Soil Health and Climate Change*, Springer, Heidelberg, Germany.

Becerra, P., González-Rodríguez, P., Smith-Ramírez, C. and Armesto, J. J. (2011) Spatiotemporal variation in the effect of herbaceous layer on woody seedling survival in a Chilean mediterranean ecosystem, *Journal of Vegetation Science*, vol 22, pp. 847–855.

Benayas, J. J., Bullock, J. and Newton, A. (2008) Creating woodland islets to reconcile ecological restoration, conservation, and agricultural land use, *Frontiers in Ecology and the Environment*, vol 6, pp. 329–336.

Blumler, M. A. (2005) Three conflated definitions of Mediterranean climates, *Middle States Geographer*, vol 38, pp. 52–60.

Bond, W. J. and van Wilgen, B. W. (1996) *Fire and Plants*, Chapman & Hall, London, UK.

Bozzolo, F. H. and Lipson, D. A. (2013) Differential responses of native and exotic coastal sage scrub plant species to N additions and the soil microbial community, *Plant and Soil*, vol 371, pp. 37–51.

Chirino, E., Vilagrosa, A., Cortina, J., Valdecantos, A., Fuentes, D., Trubat, R., Luis, V. C., Puertolas, J., Bautista, S., Baeza, J., Peñuelas, J. L. and Vallejo, V. R. (2009) Ecological restoration in degraded drylands: the need to improve the seedling quality and site conditions in the field, in S.P. Grossberg (ed.), *Forest Management*, Nova Publisher, New York, NY.

Cortina, J., Amat, B., Castillo, V., Fuentes, D., Maestre, F. T., Padilla, F. M. and Rojo, L. (2011) The restoration of vegetation cover in the semi-arid Iberian southeast, *Journal of Arid Environments*, vol 75, pp. 1377–1384.

Cowling, R. M., Rundel, P. W., Lamont, B. B., Kalin Arroyo, M. and Arianoutsou, M. (1996) Plant diversity in mediterranean-climate regions, *Trends in Ecology and Evolution*, vol 11, pp. 362–366.

Delitti, W., Ferran, A., Trabaud, L. and Vallejo, V. R. (2005) Effects of fire recurrence in *Quercus coccifera* L. shrublands of the Valencia region (Spain): I. Plant composition and productivity, *Plant Ecology*, vol 177, pp. 57–70.

Duguy, B., Paula, S., Pausas, J. G., Alloza, J. A., Gimeno, T. and Vallejo, V. R. (2013) Effects of climate and extreme events on wildfire regime and their ecological impacts, in A. Navarra and L. Tubiana (eds), *Regional Assessment of Climate Change in the Mediterranean, Volume 2: Agriculture, Forests and Ecosystem Services and People,* Springer, Dordrecht, the Netherlands.

Fuentes, E. R., Jaksic, F. M. and Simonetti, J. A. (1983) European rabbits versus native rodents in central Chile: effects on shrub seedlings, *Oecologia*, vol 58, pp. 411–414.

Fuentes, E. R., Otaiza, R. D., Alliende, M. C., Hoffmann, A. J. and Poiani, A. (1984) Shrub clumps of the Chilean matorral vegetation: structure and possible maintenance mechanisms, *Oecologia*, vol 62, pp. 405–411.

Fuentes-Castillo, T., Miranda, A., Rivera-Hutinel, A., Smith-Ramirez, C. and Holmgren, M. (2012) Nucleated regeneration of semiarid sclerophyllous forests close to remnant vegetation, *Forest Ecology and Management*, vol 274, pp. 38–47.

Fulé, P. Z. (2008) Does it make sense to restore wildland fire in changing climate?, *Restoration Ecology*, vol 16, pp. 526–531.

Gaertner, M., Richardson, D. M. and Privett, S. J. (2011) Effects of alien plants on ecosystem structure and functioning and implications for restoration: insights from three degraded sites in South African Fynbos, *Environmental Management*, vol 48, pp. 57–69.

Gaertner, M., Holmes, P. M. and Richardson, D. M. (2012a) Biological invasions, resilience and restoration', in J. van Andel and J. Aronson (eds), *Restoration Ecology: The New Frontier*, Blackwell, Oxford, UK.

Gaertner, M., Nottebrock, H., Fourie, H., Privett, S. D. J. and Richardson, D. M. (2012b) Plant invasions, restoration, and economics: Perspectives from South African fynbos, *Perspectives in Plant Ecology, Evolution and Systematics*, vol 14, pp. 341–353.

Gardner, J. H. (2001) Rehabilitating mines to meet land use objectives: bauxite mining in the jarrah forest of Western Australia', *Unasylva*, vol 207, pp. 3–8.

Grant, C. J. and Koch, J. (2007) Decommissioning Western Australia's first bauxite mine: co-evolving vegetation restoration techniques and targets, *Ecological Management and Restortation*, vol 9, pp. 92–105.

Hechenleitner, P., Gardner, M., Thomas, P., Echeverría, C., Escobar, B., Brownless, P. and Martínez, P. (2005) *Plantas amenazadas del centro-sur de Chile*, Trama Impresores, Chile.

Holmes, P. M. (2001) Shrubland restoration following woody alien invasion and mining: effects of topsoil depth, seed source and fertilizer addition, *Restoration Ecology*, vol 9, pp. 71–84.

Holmes, P. M. (2002) Depth distribution and composition of seed-banks in alien-invaded and uninvaded fynbos vegetation, *Austral Ecology*, vol 27, pp. 110–120.

Holmes, P. M. and Richardson, D. M. (1999) Protocols for restoration based on recruitment dynamics, community structure and ecosystem function: perspectives from South African Fynbos, *Restoration Ecology*, vol 7, pp. 215–230.

Holmes, P. M., Richardson, D. M., van Wilgen, B. W. and Gelderblom, C. (2000) Recovery of South African fynbos vegetation following alien woody plant clearing and fire: implications for restoration', *Austral Ecology*, vol 25, pp. 631–639.

Holmgren, M., Aviles, R., Sierralta, L., Segura, A. M. and Fuentes, E. R. (2000) Why have European herbs so successfully invaded the Chilean matorral? Effects of herbivory, soil nutrients and fire. *Journal of Arid Environments*, vol 44, pp. 197–211.

Hopper, S. D. (2009) OCBIL theory: towards an integrated understanding of the evolution, ecology and conservation of biodiversity on old, climatically buffered, infertile landscapes, *Plant and Soil*, vol 322, pp. 49–86.

Horton, T. R., Bruns, T. D. and Parker, V. T. (1999) Ectomycorrhizal fungi associated with *Arctostaphylos* contribute to *Pseudotsuga menziesii* establishment, *Canadian Journal of Botany*, vol 77, pp. 93–102.

Keeley, J. E. and Davis, F. (2007) Chaparral, in M. G. Barbour, T. Keeler-Wolf and A. Schoenherr (eds), *Terrestrial Vegetation of California*, 3rd edition, University of California Press, Berkeley, CA.

Keeley, J. E. and Fotheringham, C. J. (2001) The historic fire regime in southern California shrublands, *Conservation Biology*, vol 15, pp. 1536–1548.

Keeley, J. E. and Keeley, S. C. (1984) Postfire recovery of California coastal sage scrub, *American Midland Naturalist*, vol 111, pp. 105–117.

Keeley, J. E., Bond, W. J., Bradstock, R. A., Pausas, J. G. and Rundel, P. W. (2012) *Fire in Mediterranean Ecosystems. Ecology, Evolution and Management*, Cambridge University Press, New York, NY.

Koch, J. M. (2007) Alcoa's mining and restoration process in South-Western Australia, *Restoration Ecology*, vol 15, pp. S11–S16.

Kovats, R. S., Valentini, R., Bouwer, L. M., Georgopoulou, E., Jacob, D., Martin, E., Rounsevell, M. and Soussana, J.-F. (2014) Europe, in V. R. Barros, C. B. Field, D. J. Dokken, M. D. Mastrandrea, K. J. Mach, T. E. Bilir, M. Chatterjee, K. L. Ebi, Y. O. Estrada, R. C. Genova, B. Girma, E. S. Kissel, A. N. Levy, S. MacCracken, P. R. Mastrandrea and L. L. White (eds), *Climate Change 2014: Impacts, Adaptation, and Vulnerability. Part B: Regional Aspects. Contribution of Working Group II to the Fifth Assessment Report of the Intergovernmental Panel on Climate Change*, Cambridge University Press, Cambridge, UK.

Kribeche, H., Bautista, S., Chirino, E., Vilagrosa, A. and Vallejo, V. R. (2012) Effects of landscape spatial heterogeneity on dryland restoration success: the combined role of site conditions and reforestation techniques in southeastern Spain, *Ecologia Mediterranea*, vol 38, pp. 5–18.

Lambers, H. (ed.) (2014) *Plant Life on the Sandplains in Southwest Australia, a Global Biodiversity Hotspot*, University of Western Australia Publishing, Crawley, Australia.

Lambers, H., Raven, J. A., Shaver, G. R. and Smith, S. E. (2008) Plant nutrient-acquisition strategies change with soil age, *Trend in Ecology and Evolution*, vol 23, pp. 95–103.

Lambers, H., Shane, M. W., Laliberte, E., Swarts, N. D., Teste, F. and Zemunik, G. (2014) Plant mineral nutrition', in H. Lambers (ed.) *Plant Life on the Sandplains in Southwest Australia, a Global Biodiversity Hotspot*, University of Western Australia Publishing, Crawley, Australia.

Laughlin, D. C. (2014) Applying trait-based models to achieve functional targets for theory-driven ecological restoration, *Ecology Letters*, vol 17, pp. 771–784.

Lloret, F., Pausas, J. G. and Vilà, M. (2003) Response of Mediterranean plant species to different fire regimes in Garraf Natural Park (Catalonia, Spain): field observations and modelling predictions, *Plant Ecology*, vol 167, pp. 223–235.

Moreno, J. M., Vallejo, V. R. and Chuvieco, E. (2013) Current fire regimes, impacts and the likely changes – VI: Euro Mediterranean, in J. G. Goldammer (ed.) *Vegetation Fires and Global Change – Challenges for Concerted International Action. A White Paper Directed to the United Nations and International Organizations (Global Fire Monitoring Center (GFMC)*, Kessel Publishing House, Remagen-Oberwinter, Germany.

Mucina, L. and Dobrowolski, M. (2015) Post-mining restoration of species-rich kwongan shrublands: The time of paradigm change is overdue, in Y. Pung, L. Cronin and N. Bowman (eds), *Proceedings of the Third Workshop on Australian Mine Rehabilitation, Adelaide, South Australia. 18–20 August 2015*, JKTech, Brisbane, Australia.

Mucina, L., Laliberté, E., Thiele, K. R., Dodson, J. R. and Harvey, J. (2014) Biogeography of kwongan: origins, diversity, endemism, and vegetation patterns, in H. Lambers (ed.), *Plant Life on the Sandplains in Southwest Australia, a Global Biodiversity Hotspot*, University of Western Australia Publishing, Crawley, Australia.

Mucina, L. and Wardell-Johnson, G. (2011) Landscape age and soil fertility, climatic stability, and fire: beyond the OCBIL framework. *Plant and Soil*, vol 341, pp. 1–23.

Myers, N., Mittermeier, R. A., Mittermeier, C. G., da Fonseca, G. A. B. and Kent, J. (2000) Biodiversity hotspots for conservation priorities, *Nature*, vol 403, pp. 853–858.

Ovalle, C., Aronson, J., del Pozo, A. and Avendaño, J. (1999) Restoration and rehabilitation of mixed espinales in central Chile: 10-year report and appraisal, *Arid Soil Research and Rehabilitation*, vol 13, pp. 369–381.

Radloff, F. G. T., Mucina, L. and Snyman, D. (2013) The impact of native large herbivores and fire on the vegetation dynamics in the Cape renosterveld shrublands of South Africa: insights from a six-yr field experiment, *Applied Vegetation Science*, vol 17, pp. 456–469.

Raimondo, D., Von Staden, L., Foden, W., Victor, J. E., Helme, N. A., Turner, R. C., Kamundi, D. A. and Manyama, P. A. (2009) *Red List of South African Plants*, South African National Biodiversity Institute, Pretoria.

Rebelo, A.G., Boucher, C., Helme, N., Mucina, L., Rutherford, M. C., Smit, W. J., Powrie, L. W., Ellis, F., Lambrechts, J. J., Scott, L., Radloff, F. G. T., Johnson, S. D., Richardson, D. M., Ward, R. A., Procheş, S. M., Oliver, E. G. H., Manning, J. C., Jürgens, N., McDonald, D. J., Janssen, J. A. M., Walton, B. A., Le Roux, A., Skowno, A. L., Todd, S. W. and Hoare, D. B. (2006) Fynbos biome, in L. Mucina and M. C. Rutherford (eds), *The Vegetation of South Africa, Lesotho and Swaziland*, South African National Biodiversity Institute, Pretoria, South Africa.

Rebelo, A. G., Holmes, P. M., Dorse, C. and Wood, J. (2011) Impacts of urbanization in a biodiversity hotspot: conservation challenges in Metropolitan Cape Town, *South African Journal of Botany*, vol 77, pp. 20–35.

Reid, S. and Armesto, J. J. (2011) Interaction dynamics of avian frugivores and plants in a Chilean Mediterranean shrubland, *Journal of Arid Environments*, vol 75, pp. 221–230.

Reid, T. S. and Murphy, D. D. (1995) Providing a regional context for local conservation action, *BioScience*, vol 45 (Supplement: Science and Biodiversity Policy), pp. S84–S90.

Roche, S., Koch, J. M. and Dixon, K. W. (1997) Smoke enhanced seed germination for mine rehabilitation in the southwest of Western Australia, *Restoration Ecology*, vol 5, pp. 191–203.

Schulz, J. J., Cayuela, L., Rey Benayas, J. M and Schröder, M. (2010) Factors influencing vegetation cover change in Mediterranean central Chile (1975–2008), *Applied Vegetation Science*, vol 14, pp. 571–582.

Temperton, V. M., Hobbs, R. J., Nuttle, T. and Halle, S. (eds) (2004) *Assembly Rules and Restoration Ecology: Bridging the Gap between Theory and Practice*, Island Press, Washington, DC.

Valdecantos, A., Baeza, J. and Vallejo, V. R. (2009) Vegetation management for promoting ecosystem resilience in fire-prone Mediterranean shrublands, *Restoration Ecology*, vol 17, pp. 414–421.

Valdecantos, A., Fuentes, D., Smanis, A., Llovet, J., Morcillo, L. and Bautista, S. (2014) Effectiveness of low-cost planting techniques for improving water availability to *Olea europaea* seedlings in degraded drylands, *Restoration Ecology*, vol 22, pp. 327–335.

Vallejo, V. R. and Alloza, J. A. (2015) Postfire ecosystem restoration, in D. Paton (ed.) *Wildfire Hazards, Risks, and Disasters*, Elsevier, Amsterdam, the Netherlands.

Vallejo, V. R., Allen, E. B., Aronson, J., Pausas, J. G., Cortina, J. and Gutiérrez, J. R. (2012a) Restoration of Mediterranean-type woodlands and shrublands, in J. van Andel and J. Aronson (eds), *Restoration Ecology: The New Frontier*, 2nd edition, Blackwell Publishing, Malden, MA.

Vallejo, V. R., Arianoutsou, M. and Moreira, F. (2012b) Fire ecology and post-fire restoration approaches in Southern European forest types, in F. Moreira, M. Arianoutsou, P. Corona and J. De Las Heras (eds), *Post-fire Management and Restoration of Southern European Forests*, Springer, Dordrecht, the Netherlands.

Vallejo, V. R., Smanis, A., Chirino, E., Fuentes, D., Valdecantos, A. and Vilagrosa, A. (2012c) Perspectives in dryland restoration: approaches for climate change adaptation. *New Forests*, vol 43, pp. 561–579.

Van de Wouw, P., Echeverría, C., Rey Benayas, J. M. and Holmgren, M. (2011) Persistent *Acacia* savannas replace Mediterranean sclerophyllous forests in South America, *Forest Ecology and Management*, vol 262, pp. 1100–1108.

Van Wilgen, B. W., Forsyth, G. G., Le Maitre, D. C., Wannenburgh, A., Kotzé, J. D. F, van den Berg, E. and Henderson, L. (2012) An assessment of the effectiveness of a large, national-scale invasive alien plant control strategy in South Africa, *Biological Conservation*, vol 148, pp. 28–38.

Vasey, M. C., Parker, V. T., Holl, K. D., Loik, M. E. and Hiatt, S. (2014) Maritime climate influence on chaparral composition and diversity in the coast range of central California, *Ecology and Evolution*, vol 4, no 18, pp. 3662–3674.

Wilson, J. R., Gaertner, M., Griffiths, C. L., Kotze, I., Le Maitre, D. C., Marr, S. M., Picker, M. D., Spear, D., Stafford, L., Richardson, D. M., van Wilgen, B. W. and Wannenburgh, A. (2014) Biological invasions in the Cape Floristic Region: history, current patterns, impacts, and management challenges, in N. Allsopp, J. E. Colville and G. A. Verboom (eds), *Fynbos: Ecology, Evolution and Conservation of a Megadiverse Region*, Oxford University Press, Oxford.

Yelenik, S. G., Stock, W. D. and Richardson, D. M. (2004) Ecosystem level impacts of invasive *Acacia saligna* in the South African Fynbos, *Restoration Ecology*, vol 12, pp. 44–51.

14

ALPINE HABITAT CONSERVATION AND RESTORATION IN TROPICAL AND SUB-TROPICAL HIGH MOUNTAINS

Alton C. Byers

Introduction

Alpine ecosystems, the 'land above treeline', cover three per cent of the Earth's land surface and contain over 10,000 species of plants (Körner 1999), which ranks them among the most biodiverse per unit area of any ecosystem in the world. In sub-tropical and tropical high mountains of the world, they are of critical importance as sources of highly valuable medicinal and aromatic plants (Olsen and Larsen 2003; Buntaine *et al.* 2007; Byers *et al.* 2014); of freshwater supplies for millions of people living downstream (Bandyopadhay *et al.* 1997); for livestock grazing (Stevens 1993; Nagy and Grabherr 2009); extractive industries such as mining (Fox 1997); and as adventure tourism destination sites (Price and Kohler 2013). They are also among the most fragile ecosystems on Earth, with thin soils, cold environments, and slow-growing

Figure 14.1　Upper Imja Khola alpine ecosystem, Sagarmatha (Everest) National Park, Khumbu, Nepal

vegetation that is highly susceptible to disturbances as a result of turf mining, overgrazing, and the harvesting of shrubby vegetation for fuel (Byers 2005; Nagy and Grabherr 2009). In all cases, once the 'geomorphic glue' of alpine shrubs and protective alpine turf is removed from the thin alpine soils, rapid accelerations of mass wasting and soil loss can occur that can take decades, if not centuries, to heal (Körner 1999; Byers 2005, 2013).

Alpine ecosystems throughout the world have been heavily impacted by human activities, both historically (e.g. the systematic burning and removal of shrub juniper and dwarf rhodo-dendron to increase pasture area in the Himalayas, including the impacts of grazing; Byers 2013; Byers *et al.* 2014) and more recently with the exponential growth of mountaineering, trekking, and adventure tourism (Byers 1996, 2005; Nagy and Grabherr 2009). During the past decade, the high prices and global demands for various medicinal plants has led to the influx of tens of thousands of harvesters per year with heavy impacts upon the soils, vegetation, and wildlife (Byers *et al.* 2014). Collectively, it is probably safe to say that a majority of alpine ecosystems within the tropics and sub-tropics have been heavily modified, disturbed, and impacted by humans and cattle for thousands of years, processes now exacerbated by unregulated adventure tourism, globalization, new markets for medicinal plants, and climate change.

The following chapter presents a discussion of alpine ecosystems and restoration methods in tropical and sub-tropical high mountains of the world. Tropical mountains occur between the equatorial zones of 20° N and S and include the mountains of southern Mexico, Mesoamerica, the northern and central Andes, East African Mountains, and the New Guinea and southeast Asian Islands (Rhoades 2007: 65). Subtropical mountains lie within 20° and 40° N and S latitude and include the Karakorums, much of the Himalaya, southern Rockies, Spanish Sierra Nevada, the Atlas Mountains, the Zagros and Balkan ranges, the Bolivian and Argentinean Andes, and other ranges sharing a Mediterranean type of climate (*ibid.*: 64). Most of the countries contained within these two zones are considered to be lesser developed or developing countries, and with the exception of recent mining activity in the Andes exhibit fewer of the mechanized forms of disturbance to their alpine ecosystems common to mid lati-tude mountains (e.g. through mining and drilling, ski resorts, mechanized four-wheel travel) as opposed to traditional and recreational forms of disturbance (e.g. grazing, trekking and climb-ing impacts). As we shall see, restoration methods tested in tropical and sub-tropical mountains to date also tend to focus less on the use of direct prescriptions (e.g. re-seeding, soil fertiliza-tion, turf transplanting, biodegradable geotextiles; Urbanska and Chambers 2008) and more on community-determined behaviour and land use changes. Likewise, the most successful approaches in tropical and sub-tropical mountains have consistently been science-based and community-driven, as opposed to the science-based, government-financed and driven approaches of the west. Restoration methods and lessons learned are then presented, followed by case studies of successful alpine conservation projects in Nepal, Peru, and east Africa.

High-elevation alpine ecosystems: clarifying the terminology

In Europe, New Zealand, and Japan, 'alpine' commonly refers to entire mountain ecosystems, i.e. it includes valleys, forests, farmland, and pastures. In biogeographical terms, however, the alpine life zone is less inclusive and in much of the northern hemisphere it is considered to be confined to vegetation above the upper treeline, i.e. between the upper treeline and nival zone dominated by rock, ice, and snow. By this definition, the term is usually applicable to the mid-to high latitudes, but has been called 'unsuitable' for tropical and arid mountains (Troll 1988: 51) where it is often replaced by terms such as 'high-Andean' or 'afro-montane'. For example, native *Polylepis* forests in Peru, the highest growing trees in the world, ascend from a lower limit of 3,600 m to over 4,900 m in elevation, or nearly to the zone of permanent snow and ice, with no distinctive 'treeline' present (Körner 1999: 85). Furthermore, the intermingling of subalpine with alpine species further contributes to the challenges of explicitly defining alpine conditions or 'zones': plants common in lower elevations are also found above local treeline, making the distinction between subalpine and alpine difficult if defined strictly by species composition.

'Tundra' ('treeless plain') originally referred to high latitude arctic areas north of the timberline (Hadley *et al.* 2013). Although high latitude arctic and high altitude alpine ecosystems share many of the same species, they become increasingly dissimilar with increasing latitude. Cronin (1979) discourages the comparisons of vegetation belts on mountains with latitudinal changes in vegetation by pointing out that arctic environments are dominated by severe cold as a result of the Earth's shape and tilt of the axis, whereas the atmosphere of alpine ecosystems is modified by decreasing atmospheric pressure – the higher one goes, the less air pressure and density, and thus less warming of the air mass itself. Likewise, high latitude species are adapted to large *seasonal* temperature changes and stress (i.e. long, cold winters and short summers), whereas tropical and sub-tropical alpine species are adapted to large *diurnal* temperature stresses (i.e. large temperature differences between day and night).

Types of disturbance and management challenges

Urbanska and Chambers (2008) note that most high elevation ecosystems throughout the word have five things in common: (i) comparatively short growing seasons and unpredictable weather patterns, (ii) cold environments, (iii) restricted water availability, both physical and physiological, (iv) wind, and (v) potentially high radiant energy fluxes. Soils are young, thin, and often capped by a layer of alpine turf that, once broken or disturbed, exposes the soils to wind, water, gravity, and accelerated erosion and mass wasting processes. Plants have adapted to the cold, windy environment by having most of their root systems and biomass underground (Hadley *et al.* 2013); ground-hugging, low growth forms (Hadley *et al.* 2013; Cronin 1979); and waxy or hairy covering on leaves to help trap and store heat. Cronin (*ibid.*: 115) likens the alpine environment of the eastern Himalaya to the lunar landscape, with 'wide fluctuations in temperature, an attenuated atmosphere, intense solar radiation, and extreme aridity'. Alpine species and ecosystems in tropical and sub-tropical mountains are somewhat less sensitive to disturbances than those in the high latitudes, since the growing seasons are comparatively much longer (i.e. up to five months in the Mt. Everest region, and less than four weeks on Mt. Rainier).

Alpine ecosystems in tropical and sub-tropical mountains, although resilient under natural conditions, are nevertheless extremely sensitive to anthropogenic disturbances that occur in the course of livestock grazing, tourism infrastructure development, recreation, medicinal plant harvesting, and mining. The subalpine ecosystems and forests below can also be heavily impacted as a result of fuelwood harvesting, overgrazing, and contaminated water from mine tailings. Several of the more harmful uses of alpine ecosystems in the world's high mountain landscapes are discussed in the following text.

Fuelwood harvesting

One of the most common impacts of human usage of the alpine, and one normally and directly linked to adventure and trekking tourism, is the removal of slow-growing alpine shrubs and cushion plants by lodge owners, porters, and climbing/trekking parties for use as fuel (Stevens 1993; Byers 2001, 2005). Until recently, climbing expeditions in the Mt. Everest region of

Nepal and Tibet burned tons of shrub juniper annually as a *puja* (prayer or ceremony) requesting the gods for safe climbing conditions (J. Reinhard, pers. comm., 2015). In the absence of effective conservation and management systems, lodge owners in particular turn to the most convenient and free sources of local fuel, which happens to be the low and slow-growing shrubs found in their alpine ecosystems. One cross section of a juniper shrub harvested near the seasonal village of Chukkung, for example, showed a diameter of 5.5 cm and age of 157 years (Byers 2005). The shrubs are usually cut near the base, exposing normally protected soils and wildlife habitat (Figure 14.2). Roots and remaining woody material are often dug out as well, breaking up the alpine turf that protects the shallow and fragile inceptisols of the typical alpine environment (Price and Harden 2013). Mass wasting of these newly created 'turf islands' results, as well as accelerations of soil loss (Byers 1987, 2005; Körner 1999, 2002).

Figure 14.2 Soils exposed as a result of shrub juniper harvesting

Turf cutting

Other forms of disturbance include turf cutting (Byers 1987, 2005; Olsen and Larsen 2003), usually for use in the construction of rock walls as well as for flooring in the outdoor patios of lodges. If the turf is cut from relatively flat areas and left undisturbed, re-colonization by plants can occur within a matter of several years. If cut from a hillslope, however, the breakage of the protective turf, gravity, and exposure of fragile soils can result in increased soil loss, gully formation, and other essentially irreversible forms of mass wasting (Figure 14.3).

Pack animals

The increased numbers of pack animals in the Everest region (i.e. *yak* and *dzopio*; see Byers 2005, 2011) to accommodate growing numbers of tourists can result in soil compaction, overgrazing, and degradation of alpine pastures in general. Corridors of disturbance (i.e. swaths of highly disturbed alpine ecosystems several metres on either side of a popular trail) commonly form as well. In the Pisco and other valleys of Huascaran National Park, Peru, donkeys are used to transport climbers' gear to *refugios* or mountaineering lodges in the vicinity, for climbing objectives (Figure 14.4). Cattle also graze in the park up to altitudes of 5,000 m or more, but locals claim that the impacts of donkeys are much greater because their sharper hooves dislodge and displace soils and vegetation. Heavy grazing on alpine soils and vegetation has also been documented in Scandinavia, the High Atlas, South African Drakensberg, Tibet, and Bolivian Altiplano (Nagy and Grabherr 2009: 294–295).

Figure 14.3 Removal of protective alpine turf, whether by 'mining' for use in patio floors, overgrazing, or shrub juniper harvesting, can expose the thin and fragile alpine soils and accelerate soil erosion processes

Figure 14.4 Alpine cushion plants in the vicinity of the Pisco Refugio, Huascaran National Park, Peru. They are particularly susceptible to damage by donkeys left to graze once the climber or trekker's gear has been off-loaded

Improper solid and human waste disposal

The accumulation of solid waste at camping sites, basecamps, and high camp regions has been a chronic problem of alpine ecosystems since mountaineering first became popular in Europe in the mid–1800s, accelerating steadily with the exponential growth of trekking and mountaineering during the past 30+ years (Byers *et al.* 2011; Goldenberg 2011). In the Everest region, the growing presence of landfills and human waste disposal pits in the vicinity of villages along the main trekking routes (Rogers and Aitchison 1998; Byers *et al.* 2011; Goldenberg 2011) poses a growing health and safety concern for humans and livestock alike, as the landfills and leaking septic systems contaminate freshwater supplies and increase the incidence of gastrointestinal diseases among tourists and local people (Manfredi *et al.* 2010; McDowell *et al.* 2013). The problem is becoming more acutely felt in many popular trekking/mountaineering destinations, where traditional freshwater sources and springs are disappearing as a result of changing precipitation and snowfall patterns (Manfredi *et al.* 2010; Byers 2013). It is not, however, strictly limited to tourists – harvesters of the highly valuable yarsugumba ('caterpillar fungus') in alpine ecosystems throughout the central and eastern Himalayas reportedly leave behind massive amounts of garbage each year (e.g., Winkler 2008; Weckerle *et al.* 2008), in addition to negatively impacting vegetation, wildlife, and wildlife habitat (Cox 2008).

Medicinal plant harvesting

Alpine ecosystems are the home to many dozens of medicinal and aromatic plants that in recent years have developed a growing international market. An archetypical example is the case of *yarsugumba* (*Ophiocordyceps sinensis*; caterpillar fungus). Translated as 'summer grass, winter worm' in Nepali, *yarsugumba* is a parasitic complex formed by the relationship of the fungus *Ophiocordyceps sinensis* with the larval stage of several species of moth of the genus *Thitarodes*, known as the 'ghost moth'. *Yarsugumba* is one of the most valuable medicinal fungi in the world, an alleged aphrodisiac that can ultimately command prices of over $50,000 per kilogram in Chinese markets. Today, tens of thousands of harvesters per year live and collect *yarsugumba* in the alpine regions of Nepal and Tibet between April and July, in valleys that historically have seen only a handful of herders and brief visits from pilgrims and trekkers. Impacts in the form of poaching and decreased wildlife populations, de-vegetation of alpine hillslopes, solid and human waste problems can occur as a result (Byers *et al.* 2014).

Climate change

Alpine plants and animals may be adapted to extreme environments, but as the climate becomes warmer they are poor competitors against more aggressive or generalist species from lower altitudes that are now migrating upward. The sound stewardship, conservation, and restoration of alpine ecosystems are most likely the most promising buffers against these and other impacts of climate change, since a continuous (as opposed to disturbed or degraded) vegetative cover cools the environment, intercepts rainfall, promotes better water infiltration, reduces soil loss, and discourages invasions from lowland species through ecosystem integrity.

Ineffective fee systems

Most parks in the tropics and sub-tropics charge entrance fees, and many charge additional fees for mountain climbing. The Government of Nepal, for examples, collects US$2 million in Everest expedition fees per year and an unknown amount for the 'trekking peaks', or those generally between 5,650 and 6,500 m in elevation. The cost is between US$350 and US$500 per climber or climbing group of one to four people. These fees, however, are deposited in a central government fund and are not normally available for conservation projects.

Disengaged stakeholders

In 2004, The Mountain Institute (TMI) and American Alpine Club (AAC) initiated a promising start in the conservation of the world's alpine ecosystems by establishing the Alpine Conservation Partnership (Byers 2007, 2008). Although highly successful alpine conservation and restoration pilot projects were implemented by the Partnership in the Everest/Makalu regions of Nepal and the Cordillera Blanca region of Peru, securing the active involvement of international climbing clubs, conservation organizations, trekking and climbing service providers, and the outdoor retail industry in follow-on activities fell short of expectations for reasons that are not entirely understood. Clearly, finding new and creative ways to engage all relevant stakeholders will be critical to preserving of the world's alpine ecosystems for the use and enjoyment of generations to come. Continued awareness building among adventure tourists, service providers, governments, and other stakeholders would appear to be the first step in accomplishing this intuitively logical, but thus far evasive, task.

Field sampling methods and approaches

Urbanska and Chambers (2008) write that 'recognition of the differences that exist in the spatial and temporal scales as well as severity of disturbance is critical for determining the appropriate restoration approach'. Understanding the disturbance history of a region is critical to the development of meaningful and effective restoration approaches. We therefore advocate an integration of the physical and social sciences in the interpretation of a disturbed alpine ecosystem that can include the following steps:

1   Develop trust with local governments, the national park, and/or local communities that includes clarity of project objectives, roles and responsibilities, funding availability, and memoranda of understanding (the latter in the interests of project continuation in the event of staff transfer).
2   Work with local and indigenous experts to understand the social, historical, ecological, cultural, and livelihood characteristics of the place and its current condition.
3   Collect landcover data (percentage bare ground, shrub, herbaceous, rock, cut juniper) for the study site.
4   Determine and map the extent of landcover change (high altitude vegetation and glacial) within the study sites.
5   Determine the timing of change through primarily physical methods (repeat photography and satellite data time series vegetation/glacial mapping and change detection), combined with ground truth sampling, vegetation age analyses (e.g. ring counts), and oral testimony.
6   Identify the probable underlying causes of change within the study sites, including political, economic, subsistence, tourism development, social, and cultural factors, through interviews, observations, and interpretations of landcover change data.
7   Evaluate the conservation practices and performance of local communities and central government protected area authorities, if any, based on correlating findings on ecological change with the geographies of natural resource management and conservation responsibilities and efforts.
8   Use the insights from this spatial analysis and from interviews with indigenous residents and protected area officials and staff to analyse the factors which may have helped or hindered effective action.
9   Based on project results, propose prospective preventative, restorative, and management recommendations that could help minimize unsustainable land use processes.

Restoration methods and approaches

Formal relations and memoranda of understanding

Trust and formal memoranda of understanding (MOU) with local governments, park authorities, and stakeholders must be established prior to commencement of any community consultation process in order to maximize the potential for a project's success, effectively mainstream identified priorities into available funding opportunities, and avoid misunderstandings regarding roles, responsibilities, and funding. Even with the full endorsement of a current park warden, a formal MOU should be in place in the event of the warden's transfer and replacement with someone with little interest in the project. Consultations with religious leaders should also be held to secure their good faith and endorsement of any proposed intervention or programme, as they are normally held in the highest esteem by local people and considered to be an integral part of any community development or conservation project.

Build local leadership

Seek the active involvement of local NGOs, lodge owners' associations, trekking and climbing outfitters, youth groups, the national park, and others in the design and implementation of the project. Train interested stakeholders in field and data analysis methods, actively incorporating their own insights regarding historical and contemporary use of a region's alpine ecosystem. All data collected during the research phase should be freely shared, particularly with national parks and other government agencies charged with conservation or natural resource management. Include field assistants and participants as co-authors in any resultant peer-reviewed and popular publications to enhance their own professional growth and outreach.

Mainstream plans into existing sources of funding

Given the scarcity of resources, final project workplans and budgets should be mainstreamed into existing sources of funding, such as national park management plans, local government development plans, or on-going conservation projects funded by other donors. This is a process that begins with the establishment of formal relationships as discussed previously.

Incentives and alternatives

As mentioned previously, alpine ecosystem restoration methods in the tropics and sub-tropics are largely linked to the removal or lessening of stressors concurrent with some level of behaviour change. Once the root problems have been determined and shown to be deleterious, alternatives and incentives need to be provided for communities to realistically change their current practices. Cost-effective fuelwood substitutes, incentives to recycle or discontinue the use of plastic bottles, and clear benefits of conserving and protecting an ecosystem need to be provided.

Banning the harvesting and burning of woody shrubs

Alpine shrubs can take hundreds of years to grow several centimetres in diameter (Byers 2005), and alpine cushion plants hundreds of years to grow millimetres (Cronin 1979). A continuous cover of alpine vegetation also binds thin and fragile soils, provides habitat for other plants and wildlife, promotes rainwater infiltration and downstream freshwater supplies, and cools the environment during a time of rapid climate change. The indiscriminate destruction of alpine vegetation for economic purposes is not only unacceptable, but illegal in all of the world's high mountain parks. Most community groups will readily acknowledge the problem once quantified and demonstrated, at which point viable alternatives need to be discussed.

Fuel and energy substitutes

Substituting kerosene and propane for woody vegetation as fuel at present is the most effective short-term solution to the problem. However, because kerosene and propane are fossil fuels, expensive and dangerous to transport, and often plagued by unreliable supplies, they should not be considered as best practices. Rather, alternative sources of fuel, particularly hydropower, offer the most promising sources of energy for stoves and cooking. Today, solar water heaters, battery chargers, and lights are used extensively in the Mt. Everest alpine region of Nepal (see Byers *et al.* 2011).

Porter rest houses

In addition to lodges and adventure tourists, porters can sometimes be part of the problem if not provided with adequate food, shelter, and clothing by their trekking or climbing agencies. In the absence of these basic camping needs, porters in the Everest region were often forced to spend the night in caves or makeshift shelters, burning whatever fuel happened to be handy and causing considerable damage to various high altitude camping sites over the years. The construction of porter shelters as part of an alpine conservation plan can remove much of the stress by providing them with a warm place to sleep and eat. Better yet, trekking companies need to better provide for their porters in the form of adequate shelter and high altitude clothing, which thankfully now appears to be becoming standard policy in some high mountain countries.

Restoration exclosures

Cattle and livestock exclosures, commonly 1 ha in size, are often an effective method of removing a primary source of disturbance that in turn allows a hillslope to recover, re-vegetate, and stabilize. They can also become a before-and-after learning and educational tool in demonstration of the recovery potential of a subalpine, forest, and grassland region, as well as a source of pride for local communities eager to showcase their effort's success. Exclosures can also be quite successful in alpine regions, if the alpine turf and soils have not been too heavily damaged. Care also needs to be taken that any species introduced into the exclosure for re-vegetation purposes is both native as well as the correct species (e.g. subalpine juniper species will not necessarily grow in upper alpine regions). Exclosures need to be maintained and protected against cattle entering through breaks in the fence or other forms of disturbance, otherwise the progress made in vegetation recovery may be erased within a day. Finally, it is recommended that a series of exclosures be constructed throughout the project region at different altitudes, aspects, and exposures for comparative analysis purposes.

Conservation contracts

In the Aquia region due east of Huascaran National Park, assistance was provided to improve the quality of local community pastures which resulted in an increased quality and volume of milk production. The arrangement was in exchange for the community's active involvement in the restoration of biodiversity-rich *Polylepis* forests in the upper valley – a 'conservation contract' approach that resulted in both groups achieving their desired objectives.

Visitor education

Awareness of a problem, its root causes, and prospective remedial solutions are key to reversing any negative usage trend in high mountain ecosystems. Climbers, trekkers, and other adventure tourists often have little idea that their actions or mere presence may be having deleterious impacts, such as the fact that their tea was heated using shrub juniper that took 200 years to reach a diameter of 5 cm. They may leave a high altitude basecamp after a climb with the full assurance of their outfitters that all garbage will be collected, which may or may not happen. Education and awareness building is key to reversing such trends and behaviours, both for client and service provider alike.

Case studies

Case study 1: tourism impacts in the Upper Imja Khola Valley of the Sagarmatha (Everest) National Park, Nepal

During the past several decades, adventure tourism has grown dramatically throughout the high mountain world. This is particularly true in the Mt. Everest region of Nepal, which in the late 1960s experienced only a handful of western tourists. Tourism has now grown to more than 35,000 per year, not including support staff which could increase the total at least twofold. Since such large numbers of visitors could be expected to have impacts on fragile alpine ecosystems, a study was conducted in 2001 to ascertain what precisely these impacts might be (Byers 2005). As part of the study, 25 systematic, belted transects and 225 sampling quadrats (25 m^2 each) were established between the altitudes of 4,200 m and 5,300 m in the Gokyo and upper Imja alpine valleys of the park (*ibid*.). In the upper Imja valley, results showed an average cover of 29 per cent shrub, 34 per cent herbaceous, 2 per cent detritus, 14 per cent rock, and 22 per cent bareground. 77 per cent of the plots in the upper Imja contained shrub juniper, of which 35 per cent had been cut and removed for fuelwood by local tourist lodges in Pheriche, Dingboche, Chukung, Dugla, Lobuche, Dragnag, Gokyo Peak, Ama Dablam basecamp, and Kala Patar (disturbance in Gokyo was somewhat less because of the region's fewer visitors and lodges). The removal of shrub juniper, dwarf rhododendron, and slow-growing cushion plants for fuel was actively de-stabilizing hillslope integrity and was most certainly a contributing factor to the very high rates of soil loss quantified during an earlier slope process study (Byers 1987, 2005). Rather than the highly publicized garbage and Everest basecamp cleanups representing the major adventure tourism threat to the park, an insidious but growing degradation of its alpine ecosystems was occurring that showed signs of acceleration in the face of continued lodge construction and growing numbers of trekkers, climbers, and porters.

When consulted, local people expressed concern over the problem but were not clear on what alternatives existed, or what exactly to do. Thanks to progressive thinking on the part of a number of lodge owners, businessmen, and businesswomen, however, a Khumbu Alpine Conservation Council (KACC) was formed (Figure 14.5), initially supported through a partnership between The Mountain Institute and the American Alpine Club. The KACC's first action was to prohibit the yearly harvesting of an estimated 70,000 kilograms of slow-growing shrub juniper (estimated at 2000 *doko* baskets per year at 35 kilograms per basket = 70,000 kilograms), followed by the establishment of a kerosene depot as a fuel substitute. A porter's shelter was restored at Lobuche that provided hot food and a warm place to sleep for porters, thus decreasing their demands for fuelwood at the end of the day. Burning of shrub juniper at the Everest basecamp for the previously described climbing *pujas* was banned next, followed by

Figure 14.5 The Khumbu Alpine Conservation Council, Dingboche village, Nepal

alternative energy trials and construction of a 0.5 ha demonstration exclosure above the village of Dingboche. A KACC office and community centre was built with educational alpine conservation displays, and the work of the KACC was endorsed by the highly respected Tengboche Rimpoche which provided additional credibility and prestige to the project.

The harvesting of shrub juniper was technically illegal in Nepal's national parks when the Imja and Hinku valley studies began. However, there was no enforcement of the law prior to the establishment of the KACC in 2004 and Mera Alpine Conservation Group in 2007 for reasons that are not entirely understood. Regardless, and following the Khumbu Alpine Council's decision to ban the harvesting of most shrub juniper,[1] stockpiles outside of the typical tourist lodge vanished within one season, the burning of juniper now replaced by kerosene and Indian-style primus stoves. Perhaps equally impressive was the Council's ban a year later on the burning of tons of shrub juniper at the Everest base camp, a part of every climbing expedition's *puja* (prayer) ceremony since the first attempts to climb the mountain by the Swiss in 1952.

The KACC remains a vibrant organization as of this writing. However, even after 10 years there is little change in the condition of hillslopes above Dingboche village (Figure 14.5) because of the extremely high levels of disturbance that the original alpine turf, shrub juniper, and soils experienced at some point in the recent past. Other alpine hillslopes throughout the upper Imja and Gokyo valleys appear to be in better condition, and it is likely that with rest the continuous cover of shrub juniper, grass, and forbs that greeted Sir Edmund Hillary during his 1951/952 reconnaissance trips will return once again (Hillary 1999). For Dingboche and similar regions, however, it could take hundreds of years to see any actual change, and high-lights the fact that prevention of disturbance in an alpine ecosystem is far better than attempts at restoration

Case study 2: Kilimanjaro

Contrary to popular accounts, the Marangu ('Coca-Cola') trail to the summit of Africa's high-est mountain is in relatively good condition in spite of the 200,000 visitors who attempt to climb to the mountain each year (climbers plus porters and support staff). The 'huts' at each site are reminiscent of a small village, with tastefully constructed A frames for tourists, lodges for porters, kitchens, storage, and flush toilets at Mandara and Horombo (the septic systems appeared to be fine when visited in 2007, although the toilets at Mandara needed some minor plumbing repairs). Garbage is cleaned up on the trail by garbage patrols, and packed out from each hut 'village' once 100 kilograms accumulate. The entire trail from gate to Kibo is well maintained with drainage ditches and breaks identical to those found on the best trails in US national parks. 'Corridor' impacts such as those found on the Everest trek or Huascaran National Park basecamp approaches, are absent since no pack animals are used, only porters who walk single file on the established trail.

At Kibo huts (the last prior to the final ascent to the summit), however, the existing toilet structures resemble huge Sherpa-style *charpis* which are emptied once a year by hand and shovel, with the waste buried a few metres downslope in the thin alpine soils. This practice is unacceptable from a health and environmental point of view, and once again argues for the need to accelerate the development of reliable high-altitude waste management systems, such as solar composters (see Byers *et al.* 2011).

In summary, although Kilimanjaro tourism continues to suffer from a range of social and managerial problems (e.g. limited rescue and medical facilities, poorly clothed and equipped porters), the trails and ecosystems throughout the climb are in good condition and a testimony to proper physical maintenance and stewardship.

Case study 3: grazing impacts in the Ishinca Quebrada of the Huascaran National Park, Peru

Between 2001 and 2008, The Mountain Institute developed a pasture improvement programme in the lower altitude village of Collón, Ishinca valley, with the following three objectives:

- improve knowledge of the status of grassland uses in the alpine regions of Ishinca valley;
- develop conservation agreements with the community sectors of Pashpa and Collon with the objective of improving ecosystems within Ishinca valley; and
- reduce the pressures of grazing upon the Ishinca alpine zone by constructing a series of lower altitude, grassland restoration exclosures with superior pasturage.

The initiative was based upon the hypothesis that pasture improvement with the park's inhab-ited buffer zone would induce cattle owners to reduce their numbers of cattle that grazed the *puna*, or alpine grasslands, inside of the national park (Byers 2009, 2011). Five cattle-proof exclosures (1 ha each) were constructed in the vicinity of Collon in 2000, used primarily for community cattle grazing but also rented to *arrieros* for the grazing of their donkeys, llamas, and horses used in the tourist trade. Between 2001 and 2007, an additional 40 exclosures were constructed by the village of Collon using their own resources, and 15 more were constructed in 2008 as part of The Mountain Institute and the American Alpine Club's Alpine Conservation Partnership project (Byers 2008). In the early 2000s, 100 of the total 140 cattle that normally free ranged the upper Ishinca alpine zone were removed to Collon, presumably because of better grazing conditions, leaving 40 of the original and the occasional *chucaro* (feral cattle) behind. Annual burning of the upper Ishinca alpine pastures was also discontinued at this time.

On 31 May, 2009 two of the eight transects from the 2001 study were re-sampled. Transects 1 and 2, located east and about 0.5 km up-valley of the Ishinca Refugio, ranged in elevation from 4,360 m to 4,802 m and 4,350 m and 4,670 m respectively, both with predominantly south-facing exposures. GPS allowed for the re-location of each original central plot to within ±10 m. Because the coordinates of the other two randomly sited plots were not recorded, variable data from all three of the 2001 plots were averaged to establish a single average value for the particular stratification. 2009 data were derived from re-sampling the central plot as re-located by the GPS. All variables recorded during the 2001 field work were included in the 2009 sample.

Between the altitudes of 4,360 m and 4,775 m, results showed that bare ground had decreased between 70 and 100 per cent, and grass cover had increased between 39 and 69 per cent during the interim period. Herbaceous cover decreased between 25 and 63 per cent, presumably replaced by grass. Per cent terracette cover was reduced by nearly 100 per cent, no evidence of burning was found, and overland flow processes, a feature of concern in 2002, were now practically absent. The exception to this overwhelmingly positive trend was the lower stratification at an altitude of 4,360 m, the interface between alluvial plain hillslope, where bare ground increased by 19 per cent and grass cover decreased by 19 per cent. This was attributed in 2002, as well as in 2009, to the impacts of burros and horses used to carry tourist equipment to the basecamp and Refugio.

In Transect 2, between the altitudes of 4,390 m and 4,670 m, bare ground decreased nearly 100 per cent for all stratifications, grass cover increased from 64 to 153 per cent, and herbaceous cover again decreased significantly (although the average 2001 value was only 8 per cent ground cover). Per cent terracette cover decreased from between 73 and 100 per cent, no evidence of burning was found, and overland flow processes were now absent. Again the exception occurred at the 4,350 m stratification, where bare ground increased by 36 per cent and grass cover decreased by 67 per cent, again attributed to pack animals left to free range at the end of the journey from Collon to the Refugio.

Based on the comparison of groundcover variables from two transects sampled in 2002 and again in 2009, pasture and geomorphic conditions improved dramatically within the hillslope re-sampled that can most likely be attributed to the (a) reduction of cattle numbers and (b) cessation of annual burning that occurred during the interim period. The phenomenon demonstrates the resilience of even heavily damaged alpine ecosystems once protective measures are established, be it the banning of shrub juniper harvesting for expedition and lodge fuelwood, as in the Mt. Everest region, or the reduction of cattle in the Ishinca valley of Peru.

Conclusion

Alpine conservation and restoration methods in tropical and sub-tropical mountains to date tend to focus less on the use of direct prescriptions (e.g. re-seeding, soil fertilization) and more on community-determined behaviour and land use changes. Science-based evidence is a key component to the eventual reversal of negative trends, as local communities tend to appreciate the opportunity to incorporate science into their decision-making processes. Trust needs to be established between the field practitioner and all stakeholders involved – local communities, entrepreneurs, governments – prior to the community consultation process in the interests of clarity, realistic expectations, and accessing project funding opportunities. Once established, significant improvements in an ecosystem's management can occur within relatively short periods of time and at minimal cost.

For each case study presented earlier in the chapter, the key ingredients to project success were linked to four interrelated factors. Fundamental was (a) the science-based approach to the understanding of alpine disturbance problems, which in partnership with local stakeholders enable each project to (b) reduce the level of key threats, (c) achieve the desired ecosystem and social/behavioural result, and (d) ensure that long-term conservation mechanisms were in place.

In close partnership with local stakeholders, the reduction of threats and achievement of desired social and environmental results depended on the provision of viable alternatives or incentives to existing systems. Experiences will differ community by community, but the case studies shown previously also provide examples of how incorporating, linking, and quantifying livelihood improvement components within the conservation project or interventions can result in a higher level of success than just a conservation project alone. 'Long-term conservation mechanisms' can include the development of conservation easements, memoranda of understanding, new governmental regulations, new protected area status, and/or the establishment of local user groups or conservation committees dedicated to the improved management of the natural resource in question. The main challenge for the next generation of international alpine conservationists, however, will be the determination of ways in which to actively and meaningfully engage the adventure tourism, conservation, and retail industry communities in the actual conservation of alpine ecosystems through behavioural, financial, and advocacy means.

Note

1   Small amounts of shrub juniper leaves are burned as incense by households and monasteries as part of the morning *puja*, but the amounts are nothing compared to the thousands of tons burned for fuel in previous days.

References

Bandyopadhay, J., Rodda, J., Kattelmann, R., Kundzewicz, W. and Kraemer, D. 1997. Highland waters – a resource of global significance. In B. Messerli and J. D. Ives, (eds), *Mountains of the World: A Global Priority*. New York: Parthenon Publishing Group. pp. 131–156.

Buntaine, M. T., Mullen, R. B. and Lassoie, J.P. 2007. Human use and conservation planning in alpine areas of northwestern Yunnan, China. *Environment, Development and Sustainability* 9: 305.

Byers, A. C. 1987. Landscape change and man-accelerated soil loss: the case of the Sagarmatha (Mt. Everest) National Park, Khumbu, Nepal. *Mountain Research and Development* 7(3): 209–216.

Byers, A. C. 1996. Historical and contemporary human disturbance in the upper Barun valley, Makalu-Barun National Park and Conservation Area, east Nepal. *Mountain Research and Development* 16(3): 235–247.

Byers, A. C. 2001. *An Assessment of Contemporary Impacts on Alpine Ecosystems in the Sagarmatha (Mt. Everest) National Park, Nepal*. Final Report to the National Geographic Society, Committee for Research and Exploration. Washington, DC: Mountain Institute.

Byers, A. C. 2005. Contemporary human impacts on alpine landscapes in the Sagarmatha (Mt. Everest) National Park, Khumbu, Nepal. *Annals of the Association of American Geographers* 95(1): 112–140.

Byers, A. C. 2007. The Alpine Conservation Partnership: a global initiative to protect and restore alpine ecosystems. *American Alpine Journal* Winter: 121–123.

Byers, A. C. 2008. The Alpine Conservation Partnership: Saving Mountain Ecosystems, Building Resilience to Climate Change. *Alpine Journal* Fall: 210–220.

Byers, A. C. 2009. A comparative study of tourism impacts on alpine ecosystems in the Sagarmatha (Mt. Everest) National Park, Nepal and the Huascarán National Park, Peru. In J. Hill and T. Gale (eds), *Ecotourism and Environmental Sustainability*. London: Ashgate.

Byers, A. C. 2011. *Recuperacion de pastos alpinos en el valle de Ishinca, Parque Nacional de Huascaran, Peru: Implicaciones para la conservacion, las communicades y el cambio climatico.* Documento de Trabajo no. 1. Huaraz: Instituto de Montana.

Byers, A. C. 2013. The nature of Everest. In C. Anker (ed.), *The Call of Everest.* Washington, DC: National Geographic Society.

Byers, A. C., Culhane, T., Marcinkowski, D., Vaidya, S. and Howe, C. 2011. *Protecting and Restoring the World's Alpine Ecosystems through NGS Explorer Collaboration and Exchange: The Role of Awareness Building, Improved Conservation Practices, and High Altitude Alternative Energy Development.* Interim Report to the National Geographic Society-Blackstone Challenge Grant Program. Washington, DC: The Mountain Institute.

Byers, A., Byers, E. and Thapa, D. 2014. *Conservation and Restoration of Alpine Ecosystems in the Upper Barun Valley, Makalu-Barun National Park, Nepal.* Final Report. Elkins, WV: The Mountain Institute, Science and Exploration Program.

Cox, J.H. 2008. Uncontrolled exploitation of yarsugumba *Ophiocordyceps sinensis* in Rukum and Dolpa districts, Nepal: Observations in May–June 2007 and a suggested course of action. Kathmandu: The Mountain Institute and Department of National Parks and Wildlife Conservation. Report.

Cronin, E. 1979. *The Arun: A Natural History of the World's Deepest Valley.* Boston, MA: Houghton Mifflin Company.

Fox, D. J. 1997. Mining in mountains. In B. Messerli and J. D. Ives (eds), *Mountains of the World: A Global Priority.* New York: The Parthenon Publishing Group.

Goldenberg, S. 2011. Himalayas in danger of becoming a giant rubbish dump. *Guardian*, 12 September.

Hadley, K., Price, L. and Grabherr, G. 2013. Mountain vegetation. In M. Price, A. Byers, D. Friend and T. Kohler (eds), *Mountains: Physical and Human Dimensions.* Berkeley, CA: University of California Press.

Hillary, E. 1999. *View from the Summit.* London: Transworld Publishers.

Körner, C. 1999. *Alpine Plant Life.* Heidelberg: Springer.

Körner, C. 2002. Mountain biodiversity, its causes and function: an overview. In C. Körner and E. Spehn (eds), *Mountain Biodiversity: A Global Assessment.* London: Parthenon.

Manfredi, C., Flury, B., Vivlano, G., Thakuir, S., Khanal, S., Jha, R., Maskey, R., Kayastha, R., Kafle, K., Bhochhlbhoya, S., Ghimire, N., Shrestha, B., Chaudhary, G., Giannino, F., Carteni, F., Mazzoleni, S., Salerno, F. 2010. Solid waste and water quality management models for Sagarmatha National Park and Buffer Zone. *Mountain Research and Development* 30: 2, 127–142.

McDowell, G., Ford, J., Lehner, B., Berrang-Ford, L. and Sherpa, A. 2013. Climate-related hydrological change and human vulnerability in remote mountain regions: a case study from Khumbu, Nepal. *Regional Environmental Change* 13(2): 299–310.

Nagy, L. and Grabherr, G. 2009. *The Biology of Alpine Habitats.* New York: Oxford University Press.

Olsen, C. S. and Larsen, H. O. 2003. Alpine medicinal plant trade and Himalayan mountain livelihood strategies. *Geography Journal* 169 PN 3, 243–254.

Price, L. W. and Harden, C. P. 2013. Mountain soils. In M. F. Price, A. C. Byers, D. A. Friend, T. Kohler and L. W. Price (eds), *Mountain Geography: Physical and Human Dimensions.* Berkeley, CA: University of California Press.

Price, M. and Kohler, T. 2013. Sustainable mountain development. In M. Price, A. Byers, D. Friend and T. Kohler (eds), *Mountains: Physical and Human Dimensions.* Berkeley, CA: University of California Press.

Rhoades, R. 2007. *Listening to Mountains.* Dubuque: Kendall/Hunt Publishing Company.

Rogers, P. and Aitchison, J. 1998. *Towards Sustainable Tourism in the Everest Region of Nepal.* Kathmandu: IUCN.

Stevens, S. F. 1993. *Claiming the High Ground: Sherpas, Subsistence, and Environmental Change in the Highest Himalaya.* Berkeley, CA: University of California Press.

Troll, C. 1988. Comparative geography of high mountains of the world in the view of landscape ecology: a development of three and a half decades of research and organization. In N. Allan, G. Knapp and C. Stadel (eds), *Human Impact on Mountains.* Totowa, NJ: Rowman & Littlefield.

Urbanska, K. and Chambers, J. 2008. High elevation ecosystems. In M. Perrow and A. Davy (eds), *Handbook of Ecological Restoration.* Cambridge: Cambridge University Press.

Weckerie, C., Yang, Y., Huber, F., and Li, Q. 2010. People, money, and protected areas: the collection of the caterpillar mushroom *Ophiocordyceps sinensis* in the *Baima Xueshan* Nature Preserve, Southwest China. *Biodivers Conserv* 19: 2685–2698.

Winkler, D. 2008. Yartsa Gunbu (*Cordyceps sinensis*) and the fungal commodification of the rural economy in Tibet AR. *Economic Botany* 62(3): 291–305. doi:10.1007/s12231-008-9038-3.

15

RESTORATION OF RIVERS AND STREAMS

Benjamin Smith and Michael A. Chadwick

A brief history of degradation

Rivers around the world have been consistently exploited and modified to meet human needs. There is evidence for regulation of the Yellow River in China as long ago as 2000 BC, both Roman and Chinese civilizations used weirs and other structures to improve rivers for fisheries, and in the UK, the Domesday Book of 1086 records significant numbers of water mills and river embankments (Brookes 1988; Downs and Gregory 2004). European settlers in the United States and Australia rapidly altered rivers on a large scale. For example, clearing of native vegetation for agriculture in Australia frequently caused transitions in river type (Brierley and Fryirs 2005). In the Unites States, historical accounts of North American fisheries depicted how human settlements and their subsequent effects including dams, mills, boat passage and effluent from towns and cities among others 'destroy the fisheries of estuaries and rivers far more rapidly than the hook, net, or seine of the fisherman' (Gesner 1859: 289).

Following on from the industrial revolution there was rapid acceleration in the scale and intensity of negative impacts to aquatic systems and rivers (Newson 1992). Rivers were viewed as resources to be exploited or hazards to be controlled with little regard for ecological deterioration (Brookes 1988). Direct modification of river channels, via culverting, stream burial, damming, straightening, diversion and channelization heavily altered the morphology of the channel (Downs and Gregory 2004). At the same time, impacts on water quality primarily via discharges of untreated or minimally treated sewage, industrial effluent and run-off from agricultural land devastated aquatic ecosystems – with iconic rivers such as the Thames declared biologically dead (Brookes 1988). In addition to direct impacts on the river channels, extensive modification of catchment land-use, primarily for agricultural and urban development, severely altered the hydrological regime and sediment supply of large numbers of rivers (Allan 2004). Increases in fine sediments, nutrients such as nitrogen and phosphorus, and other more specific pollutants such as hydrocarbons and insecticides increased the risk of eutrophication and colmation of the streambed (i.e. clogging with fine sediment) and severely reduced the diversity and richness of aquatic species (*ibid.*).

The impacts of urbanization in particular occur at extremely low levels of urban land cover, and result in urban streams and rivers which are in general homogenous channels with simplified, generalist ecosystems (Paul and Meyer 2001). The term Urban Stream Syndrome (Walsh *et al.* 2005a) has been coined to describe the broad assemblage of changes associated with urban land cover, including channelization and subsequent lack of both lateral and vertical connectivity, elevated levels of nutrients and pollutants, and decreased longitudinal connectivity as a result of culverts and weirs, changes to sediment supply, and an altered hydrological regime (Paul and Meyer 2001; Walsh *et al.* 2005a; Wenger *et al.* 2009).

As a result the proportion of rivers in Europe, the UK and North America which can be considered to be in a natural condition is very small, with estimates of less than 5 per cent for the US and Europe (Graf 2001; Mant *et al.* 2012). This chapter focuses on restoration in developed countries, however, it is worth noting that although developing countries in general retain a larger proportion of their river networks in good condition, the pace of degradation is rapid.

The rise of river restoration

There is no single definition of river restoration, or even agreement as to the appropriateness of restoration as a term compared to rehabilitation or enhancement (Downs and Gregory 2004). Figure 15.1, adapted from Bradshaw (1996), shows the conceptual difference between restoration, rehabilitation, and enhancement or ecosystem creation. Restoration in its strictest sense aims for the complete return to a pre-disturbance state, focusing on the return to reference conditions. Rehabilitation is a more nuanced definition of restoration as a process of recovery (*ibid.*), while river enhancement, or the creation of novel ecosystems, can be seen as a type of ecological engineering in which rivers are managed to fulfil certain objectives set by society, for example denitrification (Palmer *et al.* 2014a).

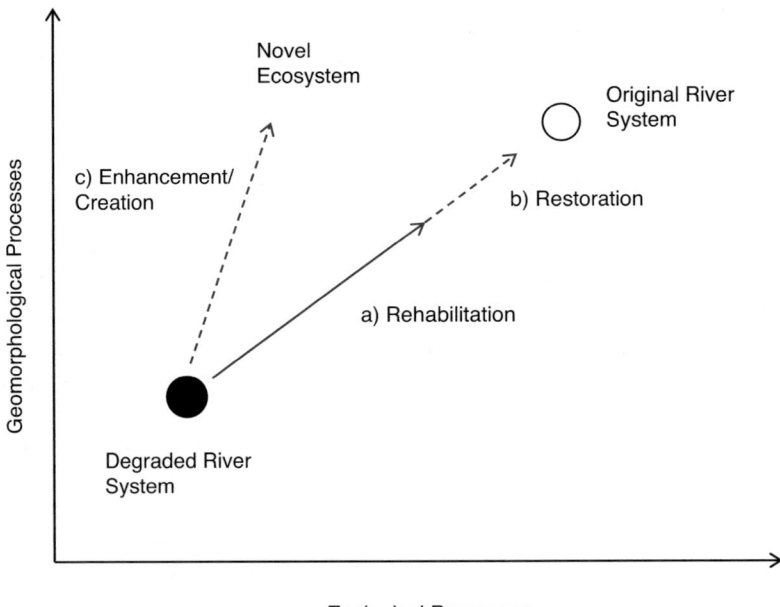

Figure 15.1 The conceptual difference between restoration, rehabilitation and enhancement. Enhancement is a form of ecological engineering and follows a pathway of improvement which does not target a return to the original ecosystem

Source: based on Bradshaw (1996: 4)

Although there are some early examples of awareness of the poor condition of rivers and the need for some form of restoration (see for example Hager and Barrett 1866), the modern field of river restoration began in earnest in the 1970s and 1980s (Downs and Gregory 2004). Work by Keller in the 1970s in the United States outlined efforts to increase hydrological and geomorphological diversity in channelized reaches, while in the 1980s restoration projects were also beginning across Europe and Japan (Keller 1978; Mant *et al.* 2012).

River restoration can be viewed as a phase of resource reconstruction following initial exploitation and degradation, and emerged as changing priorities within developed societies created the conditions for the rise of environmentalism (Wohl *et al.* 2008). Figure 15.2 illustrates the evolution of different phases in river degradation and restoration, following the model of an environmental Kuznets curve (Carson 2010). The economic transformation away from heavy industry and towards a service-oriented economy, coupled with increased environmental awareness from the 1970s onwards, has allowed space for improvement in many river systems. Deintensification of both catchment and channel use, whether deliberate or as a side-effect of other socio-economic trends, appears to have created a turning point in the pattern of degradation previously seen.

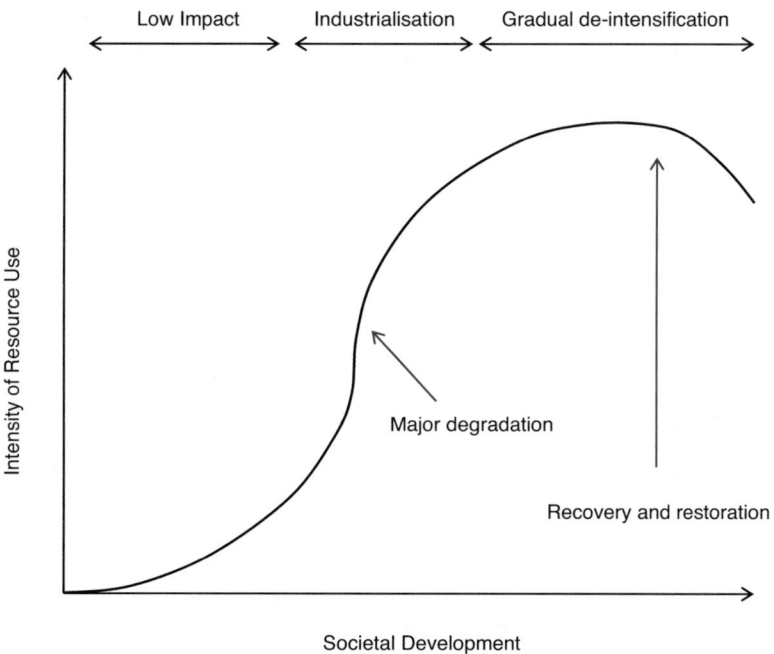

Figure 15.2   The relationship between societal development, the intensity of resource use and the pattern of degradation of river systems over time

Principles for restoration

Restoration is carried out for a variety of reasons, including, *inter alia*, fisheries improvement, meeting ecological objectives, climate change adaptation, natural forms of flood defence, aesthetic and recreational goals, and as a form of community-building (Smith *et al.* 2014). Over time, there has been a general shift in restoration projects from early small-scale projects focussed on fisheries and natural approaches to erosion control, to larger catchment-scale projects seeking to make improvements for the entire ecosystem (*ibid.*). There is also significant variation in the priorities for restoration between countries. In Australia and South Africa restoring environmental flows is the main focus, in Japan restoring fish passage dominates, while in Europe the Water Framework Directive is the main driver for restoration but significant variations remain, from fisheries restoration in Ireland, to flood defence in the Czech Republic (*ibid.*). A central focus in the United States, driven by the Clean Water Act, has been improving water quality, however, there are large differences between States, from urban restoration in Maryland, to fisheries in the Pacific Northwest (Palmer *et al.* 2007).

Although river restoration in developing countries is in general much less of a priority (Capps *et al.* 2016), with an understandable focus on basic issues of water and sanitation, it is important to note that restoration is not limited to Western Europe, the US and Australia. Restoration projects can be found in a range of countries, including, *inter alia,* Colombia, Ecuador, Estonia, South Korea, South Africa, China and Puerto Rico (Smith *et al.* 2014).

Despite large differences in the motivation for restoration, the last 25 years have seen the development of a set of overarching principles for how restoration should be carried out (see for example Boon 1992; Petts and Amoros 1996; Palmer *et al.* 2005; Beechie *et al.* 2010). Key principles which have emerged from the literature over the last 25 years are summarized later in the chapter, spanning issues of process and scale, but also now extending into the importance of social elements of restoration, and in particular stakeholder participation.

Process-based restoration

Following geomorphological principles, it is now widely accepted that restoration should focus on river processes rather than attempting to create a static channel form (Clarke *et al.* 2003; Brierley and Fryirs 2005). A focus on the geomorphological, hydrological and ecological processes needed to create a self-sustaining river system, such as sediment fluxes, nutrient cycling and a natural hydrological regime minimizes the need for repeated intervention and maintenance (Palmer *et al.* 2005; Beechie *et al.* 2010). Restoring processes allows the river to make its own adjustments and recovery in response to current and changing boundary conditions.

The emphasis on process-based restoration is a response to the many restoration projects which focussed on the creation of stable channel forms, often based on 'cookbook' type approaches such as Rosgen's Natural Channel Design (Rosgen 1994), which have been applied without appropriate knowledge of the local conditions (Kondolf 1995; Palmer et al. 2010). These designs are often inappropriate for catchment hydrological and geomorphological conditions, with subsequent problems with fine sedimentation or the washing out of structures during flood flows (Palmer *et al.* 2005).

Kondolf *et al.* (2001) exemplify this through a case study of Uvas Creek, California. The Creek rises in the forested mountains of central California, in a region with high variability in seasonal rainfall and flood peaks. This historically braided stream was restored in 1995 as a single meandering channel, based on misassumptions about the type of channel appropriate for the site, using a channel classification scheme. The new channel was held in place with boulders, with channel stability prioritized. The new channel lasted three months before being washed out by a small flood, after which the channel returned to a braided state. Despite consensus that restoration should be process-based, the re-creation of form is still widespread (Wohl *et al.* 2008).

Restoration at a catchment scale

The importance of catchment-scale processes in shaping the behaviour of river systems is a central tenet of fluvial geomorphology. Rivers are seen as part of a spatially nested system in which large-scale processes operating at the catchment scale set the boundary conditions for reach-scale behaviour (Frissel *et al.* 1988; Brierley and Fryirs 2005). Catchment geology, shape and land use mediate regional climate and shape sediment availability and discharge regime (Knighton 1998). Channel morphology is in turn determined by the interaction between sediment and discharge regimes and will vary according both to position within a catchment and the dominant catchment characteristics (Brierley and Fryirs 2005).

It follows that there is wide recognition that restoration should ideally take place at the catchment scale in order to address the underlying stressors affecting the river (Wohl *et al.* 2005; Hermoso *et al.* 2012). If restoration is to be effective then it must, therefore, take place at a scale which is large enough to maintain habitats and biophysical processes necessary for the functioning of the system.

The River Styles framework developed by Brierley and Fryirs (2005) in an Australian context explicitly integrates the catchment perspective into protocols for restoration planning. It takes a geomorphological approach to understanding the catchment and its history and how this influences river behaviour and trajectories of change and in particular, it emphasizes the need for individual reach-based restoration activities to be carried out in the context of a larger vision for the catchment and understanding of linkages between the different components and sub-systems (Brierley and Fryirs 2005). The explicit focus on the catchment provides a useful template for restoration at the requisite scale but achieving catchment-based restoration is difficult in reality. While successful examples of catchment-scale restoration exist, the administrative and socio-economic complexities involved in restoration at a catchment scale mean that the majority of projects remain reach-scale in nature (Hermoso *et al.* 2012; Palmer *et al.* 2014b).

Addressing the right stressors

Many authors stress the failure of restoration projects to address the main causes of degradation (e.g. Beechie *et al.* 2010; Kail *et al.* 2012). Because it is often practical to accomplish, reach-scale restoration tends to be common. Since this scale tends to focus on improving in-channel habitat, larger-scale problems such as sediment supply, water quality or the hydrological regime, are not addressed (Roni *et al.* 2008; Beechie *et al.* 2010). Kail *et al.* (2012) emphasize the dominant role of water quality and catchment land use as limiting factors constraining small-scale restoration efforts, while Sundermann *et al.* (2011) stress the importance of connectivity to colonizing populations as a factor limiting ecological recovery. In this context there is a call for clearer assessment of the pathways of ecosystem degradation in each case and what can be done to address these, rather than assuming that ecological improvement will necessarily follow habitat improvement.

Participation and multiple stakeholders

Community participation is now recognized as an essential element in restoration projects (Gregory *et al.* 2011). Different stakeholders may have different views of what a restoration project should achieve, and including these in project design is an important part of ensuring acceptance and long-term sustainability (Wohl *et al.* 2005). Petts (2006) notes the importance of facilitating a process in which all parties can learn from each other and contribute to the design of projects, thus increasing local ownership and commitment to the restoration. Advocates of strong participatory processes in river restoration note that in highly modified systems, there may be significant education, well-being and community-building benefits which accrue through projects, even where ecological recovery is missing (e.g. McDonald *et al.* 2004).

Case study: natural and cultural perspectives on river management on the River Wharfe, UK

McDonald *et al.* (2004) illustrate the difficulties involved using catchment-scale approaches and addressing the right stressors where multiple stakeholders are involved through a detailed case study of a river restoration project on the River Wharfe in the UK.

The River Wharfe is an upland gravel-river in the north of England, which has experienced problems with bed aggradation, and subsequent erosion and increased flood risk. Analysis showed that changes to catchment land use, combined with climatic change, had led to increased sediment supply, and were the root causes of bed aggradation.

In the 1980s new flood embankments were put in place, alongside a gravel trap, in order to address the problem, however, the trap filled up rapidly, and was ineffective by the mid-1990s. Aggradation of the river bed continued, leading to increased lateral erosion, and higher flood risk. In the late 1990s, the Upper Wharfedale Best Practice Project (UWBPP) was instigated to bring together the multiple stakeholders involved in managing the river, with the goal of developing best practice in upland land-water management. The local stakeholder context was complex, and included: local landowners, the Environment Agency (responsible for flood risk), the Department of the Environment, Food and Rural Affairs, the Forestry Commission, the Yorkshire Dales National Park, NGOs such as the National Trust, local parish councils and academics.

A detailed model of the system was developed, highlighting the drivers of sediment delivery, catchment-scale processes, and distinct zones of aggradation and erosion. The model noted that any measures to increase local transport capacity ran the risk of simply causing additional problems with downstream deposition.

Management options were developed through sustained engagement between technical teams, local representatives and the project steering committee, before being presented to the community as a whole. These ranged from hard engineering to keep the channel in place, to managed retreat in certain areas, or working with the river's natural processes to manage flood and erosion risk.

The development of the management options acknowledged that the root causes were linked to catchment hydrology and associated sediment delivery, however, the team noted that these were difficult to address given the management complexities involved, and community views about their cultural landscape. As such the options presented did not focus on management approaches which were deemed unrealistic given the management context.

The chosen option incorporated bank protection for one side of the river, but decreasing bank protection on the other side to allow some degree of river migration, along with removal of the previous gravel trap. Although this allows some floodplain building on the left bank, a key element in the community meeting where the proposal was chosen was to protect the land of riparian landowners in the area, representing a prioritization of river stability over dynamic processes and ecological objectives.

The end restoration project did not address the root causes of excess sediment and subsequent aggradation and erosion, but rather was driven by local concerns over the protection of pasture land and the stability of the river. A key element in the consideration of the scheme was the protection of the culturally influenced character of the local environment. The authors note that while the community stakeholders accepted the role of natural dynamics in the system, this was not enough to overcome a strongly ingrained desire for control. Crucially, they noted that the end project was a hybrid of technical and stakeholder perspectives, or natural and cultural perspectives.

The final design is not the technical approach which would have been taken if the project had been designed in isolation, and doesn't address root causes at the catchment scale. It is, however, an exemplary case study of how to manage tensions between different views of what a restoration project should achieve, and resulted in a project which was an acceptable compromise. The authors also note the importance of the project in incrementally changing perspectives on what river management can look like, and thus potentially setting the scene for more ecologically informed restoration at a later date. This is a point which is expanded on in an urban context in a 2016 paper highlighting the interactions between nature and culture in degraded urban systems (Smith *et al.* 2016).

Monitoring

The monitoring and appraisal of restoration has traditionally been poor, despite early calls to do so in the literature (e.g. Brookes 1988). In their assessment of river restoration projects in the US, for example, Bernhardt *et al.* (2005) found just 10 per cent were monitored. This lack of systematic monitoring has limited the potential to learn from past practice and test the application of ecological theories. Morandi *et al.* (2014), in an assessment of restoration projects in France, highlight another interesting element of the importance of monitoring, noting that projects with the weakest evaluation strategies reported the best outcomes. This suggests that poor monitoring not only limits the opportunity for learning, but also biases reporting of outcomes.

While the lack of monitoring and evaluation has hampered understanding of the geomorphological, hydrological and ecological response to restoration projects, there is evidence that the situation is improving, with Palmer *et al.* (2014b) reporting more projects with monitoring in place than during previous meta-analyses. In addition in Europe, reporting requirements under the Water Framework Directive should significantly improve post-project appraisal.

Citizen science, and the proliferation of cheap technologies for remote sensing such as small unmanned aerial vehicles, or 'drones', also provide the potential for cheap and comprehensive monitoring of restoration projects. Engaging community groups in the monitoring and evaluation group from the beginning, through repeated photography, or the provision of a drone with cheap sensors for repeated transects of a site, could greatly increase data on post-project dynamics and response. If integrated into projects in a coordinated way, the potential exists to greatly increase understanding of the response of different types of system to restoration interventions.

Challenges for river restoration

Despite its rapid growth and growing role as a tool for environmental management, river restoration faces major challenges – a common theme no matter if it is restoration of rivers, wetlands or terrestrial systems. These include persistent critiques such as the widespread lack of monitoring and evaluation, inappropriate and unsustainable designs, and poor evidence for ecological improvement following projects (e.g. Downs and Kondolf 2002; Bernhardt *et al.*

2005; Hermoso *et al.* 2012). In addition, emerging challenges such as how to ensure restoration projects are adaptive in the context of climate change, and how to reconcile ecological and social objectives for restoration projects must also be addressed (Wilby *et al.* 2010; Gilvear *et al.* 2013).

Limited evidence for success

Despite an overall lack of data on restoration outcomes, there is now a growing body of literature focussing on the evaluation of restoration projects, much of which questions their effectiveness (e.g. Roni *et al.* 2008; Palmer *et al.* 2010). In particular, a strong critique of the idea that ecological improvement will always follow habitat improvement has developed.

Various meta-analyses have been carried out to assess the effectiveness of different types of restoration project. In a global review of 345 studies, Roni *et al.* (2008) found that the effectiveness of different types of restoration technique was largely dependent on the suite of stressors in each project. Although they found instances of success, and certain techniques such as the removal of barriers to fish passage to be generally effective, the main finding was that responses were highly variable. Similarly, Stewart *et al.* (2009) in a review of restoration for salmonid species noted the unclear and highly variable response of populations to in-stream restoration structures depending on type of structure and local conditions. In a meta-analysis of the effects of restoration on macroinvertebrates, Miller *et al.* (2010) found mixed results depending on location, limiting factors and project design, with connectivity to regional species a crucial factor. Palmer *et al.* (2010) noted only two out of 78 studies reviewed showed an increase in invertebrate richness corresponding to the increase in habitat diversity through restoration. The main finding appears to be that there is no evidence that habitat heterogeneity is the main factor affecting invertebrate richness, in particular in a restoration context.

Subsequently, Palmer *et al.* (2014b) found that for projects using techniques to improve channel hydromorphology, or in-stream habitat, only 16 per cent resulted in any biological improvement. Restoration of the riparian zone exhibited a much stronger response, with ecological improvement of the river ecosystem found in 69 per cent of cases. The picture which emerges from these meta-analyses is that some techniques can be shown to be successful under certain circumstances, but that evidence for more widespread ecological improvement following restoration is limited.

Meeting legislative requirements – the Water Framework Directive

River restoration in Europe is widely acknowledged as a key tool in River Basin Management Plans (RBMPs) and programmes of measures associated with Water Framework Directive targets. This is highlighted by analysis showing that 96 per cent of RBMPs contain hydromorphological restoration measures aimed at achieving good ecological status (EEA 2012). Unfortunately, there is limited evidence to suggest that these restoration measures will be sufficient. As the EEA note, 'It is generally not clear how the proposed hydromorphological measures are expected to contribute to the improvement of ecological status or potential' (*ibid.*: 82).

There has been little work assessing the effectiveness of restoration projects in relation to the WFD. A notable exception is Haase *et al.* (2012), who evaluated 24 German restoration projects. While they found that projects were successful in increasing hydromorphological diversity, the biological response to restoration was weak. Following restoration, seven projects improved in overall WFD status compared to their reference sites, one decreased and 16 stayed the same, while only four of the 24 sites achieved Good Ecological Status overall. Both the EEA (2012) and Haase *et al.* (2012) noted that more evidence is needed that restoration can achieve its objectives in RBMPs. Wider trends in improving water quality are encouraging, however, significant improvements are required if WFD targets are to be met, even on an extended timescale (EEA 2012).

Blending the social and ecological: an ecosystem services perspective

Restoration is increasingly being framed from an ecosystem services perspective (e.g. Gilvear *et al.* 2013). A distinction should be made here between projects which are implemented in a traditional restoration context but then evaluated using an ecosystem services approach (e.g. the Mayes Brook restoration in London; Environment Agency 2011), and those which are designed specifically to target and improve one or more ecosystem services (e.g. nutrient cycling – see examples in Palmer *et al.* 2014a). The latter approach is a form of ecological engineering focussing on creating novel ecosystems specifically to benefit society, whereas the former retains a more broadly based ecosystem approach.

There are issues associated with both approaches. In urban areas certainly, and potentially more rural locations as well, evaluation of the costs and benefits of restoration using an ecosystem services approach shows the dominance of social and cultural benefits over ecological benefits. For example, the Mayes Brook restoration project in London showed the annual value of supporting and regulatory services is £59,000 whereas the value of cultural services is £820,000 annually (Environment Agency 2011). While highlighting the importance of the social and cultural dimension of river restoration, there is a danger that if an economic perspective dominates then planners will decide that ecological considerations are not important and design projects for social gain only.

An ecological engineering approach, focussing on enhancing certain ecosystem services may take a narrow perspective on what is needed in the restoration project. If the goal is to enhance nutrient retention, then there is no need to also attempt to try to restore communities of macroinvertebrates or fish (Palmer *et al.* 2014a). There is also little reason, *a priori*, to restore dynamic natural processes if the same effect can be achieved with engineered structures. While this approach may be appropriate in highly degraded locations where there is little potential for ecosystem improvement, it also may marginalize those elements of a river which do not provide a direct 'service' of interest (*ibid.*).

Reconciling social and ecological objectives remains a challenge for restoration, particularly where trade-offs are needed. There is potential for an ecosystem services approach to be effective in this regard, with its explicit focus on both social and ecological considerations, however more work is required to ensure that a balance can be found (Smith *et al.* 2014).

Climate change

Measures to adapt the freshwater environment to climate change have been limited, despite widespread recognition of the problem (Wilby *et al.* 2010). The potential clearly exists for restoration to play a key role in adapting to climate change – both in terms of shielding ecosystems from adverse effects (e.g. increasing shading to reduce the effects of temperature rises; Arnell *et al.* 2015), and in terms of protecting society (e.g. natural flood storage; EEA 2012). Although the links between restoration and climate change are frequently made there are very few projects explicitly addressing climate change (see Environment Agency 2011 for a review of the Mayes Brook 'climate change park' project, and Thomas *et al.* 2015 for a project assessing how riparian vegetation can protect salmonids from rising temperatures). Where restoration is thought of in relation to climate change adaptation the majority of references are to its potential in flood management.

Table 15.1 highlights potential restoration measures which could be used to support climate change adaptation. The greatest focus has been on the potential for riparian restoration to counter rising water temperatures, and flood storage measures to reduce the impact of future flooding (Arnell *et al.* 2015). Hydromorphological restoration can also play an important role, for example through the narrowing of over-widened channels (Natural England and RSPB 2014). In addition to specific restoration measures it is acknowledged that removing current constraints on freshwater systems will increase their ability to cope with climate change (Wilby *et al.* 2010; Natural England and RSPB 2014). In particular the removal of barriers to species dispersal, and the restoration of natural processes are expected to increase overall ecosystem resilience.

In order to do more than pay lip-service to climate change, however, there will need to be a concerted effort to include changes in the hydrological and thermal regimes into project planning. The significant variation found in catchment responses to both temperature and hydrological changes (Arnell *et al.* 2015) suggests that approaches using average changes at the national level are unlikely to be robust. Where available, information on likely changes in different catchments should be available to restoration practitioners, and if necessary translated into a form which is more easily accessible.

Table 15.1 Restoration measures to support climate change adaptation

Climate change impact	Restoration measures
Higher water temperatures	Riparian restoration to increase shading. Morphological restoration to narrow and deepen overwide channels.
Reductions in flow	Narrowing of over-widened channels to concentrate flow. Riparian restoration increases shading and reduces likelihood of algal blooms. Increase water retention in upper catchments.
Increase in flooding	Floodplain reconnection. Afforestation and in upland increased water retention in upper catchments.
Urban heatwaves	Riparian restoration enhances local cooling effects from urban rivers.
Increase in extreme rainfall events	Slowing water in upland areas will decrease peak flows. Sustainable urban drainage systems will decrease the pulse of pollutants into urban rivers following intense rainfall.
Favourable conditions for invasive species	Restoration which enhances overall ecosystem resilience will improve conditions for native species.

Future directions

There is a tension in river restoration between the principles for effective projects outlined before in this chapter, and the reality of restoration projects as they are planned and implemented in practice. A recurring theme in the literature around restoration failure is that projects fail because they do not take these principles into account, with the implicit assumption that what is needed is more information, more communication between scientists and practitioners, and further protocols for restoration. While this may be true in part, it ignores the constraints present in many restoration projects.

The majority of projects are carried out opportunistically, with land availability being a principle driver of project location (Palmer 2009). As Palmer notes: 'For most projects, sites are selected based on land availability even though this does little to ensure project success, and in some cases results in the selection of sites that are clearly suboptimal' (*ibid.*: 6).

Equally, restoration at a catchment scale very quickly becomes so complex from an administrative and political perspective, with the need to manage relationships with multiple land-owners, local authorities and other government agencies that in most cases it is unrealistic (McDonald *et al.* 2004). This is particularly the case in highly urban catchments, where current restoration best practice emphasizes the need for distributed stormwater management systems (Walsh *et al.* 2005b), which in most cases would need to be retrofitted to the existing urban environment.

The reality is that the choice of sites for restoration projects is not taken in a systematic way, and is driven more by social and economic considerations than ecological or geomorphological principles. They therefore face a suite of constraints driven by their 'sub-optimal' locations and difficult boundary conditions. The failure of many restoration projects to adhere to the principles outlined for restoration at a theoretical level is not necessarily due to a lack of information, but is intrinsic to the messy realities of site selection and project implementation. The diversification of objectives for restoration, and implementing groups, increases the likelihood of restoration in locations which face difficult constraints.

Increasing the effectiveness of many restoration projects may therefore rest on the ability to design and plan interventions which achieve ecological objectives in constrained situations. Is it possible, for example, to address water quality issues in urban streams through in-stream purification systems, rather than distributed drainage? Targeting research into what can be achieved in locations which are typical of restoration projects should therefore be an important area of focus.

References

Allan, D. J. (2004) Landscapes and riverscapes: the influence of land use on stream ecosystems. *Annual Review of Ecology, Evolution, and Systematics* 35: 257–284.

Arnell, N. W., Halliday, S. J., Battarbee, R. W., Skeffington, R. A., and Wade, A. J. (2015) The implications of climate change for the water environment in England. *Progress in Physical Geography* 39(1): 93–120.

Beechie, T. J., Sear, D. A., Olden, J. D., Pess, G. R., Buffington, J. M., Moir, H., Roni, P. and Pollock, M. M. (2010) Process-based principles for restoring river ecosystems. *BioScience* 60(3): 209–222.

Bernhardt, E. S., Palmer, M. A., Allan, J. D. and the National River Restoration Science Synthesis Working Group (2005) Restoration of US rivers: a national synthesis. *Science* 308: 636–637.

Boon, P. J. (1992) Essential elements in the case for river conservation. In P. J. Boon, P. Calow and G. E. Petts (eds), *River Conservation and Management*, pp. 11–34. Chichester: John Wiley & Sons.

Bradshaw, A. D. (1996) Underlying principles of restoration. *Canadian Journal of Fisheries and Aquatic Science* 53(Suppl. 1): 3–9.

Brierley, G. J. and Fryirs, K. A. (2005) *Geomorphology and River Management: Applications of the River Styles Framework*. Oxford: Blackwell.

Brookes, A. (1988) *Channelized Rivers: Perspectives for Environmental Management*. Chichester: John Wiley & Sons.

Capps, K. A., Bentsen, C. N. and Ramírez, A. (2016) Poverty, urbanization, and environmental degradation: urban streams in the developing world. *Freshwater Science* 35(1): 429–435.

Carson, R. T. (2010) The environmental Kuznets curve: seeking empirical regularity and theoretical structure. *Review of Environmental Economics and Policy* 4(1): 3–23.

Clarke, S. J., Bruce-Burgess, L. and Wharton, G. (2003) Linking form and function: towards an eco-hydromorphic approach to sustainable river restoration. *Aquatic Conservation: Marine and Freshwater Ecosystems* 13: 439–450.

Downs, P. W. and Gregory, K. J. (2004) *River Channel Management: Towards Sustainable Catchment Hydrosystems*. London: Hodder.

Downs, P. W. and Kondolf, M. G. (2002) Post-project appraisals in adaptive management of river channel restoration. *Environmental Management* 29(4): 477–496.

EEA (2012) European Waters – assessment of status and pressures. EEA Report no. 8/2012. Retrieved from www.eea.europa.eu/publications/european-waters-assessment-2012.

Environment Agency (2011) *The Mayes Brook Restoration in Mayesbrook Park, East London: An Ecosystem Services Assessment*. Report for the Environment Agency. Retrieved from www.theriverstrust.org/projects/water/Mayes%20brook%20restoration.pdf.

Frissel, C. A., Liss, W. J., Warren, C. E. and Hurley, M. D. (1988) A hierarchical framework for stream habitat classification: viewing streams in a watershed context. *Environmental Management* 10(2): 199–214.

Gesner, A. (1859) North American fisheries. *Journal of the American Geographical and Statistical Society* 1(10): 288–296.

Gilvear, D. J., Spray, C. J. and Casas-Mulet, R. (2013) River rehabilitation for the delivery of multiple ecosystem services at the river network scale. *Journal of Environmental Management* 126: 30–43.

Graf, W. L. (2001) Damage control – restoring the physical integrity of America's rivers. *Annals of the Association of American Geographers* 91(1): 1–27.

Gregory, C., Fisher, K., Brierley, G. and Clifford, N. (2011) Approaches to participation in sustainable river management: a comparative analysis of contemporary practices in Europe and New Zealand. *The International Journal of Environmental, Cultural, Economic and Social Sustainability* 7(4): 85–107.

Haase, P., Hering, D., Jähnig, S. C., Lorenz, A. W. and Sundermann, A. (2012) The impact of hydromorphological restoration on river ecological status: a comparison of fish, benthic invertebrates, and macrophytes. *Hydrobiologia* 704(1): 475–488.

Hager, A. D. and Barrett, C. (1866) *Report of Commissioners Relative to the Restoration of Sea-Fish to the Connecticut River and its Tributaries*. Montpelier: Freeman Steam Printing Establishment.

Hermoso, V., Pantus, F., Olley, J. O. N., Linke, S., Mugodo, J. and Lea, P. (2012) Systematic planning for river rehabilitation: integrating multiple ecological and economic objectives in complex decisions. *Freshwater Biology* 57(1): 1–9.

Kail, J., Arle, J. and Jähnig, S. C. (2012) Limiting factors and thresholds for macroinvertebrate assemblages in European rivers: empirical evidence from three datasets on water quality, catchment urbanization, and river restoration. *Ecological Indicators* 18: 63–72.

Keller, E. A. (1978) Pools, riffles, and channelization. *Environmental Geology* 2(2): 119–127.

Knighton, D. (1998) *Fluvial Forms and Processes*. London: Edward Arnold.

Kondolf, G. M. (1995) Five elements for effective evaluation of stream restoration. *Restoration Ecology* 3(2): 133–136.

Kondolf, G. M., Smeltzer, M. W. and Railsback, S. F. (2001) Design and performance of a channel reconstruction project in a coastal California gravel-bed stream. *Environmental Management* 28(6): 761–776.

Mant, J., Janes, M., Hammond, D. and Gill, A. (2012) In J. Van Andel and J. Alonson (eds), *Restoration Ecology: The New Frontier*, pp. 214–232. Chichester: Wiley-Blackwell.

McDonald, A., Lane, S. N., Haycock, N. E. and Chalk, E. A. (2004) Rivers of dreams: on the gulf between theoretical and practical aspects of upland river restoration. *Transactions of the Institute of British Geographers* 29: 257–281.

Miller, S. W., Budy, P., and Schmidt, J. C. (2010). Quantifying macroinvertebrate responses to in stream habitat restoration: applications of meta-analysis to river restoration. *Restoration Ecology* 18(1): 8–19.

Morandi, B., Piégay, H., Lamouroux, N. and Vaudor, L. (2014) How is success or failure in river restoration projects evaluated? Feedback from French restoration projects. *Journal of Environmental Management* 137: 178–188.

Natural England and RSPB (2014) *Climate Change Adaptation Manual: Evidence to Support Nature Conservation in a Changing Climate*. Natural England Report NE546.

Newson, M. D. (1992) River conservation and catchment management: a UK perspective. In P. J. Boon, P. Calow and G. E. Petts (eds), *River Conservation and Management*, pp. 385–396. Chichester: John Wiley & Sons.

Palmer, M. A. (2009) Reforming watershed restoration: science in need of application and applications in need of science. *Estuaries and Coasts* 32(1): 1–17.

Palmer, M. A., Bernhardt, E., Allan, J. D. and the National River Restoration Science Synthesis Working Group (2005) Standards for ecologically successful river restoration. *Journal of Applied Ecology* 42: 208–217.

Palmer, M., Allan, J. D., Meyer, J. and Bernhardt, E. S. (2007) River restoration in the 21st century: data and experiential knowledge to inform future efforts. *Restoration Ecology* 15(3): 472–481.

Palmer, M. A., Menninger, H. L. and Bernhardt, E. (2010) River restoration, habitat heterogeneity and biodiversity: a failure of theory or practice? *Freshwater Biology* 55(Suppl 1): 205–222.

Palmer, M. A., Filoso, S. and Fanelli, R. M. (2014a) From ecosystems to ecosystem services: stream restoration as ecological engineering. *Ecological Engineering* 65: 62–70.

Palmer, M. A., Hondula, K. L. and Koch, B. J. (2014b) Ecological restoration of streams and rivers: shifting strategies and shifting goals. *Annual Review of Ecology, Evolution and Systematics* 45: 247–269.

Paul, M. J. and Meyer, J. L. (2001) Streams in the urban landscape. *Annual Review of Ecology, Evolution and Systematics* 32: 333–365.

Petts, J. (2006) Managing public engagement to optimize learning: reflections from urban river restoration. *Human Ecology Review* 13(2): 172–181.

Petts, G. and Amoros, C. (1996) *Fluvial Hydrosystems*. London: Chapman & Hall.

Roni, P., Hanson, K. and Beechie, T. (2008) Global review of the physical and biological effectiveness of stream habitat rehabilitation techniques. *North American Journal of Fisheries Management* 28: 856–890.

Rosgen, D. L. (1994) A classification of natural rivers. *Catena* 22(3): 169–199.

Smith, B., Clifford, N. J. and Mant, J. (2014) The changing nature of river restoration. *WIREs Water* 1: 249–261.

Smith, R. F., Hawley, R. J., Neale, M. W., Vietz, G. J., Diaz-Pascacio, E., Herrmann, J., Lovell, A. C., Prescott, C., Rios-Touma, B., Smith, B. and Utz, R. M. (2016) Urban stream renovation: incorporating societal objectives to achieve ecological improvements. *Freshwater Science* 35(1), 364–379.

Stewart, G. B., Bayliss, H. R., Showler, D. A., Sutherland, W. J., and Pullin, A. S. (2009) Effectiveness of engineered in-stream structure mitigation measures to increase salmonid abundance: a systematic review. *Ecological Applications* 19(4): 931–941.

Sundermann, A., Stoll, S. and Haase, P. (2011) River restoration success depends on the species pool of the immediate surroundings. *Ecological Applications* 21(6): 1962–1971.

Thomas, S. M., Griffiths, S. W. and Ormerod, S. J. (2015) Adapting streams for climate change using riparian broadleaf trees and its consequences for stream salmonids. *Freshwater Biology* 60(1): 64–77.

Walsh, C. J., Roy, A. H., Feminella, J. W., Cottingham, P. D., Groffman, P. M. and Morgan, R. P. (2005a) The urban stream syndrome: current knowledge and the search for a cure. *Journal of the North American Benthological Society* 24(3), 706–723.

Walsh, C. J., Fletcher, T. D. and Ladson, A. R. (2005b) Stream restoration in urban catchments through redesigning stormwater systems: looking to the catchment to save the stream. *Journal of the North American Benthological Society* 24(3), 690–705.

Wenger, S. J., Roy, A. H., Jackson, C. R., Bernhardt, E. S., Carter, T. L., Filoso, S., Gibson, C. A., Hession, W. C., Kaushal, S. S., Martí, E. and Meyer, J. L. (2009) Twenty-six key research questions in urban stream ecology: an assessment of the state of the science. *Journal of the North American Benthological Society* 28(4): 1080–1098.

Wilby, R. L., Orr, H., Watts, G., Battarbee, R. W., Berry, P. M., Chadd, R., Dugdale, S. J., Dunbar, M. J., Elliott, J. A., Extence, C., Hannah, D. M., Holmes, N., Johnson, A. C., Knights, B., Milner, N. J., Ormerod, S. J., Solomon, D., Timlett, R., Whitehead, P. J. and and Wood, P. J. (2010) Evidence needed to manage freshwater ecosystems in a changing climate: turning adaptation principles into practice. *Science of the Total Environment* 408(19): 4150–4164.

Wohl, E., Angermeier, P. L., Bledsoe, B., Kondolf, G. M., MacDonnell, L., Merritt, D. M., Palmer, M. A., Poff, N. L. and Tarboton, D. (2005) River restoration. *Water Resources Research* 41: W10301.

Wohl, E., Palmer, M. and Kondolf, G. M. (2008) River management in the United States. In G. J. Brierley and K. A. Fryirs (eds), *River Futures: An Integrative Scientific Approach to River Repair*, pp. 174–200. Washington, DC: Island Press.

16

LAKE RESTORATION

Erik Jeppesen, Martin Søndergaard and Zhengwen Liu

Introduction

Why lake restoration?

Eutrophication following exposure to high nutrient loading has been the key environmental problem in lakes and reservoirs world-wide for many decades. Phosphorus (P) is invariably considered the most critical nutrient for the ecological state of lakes as well as any restoration attempts, though recent studies suggest that nitrogen (N) is of higher importance than previously assumed (Moss *et al.* 2013). The consequences of eutrophication are increased biomass of phytoplankton, resulting in turbid water, blooms of (often toxic) cyanobacteria and loss of biodiversity.

Exposure to excessive nutrients and the resulting eutrophication entail higher fish predation on invertebrates due to higher fish biomass and a shift towards dominance of coarse fish at the expense of fish-eating fish. This leads to enhanced predation on efficient large-bodied zooplankton grazers, such as waterfleas, and thereby to less filtering of phytoplankton. Moreover, many coarse fish stir up sediment when searching for food, creating higher turbidity, greater nutrient recycling and less light, diminishing the growth of submerged macrophytes. As submerged macrophytes disappear or their abundance considerably declines, the food sources and feeding habitats of aquatic birds and invertebrates become impoverished.

The vast majority of lake restoration projects aim to combat eutrophication (Cooke *et al.* 2005; Lathrop 2007; Søndergaard *et al.* 2007; Moss *et al.* 1996). However, lake restoration has also been used for other purposes, for instance to counteract acidification in soft water lakes by reducing the Sox loading and liming (Gunn and Sandöy 2003). However, here eutrophication is the main subject of our interest.

First solve the external problem

The first step in restoring lakes from eutrophication is to reduce the external nutrient input from waste water or other point sources by improved treatment or diversion of sewage. In agricultural areas, diffuse sources of nutrients are a key factor of eutrophication, and often major actions have to be taken to reduce the input from cultivated fields to freshwaters.

A multi-faceted approach is often needed to reach a sufficiently low external nutrient loading to lakes to achieve clear water conditions. Approaches include P stripping and occasionally

N removal at sewage works, sewage diversion, increased use of phosphate-free detergents, establishment of regulations concerning animal fertilizer storage capacity, fertilizer application practices, fertilization plans and green cover in winter. In addition, various measures can be implemented to enhance the nutrient retention and N loss capacity in lake catchments, better agricultural practices with less use and loss of fertilizers, re-establishment of wetlands, stabilization of river banks to reduce erosion, re-establishment of a natural riparian zone and flooding of riverine areas.

A number of simple empirical models have been developed to predict total phosphorus (TP) and total nitrogen (TN) concentrations in lakes based on information on external TP and TN loading and lake characteristics. Supplemented with empirical models between TP or TN and ecological state variables, such models may act as useful tools for setting critical loading targets for the different types of lakes (OECD 1982; Jeppesen *et al.* 2005).

A reduction of the external nutrient loading to a lake is sometimes termed 'lake restoration', but most frequently lake restoration is – as here – interpreted as within-lake physical, chemical and biological methods to restore a healthy ecosystem.

Focus on the within-lake problem

After a reduction of the external nutrient loading the state of the lake may not necessarily be satisfactory, either due to insufficient nutrient reduction, implying inadequate nutrient limitation of the phytoplankton, or chemical or biological resistance within the lake, preventing or delaying improvement. Chemical resistance is caused by P release – most prominently during summer in shallow lakes and seasonal overturns in deep, summer-stratified lakes – from a pool accumulated in the sediment when loading was high. Depending on the loading history and release mechanisms, this internal P loading typically persists for 10–15 years or even longer after the loading reduction until new equilibrium conditions are established (Jeppesen *et al.* 2005). Biological resistance may especially be attributed to (1) slow response of the fish community, maintaining the dominance of coarse fish, or (2) delayed recolonization of submerged macrophytes, which particularly in shallow lakes are important for maintaining stable clear water conditions (Hansson *et al.* 1998; Jeppesen *et al.* 2012).

Reinforcing recovery

To reinforce lake recovery, numerous physical-chemical and biological restoration methods have been developed. These methods aim either to reduce nutrient availability for phytoplankton (bottom-up effects) or increase the grazing pressure on phytoplankton (top-down effects), both of which may reduce the phytoplankton biomass. The key methods are described in Table 16.1.

Physical-chemical restoration methods

Various physical-chemical methods have been used to reduce P release from the sediment. These include sediment removal and various chemical treatments of the sediment or oxidizers (oxygen, nitrate) to increase the P retention capacity in the lake.

Sediment removal

Sediment removal is an efficient but relatively costly way to reduce an internal loading problem. Dredging may also serve to deepen a lake that is gradually filling in. Due to the potential

Table 16.1 Commonly used lake restoration methods to overcome eutrophication

Methods	Main principles	Possible problems
Physical-chemical methods		
Sediment dredging	Phosphorus-rich sediment is removed to reduce the internal loading of phosphorus.	Relatively expensive method.
Dilution/flushing	The lake is flushed with nutrient-poor water to decrease phosphorus concentrations in the lake water.	Depends on the availability of large amounts of nutrient-poor water.
Water-level manipulation	Increased water level decreases the risk of resuspension of sediment in shallow lakes. Decreased water level increases the growth potential of submerged macrophytes.	Not always easy to manipulate water level or obtain permission to do so.
Hypolimnetic withdrawal	To increase the outflow of phosphorus, the lake water outflow is redirected through a pipe to the hypolimnion where phosphorus accumulates during summer stratification in deep lakes.	The outlet of nutrient-rich water may create problems for downstream waterbodies.
Hypolimnetic oxygenation	Oxygen (or air) is fed to the hypolimnion of deep lakes to increase redox conditions and resorption of phosphorus.	Long-term treatment (>10 years) is often needed; yet, permanent effects are difficult to obtain.
Phosphorus inactivation	Salts of aluminum, iron, calcium or other phosphorus-binding agents are added to the water and/or the sediment to decrease the internal loading of phosphorus.	Aluminum is toxic at low and high pH, hampering its use in soft-water lakes.
Nitrate addition	Nitrate is added to the sediment to oxidize the organic matter, decrease the future oxygen demand and increase phosphorus resorption.	Nitrate addition is problematic in nitrogen-limited lakes.
Biological methods		
Removal of zooplankton-eating fish	Zooplankton- and/or benthic invertebrate-eating fish are removed to increase the number of large zooplankton and their grazing on phytoplankton.	Permanent effects may be difficult to obtain.
Stocking of predatory fish	Fish-eating fish are added to decrease the number of zooplankton-eating fish and augment zooplankton numbers and phytoplankton grazing.	Correct timing of stocking is crucial to achieve the desired effects.
Stocking with herbivorous fish	Plant-eating fish (grass carps) are added to reduce excessive growth of submerged macrophytes.	Phytoplankton takes over instead of macrophytes and creates turbid water.
Macrophyte transplantation and protection	Submerged macrophytes are transplanted and protected from plant-eating birds or fish to maintain high macrophyte coverage.	Relatively large areas of protected macrophytes are needed.
Macrophyte harvest	Nuisance growth of macrophytes is prevented to improve conditions for boating and fishing.	At favourable growth conditions, the macrophytes will soon reestablish.

Table 16.1 continued

Methods	Main principles	Possible problems
Introduction of mussels	Mussels are introduced to increase filtration of the water and create clearer water.	Often zebra mussels (*Dreissena polymorpha*) are used, but this species is also invasive and may reach very high densities with negative biological and physical consequences.
Combined methods		
Chemical-physical	Oxygenation and chemical treatment of sediment	
Combined chemical	Polyaluminium chloride and Phoslock	
Combined biological	Fish removal, stocking of piscivorous fish and transplantation of plants	
Bio-chemical	Fish removal and chemical treatment of the sediment	
Bio-physical	Fish removal and oxygenation of the hypolimnion	

Source: modified from Søndergaard *et al.* (2012)

presence of toxic substances (such as various heavy metals) in the sediment, a key problem is how to dispose of the sediment. Other concerns are how to obtain sufficient storing capacity of the sediment, disturbance of wildlife during the process and release of toxic substances to the lake and/or its outlets. Also, redistribution of sediment during the dredging period is important to consider and prevent, when possible. Typically, the upper 20–60 cm is removed, sometimes differentiated so that more sediment is removed from 'hot spot' areas.

Removal of sediment has in many cases led to an immediate and substantial reduction of the internal P loading. However, long-term success has frequently been hampered by a still too high external loading. Sediment removal is probably the most reliable method to reduce internal loading as it permanently removes the loading-triggering factors. For practical reasons, dredging is most useful in shallow lakes and reservoirs, in the latter preferably combined with water level draw-down during the process to reduce costs (Annadotter *et al.* 1999; Cooke *et al.* 2005).

Hypolimnetic withdrawal

This method aims at removing water from the hypolimnion (bottom water) where P and N accumulate during summer stratification. Typically, a pipe is installed in the deepest part of the lake with an outlet downstream below the lake level, enabling the pipe to act as a siphon. The method has often generated valuable results. In most cases, hypolimnion TP decreases and so does the depth of the anoxic layer, leading to reduced internal loading. An important drawback of the method is, however, that the effluents typically pollute downstream systems (high loading of total P, ammonia, low oxygen concentrations and, occasionally, an obnoxious smell of hydrogen sulphide). Another problem is that hypolimnetic withdrawal may weaken the thermal stratification, with potential release of nutrients from the bottom water to the surface water where the algal growth occurs. The method is particularly easy to apply in reservoirs where withdrawal can be established at the dam, but it may also be useful in summer-stratified lakes (Nürnberg 2007; Lathrop 2007; McDonald *et al.* 2004).

Artificial deep water mixing

In deep and stratified lakes, higher nutrient elimination during the stagnation period can be achieved using artificial mixing or destratification so that deep waters with high nutrient contents are mixed with the surface water and exported with the outflow at the surface. Moreover, deep mixing may lead to lower phytoplankton biomass, as the time in dark water reduces phytoplankton growth, and to fewer cyanobacteria. The method most often applied for destratification is introduction of compressed air by boreholes in pipes inserted horizontally above the lake bottom. The air-water mixture has a lowered specific weight, causing a rising water curtain, which destroys stratification, and ideally the lake remains fully circulated the entire year.

The results with this method have been variable and improvements have typically been inadequate and of short-term duration. Intuitively, it would be most easy to keep the system mixed in warm lakes where the temperature difference is smaller and the strength of the stratification weaker. Unless a sufficient amount of nutrients is transported out of the lake, the lakes typically return to the turbid state in the year after mixing is ended (Cooke *et al.* 2005).

Oxidation of hypolimnion

This method employs addition of oxidizers to the bottom water to improve resorption of phosphate to iron in the sediment and thereby reduce internal P loading. Most often oxygen is used, but alternatively electron acceptors such as nitrate can be applied. Oxygenation may also serve to improve the living conditions for fish and invertebrates as well as enhance coupled nitrification-denitrification, thereby enhancing nitrogen removal.

Oxygen is added either as pure oxygen or as atmospheric air and is typically injected via a number of diffusers creating fine oxygen or air bubbles in the deepest part of the lake. For nitrate, a liquid solution is added by stirring it into the upper sediment layer or by injecting it into the hypolimnion.

Once again, the results obtained have been variable. While some oxidation experiments have clearly led to lower accumulation of phosphate and reduced elements (e.g. ammonium) in the bottom water, others have produced no change in internal loading. This difference may reflect variation in the efficacy of the oxygenation and in the internal pool of organic matter and ammonia. Intuitively it would also be more difficult to get successful results in warm lakes as higher temperatures in the hypolimnion will enhance oxygen consumption and also oxygen concentrations, saturation concentrations being lower in warm lakes. Hypolimnetic oxygenation must be continued for many years to maintain the effects. Oxygenation increases the decomposition of organic matter in the hypolimnetic sediment and potentially increases the pool of mobile phosphorus that may be released when the oxygenation eventually stops. There is a risk that the method may break or destabilize the thermal stratification, not least in warm lakes where the stratification typically is weaker (Gächter and Wehrli 1998; Liboriussen *et al.* 2009).

Water level alterations

Water level management has been used extensively to enhance submerged macrophyte growth and thereby improve the habitat for waterfowl and water quality, and further to promote game fishing. The ultimate regulation is a complete draw-down, which may facilitate a shift to clear-water conditions in nutrient-rich turbid lakes, at least in the short term as drying out may consolidate the sediment. Moreover, fish kill mediated by the draw-down enhances zooplankton grazing on phytoplankton, leading to improved water clarity and thus better growth conditions for submerged macrophytes and food resources for waterfowl.

Changes in water level may also influence lakes indirectly by affecting fish recruitment, not least in reservoirs. Lack of flooding of marginal meadows in spring has been suggested as an important factor for poor recruitment of pike (*Esox lucius*) in regulated lakes. Short-term partial draw-down has been used to improve game fishing since it enhances the biomass and size of predatory fish at the expense of planktivorous and benthivorous fish, either through the increased predation risk due to the lower water table or drying of fertilized eggs. In shallow lakes, water level draw-down has not always resulted in improved water quality due to more frequent episodes of wind-induced sediment resuspension, enhancing lake water turbidity. This potential lack of improvement is highest in large lakes that are more exposed to wind stress. Water level regulation as a restoration tool may be particularly useful in small shallow lakes and in reservoirs (Cooke *et al.* 2005; Coops *et al.* 2003; Chow-Fraser 2005; Jeppesen *et al.* 2015).

Aluminium, iron, calcium and Phoslock treatment

Geoengineering aims at manipulating biogeochemical processes in lakes to improve their ecological state. It most commonly focuses on a fast reduction of the phosphorus concentration in the water column and a decrease of the internal P release from the sediment, leading to reduction of phytoplankton biomass and, in shallow lakes, a shift to a macrophyte-dominated state. The method aims at supplying new resorption sites for P in the lakes and is commonly applied to the inflows, surface waters, bottom waters or sediments using a range of slurry injection and spraying equipment. When added to the water column, the materials strip out dissolved phosphorus as they sink to the sediments where they may continue to reduce or prevent diffusion of dissolved phosphorus from the sediment to the water.

Phosphate (inorganic P) absorbs readily to calcite and hydroxides of oxidized iron and aluminum. Phosphate precipitation with calcite has been used in hardwater lakes, but the method is somewhat unpredictable because pH often drops below 7.5 in the sediment at which level calcite is dissolved (Cooke *et al.* 2005).

Of the two hydroxides, iron has the highest affinity for phosphate, and P release from oxic sediment surfaces is often controlled by iron when present in a molar ratio higher than 8:1 relative to phosphorus. However, iron is redox sensitive and typically the oxic sediment surface layer is reduced in summer when mineralization is high, which may result in release of phosphate from the sediment to the water. In deep stratified lakes this can be seen as an accumulation of phosphate in the anoxic hypolimnion during summer.

In contrast, aluminium hydroxide ($Al(OH)_3$) is stable and independent of redox conditions. Aluminum is usually added as alum (Al_2SO_4) or poly-Al chloride. During treatment, a fresh aluminium hydroxide floc is formed within hours, and when the floc sinks to the bottom this efficiently strips the lake water for phosphate and suspended solids. Silicate and humic acids may interfere with the binding of P, and ageing of the aluminium floc reduces the phosphate-binding capacity. The 'pH window' where aluminium treatment can be used ranges between 6 and 8.5 due to formation of soluble and toxic hydroxides at lower or higher levels. Therefore, aluminium should not be applied to lakes with the risk of high or low pH exposure of sediments and is not recommended for use in lakes with an alkalinity below 1 meq l^{-1} and in lakes where pH can be high due to, for example, high primary production (Cooke *et al.* 2005).

Of the three chemicals listed here, treatment with aluminium is the most widely used chemical restoration method and has been employed in more than 200 lakes worldwide. It is also a very cost-efficient solution compared to sediment dredging. Aluminium should be dosed in a molar ratio of at least 10:1 relative to the pool of potentially mobile P in the lake sediment and the water column (Søndergaard *et al.* 2007).

A number of other sediment-capping materials have emerged recently. One of these is Phoslock, which is lanthanum-modified clay and has an advantage over aluminium because it is not toxic. In soft-water lakes where aluminium is not safe to use, Phoslock may therefore act as a possible substitute. Here phosphate is bound irreversibly (depending on the conditions prevailing in the lake sediment) to lanthanum which is adsorbed to the clay matrix (Box 16.1). The downside of the method is the high cost of Phoslock compared with aluminium. The recommended dose of Phoslock is 100 g per gram of phosphorus to be retained. Other geoengineering materials used to reduce phosphorus concentrations in the water column include modified local soil materials such as Fe-modified bentonite or modified zeolite (Mackay *et al.* 2014).

Results of sediment capping with different materials vary. Up to 33 per cent of Al treatments have been unsuccessful or effective for only one or two years. Sometimes this can be attributed to an insufficiently low external nutrient loading or a too low Al dosing, but often the reasons are not clear. The effects of Phoslock treatment have been highly site-specific but often with significant effects on annual P concentrations and some effects also on chlorophyll a and Secchi depth (Lürling and Oosterhout 2013).

Biological restoration methods

For the past 20–30 years a number of biological methods have been developed.

Fish manipulation

One method of fish manipulation is to reduce the density of coarse fish, preferably by at least 75–80 per cent during a one- to two-year period. An alternative or supplementary method is stocking of fish-eating fish. A simple and feasible strategy of fish removal is to catch non-moving fish with active gear and actively moving fish with passive gear using information on the seasonal behaviour of fish, such as spawning or foraging migration and shoaling of the target species. In addition, ice (when present) fishing during winter, when fish may aggregate near the bottom, and removal of fish seeking winter refuge in adjacent streams have also been used. A more drastic method is to apply a piscicide (typically rotenone) treatment to part of the lake and/or its tributary streams. Rotenone affects all fish species, as well as invertebrates, and thus requires ethical considerations, and legal permissions may be difficult to obtain in some countries. Nonetheless, rotenone treatment has been used to regulate fish communities for sport fishing and to eliminate invasive species (Cooke *et al.* 2005).

Dramatic and cascading effects are generally achieved in eutrophic lakes from efficient fish reduction in the form of reduced phytoplankton biomass, less suspended matter, dominance by large-sized zooplankton and improved lake water transparency. Fish manipulation may cascade to nutrients as well. A 30–50 per cent reduction in lake TP concentration has been recorded in the most successful fish manipulation experiments (Box 16.2). So far, the long-term perspectives are less promising, though. A gradual return to the turbid state and higher abundance of coarse fish after five to ten years have been reported in many cases; however long-term or a permanently lower abundance of some coarse fish, like bream (*Abramis brama*) and carp

Box 16.1 Chemical treatment of Lake Rauwbraken, the Netherlands

A combined treatment of 2 tonnes PAC (poly aluminium chloride) and 18 tonnes Phoslock in Lake Rauwbraken, the Netherlands (4 ha, max depth 15 m, stratified in summer), in April 2008, effectively sedimented a developing bloom of the cyanobacteria *Aphanizomenon flos-aquae*. The average chlorophyll a concentration in the two years prior to this Flock and Lock treatment was 19.5 µg L^{-1} and was as low as 3.7 µg L^{-1} in the years following the treatment. The combined treatment effectively reduced the amount of total phosphorus (TP) in the water column from, on average, 169 µg P L^{-1} before the application to 14 µg P L^{-1} after the treatment. Based on mean summer chlorophyll a and TP concentrations, the lake shifted from a eutrophic-hypertrophic state to an oligo-mesotrophic state. From directly after the treatment in April 2008 until and including 2013, the lake remained in an oligo-mesotrophic clear water state where TP was reduced to less than 10 per cent of the pre-treatment level. This result shows that relatively small, isolated, stratifying lakes can be restored by targeting both water column and sediment P using a combination of flocculent and solid phase P-sorbent (Lürling and Oosterhout 2013).

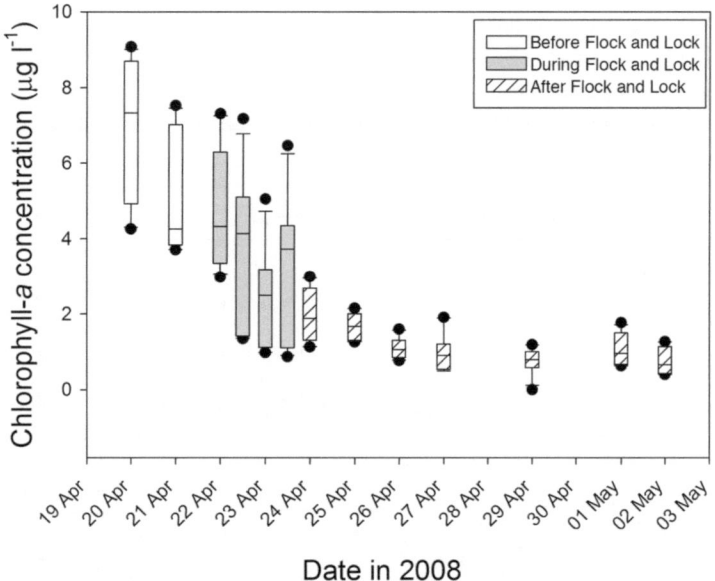

Figure 16.1   Chlorophyll a concentrations in Lake Rauwbraken, the Netherlands, in April/May 2008 around the combined flocculent PAC and lanthanum modified clay Phoslock (Flock and Lock) application. White boxes indicate the period before the Flock and Lock application, light grey boxes during treatment and dashed boxes after the treatment. The boundary of the boxes closest to zero indicates the 25th percentile, lines within the boxes mark the medians and the boundary of the box farthest from zero shows the 75th percentile. Whiskers above and below the boxes indicate the 90th and 10th percentiles, while the dots above and below the boxes represent the 95th and 5th percentiles

Source: Lürling and Oosterhout (2013)

Box 16.2 Fish manipulation in Lake Engelsholm, Denmark

Lake Engelsholm was subjected to nutrient loading reduction and showed delayed recovery. Therefore, restoration by biomanipulation was conducted in 1992–1994. Nineteen tonnes of cyprinid fishes were removed and their estimated biomass subsequently decreased from 675 to 150–300 kg ha^{-1}. The biomanipulation led to a substantial reduction of phytoplankton chlorophyll a, TP and TN, as well as an increase in Secchi depth. Submerged macrophyte coverage has generally been very low, both before and during the early years following the restoration. However, during 2007–2010 coverage increased gradually from 2 to 12 per cent (Jeppesen *et al.* 2012).

Change Point Analysis identified a shift in phytoplankton biomass trends in July 1993, immediately after the fish biomanipulation. There was a tenfold decrease in both median and minimum phytoplankton biomass (10th percentile) following the shift. Time series analysis revealed no trend in phytoplankton biomass before biomanipulation and a negative trend afterwards. Mean within-year variation in phytoplankton biomass almost doubled after the shift. Both median phytoplankton richness and evenness increased after the shift, whereas within-year variation increased for richness but decreased for evenness, indicating a year-round higher evenness of phytoplankton. Following biomanipulation, phytoplankton composition shifted from year-round dominance of cyanobacteria to, first, stronger seasonal succession among chlorophytes, cyanobacteria and diatoms and, next, higher dominance of the remaining groups.

There were also marked changes in zooplankton in the lake. Median zooplankton biomass as well as its mean within-year variation decreased after biomanipulation. Zooplankton richness and its within-year/interannual variation increased slightly with a positive trend after the shift (Jeppesen *et al.* 2012).

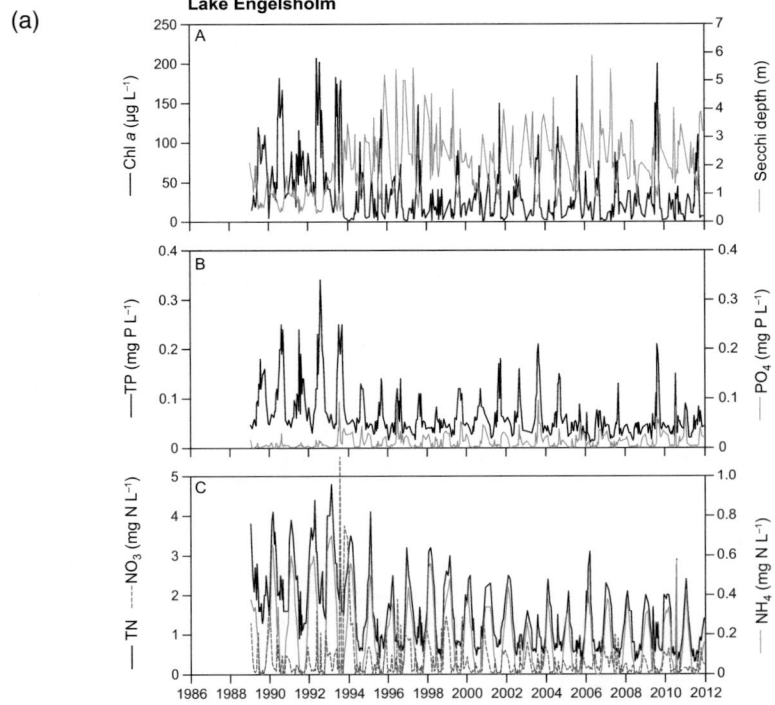

Figure 16.2

(b)

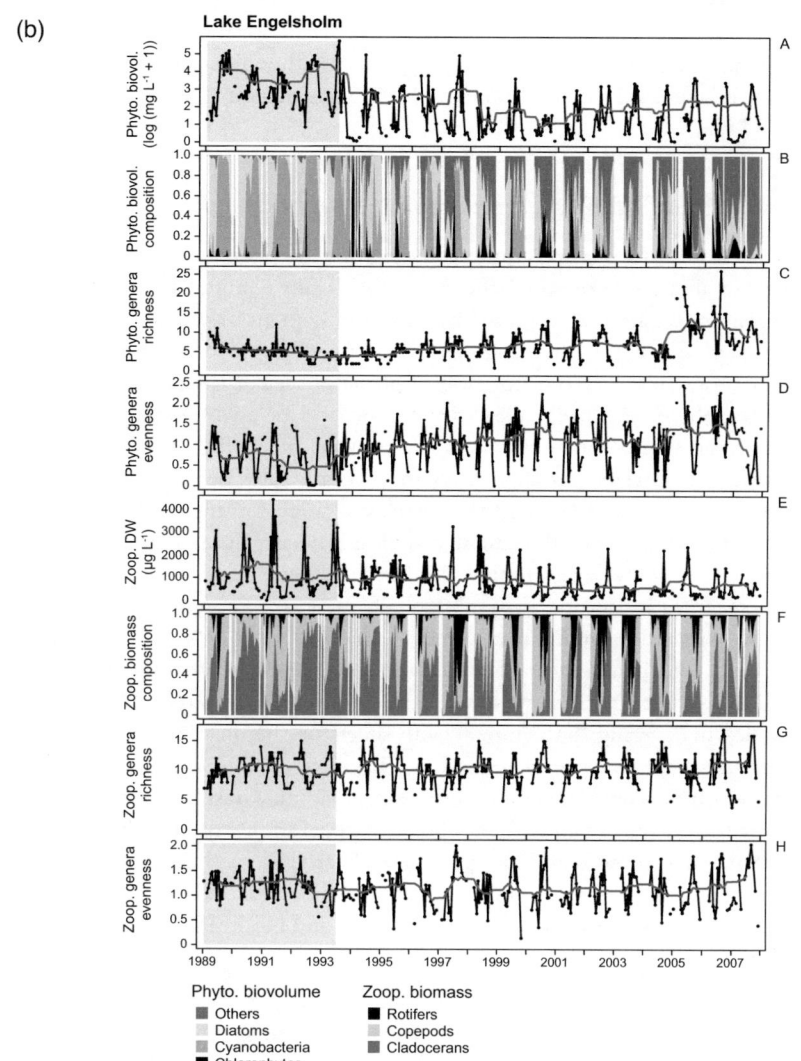

Lake Engelsholm

Phyto. biovolume
- Others
- Diatoms
- Cyanobacteria
- Chlorophytes

Zoop. biomass
- Rotifers
- Copepods
- Cladocerans

Figure 16.2 (a) Time series of chlorophyll a and Secchi depth (A), total phosphorus and ortho-phosphate (B), and total nitrogen, nitrate and ammonia (C) in Lake Engelsholm, Denmark, from 1989 to 2010. Fish removal was conducted in 1992–1993. (b) Time series for phytoplankton biomass (log-transformed, A), phytoplankton biomass composition across four main groups (chlorophytes, cyanobacteria, diatoms and others, B), phytoplankton genera richness (C) and evenness (D), as well as zooplankton biomass (E), zooplankton biomass composition (F), zooplankton genera richness (G) and evenness (H) in Lake Engelsholm, Denmark. Grey and white backgrounds indicate the shift in the phytoplankton biovolume trend identified by change point analysis. Lines in scatter plots denote yearly running means excluding winter samples. Lines in scatter plots and polygons in composition plots were left unconnected if two samples were more than 30 days apart

Source: Jeppesen *et al.* (2012)

(*Cyprinus carpio*), has occurred, resulting in improved water clarity in the long term. Repeated fish removal either annually ('maintenance fishing') or more drastically every 5–10 years may be needed to maintain high water quality, but experiences (though scarce) indicate that less effort is needed in the repeated fishing (Benndorf 1995; Søndergaard *et al.* 2008; Jeppesen *et al.* 2012).

Protection of submerged macrophytes and transplantation

Construction of exclosures to protect submerged macrophytes against waterfowl grazing has been employed as an alternative or supplementary tool to fish manipulation. The exclosures enable the macrophytes to grow in a grazer-free environment from where they may spread seeds, turions or plant fragments, thus augmenting colonization. Moreover, they serve as a daytime refuge for zooplankton. Transplantation of plants or seeds is an alternative method.

Submerged macrophytes stabilize a clear water state and re-establishment of submerged macrophytes is therefore a goal of most restoration projects in shallow lakes; however, dense plant beds appearing in nutrient-enriched lakes may occasionally be considered a nuisance since they impede navigation and reduce the recreational value for anglers. Moreover, excessive growth of invading species, like the Eurasian milfoil, *Myriophyllum spicatum*, *Pistia stratiotes* or *Eichhornia crassipes* in many lakes in the US, South America and Africa, or the North American *Elodea canadensis* in Europe, may substantially alter lake ecosystems and constitute a serious threat to the native flora and fauna.

Methods to combat such nuisance plant growth are manual harvesting, introduction of specialist phytophagous insects such as weevils or herbivorous grass carp (*Ctenopharyngon idella*), water level draw-down, covering the sediment with sheets or chemical treatment with herbicides. Often, harvesting and water level draw-down have only a temporary effect because of fast regrowth of the plant community and high external loading. Grass carp may have a strong effect on plant growth and are currently used in many parts of the world to reduce macrophyte abundance, but a shift to a turbid state is a typical side effect. The method should therefore be used with caution. Moreover, before planning plant removal one has to bear in mind that these plants generally have a positive effect on lake water clarity and biodiversity (Bakker *et al.* 2013; Lauridsen *et al.* 2003).

Introducing invertebrate filtrators

In some cases mussels have been introduced to increase the filtration of the water and create clear water conditions. Zebra mussels (*Dreissena polymorpha*) have been suggested as they may greatly increase water transparency where they colonize. The zebra mussel is, however, an invasive species and may reach very high densities with negative biological and physical consequences. It has been shown that adult zebra mussels can use cyanobacteria as food from the water column, irrespective of the size, shape, form and toxicity of these phytoplankton species. Mussels even seem to prefer cyanobacteria over other phytoplankton groups and detritus. Zebra mussels assimilate cyanobacterial toxins but only to a limited extent, the quantity being, however, high enough to produce liver damage in diving ducks, the main mussel predators. Cage cultures outside the mussel production period have been proposed as an alternative as it allows control of the population density (Gulati *et al.* 2008).

Introduction of large native mussel species (*Anodonta*, *Unio*) may, however, be considered as a better option in case they have been eliminated during the eutrophication period.

Combined methods

Most restoration attempts have shown only temporary effects; thus, it seems uncertain whether the effects will last in the long term. This suggests that a combination of methods may prove to be more useful (Table 16.1). The hypothesis is that a combined treatment may be more effective than individual application of methods due to synergistic effects. Combined methods may therefore potentially reduce the costs of restoration and perhaps the need for subsequent intervention (Jeppesen *et al.* 2012).

Various chemical methods may also be combined (see, for instance, the example in Box 16.1). Another approach is to combine different biomanipulation methods, such as fish removal, stocking of piscivores and transplantation of plants. This approach has been used in several South Chinese lakes, the rationale being that due to high temperatures (subtropical and tropical climate) and consequently high fish recruitment, it is impossible to obtain a top-down control effect by zooplankton on phytoplankton as in temperate lakes. Only bottom-up effects can lead to clear water conditions: fish removal and control by the added piscivores lead to reduced fish-induced resuspension and nutrient release from the sediment, and enhanced abundance of macrophytes through transplantation reduces nutrients and wind-induced resuspension, and chemical warfare is being waged against the phytoplankton (Box 16.3).

Box 16.3 Biomanipulation in Huizhou West Lake, China

Huizhou West Lake is a shallow lake in tropical China with a surface area of about 1.6 km^2 and is divided into several basins connected through waterways. The lake was dominated by submerged macrophytes before the 1960s. Following the initiation of fish aquaculture in the lake during the 1970s and increased waste water input, the lake became eutrophic and submerged macrophytes have been absent since the 1980s. In spite of restoration efforts, including effluent diversion and sediment removal, the lake has remained eutrophic and turbid. Furthermore, fish stockings have occasionally been conducted, the fish community being dominated by omnivorous and benthivorous species including Nile tilapia (*Oreochromis niloticus*), common carp and Crucian carp (*Carassius auratus* (L.)). In order to improve water quality, a large-scale biomanipulation experiment was conducted in a 1 ha basin of the lake at the end of 2004. The biomanipulation included fish removal followed by submerged macrophyte transplantation and stocking of piscivorous fish. After isolating the area from the rest of the lake, the water level was lowered to roughly 60 cm and roughly 200 kg ha^{-1} fish, including tilapia, common carp, Crucian carp, mud carp (*Cirrhina molitorella* Cuvier et Valenciennes), silver carp (*Hypophthalmcihthys molitix* (Valenciennes)) and bighead carp (*Aristichthys nobilis* (Richardson)), were removed over a two-month period. Submerged macrophytes, including *Hydrilla verticillata* Royle, *Vallisnaria natans* L. and *Myriophyllum spicatum* L., were planted and coverage reached 60 per cent in May–June 2005 and > 80 per cent in summer 2006. Piscivorous fish, snakehead (*Channa argus* Cantor) and mandarin fish (*Siniperca chuatsi* Basilewsky), were stocked after macrophyte transplantation (Liu *et al.* 2014; Jeppesen *et al.* 2012).

Annual mean TN and TP decreased substantially after the biomanipulation. Fish removal likely reduced sediment resuspension, while submerged macrophytes protected the sediment from resuspension, and macrophytes and benthic algae may have reduced the nutrient release from the sediments. Accordingly, annual mean algal biomass (expressed as the pigment Chl a) decreased markedly in the restored area and suspended solids declined as well (reduction of both inorganic suspended solids and phytoplankton in the restored part). Densities of planktonic crustaceans, both cladocerans and copepods, were extremely low in both the restored and unrestored areas, suggest-

---

ing that zooplankton grazing played a minor role in controlling phytoplankton in the restored area. So, the biomanipulation in this tropical shallow lake did not increase the number of zooplankton grazers as seen in north temperate lakes, likely reflecting high predation by the abundant small fish, including fish fry. Many fishes can spawn several times a year in the tropical region, including species such as tilapia and Crucian carp which are abundant in the lake (Z. Liu, unpublished data).

This restoration demonstrates that fish removal and transplantation of submerged macrophytes can restore a tropical eutrophic lake to a clear water state via enhanced bottom-up control and reduced sediment resuspension, even without increasing zooplankton grazing on phytoplankton.

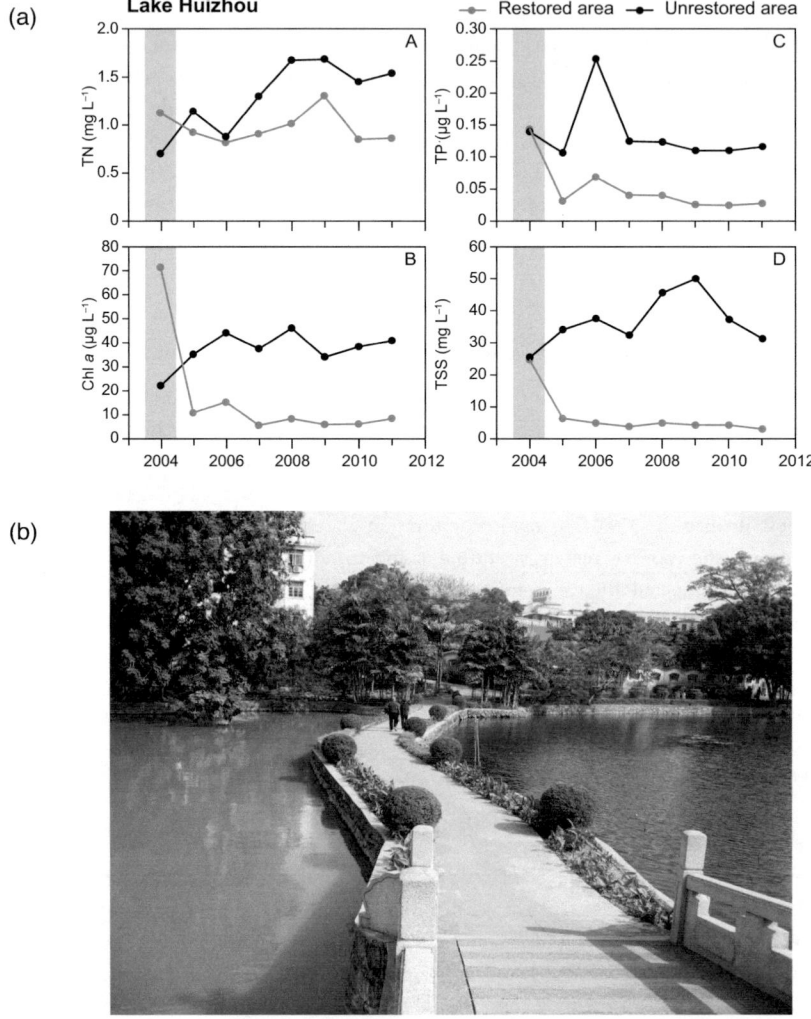

Figure 16.3   (a) Comparison of levels of total nitrogen (A), chlorophyll a (B), total phosphorus (C) and total suspended solids (D) in the restored and unrestored area in Huizhou West Lake, China. The highlighted period in grey shows pre-restoration levels. (b) Unrestored (left) and restored area of Huizhou West Lake

Source: Jeppesen *et al.* (2012). Photo: Zhengwen Liu.

Biomanipulation may also be combined with chemical treatment of the water in order to precipitate phosphorus (Table 16.1, Box 16.4). The resulting higher water clarity may potentially reinforce recovery by altering the top-down control of zooplankton by fish through a trophic and/or a behavioural cascade.

Biomanipulation can also be combined with oxygenation of the hypolimnion (deep lakes), which may result in reduced internal loading; however, it may also affect the fish community and eventually the entire pelagic ecosystem. If oxygen is available in the hypolimnion, the conditions for macroinvertebrates improve. In addition, piscivorous fish species may be favoured. Some piscivores fish like perch (*Perca fluviatilis*) may obtain better foraging possibilities in the benthos, allowing them to pass from consuming zooplankton through a macroinvertebrate-eating stage before becoming piscivores (pass the 'macroinvertebrate bottleneck'), facilitating higher control of planktivorous fish. A behavioural cascade is also expected if water clarity increases in the epilimnion as it enhances grazing and thus reduces sedimentation and thereby the need for oxygenation in the hypolimnion. Such an approach has been successfully applied in deep Lake Furesø, Denmark (Jeppesen *et al.* 2012).

Conclusions and perspectives

Lake restoration can be a powerful tool to improve lake water quality, but for restoration and lake recovery to be successful the external nutrient loading must first be reduced sufficiently and, when needed, combined with in-lake measures. An important prerequisite is to have prior knowledge of the lake and its catchment before planning and conducting the intervention. Focus should in particular be directed at the external nutrient loading – if this is too high, long-lasting effects of restoration cannot be expected, and the lake will sooner or later return to pre-intervention conditions. For shallow lakes phosphorus equilibrium concentrations below 0.05 mg P L^{-1} have been suggested, and in deeper lakes concentrations below 0.02 mg P L^{-1} may be required.

So far, most experiences with lake restoration come from northern temperate regions; however, in recent years lake restoration projects have been undertaken increasingly also in warm countries. Knowledge from temperate lakes cannot readily be transferred to Mediterranean, subtropical and tropical lakes as these differ in many aspects from temperate lakes, for instance by having (1) faster nutrient cycling, (2) stronger predation pressure on zooplankton, being dominated by numerous small fish with fast recruitment, (3) higher risk of dominance by cyanobacteria owing to the warmer climate and higher abundance of nuisance floating plants, and (4) less seasonality with high biological activity throughout the year. New or modified methods for such lakes are currently under development and need further consideration. Generally, most reported restoration initiatives have led to improvement. However, the longevity of the improvements varies considerably, from a few months to more than ten years. Unfortunately, only few restoration projects exist that document effects for more than a few years following intervention, making it difficult to elucidate the longevity of the restorations.

In the future, lake restoration may become even more relevant because of an increasing demand for clean drinking water and increasing farming intensity, and with it intensified nutrient loading, due to human population growth. In developing countries, enhanced use of fertilizers and introduction of sewage systems may further increase the nutrient loading. Climate change may affect both temperature and precipitation patterns throughout the world, further reinforcing eutrophication and the risk of toxic algal blooming, and the demand for successful lake restoration techniques will therefore grow.

Box 16.4 Dual treatment: Chemical treatment and fish manipulation in Lake Kollelev, Denmark

Shallow Lake Kollelev is an example of combined chemical and biological treatment. The lake is divided into three basins connected by channels. Until 1998 the lake received waste water or storm water with overflow, but after waste water was diverted the lake remained hypertrophic. Different in-lake measures were applied in the period 1999–2005 to improve water clarity. Iron addition to two of the basins in 1998 had little effect. Biomanipulation including cyprinid removal and perch stocking in all basins in 1999 was also ineffective, while aluminium treatment of basin 1 and 2 in 2003 immediately resulted in lower lake water TP. Meanwhile, no improvements in Secchi depth were observed in any of the basins. A new biomanipulation effort undertaken in all three basins resulted in an immediate and strong improvement of water clarity in the aluminium-treated basins but a much less pronounced and only gradual improvement of Secchi depth in the untreated basin, coinciding with a gradual decrease in TP. This case study indicates that only the combined treatment with aluminium (bottom-up control) and biomanipulation (top-down control) ensured rapid improvement in water clarity. However, as this experiment was not replicated, it is possible that other factors may be involved, and controlled follow-up experiments are thus needed before any firm conclusions can be drawn (Jensen *et al.* 2015).

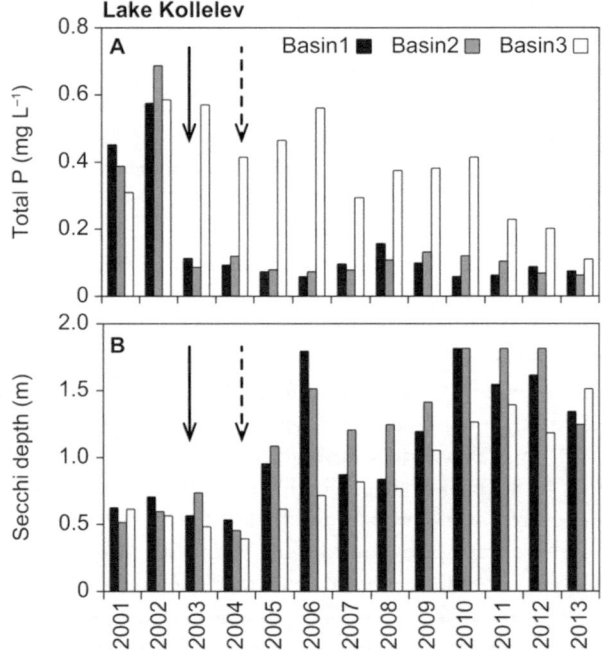

Figure 16.4   Average summer values for lake water TP (A) and Secchi depth (B) for the three basins in Lake Kollelev, Denmark. Solid arrows indicate time of Al treatment. Dotted arrows show time of biomanipulation

Source: Jensen *et al.* (2015).

References

Annadotter, H., Cronberg, G., Aagren, R., Lundstedt, B., Nilsson, P. A. and Strobeck, S. (1999) Multiple techniques for lake restoration. *Hydrobiologia*, vol. 395, pp. 77–85.

Bakker, E. S., Sarneel, J. M., Gulati, R. D., Liu, Z. and Van Donk, E. (2013). Restoring macrophyte diversity in shallow temperate lakes: biotic versus abiotic constraints. *Hydrobiologia*, vol. 710, pp. 23–37.

Benndorf, J. (1995) Possibilities and limits for controlling eutrophication by biomanipulation. *Internationale Revue der gesamten Hydrobiologie*, vol. 80, pp. 519–534.

Chow-Fraser, P. (2005) Ecosystem response to changes in water level of Lake Ontario marshes: lessons from the restoration of Cootes Paradise Marsh. *Hydrobiologia*, vol. 539, pp. 189–204.

Cooke, G. D., Welch, E. B., Peterson, S. A. and Nicholson. S. A. (2005) *Restoration and management of lakes and reservoirs*. Boca Raton, FL: CRC Press.

Coops, H., Beklioğlu, M. and Crisman, T. L. (2003) The role of water-level fluctuations in shallow lake ecosystems: workshop conclusions. *Hydrobiologia*, vol. 506, pp. 23–27.

Gächter, R. and Wehrli, B. (1998) Ten years of artificial mixing and oxygenation: No effect on the internal P loading of two eutrophic lakes. *Environmental Science and Technology*, vol. 32, pp. 3659–3665.

Gulati, R. D., Pires, L. M. D. and Van Donk, E. (2008) Lake restoration studies: Failures, bottlenecks and prospects of new ecotechnological measures. *Limnologica*, vol. 38, pp. 233–247.

Gunn, J. M. and Sandöy, S. (2003) Introduction to the Ambio special issue on biological recovery from acidification: Northern Lakes recovery study. *Ambio*, vol. 32, pp. 162–164.

Hansson, L.-A., Annadotter, H., Bergman, E., Hamrin, S. F., Jeppesen, E., Kairesalo, T., Luokkanen, E., Nilsson, P.-Å., Søndergaard, M. and Strand, J. (1998) Biomanipulation as an application of food-chain theory: constraints, synthesis, and recommendations for temperate lakes. *Ecosystems*, vol. 1, pp. 558–574.

Jensen, H. S., Reitzel, K. and Egemose, S. (2015) Evaluation of aluminum treatment efficiency on water quality and internal phosphorus cycling in six Danish lakes. *Hydrobiologia*, vol. 751, pp. 189–199.

Jeppesen, E., Søndergaard, M., Jensen, J. P., Havens, K., Anneville, O., Carvalho, L., Coveney, M. F., Deneke, R., Dokulil, M., Foy, B., Gerdeaux, D., Hampton, S. E., Kangur, K., Köhler, J., Körner, S., Lammens, E., Lauridsen, T. L., Manca, M., Miracle, R., Moss, B., Nõges, P., Persson, G., Phillips, G., Portielje, R., Romo, S., Schelske, C. L., Straile, D., Tatrai, I., Willén, E. and Winder, M. (2005) Lake responses to reduced nutrient loading – an analysis of contemporary long-term data from 35 case studies. *Freshwater Biology*, vol. 50, pp. 1747–1771.

Jeppesen, E., Søndergaard, M., Lauridsen, T. L., Davidson, T. A., Liu, Z., Mazzeo, N., Trochine, C., Özkan, K., Jensen, H. S., Trolle, D., Starling, F., Lazzaro, X., Johansson, L. S., Bjerring, R., Liboriussen, L., Larsen, S. E., Landkildehus, F. and Meerhoff, M. (2012) Biomanipulation as a restoration tool to combat eutrophication: recent advances and future challenges. *Advances in Ecological Research*, vol. 47, pp. 411–487.

Jeppesen E., Brucet, S., Naselli-Flores, L., Papastergiadou, E., Stefanidis, K., Nõges, T., Nõges, P., Attayde, J. L., Zohary, T., Coppens, J., Bucak, T., Menezes, R. F., Freitas, F. R. S., Kernan, M., Søndergaard, M., and Beklioğlu, M. (2015) Ecological impacts of global warming and water abstraction on lakes and reservoirs due to changes in water level and salinity. *Hydrobiologia*, vol. 570, pp. 201–227.

Lathrop, D. (ed.) (2007) Long-term perspectives in lake management. Special issue of *Lake and Reservoir Management*.

Lauridsen, T. L., Sandsten, H. and Møller, P. H. (2003) The restoration of a shallow lake by introducing *Potamogeton* spp. The impact of waterfowl grazing. *Lakes and Reservoirs: Research and Management*, vol. 8, pp. 177–187.

Liboriussen, L., Søndergaard, M., Jeppesen, E., Thorsgaard, I., Grünfeld, S., Jakobsen, T. S. and Hansen, K. (2009) Effects of hypolimnetic oxygenation on water quality: results from five Danish lakes. *Hydrobiologia*, vol. 625, pp. 157–172.

Liu, Z. W., Zhong, P., Zhang, X., Ning, J., Larsen, S. E. and Jeppesen, E. (2014) Successful restoration of a tropical shallow eutrophic lake: strong bottom-up but weak top-down effects recorded. In: G. Kattel (ed.), *Australia–China Wetland Network Research Partnership: Proceedings of the Australia–China Wetland Network Research Partnership Symposium, Nanjing, China*, pp. 78–86. Ballarat, Australia: Federation University Australia.

Lürling, M. and van Oosterhout, F. (2013) Controlling eutrophication by combined bloom precipitation and sediment phosphorus inactivation. *Water Research*, vol. 47, pp. 6527–6537.

Mackay, E. B., Maberly, S. C., Pan, G., Reitzel, K., Bruere, A., Corker, N., Douglas, G., Egemose, S., Hamilton, D., Hatton-Ellis, T., Huser, B., Li, W., Meis, S., Moss, B., Lürling, M., Phillips, G., Yasseri, S., and Spears, B. M. (2014) Geoengineering in lakes: welcome attraction or fatal distraction? *Inland Waters*, vol 4, pp. 349–356.

McDonald, R. H., Lawrence, G. A. and Murphy, T. P. (2004) Operation and evaluation of hypolimnetic withdrawal in a shallow eutrophic lake. *Lake and Reservoir Management*, vol. 20, pp. 39–53.

Moss, B., Madgwick, J. and Phillips, G. (1996) *A guide to the restoration of nutrient enriched shallow lakes.* Bristol: Environment Agency.

Moss, B., Jeppesen, E., Søndergaard, M, Lauridsen, T. L. and Liu, Z. W. (2013) Nitrogen, macrophytes, shallow lakes and nutrient limitation – resolution of a current controversy? *Hydrobiologia*, 710: 3–21.

Nürnberg, G. K. (2007) Lake responses to long-term hypolimnetic withdrawal treatments. *Lake and Reservoir Management*, vol. 23, pp. 388–409.

OECD (1982) *Eutrophication of waters: monitoring, assessments and control.* Paris: Organisation for Economic Co-operation and Development.

Søndergaard, M., Jeppesen, E., Lauridsen, T. L., Skov, C., Van Nes, E. H., Roijackers, R., Lammens, E. and Portielje, R. (2007). Lake restoration: successes, failures and long-term effects. *Journal of Applied Ecology*, vol. 44, 1095–1105.

Søndergaard, M., Liboriussen, L., Pedersen, A. R. and Jeppesen, E. (2008) Lake restoration by fish removal: Short and long-term effects in 36 Danish lakes. *Ecosystems*, vol. 11, pp. 1291–1305.

Søndergaard, M., Jeppesen, E. and Jensen, H. S. (2012) Lake restoration. In L. Bengtson, R. W. Herschy and R. W. Fairbridge (eds), *Encyclopedia of Lakes and Reservoirs*, pp. 455–458. Berlin: Springer.

17

RESTORATION OF FRESHWATER WETLANDS

Paul A. Keddy

Introduction

All life needs water. Therefore, wetlands have always influenced humans, and been influenced by humans in return. Early agricultural civilizations first arose along the edges of rivers in the fertile soils of floodplains. Wetlands also produce many services for humans – along with fertile soils for agriculture, they provide food such as fish and water birds, and, of course, fresh water. Additionally, wetlands have other vital roles that are less obvious. They produce oxygen, store carbon, and process nitrogen. Since wetlands form at the interface of terrestrial and aquatic ecosystems, they possess features of both. They are often overlooked in standard books, since terrestrial ecologists focus on drier habitats, while limnologists focus on deeper water. Shallow water, and seasonally flooded areas, fall comfortably into neither category. All wetlands share one causal factor: flooding. Hence, any discussion of wetland ecology has to place a primary focus on getting the water right (Keddy 2010; Middleton 2002; Pierce 2015). While wetlands may be highly variable in appearance and species composition, flooding produces distinctive soil processes and adaptations of the biota. Thus wetlands and water are inseparable.

Two general obstacles must be met in coming to grips with the scientific literature for wetlands in general, and for wetland restoration in particular. First, much of the work on wetlands is scattered across ecological journals and may not even appear under key word searches for wetland; instead, material may appear under a term such as bog, fen, shoreline, lake, floodplain, pothole, playa, peatland, or mire (or a dozen other terms). This problem is compounded when you add in the names used to describe wetlands in other human languages. Second, this discipline seems to have attracted a large number of conference symposia, the findings of which are recorded often in expensive books with a haphazard collection of papers, written by a haphazard collection of people, with no unifying theme whatsoever except that all deal with wet areas. One can easily be exhausted by an accumulated array of examples that seem to have few general principles. Hence, the need is pressing for a few general principles to guide restoration. In this chapter I will focus on general causal factors and their relative importance. This framework applies across wetland types and across biogeographic regions. The framework focuses upon the pool of species available, and the filters that control their relative composition, an approach which is sometimes termed assembly rules or trait-based assembly rules (Weiher and Keddy 1999).

I will first briefly introduce you to some basic information: what a wetland is, the kinds of wetlands that exist, and some key processes that occur within them. Then I will turn to causal factors. Flooding creates wetlands, so it receives a full section. Then I will consider how nutrient availability modifies wetlands. As a third key factor, I will consider the role of natural disturbances, and how they counterbalance competition and succession to produce a diversity of wetland types in a landscape. As Figure 17.1 shows, any particular wetland exists at a dynamic equilibrium set by the relative impacts of these three general processes. If one becomes predominant, the wetland will shift in area, composition, and ecological services. In the most general sense, restoration can be viewed as re-establishing the natural balance among these forces. There are two cautions. First, the relative importance of these factors differs significantly among wetland types: you cannot manage or restore a fen like you would an alluvial forest. There is no one size fits all! Second, each specific location will have additional causal factors, such as salinity, competition, herbivory, or roads. However, as Table 17.1 suggests, if you think about the problem of restoration in terms of causal factors, the first few are likely the most important. If you get these right, you can address the other factors on a case by case basis.

The kinds of wetlands

Wetlands are inherently variable. Consider that the term wetland applies equally to a coastal mangrove swamp, a beaver pond, a forested floodplain, and a wet prairie. Is there some natural

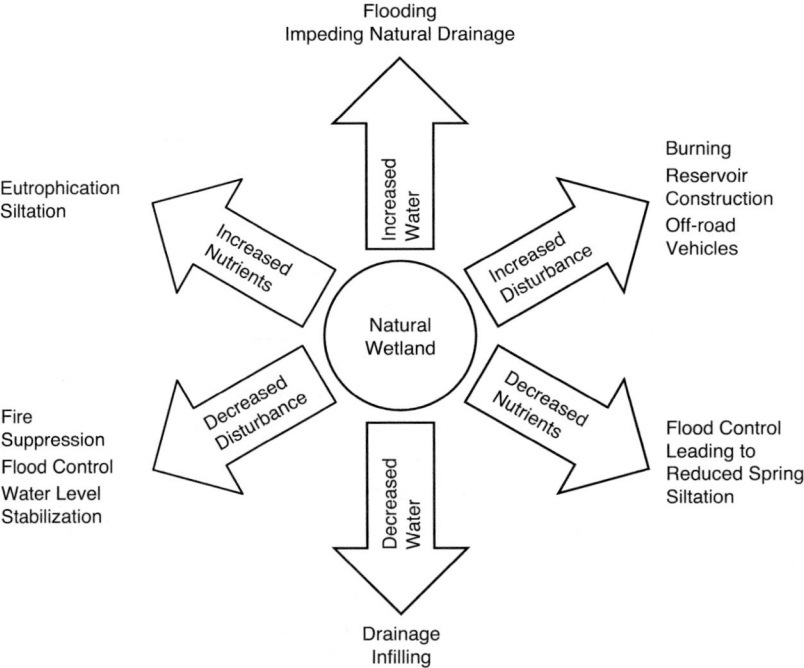

Figure 17.1 Any particular wetland exists at a dynamic equilibrium set by the relative impacts of these three general processes: flooding, fertility, and natural disturbance

Source: Keddy (1983)

Table 17.1 The estimated relative importance of environmental factors that determine the properties of wetlands. These can be considered the key filters for assembling wetlands from species pools

Environmental factor	Relative importance (%)
hydrology	50
fertility	15
salinity	15
disturbance	15
competition	<5
herbivory	<5
burial	<5

Source: Keddy (2010)

way to sort them into similar types? Each type can be visualized as a particular set of plant and animal associations that recur. This recurrence probably means that the same causal factors are at work. Unfortunately, the search for patterns is complicated by the terminology for describing wetlands that varies both among human societies, and among their scientific communities. Thus one finds an abundance of words used to describe wetlands – bog, bayou, carr, fen, flark, hochmoore, lagg, marsh, mire, muskeg, swamp, pocosin, pothole, quagmire, savannah, slob, slough, swale, turlough, yazoo – in the English language alone!

To keep the terminology simple, we will begin with four types of wetland:

1 *Swamp:* A wetland that is dominated by trees that are rooted in hydric soils, but not in peat. Examples include the tropical mangrove swamps (mangal) of Bangladesh and bottom-land forests in floodplains of the Amazon River in Brazil.

2 *Marsh:* A wetland that is dominated by herbaceous plants that are usually emergent through water and rooted in hydric soils, but not in peat. Examples include cattail (*Typha augustifolia*) marshes around the Great Lakes and reed (*Phragmites australis*) beds around the Baltic Sea.

3 *Bog:* A wetland dominated by *Sphagnum* moss, sedges, Ericaceous shrubs, or evergreen trees rooted in deep peat with a pH less than 5. Examples include the blanket bogs which carpet mountainous areas of the Himalayas, and the vast peatland of the West Siberian Lowland in central Russia, as well as bogs in southern South America.

4 *Fen:* A wetland that is usually dominated by sedges and grasses rooted in shallow peat, often with considerable ground water movement, and with pH greater than 6. Examples can be found within the extensive peatlands of northern Canada and Russia, as well as in smaller seepage areas throughout the temperate zone.

Two other wetland types could be added to these four.

5 *Wet meadow:* A wetland dominated by herbaceous plants rooted in occasionally flooded soils. Temporary flooding excludes terrestrial plants and swamp plants, but drier growing seasons then produce plant communities typical of moist soils. Examples would include wet prairies along river floodplains, or herbaceous meadows on the shorelines of large lakes. These habitats often have inordinately high plant diversity, and are one of the first habitats to be lost when dams and levees are constructed along rivers.

6   *Shallow water or aquatic:* A wetland community dominated by truly aquatic plants growing in and covered by at least 25 cm of water. Examples include the littoral zones of lakes, bays in rivers and the more permanently flooded areas of prairie potholes.

So, if you are going to restore a wetland, an obvious and essential first question is this: what kind of wetland are you trying to create? Of course, within each of these six categories there are thousands of subgroups depending upon which ecoregion you are in. If you are beginning a wetland restoration project, you must find the wetland classification that is applicable to your ecoregion. Once you locate an appropriate regional system, you will want to familiarize yourself with important causal factors that produce this array of wetlands. To put it into a global context, you may wish to refer to larger scale classification schemes such as those found in Vitt (1994) or Gopal *et al.* (1990).

Restoration needs

Overall, the past few centuries have seen major losses in wetland area around the globe. Hence, a first priority is to restore wetland area. This requires an understanding of why wetlands have disappeared. The most obvious cause is drainage ditches. Too often, wetlands are drained for agriculture or urbanization. In such cases, the primary tool for restoration is to plug or back-fill drainage ditches. In other cases, wetlands have been lost through the deliberate construction of levees or dykes to obstruct the natural flow of water through the site and replace it with a polder. In this case outright removal of the dyke will restore wetlands.

In some landscapes wetlands will need to be reconstructed by physically creating depressions and obstacles to water flow. This allows much more precise control over topography and hydrology. However, the cost per restored acre is likely to be much higher. Here, important issues include (1) determining the availably of water to maintain the wetland (a wetland hydrograph is advised), (2) constructing the basin to create appropriate water levels and gradients (sub grading, see Pierce 2015), and (3) ensuring the availability of the right species pool, either through natural sources, added seeds, or outright planting.

An equally important target is restoring wetland composition. Often degraded wetlands become dominated by a few fast-growing dominant species of grass, or of the genus *Typha*, along with a few common species of amphibians and birds. While this may qualify as a wetland, it may not contribute to maintaining biological diversity. A large portion of the world's rare and endangered species require wetlands, and if we do not recreate the natural wetlands that once occurred in our landscapes, we will lose large numbers of wetland species. Examples you ask? The giant ibis (seasonal wet meadows in northern Cambodia); the Basra reed warbler (marshes of the Tigris-Euphrates). The eastern prairie fringed-orchid (in fens and wet prairies of North America); the Venus flytrap (coastal bogs in the Carolinas); the southern corroboree frog (*Sphagnum* bogs in subalpine woodlands in eastern Australia); the Mekong giant catfish (Mekong River in southeast Asia). For a full list of species at risk, and their habitats, consult the IUCN Red List of Threatened Species (www.iucnredlist.org). The IUCN estimates that more than 125,000 known species depend upon freshwater wetlands, including 15,000 species of fish and 5600 species of odonata.

The important point, then, is that it is not enough to restore wetland area, but one must set meaningful targets for species composition to provide habitat for the full array of wetland plants and animals. This means that restoration must consider not only regionally common wetland species, but also the ones unique to each of the world's ecological regions. According to Olson *et al.* (2001) there are a total of 867 such ecoregions, nested within 14 biomes and 8 biogeographic realms (for an online version of this map consult www.worldwildlife.org/

science/wildfinder). The first restoration challenge is to set an appropriate target for the desired species composition. This requires careful consideration of the ecoregion in which you are working, ecological states, and the tools available for restoring key environmental factors. Once the targets are set, one needs a monitoring program to measure success, and adaptive management to correct any mistakes (Keddy 2010: 373–376). These steps are summarized for your convenience in Table 17.2.

The importance of flooding and hydroperiod

Flooding makes wetlands. The conspicuous zonation of wetland plants within wetlands (Figure 17.2) shows just how important flood duration is to wetland plants. The causes of such zonation are complicated, and in part arise from reduced oxygen levels in the soil. These changes

Table 17.2 Four steps in the plan for restoring a wetland, with some guiding questions

Step	Questions
1. Set a target for species composition	What was the original array of wetland types in the landscape?
	What was the original array of gradients?
	What were the original key factors (filters)?
	What rare and significant species could serve as indicators?
	What was the natural landscape really like when human populations were lower?
	What was the original pool of species? If you can't answer such simple questions, you need to do more homework on environmental history.
2. Determine the key causal factors	What is the projected maximum water depth at a set of locations?
	What is the projected seasonal variation?
	What is the projected decadal variation?
	What is the target value for N and P?
	What other key factors must be considered (fire? herbivores? salinity?)
	How will you ensure that these factors create a biologically significant wetland with natural gradients in species composition, as opposed to a circular wet hole with cattails and a few ducks?
3. Decide how each key factor can be created or maintained	To what extent can you work with nature?
	Do you need to first recontour the site to enhance natural gradients?
	Are there existing obstacles to natural seasonal flows?
	Are there existing channels that remove too much water?
	Consider dykes with water control structures to be a last resort.
	Artificial structures are expensive to build, expensive to run, and they will eventually fail unless given continual maintenance. For this reason, consider gently sloping berms rather than steeply angled dykes.
4. Plan for adaptive management	What key factors will be monitored?
	What species will serve as indicators of desired conditions?
	Who will do the monitoring?
	How long will monitoring continue?
	Who will store the data and write updates?
	Who will make the adaptive changes, if any?
	In very few cases, if any, will it be acceptable to build it, walk away, and hope for the best. This, like children without a father, is still far too common, and increases the onus to get it right.

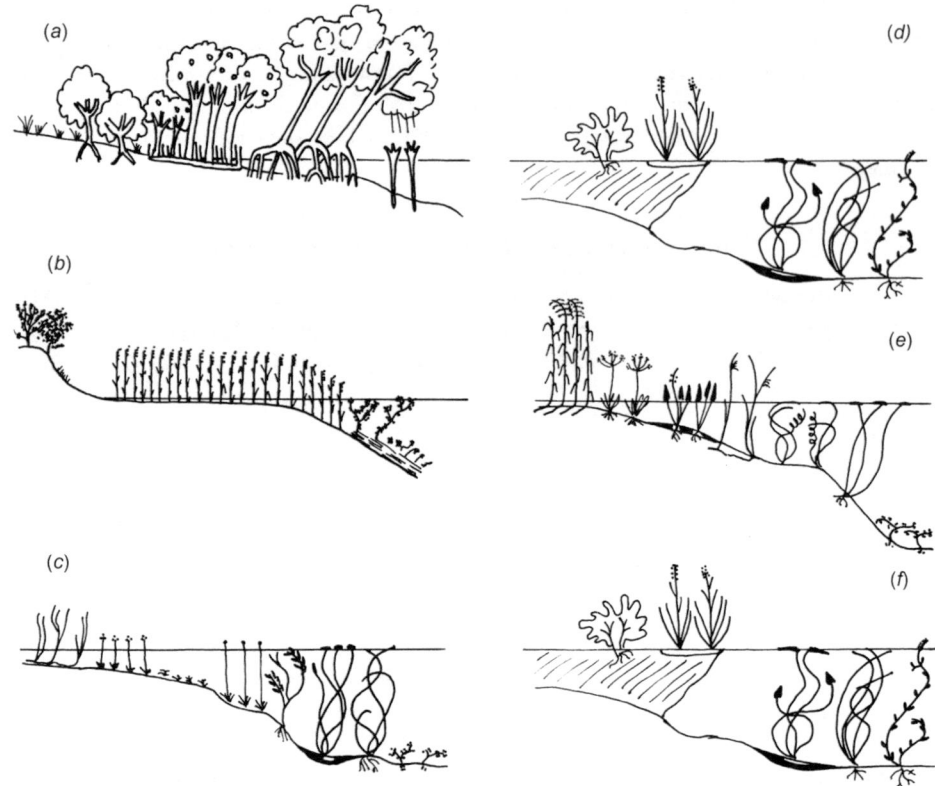

Figure 17.2 Flooding is the primary factor that produces wetlands, and the factor that controls much of the variation seen within wetlands. Examples include (a) mangroves along ocean coasts, (d) pools in northern peatlands, and (b, c, e, f) shorelines of lakes and rivers. The species names will change depending on the biogeographic region, but the wide occurrence of zonation emphasizes the overwhelming importance of getting the water right. Indeed, the wider the range of water levels, the more kinds of plants

Source: Keddy (2010)

are generally described in Keddy (2010) and Mitsch and Gosselink (2015). Hence, plants and animals have to adapt to reduced oxygen levels. The presence of distinctive plants with channels for transmitting oxygen from the atmosphere to the roots (aerenchyma) is a defining characteristic of wetlands. Aquatic plants offer the most extreme case of plants adapted to flooding (Sculthorpe 1985).

It is easy to think about zonation as resulting from some sort of mean water level, but in wetlands, the fluctuations in water level may be just as important as the mean. High spring flooding makes extensive areas of wetlands along the shores of lakes, and in many other kinds of depressions. Nearly every wetland in the world has water level fluctuations. Along the Amazon these may exceed 10 m within a year (Junk 1993). In large lakes like the Great Lakes, fluctuations may extend over 10 m over a period of decades (Keddy and Reznicek 1986; Wilcox 2012). These natural cycles must be considered in any wetland restoration project. In other books, such as Middleton (2002), this is described as 'flood pulsing'. Hughes (2003) explores how the restoration of spring floods in rivers is necessary for restoring ecological health to wetlands and watersheds. At smaller scales, where one is working with a single basin rather than a watershed, it is necessary to construct a wetland hydrograph to ensure that enough water is available to maintain desired water levels, and rather more engineering may be involved (Pierce 2015).

Trying to restore water levels is always the first step in wetland restoration. But it also brings you face to face with human intransigence. You can say it a hundred times and write books on the topic – yet people will express shock and dismay that their floodplain property is flooded in the spring, and they will equally complain about low water levels in the summer make it inconvenient to use their boat docks. They will also complain when some authority tells them they cannot build a house or factory in a flood-prone area, expecting, of course, that if anything does happen, an insurance company or government will pay for the damage. Yet, so long as snow melts in the spring and rainy seasons arrive, water levels in rivers will have high periods. A major impact humans have had on wetlands is the systematic disruption of such flood peaks in watersheds around the world (Nilsson *et al.* 2005). The importance of flood pulsing is now well documented, yet no doubt individuals will continue to think that rivers and lakes should have stable levels so they can build their houses wherever they care – alas, excellent science does not seem to provide an antidote to ignorance.

As an example of the challenges that lie ahead, consider the Tigris–Euphrates. It was one of the earliest centres of human civilization. Over the last century 32 enormous dams have been constructed, with eight more under construction and 13 more planned (Partow 2001; Lawler 2005). One of the largest dams is Turkey's Ataturk Dam. The cumulative effect of these dams allows storage of five times the volume of the entire flow of the Euphrates! The downstream effects on Mesopotamian marshes have been catastrophic. The area of marsh in the early 1970s was some 8,900 km^2 (about the original size of the Everglades), but had shrunk to 1,296 km^2 by 2000.

The importance of nutrients

Two elements, nitrogen and phosphorus, control rates of primary production in wetlands, and they also determine species composition. Alluvial floodplains and deltas usually have high production, as nutrients are carried in by spring flood waters, and these nutrients accumulate in sediment. Here one finds some of the highest rates of primary production in the world, in excess of 1000 gm^2 yr^{-1} (Keddy 2010: Fig. 11.1). This often translates directly into animals, particularly fish (Welcomme 1979). It is difficult to generalize whether it is nitrogen or phosphorous that limits growth (Verhoeven *et al.* 1996). Nutrients are not necessarily beneficial. In shallow water nutrients can generate algal blooms with negative consequences on marsh and aquatic vegetation, while at larger scales, entire lakes or estuaries may become so nutrient enriched that the resulting decay consumes oxygen, producing 'dead zones' (Turner and Rabelais 2003). The Gulf of Mexico, Chesapeake Bay, and the Baltic Sea are well-known examples of this phenomenon. Other types of wetlands, such as peatlands and shorelines, may have very low levels of available nutrients. Distinctive and rare wetland species often occupy these nutrient-deficient wetlands (Keddy 2010): the rare biota of the New Jersey Pine Barrens (Zampella *et al.* 2006) and the Everglades (Davis and Ogden 1994) are classic examples.

Hence, it may be useful to visualize wetlands arrayed along a nutrient gradient. At one end, infertile wetlands have many rare and unusual species. In these cases, the challenge is to maintain low nutrient levels to protect the unusual biota. At the other extreme, fertile wetlands, the challenge may be to maintain existing elevated nutrient levels, particularly those associated with spring flood pulses, and wisely manage the sustainable harvest of wildlife. Since eutrophication is a now a global process (with nutrients being released from burning coal, eroding uplands, agriculture, and sewage), we may expect infertile wetlands, and their associated biota, to become increasingly scarce in the future (Turner and Rabelais 2003; Keddy 2016). Dead zones, in contrast, may become more common.

In general, erosion, agriculture, and cities add nutrients to water courses, and hence to wetlands. In most cases, restoring a wetland will require minimizing the input of nutrients. This raises another problem: it is easy to add nutrients to wetlands; it is hard to remove them. Thus, one should err on the side of caution. If one is rebuilding a wetland basin to create a new wetland, the use of fertile topsoil as a substrate should likely be avoided.

There is a more general context for considering nutrients in wetlands. Most natural wetlands have fertility gradients, with some areas being fertile, productive, and dominated by nutrient-demanding species such as *Typha* spp. Other areas of the wetland, or nearby wetlands, may have lower levels of nutrients. They may contain species known as stress tolerators, with inherently slow growth and evergreen foliage (Keddy 2010). A particularly good indicator for such conditions is carnivorous plants (which compensate for low soil nutrients by capturing invertebrates) and orchids (which compensate for low soil nutrients with mycorrhizae). If you look at the natural fertility gradients in any particular landscape, you can often see evidence of centrifugal organization (Figure 17.3). There is one core habitat dominated by large fast-growing canopy-forming species that are likely competitive dominants. There are many other kinds of peripheral habitats with distinctive features such as low N, low P, recurring disturbance, and recurring drought, that have relatively uncommon species. Although each of these habitats may be uncommon, in total, they often have a large proportion of the biological diversity in a landscape. Hence, any planned restoration should consider nutrient gradients, and where possible, maintain natural gradients. Since, it is the peripheral habitats that are often most at risk in a landscape, particular attention needs to be given to maintaining existing peripheral habitats, and, if possible, constructing new ones.

Other causal factors

For each particular wetland, there is a hierarchy of causal factors. The challenge for a scientist or a manager is to identify these causal factors and to determine which ones are the most important at a specific site. Two factors of overriding importance, flooding and nutrients, have already been discussed. Superimposed upon these is a long list of other factors including: disturbance, competition, herbivory, roads, and burial. Here we will consider just four beyond flooding and fertility:

1 *Salinity* is a very important factor near coastlines, with species and communities arranged along salinity gradients created by freshwater inputs (Keddy 2010; Mitsch and Gossleink 2015).
2 *Herbivores* can have a major impact. The impacts of muskrats in marshes provides a classic case in which high population densities of herbivores can lead to almost total loss of aboveground vegetation (Keddy 2010). Such top-down effects are becoming better understood; when humans remove the top carnivores (such as crabs or alligators), the effects can be dramatic (Silliman *et al.* 2009).
3 *Fire* can occur during drought. Fire in the Everglades (White 1994) is a classic example; here, fire not only removes plant biomass, but it can even remove peat, thereby producing new areas of open water during the next wet period.
4 *Roads* can have a significant effect upon the biota of wetlands in populated regions. Not surprisingly, road density is a rather good surrogate for the overall impacts of humans in

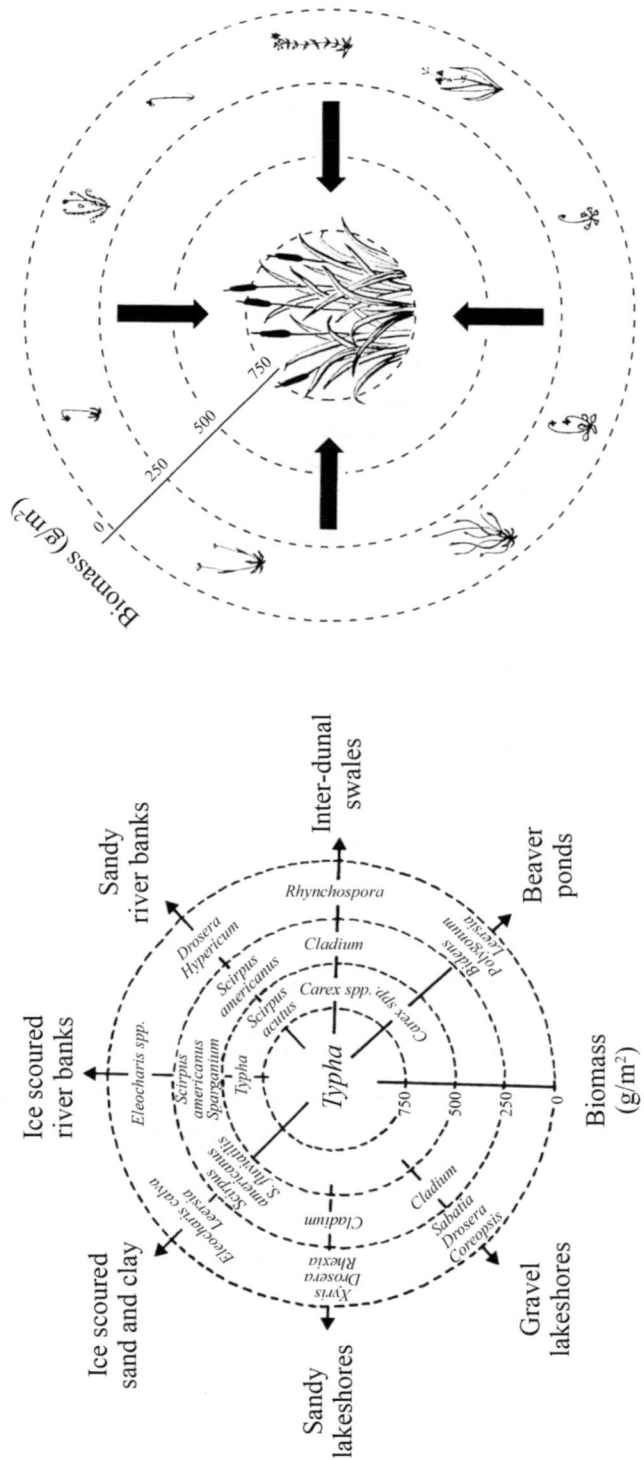

Figure 17.3 Wetlands contain multiple gradients, not just the water depth gradients shown in Figure 17.2. The centrifugal model shows how many different gradients and vegetation types generate plant diversity in wetlands. The centre (core habitat) has high biomass (>750 gm²) and is usually dominated by a few large canopy-forming rhizomatous plants. The edges (peripheral habitats) have low biomass and distinctive species that are restricted to unusual habitats. Here is where a wide array of rosette plants, isoetids, orchids and carnivorous plants often occur. The left shows some of the presumed gradients in wetlands found along lakes and rivers in eastern North America. The right is simplified to show the general changes in growth form likely to be found in any herbaceous wetland. The black arrows show the effects of eutrophication: the loss of peripheral habits and dominance by competitive dominants such as *Typha* and *Phragmites*

Sources: left image, Moore *et al.* (1989); right image, Keddy (2010)

the landscape (Houlahan *et al.* 2006). One sometimes sees road networks being built to carry out restoration; they should be avoided when possible.

The most important point when reading about these other causal factors is to keep them in perspective. In each wetland, some are very important while others are less important. Here is a case where wetland ecology is contingent: it is essential to know not only the important general factors that create a wetland, but also how these are modified by local circumstances and other causal factors. While reading the literature, one should make a concerted effort to rank other causal factors in order of relative importance.

Examples

In this section I will look at a small set of examples, arrayed along one axis: the degree of human intervention required, and, perhaps more the point, the cost of the intervention. I have a preference for simple and inexpensive methods. Partly this is a philosophical position: that I prefer to work with nature and natural forces in general, rather than trying to replace them with concrete and steel. Partly this is because my experience has led me to mistrust the ability of humans to manage large complicated engineering projects. And mostly, it is practicality: the less a restoration programme costs, the more likely it is to be implemented. However, I will indeed end with giant engineering projects that illustrate large-scale restoration with an abundance of concrete and steel.

Sometimes it is necessary to state (and restate) the obvious. With regard to wetland restoration, I need to remind you that the best option is to avoid the need for restoration in the first place. In a wisely-managed landscape, natural forces will generate biological diversity and ecological services with minimal human cost or oversight. Hence, our first rule might well be a sort of Hippocratic oath: dig no ditches or canals, erect no levees or dykes. This will obviate the need for future restoration. Alas, even if all such obscene practices were halted tomorrow, we would still have vast areas that already need restoration. In many cases, the wetlands that remain in a landscape are not only much smaller than they once were, but their composition has been greatly altered. Thus our challenge is to restore the original area and the original variety of wetland types. Some examples follow. Much remains to be done.

Low-tech examples: dealing with drainage ditches

Beaver ponds in the Canadian Shield

In the early 1800s, large numbers of settlers were brought to southern Canada from the United Kingdom. New townships were surveyed into large squares with straight roads dividing the land into rectangular lots. In order to grow their own food, these settlers had two main tasks: clear the forests and drain the wetlands. At the same time, many large species of mammals including caribou, elk, moose, and fisher were extirpated. By the time of the First World War, much of the upland area had been deforested and most wetlands had been drained either for pastures or crop production. The rocky land of the Canadian Shield however, was not well suited for mechanized agriculture, and many of the least productive farms were abandoned. This abandoned land received limited use, mostly for hunting, trapping, and logging. There was no plan for restoration, simply abandonment. But then beaver populations began to recover and by 1990 beavers had plugged many of the drainage ditches and created ponds and wetlands (Keddy 2010: 367–369). Wetland species began to recover. Other mammals such as fishers, otters, and muskrats became more common. Great blue herons and waterfowl returned to nest.

Osprey fished in the larger ponds. Snapping turtles, painted turtles and Blanding's turtles were frequently sighted. As beaver colonies collapsed from lack of food, water levels fell, and a natural cycle of flooding and seed bank regeneration was re-established.

I include this example because it is very familiar to me: my house now overlooks one of those beaver ponds. But, more importantly, the example illustrates how effective it is to simply plug drainage ditches. Beavers do it free. To complete the story, my wife and I bought several of those old farms as they became available, starting in 1975 when we borrowed money for the first hundred acres. Recently we donated a mixture of land and development rights to the Mississippi Madawaska Land Trust (www.mmlt.ca), which will protect nearly a square mile of forest and wetlands in perpetuity.

This is not to say beavers are a magical solution. They have costs, and they may generate new restoration challenges for the coming generations. Beavers need trees to construct dams, and the surrounding forests are strongly shaped by beaver cutting, which tends to shift composition away from deciduous trees toward coniferous trees. Beavers have been so effective at constructing ponds that they have all but eliminated natural seepage areas, streamside wet-meadows, and small streams. Future management may require control of beaver populations to protect these locally uncommon wetland habitats.

The Great Fen in England

The English fens are a good example to consider, because we have a long history of human activity there, and more than a century of efforts at restoration to consider. The Woodwalton Fen occurs in a flat area of eastern England. Descriptions of the fen go back to the Domesday survey of 1086; recall that, after England was conquered by Norman armies, this list was needed for the disposition of new land and other plunder. Thereafter is a period of decline from drainage and over-hunting. I have described these events in *Wetland Ecology* (Keddy 2010: 411–412), and for a longer essay you may read Sheail and Wells (1983). By the late 1890s, most of what remained was 'a dreary flat of black arable land, with hardly a jack snipe to give it a charm and characteristic attraction'. In 1910, 137 hectares were purchased as a nature reserve, but owing to the falling water table, the fen continued to deteriorate and was invaded by woody plants. Thereafter, restoration activities mostly focused upon blocking drainage ditches, and in one case, in 1935, using a portable pump to try to raise the water table during a drought. In 1972 a clay-cored bank was constructed to try to reduce the percolation of water out of the reserve. More recently, another relatively natural remnant of 256 ha has been acquired as the Holme Fen National Nature Reserve. Woodwalton and Holme will now become core areas within a 3000 ha restored wetland. The two problems of low water tables and high nutrient inputs will continue as challenges. You can read more about this under the title of The Great Fen Project (www.greatfen.org.uk). The section on restoration states:

> The Great Fen has inherited a complex and efficient network of drains, dykes and ditches whose primary purpose has been to get water away from the arable farmland as quickly as possible. Generations of farmers have deepened and straightened field ditches, and as a result, the peat fields rarely have any of the standing water that can be seen in other parts of the country after heavy rainfall. But now a major aim of the project is to retain water, rather than to drain it away.

For more on fen restoration elsewhere, you can consult Lamers *et al.* (2015). For the restoration of peat bogs, you can find useful practical instructions in Quinty and Rochefort (2003).

Larger-scale restoration

Levees, dykes, and canals in the Danube Delta

Restoration ecologists may also be challenged with larger tracts of dysfunctional landscapes. Even here, however, the principal causes may be obvious: drainage ditches and dykes. Consider the Danube River Delta in the Black Sea, which at 800,000 ha, is the largest in Europe (Gastescu 1993). The natural hydrology of this European waterway has been greatly altered – over 700 dams and weirs have been built along the river and its tributaries. The delta in the Black Sea has therefore been shrinking from lack of sediment. In addition, the delta has been criss-crossed with more than 1700 km of dredged canals. In the mid-1980s the communist dictator Nicolae Ceauşescu decreed that large areas of the delta should be transformed into agricultural land (Simons 1997). He sent 6000 men to build dikes, pump the land dry, and convert it into grain fields. Tataru Island, for example, was half drained and the local forest service had to supply 1000 m³ of wood, 3 tonnes of meat, 700 kg of honey, 3000 muskrats, and 0.5 tonnes of medicinal plants to the state every year. The challenge of repairing their damage remains.

One relatively easy way to restore habitat along rivers is simply to remove, or breach, the levees. In autumn 2003, for example, some 6 km of levee that surrounded the aforementioned Tataru Island were removed, restoring natural flooding, and therefore in 2004 the Danube again flowed freely over the island. In 1994 and 1996, levees were also opened in two former agricultural polders, Babina (2100 ha) and Cernovca (1560 ha), in Romania (Schneider *et al.* 2008). Seventeen major floodplain restoration sites have been identified along the Danube, as part of a larger plan to re-create a green corridor along the river (World Wildlife Fund 1999).

Rebuilding landscape contours with constructed wetlands

In some watersheds, the landscape has been so transformed by dykes, levees, ditches, fill, highways, canals, and cities that it is necessary to physically create or at least re-shape the land before flooding. This physical shaping has costs. There is the cost of the equipment, and the engineering planning. There is also the cost of harm done to remnant ecosystems during the reshaping. Balanced against this are the benefits of being able to construct a desirable set of contours with complex gradients, and the ability to control the substrate type.

As an example, consider the set of constructed wetlands in the south central United States described in Pierce (2015). He describes five steps in building such as constructed wetland:

1 Defining goals and preparing plans.
2 Defining the hydrogeomorphic setting.
3 Preparing a quantitative description of the hydrologic regime.
4 Developing a substrate and subgrade management plan.
5 Preparing a planting plan.

In general, Pierce concludes that the failure to develop a predictive model for the hydrologic regime is one of the most common failings. Without the appropriate water levels, one does not end up with a desirable wetland, or a wetland at all. Hence, it is important to know the water inputs and the water outputs, and to incorporate them into a wetland hydrograph. This advice comes from decades of practical experience in constructed wetlands. It is reassuring that I have quite independently suggested (Table 17.1) that about the half the variation we see in wetlands is caused by differences in water characteristics.

Let me say more about the potential merits of constructed wetlands. One of the real advantages of a constructed wetland is the ability to make new wetlands. This is a step up from refilling existing depressions and channels. It also provides an opportunity to make kinds of wetlands that have all but vanished from local landscapes. In my experience, fens, seepage areas, and wet meadows are particularly vulnerable to being lost from landscapes. I suggest that many of these can be considered peripheral habitats (see Figure 17.3) that likely supported much of the plant diversity in the original landscape. Constructed wetlands, then, may allow the creation of not just common wetland types, but some of the rare and more locally significant types that will further enhance biodiversity. It is easy to think only in terms of the single site at hand, that is, the particular plot of land designated for a constructed wetland. But the planning process really asks us to consider the surrounding landscape as a whole. What was the original mixture of wetlands in the landscape? What were the natural gradients and causal factors? Which kinds of wetlands and kinds of species were rare, and which were common? Given the regional context, what kind of wetland would provide the greatest number of services?

This is where wetland construction grades into the entire topic of landscape conservation. Each particular wetland will have a regional context where, in many cases, there will be core protected areas, buffer zones, and ecological corridors (Noss and Cooperrider 1994). Your restored wetland may therefore become part of the regional network of conservation lands. Constructed wetlands may allow us to enhance all three components of protected area systems: restoring core areas, enhancing the quality of buffer areas, and expanding the network of corridors. This is where we can learn from the concept of biosphere reserves as developed by UNESCO. Biosphere reserves contain core areas with high ecological value, are surrounded by a buffer zone, and include management plans to maximize human benefits while minimizing human damage. As of 2015 there are 651 biosphere reserves; there is an interactive map at www.unesco.org/mabdb/bios1-2.htm. Many are familiar for their wetlands: examples include the Doñana (Spain), the Pantanal (Brazil), the Danube Delta (Romania and Ukraine), and the Sundarbans (Bangladesh and India).

Really large-scale restoration

The Everglades in Florida

It is impossible to write about wetland restoration without saying something about the Everglades. It is an extreme case which provides a context for many other projects. Comprehensive Everglades Restoration Plan (CERP) is priced at more than 8 billion US dollars. There is an ongoing flood of reports and scholarly papers; one of the main planning documents exceeds 4000 pages! Pages on the Everglades are likely being written faster than you can read them. So, what can I say in a few short paragraphs? I intend to avoid a long description of the Everglades and CERP, except for some references to guide your reading. I will try to extract a few general lessons for younger practitioners from these early years of CERP. These lessons relate primarily to nutrients, and to plant diversity in natural wetlands.

First, the Everglades themselves. They were once a vast rain-fed wetland, with extremely low nutrient levels, and steady flow from north to south, producing a distinctive sedge-dominated vegetation type adapted to wet infertile conditions (Davis and Ogden 1994). The slow but steady flow of water, combined with extremely low nutrients, and drier periods with fire, appear to have been the main environmental factors that created and maintained the system (recall Figure 17.1). Drainage began in the 1880s. Humans were principally concerned with water, extracting it for growing cities, or to create drier conditions for agriculture and urbanization.

The battles over land development, drainage, and irrigation were legendary and include many stories of political intrigue and outright corruption (Grunwald 2006). As the Everglades began to change, populations of wading birds declined. The area of natural wetland began to shrink.

A first general lesson from the Everglades is the importance of low nutrient levels to successful restoration. Phosphorus concentrations across most of the Everglades were likely as low as 4 to 10 pg/l and loading rates averaged less than 0.1 g P/m^2/year. This means that many of the species in the Everglades could be termed stress tolerators with particular life history traits associated with low nutrient levels, such as evergreen plants and carnivorous plants. So, here is one lesson to draw to your attention: nutrients really do matter. Of course you have to get the water right for restoration. A huge network of canals, berms, and water control structures is intended to recreate the natural surface flow from south to west, and into Everglades National Park. But a flow of nutrient-rich water will simply increase the degradation, converting a rich mixture of stress-tolerant plants into a cattail-dominated wetland. Hence, a second objective of CERP is to reduce nutrient concentrations in the water to below 10 μg l^{-1} phosphorus. Recall that natural rainwater has minimal phosphorous, since it has come from evapotranspiration. Once such distilled water begins to flow across ground, nutrients accumulate, and if farmers are pouring phosphorus into their fields, the water will quickly become contaminated with high levels of P. In an attempt to deal with this, enormous (18,000 ha) treatment wetlands (STAs or stormwater treatment areas) have been constructed to reduce nutrient levels in runoff before this water enters the Everglades (Sklar *et al.* 2005). The general idea is that plants in the treatment ponds will extract enough phosphorus to ensure that the runoff will cause less harm to the Everglades. This, in my opinion, is one of the great untested assumptions in CERP. It is true that aquatic plants can remove phosphorus from water. But surely there are lower limits to the physiological capacity of plants to remove phosphorus – the original 4 to 10 pg/l is a very low level indeed.

A second general lesson is the importance of scale. That is, we need simple models to help us think, but they should not blind us to the wild diversity of wild nature. If you look at conceptual diagrams for the Everglades, they usually involve less than ten vegetation types (Figure 17.4). These ten types include sloughs, tree islands, and mangrove swamp. When one is managing an area the size of the Everglades, it is of course necessary to simplify the vegetation for some kinds of management. But, it is easy for engineers and zoologists to then begin to believe that there really are only ten or so vegetation types. And since many of these are dominated by just a few plant species, it is easy to begin to think that managing the Everglades means managing about 20 or so plant species. In fact, the vegetation of the Everglades was a rich mixture of species, including, as just one example, calcareous wet prairies maintained by fire (Orzell and Bridges 2006). There was high plant diversity, with more than 100 species per 1000 m^2. These habitats graded into different kinds of seasonally wetted rocklands and savannas. Thus the gradient structure in species composition was extremely complex. And these large numbers of plants rarely show up in Everglades models. Although Figure 17.4 is a classic, it risks becoming a problem if it replaces reality rather than illuminating it. That is, if the vast biological diversity of 'wet prairies' ends up being treated as one box with a couple of dominants, there is significant risk of losing much of the original diversity. Indeed, much of the plant knowledge in the Everglades relates to just a few wetland plants, particularly sawgrass and cattails. It might be helpful to have more information on vegetation gradients, indicator species, the ecology of stress tolerators, and the structure of those wet prairies, which are among some of the most speciose herbaceous vegetation types in the world. Here is where historical and palaeoecological information may help set restoration targets (Riedinger-Whitmore 2015).

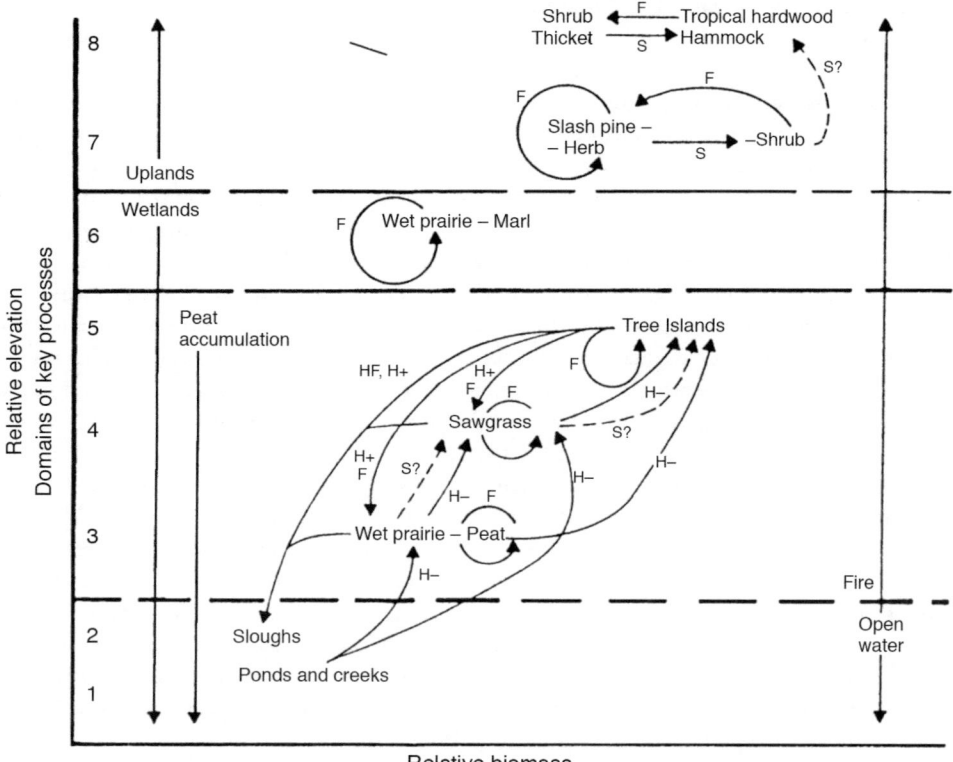

Figure 17.4 A classic illustration showing how the vegetation of the Everglades results from a few key factors. The vertical axis is elevation, which is controlled not only by the underlying topography, but by the accumulation of peat. The wettest sites have herbaceous vegetation in pools or sloughs with seasonal flooding. If enough peat accumulates, the herbaceous wetlands become tree islands (upper right). Succession then slowly moves the system from left to right. Fires move the system the other direction, from right to left. Light fires mostly change species composition, while more severe fires can create new shallow water sloughs (lower left). Superimposed upon this is a third controlling factor: nutrients. The very low phosphorous levels control both the kinds of plants found, and the rate at which sites recover from fire

Source: White (1994)

Other examples of and lessons from large-scale restoration

Much more could be written about large-scale restoration. Big scales have two potential problems. First, the stakes are bigger. Mistakes can have much bigger consequences. This is why we must get the science right. In some cases, I am far from impressed. Doyle and Drew (2008) have described five case studies of large-scale ecosystem restoration in the United States. To judge from work I have reviewed, it is easy to get the impression that teams of engineers are trying to build models of wetlands with minimal input from the science of plant ecology. We should not be reinventing the wheel. Existing knowledge about plant life history strategies, environmental gradients, succession, and pools and filters should be used, not ignored. A workshop of engineers and vertebrate ecologists, however well-intentioned, cannot reinvent a discipline they do not understand. Such oversights not only raise the costs, but they reduce the probability of success. The existence of this disconnect is readily apparent to anyone who understands plant ecology and then reads the reports and papers.

Second, the larger the scale, the more money and the greater the opportunity for abuse. Mark Twain may have said it best more than a century ago, opining that everyone disagreed what should be done about flooding along the Mississippi – but they all agreed it would take lots of federal money. Greed for federal money often over-rides scientific interests. While working in the Manchac Swamp in coastal Louisiana (Keddy *et al.* 2007) I saw distressing examples of money for restoration being squandered by administrators. Serious meetings about planning for the future of the coastal wetlands were lightly attended. But at the suggestion of federal money being available (an RFP, request for proposal), the room would packed, often with even one or more university deans present to monitor the scene. Our university received several million dollars for ecological restoration of the Manchac Swamp and for enhancing our field station. Much of it was handed out to biologists who knew or cared little about restoration. Typically, a microbial ecologist (said to be knowledgeable about plastic decomposition) announced loudly at a meeting 'This is just federal pork and I want my cut.' He got not just one, but several prime cuts, including a new boat. It became readily apparent that the vast scientific literature on restoration, community ecology, ecosystem resilience, and ecosystem health could be safely ignored, except in titles for the grant proposals. If you want an indicator for the consequences of the Manchac restoration money, you might be better to look at the participants: the size of their pick-up trucks, the upgrades to their houses, and the quality of alcohol consumed therein. All these improved markedly. The swamp did not. Without action, it may stay an anthropogenic marsh, degrade into brackish water, or even, as the climate warms, become a mangrove swamp (Keddy *et al.* 2007). More money won't help unless it is wisely spent. You can spend a lot of money on helicopters and airboats, and accomplish nothing.

I will not bore you with other stories: trainloads of rock being dumped in the swamp to hold back flood waters (one could mention King Canute but no one knows about him anymore), studies on ecosystem 'health' with minimal understanding of the environmental history of the region, new construction in the very areas flooded by hurricane Katrina, the Deepwater Horizon oil spill of 2010, or rooms of Louisiana residents chanting 'Drill baby drill!' Yes, large areas of the state are just above sea level, and yes drilling for oil will cause the land to subside, and yes burning it will cause the sea to rise, but apparently these are unwelcome facts to be ignored.

Such irritations do raise a deeper question for younger scholars to consider. What would you have done as a wetland ecologist in the Danube Delta in Romania during the 1960s? Or in the Mesopotamian wetlands in Iraq during the 1980s? Or, for that matter, in the Manchac Swamp in Louisiana in 2000? There is no easy answer. If you participate and do good work, it may simply be used as camouflage to hide the much larger body of bad work. If you walk away, there may be no one to document the waste and abuse, or to insist on at least minimal standards of scientific credibility.

These situations remind me of the dire story of the destruction of the forests of Easter Island (Wright 2004). The task of restoration is a challenging one, requiring a knowledge of wetland ecology (causal factors in wetlands), community ecology (pools and filters), and environmental history (recall Table 17.2). But the biggest challenge may be managing our own species. It appears that greed, cronyism, and corruption can at times overwhelm our better nature. How else can one explain Easter Island? Were there public meetings where the islanders chanted 'Log baby log?' I raise these unhappy topics in this handbook because there is a great risk for young restoration ecologists that they will be trampled in the rush for money by those far less qualified and even wilfully ignorant of the field of restoration and the science of ecology altogether. This is an unhappy reality, and while I once expected it to recede with time and education, I am now more inclined to think of it as an inherent part of human nature.

Conclusion

We have come a long way from Figure 17.1. It is time to remind you to follow the four steps in Table 17.2. Learn about the environmental history of your project area. Get the water right. Get the nutrients right. Do the very best science you can. Plan your restoration work with the highest aspirations for success.

Oh yes, while I am dispensing advice, let me say one more thing before I return to the forest. Instead of spending your weekends playing sports or hanging out in bars or mowing your lawn, get a canoe and get to know your wetland personally. Frogs and egrets and alligators and even dragonflies all have something useful to say, if you get to know them on their own terms, and if you take the time to listen to them, over the orchestrated din of organized sports, academic infighting, cronyism, and pork barrel politics. The better you know your wetland, and its many inhabitants, the greater your probability of success in restoration. This may not make you rich, but it should give you a life worth living.

References

Davis, S. M. and J. C. Ogden (eds). 1994. *Everglades: The Ecosystem and Its Restoration*. Delray Beach, FL: St Lucie.

Doyle, M. and C. A. Drew (eds). 2008. *Large-Scale Ecosystem Restoration: Five Case Studies from the United States*. Washington, DC: Island Press

Gastescu, P. 1993. The Danube Delta: geographical characteristics and ecological recovery. *Earth and Environmental Science* 29: 57–67.

Gopal, B., J. Kvet, H. Löffler, V. Masing and B. C. Patten. 1990. Definition and classification. Pp. 9–15 in B.C. Patten (ed), *Wetlands and Shallow Continental Water Bodies*. Vol. 1, *Natural and Human Relationships*. The Hague: SPB Academic Publishing.

Grunwald, M. 2006. *The Swamp: The Everglades, Florida, and the Politics of Paradise*. New York: Simon & Schuster.

Houlahan, J. E., P. A. Keddy, K. Makkay and C. S. Findlay. 2006. The effects of adjacent land use on wetland plant species richness and community composition. *Wetlands* 26.1: 79–96.

Hughes, F. M. R. (ed.). 2003. *The Flooded Forest: Guidance for Policy Makers and River Managers in Europe on the Restoration of Floodplain Forests*. Cambridge: Department of Geography, University of Cambridge.

Junk, W. J. 1993. Wetlands of tropical South America. Pp. 679–739 in D. F. Whigham (ed), *Wetlands of the World I*. Dordrecht, the Netherlands: Kluwer Academic Publishers.

Keddy, P. A. 1983. Freshwater wetlands human induced changes: Indirect effects must also be considered. *Environmental Management* 4: 299–302.

Keddy, P. A. 2010. *Wetland Ecology: Principles and Conservation*. 2nd edn. Cambridge: Cambridge University Press.

Keddy, P. A. 2016. *Plant Ecology*. 2nd edn. Cambridge: Cambridge University Press.

Keddy, P. A. and A. A. Reznicek. 1986. Great Lakes vegetation dynamics: the role of fluctuating water levels and buried seeds. *Journal of Great Lakes Research* 12: 25-36.

Keddy, P. A., D. Campbell, T. McFalls, G. Shaffer, R. Moreau, C. Dranguet and R. Heleniak. 2007. The wetlands of lakes Pontchartrain and Maurepas: past, present and future. *Environmental Reviews* 15: 1–35.

Lamers, L. P. M., ,Vile, M. A., Grootjans, A. P., Acreman, M. C., van Diggelen, R., Evans, M. G., Richardson, C. J., Rochefort, L., Kooijman, A. M., Roelofs, J. G. M. and Smolders, A. J. P. 2015. Ecological restoration of rich fens in Europe and North America: from trial and error to an evidence-based approach. *Biological Reviews* 90: 182–203.

Lawler, A. 2005. Reviving Iraq's wetlands. *Science* 307: 1186–1189.

Middleton, B. A. (ed.). 2002. *Flood Pulsing in Wetlands: Restoring the Natural Hydrological Balance*. New York: Wiley.

Mitsch, W. J. and J. G. Gosselink. 2015. *Wetlands*. 5th edn. Hoboken, NJ: Wiley.

Moore, D. R. J., P. A. Keddy, C. L. Gaudet and I. C. Wisheu. 1989. Conservation of wetlands: Do infertile wetlands deserve a higher priority? *Biological Conservation* 47: 203–217.

Nilsson, C., C. A. Reidy, M. Dynesius and C. Revenga. 2005. Fragmentation and flow regulation of the world's large river systems. *Science* 308: 405–408.

Noss, R. F. and A. Y. Cooperrider. 1994. *Saving Nature's Legacy*. Washington, DC: Island Press.

Olson, D. M., Dinerstein, E., Wikramanayake, E. D., Burgess, N. D., Powell, G. V. N., Underwood, E. C., D'amico, J. A. Itoua, I., Strand, H. E., Morrison, J. C., Loucks, C. J., Allnutt, T. F., Ricketts, T. H., Kura, Y., Lamoreux, J. F., Wettengel, W. W., Hedao, P. and Kassem, K. R. 2001. Terrestrial ecoregions of the world: a new map of life on Earth. *Bioscience* 51: 933–938.

Orzell, S. L. and E. Bridges. 2006. Floristic composition and species richness of subtropical seasonally wet *Muhlenbergia sericea* prairies in portions of central and south Florida. Pp 136–175 in R. Noss (ed), *Land of Fire and Water: The Florida Dry Prairie Ecosystem*. DeLeon Springs, FL: Painter Printing Company.

Partow, H. 2001. *The Mesopotamian Marshlands: Demise of an Ecosystem, Early Warning and Assessment Technical Report*. Nairobi, Kenya: United Nations Environment Programme.

Pierce, G. J. 2015. *Wetland Mitigation: Planning Hydrology, Vegetation, and Soils for Constructed Wetlands*. Glenwood, NM: Wetland Training Institute.

Quinty, F. and L. Rochefort. 2003. *Peatland Restoration Guide*. 2nd edn. Quebec, Quebec: Canadian Sphagnum Peat Moss Association and New Brunswick Department of Natural Resources and Energy.

Riedinger-Whitmore, M. A. 2015. Using palaeoecological and palaeoenvironmental records to guide restoration, conservation and adaptive management of Ramsar freshwater wetlands: lessons from the Everglades, USA. *Marine and Freshwater Research* 67: 707–720.

Schneider, E., M. Tudor, and M. Staras (eds). 2008. *Evolution of Babina Polder after Restoration Works*. Frankfurt am Main, Germany: WWF Germany, Department of Water and River Basin Management at the University of Karlsruhe, and Danube Delta National Institute for Research and Development.

Sculthorpe, C. D. 1985. *The Biology of Aquatic Vascular Plants*. Königstein, Germany: Koeltz Scientific.

Sheail, J. and T. C. E. Wells, 1983. The Fenlands of Huntingdonshire, England: a case study in catastrophic change. Pp. 375–93 in A. J. P. Gore (ed.) *Ecosystems of the World, vol. 4B, Mires: Swamp, Bog, Fen and Moor – Regional Studies*. Amsterdam: Elsevier.

Silliman, B. R., E. D. Grosholz, and M. D. Bertness (eds). 2009. *Human Impacts on Salt Marshes: A Global Perspective*. Berkeley, CA: University of California Press.

Simons, M. 1997. Big, bold effort revives the Danube wetlands. *The New York Times*, 19 October, pp. 1, 8.

Sklar, F. H., Chimney, M. J., Newman, S., McCormick, P., Gawlik, D., Miao, S., McVoy, C., Said, W., Newman, J., Coronado, C., Crozier, G., Korvela, M. and Rutchey, K. 2005. The ecological–societal underpinnings of Everglades restoration. *Frontiers in Ecology and the Environment* 3: 161–169.

Turner, R. E. and N. N. Rabelais. 2003. Linking landscape and water quality in the Mississippi River Basin for 200 years. *BioScience* 53: 563–572.

Verhoeven, J. T. A., W. Koerselman and A. F. M. Meuleman. 1996. Nitrogen- or phosphorus-limited growth in herbaceous, wet vegetation: Relations with atmospheric inputs and management regimes. *Trends in Ecology and Evolution* 11: 494–497.

Vitt, D. 1994. An overview of factors that influence the development of Canadian peatlands. *Memoirs of the Entomological Society of Canada* 169: 7–20.

Weiher, E. and P. Keddy (eds). 1999. *Ecological Assembly Rules: Perspectives, Advances, Retreats*. Cambridge: Cambridge University Press.

Welcomme, R. L. 1979. *Fisheries Ecology of Floodplain Rivers*. London: Longman.

White, P. S. 1994. Synthesis: Vegetation pattern and process in the Everglades ecosystem. Pp. 445–460 in S. Davis and J. C. Ogden (eds), *Everglades: The Ecosystem and its Restoration*. Delray Beach, FL: St Lucie.

Wilcox, D. A. 2012. Great Lakes coastal marshes. Pp. 173–188 in D. P. Batzer and A. H. Baldwin (eds), *Wetland Habitats of North America*. Berkeley, CA: University of California Press.

World Wildlife Fund. 1999. *Evaluation of Wetlands and Floodplain Areas in the Danube River Basin: Final Report*. Sofia, Bulgaria: WWF Danube–Carpathian Programme, and Rastatt, Germany: WWF Auen-Institut.

Wright, R. 2004. *A Short History of Progress*. Toronto: Anansi Press

Zampella, R. A., J. F. Bunnell, K. J. Laidig and N. A. Procopio. 2006. Using multiple indicators to evaluate the ecological integrity of a coastal plain stream system. *Ecological Indicators* 6: 644–663.

18

SALT MARSHES

David M. Burdick and Susan C. Adamowicz

Tidal marsh systems

Definition and distribution

Salt marshes are wetlands dominated by salt-tolerant herbaceous plants (halophytes) that typically develop in depositional environments subject to tidal inundation. We limit discussion to tidally influenced systems because processes associated with hydrology are critical for restoration planning and success. Tidal marshes (hereafter salt marshes) transition from salt to brackish and fresh as fresh water contributions increase. Found along the margins of all continents except Antarctica, salt marshes typically are limited to latitudes greater than 25° and can extend into arctic coastlines. Their abundance and extent at any particular locale depends upon the physical exposure and slope of the shoreline, with extensive marshes along flat trailing edges of continental plates, especially landward of barrier beach systems; whereas narrow, fringing marshes are typically found along steep shorelines (Mitsch and Gosselink 2000; Morgan *et al.* 2009). Salt marshes embody several ecological functions that are important to humans as ecosystem services, which range from support of coastal food webs and biodiversity to storm protection and carbon storage.

Complete and accurate figures of salt marsh distribution worldwide are unavailable, with estimates for the US, Canada and Europe at 22,000 km² (Mcleod *et al.* 2011) and globally at 37,500km² (Halpern *et al.* 2012). The area of salt marsh loss is much more difficult to estimate because many losses occurred before remote sensing tools were developed (e.g. San Francisco Bay; US Atlantic and Gulf coasts; southeast England) and it is not always clear how human management has contributed to losses (e.g. levee and diversion construction throughout the Mississippi Delta). Regional studies in the US and Canada have shown about 37 per cent of tidal marsh area has been lost in New England, 93 per cent lost in California and 64 per cent lost in the Canadian Maritimes (Bromberg Gedan *et al.* 2009), with about 20 per cent of the remaining marsh area currently being degraded by tidal restrictions (Bromberg and Bertness 2005). Much of the remaining marsh has been compromised by direct impacts like ditching and pollution and indirect impacts from a variety of stressors, including excess nitrogen, invasive species and climate change. A sample of twelve estuaries of the world (most in NA) averaged 65 per cent loss of salt marsh area (Lotze *et al.* 2006). Because of their limited distribution, historic losses and high ecosystem service values, salt marshes are important and common targets of restoration efforts in the US, Canada and Europe. By restoration we mean re-establishing natural processes needed to support functions, especially self-maintenance. Further, threats associated with climate change, particularly increased rates of sea level rise, create an urgency for marshes to be restored as well as wariness of resources to be spent on potentially short-lived benefits.

Tidal marsh development and response to sea level rise

The current paradigm of formation and development of salt marshes framed by Redfield (1972) has been used and modified by a generation of scientists, most recently to explain marsh growth (and loss) with recent acceleration of sea level rise (Morris *et al.* 2002; Kirwan *et al.* 2010). Generally, where salt marsh vegetation colonizes unvegetated intertidal elevations, plants interact with floodwaters to build to an elevation that approximates mean high tide. Under conditions of slow sea level rise, marshes may contract or expand along their seaward edge slowly, depending upon sediment supply and exposure to waves and ice (Figure 18.1). Marsh plants that dominate low marsh habitats may be replaced by competitively superior ones of the high marsh (Levine *et al.* 1998) if the marsh builds more quickly than SLR. As seen for many tidal marshes around the world, large expanses of high marsh can result, extending from the mean high tide landward to the spring high tide line.

With accelerating SLR, low marsh can replace high marsh communities. High marsh, in turn, can migrate inland, eventually replacing uplands, especially along coasts with low slopes. Under stable or slow SLR (0-5 mm/yr), marshes seem to persist unless subjected to disturbances that disrupt their negative feedback system – a dynamic equilibrium between building

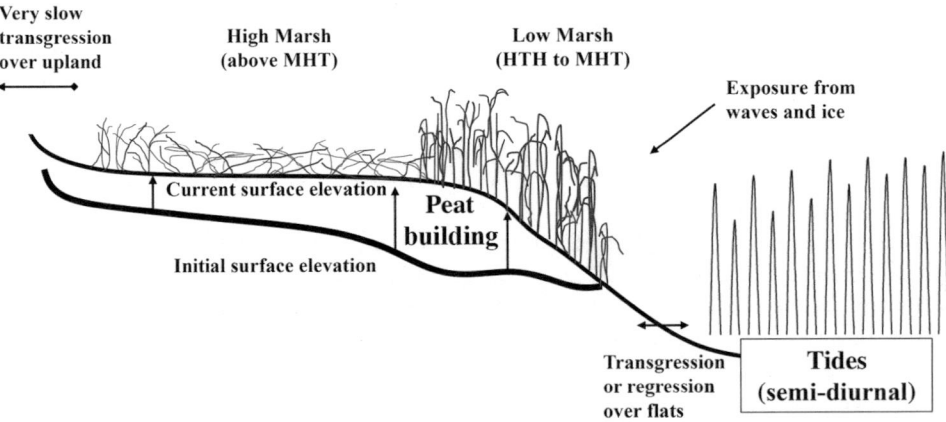

Figure 18.1 Conceptual model of interactions promoting marsh persistence and loss. As tidal waters flow over the marshes, plants slow floodwaters and suspended sediment drops out or is captured on leaves while peat decomposition is slowed by hypoxic conditions. Storms can be important sedimentation events, as are ice deposits in colder climates. Under slow to moderate sea level rise (1–5 mm/yr) shown here, marshes can maintain relative elevation through accretion of inorganic sediments and organic matter storage; falling sea level results in draining, oxidation of peat, and subsidence. Half tide height (HTH) and mean high tide (MHT) bound the lower and upper low marsh, respectively

and destructive processes (Morris *et al.* 2002; Burdick and Roman 2012). In simplest terms, marshes can build in elevation with sediment addition and peat formation, both of which require tidal flooding to supply suspended sediments and prevent oxidation of the accumulating organic matter. Over the past 4–6,000 years, the slow increase in sea level (1 mm/yr) has allowed marshes with limited physical exposure and relatively high sediment supplies to build and expand throughout their geographic range. We can think of these systems as self-maintaining, with the halophytes and tides interacting to maintain elevation and ecological functions under dynamic conditions (Figure 18.1).

Call for restoration

Certainly the need for conservation and restoration of salt marshes today is as urgent and even more global in appeal (Bromberg Gedan *et al.* 2009) than when salt marsh losses were first called to our attention by Teal and Teal in 1969. At that time most impacts were direct, caused by conversion to agriculture, filling for development, filling/dredging for overland and marine transportation, and discharge of pollutants. Across much of the developing world impacts are similar, but in developed areas many direct impacts are prohibited or regulated and indirect impacts (e.g. tidal restriction and invasive species) are becoming more important threats. Further, many threats are now global rather than local in nature (e.g. human-induced sea level rise) and may require large-scale regional programmes to address them (Silliman *et al.* 2009).

While undeveloped habitats could be engineered using biological and physical processes to create new tidal marshes, restoration of marshes degraded by direct and indirect impacts can be pursued with lower costs and better outcomes without the need to convert other valuable habitats. These sites represent a large portion of existing marshes and should be restored quickly, especially in cases where further human development would prevent restoration (e.g. see Reiner 2012), or changing climate and physical conditions could interact with human impacts to prevent restoration success or interfere with self-maintenance of restored systems. A prime example of this is Blackwater National Wildlife Refuge in Chesapeake Bay on the eastern shore of Maryland. Blackwater marshes have been affected by an historic hand-dug channel, prescribed burns, water withdrawals, geologic subsidence and sea level rise. The once extensive tidal marshes have converted to open water locally known as 'Blackwater Lake'.

Human impacts to tidal marshes and restoration approaches

Human impacts to tidal marshes have been discussed extensively in a recent volume (Silliman *et al.* 2009). Occupation of coastal areas by humans with agrarian lifestyles has led to diking of salt marshes and conversion to agriculture. Some impounded sites were left unimproved as fresh water non-tidal wetlands, whereas sediment was used as fill in others to reduce flooding. Increasing development led to filling of marshlands (e.g. Boston: Teal and Teal 1969; Bromberg and Bertness 2005), and building earthen dikes as coastal defences (e.g. western Europe: Pethick 2002; Wolters *et al.* 2005). Transportation needs led to corridors built across marshes (fill for overland; dredge for vessel passage). Both filling and dredging directly destroyed tidal marshes, but remaining marshes continue to be degraded by a variety of indirect impacts – those that interfere with marsh processes. For example, hydrologic impacts include the widespread practice of reducing or eliminating tides with undersized culverts, physical alterations in attempts to control mosquito production (ditching, open marsh water management) and waterfowl habitat improvements including creation of impoundments, pool excavation and ditch plugging.

Restoring tidal exchange

Human impacts to tidal hydrology may be unintentional (undersized culverts and bridges, etc.) or intentional (drainage, berm and dike barriers for agriculture, storm protection dikes, impoundments for wildlife, salt production, etc.). Regardless of intent, hydrologic barriers restrict tidal flooding, leading to freshening of the upstream marsh, invasion by plant species less tolerant of saltwater, oxidation of organic sediments and subsidence (Roman *et al.* 1984; Portnoy 1999; Adamowicz and O'Brien 2012). Some marsh elevations have fallen by one meter or more (Frenkel and Morlan 1991; Pethick 2002; Williams and Orr 2002; Anisfeld 2012) due to diking and subsequent peat oxidation. Another key value of tidal marshes, support of coastal food webs, is substantially degraded by tidal restrictions. Clearly areas with little to no flooding cannot support fish, but even impounded sites that have water and support small resident fish do not allow uninhibited fish passage (Eberhardt *et al.* 2011) and are less valuable habitat (Dibble and Meyerson 2012). Restoration of tidal flows, however, can both improve fish habitat (*ibid.*) and support passage (Eberhardt *et al.* 2011).

Restoration of tidal hydrology can be achieved by expanding a restricting inlet (bridge or culvert) or removal of a berm or dike to allow for full tidal flows. Many such projects have been carried out in the past 30 years because they are often straightforward from an engineering perspective and provide a better return on investment in terms of area restored compared to more intensive restoration techniques. In Western Europe, dike removal, termed de-embankment or managed realignment, is not always popular with the public (Musch 2011). It is, however, seen as a way to provide habitat benefits and adapt to sea level rise as salt marsh restoration will provide less expensive storm protection than dike maintenance (Wolters *et al.* 2005).

Coastal states in New England and elsewhere have developed entire programmes focused upon restoring tidal exchange to marshes (Roman and Burdick 2012). Unfortunately due to greater costs associated with replacing larger spans or completely removing barriers, and the potentially increased risk that full tidal exchange may damage homes and infrastructure landward of the restriction, full tidal restoration is rarely, if ever, achieved. Fortunately, provision of 80 per cent or more of potential tidal exchange has been found to reinitiate marsh carbon storage and building processes even as it inhibits spread of less salt-tolerant invasive plants (Konisky *et al.* 2006). Whereas complete dike removal may be the best solution from an ecological perspective, partial removal costs less and is usually pursued, even though surface flows continue to be restricted and flood levels are lower than unrestricted marshes.

Some projects to restore tidal exchange are large and expensive and/or controversial due to high population densities, local concerns over past and current wetlands, and the nature of the tidal restoration approach – reintroducing tides to an entire wetland. In these cases hydrologic models are needed to size the openings (MacBroom and Schiff 2012) and engineering may be required to design a water control structure and fortify abutments. In addition, active management may be needed to promote marsh development and avoid or reduce undesirable results. Such projects typically feature predictive models that are required by the public to show the likely results of tidal reintroduction. Results of the management action also need to be examined using pre- and post-restoration monitoring to demonstrate to local stakeholders whether ecological expectations were met.

Where dikes have been maintained and marshes drained for agriculture, surface elevation losses may be so great that complete tidal restoration will drown all vegetation, and even low marsh plants will be unable to return. In these cases, partial tidal restoration may allow some sedimentation and peat building and a sequenced approach like that used in Hatches Harbor (Smith *et al.* 2009) could restore greater tides in the future. Others call for filling to a grade that will support vegetation and marsh development (Cornu and Sadro 2002; Williams and Orr 2002). Standard hydrologic models are available to predict tidal flooding based on a variety of inlet designs and flow control structures (McMillan 2011; MacBroom and Schiff 2012). Ecological models are available to help predict marsh outcomes based on potential tide heights and elevation ranges for local marsh habitats (Craft *et al.* 2009; Konisky 2012).

While restoration of undeveloped farmland back to tidal marsh is fairly straightforward, there are several issues that often arise with tidal restoration. Foremost among these is the development of residences, businesses and infrastructure (roads, utilities, etc.) that could become susceptible to flooding with re-establishment of tidal flow. In these cases, partial tidal restoration may be chosen as the best alternative and can be achieved by correct sizing of a culvert or use of self-regulating or human-activated tide gates. Increasing tidal range and influence will increase soil salinity and reduce peat oxidation, but results with tide gates have been mixed (Reiner 2012; Adamowicz and O'Brien 2012) and ultimately may not allow the system to respond to sea level rise. Thus, partial tidal restoration may provide ecological benefits and increase a system's resilience over the short term, but still may not allow long-term survival. Many sites in Europe have only partial tidal restoration, with the use of self-regulating tide gates gaining recognition as a restoration tool (Wolters *et al.* 2005; ABPmer 2014).

Collision of ecological and social expectations can occur with tidal restorations. Freshwater mosquito populations, which are seasonal breeders, are replaced by saltwater species that breed on spring tide intervals throughout the warmer months. If forests have developed within drained areas, restoring tidal flow will kill trees, which will then fall over, creating small pools for mosquito production. Exposure of impounded sites may release foul odours for several weeks. The public may not be aware of such potential changes and management of expectations is an important component of tidal restoration.

Fill removal

Because the elevation of large expanses of high marsh lies just a few feet below the highest astronomical tides, tidal marshes have provided an easy target for landowners and local governments wanting to increase the area of usable upland. Throughout the twentieth century thousands of hectares of marshlands were filled for development across the US (Bromberg and Bertness 2005; Bromberg Gedan *et al.* 2009) and Europe (e.g. Guadalquivr Estuary, Spain; Gallego Fernández and García Novo 2007). Filling of marshlands was especially true in port cities where dredging to create deep navigational channels provided a ready source of fill.

Restoration of a filled tidal marsh typically begins with historical investigation of when and why the area was filled. Toxicity of the fill needs to be considered as well as presence of invasive species propagules when locating a site to receive the sediments. In addition, rising sea levels should be considered when the tidal marsh is designed (McMillan 2011). Thus, elevations for vegetated areas and creeks should be planned for a minimum life expectancy for explicitly stated sea levels projected at least 50 years into the future. Resilience to rising sea levels can be designed into projects by creating areas where marshes may migrate in the future (e.g. tidal freshwater marsh or recreational open space) as marsh seaward edges are lost to rising tides.

Large unvegetated areas may need to be planted if local seed sources are absent or if expectations require accelerated development of vegetation, especially where invasive plants have been removed (see following text). Some planting to establish seed sources can be valuable and accomplished through planting projects with local schools to develop authentic educational experiences. Plantings can also be used to increase diversity to enhance marsh functions (Callaway *et al.* 2003) or deter invasives (Peter and Burdick 2010).

Invasive species and trophic cascades

One might think that the relatively harsh physical conditions encountered in salt marshes would themselves preclude invasions by non-native plants and animals, but the adaptations of a species on one coast may serve it well when transported to a new coast. Hence, *Phragmites australis* (Eurasian variety) is a major concern among managers in the US (Hazelton *et al.* 2014), while in Europe managers are concerned about its dieback (Brix and Cizkova 2001). In contrast, *Spartina alterniflora*, a foundation of East Coast marshes, has invaded the Pacific coasts of North America and Asia (Strong and Ayers 2009). In addition to *S. alterniflora*, *Lepidium latifolium* (perennial pepperweed) has stimulated restoration efforts on the west coast of North America (Powers and Boyer 2013). In southwestern Spain, South American *Spartina densiflora* has been removed and native *S. maritima* planted to restore tidal marshes (Curado *et al.* 2014). Approaches to control or eradicate invasive plants in marshes include hand-pulling, mowing, burning, flooding with fresh or saltwater, covering, excavation, grazing, biocontrol and herbicide. Eradication is rarely achieved for the most tenacious invasives (e.g. *P. australis*), but restoration can be successful using herbicide where a thoughtful programme of monitoring and treatment has been continued over several years and/or when used in conjunction with other methods, especially a change in processes (e.g. tidal restoration to increase salinity; Teal and Peterson 2009; Hazelton *et al.* 2014).

The current paradigm of marsh development and expansion has served us well for many years, as has our concept of how marshes provide trophic support for coastal ecosystems, primarily through the detrital food web. However, several cogent examples have shown that trophic interactions, heretofore considered of minor importance, may indeed be important, especially at large scales (Silliman *et al.* 2009; Altieri *et al.* 2012). These cases often have an anthropogenic basis and thus are candidates for restoration. As pointed out by Teal and Peterson (2009), impacts from invasive European snails and green crabs on US Atlantic coast marshes are difficult to assess and perhaps more difficult to control as part of a restoration effort.

Some of these problematic organisms are exotic, but others are native. Large aggregations of migrant waterfowl (snow geese) have been assigned blame for destroying or modifying large areas of coastal marshes in Hudson Bay (Jefferies *et al.* 2006) and Texas (Miller *et al.* 1997). Destructive impacts from native muskrat and exotic nutria have been described as 'eat outs' in marshes along the Gulf of Mexico (Ford and Grace 1998). Similarly, a native snail (*Littorina irrorata*) and crab (*Sesarma reticulatum*) have been implicated in marsh die-off episodes in South Carolina and Massachusetts (Silliman and Zieman 2001; Altieri *et al.* 2012; but see Alber *et al.* 2008). Loss of top predators (fish and herons) feeding on these nocturnal crabs may be responsible, requiring thoughtful approaches to restoration (e.g. restore predators or change herbivore behaviour; Powers and Boyer 2013).

Ditches and their legacy

Initially, farmers of salt hay (*Spartina patens*) dug ditches to drain portions of the marsh for access and to increase production, but early in the twentieth century mosquitoes were identified as the vector for malaria and populations of salt marsh mosquitoes were targeted by digging miles of drainage ditches through marshes. Seventy years later, these ditches have lowered high marsh elevations and their spoil piles have impounded marsh sections, both of which lead to loss of resilience as sea levels rise (Raposa *et al.* 2015). Managers tried to plug ditches to remove their impact, but this led to impoundments with loss of vegetation, peat integrity and elevation (Vincent *et al.* 2013; Elsey-Quirk and Adamowicz 2016). A more recent approach is to fill ditches from the bottom up to reduce peat oxidation and loss. At the end of the growing season, aboveground portions of high marsh plants are cut, placed in the ditches and secured with twine so the ditches shallow as plants collect sediments (Wilson and Adamowicz, unpublished data). To drain the inter-spoil impoundments, some managers are installing shallow drainage paths, called runnels, to allow plants to recolonize and rebuild peat (Raposa *et al.* 2015).

Sea level rise and erosion

During periods of slow sea level rise (1 to 5 mm/yr) over the past several thousand years, marshes have expanded into shallow bays and gently sloped uplands. Accelerated climate change has led to changing conditions including increased water and air temperatures, increased CO_2 levels, higher water levels, changes in amounts and patterns of precipitation, and an increase in storms or climatic events (e.g. drought). Such changes have altered ecosystem interactions that might benefit or degrade marshes. Other large-scale actions have increased (deforestation; watershed development) or decreased (reforestation, damming rivers, armouring shorelines, dredging, stabilizing inlets and preventing inlet movement and overwash of barrier systems) sediment supply needed for marsh elevation growth. Some changes have accelerated marsh expansion (breakwaters or causeways that reduce physical exposure). Increasing temperatures, varying salinity and increased supply of nitrogen all may increase decomposition of peat (Portnoy 1999; Kirwan and Blum 2011; Deegan *et al.* 2013) that would decrease marsh elevations and their ability to vertically accrete through peat storage. However, sea level rise, with estimates ranging from 3 to 20 mm/yr and the possibility of rapid increases from the collapse of the western Antarctic ice shelf (Nichols and Cazenave 2010), makes the widespread loss of tidal marshes seem inevitable. Although we do not suggest restoration can occur today in preparation for future impacts, we can increase resilience of impacted marshes to sea level rise and we can design restoration projects to account for higher sea levels (McMillan 2011).

In Louisiana, subsidence associated with natural and human actions (including extraction of fossil fuels) has resulted in substantial areas of marsh loss (Penland *et al.* 1990). Approaches to restore lost marshes have included thin layer deposition, beneficial use of dredge spoil, and creation of diversions that direct sediment-laden waters from the Mississippi River to areas of rapid marsh loss. The process is termed sediment recharge and has been used in other locations. For example, several US National Wildlife Refuges are using this technique to increase marsh resilience following Hurricane Sandy.[1] Sixteen projects have reused dredged sediment to reduce erosion and stabilize marshes in southern England (ABPmer 2014).

In regions without severe subsidence, some marshes are showing losses in vegetation that have been attributed to sea level rise (e.g. New England; Smith 2009; Raposa *et al.* 2015) and the previously mentioned techniques may be useful approaches as may the living shorelines concept where restoration includes some sort of artificial erosion control at the marsh edge (Needelman *et al.* 2012). Our evolving model of marsh development has recently been modified to include processes related to expansion and erosion of the seaward edge of the marsh (Fagherazzi *et al.* 2012). Marsh restoration projects lacking an erosion-resistant edge of peat may benefit from installation of a seaward sill (rock, shellfish reef, coir logs) that protects the marsh (Gittman *et al.* 2014; Restore America's Estuaries 2015).[2]

Examples of restoration approaches

Restoration practitioners would do well to consider tidal marshes as poised systems that exist within an envelope of physical conditions and where physical and biological processes interact over centuries to develop low marsh and high marsh expanses intersected by a system of tidal creeks. Human activities that change conditions or disrupt the interchange of processes can have long-term negative impacts – resulting in changes in marsh structure and function that are both symptoms of impacts and a signal that such systems can (and should) be restored. Each restoration could be viewed as an experiment, and if an experimental approach is embraced and the results shared, we can learn from each project. What follows is a series of short case studies that exemplify different restoration approaches and have contributed to our knowledge of tidal marsh restoration.

Hydrologic restoration

Restoring tidal hydrology to systems with diminished tidal regime has been practised success-fully at large and small scales around the US and Canada (Cornu and Sadro 2002; Williams and Orr 2002; Teal and Peterson 2009; Konisky *et al.* 2006). A guidance manual has been produced recently by NOAA that includes a variety of valuable resources, including tool kits and exam-ples of projects in the southeastern US (Craig *et al.* 2010). Examples of hydrologic restoration from the northeastern US and Canada, including projects with self-regulating tide gates, are featured in Roman and Burdick (2012).

In northwestern Europe extending from Spain to Denmark, removing dikes and seawalls to allow marsh rebuilding (managed realignment or de-embankment) has developed into a signif-icant restoration approach (ABPmer 2014). Some projects proposed for natural protection of shorelines and restoration of valuable wildlife habitat are proactive; however, others are proposed as compensation for losses elsewhere and may involve only partial tidal restoration (Musch 2011). Examples of de-embankment and other European approaches can be found in Wolters *et al.* (2005), Musch (2011) and ABPmer (2014).

Removing fill to re-establish tidal elevations, hydrology and native vegetation

Broad Meadows Marsh in Quincy, MA (USA) was used to dispose of material dredged from the adjacent Town River. This type of marsh destruction was common in the coastal wetlands across the nation (Bromberg Geden and Silliman 2009). In Quincy, 45 ha of salt marsh were filled with 1.8–3.0 metres of sediment from dredge projects in 1938 and 1956. Since then, 15 ha of the site were developed for municipal purposes (school, public works yard, ice hockey rink) thus removing those areas as potential restoration sites. In 2004, the Town of Quincy, the US Army Corps of Engineers (USACE), the Neponset River Watershed Association and other partners designed a project that removed over 298,000 cubic meters of dredged material, restoring 13 ha of salt marsh, 2 ha of wet meadow grasses and 9 ha of coastal grasslands (Gendron 2013). The work was partially funded by the USACE.[3]

Remediating the legacy of ditch effects

Ditching for agricultural and other purposes in North American tidal marshes dates back to European settlement. While Colonial ditching was limited, twentieth-century ditching was more pervasive and ultimately motorized so that by 1950, Bourne and Cottam (1950) concluded that nearly 90 per cent of coastal marshes from Maine to Virginia had been ditched. In response to ditch legacy effects, integrated marsh management (IMM) was implemented at Wertheim National Wildlife Refuge in Long Island, NY. This project completely filled some ditches, increased sinuosity on others and created pools in hummocky mosquito breeding locations to reduce larval habitat and to provide refugia for their fish predators. As a result, invasive *Phragmites* was reduced, native vegetation increased cover, fish abundance and diversity increased, as did avian species (Rochlin *et al.* 2012).

Removing invasive and restoring native halophytes

Restoration efforts that focus upon eradication of invasive plants seem to be poorly monitored and reported in the literature, with most publications centred on a new or modified method to weaken or eradicate them. A review of US control methods for Eurasian *Phragmites australis* found few studies presented long-term data for native plant community responses to invasive control and called for planning and control efforts at larger scales (Hazelton *et al.* 2014). Similarly for *Spartina* control along Pacific Coast estuaries of North America, short-term, but not long-term efficacy is reported, (e.g. do plants return after a number of years?) (Hedge *et al.* 2003). The authors' call for large-scale planning and funding for *Spartina* control was heeded. A programme managed by the Washington State Department of Agriculture documents (in 15 years of annual reports) the removal of large infestations but also the occurrence of almost 20,000 single and small clones of *Spartina* plants spread widely across 3,000 miles of coastline.[4]

Increasing elevation

There are few processes that can rapidly increase marsh elevation. The most common method is an application of sediment material of sufficient depth to reach a desired elevation with respect to a local tidal datum. Since 1992, the US Army Corps of Engineers has been pursuing the beneficial use of dredged sediments,[5] which provides opportunities across the US to partner with salt marsh restoration projects.

Just outside of New York City, salt marshes in Jamaica Bay have been eroding for decades. Concerned about the accelerated loss of salt marshes (at 16–20 ha/yr in 2001; Gateway NRA 2001; Hartig *et al.* 2002), a broad partnership sought to restore Big Egg Marsh through the use of locally dredged material. Unanticipated difficulties included grazing and disturbance of new vegetation by wild geese and horseshoe crabs (Frame 2006). The effort was judged enough of a success (Rafferty *et al.* 2011) that it was expanded to use dredged material from harbour deepening projects (478,000 m³ from 2012–2013) at five other sites within Jamaica Bay to produce an additional 61 ha of restored marsh. Costs of this technique are significant enough ($773,000 per ha for one project; Rafferty *et al.* 2011) to limit its availability to large budget projects.[6]

Summary and future outlook

Restoration of salt marshes is neither inexpensive nor immediate; it *begins* when the construction team has completed re-establishing conditions so that tidal floodwaters interact with the sediments, plants and animals as they do in natural systems. Regardless of the restoration approach, setting up the system with appropriate physical and biological processes that will help develop a productive, self-maintaining and resilient system is our creed for tidal marsh restoration.

At the close of the twentieth century, salt marshes were seen as long-lived relatively stable ecosystems, requiring only the removal of local anthropogenic stressors to set them on a trajectory toward mimicking pre-agrarian settlement conditions. As restoration ecologists, we view such restoration goals, if even achievable, no longer desirable. We recognize that coastal salt marshes are dynamic systems that will need to adapt to be resilient in the face of many stressors operating at local to global scales (development, eutrophication, invasive species, reduced sediment inputs, climate change, etc.) if they are to be self-sustaining.

Reducing the number and severity of stressors will aid restoration efforts and decrease the amount of restoration necessary in the future if salt marshes are expected to continue to provide us with ecosystem services. For example, curtailing fossil fuel use and preserving marsh peat both serve to curb CO_2 emissions and slow sea level rise. Re-establishing sediment supply to marshes extends their existence under conditions of rapid sea level rise, but much of our current marsh area will be drowned and provision for marsh migration onto gently sloping uplands will be a challenge, especially along highly developed coasts. Experimentation with new methods and approaches to tidal marsh restoration are being piloted in the US (USFWS) and Europe (Theseus Project).[7]

To ensure the continued existence of salt marshes and their ecosystem services for future generations, including diversity of marsh-dependent plants and animals, regional approaches must be undertaken by broad partnerships of governmental agencies, universities and conservation organizations. Regional efforts should identify current marshes (and potential future marshes under different climate change scenarios) that are critical to the survival of dependent species and to the resiliency of coastal communities. These locations should become focal areas for future restoration efforts to improve marsh resilience and enable marsh migration.

Notes

1　See www.fws.gov/hurricane/sandy/storymap.
2　See also www.habitat.noaa.gov/restoration/techniques/livingshorelines.html.
3　See www.mvr.usace.army.mil/BusinessWithUs/OutreachCustomerService/EcosystemRestoration.aspx.
4　See http://agr.wa.gov/PlantsInsects/Weeds/Spartina/default.aspx.
5　See www.mvr.usace.army.mil/BusinessWithUs/OutreachCustomerService/EcosystemRestoration.aspx.
6　See www.nan.usace.army.mil/Missions/CivilWorks/ProjectsinNewYork.aspx.
7　See www.fws.gov/hurricane/sandy/storymap and www.theseusproject.eu.

References

ABPmer Online Marine Registry. 2014. Database of international shoreline adaptation projects (updated 30 July 2014). Retrieved from www.omreg.net (accessed 22 November 2015).

Adamowicz, S. C., K. M. O'Brien. 2012. Drakes Island tidal restoration, Science, community and compromise. In Roman, C. T. and D. M. Burdick (eds) *Tidal Marsh Restoration: A Synthesis of Science and Practice* pp. 315–332. Washington, DC: Island Press.

Alber, M., E. M. Swenson, S. C. Adamowicz, I. A. Mendelssohn. 2008. Salt Marsh Dieback: An overview of recent events in the US. *Estuarine Coastal and Shelf Science* 80: 1–11.

Altieri, A. H., M. D. Bertness, T. C. Coverdale, N. C. Herrmann, C. Angelini. 2012. A trophic cascade triggers collapse of a salt-marsh ecosystem with intensive recreational fishing. *Ecology* 93: 1402–1410.

Anisfeld, S. 2012. Biogeochemical responses to tidal restoration. In Roman, C. T. and D. M. Burdick (eds) *Tidal Marsh Restoration: A Synthesis of Science and Practice* pp. 39–58. Washington, DC: Island Press.

Bourne, W. S., C. Cottam. 1950. *Some Biological Effects of Ditching Tidewater Marshes*. Washington, DC: US Department of the Interior, US Fish and Wildlife Service.

Brix, H., H. Cizkova. 2001. Introduction: *Phragmites*-dominated wetlands, their functions and sustainable use. *Aquatic Botany* 69: 87–88.

Bromberg, K. D., M. D. Bertness. 2005. Reconstructing New England salt marsh losses using historical maps. *Estuaries* 28: 823–832.

Bromberg Gedan, K., B. R. Silliman. 2009. Patterns of salt marsh loss within regions of North America. In Silliman, B. R., *et al.* (eds) *Human Impacts to Salt Marshes: A Global Perspective* pp. 253–265. Berkeley, CA: University of California Press.

Bromberg Gedan, K., B. R. Silliman, M. D. Bertness. 2009. Centuries of human-driven change in salt marsh ecosystems. *Annual Review of Marine Sciences* 1: 117–141.

Burdick, D. M., C. T. Roman. 2012. Salt marsh responses to tidal restriction and restoration. A summary of experiences. In Roman, C. T. and D. M. Burdick (eds) *Tidal Marsh Restoration: A Synthesis of Science and Practice* pp. 373–382. Washington, DC: Island Press.

Callaway, J. C., G. Sullivan, J. B. Zedler. 2003. Species-rich plantings increase biomass and nitrogen accumulation in a wetland restoration experiment. *Ecological Applications* 13: 1626–1639.

Cornu, C. E., S. Sadro. 2002. Physical and functional responses to experimental marsh surface elevation manipulation in Coos Bay's South Slough. *Restoration Ecology* 10: 474–486.

Craft, C., J. Clough, J. Ehman, S. Joye, R. Park, S. Pennings, H. Guo, M. Machmuller. 2009. Forecasting the effects of accelerated sea-level rise on tidal marsh ecosystems. *Frontiers in Ecology and the Environment* 7: 73–78.

Craig, L., K. McCraken, H. Schnabolk, B. Ward. (eds) 2010. *Returning the Tide: Tidal Hydrology Restoration Guidance Manual.* Silver Spring, MD, NOAA Restoration Center and NOAA Coastal Services Center.

Curado, G., A. E. Rubio-Casal, E. Figueroa, J. M. Castillo. 2014. Plant zonation in restored, nonrestored, and preserved *Spartina maritima* salt marshes. *Journal of Coastal Research* 30: 629–634.

Deegan L., D. S. Johnson, R. S. Warren, B. J. Peterson, J. W. Fleeger, S. Fagherazzi, W. M. Wolheim. 2013. Coastal eutrophication as a driver of salt marsh loss. *Nature* 490: 388–392.

Dibble K. L., L. A. Meyerson. 2012. Tidal flushing restores the physiological condition of fish residing in degraded salt marshes. *PLoS ONE* 7(9): e46161.

Eberhardt, A. L, D. M. Burdick, M. Dionne. 2011. The effects of road culverts on nekton in New England salt marshes: Implications for tidal restoration. *Restoration Ecology* 19: 776–785.

Elsey-Quirk, T., S. Adamowicz. 2016. Influence of physical manipulations on short-term salt marsh morphodynamics: Examples from the north and mid-Atlantic coast, USA. *Estuaries and Coasts* 39(2): 423–439.

Fagherazzi, S., M. L. Kirwan, S. M. Mudd, G. R. Guntenspergen, S. Temmerman, A. D' Alpaos, J. van de Koppel, J. M. Rybczyk, E. Reyes, C. Craft, J. Clough. 2012. Numerical models of salt marsh evolution: Ecological, geomorphic, and climatic factors. *Reviews of Geophysics* 50: RG1002.

Ford, M. A., J. B. Grace. 1998. Effects of vertebrate herbivores on soil processes, plant biomass, litter accumulation and soil elevation changes in a coastal marsh. *Journal of Ecology* 86: 974–982.

Frame, G. W., M. K. Mellander, D. A. Adamo. 2006. Big Egg Marsh experimental restoration in Jamaica Bay, New York. In Harmon, D. (ed.) *People, Places, and Parks: Proceedings of the 2005 George Wright Society Conference on Parks, Protected Areas, and Cultural Sites* pp. 123–130. Hancock, MI: The George Wright Society.

Frenkel, R. E., J. C. Morlan. 1991. Can we restore our salt marshes? Lessons from the Salmon River, Oregon. *Northwest Environmental Journal* 7: 119–135.

Gallego Fernández, J. B., F. García Novo. 2007. High-intensity versus low-intensity restoration alternatives of a tidal marsh in Guadalquivir estuary, SW Spain. *Ecological Engineering* 30: 112–121.

Gateway NRA. 2001. *The Jamaica Bay Blue Ribbon Panel on Marsh Loss and Coastal Sea Level Rise: A Future Agenda for Mitigation and Pilot Investigations.* Final Report, July. Staten Island, NY: Gateway National Recreation Area.

Gendron, W. C. 2013. Broad Meadows marsh restoration. US ACOE, New England District. Retrieved from www.nae.usace.army.mil/Missions/ProjectsTopics/broadmeadows.aspx (accessed 3 August 2015).

Gittman, R. K., A. M. Popowich, J. F. Bruno, C. H. Peterson. 2014. Marshes with and without sills protect estuarine shorelines from erosion better than bulkheads during a category 1 hurricane. *Ocean & Coastal Management* 102: 94e102.

Halpern, B. S., C. Longo, D. Hardy, K. L. Mcleod, J. F. Samhouri, S. K. Katona, K. Kleisner, S. E. Lester, J. O'Leary, M. Ranelletti, A. A. Rosenberg, C. Scarborough, E. R. Selig, B. D. Best, D. R. Brumbaugh, F. S. Chapin III, L. B. Crowder, K. L. Daly, S. C. Doney, C. Elfes, M. J. Fogarty, S. D. Gaines, K. Jacobsen, L. B. Karrer, H. M. Leslie, E. Neeley, D. Pauly, S. Polasky, B. Ris, K. St. Martin, G. S. Stone, U. R. Sumaila, and D. Zeller. 2012. An index to assess the health and benefits of the global ocean. *Nature* 488: 315–320.

Hartig, E. K., V. Gornitz, A. Kolker, F. Mushacke, D. Fallon. 2002. Anthropogenic and climate-change impacts on salt marshes of Jamaica Bay, New York City. *Wetlands* 22: 71–89.

Hazelton E. L. G., T. J. Mozdzer, D. M. Burdick, K. M. Kettenring, D. F. Whigham. 2014. *Phragmites australis* management in the United States: 40 years of methods and outcomes. *AoB Plants* 6: plu001.

Hedge, P., L. K. Kriwoken, K. Patten. 2003. A review of *Spartina* management in Washington State, US. *Journal of Aquatic Plant Management* 41: 82–90.

Jefferies, R. L., A. P. Jano, K. F. Abraham. 2006. A biotic agent promotes large-scale catastrophic change in the coastal marshes of Hudson Bay. *Journal of Ecology* 94: 234–242.

Kirwan, M. L., L. K. Blum. 2011. Enhanced decomposition offsets enhanced productivity and soil carbon accumulation in coastal wetlands responding to climate change. *Biogeosciences* 8: 987–993.

Kirwan, M. L., G. R. Guntenspergen, A. D'Alpaos, J. T. Morris, S. M. Mudd, S. Temmerman. 2010. Limits on the adaptability of coastal marshes to rising sea level. *Geophysical Research Letters* 37: L23401.

Konisky, R. A. 2012. Role of simulation models in understanding the salt marsh restoration process. In Roman, C. T. and D. M. Burdick (eds) *Tidal Marsh Restoration A Synthesis of Science and Practice* pp. 253–276. Washington, DC: Island Press.

Konisky, R. A., D. M. Burdick, M. Dionne, H. A. Neckles. 2006. A regional assessment of salt marsh restoration and monitoring in the Gulf of Maine. *Restoration Ecology* 14: 516–525.

Levine, J. M., J. S. Brewer, M. D. Bertness. 1998. Nutrients, competition and plant zonation in a New England salt marsh. *Journal of Ecology* 86: 285–92.

Lotze, H. K., H. S. Lenihan, B. J. Bourque, R. H. Bradbury, R. G. Cooke, M. C. Kay, S. M. Kidwell, M. X. Kirby, C. H. Peterson, J. B. C. Jackson. 2006. Depletion, degradation, and recovery potential of estuaries and coastal seas. *Science* 312: 1806–1809.

MacBroom, J. G., R. Schiff. 2012. Predicting the hydrologic response of salt marshes to tidal restoration. In Roman, C. T. and D. M. Burdick (eds) *Tidal Marsh Restoration: A Synthesis of Science and Practice* pp. 13–38. Washington, DC: Island Press.

Mcleod, E., G. L. Chmura, M. Bjork, S. Bouillon, C. M. Duarte, C. Lovelock, R. Salm, W. Schlesinger, B. Silliman. 2011. A blueprint for blue carbon: Toward an improved understanding of the role of vegetated coastal habitats in sequestering CO_2. *Frontiers in Ecology and the Environment* 9: 552–560.

McMillan, H. 2011. *Planning for Sea Level Rise in the Northeast: Considerations for the Implementation of Tidal Wetland Habitat Restoration Projects.* Gloucester, MA: NOAA Restoration Center, Northeast Region.

Miller, D. L., F. E. Smeins, J. W. Webb, M. T. Longnecker. 1997. Regeneration of *Scirpus americanus* in a Texas coastal marsh following lesser snow goose herbivory. *Wetlands* 17: 31–42.

Mitsch W. J., J. G. Gosselink. 2000. *Wetlands*, 3rd edn. New York: John Wiley & Sons.

Morgan, P. A., D. M. Burdick, F. T. Short. 2009. The functions and values of fringing salt marshes in northern New England, USA. *Estuaries and Coasts* 32: 483–495.

Morris, J. T., P. V. Sundareshwar, C. T. Nietch, B. Kjerfve, D. R. Cahoon. 2002. Responses of coastal wetlands to rising sea level. *Ecology* 83: 2869–2877.

Musch, O. 2011. Development of a European site selection tool for managed realignment. Erasmus Mundus MSc Programme thesis, University of Southampton, UK.

Needelman, B. A., S. Crooks, C. A. Shumway, J. G. Titus, R. Takacs, J. E. Hawkes. 2012. *Restore-Adapt-Mitigate: Responding to Climate Change Through Coastal Habitat Restoration.* Washington, DC: Restore America's Estuaries.

Nichols, R. J., A Cazenave. 2010. Sea-level rise and its impact on coastal zones. *Science* 328: 1517–1520.

Penland, S., H. H. Roberts, S. J. William, H. Sallenger Jr, D. R. Cahoon, D. W. Davis, C. G. Groat. 1990. Coastal land loss in Louisiana. *Gulf Coast Association of Geological Societies Transactions* 40: 685–699.

Peter, C. R., D. M. Burdick. 2010. Can plant competition and diversity reduce the growth and survival of exotic *Phragmites australis* invading a tidal marsh? *Estuaries and Coasts* 33: 1226–1236.

Pethick, J. 2002. Estuarine and tidal wetland restoration in the United Kingdom: Policy versus practice. *Restoration Ecology* 10: 431–437.

Portnoy J. W. 1999. Salt marsh diking and restoration: Biogeochemical implication of altered wetland hydrology. *Environmental Management* 24: 111–120.

Powers, S. P., K. E. Boyer. 2013. Chapter 22: Marine restoration ecology. In Bertness, M. D. *et al.* (eds) *Marine Community Ecology and Conservation.* Sunderland, MA: Sinauer Associates.

Rafferty, P., J. A. Castagna, D. Adamo 2011. Building partnerships to restore an urban marsh ecosystem at Gateway National Recreational Area. *Park Science* 27: 34–41.

Raposa, K. B., R. L. Weber, M. Cole Ekberg, W. Ferguson. 2015. Vegetation dynamics in Rhode Island salt marshes during a period of accelerating sea level rise and extreme sea level events. *Estuaries and Coasts* doi: 10.1007/s12237-015-0018-4

Redfield, A. C. 1972. Development of a New England salt marsh. *Ecological Monographs* 42: 201–237.

Reiner, E. L. 2012. Restoration of tidally restricted salt marshes at Rumney Marsh, Massachusetts. In Roman, C. T. and D. M. Burdick (eds) *Tidal Marsh Restoration A Synthesis of Science and Practice* pp. 355–370. Washington, DC: Island Press.

Restore America's Estuaries. 2015. *Living Shorelines: From Barriers to Opportunities.* Arlington, VA.

Rochlin, I., M. J. James-Pirri, S. C. Adamowicz, M. E. Dempsey, T. Iwanejko, D. V. Ninivaggi. 2012. The effects of integrated marsh management (IMM) on salt marsh vegetation, nekton, and birds. *Estuaries and Coasts* 35: 727–742.

Roman, C. T. and D. M. Burdick (eds). 2012. *Tidal Marsh Restoration: A Synthesis of Science and Practice.* Washington, DC: Island Press.

Roman, C. T., W. A. Niering, R. S. Warren. 1984. Salt marsh vegetation change in response to tidal restrictions. *Environmental Management* 8: 141–149.

Silliman B. R., J. C. Zieman. 2001. Top-down control of *Spartina alterniflora* production by periwinkle grazing in a Virginia salt marsh. *Ecology* 82: 2830–2845.

Silliman, B. R., E. D. Groshholz, M. D. Bertness. 2009. Salt marshes under global siege. In Silliman, B. R., *et al.* (eds) *Human Impacts to Salt Marshes: A Global Perspective* pp. 391–398. Berkeley, CA: University of California Press.

Smith, S. M. 2009. Multi-decadal changes in salt marshes of Cape Cod, MA: Photographic analyses of vegetation loss, species shifts, and geomorphic change. *Northeastern Naturalist* 16: 183–208.

Smith, S. M., C. T. Roman, M. James-Pirri, K. Chapman, J. Portnoy, E. Gwilliam. 2009. Responses of plant communities to incremental hydrologic restoration of a tide-restricted salt marsh in southern New England (Massachusetts, USA). *Restoration Ecology* 17: 606–618.

Strong, D. R., D. R. Ayers. 2009. *Spartina* introductions and consequences in salt marshes. Pp. 3–22 in Silliman, B. R. *et al.* (eds) *Human Impacts to Salt Marshes: A Global Perspective.* Berkeley, CA: University of California Press.

Teal, J. M., S. Peterson. 2009. The use of science in the restoration of Northeastern US salt marshes. In Silliman, B. R. *et al.* (eds) *Human Impacts to Salt Marshes: A Global Perspective* pp. 267–283. Berkeley, CA: University of California Press.

Teal, J., M. Teal. 1969. *Life and Death of the Salt Marsh.* New York: Ballantine Books.

Vincent, R. E., D. M. Burdick, M. Dionne. 2013. Ditching and ditch-plugging in New England salt marshes: Effects on hydrology, elevation and soil characteristics. *Estuaries and Coasts* doi 10.1007/s12237-012-9583-y.

Williams, P. B, M. K. Orr. 2002. Physical evolution of restored breached levee salt marshes in the San Francisco Bay Estuary. *Restoration Ecology* 10: 527–542.

Wolters, M., A. Garbutt, J. P. Bakker. 2005. Salt-marsh restoration: Evaluating the success of de-embankments in north-west Europe. *Biological Conservation* 123: 249–268.

19

OYSTER-GENERATED MARINE HABITATS

Their services, enhancement, restoration and monitoring

Loren D. Coen and Austin T. Humphries

Introduction

The focus of this chapter is on reef-forming native and non-native bivalves (oysters primarily), where restoration efforts for non-extractive services have been directed since the 1990s (Luckenbach *et al.* 1999). Coastal estuarine habitats are recognized as some of the most productive and important aquatic ecosystems worldwide, providing foraging, nursery, and spawning habitats for ecologically and economically important organisms. Simultaneously they are also some of the most heavily degraded ecosystems on the planet (Lotze *et al.* 2006).

In 2000, almost 53 per cent of the inhabitants of the USA resided in coastal areas (including Great Lakes) that make up just 17 per cent by area, with numbers expected to reach over 24 per cent by 2025 (UN 2012). Along with similar increases worldwide, and as a consequence of anthropogenic activities (e.g. eutrophication and related effects), stressors, both natural and human-related (food webs, introduction of non-native or exotic species) will continue to grow (Ruesink *et al.* 2005; Coen and Bishop 2015 and references therein) both near and offshore. This degradation requires that we increase our efforts to restore and enhance these key habitats, in part because of their ecosystem services (MEA 2005).

Bivalve harvesting for human consumption has a long history worldwide and it was not until perhaps the nineteenth century that closed harvesting seasons were truly implemented and enforced (Beck *et al.* 2011; zu Ermgassen *et al.* 2012 and citations therein). Oysters, like lobsters, were often not considered a high value resource, with harvesting not just for consumption, but also for their shell use (Luckenbach *et al.* 1999; Coen and Grizzle 2016). Today, there are very few remaining wild shellfish reefs that produce oysters for direct harvesting without fisheries enhancement (reviewed in Beck *et al.* 2011). Most USA oyster fisheries were primarily subtidal (Figure 19.1J) found in estuaries like the Chesapeake (Maryland and Virginia) and Delaware Bays (Delaware and New Jersey) and throughout the Gulf of Mexico (Florida to Texas) (Luckenbach *et al.* 1999; NRC 2004; zu Ermgassen *et al.* 2012). In contrast, in the western Atlantic intertidal oyster (*C. virginica*) reefs dominate, including the seaside of VA, parts of NC and FL, SC, GA, USA and portions of the other four Gulf of Mexico states (Figure 19.1C–F; ASMFC 2007). Presently as much as 85 per cent of historical intertidal and subtidal oyster

Figure 19.1   Composite figure of different intertidal and subtidal bivalve-generated habitats. (A) Dense
pen shell aggregation in an intertidal seagrass bed in Dubai. (B) Subtidal pen shell
aggregations of *Atrina zelandica*, in New Zealand. (C) *Crassostrea virginica* restored fringing
intertidal oysters reefs adjacent to salt marsh, in South Carolina, USA. (D) Natural oyster (*C.
virginica*) patch reefs in South Carolina, USA. (E) Natural oyster (*C. virginica*) recruitment
onto red mangrove prop roots in Florida, USA. (F) Natural oyster (*C. virginica*) intertidal reef
below mangroves in Florida, USA. (G) *Modiolus modiolus* subtidal mussels assemblages in St.
Joe Bay, Florida, USA. (H) Suspended oyster culture in shrimp ponds near Charleston, South
Carolina, USA. (I) Suspended and bottom planted racks and cages for bivalve oyster
aquaculture, northwest coast of USA. (J) Subtidal restored *C. virginica* oyster reef

Sources: (A) R. Grizzle, UNH, Durham, NH, USA; (B) S. Thrush, University of Auckland, New Zealand; (C) J. Monck,
SCDNR; (D–F) L. Coen; (G) B. Peterson, SUNY, Stony Brook, NY; (H) B. Cox, Island Fresh Seafood, South
Carolina, USA; (I) A. Suhrbier, Pacific Shellfish Institute, Washington, USA; (J) R. Lipcius, VIMS, Gloucester Pt.,
VA, USA

reefs have been lost worldwide, suggesting that oysters in particular may be the most imperilled nearshore estuarine biogenic habitat (Beck *et al.* 2011). Oyster reefs are much like those
formed by corals with only a veneer of living animals, and an extensive foundation composed
of dead skeletons above and below the sediment surface. Thus, harvesting can easily remove
millennia of reef growth in decades or centuries (Beck *et al.* 2011).

Ecosystem services and foundation species

Bivalve habitat types include aggregations, beds, and reefs generated by oysters, clams, and
mussels (Figure 19.1). These filter-feeding species are vital components of coastal ecosystems
providing ecosystem services including:

1 enhanced wildstock populations (consumptive uses);

2 improved water quality (clarity), nutrient sequestration/denitrification (potentially decreasing harmful algal blooms or HABs) and hypoxia;

3 habitat for associated organisms (finfish and invertebrates) and secondary production; and

4 enhancement of adjacent habitat/shoreline (often vegetated) by stabilization reducing erosion.

Shell mounds (middens) created by indigenous people are found in nearly all coastal areas worldwide (Alleway and Connell 2015), where most bivalves being important fisheries at one time (Kirby 2004). Cultural services must also include jewelery, trade, and aesthetic values (MEA 2005).

Increased foraging and sheltering habitat for vertebrate and invertebrate organisms, including finfish, birds (at exposure) and even mammals, is the most well studied service in natural and restored oyster habitats (Coen *et al.* 1999b, 2007; Baggett *et al.* 2014; Coen and Bishop 2015). Reef-associated organisms can rival seagrasses and other habitats (salt marshes or mangroves) for associated resident (present even during low tide), and transient (absent during reef exposure) organisms (Figure 19.1) and oyster reefs support much greater animal abundances than surrounding unstructured sand/mud habitats (Coen *et al.* 1999b, 2007, 2011; Coen and Grizzle 2016 and citations therein).

USA oyster habitats have been shown to support a diverse suite of resident and transient species. For example, >300 micro- to macroscopic species were found in the 1960s by H. W. Wells (1961) in NC. More recently, over 75 resident and 59 transient macroscopic species were collected on SC, intertidal reefs by Coen and colleagues (Coen *et al.* 1999a,b; Coen and Grizzle 2016 and citations therein), and over 100 species were on reefs in LA (Humphries and La Peyre 2015). Oyster aquaculture can also play many parallel roles (Dumbauld *et al.* 2009; Coen *et al.* 2011).

Of late, research has focused also on quantifying one or more ecosystem functions in economic terms (Peterson *et al.* 2003; Grabowski and Peterson 2007; Grabowski *et al.* 2012; Humphries and La Peyre 2015), with Grabowski *et al.* (2012) estimating oyster reefs services valued at >$99,000/hectare/year once restored. While it is difficult to capture one or all of the ecosystem services provided by bivalve habitats (Grabowski *et al.* 2012), it is clear that they are exceedingly important (Peterson *et al.* 2003; Baggett *et al.* 2014; Barbier *et al.* 2014; Powers and Boyer 2014).

In a meta-analysis of data from USA, zu Ermgassen and colleagues (2012, 2013) showed that losses of oyster habitat have resulted in lost filtration capacity by ~85 per cent having both direct and indirect consequences for estuarine health, through increased water residence times and lost estuarine processing of nutrient loads as filtration capacity scales with oyster population densities and size (Luckenbach *et al.* 1999; La Peyre *et al.* 2014). Filtration by bivalves can impact water quality and clarity by removing particulate matter (or seston) from overlying waters (Dame 1996; Luckenbach *et al.* 1999). Oysters ingest material and consume particles and then reject or ingest bound in mucus, then depositing them onto the sediment surface as faeces and pseudofaeces. This process can contribute to improved water quality (clarity) when oysters are sufficiently dense and the overlying water column is relatively shallow (Dame 1996; Newell 2004; and see Coen and Grizzle 2016 and papers cited within).

The role of reefs as nutrient sinks in estuaries has only recently been quantified and better understood, but results indicate it is significant when compared to other important habitats (Kellogg *et al.* 2014; Smyth *et al.* 2015). Oyster reefs can assimilate nutrients into tissues or shell, and augment subtidal denitrification along the sediment–water interface (reviewed in Kellogg *et al.* 2014), with net denitrification rates from 30–56 $g\,N\,m^{-2}\,y^{-1}$ and shell and tissue assimilation from 0.2 per cent and 9.3 per cent N g^{-1} dw, respectively (Kellogg *et al.* 2014). However, not all studies show significant differences between denitrification of intertidal oyster reefs and surrounding habitats (Smyth *et al.* 2015).

Bivalve reefs can function also as living breakwaters – often called 'living shorelines' (LS)[1] – or erosion reducers reducing waves and increasing sedimentation or decreasing sediment resuspenson around a LS structure (Figures 19.1C,F, 4F; Walles *et al.* 2015). In fact, oyster reefs facilitate sediment accretion by as much as 6.3 cm annually (Meyer *et al.* 1997). However, this is not the same as reducing erosion which is of particular importance when ameliorating loss of landward marsh habitat. Evidence that reefs buffer shorelines from waves (wind- or tidally generated) is equivocal (Piazza *et al.* 2005; Scyphers *et al.* 2011; La Peyre *et al.* 2014; Coen and Grizzle 2016), although the consensus is that effects are generally positive, but context-dependent (La Peyre *et al.* 2015). To be beneficial, habitat accretion must also exceed sea level rise (Rodriguez *et al.* 2014; Walles *et al.* 2015, 2016), and there may be a threshold where reefs cannot exist as larger storms exceed some upper ceiling (but see Walters *et al.* 2007).

Non-native foundation oyster species

Many oysters once supported significant commercial fisheries, but many are currently at <1–10 per cent of historical levels in various estuaries such as the Chesapeake Bay, USA (Beck *et al.* 2011). Worldwide, the Japanese or Pacific oyster (*Crassostrea gigas*) has been introduced by accident or design as an alternative species (Ruesink *et al.* 2005). In some areas its use as an engineer is being encouraged given its potential reef-building capacity and resistance to native diseases (Walles *et al.* 2015). Non-native species introductions, either through direct or accidental introductions, were first mentioned in Elton's 1958 seminal work on invasive species, and are having complex (positive, negative, or even neutral) impacts in many estuaries throughout the world where native species have declined (reviewed in NRC 2004; Ruesink *et al.* 2005; Coen and Bishop 2015). Introductions and expansion of Japanese or Pacific oyster, *Crassostrea gigas*, have had mixed results.

How non-native species contribute to novel ecosystems is something being hotly debated (see Coen and Bishop 2015 and citations therein). Introduced oyster species are transforming the landscape in many novel ways (Smaal *et al.* 2005). In the Wadden Sea for example, the invasion of the Pacific oyster *C. gigas* has caused major habitat shifts from the formerly dominant native bivalves such as blue mussel, *Mytilus edulis,* the native flat oyster, *Ostrea edulis,* and cockles, which formed dense beds to intertidal oyster reefs (Smaal *et al.* 2005; Coen and Grizzle 2016). The consequences for native benthic communities, mussel-eating invertebrates, and other higher food web vertebrate consumers (i.e. birds) have yet to be resolved. In contrast, in the Netherlands, these non-native ecosystem engineers are reducing erosion and adding novel intertidal habitat (Smaal *et al.* 2005; Walles *et al.* 2015). In Australia, no obvious negative consequences have been observed with *C. gigas* (Wilkie *et al.* 2012). Whether introduced biogenic-forming species will continue to thrive, given novel diseases and native and introduced predators, will be of interest to ecologists long into the future (Coen and Bishop 2015; Walles *et al.* 2015; Coen and Grizzle 2016).

Enhancement and restoration of oyster reef habitats

No matter what area or species, it is clear that bivalve restoration will need to be scaled up significantly from 1–10 hectares to hundreds soon (e.g. the Chesapeake Bay or Gulf of Mexico supported by Deep Water Horizon oil spill funding). Central to successful enhancement and restoration efforts is an understanding of how the target species interacts with its biophysical environment.

For example, oysters are found both intertidally and subtidally, in both nearshore and estuarine waters worldwide (Galtsoff 1964). Intertidal oysters form isolated reefs away from shorelines (patch reefs) or bordering vegetated (salt marsh or mangrove) shorelines in tidal creeks, rivers, sounds, and bays (fringing reefs) (Galtsoff 1964; ASMFC 2007; Coen and Grizzle 2016). Shallow (<10 m) reef-forming oysters create different individual and habitat morphologies in response to various environmental and biological forces (*ibid.*). To examine bivalve habitat-forming morphologies, the following categories have been suggested:

1 reef-forming (e.g. genus *Crassostrea*);
2 aggregation–forming (e.g. *Ostrea* spp.); and
3 shell-accumulating (many scallop and clam) species (Figure 19.1; ASMFC 2007; Coen and Grizzle 2016).

Oyster biology as it relates to restoration

Regardless of form, most bivalves are filter-feeders, with the exception of some specialists (Coen and Bishop 2015; Coen and Grizzle 2016), such that a sufficient supply of seston (or suspended food and sediment) is needed to support viable populations (Dame 1996). Interestingly, several studies have suggested, based on remote imagery, that intertidal oyster reef organization is not random, but rather potentially self-organizing, paralleling observations of European mussels (van de Koppel *et al.* 2008). For subtidal oysters in the Gulf of Mexico, habitat suitability index (HSI) models have incorporated biotic and abiotic factors to help assist with the site selection for restoration efforts, as well as how oyster populations thrive under different biological and physical scenarios (Pollack *et al.* 2012; Soniat *et al.* 2013; La Peyre *et al.* 2015).

The life cycle of oysters consists of a larval (either a crawling or planktonic) phase prior to settlement and a sedentary, or sessile, adult phase. Once having settled (cemented) on the required hard substrate, adults cannot relocate. Chemical clues are also involved in reef aggregation (Kennedy *et al.* 1996). Successful recruitment of oyster larvae and post-larvae over successive years (multiple age classes) is fundamental to reef persistence and expansion (Coen and Luckenbach 2000; Luckenbach *et al.* 2005).

Healthy populations require the renewal of hard substrate, as both harvesting and natural mortality, burial and shell degradation (i.e. taphonomic processes) remove reef substrate. In the past, buried shell was dredged from estuary bottoms, but this has been discontinued in many areas. When shell is unavailable, then alternative materials such as fossil shell, limestone, granite or recycled manmade materials (concrete) can provide the necessary substrate for reef construction. Presently and into the future, increasing restoration efforts (hundreds hectares) will require much more shell (i.e. alternative materials) than is currently available.

The quantity and quality of food can be directly correlated with water quality (e.g. nutrients, salinity, temperature, and hydrology; Pollack *et al.* 2012), with oysters thriving in highly turbid waters. Many intertidal *C. virginica* reefs (western Atlantic and Gulf of Mexico) (Figure 19.1C–F) can thrive where salinities are higher than subtidal ranges (15–30 ppt vs. 10–15 ppt), and sediments are quite fine (Galtsoff 1964; Bahr and Lanier 1981).

Today, aquaculture has taken over as the primary method for producing oysters for consumption (Figure 19.1H–I; Shumway 2011). Worldwide, many use a hybrid approach where oyster larvae are set onto deployed substrates and then harvested or relayed to better areas or seed beds for grow-out (e.g. LA, USA). Elsewhere, including the Pacific coast of the USA, several non-native oysters (especially the Pacific oyster, *C. gigas*) have been introduced, requiring intensive labour and capital (e.g. land-based hatcheries) (Shumway 2011 and citations therein).

Need for restoration

Whether enhancing or restoring (creating) these important bivalve biogenic habitats it should be noted that the approach is quite different from those employed in the recovery of other non-foundation (and mobile) species such as finfish or crustaceans (Coen and Luckenbach 2000; Coen *et al.* 2011). Past oyster restoration efforts generally focused only on recovering lost or impaired oyster fisheries (extractive services). The causes for the decline of many bivalves are manifest and include more than one reason (e.g. over-harvesting, pollution and related impacts, habitat destruction, and native and non-native oyster diseases; Lenihan *et al.* 1999; Coen and Luckenbach 2000; Kirby 2004; NRC 2004).

As mentioned earlier in this chapter, early European settlers into the Americas were able to harvest abundant shellfish in relatively shallow estuarine waters. However, by the late nineteenth and early twentieth centuries, most of these populations were significantly depleted or driven to near local extinction (Kirby 2004; Beck *et al.* 2011). Elsewhere (Europe, Asia) and even earlier, many native oyster populations were extirpated (Beck *et al.* 2011; Alleway and Connell 2015).

Today, native and non-native species are receiving attention either through enhancement or restoration, especially in the USA for the non-extractive ecosystem services of two native species, the Eastern oyster (*Crassostrea virginica*, Atlantic and Gulf of Mexico), and the greatly depleted Olympia oyster (*Ostrea lurida,* Pacific coast of North America). Similarly, but lagging those USA and Canadian efforts are plans to restore the native flat oyster (*Ostrea edulis*) in Europe and the UK. However, diseases and depleted local populations may make their return difficult (Coen and Bishop 2015).

Because of the numerous ecosystem services mentioned previously, bivalves are being increasingly protected, enhanced, or restored in greater numbers. A major requirement before enhancement or restoration occurs is to assess their current status and to eventually conduct triage assessments determining where to focus limited resources (NRC 2017). This typically requires that one map, assess, and quantify the extent and health of oyster reefs, and then input the data into a GIS geodatabase for long-term use in assessing status and change.[2]

Also critical to any restoration assessment are: (1) the inclusion of explicit goals and objectives; (2) consistent approaches, related metrics, and success criteria; (3) rigorous designs for monitoring natural (including replicated constructed, reference, and/or control areas) and restored habitats; and (4) monitored over a sufficiently appropriate time period (Coen *et al.* 2004; SER 2004; Baggett *et al.* 2014; NRC 2017). These should be developed a priori and be appropriate to the proposed effort, including the distinction as to whether one is restoring intertidal, shallow subtidal, or deeper (>5 m MLW) habitats. The importance of population connectivity (metapopulations) also needs to be considered as part of the effort.

Aquaculture is also having an increasing role for bivalve sustainability, not just for seafood (Dumbauld *et al.* 2009; Shumway 2011 and chapters within; Coen and Bishop 2015). A recent survey (Shinn *et al.* 2015) of the top 69 aquatic (brackish to marine waters) cultured species found that molluscan aquaculture leads animal production by tonnage (31.7 per cent overall). Of special concern, molluscan aquaculture, as compared to finfish, is always concentrated relatively close to the coastal zone (including estuaries), an area concurrently being heavily impacted by human populations and related development (Coen *et al.* 2011; Coen and Bishop 2015 and papers cited therein). While we know a great deal about shellfish populations in North America and Europe, the status and potential trends in data-poor areas such as Africa, South America, and southeastern Asia are even less clear.

Although bivalve restoration for other services has seen a recent surge outside of North America, such as in Europe and the Pacific from China to New Zealand, we still know relatively little about their successes or failures, in part due to a lack of published reports or extended duration (>10 years). For instance, over 120 km of intertidal oyster reef has been created in the Yangtze River estuary, China – an extremely large-scale effort by most accounts, but only a limited portion of the results have been reported to date (Quan *et al.* 2013 and citations therein). Japan has a long history of enhancing shellfish populations for consumption using aquaculture and Australia and New Zealand have only recently begun to recognize the need for restoration of native oyster habitat (e.g. Alleway and Connell 2015; Gillies *et al.* 2015). Elsewhere, sea level rise is making living shoreline efforts with oysters a potentially important approach for the shorter-term impacts in places like Bangladesh (Bilkovic *et al.* 2017).

As mentioned earlier many bivalve species (e.g. clams, scallops, mussels) can relocate as conditions become unfavourable (low dissolved oxygen, sedimentation), reef-forming species cannot (Kennedy *et al.* 1996). This sedentary life makes oysters susceptible to any environmental stressors (oiling and burial from dredging, hurricanes; Coen and Bishop 2015), and must be considered when planning restoration projects as part of site selection (Coen and Grizzle 2016).

Critical steps for restoration-related efforts

For the purposes of our discussion below, we define restoration as: 'the process of establishing or reestablishing a habitat that in time can come to closely resemble a natural condition in terms of structure and function' (Coen and Luckenbach 2000; Peterson *et al.* 2003; Grabowski and Peterson 2007).

Site selection concepts, historical information, and related concepts

Site selection may be the single-most important set of factors for determining the success of an enhancement or restoration project. Derived from a workshop of restoration practitioners in 2004, site selection parameters were assessed and then ranked (Table 19.1) (Coen and Luckenbach 2000; Coen *et al.* 2004; ASMFC 2007; Brumbaugh and Coen 2009; Baggett *et al.* 2014). Responses were by reef type (intertidal or subtidal) for *C. virginica*. Some recommended establishing restored reefs only in areas (reef footprints) where oyster populations existed historically. These can typically be determined from navigation charts, past historical surveys, state mapping efforts, or published fishing records[3] (Baggett *et al.* 2014), while others recommended using habitat suitability index (HSI) models to determine optimal site conditions, potentially ensuring longer-term sustainability (Pollack *et al.* 2012; La Peyre *et al.* 2015 and references therein; but see Coen and Bishop 2015; NRC 2017).

Briefly here we describe the most important site selection criteria for intertidal and subtidal restoration focusing on physical and biological site traits. Highest ranked parameters (Table 19.1) are by reef type. These parameters and monitoring logistics are crucial to a project's success and associated costs. Later adaptive management (including as a means of tweaking a project or understanding why a particular effort worked or failed) need to be based on sufficient project monitoring whose aim is to ensure a long-lasting, positive result (Coen and Luckenbach 2000; Coen *et al.* 2004; Baggett *et al.* 2014; NRC 2017).

Table 19.1 A summary of ranked site selection parameters for intertidal and subtidal oyster reef restoration modified from Coen *et al.* (2004) based on responses to a questionnaire circulated among attendees, other active restoration practitioners, and oyster fisheries managers working with *Crassostrea virginica* in the USA

Subtidal	Ranking	Intertidal
Physical parameters		
Reef depth	1	Primary underlying substrate
Primary substrate for planting	2	Mean salinity
Substrate firmness for planting	3	Substrate firmness for planting
Water quality	4	Siltation/sedimentation
Mean salinity	5	Potential reef height relative to MLW
Elevation off bottom	6	Water quality
Siltation/sedimentation	7	Runoff from adjacent land
Current flow rate	8	Current flow rate
Reef orientation	9	Bank slope to be planted
Channel depth (lowest tide) logistics for construction	10	Width of intertidal zone for planting
Runoff from adjacent land	11	Erosion potential
Erosion potential if shallow	12	Fetch (wind wave exposure)
Fetch (wind wave exposure)	13	Channel width and depth for (lowest tide) logistics for construction
	14	Reef orientation
Biological parameters		
Disease (MSX, Dermo)	1	Recruitment of oysters (larval supply)
Recruitment of oysters (larval supply)	2	Disease (MSX, Dermo)
Predation	3	Fouling communities
Proximity to extant oyster populations	4	Food quantity and quality
Food quantity and quality	5	Predation
Fouling communities	6	Proximity to extant oyster populations

Also critical for a given project's success are:

1 scale of restoration footprint (Figures 19.2–19.4; hundreds of metres to hundreds of hectares);
2 reef type (intertidal or subtidal) and nearby sources of larvae;
3 relevant permitting details (e.g. adjacent habitats, species of concern, materials, direct and indirect effects, attractive nuisance effects, signage, etc.); and
4 pertinent site staging logistics (e.g. proximity to boat ramps, staging of personnel for volunteer labour, access by boat or barge, etc.).

Project goals and related objectives significantly affect *all* of the above-mentioned concepts. A given project's scale (Figures 19.2–19.4) also influences sampling design, metrics, replication, duration, etc. and should not be given short shrift in the 'costing-out' of a given restoration project (Coen *et al.* 1999a, 2004; Kennedy *et al.* 2011; Baggett *et al.* 2014; NRC 2017). We cannot emphasize enough one's attention to the details discussed here and are not able to describe most in any great detail here.[4] The above-mentioned designs and sampling must be considered upfront prior to the initiation of any given restoration effort or else their later inclusion will be of limited value.

Figure 19.2 (A & B) Examples of different oyster reef restoration techniques using loose unaggregated shell; (A) Barge with shell; (B) Barge deploying shell using One Ton Bags™; (C) Trianglular metal rebar structures filled with shell contained within, Mad Island, TX., USA. (Reefblk™); and (D) Reef balls made from composite cement substrate before and after deployment with recruited oysters in Tampa, FL, USA

Sources: (A) M. Berrigan, Department of Agriculture and Consumer Services; (B) see www.onetonbag.com/. R. Konisky, NH TNC, USA; (C) TNC; (D) Tampa BayWatch

Site selection attributes

Subtidal/intertidal reef type, height, and depth

Constructing subtidal reefs with sufficient vertical relief (height above the sediment's surface, cm to 0.5 m or more in some cases) can reduce the negative effects of sedimentation and dissolved oxygen, while enhancing local flow (Lenihan 1999; Schulte *et al.* 2009), including the use of taller vertical areas of material (cultch), interspersed with lower relief areas. On a smaller spatial scale this is often termed rugosity which can also enhance flow and reef complexity with recruitment, and thus reef use by associated organisms (Coen *et al.* 1999b; Coen and Luckenbach 2000; Baggett *et al.* 2014).

A suggested minimum value or range for subtidal reefs encountering low DO is ~0.5–1 m (Lenihan 1999; Gregalis *et al.* 2009). However, increasing the vertical reef height beyond >15 cm post-construction yields a concomitant increase in material and cost that many projects cannot afford. For intertidal reefs, a minimum has yet to be established, but 7–15 cm should be a minimum post-construction height after settling and cultch dispersal, especially where boat wakes, soft sediments, and sloping shorelines (often with high fetches and waves) affect pre-aggregation (i.e. cementing by oysters, etc.), reef longevity, and ultimately success (but see Rodriguez *et al.* 2014 and Byers *et al.* 2015 related to sea level rise and tidal range).

Figure 19.3 Scale of restoration and enhancement efforts. These include: (A) *C. virginica* shell delivery (15.3 m³ or 20 yd³) for SCDNR restoration project in SC, USA; (B) small- to medium-scale shell bagging (1000s) using polypropylene mesh with community volunteers; (C) community restoration effort bagging (mesh) shell or alternative materials for intertidal reef constructionn; (D) use of articulated concrete blocks (SHOREBLOCK™ SD Series, see www.shoretec.com/shoreblock-sd.php) for capping of sediments using geotextile material and blocks as hard substrate at the SC Aquarium and Ft. Sumter U.S. Park Service facility; (E) larger scale intertidal planting of loose oyster shell onto reefs fringing salt marsh using barge and high pressure water cannon at high tide in South Carolina, USA; (F) shell (mesh) bags deployed at a relatively large reef construction site near Charleston, South Carolina, USA; (G) moving shell and other materials for intertidal restoration experiments using barge and buckets at Cape Romain, NWR, South Carolina, USA; (H) planted shell exposed at low tide for intertidal habitat enhancement using barge and crane for TNC project in Long Creek, Virginia, USA

Sources: (A–C) L. Coen; (D) T. Effinger, SCANA; (E–G) L. Coen; (H) E. Moleen

Figure 19.4   Restoration of oyster reef habitats. (A) Post-construction view of one of three replicate
10 m^2 intertidal oyster shell bag (100 total) reefs constructed in Cape Romain National
Wildlife Refuge (NWR), South Carolina, USA; (B) Loading of loose oyster shell onto
truck with front end loader for later intertidal oyster reef planting, SCDNR, South
Carolina, USA; (C) Close-up of intertidal oysters (*C. virginica*) recruited onto loose shell
growing through polypropylene mesh in ACE Basin NERR, South Carolina, USA;
(D) Intertidal to shallow subtidal oyster shell 'mats' submerged in Indian River Lagoon,
Florida, USA. Mats composed of ~36 shells, attached to a small mesh mat with zipties.
Individual mats later attached to each other, forming a large quilt-like reef structure;
(E) Large intertidal shell bag reef (~124 m^2) adjacent to the South Carolina Aquarium,
Charleston, South Carolina, USA; (F) Large shell bag reefs (total ~340 m, acreage,
~0.22 ha) constructed by TNC at Coffee Island, Alabama, USA; (G) Large-scale shell
barge planting of subtidal reefs by FDACS in Apalachicola, Florida, USA. Note the
surrounding sediment plume; (H) Aerial image of replicate shallow subtidal reefs built by
Dauphin Island Sea Lab and TNC that are also referred to as 'Living Shorelines', Pelican
Point, Alabama, USA; and (I) Tens of thousands of oyster shell bags in consolidated bundles
filled using automated bagging machine and to be deployed in coastal Alabama, USA as
part of the NOAA-ARRA large scale efforts

Sources: (A–C) L. Coen; (D) A. Birch, TNC; (E) L. Coen; (F) B. Maynor Young, TNC; (G) M. Berrigan, FDACS; (H) S.
St. John; and (I) J. DeQuattro, TNC

Sediment dynamics (reef siltation/sedimentation)

In areas receiving high suspended sediment loads, deployed reef base substrates can typically
experience poor recruitment success through fouled or covered substrates and high post-settlement mortality through burial (Coen and Grizzle 2016). Subsidence, flow, and sedimentary
processes also decrease the post-construction overall reef footprint through time (Coen *et al.*
2004; Baggett *et al.* 2014; Coen and Grizzle 2016). Sedimentation is often greater at a reef's

base where water currents are often slowest and finer particles tend to settle out (Lenihan 1999; Coen *et al.* 2004). As mentioned before, local conditions also influence sediment characteristics (grain size, depth, and load). Construction using heavier (denser) materials such as granite, fossil shell or recycled cement increase sinking rates, especially where sediments are finer (softer) requiring additional material to attain a acceptable final vertical reef height and avoidance of later costly adaptive management (Brumbaugh and Coen 2009).

Reef height relative to mean low water/width of intertidal zone

Oyster reef aerial exposure can greatly influence oyster reproduction, disease susceptibility, and responses to other anthropogenic stressors (Kennedy *et al.* 1996; Coen and Bishop 2015; Walles *et al.* 2015). Intertidal placement of reef base material relative to MLW (mean low water) determines the duration of exposure and likelihood of survival, especially in warm temperate to subtropical areas. In the western Atlantic, intertidal oysters in the USA typically occur from below MLW to about 1 m above MLW (Bahr and Lanier 1981), but this varies by site (latitude, shade from adjacent habitats, predominant solar orientation), tidal range, and type (diurnal vs. semi-diurnal) and other variables (fetch, waves, wind driven or boat wakes, 'rewetting' oysters during exposure (Walles *et al.* 2016; D. Bushek and L. Coen, personal observation).

Underlying substrate type and deployment

Existing substrate(s) type for reefs to be constructed requires sandy to muddy sand to mud (intertidal often) as a foundation for placing hard substrate or 'cultch'. Often larger projects use coarser natural or recycled material (granite or limestone or recycled cement) capped with oyster or fossil shell if in short supply (Luckenbach *et al.* 1999; Coen and Luckenbach 2000; ASMFC 2007; Powers and Boyer 2014). This is especially true as shell is limiting across the USA with increasing number and size of reef projects (>1–10 to hundreds hectares). One method using plastic mesh bags filled with material (Figures 19.3B–D, 4A,E,F,I; Baggett *et al.* 2014) or an under- or overlayments of geomeshes are used to reduce sinking or to retain shelly (or alternative) material in shallow areas where waves, wakes, and significant erosion occur. They are also easier to transport to and from the field (for assessment) when community volunteer restoration efforts are employed (Brumbaugh and Coen 2009).

Sediments consisting of high silt/clay percentages should be avoided wherever possible as sedimentation, siltation, and burial (subsidence) can be quite problematic for restoration efforts, requiring more material to attain an equivalent final reef height versus areas with coarser sediments. This can be assessed initially as the first part of site selection process to minimize failure. Reef footprints (area and vertical height above the surrounding sediment) through time need to be assessed as part of a longer-term monitoring and adaptive management programme (Table 19.1; Coen *et al.* 2004; Baggett *et al.* 2014; NRC 2017).

Dissolved oxygen and temperature

Reduced levels of dissolved oxygen (<2 mg L^{-1}), especially for extended durations of days to weeks, can cause significant oyster and associated species mortality (Lenihan and Peterson 1998; Lenihan 1999; ASMFC 2007). Hypoxia (<4 mg L^{-1}) often results in mobile species relocating from these areas to those with higher DO. Subtidal oyster reefs commonly can occur <5–10 m (MLW) of water below the surface. Historically they occurred deeper, perhaps given fewer hypoxic events. Natural subtidal reefs were often shallow (~1–5 m) and sufficiently elevated off the bottom so that oysters experienced fewer low oxygen (hypoxic to anoxic) conditions. DO varies with tide, and time of day, although in tidal creeks, low levels often occur nocturnally.

Intertidally, oysters encounter a suite of unique physiological challenges as they are exposed often to very high summer temperatures (e.g. in the southeastern USA, these can exceed 44°C internally; L. Coen and P. Lara, personal observation; Bahr and Lanier 1981; Walles *et al.* 2016) or very low winter freezing temperatures (e.g. north of VA; Coen *et al.* 2007; Coen and Bishop 2015). DO is less of a problem for intertidal reefs and associated mobile organisms as they must either move as tides fall or remain on exposed reefs 1–2 times per day (Coen *et al.* 1999a). Regardless, sessile intertidal organisms are aerially exposed daily and many (bivalves) can respire so that low DOs are less of a problem if they occur for <6–12 h d^{-1}.

Salinity

Prolonged exposure to low salinities (< 5 ppt or psu) can reduce feeding, growth, and repro-duction, while also reducing observed negative effects of disease and predation (Figure 19.2; Coen and Bishop 2015). The oyster protozoan pathogens, *Perkinsus marinus* (Dermo disease), and *Haplosporidium nelsoni* (or MSX) are generally intolerant of salinities <10 ppt (summarized in Coen and Bishop 2015). For subtidal populations especially, higher temperatures and salini-ties (> 15–34 ppt) tend to cause increased disease levels and even epizootics (Coen and Bishop 2015 and references therein). Restoration located in close proximity to highly fluctuating or longer duration freshwater inflows can be affected significantly. Subtidal predators often greatly increase with higher (>15 ppt) salinities. Smaller species (flatworms) may feed on earlier post-settlement oyster stages (spat) and larger predators (gastropods, finfish, crabs; Figure 19.2) (Kennedy *et al.* 1996; Newell *et al.* 2000; Luckenbach *et al.* 1999; ASMFC 2007). The possible benefits of sites near freshwater inflows may be counter-balanced by increased oyster mortality either directly via osmotic stress or indirectly from sedimentation (Coen *et al.* 1999a, 2004; La Peyre *et al.* 2009) ultimately reducing the habitat value of both natural and restored reefs (Coen *et al.* 2004; ASMFC 2007).

Predators and competitors

Predators can have significant effects on both oysters (spat to adults), as well as on other reef-associated species. Predators range from smaller species such as flatworms (genus *Stylochus* spp. can feed on small, early post-settlement spat recruits; Newell *et al.* 2000), to larger invertebrate species such as other molluscs (gastropods), echinoderms (starfish), crabs, and vertebrates (finfish) (White and Wilson 1996). Many stenohaline predators only live in higher salinity envi-ronments (e.g. gastropod drills, echinoderms, rays), but are greatly diminished when salinities are relatively low (< 10–15 ppt). At most *C. virginica* sites, decapod crabs such as xanthids (e.g. *Panopeus* spp., *Eurypanopeus* spp., *Menippe* spp.) or portunids (genus *Callinectes* spp.) can cause significant oyster mortality, as well as other reef-associated species (White and Wilson 1996; Baggett *et al.* 2014). Oyster predators also include vertebrates spanning the gambit from finfish (including rays) when submerged, to raccoons and birds who feed on intertidal reef habitats when exposed.

Competitors such as fouling organisms (barnacles, tunicates, sponges, and even other bivalve mussels) may attach to oyster shell and limit free-space available for settlement of oyster larvae (Luckenbach *et al.* 2005; Brumbaugh *et al.* 2006; Brumbaugh and Coen 2009; Baggett *et al.* 2014). These can include both native, as well as non-native species (Kennedy *et al.* 1996; Luckenbach *et al.* 1999; NRC 2004; ASMFC 2007; Coen and Bishop 2015).

Hydrology and food

Sites with greater water (current) flows are often associated with greater oyster survival and faster growth (Lenihan 1999). Oyster shape, unlike many other molluscs, is quite plastic, with flow, sediments, and exposure affecting overall size, shell shape, and thickness (Galtsoff 1964). Flows ranging from 156–260 cm sec^{-1} are often associated with enhanced growth (Lenihan 1999; Coen *et al.* 2004). Oysters filter seston (phytoplankton, resuspended benthos, and other organic particles) from the water column (Kennedy *et al.* 1996) so an adequate food concentration is necessary for growth and survival. Currents not only deliver seston, but also carry away silt, oyster pseudofaeces, and faeces from reefs (Dame 1996). Water column chlorophyll *a* levels can be used as a proxy for food quantity. The quality of food and the potential filtering effects of developing reefs is more difficult to quantify, but it can be done (zu Ermgassen *et al.* 2013). Circulation patterns should be examined (along with appropriate modelling, see Kim *et al.* 2013) during peak spawning when larvae are recruiting (Southworth and Mann 1998).

Based on limited efforts, we know that lab-reared oysters with reduced (<4 cm sec^{-1}) flows have slower growth and greater mortality versus those reared under higher flow rates (7–20 cm sec^{-1}). Cholorophyll *a* concentrations greater than 30 mg m^{-3} result in rapid oyster growth.

Oyster recruitment potential (larval supply)

The availability of adequate bivalve (oyster spat) larval supply is critical for successful reef restoration. Otherwise as mentioned elsewhere the cost of the effort will be greatly increased. As part of site selection for any restoration effort, one or more oyster recruitment seasons (years) should be assessed prior to construction to ensure the availability of oyster recruits is sufficient (Coen and Luckenbach 2000; Baggett *et al.* 2014; NRC 2017).

One way to assess where one's sites might reside on the continuum between recruit- to substrate-limited sites is through the assessment of post-settlement recruitment and survival. The deployment of settlement plates, shell strings, vertical cylinders, and containers filled with various materials can be used to assess larval supply, growth, and ultimately juvenile to adult densities (reported in m^{-2}; Coen *et al.* 1999a; Baggett *et al.* 2014).

One easy, low-cost method to assess recruitment potential (larval supply) for shallow subtidal and intertidal projects is to deploy large mesh-covered replicate plastic trays filled with appropriate material at sites under consideration prior to the recruitment season (generally May to October). These can be placed on subtidal or intertidal reefs or bottoms and shorelines without hard substrates or for uniformity (age, timing, and material) to assess colonization of substrates before and even during ongoing restoration projects (Coen *et al.* 1999a, 2004; Luckenbach *et al.* 1999; Brumbaugh *et al.* 2006; Brumbaugh and Coen 2009; Baggett *et al.* 2014). They can be collected and assessed sometime in the fall to winter or even the spring of the following year. By assessing live (and dead) oysters and their relative sizes (and frequencies and densities m^{-2}), one can gain a lot of information prior to investing significant resources, especially for large-scale restoration projects (Baggett *et al.* 2014).

However, in some areas (e.g. New York Hudson River estuary), larval supply and associated recruitment is so low that costly alternatives such as: (1) deploying spat-on-shell (spat set on shell or other substrate in hatchery and grown-out in a field nursery); (2) alternatively, though more costly, small single juvenile oysters (< 10–15 mm shell height, SH), can be added to reefs to jump start restoration efforts (Baggett *et al.* 2014; Lodge *et al.*, unpublished data).[5]

Proximity to natural (extant) oyster reefs

Larval supply is critical for the survival and development of restored reefs (Brumbaugh and Coen 2009). It is also invaluable for cost-effective restoration efforts, especially at larger scales. As broadcast spawners, members of the genus *Crassostrea* are sequential protandric hermaphrodites, with external fertilization. Larvae are then carried by local currents for up to two weeks until they are ready (competent) to settle on an appropriate hard substrate. For the genus *Crassotrea*, recruitment can occur into areas without any nearby (many km) reefs given that the requisite hard substrate is present first and relatively clean as many sites may be substrate- vs. recruit-limited (Brumbaugh and Coen 2009). Other genera (*Ostrea* spp., west coast of North America, Europe, IWP-Asia) are brooders undergoing internal fertilization and releasing juveniles rather than gametes for external fertilization. By placing material into areas without prior substrate, novel reefs can arise where did not occur formerly. Settlement behaviour is also mediated chemically (Kennedy *et al.* 1996 and references therein) such that newly settling oysters settle gregariously onto existing reefs with live oysters. Under certain conditions, natural recirculation patterns (retention estuaries) may enable existing reefs to resupply native or constructed reefs with new recruits (Southworth and Mann 1998).

Disease

Oyster diseases worldwide are diverse (Coen and Bishop 2015), but *C. virginica* usually refers to the presence of either Dermo (occurs from the Gulf of Mexico to Maine), whereas MSX has been observed from Florida to Canada (Kennedy *et al.* 1996 and references therein).[6] Infection by these two pathogens causes reduced growth rates, and ultimately death in areas with appropriate salinities where disease prevalence is high; these subtidal areas should be avoided where possible as potential restoration sites (Coen and Bishop 2015).

Explicit goals and related objectives for restoration

As mentioned previously any oyster restoration effort should be well-planned with clear goals that represent desired outputs and ecosystem services specified *a priori* (Coen and Luckenbach 2000; Kennedy *et al.* 2011; Baggett *et al.* 2014). Given limited space here, we refer the reader to recent reviews such as Coen *et al.* (2004), Kennedy *et al.* (2011), Baggett *et al.* (2014, 2015), La Peyre *et al.* (2014), and Powers and Boyer (2014). An NRC (2017) review addresses pre-construction and post-construction monitoring and related issues. All of the above discuss explicit goals which represent specific ecosystem services, individual objectives and related metrics, and associated success criteria to be sampled to achieve these.

However, because these ecosystems are quite complex, *measuring all variables is not feasible*. Hence, selection of which variables provide the most value for a given cost for assessing oyster reef restoration progress requires a clear understanding of appropriate protocols, and a clear knowledge of suitable sampling techniques. However, the selection of a limited suite of either general, basic, or as some have referred to them 'universal metrics and variables', should be required for *all* projects using comparable methods to assess performance (maximize success or understand failure), and later, if required, apply appropriate adaptive management to get back on track restoration project positive trajectories (reviewed in Coen and Luckenbach 2000; Kennedy *et al.* 2011; Baggett *et al.* 2014; NRC 2017).

Suffice it to say that these universal metrics and universal environmental variables and Restoration Goal-Based Metrics have been put forth in Baggett *et al.* (2014) and are being used more and more by practitioners across the USA for *Crassostrea* spp. and *Ostrea* spp.[7] Below we discuss very briefly some goals and objectives that provide support for documenting specific ecosystem services as part of potential oyster restoration goals and related success.

Summary and future considerations

Future oyster restoration/enhancement projects need to consider the lessons learned from previous attempts, whether viewed as successes or failures, and then these need to be applied *a priori* or within an adaptive management framework (NRC 2017). Below we consider a few of the lessons learned and what future efforts could be implemented to increase the probability of a given project's success.

One of the key differences between a successful and non-successful restoration effort is adequate (post-construction reef footprint and vertical relief) hard substrate for settlement of oyster spat. Note simulation modelling is getting better at predicting some of the critical processes as they relate to successful restoration (e.g. Kim *et al.* 2013).

Another fundamental result of early subtidal restoration efforts is that where dissolved oxygen is a potential challenge, the use of higher relief reefs, mimicking historical natural ones, has proven to be more successful than low or very low relief reefs (Lenihan and Peterson 1998; Lenihan 1999; Woods *et al.* 2005; Breitburg *et al.* 2009; Coen and Luckenbach 2000). This has been shown in the Gulf of Mexico (Gregalis *et al.* 2009), the southeastern USA (Lenihan 1999), as well as the mid-Atlantic USA (Luckenbach *et al.* 1999; Schulte *et al.* 2009).

Shell budgets for subtidal oyster reefs in the northeastern USA have been calculated and used to assess reef shell trajectories and the likelihood of longer-term restoration success (Powell *et al.* 2006, 2011; Waldbusser and Salisbury 2014; Casas *et al.* 2015; Ekstrom *et al.* 2015). Also concurrent multiple habitat restoration can maximize the overall success if the association of adjacent habitats is potentially positive (Milbrandt *et al.* 2015).

In areas where larval supply is limited (e.g. Hudson River Estuary), shell with small already set oysters, either from hatcheries or from redeployed field sets, can be one approach to increase success (Coen and Luckenbach 2000; Coen *et al.* 2004; Brumbaugh *et al.* 2006; Baggett *et al.* 2014). Once the oysters reach a refuge size (often 2–4 cm SH) or shell thickness, effectively reducing predation losses, they can be added onto reefs loosely with the deployed shell and seeded directly onto newly constructed reefs. However, with larger deployed oyster size, comes an obvious trade-off as hatchery/nursery costs also increases significantly with this increasing size.

Intertidal evaluations of natural oyster reef changes and restoration success can be more easily assessed using a number of approaches (Coen *et al.* 1999a; Luckenbach *et al.* 2005; ASMFC 2007; Powers *et al.* 2009; Baggett *et al.* 2014 and references therein). For a large number of restoration footprints over a range of post-construction ages, Powers *et al.* (2009) reassessed NC reefs and determined that intertidal success was much greater than for subtidal reefs. However, this finding may be confounded by a number of potential methodological problems by time and parameter selection.

A lot more work needs to be done with regard to the success of reef restoration beyond the normal funding cycle of 1–3 years perhaps (Brumbaugh and Coen 2009; Kennedy *et al.* 2011; Baggett *et al.* 2014; NRC 2017). The large-scale 2009 American Recovery and Reinvestment Act (NOAA ARRA) oyster reef-related projects across the Gulf of Mexico and eastern USA may provide some of these answers, but initial funding was limited for 12–18 months, a period too short to assess success.

Knowing the status and condition (for trend analyses) of relevant habitats to be restored are critical for placement and extent of construction.[8] In many areas, major efforts have taken place with new imagery and related mapping (ASMFC 2007; SCDNR 2008)[9] or are under way (e.g. RESTORE funding for the Gulf of Mexico post DWH spill) to assess natural reefs for triaging recovery efforts (e.g. Harris Creek Oyster Restoration Tributary Plan, MD, USA).[10]

There are many parallels in the services rendered by farmed and natural reef restoration approaches (Dumbauld *et al.* 2009; Coen *et al.* 2011). It should be noted that worldwide, aquaculture is playing an ever increasing role in the supply of bivalve molluscs (Dumbauld *et al.* 2009; Shumway 2011; Shinn *et al.* 2015). Of late it has value as a potential tool for documenting also other non-consumptive ecosystem services for coastal habitats worldwide (Coen *et al.* 2007, 2011; Grabowski and Peterson 2007; Brumbaugh and Coen 2009; Beck *et al.* 2011; Powers and Boyer 2014). Some have gone on to even suggest that mussel and potentially other bivalve (e.g. oyster) aquaculture may provide a mechanism for reducing eutrophication (Shumway 2011). However, not all of the aquaculture impacts are positive (Dumbauld *et al.* 2009; Coen *et al.* 2011; Coen and Bishop 2015).

Despite the loss or dramatic decline of many habitat-generating bivalve species worldwide (Beck *et al.* 2011), these species face even greater obstacles to their recovery, let alone continued existence. Current and future alteration of 'typical' conditions by: (1) climate change and associated sea level rise; (2) ocean and nearshore acidification and related changes in pH; (3) rising temperatures impacting diseases and species ranges; (4) rainfall and salinity shifts; (5) increasing hypoxic areal extent and duration for subtidal populations; (6) continued coastal development and related vegetated habitat losses; (7) introduction of non-native competitors, predators, diseases; and (8) changes in native diseases will create natural habitat winners and losers, while impacting restoration outcomes (discussed in Coen and Bishop 2015; NRC 2017).

Recently oyster reefs were included as one of nine potentially important nearshore habitats that can potentially protect coastal communities and infrastructure (Arkema *et al.* 2013). Because bivalve habitats are potentially one of the most important nearshore habitats that have the ability to feed populations, as well as helping to protect coastal communities and infrastructure (*ibid.*), it is critical that we increase the scale of our restoration efforts. Not only do we need to increase local capacity to perform such habitat restoration, but also figure out how to extend monitoring while effectively scaling-up restoration efforts.

Enhancing or restoring of bivalve populations for direct harvesting or delivery of their numerous ecosystem services will require a great deal more research on questions related to diseases, genetics, and scaling-up related to practical construction relevant to management decisions (Lotze *et al.* 2006; Powers and Boyer 2014; Coen and Bishop 2015). Novel environmental perturbations, along with declining wildstocks (Lotze *et al.* 2006) will be especially challenging as local, state, and federal budgets continue to shrink, along with many agencies being forced to ignore the effects of climate change and sea level rise, especially in the USA, and more recently in other countries such as Australia and Canada (Coen and Bishop 2015).

Notes

1   See www.oyster-restoration.org/living-shorelines.
2   Cf. www.dnr.sc.gov/GIS/descoysterbed.html.
3   E.g. for SCDNR 2008 see www.oyster-restoration.org/oyster-restoration-research-reports.
4   But see www.oyster-restoration.org.
5   See www.hudsonriver.org/?x=orrp.
6   See also www.dfo-mpo.gc.ca/science/aah-saa/diseases-maladies/index-eng.html as updated.
7   Handbook available at www.oyster-restoration.org (see 'Restoration Practices' dropdown).
8   See www.dnr.sc.gov/GIS/descoysterbed.html.

9   See www.oyster-restoration.org/oyster-restoration-research-reports.
10   See www.oyster-restoration.org.

References

Alleway, H. K., and S. D. Connell, 2015. Loss of an ecological baseline through the eradication of oyster reefs from coastal ecosystems and human memory. *Conserv. Biol.* 29:795–804.

Arkema, K. K., G. Guannel, G. Verutes, S. A. Wood, A. Guerry, M. Ruckelshaus, P. Kareiva, M. Lacayo, and J. M. Silver, 2013. Coastal habitats shield people and property from sea-level rise and storms. *Nature Clim. Change* 3:913–918.

ASMFC, 2007. *The importance of habitat created by shellfish and shell beds along the Atlantic coast of the US,* prepared by L. D. Coen, and R. Grizzle, with contributions by J. Lowery and K.T. Paynter, Jr.

Baggett, L. P., S. P. Powers, R. Brumbaugh, L. D. Coen, B. DeAngelis, J. Green, B. Hancock, and S. Morlock, 2014. *Oyster habitat restoration: monitoring and assessment handbook.* The Nature Conservancy, Arlington, VA, USA. Retrieved from www.oyster-restoration.org/wp-content/uploads/2014/01/Oyster-Habitat-Restoration-Monitoring-and-Assessment-Handbook.pdf.

Bahr, L. M. and W. P. Lanier, 1981. *The ecology of intertidal oyster reefs of the South Atlantic Coast: a community profile.* US Fish Wildl. Serv. Program FWS/OBS/–81/15.

Barbier, E. B., H. M. Leslie, and F. Micheli, 2014. Services of marine ecosystems: a quantitative perspective. Pages 403–425 in M. D. Bertness, B. J. Silliman, and J. J. Stachowicz (eds), *Marine community ecology and conservation.* Sinauer Associates, Sunderland, MA.

Beck, M. W., R. D. Brumbaugh, L. Airoldi, A. Carranza, L. D. Coen, C. Crawford, O. Defeo, G. J. Edgar, B. Hancock., M. C. Kay, H. S. Lenihan, M. W. Luckenbach, C. L. Toropova, G. Zhang, and X. Guo, 2011. Oyster reefs at risk and recommendations for conservation, restoration and management. *BioScience* 61:107–116.

Bilkovic, D. M., M. M. Mitchell, M. K. La Peyre, and J. D. Toft, eds., 2017. *Living shorelines: the science and management of nature-based coastal protection.* CRC Marine Science Series, CRC Press, Taylor & Francis Group.

Breitburg, D. L., L. Hondorp, W. Davias, and R. J. Diaz, 2009. Hypoxia, nitrogen and fisheries: integrating effects across local and global landscapes. *Ann. Rev. in Mar. Sci.* 1:329–350.

Brumbaugh, R. D. and L. D. Coen, 2009. Contemporary approaches for small-scale oyster reef restoration to address substrate versus recruitment limitation: a review and comments relevant for the Olympia oyster, *Ostrea lurida* (Carpenter, 1864). *J. Shellfish Res.* 28:147–161.

Brumbaugh, R. D., M. W. Beck, L. D. Coen, L. Craig and P. Hicks, 2006. *A practitioners' guide to the design and monitoring of shellfish restoration projects: an ecosystem services approach.* MRD Educational Report no. 22. The Nature Conservancy, Arlington, VA. Retrieved from http://www.habitat.noaa.gov/pdf/tncnoaa_shellfish_hotlinks_final.pdf.

Byers, J. E., J. H. Grabowski, M. F. Piehler, A. R. Hughes, H. W. Weiskel, J. C. Malek, and D. L. Kimbro, 2015. Geographic variation in intertidal oyster reef properties and the influence of tidal prism. *Limnol. Oceanogr.* 60:1051–1063.

Casas, S. M., J. La Peyre, and M. K. La Peyre, 2015. Restoration of oyster reefs in an estuarine lake: population dynamics and shell accretion. *Mar. Ecol. Prog. Ser.* 524:171–184.

Coen, L. D., and M. J. Bishop, 2015. The ecology, evolution, impacts and management of host–parasite interactions of marine molluscs. *J. Invertebr. Pathol.* 13:177–211.

Coen, L. D., and R. E. Grizzle, 2016. Bivalve molluscs. Pages 89–109 in M. Kennish (ed.), *Encyclopedia of estuaries.* Springer, Dordrecht.

Coen, L. D. and M. W. Luckenbach, 2000. Developing success criteria and goals for evaluating oyster reef restoration: ecological function or resource exploitation? *Ecol. Eng.* 15:323–343.

Coen, L. D., D. M. Knott, E. L. Wenner, N. H. Hadley, and A. H. Ringwood, 1999a. Intertidal oyster reef studies in South Carolina: design, sampling and experimental focus for evaluating habitat value and function. In M. W. Luckenbach, R. Mann, and J. A. Wesson (eds), *Oyster reef habitat restoration: a synopsis and synthesis of approaches,* 131–156. VIMS Press, Gloucester Point, VA.

Coen, L. D., M. W. Luckenbach, and D. L. Breitburg, 1999b. The role of oyster reefs as essential fish habitat: a review of current knowledge and some new perspectives. Pages 438–454 in L. R. Benaka (ed.), *Fish habitat: essential fish habitat and rehabilitation.* American Fisheries Society, Symposium 22, Bethesda, MD.

Coen, L., K. Walters, D. Wilber, and N. Hadley, 2004. *A South Carolina Sea Grant report of a 2004 workshop to examine and evaluate oyster restoration metrics to assess ecological function, sustainability and success results and related information*. SC Sea Grant Publication, Charleston, SC. Retrieved from www.oyster-restoration.org/wp-content/uploads/2012/06/SCSG04.pdf.

Coen, L. D., R. D. Brumbaugh, D. Bushek, R. Grizzle, M. W. Luckenbach, M. H. Posey, S. P. Powers, and G. Tolley, 2007. As we see it: a broader view of ecosystem services related to oyster restoration. *Mar Ecol. Prog. Ser.* 341:303–307.

Coen, L. D., B. R. Dumbauld, and M. L. Judge, 2011. Expanding shellfish aquaculture: a review of the ecological services provided by and impacts of native and cultured bivalves in shellfish-dominated ecosystems. Pages 239–295 in S. E. Shumway (ed.), *Shellfish aquaculture and the environment*, Wiley-Blackwell, NY.

Dame, R., 1996. *Ecology of marine bivalves: an ecosystem approach*. CRC Marine Science Series, Boca Raton, FL.

Dumbauld, B. R., J. L. Ruesink, and S. S. Rumrill, 2009. The ecological role of bivalve shellfish aquaculture in the estuarine environment: a review with application to oyster and clam culture in West Coast (USA) estuaries. *Aquaculture* 290:196–223.

Ekstrom, J., L. Suatoni, S. Cooley, L. Pendleton, G. G. Waldbusser, J. Cinner, J. Ritter, C. Langdon, R. van Hooidonk, D. Gledhill, K. Wellman, M. Beck, L. Brander, D. Rittschof, C. Doherty, P. Edwards, R. Portela, 2015. Vulnerability and adaptation of US shellfisheries to ocean acidification. *Nature Clim. Change* 5:207–214.

Elton, C., 1958. *The ecology of invasions by animals and plants*. Methuen and Co., London.

Galtsoff, P. S., 1964. Morphology and structure of shell. The American oyster, *Crassostrea virginica* Gmelin. *Fish. Bull.* 64:16–47.

Gillies, C.L., C. Creighton, and I.M. McLeod, eds., 2015. Shellfish reef habitats: a synopsis to underpin the repair and conservation of Australia's environmentally, socially and economically important bays and estuaries. Report to the National Environmental Science Programme, Marine Biodiversity Hub. Centre for Tropical Water and Aquatic Ecosystem Research (TropWATER) Publication, James Cook University, Townsville, 68pp.

Grabowski, J. H., and C. H. Peterson, 2007. Restoring oyster reefs to recover ecosystem services. Pages 281–298 in K. Cuddington, J. E. Byers, W. G. Wilson, and A. Hastings (eds), *Ecosystem engineers: concepts, theory and applications*. Elsevier/Academic Press, Amsterdam, the Netherlands.

Grabowski, J. H., R. D. Brumbaugh, R. F. Conrad, A. G. Keeler, J. J. Opaluch, C. H. Peterson, M. F. Piehler, S. P. Powers, and A. R. Smyth, 2012. Economic valuation of ecosystem services provided by oyster reefs. *BioScience* 62:900–909.

Gregalis, K. C., M. W. Johnson and S. P. Powers, 2009. Restored oyster reef location and design affect responses of resident and transient fish, crab, and shellfish species in Mobile Bay, Alabama. *T. Am. Fish. Soc.* 138:314–327.

Humphries, A. T., and M. K. La Peyre, 2015. Oyster reef restoration supports increased nekton biomass and potential commercial fishery value. *PeerJ* 3:e1111.

Kellogg, M. L., A. R. Smyth, M. W. Luckenbach, R. H. Carmichael, B. L. Brown, J. C. Cornwell, M. F. Piehler, M. S. Owens, D. J. Dalrymple, and C. B. Higgins, 2014. Invited feature: use of oysters to mitigate eutrophication in coastal waters. *Estuar. Coast. Shelf S.* 151:156–168.

Kennedy, V. S., R. I. E. Newell, and A. F. Eble (eds), 1996. *The eastern oyster:* Crassostrea virginica. Maryland Sea Grant, College Park, MD.

Kennedy, V. S., D. L. Breitburg, M. C. Christman, M. W. Luckenbach, K. Paynter, J. Kramer, K. G. Sellner, J. Dew-Baxter, C. Keller, and R. Mann, 2011. Lessons learned from efforts to restore oyster populations in Maryland and Virginia, 1990 to 2007. *J. Shellfish Res.* 30:719–731.

Kim, C.-K, K. Park, and S. P. Powers, 2013. Establishing restoration strategy of eastern oyster via a coupled biophysical transport model. *Restor. Ecol.* 21:353–362.

Kirby, M. X., 2004. Fishing down the coast: historical expansion and collapse of oyster fisheries along coastal margins. *P. Natl. Acad. Sci. USA* 101:13096–13099.

La Peyre, M. K., B. Gossman, and J. F. La Peyre, 2009. Defining optimal freshwater flow for oyster production: effects of freshet rate and magnitude of change and duration on Eastern oysters and *Perkinsus marinus* infection. *Estuaries Coasts* 32:522–534.

La Peyre, M. K., J. N. Furlong, L. A. Brown, B. P. Piazza and K. Brown, 2014. Oyster reef restoration in the northern Gulf of Mexico: extent, methods and outcomes. *Ocean. Coast. Manage.* 89:20–28.

La Peyre, M. K., K. Serra, T. A. Joyner, and A. Humphries, 2015. Assessing shoreline exposure and oyster habitat suitability maximizes potential success for sustainable shoreline protection using restored oyster reefs. *PeerJ* DOI 10.7717/peerj.1317.

Lenihan, H. S., 1999. Physical-biological coupling on oyster reefs: how habitat structure influences individual performance. *Ecol. Monogr.* 69:251–275.

Lenihan, H. S., and C. H. Peterson, 1998. How habitat degradation through fishery disturbance enhances impacts of hypoxia on oyster reefs? *Ecol. Applic.* 8:128–140.

Lenihan, H. S., F. Micheli, S. W. Shelton, and C. H. Peterson, 1999. The influence of multiple environmental stressors on susceptibility to parasites: an experimental determination with oysters. *Limnol. Oceanogr.* 44:910–924.

Lotze, H. K., H. S. Lenihan, B. J. Bourque, R. H. Bradbury, R. G. Cooke, M. C. Kay, S. M. Kidwell, M. X. Kirby, C. H. Peterson, and J. B. C. Jackson, 2006. Depletion, degradation, and recovery potential of estuaries and coastal seas. *Science* 312:1806–1809.

Luckenbach, M. W., R. Mann, and J. A. Wesson (eds), 1999. *Oyster reef habitat restoration. a synopsis and synthesis of approaches*. Virginia Institute of Marine Science Press, Gloucester Point, VA.

Luckenbach, M. W., L. D. Coen, P. G. Ross, Jr., and J. A. Stephen, 2005. Oyster reef habitat restoration: relationships between oyster abundance and community development based on two studies in Virginia and South Carolina. *J. Coastal Res.* Special Issue 40:64–78.

MEA, 2005. *Ecosystems and human well-being: synthesis*. Island Press, Washington, DC.

Meyer, D. L., E. C. Townsend, and G. W. Thayer, 1997. Stabilization and erosion control value of oyster cultch for intertidal marsh. *Restor. Ecol.* 5:3–99.

Milbrandt, E. C., M. Thompson, L. D. Coen, R. E. Grizzle, and K. Ward, 2015. A multiple habitat restoration strategy in a semi-enclosed Florida embayment, combining hydrologic restoration, mangrove propagule plantings and oyster substrate additions. *Ecol. Eng.* 3:394–404.

Newell, R. I. E. 2004. Ecosystem influences of natural and cultivated populations of suspension-feeding bivalve mollusks: a review. *J. Shellfish Res.* 23:51–61.

Newell, R. I. E., G. S. Jr. Alspach, V. S. Kennedy, and D. Jacobs, 2000. Mortality of newly metamorphosed eastern oysters (*Crassostrea virginica*) in mesohaline Chesapeake Bay. *Mar. Biol.* 136:665–676.

National Research Council (NRC) 2004. *Nonnative oysters in the Chesapeake Bay*. National Academies Press, Washington, DC.

National Research Council (NRC), 2017. *Effective monitoring to evaluate ecological restoration in the Gulf of Mexico*. National Academies Press, Washington, DC.

Peterson, C. H., J. H. Grabowski, and S. P. Powers, 2003. Estimated enhancement of fish production resulting from restoring oyster reef habitat: quantitative valuation. *Mar. Ecol. Prog. Ser.* 264:251–256.

Piazza, B. P., P. D. Banks, and M. K. La Peyre, 2005. The potential for created oyster shell reefs as a sustainable shoreline protection strategy in Louisiana. *Restor. Ecol.* 13:499–506.

Pollack, J. B., A. Cleveland, T. A. Palmer, A. S. Reisinger, and P. A. Montagna, 2012. A restoration suitability index model for the eastern oyster (*Crassostrea virginica*) in the Mission-Aransas Estuary, TX, USA. *PLoS ONE* 7(7):e40839. doi:10.1371/journal.pone.0040839

Powell, E. N., J. N. Kraeuter, and K. A. Ashton-Alcox, 2006. How long does oyster shell last on an oyster reef? *Estuar. Coast. Shelf S.* 69:531–542.

Powell, E. N., G. M. Staff, W. R. Callender, K. A. Ashton-Alcox, C. E. Brett, K. M. Parsons-Hubbard, S. E. Walker and A. Raymond, 2011. Taphonomic degradation of molluscan remains during thirteen years on the continental shelf and slope of the northwestern Gulf of Mexico. *Palaeogeogr. Palaeoclimatol. Palaeoecol.* 312:209–232.

Powers, S. P. and K. E. Boyer, 2014. Ch. 22. Marine restoration ecology. Pages 495–516 in M. D. Bertness, J. F. Bruno, B. R. Silliman, and J. J. Stachowicz (eds), *Marine community ecology and conservation*, Sinauer Associates, Sunderland, MA.

Powers, S. P., C. H. Peterson, J. H. Grabowski, and H. S. Lenihan, 2009. Success of constructed oyster reefs in no-harvest sanctuaries: implications for restoration. *Mar. Ecol. Prog. Ser.* 389:159–170.

Quan, W. M., L. Zheng, B. Li, and C. An, 2013. Habitat values for artificial oyster (*Crassostrea ariakensis*) reefs compared with natural shallow-water habitats in Changjiang River estuary. *Chin. J. Oceanol. Limnol.* 31:957–969.

Rodriguez, A. B., F. J. Fodrie, J. T. Ridge, N. L. Lindquist, E. J. Theuerkauf, S. E. Coleman, J. H. Grabowski, M. C. Brodeur, R. K. Gittman, D. A. Keller, and M. D. Kenworthy, 2014. Oyster reefs can outpace sea-level rise. *Nature Clim. Change* 4:493–497.

Ruesink, J. L., H. S. Lenihan, A. C. Trimble, K. W. Heiman, F. Micheli, J. E. Byers, and M. C. Kay, 2005. Introduction of non-native oysters: ecosystem effects and restoration implications. *Ann. Rev. in Ecol. Syst.* 36:643–689.

SCDNR, 2008. Final report for South Carolina's 2004–05 intertidal oyster survey and related reef restoration/ enhancement program: an integrated oyster resource/habitat management and restoration program using novel approaches, by the Marine Resources Division, SCDNR. Final Report Completed for NOAA Award No. NA04NMF4630309 December 2008. Retrieved from www.oyster-restoration.org/oyster-restoration-research-reports.

Schulte, D. M., R. P. Burke, and R. N. Lipcius, 2009. Unprecedented restoration of a native oyster metapopulation. *Science* 325:1124–1128.

Scyphers, S. B., S. P. Powers, K. L. Heck, Jr., and D. Byron, 2011. Oyster reefs as natural breakwaters mitigate shoreline loss and facilitate fisheries. *PLoS ONE* 6(8):e22396.

SER 2004. *The SER international primer on ecological restoration.* Society for Ecological Restoration International, Tucson, AZ. Retrieved from http://c.ymcdn.com/sites/www.ser.org/resource/resmgr/custompages/publications/SER_Primer/ser_primer.pdf.

Shinn, A. P., J. Pratoomyot, J. E. Bron, G. Paladini, E. E. Brooker, and A. J. Brooker, 2015. Economic costs of protistan and metazoan parasites to global mariculture. *Parasitol.* 142:196–270.

Shumway, S. E. (ed.), 2011. *Shellfish aquaculture and the environment.* Wiley-Blackwell, New York.

Smaal, A., M. van Stralen, and J. Craeymeersch, 2005. Does the introduction of the Pacific oyster *Crassostrea gigas* lead to species shifts in the Wadden Sea? Pages 277–289 in R. F. Dame and S. Olenin (eds), *The comparative role of suspension-feeders in ecosystems.* NATO Science Series IV – Earth and Environmental Sciences, Volume 47. Springer, Dordrecht, the Netherlands.

Smyth, A. R., M. F. Piehler, and J. H. Grabowski, 2015. Habitat context influences nitrogen removal by restored oyster reefs. *J. Appl. Ecol.* 52:716–725.

Soniat, T. M., C. P. Conzelmann, J. D. Byrd, D. P. Rozell, J. L. Bridevaux, K. J. Suir, and S. B. Colley, 2013. Predicting the effects of proposed Mississippi River diversions on oyster habitat quality: application of an oyster habitat suitability index model. *J. Shellfish Res.* 32:629–638.

Southworth, M., and R. Mann, 1998. Oyster reef broodstock enhancement in the Great Wicomico River, Virginia. *J. Shellfish Res.* 17:1101–1114.

UN, 2012. *World population, 2012.* United Nations, New York. Retrieved from www.un.org/en/development/desa/population/publications/pdf/trends/WPP2012_Wallchart.pdf.

Van de Koppel, J., J. C. Gascoigne, G. Theraulaz, M. Rietkerk, W. M. Mooij, and P. M. J. Herman, 2008. Experimental evidence for spatial self-organization and its emergent effects in mussel bed ecosystems. *Science* 322:739–742.

Waldbusser, G. G., and J. E. Salisbury, 2014. Ocean acidification in the coastal zone from an organism's perspective: multiple system parameters, frequency domains, and habitats. *Ann. Rev. in Mar. Sci.* 6: 221–247.

Walles, B., J. S. de Paiva, B. C. van Prooijen, T. Ysebaert, and A. C. Smaal, 2015. The ecosystem engineer *Crassostrea gigas* affects tidal flat morphology beyond the boundary of their reef structures. *Estuaries Coasts* 38:941–950.

Walles, B., F. J. Fodrie, S. Nieuwhof, O. J. D. Jewell, P. M. J. Herman, and T. Ysebaert, 2016. Guidelines for evaluating performance of oyster habitat restoration should include tidal emersion: reply to Baggett *et al. Rest. Ecol.* 24:4–7.

Walters, L. J., P. E. Sacks, M. Y. Bobo, D. L. Richardson, and L. D. Coen, 2007. Impact of hurricanes on intertidal oyster reefs in Florida: reef profiles and disease prevalence. *Fla. Sci.* 70:506–521.

Wells, H.W., 1961. The fauna of oyster beds, with special reference to the salinity factor. *Ecol. Monogr.* 31:266–329.

White, M. E., and E. A. Wilson, 1996. Predators, pests and competitors. Pages 559–580 in V. S. Kennedy, R. I. E. Newell, and A. F. Eble (eds), *The eastern oyster* Crassostrea virginica. Maryland Sea Grant College, College Park, MD.

Wilkie, E. M., M. J. Bishop, and W. A. O'Connor, 2012. Are native *Saccostrea glomerata* and invasive *Crassostrea gigas* oysters' habitat equivalents for epibenthic communities in southeastern Australia? *J. Exp. Mar. Biol. Ecol.* 420–421:16–25.

Woods, H., W. J. Hargis, C. H. Hershner, and P. Mason, 2005. Disappearance of the natural emergent 3-dimensional oyster reef system of the James River, Virginia, 1871–1948. *J. Shellfish Res.* 24:139–142.

zu Ermgassen, P. S. E., M. D. Spalding, P. Banks, B. Blake, L. Coen, B. Dumbauld, S. Geiger, J. H. Grabowski, R. Grizzle, M. Luckenbach, K. McGraw, B. Rodney, J. Ruesink, S. Powers, and R. Brumbaugh, 2012. Historical ecology with real numbers: past and present extent and biomass of an imperilled estuarine ecosystem. *Proc. R. Soc. Lond. B. Bio.* 279:3393–3400.

zu Ermgassen, P. S. E., M. D. Spalding, R. Grizzle, and R. Brumbaugh, 2013. Quantifying the loss of a marine ecosystem service: filtration by the eastern oyster in US estuaries. *Estuaries Coasts* 36:36–43.

20

ECOLOGICAL REHABILITATION IN MANGROVE SYSTEMS

The evolution of the practice and the need for strategic reform of policy and planning

Ben Brown

Background

Mangrove forest ecosystems covered 13.8 million ha of tropical shorelines in 2000 (Giri *et al.* 2011), down from 19.8 million ha in 1980 and 15.9 million ha in 1990 (FAO 2003). These losses represent about 2 per cent per year from 1980 to 1990 and 1 per cent per year from 1990 to 2000. Therefore, achieving no net loss of mangroves worldwide would require the successful restoration of approximately 150,000 ha per year, unless all major losses of mangroves ceased. Increasing the total area of mangroves worldwide towards their original extent would require an even larger effort (Lewis in Bozzano *et al.* 2014).

Unfortunately, most mangrove rehabilitation projects do not succeed in rehabilitating mangrove habitat, and consist largely of the practice of planting mangrove propagules from a single genus (*Rhizophora*) at lower intertidal elevations, inappropriate for mangrove growth. This attempt at afforestation of sub-mean sea level mudflats has been criticized recently by the IUCN Mangrove Specialist Group manifesting itself in a petition signed by 33 scientists in the Philippines to stop all such plantings in the Philippines (IUCN-MSG 2015). Survivorship of planted mangroves is reported as being low, between 10–20 per cent (Primavera and Esteban 2008), but may be much lower as monitoring efforts are frequently inadequate (Lewis 2009). Most mangrove scientists now agree that an ecological approach to mangrove rehabilitation, with a proven record of accomplishment in some parts of the world, is the way forward, yet such an approach needs promotion through advocacy and continued practice (IUCN-MSG 2015).

Ecological approaches to mangrove rehabilitation work with natural recruitment, using naturally occurring propagules as the primary source for regeneration (Field 1999), and are generally underpinned by the goal of re-establishing fully functioning ecosystems (Lewis 1990; Ellison 2000). The following section briefly covers the ecological and hydrological principles which underscore successful mangrove rehabilitation.

Key ecological and hydrological principles underscoring mangrove rehabilitation

In *The Botany of Mangroves*, Tomlinson (1986) makes the important observation that mangroves 'have clearly pronounced characteristics of pioneer species in their reproductive biology but of a mature-phase species in some aspects of their community structure.' This bodes well for mangrove rehabilitation practitioners, as in theory, mangroves should be able to colonize appropriate but un-vegetated surfaces, and then persist over time adapting to a variety of ecological changes. When undisturbed, mangrove forests will follow a general adaptive cycle (Walker *et al.* 2004) as depicted in Figure 20.1.

Starting in the upper right corner in the κ-*phase* (where species maintain their populations at the maximum carrying capacity of their environment) mangroves are at a clear advantage over nearly all other flora due to their various ranges of tolerance to tidal inundation and salinity. When an individual dies of natural causes in such a case, it is replaced by one of several mangrove species of similar environmental tolerances, unless there has been a significant change in surface elevation or other hydrological factors, as depicted in the *omega-phase* (Figure 20.1).

In this instance, *release* (Ω) was caused by a gradual lowering of surface elevation, a natural geomorphological process along this coastline of alternating white rivers (sediment laden – highland draining) and black rivers (peatland draining). The reduction in surface elevation increases the duration and frequency of tidal flooding, resulting in increased concentrations of

Figure 20.1   A visual look at the adaptive cycle along a naturally eroding mangrove coastline in Ajkwa Delta, Papua, Indonesia

Source: photos and design by Ben Brown

substances toxic to mangrove roots such as H$_2$S (McKee and Faulkner 2000). Although mangrove trees can adapt to increased soil toxicity, by oxygenating their rhizosphere, thresholds can eventually be crossed, fatal to the tree. Release is depicted in Figure 20.1 by the death and toppling of adult *Bruguiera gymnorrhiza* and *B. sexangula*, members of the mesozone which are unable to tolerate inundation frequencies exceeding approximately 30 per cent.

In a healthy system toppled adult mangroves from the mesozone may be replaced by pioneer species from the seaward zone as well as halophytic grasses tolerant to tidal inundation (Figure 20.1: *re-organization* (α) *phase*). The important factor that enables re-organization is that the 'memory' of the previous system is preserved (Walker *et al.* 2004), in this case adequate nearby sources of propagules from pioneering species. When propagules are not available for colonization (propagule limitation), renewal will not take place without assistance. It should be noted that memory of the previous system need not be exact. Biodiversity is another key factor which allows for system resilience, as redundancy of functional roles ensures that some form of colonizer will be available to continue the system. The Aijkwa River delta (Figure 20.1) is rich with redundancy, containing five *Avicennia* species/varieties (*A. marina*, *A marina* var. *eucalyptifolia*, *A. rumphiana*, *A. alba* and *A. officianalis*), three colonizing *Sonneratia* species (*S. caseolaris*, *S. alba* and *S. ovata*), and at least four halophytic grasses able to colonize the seaward zone.

The (Γ) *phase* is characterized by a high degree of growth and exploitation. In mangroves, the mechanisms of succession are not straightforward, as mangroves do not necessarily develop into a 'climax' forest in the sense that we understand it in terrestrial forests. What is clear is that a healthy, diverse, mature mangrove forest has a high degree of resilience to natural shocks and disturbances.

That being said, there are disturbances, primarily hydrological in nature and primarily caused by humans, which prevent colonization of suitable surfaces and natural secondary succession. In those instances, human assistance may be required to rehabilitate a mangrove area. To overcome human-made disturbances to secondary succession in mangroves, ecological rehabilitation has focused on hydrological repair which allows for ecological recovery. Hydrological repair requires a basic understanding of mangrove hydrology, covered in excellent depth in Mazda *et al.* (2007), and explicated below for stakeholders and practitioners.

The following four basic principles of mangrove hydrology were distilled by Lewis and Gilmore (2007).

1   Freshwater flows (surface and groundwater) moderate substrate salinities and provide nutrients for mangrove forests.
2   General distribution and dominance of mangrove forest species is controlled by the depth, duration and frequency of flooding by both the tides and freshwater sources.
3   It is relatively easy to kill mangroves by withholding freshwater drainage, or drowning them with too much water.
4   Successful hydrologic rehabilitation requires an understanding of the tidal prism associated with tidal channels at a restoration site, either existing or created.

The next section on early case studies describes the development of hydrological rehabilitation techniques beginning with early trials in the 1980s.

Early case studies and the development of ecological mangrove rehabilitation

Both early and modern day attempts at mangrove rehabilitation can be categorized by four main technical methods (Lewis and Gilmore 2007):

1   Planting/transplanting only.
2   Hydrologic rehabilitation
    –   without planting (natural regeneration)
    –   with assisted propagule dispersal
    –   with planting.
3   Major excavation/fill
    –   without planting (natural regeneration)
    –   with assisted propagule dispersal
    –   with planting.
4   Experimental (primarily in eroding shorelines).

As planting/transplanting alone is not considered an ecological mangrove rehabilitation process, the following early case studies demonstrate technical methods 2, 3 and 4 from the list above.

West Lake (500 ha), Fort Lauderdale (1986): major excavation with natural re-vegetation

Originally a freshwater ecosystem, West Lake was dredged and filled in the 1920s to create 654 housing plots. In the 1980s and 1990s the area was rehabilitated to create 607 ha of self-maintaining mangrove forests and associated intertidal and sub-tidal habitats (Figure 20.2). It was originally budgeted at $20 million for 81 ha of restoration utilizing excavation and mangrove planting (initially planned at +18cm MSL; the project design was revised based on the results of natural recruitment trials reducing costs to $6 million for 80 ha of excavation to an appropriate surface elevation (+0.9 ± 1.4 ft MSL), and the creation of a 420 ha network of tidal channels, mudflats and fish refugia (MacAdam *et al.* 1998). The project relied fully on natural recruitment, bringing back all three native species of Florida mangroves and associated intertidal vegetation. The project had a strong focus on the restoration of functional fisheries equivalent, through intentional design of fish refugia including morphologically correct tidal creeks. Post-project monitoring found no significant differences between fish populations (in terms of abundance and diversity) in reference creeks and rehabilitated sites (Lewis and Gilmore 2007).

Major lessons learned include cost reduction through appropriate planning and design and reliance on natural re-vegetation, attention to re-grading to an appropriate surface elevation, the development of an extensive tidal creek network, and the enhancement of fisheries and wildlife values through creation of mudflats and fish refugia.

Hillsborough Bay, Tampa, Florida, Cargill Fertilizer, Bayside Shoreline (1980): experimental

This mitigation project took place in the 1980s and 1990s in Florida in an industrial discharge site of phosphate fertilizer waste products that buried and killed the vegetation. Residual waste was removed prior to rehabilitation. Vertical erosion bluffs were prominent on the shoreline, preventing natural re-vegetation by mangroves and halophytic grasses. Planners determined the best method was to cut and re-grade the vertical erosion banks, creating more gently sloped intertidal and supra-tidal surfaces. The supra-tidal slopes were planted with salt-spray tolerant grass in 1990, but saline conditions from salt-spray and occasional high tides killed off the grass. In October 1990, the intertidal area was planted with smooth cordgrass (*Spartina alterniflora*) – to stabilize the slope, improve edaphic conditions and capture mangrove seedlings. The resultant salt-marsh was eventually succeeded by local mangrove species (Broome *et al.* 1992).

Figure 20.2   The proof is in the monitoring: many mangrove 'restoration' projects claim success, but few have the data to back it up. Long-term photo plots are an excellent way to rapidly convey the development of a rehabilitation site to the public. Photos (a–c) from West Lake, Florida depict (a) pre-excavation, overgrown with invasive *Casuarina*, (b) 'as-built' after excavation, re-grading and tidal creek creation and (c) time zero + 8 years. Photos (d–f) from Hillsborough Bay depict (d) re-grading, (e) planting with *Spartina* grass to hold the slope and improve edaphic conditions and (f) subsequent succession at time zero + 13 years. The graph at the bottom reveals a rapid increase in percentage cover of mangrove and halophytic grass at Cross Bayou within 12 months, and eventual succession by mangroves and 94.7 per cent vegetation cover within 60 months after rehabilitation

Source: (a–f) Lewis (2011); (g) Lewis (2004)

Important lessons learned from this project include the use of pioneering halophytic grasses, to improve and stabilize a shoreline, physically capture mangrove propagules, and improve edpahic conditions further facilitating natural recruitment and early growth of mangroves.

Cross Bayou, Pinellas Country, Florida (1999): major excavation, hydrological rehabilitation and natural regeneration

Although small (4.35 ha), this is the most well documented early ecological mangrove rehabilitation project in terms of project monitoring and reporting (Lewis 1999a, 2004). The site was rehabilitated as part of a settlement to provide restoration after a 1993 oil-spill in Tampa Bay. In March 1999, the site was cleared of all invasive plants. The surrounding mangrove forests served as a hydro-geomorphic model, guiding the removal of 5552 m³ of dredged material followed by appropriate surface elevations (Lewis *et al.* 2005). Two dendritic tidal creek systems were created and connected to the main intertidal channel at the site. *Spartina alterniflora* sprigs were planted on 0.6 m centres as a cover crop similar to the Hillsborough Bay case study (*ibid.*).

During routine monitoring, species composition, stem density, per cent cover and plant height were measured. Most notably, plant cover increased linearly from 3.7 per cent after re-grading to 94.7 per cent within five years (Figure 20.2) as mangroves succeeded the planted cordgrass. In addition, button mangrove (*Conocarpus erectus*) and endemic halophytic grass species colonized higher elevations.

An important outcome of this rehabilitation project were four guiding principles which differ slightly from the five and six-step ecological mangrove rehabilitation (EMR) processes (Lewis and Marshall 1997; Lewis 2005, 2009; Brown and Lewis 2006) that appear in most literature.

The guiding principles for ecological rehabilitation in mangroves are:

1   determine why mangroves are not naturally present in a coastal site that should support them;
2   correct the defective conditions or choose another site;
3   refer to local reference mangroves sites and measure local tidal elevations to best understand the normal topography and subtle topographic changes that control intertidal flooding, depth, duration and frequency; and
4   design the restoration to mimic the normal hydrology at a local reference site.

We continue to look at the application of these principles in the following section on the spread of ecological mangrove rehabilitation in three directions; *scaling up* in terms of project size, *scaling out* into new regions and *scaling down* through greater involvement by rural coastal communities.

Scaling up, out and down: the spread of ecological rehabilitation of mangroves

This section begins by looking at two case studies in New World (Western Hemisphere) mangroves at landscape scales, which serve as important analogues for potential future large-scale initiatives worldwide. This section continues by looking at the spread of ecological rehabilitation in Old World (Eastern Hemisphere) mangrove systems, with a sharper focus on community involvement. This section closes by considering the implications of attempting landscape-scale rehabilitation in Indonesia, which would require the application of biophysical lessons learned in the New World with the resolution of social, economic and policy issues prevalent in Indonesia and many other lower-income nations.

Scaling-up in New World mangroves

Florida: mosquito control impoundments (Indian River Lagoon), hydrological rehabilitation – natural regeneration

Throughout the twentieth century, over 16,000 ha of salt marsh and mangrove wetlands in Florida were re-engineered to maintain permanent water inundation in an effort to control mosquito breeding populations (Rey *et al.* 2011). The construction of parallel ditches and dike walls isolated coastal wetlands from adjacent estuaries, effectively controlling mosquito populations, yet unfortunately drastically reducing total coverage of mangroves and salt marsh vegetation.

Research in the 1960s and 1970s showed that flooding during only the summer breeding period was adequate to control mosquito populations, and as a result culverts were placed to reconnect tidal wetlands with adjacent estuaries. The culverts were fitted with flashboard risers in order to prevent excessive water levels (Rey *et al.* 2011). The use of dedicated pumps eliminated the need to over-flood the marsh, and the management protocols changed to flooding for mosquito control only during the summer mosquito production season, and re-establishing tidal exchange through culverts during the rest of the year. This technique is now known as Rotational Impoundment Management (RIM). These methods were responsible for between 9300–10,830 ha of coastal wetland restoration in Florida (*ibid.*).

For impoundments where mosquito control was no longer an issue further steps beyond reconnection via culverts were taken to restore mosquito control impoundments back into mangrove and salt marsh systems. These methods included strategic breaching of dike walls, and later levelling of dike walls and back-filling of perimeter ditches coupled with re-grading back to a surface elevation conducive for salt marsh and mangrove colonization.

In summary, a more natural hydrological regime has been restored to over 940 ha of previously impounded wetlands through the removal of nearly 69 km of dike. Throughout the Indian River Lagoon, the total area restored by all entities is over 1,307 ha. Of the original 16,185 ha of Indian River Lagoon impoundments, over 12,605 ha have been rehabilitated in some manner (i.e. reconnected, breached, or restored; Brockmeyer, unpublished data in Rey *et al.* 2011).

This large-scale initiative can serve as a global demonstration of an adaptive learning approach to restoring coastal wetlands. The work was funded and implemented by a variety of district, state and federal government agencies as well as other stakeholders. The effort restored a full complement of floral and faunal diversity, and ecosystem functions such as carbon sequestration and storage, storm remediation, fisheries and wildlife habitat, and water treatment and storage services.

Cienaga de Grande de Santa Marta: hydrological rehabilitation – natural regeneration

The largest example of ecological-hydrological mangrove rehabilitation in Latin America took place in the Ciénaga Grande de Santa Marta (CGSM) Delta-lagoon complex, Colombia in the mid-1990s (Rivera-Monroy *et al.* 2006, 2011; Botero and Salzwedel 1999). A mangrove die-off of 246 km^2 over a 43 year-period (1956–1999) in a 512 km^2 mangrove forest was caused by the construction of the Barranquilla-Cienaga road and infrastructure in the late 1960s. Rehabilitation cost was estimated at $40 million between 1993 and 2003.

This project focused on the hydraulic reconnection of the Clarín and Aguas Negras channels to the Magdalena River, to reduce interstitial salinities and promote natural revegetation.

Spatial analysis shows significant mangrove recovery measuring 99 km^2 between 1995 and 1999 (Rivera-Monroy *et al.* 2006). The project however, requires routine dredging of restored canals and cannot be considered self-maintaining. Failure to continue dredging beyond 2006 has led to mass die-off of much of the rehabilitated mangroves (Rivera-Monroy *et al.* 2011). This project underscores the importance of hydraulic engineering in large-scale projects, understanding an area's tidal prism and its dynamics, and designing tidal channels to self-maintain.

Scaling out, down and up in eastern hemisphere mangroves

Scaling out and down – the spread of EMR

The spread of EMR took place primarily through training workshops, held both in Florida, and in a variety of countries in Asia, Africa and Latin America including Thailand, Sri Lanka, Indonesia, India, Malaysia, Cambodia, El Salvador, Mexico, Brazil and Senegal. Initially local stakeholders had difficulty moving beyond training to implementation primarily due to the difficulty in resolving land tenure issues required to both undertake rehabilitation and ensure the longer-term protection of mangroves once established. Experience in Indonesia shows how land tenure issues were overcome, allowing for implementation of EMR at 10, 20, 30 and eventually 500 hectares in scale.

After first learning of the EMR process at a workshop in Trang, Thailand, the author returned to North Sulawesi to develop Indonesia's first EMR project in 20 ha of abandoned shrimp ponds in Tiwoho Village, Bunaken National Marine, North Sulawesi. This project was undertaken in partnership with local villagers, students from two elementary schools, the University of Sam Ratulangi (UNSRAT), Yayasan Kelola (local NGO), and the Natural Resource Conservation Agency (BKSDA). The rehabilitation design called for strategic breaching of dike walls, back-filling of artificial tidal channels and enhancement of natural tidal channels. Ten years after EMR, the site is averaging 15,763 stems/ha, representing 21 species of true mangroves including 17-metre-tall *Sonneratia alba* (Lewis and Brown 2014). The community now protects this rehabilitation site through formal village-level regulation and informally, with regular public-address system announcements on the need to protect the mangroves and to use alternate sources of firewood.

The early success of this project led to its replication in other sites in Indonesia primarily on the island of Sumatra (in Aceh, North Sumatra and Riau provinces) after the 2004 tsunami succeeded in raising global awareness of the importance of mangrove ecosystems. These sites ranged in size from 10–33 ha, taking place in disused shrimp ponds and charcoal concession logging areas with significant hydrological disturbance (Brown 2007). A hallmark of these projects was the attention to local community involvement in planning, design, implementation and monitoring in line with findings from Bosirea *et al.* (2008) and Biswas *et al.* (2009). This marked the onset of Community Based Ecological Mangrove Rehabilitation, which was being developed in parallel in Thailand (see www.mangroveactionproject.org/cbemr/blog).

Scaling up: from tens to hundreds of hectares in Indonesia, with pathways to landscape-scale rehabilitation

Shrimp pond development on Tanakeke Island, South Sulawesi, reduced mangrove coverage from 1776 ha to 580 ha from 1985 to 2000. International NGO Blue Forests (formerly MAP-Indonesia) implemented EMR in 480 ha in seven villages on Tanakeke Island with support from CIDA, OXFAM-GB (during the Restoring Coastal Livelihoods Project) and Good

Planet (during the 'Sekali Dayung' Project). Mangrove rehabilitation activities involved strategic breaching of dike walls, tidal creek creation, and routine human-assisted propagule distribution. All three year old sites have surpassed project benchmarks, which include; between 1250 and 3750 seedlings/ha growing healthy relative to reference chronoseres, and formal management agreements for each village (Brown *et al.* 2014; Figure 20.3). Supporting activities included sustainable livelihood development, formation of 'womangrove' management groups and environmental education activities with local schools.

The success of EMR on Tanakeke Island is influencing policy and practice in other locations and at the national level and is discussed next.

Implications for reforming mangrove rehabilitation policy and practice

In this section, we discuss several pathways for adoption of ecological mangrove rehabilitation practices at a variety of scales. We acknowledge that although discrete opportunities for demonstration of ecological mangrove rehabilitation should still be pursued, there is a need to go beyond opportunism and strategically reform mangrove rehabilitation policy and practice. Before discussing ways forward, a look at current, normative planting practices is necessary, to help understand obstacles hindering reform.

The problem with current 'normative' planting practices

We remind readers that evidence suggests that most attempts to rehabilitate mangroves worldwide fail completely, or fail to achieve their stated goals (Lewis 1990, 2005; Erftemeijer and Lewis 2000; Lewis and Brown 2014). The majority of these efforts involve little more than direct planting of mangroves in sub-mean sea level mudflats. This occurred and continues to occur for two main reasons:

1   Poor understanding of the ecological requirements of mangroves, and the processes which lead to their establishment and early growth.
2   Lack of political will to resolve land tenure issues in degraded mangrove areas, where rehabilitation is most warranted.

Two cases from the province of South Sulawesi, Indonesia illustrate the above-mentioned problems.

Poor understanding of ecological requirements of mangroves

A thin mangrove greenbelt (20 m wide) runs 3 km between the villages of Kurri Caddi and Kurri Lompo in Maros, South Sulawesi. This greenbelt, consisting of *Rhizophora apiculata* on the inner portion and *Avicennia alba* on the seaward side, protects 1200 ha of aquaculture ponds inland. Over the past eight years, four mangrove planting projects have been implemented on the seaward side of this greenbelt, involving the building of a 'protective' fence, rearing of *R. apiculata* and *R. mucronata* seedlings, and direct planting 100 m beyond the edge of the greenbelt in a sub-MSL mudflat. All four plantings have experienced total mortality within six months of planting. When interviewed, the forestry department extension agent in charge of site selection stated, 'Mangrove restoration is easy. There are no problems. You only need to find a muddy substrate. There are no problems' (RCL 2013). Each planting event cost approximately US$10,000–25,000, with money that is ear-marked for community development and

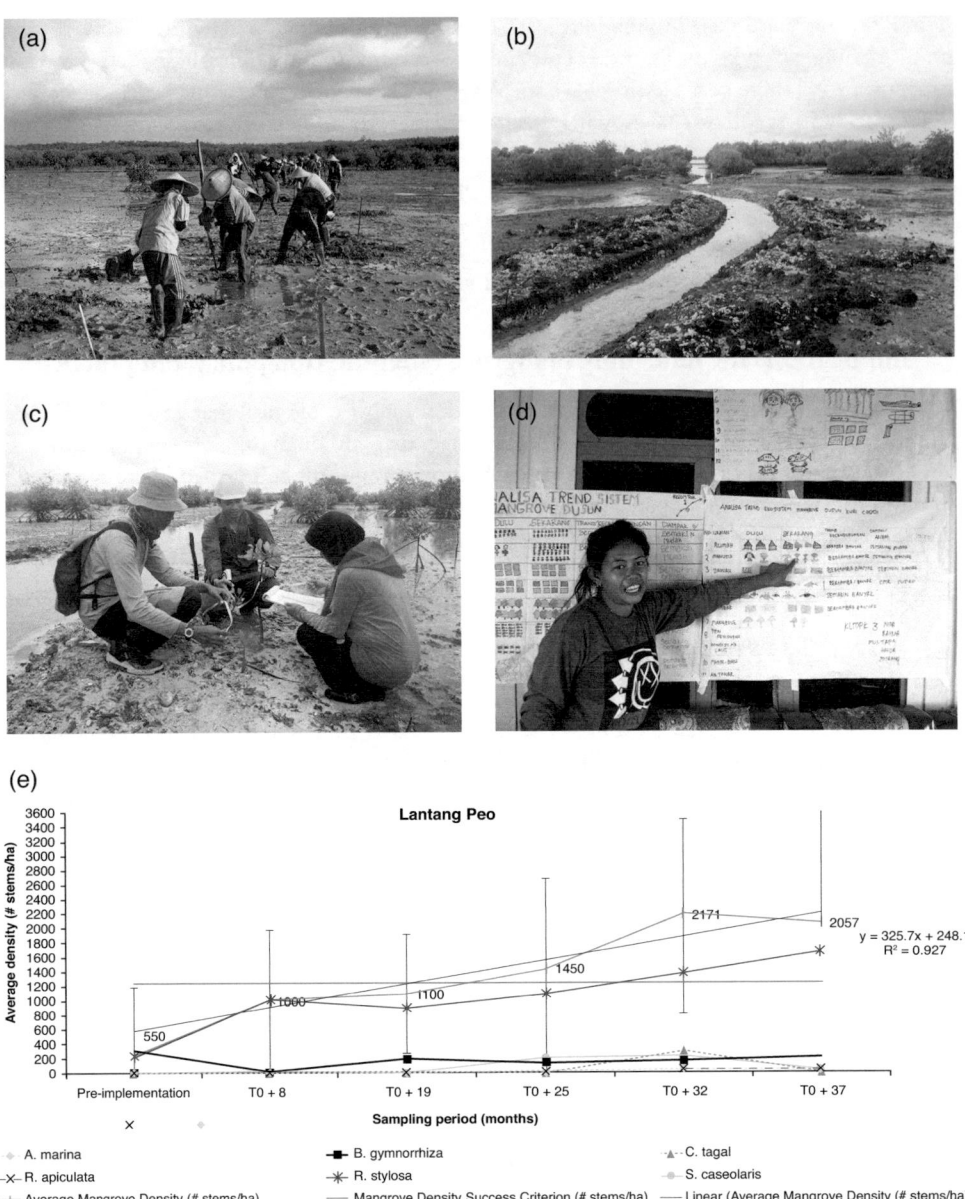

Figure 20.3 (a) Women and men from Lantang Peo Village, Tanakeke Island excavating a tidal creek to facilitate drainage as part of mid-course corrections after monitoring revealed high-instance of water logging. (b) The resultant tidal creek is self-maintaining. (c) A research team from CIFOR, Charles Darwin University, University of Hasannudin and Blue Forests monitor vegetation as well as above and below ground carbon. (d) A member of the Womangrove group presents the results of a trend analysis, to inform future management practices. (e) The graph shows recruitment data for Lantang Peo Village until TZ + 37 months. Mangrove coverage surpassed the benchmark for success (red horizontal line) before two years, and will be monitored until TZ + 60 months

Source: photos and graph courtesy of Blue Forests (Indonesian NGO)

poverty alleviation. This pattern is carried out all over Indonesia and the majority of mangrove nations around the world.

Lack of political will to resolve land tenure issues in degraded mangrove areas

The district of Barru in South Sulawesi on the Makassar Strait is a typical former mangrove area, nearly wholly converted to brackish water aquaculture ponds during the Blue Revolution. The head of the District Environmental Agency, and former head of the Forestry Department discussed the issue of mangrove planting in 2014. For the past ten years, under his leadership, his agencies have received an average of US$100,000 of federal funds per year for mangrove planting. Funds are stipulated to implement four planting projects in the district covering 400 ha planted at one-metre spacing or 4 million seedlings. This district has never attempted to restore mangroves in ex-mangrove areas, which are nearly entirely converted to shrimp ponds in his district. Plantings of *Rhizophora* spp. seedlings take place seaward of a pre-existing green-belt in sub-mean sea level, intertidal mudflats. The department head states that perhaps 0.5–2.0 ha of the 400 ha stipulated for planting will succeed in any given year, although frequently the plantings experience total mortality. Many sites are planted repeatedly. Preparation, planting and monitoring all take place within a fraction of a year. The only change in this practice over the last ten years is that his office has been asked to report odd numbers of planted mangroves and mortality to avoid questioning from the national government, so that 987,013 plants planted with 74 per cent mortality after 92 days is preferred to 1 million plants planted with 50 per cent mortality after three months.

There are 497 districts and municipal capitals in Indonesia, 324 of which are considered coastal (Working Group of Marine and Fisheries Data Arrangement 2011). Extrapolating the findings above from one district to the entire country, a total of $32,400,000 is spent annually on mangrove planting spread over 1296 planting projects with little or no increase in mangrove coverage to show for it. This fails to consider other planting projects from the Ministry of Fisheries, Climate Change Agency, CSR programmes, universities, etc.

A look at the region in general shows that events in Indonesia are likely the rule and not the exception. In neighbouring Philippines the Central Visayas Regional Project I, Nearshore Fisheries Component was provided with US$35 million from the World Bank to undertake 1000 ha of mangrove planting between 1984 and 1992. An evaluation of the success of the planting in 1995–1996 by Silliman University (de Leon and White 1999) showed that only 18.4 per cent of the 2,927,400 mangroves planted over 492 ha had survived. A separate project – funded by ADB loans and the Japanese Overseas Economic Cooperation fund totalling US$150 million called for the planting of 30,000 ha of mangroves. The project was cut short in 1995 after only 4792 ha were planted due to similar problems (Lewis 1999b).

When carried to their extreme, massive planting projects vie for recognition by the Guinness Book of World Records. In 2010, 847,257 mangroves saplings were planted in a single day, breaking an earlier record of 611,000 seedlings planted by India in 2010. This project was part of the Asian Development Bank's five-year multi-million dollar Sindh Coastal Community Development project as an effort to protect the eroding coastline. Although effective at garnering attention, the worth of these efforts needs to be called into question. Within several months 70 per cent mortality was reported for the Pakistan planting, and more recent data has not been made available (Shaikh and Tunio 2013).

In 2014, the IUCN Mangrove Specialist Group designated a working group to investigate the global track record of mangrove planting. Better data are indeed a global priority in order to advocate against this wasteful practice.

Moving forward

Capturing discrete opportunities

Not all future mangrove projects will fit into a strategic plan (one based on reform of normative practice). These discrete opportunities should continue to be taken advantage of, and used to advocate ecological mangrove rehabilitation practices until adoption of a more strategic approach is taken. A current example comes from Indragiri Hilir, Riau, Sumatra, Indonesia where rights to a 34,500 ha mangrove silvaculture concession were rescinded by the then Governor of Riau and transferred to a business friend. The subsequent concession went bankrupt, the Governor sent to jail (on various illegal forestry charges) and the un-managed concession degraded through encroachment and illegal logging. At this point 15,525 ha of the former concession is degraded requiring ecological rehabilitation and enhancement. A variety of government, non-government and industry stakeholders are currently developing a discrete plan looking at the rehabilitation of the forest and re-instatement of the concession, with the consideration that managed silviculture is a better alternative for the region than the current scenario. Rehabilitation of a timber concession is not currently considered under the national strategy, however this should not preclude the pursuance of a potentially beneficial economic and ecological recovery plan.

Strategic reform

As discussed previously, Indonesia's annual government planting programmes have generally not succeeded and in their current form represent a significant budgetary waste. In 2013, the National Mangrove Strategy of 2012 was amended, paying greater attention to assessment, appropriate restoration design, monitoring and maintenance. To move from strategy to implementation, appropriate mechanisms need to be developed with the Directorate General of Watersheds and Social Forests, the lead restoration agency within the Ministry of Forestry. This agency currently mandates that all mangrove rehabilitation activities be implemented and reported on within an annual timeframe. An alternative is to stagger rehabilitation into three phases:

(I)   assessment, appraisal and design;
(II) implementation; and
(III) monitoring and maintenance.

Using a staggered approach, although only one project phase is implemented each year at any given site, the same total number of projects will run concurrently over any given period (Figure 20.4) with a greater likelihood of success.

Such a strategy would need to be coupled with training of forestry extensionists in the methods of EMR requiring capacity building for the forestry training agency. Trained extensionists could then work with communities to develop mangrove rehabilitation plans through mechanisms like Integrated Forest Management Units (KPH), which maintain a national mandate to develop community forest management plans, including rehabilitation plans, at the village level. KPH is presently the Ministry of Forestry's primary vehicle for community-based forestry management (Nugroho et al. 2015).

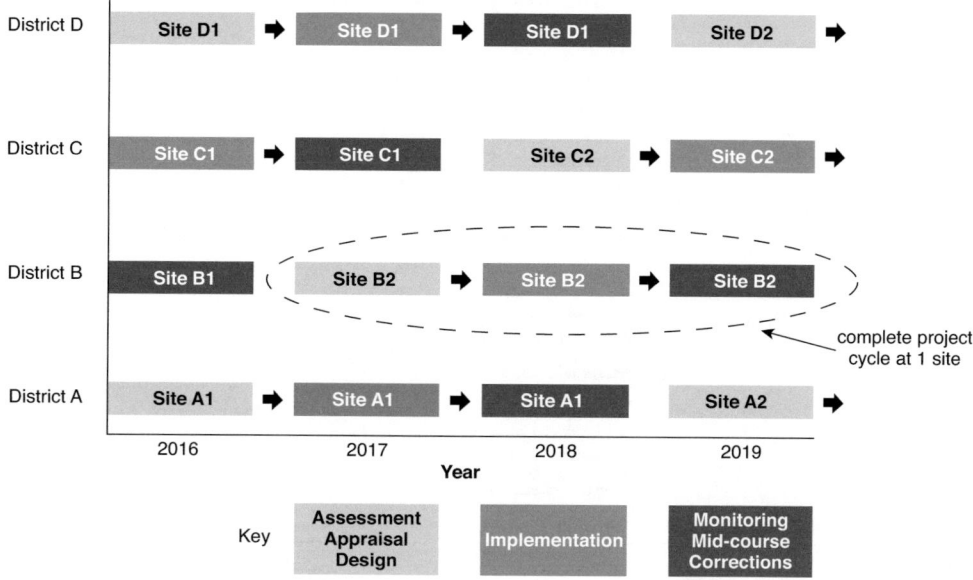

Figure 20.4 A staggered strategic approach is recommended in order to integrate ecological mangrove rehabilitation with current normative annual government planting practices

Landscape-scale assessment, site prioritization and integration

Ecological mangrove rehabilitation at the landscape scale has been attempted in the New World with examples in Florida and Columbia experiencing mixed degrees of success. The tools, skills and knowledge already exist to achieve this work on the biophysical front. The next requisite step to achieve large-scale rehabilitation in new regions are feasibility assessments, evaluating not only the biophysical feasibility of a site for recovery, but the degree to which social, political, cultural and economic factors may enable or hinder large-scale restoration.

In 2015, with institutional support from the Indonesian Ministry of Marine Affairs and Fisheries (KKP) and financial support from World Resources Institute, USAID, Green Forest Products and Dutch Sustainable Water Fund, eight rapid feasibility assessments were undertaken to gauge the potential for large-scale ecological mangrove restoration and improved collaborative management. Conceptual models (problem and objective trees; Figure 20.5) were developed for all sites followed by options analysis, rating the feasibility of identified activities and outcomes against eight criteria (Table 20.1), with higher average ratings indicating lower risk of intervention, and higher overall scores indicating a high degree of project complexity and also potential impact.

In terms of financial resources for large-scale rehabilitation, although the majority of Indonesia's mangrove restoration budget may still be allocated towards annual/routine mangrove restoration projects in coastal districts, only 5–10 per cent of the US$30 million annual mangrove planting budget would be needed to continuously implement large-scale initiatives in prioritized sites at a rate of approximately 1000 ha/yr. Government commitment

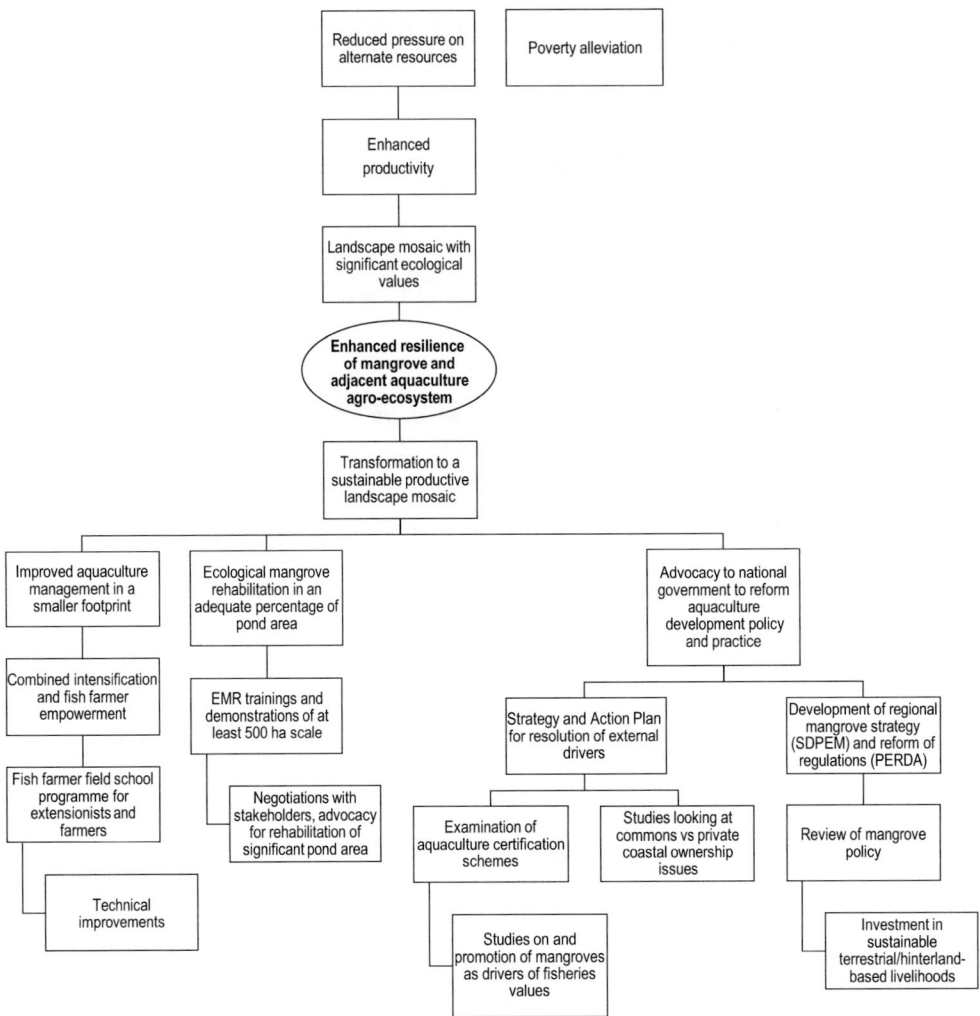

Figure 20.5　Part of an objective tree relevant to mangrove rehabilitation developed for a degrading mangrove area in Rawa Aopa Waumohai National Park, Southeast Sulawesi, Indonesia. Derived from analysis of a previously constructed problem tree, the objectives analysis depicts the ultimate programme goal in the centre of the diagram (elliptical), with the layers below derived from root causes of the main issue (degradation) and ordered in lower level purposes, objectives and activities at the bottom. Boxes above depict resolution of effects which were identified prior during problem identification

Source: USAID-IFACS (2015)

to large-scale rehabilitation will undoubtedly be matched by bi-lateral agencies, donor organizations and corporate social responsibility (CSR), enabling ecological rehabilitation to be integrated with sustainable livelihoods development, environmental education and improved governance, a strategy which has formally been included in Guiding Principles for Coastal Carbon Projects (UNEP and CIFOR 2014).

Table 20.1 Analysis for four mangrove landscapes in Indonesia

		C1	C2	C3	C4	C5	C6	C7	C8	Avg rating	Cumulative score (all activities)
Rating	1	Low	High	High	Low	None	Low	Low	Low		
	3	Med	Med	Med	Med	1	Med	Med	Med		
	5	High	Low	Low	High	>1	High	High	High		
Site A		4.13	2.57	3.80	3.80	4.73	4.43	4.17	4.07	3.97	951
Site B		4.75	3.05	3.55	3.75	5.10	5.05	4.45	5.10	3.91	696
Site C		4.46	3.46	3.69	3.69	3.92	4.00	4.15	4.46	3.67	414
Site D		3.83	2.75	4.41	4.08	4.50	4.17	4.25	3.25	3.90	375

Notes: Site A: Kubu Raya, West Kalimantan; Site B: Mahakam Delta, East Kalimantan; Site C: Rawa Aopa Watumohai, Southeast Sulawesi; Site D: Bintuni Bay, West Papua.

Scoring criteria: C1: expected conservation benefit; C2: cost; C3: social risk; C4: likelihood of success; C5: potential funding/sources; C6: available personnel; C7: experience with entailed methodologies; C8: development benefits to priority groups.

The options table above shows that Site A, Kubu Raya, has a high average rating, meaning relatively low degree of risk and also a high cumulative score (determined through ranking all project activities and outcomes against the criteria), which indicates intervention will be complex but have high potential impact. The lowest average rating (Site C, Rawa Aopa Watumohai) means project intervention carries a high risk with moderate or low impact based on the cumulative score. Although risk in Bintuni is low, potential impact of an intervention is also low, which may dissuade donors or government from supporting a programme intervention.

Reflection

To maintain current global mangrove coverage, 100,000–150,000 ha of rehabilitation would need to take place each year to keep up with rates of destruction and degradation. Clearly, mangrove rehabilitation practitioners, working with a limited pool of resources, cannot afford to proceed without improved efficiency and efficacy. The technical skills to rehabilitate whole mangrove ecosystems are available, largely relying on hydrological restoration to drive ecological recovery. Where propagules are limited, periodic propagule distribution can ensure recovery. In terms of cost, strategic breaching of dike walls in the world's vast disused and abandoned shrimp ponds may be the most efficient form of rehabilitation, while excavation, fill and finally coastal protection projects become increasingly expensive, yet necessary depending on biophysical conditions. Afforestation attempts in sub-mean sea level mudflats is clearly a dead-end and increased advocacy efforts to stop this wasteful practice are required to help shift limited resources to working solutions. Social, economic and policy issues will certainly differ from country to country, yet in general the integration of rehabilitation with management for conservation and sustainable utilization seem the clear path forward, with a high degree of participation from coastal communities in both developing and higher economic nations. Investment in the understanding of natural adaptive capacity and management so that mangrove forests do not degrade to the point of crossing thresholds which make rehabilitation exponentially more resource intensive is strongly recommended. Adaptive-collaborative management bodies should be supported to track key variables which determine the degree of resilience of a mangrove forest, and to manage their systems to maintain high total economic values, while remaining at a distance from such thresholds. This paper did not get a chance to look into future directions for ecological mangrove rehabilitation in terms of practice and research, which will need to be addressed by other authors. Demand-driven areas for research and development include identification of stressed mangroves and 'pre-emptive' hydrological rehabilitation, development of a threshold database to assist managers in avoiding serious degradation, provision of technical expertise to nations considering large-scale ecological rehabilitation, and integration of EMR with soft-engineered coastal protection measures in the face of coastal erosion, subsidence and sea level rise.

References

Biswas, S. R., A. U. Mallik, J. K. Choudhury and A. Nishat. (2009) A unified framework for the restoration of Southeast Asian mangroves – bridging ecology, society and economics. *Wetlands Ecology and Management* 17(4): 365–383.

Bosirea, J. O., F. Dahdouh-Guebasb, M. Walton, B. I. Crona, R. R. Lewis III, C. Field, J. G. Kairoa and N. Koedam. (2008) Functionality of restored mangroves: a review. *Aquatic Botany* 89(2): 251–259.

Botero, L. and H. Salzwedel. (1999) Rehabilitation of the Cienaga Grande de Santa Marta, a mangrove estuarine system in the Caribbean coast of Colombia. *Ocean Coastal and Management* 42: 243–256.

Bozzano, M., R. Jalonen, E. Thomas, D. Boshier, L. Gallo, S. Cavers, S. Bordacs, P. Smith and J. Loo. (eds) (2014) *Genetic Considerations in Ecosystem Restoration Using Native Tree Species: The State of the World's Forest Genetic Resources – Thematic Study*. Rome: FAO.

Broome, S. W., S. M. Rogers Jr. and E. D. Seneca.(1992) *Shoreline Erosion Control Using Marsh Vegetation and Low-Cost Structures*. UNC-SG-92-12. Raleigh, NC: North Carolina Sea Grant.

Brown, B. (2007) *Resilience Thinking Applied to the Mangroves of Indonesia*. Yogyakarta, Indonesia: IUCN & MAP-Indonesia.

Brown, B. and R. R. Lewis. (2006) *Five Steps to Successful Ecological Restoration of Mangroves*. Yogyakarta, Indonesia: Yayasan Akar Rumput Laut (YARL) and the Mangrove Action Project.

Brown, B., R. Fadillah, Y. Nurdin, I. Soulsby and R. Ahmad. (2014) Community Based Ecological Mangrove Rehabilitation (CBEMR) in Indonesia – from small (12–33 ha) to medium scales (400 ha) with pathways for adoption at larger scales (>5000 ha). *SAPIENS* 7(2).

De Leon, R. O. D. and White A. T. (1999) Mangrove rehabilitation in the Philippines. Pages 37–42 in W. Streever (ed.), *An International Perspective on Wetland Rehabilitation*. Dordrecht, the Netherlands: Kluwer.

Ellison, A. M. (2000) Mangrove restoration: do we know enough? *Restoration Ecology* 8(3): 219–229.

Erftemeijer, P. L. A. and R. R. Lewis III. (2000) Planting mangroves on intertidal mudflats: habitat restoration or habitat conversion? Pages 156–165 in *Proceedings of the ECOTONE VIII Seminar 'Enhancing Coastal Ecosystems Restoration for the 21st Century', Ranong, Thailand, 23–28 May 1999*. Bangkok: Royal Forest Department of Thailand.

FAO. (2003) *Status and Trends in Mangrove Area Extent Worldwide*. Forest Resources Assessment Working Paper 63. Rome: Food and Agriculture Organization of the United Nations.

Field, C. D. (1999) Rehabilitation of mangrove ecosystems: an overview. *Marine Pollution Bulletin* 37(8–12): 383–392.

Giri, C., E. Ochieng, L. Tiezen, Z. Zhu, A. Singh, T. Loveland, J. Masek and N. Duke. (2011) Status and distribution of mangrove forests of the world using earth observation satellite data. *Global Ecology and Biogeography* 20: 154–159.

IUCN-MSG. (2015) *Mangrove Specialist Group Newsletter*. Issue 1, May 22.

Lewis, R. R. (1990) Creation and restoration of coastal plains wetlands in Florida. In J. A. Kusler and M. E. Kentula (eds), *Creation and Restoration: The Status of the Science*. Washington, DC: Island Press.

Lewis, R. R. (1999a) *Time Zero Report for the Cross Bayou Mangrove Restoration Site, Pinellas County, Florida, USA*. Prepared for the Cross Bayou Project Review Group. 15 November. Tampa, FL: Florida Department of Environmental Protection.

Lewis, R. R. (1999b) Key concepts in successful ecological restoration of mangrove forests. Pages 12–32 in *Proceedings of the TCE-Workshop No. II, Coastal Environmental Improvement in Mangrove/Wetland Ecosystems, 18–23 August 1998, Ranong, Thailand*. Bangkok: Danish-SE Asian Collaboration in Tropical Coastal Ecosystems (TCE) Research and Training.

Lewis III, R. R. (2004) *Time Zero Plus 60 Months Report for the Cross Bayou Mangrove Restoration Site, Pinellas County, Florida*. Prepared for the Cross Bayou Project Review Group. 1 December. Tampa, FL: Florida Department of Environmental Protection.

Lewis, R. R. (2005) Ecological engineering for successful management and restoration of mangrove forests. *Ecological Engineering* 24: 403–418. doi:10.1016/j.ecoleng.2004.10.003

Lewis III, R. R. (2009) Methods and criteria for successful mangrove forest restoration. Pages 787–800 in G. M. E. Perillo, E. Wolanski, D. R. Cahoon and M. M. Brinson (eds), *Coastal Wetlands: An Integrated Ecosystem Approach*. Oxford: Elsevier Press.

Lewis III, R. R. (2011) Mangrove forest ecology, management and restoration. PowerPoint presentation, Restoring Coastal Livelihoods – EMR Seminar, Makassar, Indonesia.

Lewis III, R. R. and B. Brown. (2014) *Ecological Mangrove Rehabilitation: A Field Manual for Practitioners*. Mangrove Action Project–Indonesia, Canadian International Development Agency and Oxfam.

Lewis, R. R. and R. G. Gilmore. (2007) Important considerations to achieve successful mangrove forest restoration with optimum fish habitat. *Bulletin of Marine Science* 80: 823–837.

Lewis, R. R. and M. J. Marshall. (1997) *Principles of Successful Restoration of Shrimp Aquaculture Ponds Back to Mangrove Forests*. Havana, Cuba: Programa/resumes de Marcuba '97.

Lewis III, R. R., A. B. Hodgson and G. S. Mauseth. (2005) Project facilitates the natural reseeding of mangrove forests (Florida). *Ecological Restoration* 23(4): 276–277.

MacAdam, G., R. R. Lewis and M. B. Bay. (1998) Restoring West Lake: fish, shorebird habitat, tidal pools and mangroves restored in a 1500 acre mangrove preserve within densely populated Broward County, Florida. Pages 36–37 in P. J. Cannizzaro (ed.), *Proceedings of the 24th Annual Conference on Ecosystems Restoration and Creation*. Tampa, FL: Hillsborough Community College.

McKee, K. and P. Faulkner. (2000). Restoration of biogeochemical function in mangrove forests. *Restoration Ecology* 8: 247–259.

Mazda, Y., E. Wolanski and P. Ridd. (2007) *The Role of Physical Processes in Mangrove Environments: Manual for the Preservation and Utilization of Mangrove Ecosystems*. Tokyo: Terra Pub.

Nugroho, B., H. Kartodihardjo and D. R. Nurrochmat, Julijanti. (2015) *Formulasi Strategi Komunikasi Kebijakan Kehutanan: Kasus Pembangunan Kesatuan Pengelolaan Hutan [Formulation of Strategic Communication on Forestry Policy and the Development of Integrated Forestry Management Units]*. Bogor, Indonesia: IPB (Bogor Agricultural University).

Primavera, J. H. and J. M. A. Esteban. (2008) A review of mangrove rehabilitation in the Philippines: successes, failures and future prospects. *Wetlands Ecology and Management* 16(3): 173–253.

RCL. (2013) *Mangrove Journal*. Participatory video project. Makassar, Sulawesi, Indonesia: Restoring Coastal Livelihoods. See www.rcl.or.id.

Rey, J. R., D. B. Carlson and R. E. Brockmeyer Jr. (2011) Coastal wetland management in Florida: environmental concerns and human health. *Wetlands Ecology Management*, 20 November.

Rivera-Monroy, V. H., R. R. Twilley, E. Mancera, A. Alcantara-Eguren, E. Castañeda-Moya, O. Casas-Monroy, P. Reyes-Forero, J. Restrepom L. Perdomo, E. Campos, G. Cotes and E. Viloria. (2006) Adventures and misfortunes in Macondo: rehabilitation of the Cienaga Grande de Santa Marta Lagoon complex, Colombia. *Ecotropicos* 19(2): 72–93.

Rivera-Monroy, V. H., R. R. Twilley, J. E. Mancera-Pineda, C. J. Madden, A. Alcantara-Eguren, E. B. Moser, B. F. Jonsson, E. Castañeda-Moya, O. Casas-Monroy, P. Reyes-Forero and J. Restrepo. (2011) Salinity and chlorophyll a as performance measures to rehabilitate a mangrove-dominated deltaic coastal region: the Ciénaga Grande de Santa Marta–Pajarales Lagoon Complex, Colombia. *Estuaries and Coasts* 34: 1–19.

Shaikh, S. and S. Tunio. (2013) Pakistan's mangrove restoration efforts called into question. Retrieved from http://news.trust.org/item/20131203123452-q63ki.

Tomlinson, P. B. (1986) *The Botany of Mangroves*. Cambridge: Cambridge University Press.

UNEP and CIFOR. (2014) *Guiding Principles for Delivering Coastal Wetland Carbon Projects*. Nairobi: United Nations Environment Programme/Bogor: Center for International Forestry Research.

USAID-IFACS. (2015) *Rapid Feasibility Assessments Potential of Public-Private Partnerships to Drive Social, Economic and Ecological Recovery in Four Regionally Important Indonesian Mangrove Systems*. Jakarta, Indonesia: Blue Forests.

Walker, B., C. S. Holling, S. R. Carpenter and A. Kinzig. (2004) Resilience, adaptability and transformability in social–ecological systems. *Ecology and Society* 9(2): article 5. Retrieved from www.ecologyandsociety.org/vol9/iss2/art5.

Working Group of Marine and Fisheries Data Arrangement. (2011) *Marine and Fisheries Figures*. Indonesia: Center of Data, Statistics and Information, Working Group of Marine and Fisheries Data Arrangement.

21

TROPICAL SAVANNA RESTORATION

Jillianne Segura, Sean M. Bellairs and Lindsey B. Hutley

Introduction

Tropical savanna is defined as having a discontinuous C_3 woody perennial overstorey and a near-continuous herbaceous and grassy C_4 understorey. Savannas are by definition distinct from grassland (absence of or limited woody plant cover) and forest ecosystems (absence of or limited grass cover). This ecosystem occurs on nearly all major land masses within the tropical and subtropical belt around 30° north and south of the equator (Hutley and Setterfield 2008). Savannas cover 20 per cent of Earth's continental land mass and account for 30 per cent of global net primary productivity. Tropical savanna is the second largest tropical ecosystem after rainforests and occurs across the seasonal tropics of Australia, Africa and South America (Figure 21.1) with a distribution of greater than 27.6 million km² (Hutley and Setterfield 2008). The two largest uninterrupted savanna landscapes are the Australian savanna and the South American *cerrado/llanos* of Brazil, Colombia and Venezuela.

Due to their size, savannas affect global carbon, nutrient and water cycles and their frequent fires significantly influence atmospheric chemistry (van der Werf *et al.* 2010). The overarching determinants of savanna physiognomy (relative abundance of the tree and grass layer) are climate and soil type. Soil type determines availability of water and nutrients which determines the potential growth and survival of trees and grasses at a given site. Growth potential is moderated by disturbance agents, including fire, herbivory and stochastic disturbance events (Sankaran *et al.* 2005).

This chapter will examine different savanna degradation issues across the three tropical savanna regions: Africa, South America and Australia. This includes impacts from cattle grazing, clearing for cropping, alien species invasion and mining. Management and restoration efforts to reverse land degradation are described with examples from across the three continents. We identify particular challenges for restoring tropical savanna and discuss potential impacts of climate change.

Tropical savanna restoration: rehabilitation or management?

Ecological restoration has been defined as 'the process of repairing damage caused by humans to the diversity and dynamics of indigenous ecosystems' (Jackson *et al.* 1995). When reviewing

Figure 21.1   (a) *Eucalyptus*-dominated Australian tropical savanna in Howard Springs at 1100 mm mean annual rainfall. (b) South American savanna in Minas Gerais, Brazil. (c) *Pterocarpus*-dominated tropical savanna in Khaudum National Park, Namibia at 550 mm mean annual rainfall

Source: photos courtesy of (a) Lindsay Hutley, (b) Stephen Reynolds and (c) Michael Lawes

literature on the degradation of tropical savanna and efforts to restore this ecosystem, it becomes clear that there are few examples of managed restoration to a pre-existing high biodiversity state. The majority of projects focus on rehabilitation or management strategies to return the system to a desirable state. Rehabilitation works are often focused on returning degraded/transformed lands to savanna for profitability rather than restoring savanna biodiversity. Some mining projects aim to restore some degree of the original biodiversity rather than creating profitable landscapes.

Tropical savanna is often perceived as being 'unproductive' vegetation without aesthetic appeal and thus it is seen as lacking the 'importance' of other ecosystems such as tropical rainforest. In South America, government policy dictates that on any given property within the Amazon basin, 80 per cent of that property must be conserved as rainforest, however, in the dominant South American savanna type the *cerrado,* this fraction drops to 20 per cent. This is despite South American savanna being recognized for the highest floral biodiversity of any savanna (Figure 21.2; Klink and Machado 2005).

Globally, cattle grazing is a major land use in tropical savanna as the grassy understorey provides fodder. Both subsistence farming in native pasture and commercial grazing enterprises using native or improved pasture occur in tropical savanna systems. In many savanna regions, over-grazing and a reduction in fire frequency has led to increased woody growth and reduced

Figure 21.2 Comparison of (a) undisturbed South American cerrado with (b–d) areas of cerrado that have been (b) degraded through over-grazing, (c) cleared for eucalypt plantations and (d) cleared up for mining

Source: photos courtesy of Stephen Reynolds

herbaceous growth. Rehabilitation of these ecosystems to maintain grazing productivity focuses on reducing woody thickening or bush encroachment with fire being the primary tool.

Tropical savannas have been managed for thousands of years by indigenous communities using fire, thus these systems are adapted to fire. As such, an argument could be made that savannas aren't being rehabilitated so much as managed. With this in mind, the determination of whether restoration of tropical savanna is rehabilitation or management lies in its motives. These motivations will be dictated by the particular land use and value placed on native tropical savanna for this land use.

Land uses and degradation

Subsistence grazing in African and South American savannas is a tradition which spans thousands of years. These traditional practices continue in underdeveloped areas along with commercial cattle production in more developed areas. Commercial cattle production is the largest land use across African, South American and Australian tropical savanna and has had a wide range of impacts on tropical savanna. In Africa woody thickening or 'bush encroachment' is occurring as a result of over-grazing, fundamentally altering the tree:grass ratio of the savanna, leading towards a more densely wooded savanna or dry forest. Commercial cattle production in South America and Australia resulted in the introduction of improved exotic pasture which may have incorporated native species with exotic species or excluded natives altogether. African grass species such as *Andropogon gayanus* Kunth. (gamba grass) or *Melinis minutiflora* Beauv. (molasses grass) were introduced to improve pasture productivity for cattle grazing (Hoffmann *et al.* 2004; Klink and Machado 2005). These species are rapidly invading surrounding savanna where they are quickly outcompeting native species and creating hotter fires, which native grass plants and their seedbanks cannot withstand. Additionally, gamba grass can grow up to 4 m tall and is thus capable of carrying fire up into the tree canopy and increasing tree mortality thus leading to further degradation of savanna structure and composition.

Growing crops is difficult in tropical savanna due to the high seasonality of rainfall, variability in the onset of the wet season and the nutrient-poor soils, which sometimes have aluminium toxicity. However, since the 1970s areas of *cerrado* that were previously used for low-impact cattle grazing have been converted to intensive cropping (Klink and Machado 2005) with approximately 100 000 km² of *cerrado* converted to soy-bean cropping. No-till or direct-seeding with mulching have been adopted to reduce soil degradation and soil erosion (Evers *et al.* 2001). There are few examples of restoring row-crop land to tropical savanna despite 45 000 km² of land lying fallow and contributing to significant soil erosion in South America (Klink and Machado 2005).

Harvesting of native savanna trees and shrubs for fuel wood and charcoal is common in Africa and South America. Over 90 per cent of rural households in South Africa use fuel wood from savanna for cooking and heating (Shackleton *et al.* 2002), a practice that has led to land degradation, reduction of tree/shrub populations and the dominance of coppice species in African savanna (Govender *et al.* 2006). In South America a similar practice of using woody savanna trees for charcoal production for Brazil's steel industry constitutes the second greatest pressure on *cerrado* ecosystems after agriculture. Farmers clearing land for agricultural use and selling the timber were the main charcoal suppliers for the industry however, extensive eucalypt plantations now supplement ever-growing demand (Figure 21.2; Mistry 2000).

Mining in tropical savanna occupies a smaller land area than agriculture or conservation parks, however the level of restoration required is greater. There are many large gold, bauxite, uranium, copper, manganese and aluminium mines within tropical savanna regions of Australia. In the South American *cerrado* there is also a substantial amount of mining of gold, silver, limestone, copper, iron, manganese, nickel, beryllium and uranium (Mistry 2000). Mined areas need significant restoration efforts to recreate suitable topography, hydrology, soil health and vegetation, a major challenge given the area and severity of impact from this land use.

Conservation management in tropical savanna includes the iconic Kruger National Park (19,485 km²) in South Africa, which is among the oldest national parks in the world. Established in 1898 to protect game animals and preserve their habitat, Kruger National Park lies at the junction between South Africa, Swaziland and Mozambique. In the late 1960s in Kruger and other game parks, the impact of increased elephant populations (due to exclusion of poaching in these areas) on tree and shrub populations were observed and an ecological carrying capacity was set by park managers. Since this time, there have been further declines in the abundance of large trees. Adaptive management techniques seek to maintain healthy elephant populations without adverse effects to the savanna ecosystem (Owen-Smith *et al.* 2006). Furthermore, fencing of protected areas and private property (particularly pastoral land) have led to habitat fragmentation and the increasing isolation of protected areas, impacting migration patterns of large mammals and causing a reduction in their numbers (Newmark 2008).

In northern Australia, substantial areas of tropical savanna are conserved as national parks such as the World Heritage listed Kakadu National Park (19,804 km²) and Judarra/Gregory

National Park (12,800 km²). These savanna ecosystems feature intact woody cover, however they are experiencing rapid declines in small mammal populations, due in part to the migration of feral animals such as cats as well as more frequent fire regimes (Lawes *et al.* 2015). In addition, the toxic cane toad, *Rhinella marina* Linnaeus (formerly *Bufo marinus*), imported into Australia from America, has since expanded its range across savanna environments from Queensland into the Northern Territory and Western Australia (O'Donnell *et al.* 2010; Shine 2010; Urban *et al.* 2008). Restoration of critical habitat and feral animal control programmes will be key requirements for restoring the native fauna in Australian savanna (Whitehead *et al.* 2005).

Restoration, rehabilitation and management in practice

Pastoral and conservation land management

As noted, the cattle grazing industries across Africa, Australia and South America have contributed to the degradation of tropical savanna through their grazing patterns and the introduction of African grass species to improve pasture. Over-grazing of African savanna in particular has led to large areas experiencing bush encroachment from woody thickening with shrubs as opposed to the larger tree species (O'Connor *et al.* 2014). Conservation areas in Australia and South America are experiencing similar problems with grass weeds. In Africa conservation areas are being degraded through over-browsing by elephants. Here we discuss methods of managing bush encroachment, returning savanna structure after over-grazing and over-browsing, and methods for controlling weeds. In addition, this section will address methods to restore critical habitat for small mammal populations threatened by feral animals in northern Australia.

Bush encroachment following grazing

Intensive cattle grazing in savanna has contributed to large areas of woody thickening, often termed as bush encroachment (Govender *et al.* 2006; Hutley and Setterfield 2008; O'Connor *et al.* 2014). Bush encroachment of African savannas has occurred due to: the loss of megafauna populations due to hunting (pre-1900s), changed fire regimes to controlled and regular prescribed burns, through over-grazing by cattle and browsing by elephants (Chamaillé Jammes *et al.* 2007; Gil-Romera *et al.* 2010; Govender *et al.* 2006; Mistry 2000).

Intense fires can be used effectively in early stages of bush encroachment to remove woody vegetation (Mistry 2000). High intensity fires are more easily achieved in Australia savanna where mean annual late dry season burns achieve intensities of 7700 kW m⁻¹ (Andersen *et al.* 2003). In contrast, four to five years of fuel accumulation are required to achieve an average fire intensity of approximately 2300 kW m⁻¹ in Kruger National Park, South Africa. The timing of the burn (i.e. seasonality) is important to achieve high intensity fires, as fuel curing and accumulation is maximal by the late dry season resulting in high intensity fire (Govender *et al.* 2006). However, high intensity fires have been shown to increase multi-stem coppicing in African savanna, thus contributing to increased tree dominance. Fire is less effective in later stages of bush encroachment when 'bush clumps' of trees and shrubs develop (Figure 21.3). These clumps are often dominated by non-flammable species and grasses are out competed due to light competition, and the lower grass biomass reduces the fuel for fire (Mistry 2000). Govender *et al.* (2006) suggest that heterogeneous fire regimes be employed to maintain healthy savanna and reduce the dominance of low multi-stemmed trees and thus the dominance of woody plants in African savannas.

Figure 21.3   Examples of bush encroachment at varying stages in Queen Elizabeth Park, Uganda.
(a) The beginning of bush encroachment after over-grazing with bushy patches of mostly
weedy shrubs establishing under native tall *Acacia* trees. (b,c) Further encroachment and
the increasing dominance of bushy patches as fire become less effective due to increasing
dominance of non-flammable species. (d) They continue to thicken to an extent that no
grassy patches remain as observed in panel

Source: photos courtesy of Michael Lawes

Savanna structural decline following African elephant over-browsing

The African elephant (*Loxodonta africana* Blumenbach) has been referred to as an 'ecosystem
engineer' because when it is present at high population densities it has a significant effect on
vegetation structure (Chamaillé Jammes *et al.* 2007; Guldemond and Aarde 2008). This is
becoming a problem within African conservation areas where high levels of degradation are
evident due to their non-specific browsing patterns. Unlike cattle that only graze palatable
grasses, elephants will browse both grasses and soft tree leaves. This is pushing woodlands to
grasslands and sometimes to desolate land with this extreme more prevalent in fenced areas
(Guldemond and Aarde 2008). As culling of elephants is a controversial issue, alternative
management techniques are being used. A particularly effective tool is the management of artificial water supplies to contract or direct the spatial distribution of elephants. This is not
without its own problems however as the daily range of elephants is quite large such that
distance between artificial water sources becomes negligible (Chamaillé Jammes *et al.* 2007).

Weed control

Control of weeds in tropical savanna is a restoration issue as invasive plant species are contributing to declines in savanna health and structure. Australia and South America are suffering from African grass invasions whereas African savannas are exhibiting problems with Australian *Acacia* species. In both circumstances, the invasion by these plant species is altering the tree: grass ratio (Figure 21.4) and restoration of tropical savanna requires control by land managers.

Across tropical savannas in Australia and South America, the introduction of African grasses for improved pasture has led to severe ecosystem degradation. In South America, the African grass species molasses grass has been spreading in the *cerrado* region, causing widespread savanna degradation. Molasses grass can form dense mats which suppress native grass growth and tree recruitment. It also contributes to hotter fires with enhanced flame height that extend well into the tree canopy (Hoffmann *et al.* 2004; Klink and Machado 2005). These dense mats also mean that fires have a longer residence time which presents a significant threat to soil fauna and burrowing fauna species such as the capybara (Klink and Machado 2005).

Gamba grass is a highly productive perennial African grass which is impacting both Australian savanna and South American *cerrado* systems. Gamba grass is spreading from pastoral land into northern Australian conservation areas. As with molasses grass mats in the *cerrado*, high fuel production and fire severity of the invader is resulting in high rates of tree mortality (Setterfield *et al.* 2010), with low diversity, mono-specific gamba grasslands the end result of invasion (Brooks *et al.* 2010) (Figure 21.4).

Control options for these exotic grass infestations will vary according to the level of infestation present and preferred methods. In Australia, smaller infestations could be controlled using buffalo. Buffalo were introduced to northern Australia in the early 1800s as a source of food and hides for the original settlements, however when the settlements were abandoned the buffalo remained and over the next 50–60 years spread rapidly in the Alligator Rivers Region. Buffalo are now considered a pest within national parks and have led to widespread compaction and erosion of the soil, however they are perhaps the only mammal already present in Australian savanna that is capable of eating the dry, sandpaper-like leaves of gamba grass in the dry season (Petty *et al.* 2007; Alford and Schmid 2014).

Figure 21.4   Comparison of (a) healthy native tropical savanna in Australia with (b) an area of savanna heavily infested with the alien grass *Andropogon gayanus* (gamba grass) with few trees remaining

Source: photos courtesy of Natalie Rossiter-Rachor

Areas within South American and Australian savannas that have been heavily infested with African grasses to the exclusion of natives are currently abandoned as there are limited options for control and restoration. These larger infestations may require the use of fire and herbicides in order to remove the living tissue and void their seed banks (Alford and Schmid 2014). These activities will also result in the reduction of any remnant native savanna vegetation so there will be a need to introduce native trees and grasses after weed control.

Feral animal control and critical habitat restoration for Australian small mammals

In visually intact Australian savanna, there has been a dramatic decline in small native mammal populations, even in national parks such as the world heritage area Kakadu National Park (Woinarski *et al.* 2011). Twenty-two tropical marsupial species, or 30 per cent of the total, have declined since ~1970 and for two species, the decline in population size is greater than 90 per cent (Fisher *et al.* 2014). The Northern Quoll (*Dasyurus hallucatus* Gould) was once widespread across northern Australia savanna from Queensland to the Kimberley of Western Australia but regional-scale decline has been observed throughout its range and this species is now critically endangered (Woinarski *et al.* 2007). Restoration efforts are particularly focusing on the management of fire regimes, adaptation to invasive cane toads and control of introduced predators (Hill and Ward 2010). Foxes and particularly cats directly prey on the Northern Quoll and compete with it for food. There is an interaction with fire such that quolls are more vulnerable to predation after fire, due to removal of cover. Quoll populations in rocky upslope areas with more protection from predators and from fire tend to decline less than populations in flat woodland areas (Woinarski *et al.* 2011).

There are major projects implementing changes to Australian savanna fire regimes at broad landscape scales which are contributing to the improved management and restoration of healthy savanna. Methods include changing from hot fires late in the dry season to small cooler fires in the early dry season, restoring mosaic burning patterns and combining traditional and western knowledge and workforces to achieve better fire management and savanna restoration (Jacklyn and Cook 2009; Yibarbuk *et al.* 2001). Restoration of small mammal habitat requires less frequent and finer-scale burning, along with the protection of some large, infrequently burnt source areas (Andersen *et al.* 2012).

Another major threat to native fauna of Australia savanna arises from the spread of *Rhinella marina* (cane toads). The cane toad was introduced into Queensland and has spread west across north Australian savanna. This species contains strong toxins both as an adult and as a tadpole and are considered the main threat to Northern Quoll populations in central and eastern parts of northern Australia. Conservation efforts include removal of quolls to off-shore islands that are toad and cat free (Woinarski *et al.* 2007). Behavioural change in fauna can be used to reduce impacts. Northern Quolls are taught to avoid eating cane toads under captive conditions and then released back into native habitat. Captive juvenile quolls are fed a dead toad containing a nausea-inducing chemical before release and this increased their survival after release (O'Donnell *et al.* 2010).

Localized feral cat control programmes are another potential restoration tool in northern Australia. In Wongalara Wildlife Sanctuary feral cats caused local extinction of small mammal populations but in cat-proof enclosures they thrived (Frank *et al.* 2014). Another less expensive option to fencing areas is to maintain dingo populations, as dingoes compete with and harass feral cats, reducing cat activity and densities (Kennedy *et al.* 2012). Managing land to maintain ground cover through fire and grazing animal control also assists small mammals to avoid cat predation (McGregor *et al.* 2014).

Mine rehabilitation

Open-cut mining is a particularly important industry in the savanna of Australia (gold, baux-ite, manganese, uranium and iron) and the *cerrado* of Brazil (copper, iron, manganese, nickel, gold, beryllium, silver, phosphate and uranium). During the 1970s and 1980s, public pressure in Brazil and Australia resulted in improved management and increased environmental regula-tion of mining and requirements for rehabilitation of sites. Aims for restoration of mine sites vary from re-establishment of savanna vegetation suitable for inclusion in the Kakadu National Park world heritage area to requirements to stabilize the land form by re-establishing sustain-able vegetation.

The highly seasonal rainfall of savanna environments results in a range of challenges for mine site restoration. This causes several issues for restoration of vegetation on sites that are being rehabilitated after mining. The lack of vegetation cover during the initial summer wet season storms creates a high erosion risk via rapid loss of topsoil and/or burial of seeds prior to plant growth. This is particularly an issue for the large open-cut, hard-rock mines with large waste rock dumps. It is less of a problem for bauxite mines, where extensive areas are mined, but only to a shallow depth and the landform to be restored is fairly flat.

Prior to rainfall, warm humid climatic conditions occur and these can result in a relatively rapid loss of seed viability if seeds are broadcast onto a bare site during the dry season. Greater success occurs if seeds are broadcast just prior to or early in the wet season. Seed treatment is

Figure 21.5 Rehabilitated gold mine in Pine Creek, Australia
Source: photo courtesy of Sean Bellairs

generally not required for Australian savanna species, other than seed scarification of legume species, flesh removal from fleshy fruited species and storage of grass seeds (Bellairs and Ashwath 2007). For the many species that only produce high seed yields once every few years, storage of seeds in cool dry conditions is required to build up sufficient quantities of seeds for sowing.

Application of topsoil has generally been found to be favourable for restoring soil microbial activity and can provide seeds of many species to a restoration area if it is applied to a site soon after stripping, as occurs at bauxite mines in northern Australia (Corbett 1999). However, the timing of topsoil collection and subsequent management is crucial for successful savanna restoration. If the native grass seed density is too great, it can negatively affect tree establishment, however the amount of native grass seed can be managed by the timing of soil collection. If weeds are able to colonize and set seed on stockpiled topsoil, the use of this material for subsequent rehabilitation results in severe weed infestation and application of topsoil has not been recommended for some mines in northern Australia (*ibid.*).

Isolated storm events of the early wet season, followed by weeks without rainfall can result in substantial mortality of initially establishing seedlings on mining restoration sites. These effects may be mitigated by the planting of tubestock, broadcasting of seeds when the wet season is well established, contour ripping to enhance infiltration of rainfall and maximal soil moisture storage or irrigating during the season of initial establishment.

The extensive dry season of savanna environments provides conditions of potentially severe water stress to immature, establishing vegetation. Waste rock dumps of hard rock mines are characterized by substrate that is 40–60 per cent rock which has no ability to store moisture. The remaining 'soil' material can have a high sand content, also with limited ability to provide adequate plant available moisture (PAM) with high rates of hydraulic conductivity and drainage. Waste rock material at some savanna mine sites in the tropics break down to clays very quickly but this is highly variable and dependant on rock types (Corbett 1999).

It is critical to determine the depth of waste rock required on the post mining landform to support a sustainable, mature tropical savanna through the annual dry season. The hydraulic properties of waste-rock dump materials have been investigated at the Ranger mine in Australia. Ecohydrological modelling was used to determine the impact of climate variability on landform design suitability under historical climate regimes. A 4 m profile constructed of 100 per cent waste-rock was modelled without vegetation over 113 years. It was found that over the past 40 years the landform held enough water to support undisturbed tropical savanna through the dry season 75 per cent of the time (Segura *et al.* 2015). Increasing the depth of the landform or incorporating some pre-processing of spoil to separate larger particles to enable layering of coarse and fine layers would likely result in increased water-holding capacity and improve chances of restoration success (Fala *et al.* 2005). Fertilizer is typically applied to replace nutrients removed through loss of topsoil or to correct nutrient imbalances, such as very high magnesium to calcium ratios which can occur in waste rock dumps of northern Australia (Corbett 1999).

Once vegetation is established, fire and weed management are important for successful rehabilitation of the mine sites. Weeds grow vigorously during the wet season and grass weeds can create high biomass levels. When the grass weeds die off over the dry season this results in a high fuel load and the rehabilitated savanna is at risk of burning. At the Nabarlek mine site in northern Australia a combination of substantial weed establishment and hot fires killed much of the rehabilitated vegetation 12 years after rehabilitation had commenced (Bayliss *et al.* 2004). Thus control of weeds, consideration of biomass build up and establishment of fire-tolerant vegetation are all important management issues.

Role of fire in savanna restoration

Tropical savanna structure is defined by available soil moisture and fire regime. Fire regimes are important for maintaining the desired tree:grass ratio and play a part in determining the savanna boundaries with grasslands or woodland forest. Therefore fire needs to be incorporated in restoration efforts to produce a fire-tolerant and sustainable tropical savanna suitable to the region.

At an Australian bauxite mine in Gove, fire was excluded from rehabilitation areas (Cook 2012; Spain *et al.* 2006) until fire-tolerant native species reached maturity; this could take as long as 30 years. It was found that exclusion of fire resulted in the colonization of rehabilitated sites by fire-intolerant monsoon forest species, leading also to the development of frugivorous bird communities at these sites. This was considered a restoration success as native species had been re-established, yet the community that was restored is not what naturally would have formed, nor will it withstand the natural fire regimes of the area (one fire in every two years) (Cook 2012).

Excluding fire from rehabilitation sites in tropical savanna will inevitably lead to the development of large fuel loads greater than that of surrounding tropical savanna which experience frequent fires (every 2 to 6 years). A study of litter loads present at a 20-year-old rehabilitation site at the Gove bauxite mine in Australia found greater than 30 t ha^{-1} of litter present which is good for soil development but not for fire intensities (Spain *et al.* 2006). Maximum fine fuel loads found in comparable tropical savanna of Kakadu National Park were much lower at 6.3 t ha^{-1}; thus the high litter loads found at the bauxite mines rehabilitated sites have the potential to produce extremely intense fires capable of causing high tree mortalities (up to two thirds of the basal area at one site) and potentially lead to weed infestations (Cook 2012).

Predicted fire intensities at the bauxite mine in Gove have been modelled by Cook (2012). The model is based on the relationship between rehabilitation site age and litter accumulation rates and includes seasonal conditions, potential for fire spread based on mean monthly 3pm temperatures, wind speed and modelled fuel moisture. It was found that after 10 years of fire exclusion in mine rehabilitation sites, the maximum potential fire intensity was similar to the late fires in the Kapalga experiment (Andersen *et al.* 2003) and as intense as gamba grass fires. With this in mind it is important to consider management options to ensure old rehabilitation sites are burnt safely and new or young mine rehabilitation sites have fire incorporated in their management.

Management of older mining rehabilitation sites will require controlled burning planned carefully in terms of timing, method (e.g. burning against the wind) and ignition points in order to minimize potential tree mortalities. For young or new mine rehabilitation sites, fire regimes should be incorporated into the management of the sites to manage fire risk and create the required disturbance for fire-tolerant tropical savanna to establish.

Climate change impacts and savanna management

The impact of climate change on savanna will vary depending on the region. Forests could develop into savanna with increasingly long dry seasons, however increased levels of carbon dioxide favours tree growth over grasses so forests could form from savanna.

The impact of increased CO_2 on tropical savanna structure and distribution in Africa has been investigated using the Dynamic Global Vegetation Models (DGVM). One study predicted that increasing levels of atmospheric CO_2 would result in a 9 per cent reduction in area covered by African savanna between 1850 and 2100 in favour of C_3 dominated states (woodlands and forests). This large shift corresponds with the projection of a 16 per cent increase in the area of Africa covered by C_3 dominated states. The difference is made up by the predicted increases in C_3 dominated biomes in grasslands and deserts (Higgins and Scheiter 2012). Another study used a DGVM to investigate tree/grass cover interactions with historically low CO_2 and fire regimes in South Africa and found that low CO_2 favours C_4 grasses over C_3 woody components and C_4 grass fires were frequent, thus restricting sapling growth (Bond *et al.* 2003). With increased CO_2, particularly in the past 50–100 years, the survival of trees was increased likely due to increased water use efficiency at high CO_2 (Higgins and Scheiter 2012). More trees and less grass would equate to less frequent and less intense fires, thus furthering woody thickening of tropical savannas. Thus it is likely that land management and fire regimes will dictate what happens to savannas and their nearby biomes (Bond and Midgley 2012).

In South America and India, 'savannization' is occurring as a consequence of increased surface temperatures and evapotranspiration following deforestation (Salazar *et al.* 2007). In this circumstance, not only are forests being removed to directly create savanna, but the microclimate is changing leading to drier conditions more suited to savanna species. If deforestation rates continue this process will increase. With future declines in rainfall days predicted, these processes could be exacerbated, contributing to an increase in *cerrado* and *llanos* in South America despite increases in CO_2 (Hutley and Setterfield 2008; de Faria *et al.* 2015).

An Australian study investigated current woody thickening through combining a process-based model with a Bayesian network (BN) model to determine the effect of management decisions and future environmental changes, such as climate change, on savanna structure (Liedloff and Smith 2010). These models incorporated both fire and grazing as disturbance agents and climate differences were based on locality. This study found that climate (as determined by location), fire regime, initial tree population and fuel loads are more influential in the woody thickening occurring in Australian tropical savanna than grazing practices. Climate was the single greatest driver of stem density and tree basal area which suggests future change in climate will have an effect on these properties. However large time lags were observed in the effect of grazing management which could potentially be masked or altered due to floods, droughts or cyclones. The findings of this study were similar to Bond and Midgley (2012) in that when site differences are excluded, fire management is the most effective management tool for maintaining savanna structure under future climates.

Conclusions

Savannas are inherently unstable and often considered as a transitory ecological state between grassland and forest (Lehmann *et al.* 2014). The composition and distribution of savanna is constantly changing in response to natural drivers (climate variability and fire) as well as to human practices (grazing and fire management). It is important to understand that the parameters that define a healthy tropical savanna at any given site, for example a specific tree:grass ratio, may have changed historically and may change in the future due to climate shifts (rainfall), atmospheric CO_2 concentration and associated shifts in fire regimes (Scheiter and Higgins 2009). Tropical savanna supports 20 per cent of the Earth's human population and most of the world's grazing animals. Understanding savanna function is essential for improved restoration, rehabilitation and management of this significant tropical biome for future generations.

As we discussed, tropical savanna is dependent on plant available water stores over the dry season and is the first consideration for restoration. Where soil is heavily disturbed by erosion or requires reconstruction in the context of mining, it is important to ensure that further or potential erosion is controlled and that the design of the landform will hold sufficient water in the context of historical and future climatic variability. Savanna vegetation frequently occurs on sandy or skeletal soils that are readily erodible and erosion of native soils can lead to greater problems in the form of soil instability, loss of the thin top soil and propensity for weed infestations. Re-seeding native grasses and herbaceous species to stabilize the soil is essential.

In cases where there are weed infestations, the water balance and fire regimes are disturbed (Rossiter *et al.* 2003). Native species, both grass or woody, can be outcompeted for water and nutrients and the fire regime can be altered thus affecting savanna composition and structure. Current methods for removing weeds range from herbicide application to the use of fire and their combination, however research into more effective methods is ongoing (Brooks *et al.* 2010).

After available moisture, fire is the most important variable in savanna landscapes. The intensity, extent and frequency of fire determines savanna structure, in particular the tree : grass ratio, and also the species composition. Fire regimes across Africa, Australia and South America are unique, with the land use of tropical savanna an important factor. Fire is a powerful tool for savanna restoration and management and there is a growing body of research into fire management in tropical savanna. Continued integration of indigenous knowledge with western science will lead to more effective savanna fire restoration and management, critical given duel impacts of population pressure and climate change on savanna ecosystems (Bond and Midgley 2012).

Acknowledgements

The advice and input of the following persons is greatly appreciated: Mike Lawes regarding issues in African savanna and for his images, Natalie Rossiter-Rachor for information on Australian savanna weeds and her images, Stephen Reynolds for his images of South American savanna and savanna-conversions and Linda Luck for literature research support on restoration issues in South American savanna.

References

Alford, L. and Schmid, M. 2014. *Land management situation: Finniss River Aboriginal Land Trust*. Denhamia.

Andersen, A. N., Cook, G. D. and Williams, R. J. 2003. *Fire in tropical savannas: the Kapalga experiment*, New York: Springer Science and Business Media.

Andersen, A. N., Woinarski, J. C. and Parr, C. L. 2012. Savanna burning for biodiversity: fire management for faunal conservation in Australian tropical savannas. *Austral Ecology*, 37, 658–667.

Bayliss, P., Pfitzner, K. and Bellairs, S. 2004. Nabarlek revegetation project – presentations to ARRTC and Nabarlek MTC. Supervising Scientist Report 490. Department of the Environment and Heritage, Canberra.

Bellairs, S. M. and Ashwath, N. 2007. Seed biology of tropical Australian plants. In *Seeds: biology, development and ecology*, 416–427.

Bond, W. J. and Midgley, G. F. 2012. Carbon dioxide and the uneasy interactions of trees and savannah grasses. *Philosophical Transactions of the Royal Society B: Biological Sciences*, 367, 601–612.

Bond, W., Midgley, G. and Woodward, F. 2003. The importance of low atmospheric CO_2 and fire in promoting the spread of grasslands and savannas. *Global Change Biology*, 9, 973–982.

Brooks, K. J., Setterfield, S. A. and Douglas, M. M. 2010. Exotic grass invasions: applying a conceptual framework to the dynamics of degradation and restoration in Australia's tropical savannas. *Restoration Ecology*, 18, 188–197.

Chamaillé Jammes, S., Valeix, M. and Fritz, H. 2007. Managing heterogeneity in elephant distribution: interactions between elephant population density and surface water availability. *Journal of Applied Ecology*, 44, 625–633.

Cook, G. D. 2012. Fire management and minesite rehabilitation in a frequently burnt tropical savanna. *Austral Ecology*, 37, 686–692.

Corbett, M. 1999. Revegetation of mined land in the wet-dry tropics of northern Australia: a review. Supervising Scientist report 150. Department of the Environment and Heritage, Canberra.

De Faria, A., Fernandes, G. and França, M. 2015. Physiological approaches to determine the impact of climate changes on invasive African grasses in the savanna ecoregion of Brazil. *Environmental Earth Sciences*, 1–12.

Evers, G., Agostini, A. and Initiative, S. F. 2001. *No-tillage farming for sustainable land management: lessons from the 2000 Brazil study tour.* Food and Agriculture Organization of the United Nations, Rome.

Fala, O., Molson, J., Aubertin, M. and Bussière, B. 2005. Numerical modelling of flow and capillary barrier effects in unsaturated waste rock piles. *Mine Water and the Environment*, 24, 172–185.

Fisher, D. O., Johnson, C. N., Lawes, M. J., Fritz, S. A., Mccallum, H., Blomberg, S. P., Vanderwal, J., Abbott, B., Frank, A. and Legge, S. 2014. The current decline of tropical marsupials in Australia: is history repeating? *Global Ecology and Biogeography*, 23, 181–190.

Frank, A. S., Johnson, C. N., Potts, J. M., Fisher, A., Lawes, M. J., Woinarski, J. C., Tuft, K., Radford, I. J., Gordon, I. J. and Collis, M. A. 2014. Experimental evidence that feral cats cause local extirpation of small mammals in Australia's tropical savannas. *Journal of Applied Ecology*, 51, 1486–1493.

Gil-Romera, G., Lamb, H. F., Turton, D., Sevilla-Callejo, M. and Umer, M. 2010. Long-term resilience, bush encroachment patterns and local knowledge in a Northeast African savanna. *Global Environmental Change*, 20, 612–626.

Govender, N., Trollope, W. S. and Van Wilgen, B. W. 2006. The effect of fire season, fire frequency, rainfall and management on fire intensity in savanna vegetation in South Africa. *Journal of Applied Ecology*, 43, 748–758.

Guldemond, R. and Aarde, R. 2008. A meta analysis of the impact of African elephants on savanna vegetation. *The Journal of Wildlife Management*, 72, 892–899.

Higgins, S. I. and Scheiter, S. 2012. Atmospheric CO_2 forces abrupt vegetation shifts locally, but not globally. *Nature*, 488, 209–212.

Hill, B. and Ward, S. 2010. *National Recovery Plan for the northern quoll* Dasyurus hallucatus. Department of Natural Resources, Environment, The Arts and Sport, Darwin.

Hoffmann, W. A., Lucatelli, V. M. P. C., Silva, F. J., Azeuedo, I. N. C., Marinho, M. D. S., Albuquerque, A. M. S., Lopes, A. D. O. and Moreira, S. P. 2004. Impact of the invasive alien grass Melinis minutiflora at the savanna-forest ecotone in the Brazilian Cerrado. *Diversity and Distributions*, 10, 99–103.

Hutley, L. B. and Setterfield, S. 2008. Savanna. In S. E. Jorgensen and B. Fath (eds), *Encyclopedia of Ecology*. Oxford: Elsevier.

Jacklyn, P. and Cook, G. 2009. *Fire brings back country*. Darwin: Savanna Links, Charles Darwin University.

Jackson, L. L., Lopoukhine, N. and Hillyard, D. 1995. Ecological restoration: a definition and comments. *Restoration Ecology*, 3, 71–75.

Kennedy, M., Phillips, B. L., Legge, S., Murphy, S. A. and Faulkner, R. A. 2012. Do dingoes suppress the activity of feral cats in northern Australia? *Austral Ecology*, 37, 134–139.

Klink, C. A. and Machado, R. B. 2005. Conservation of the Brazilian cerrado. *Conservation Biology* 19, 707–713.

Lawes, M. J., Fisher, D. O., Johnson, C. N., Blomberg, S. P., Frank, A. S., Fritz, S. A., Mccallum, H., Vanderwal, J., Abbott, B. N. and Legge, S. 2015. Correlates of recent declines of rodents in northern and southern Australia: habitat structure is critical. *PloS ONE*, 10, e0130626.

Lehmann, C. E., Anderson, T. M., Sankaran, M., Higgins, S. I., Archibald, S., Hoffmann, W. A., Hanan, N. P., Williams, R. J., Fensham, R. J. and Felfili, J. 2014. Savanna vegetation-fire-climate relationships differ among continents. *Science*, 343, 548–552.

Liedloff, A. C. and Smith, C. S. 2010. Predicting a 'tree change' in Australia's tropical savannas: Combining different types of models to understand complex ecosystem behaviour. *Ecological Modelling*, 221, 2565–2575.

McGregor, H. W., Legge, S., Jones, M. E. and Johnson, C. N. 2014. Landscape management of fire and grazing regimes alters the fine-scale habitat utilization by feral cats. *PLoS One*, 9(10), e109097.

Mistry, J. 2000. *World savannas: ecology and human use*. Routledge, London.

Newmark, W. D. 2008. Isolation of African protected areas. *Frontiers in Ecology and the Environment*, 6, 321–328.

O'Connor, T. G., Puttick, J. R. and Hoffman, M. T. 2014. Bush encroachment in southern Africa: changes and causes. *African Journal of Range and Forage Science*, 31, 67–88.

O'Donnell, S., Webb, J. K. and Shine, R. 2010. Conditioned taste aversion enhances the survival of an endangered predator imperilled by a toxic invader. *Journal of Applied Ecology*, 47, 558–565.

Owen-Smith, N., Kerley, G., Page, B., Slotow, R. and Van Aarde, R. 2006. A scientific perspective on the management of elephants in the Kruger National Park and elsewhere: elephant conservation. *South African Journal of Science*, 102, 389–394.

Petty, A. M., Werner, P. A., Lehmann, C. E., Riley, J. E., Banfai, D. S. and Elliott, L. P. 2007. Savanna responses to feral buffalo in Kakadu National Park, Australia. *Ecological Monographs*, 77, 441–463.

Rossiter, N. A., Setterfield, S. A., Douglas, M. M. and Hutley, L. B. 2003. Testing the grass fire cycle: alien grass invasion in the tropical savannas of northern Australia. *Diversity and Distributions*, 9, 169–176.

Salazar, L. F., Nobre, C. A. and Oyama, M. D. 2007. Climate change consequences on the biome distribution in tropical South America. *Geophysical Research Letters*, 34, L09708.

Sankaran, M., Hanan, N. P., Scholes, R. J., Ratnam, J., Augustine, D. J., Cade, B. S., Gignoux, J., Higgins, S. I., Le Roux, X. and Ludwig, F. 2005. Determinants of woody cover in African savannas. *Nature*, 438, 846–849.

Scheiter, S. and Higgins, S. I. 2009. Impacts of climate change on the vegetation of Africa: an adaptive dynamic vegetation modelling approach. *Global Change Biology*, 15, 2224–2246.

Segura, J. R., Hutley, L., Bellairs, S. M. and Lu, P. 2015. Ecohydrological modelling predicts tropical savanna rehabilitation on a post-mining landform in the wet-dry tropics of northern Australia. Paper presented at World Conference on Ecological Restoration, Society for Ecological Restoration, Manchester, UK.

Setterfield, S. A., Rossiter Rachor, N. A., Hutley, L. B., Douglas, M. M. and Williams, R. J. 2010. Biodiversity research: turning up the heat: the impacts of Andropogon gayanus (gamba grass) invasion on fire behaviour in northern Australian savannas. *Diversity and Distributions*, 16, 854–861.

Shackleton, S., Shackleton, C., Netshiluvhi, T., Geach, B., Ballance, A. and Fairbanks, D. 2002. Use patterns and value of savanna resources in three rural villages in South Africa. *Economic Botany*, 56, 130–146.

Shine, R. 2010. The ecological impact of invasive cane toads (Bufo marinus) in Australia. *The Quarterly Review of Biology*, 85, 253–291.

Spain, A., Hinz, D., Ludwig, J. A., Tibbett, M. and Tongway, D. J. 2006. Mine closure and ecosystem development: Alcan Gove bauxite mine, Northern Territory, Australia. In *Proceedings of the First International Seminar on Mine Closure, Perth, Australia, 2006*, pp. 13–15.

Urban, M. C., Phillips, B. L., Skelly, D. K. and Shine, R. 2008. A toad more traveled: the heterogeneous invasion dynamics of cane toads in Australia. *The American Naturalist*, 171, E134–E148.

Van der Werf, G. R., Randerson, J. T., Giglio, L., Collatz, G., Mu, M., Kasibhatla, P. S., Morton, D. C., Defries, R., Jin, Y. V. and Van Leeuwen, T. T. 2010. Global fire emissions and the contribution of deforestation, savanna, forest, agricultural, and peat fires (1997–2009). *Atmospheric Chemistry and Physics*, 10, 11707–11735.

Whitehead, P. J., Russell-Smith, J. and Woinarski, J. C. 2005. Fire, landscape heterogeneity and wildlife management in Australia's tropical savannas: introduction and overview. *Wildlife Research*, 32, 369–375.

Woinarski, J., Pavey, C., Kerrigan, R., Cowie, I., Ward, S. and Winnard, A. 2007. Lost from our landscape: threatened species of the Northern Territory. *Pacific Conservation Biology*, 14, 184–184.

Woinarski, J. C., Legge, S., Fitzsimons, J. A., Traill, B. J., Burbidge, A. A., Fisher, A., Firth, R. S., Gordon, I. J., Griffiths, A. D. and Johnson, C. N. 2011. The disappearing mammal fauna of northern Australia: context, cause, and response. *Conservation Letters*, 4, 192–201.

Yibarbuk, D., Whitehead, P., Russell-Smith, J., Jackson, D., Godjuwa, C., Fisher, A., Cooke, P., Choquenot, D. and Bowman, D. 2001. Fire ecology and Aboriginal land management in central Arnhem Land, northern Australia: a tradition of ecosystem management. *Journal of Biogeography*, 28, 325–343.

22

RESTORATION OF TROPICAL AND SUBTROPICAL GRASSLANDS

Gerhard Ernst Overbeck and Sandra Cristina Müller

Introduction

Throughout the tropics and subtropics, human action has resulted in large-scale losses of natural vegetation, and land use change continues to occur at a staggering pace in many countries. Conservation needs of tropical forest are well established and forest restoration has become an important field of action, as well as a research field (see Chapter 23). The situation of grasslands, however, is different: even though large parts of the tropics and subtropics are covered by non-forest vegetation, these have not been the focus of much conservation and less so of restoration (e.g. Parr *et al.* 2014; Overbeck *et al.* 2015; Veldman *et al.* 2015a). Among the many reasons, these three may be the most important:

- Non-forest systems were and still are undervalued in conservation in many countries, as their presence is often considered a consequence of degradation of originally forested areas.
- Their distinct ecological features often are not sufficiently considered, in particular the role of disturbances and management.
- Recently, the necessity to mitigate climate change and increase carbon stocks has resulted in a widespread belief that planting of trees is inherently good for conservation – even in situations of afforestation, which clearly have detrimental effects for grasslands (Veldman *et al.* 2015b).

In this chapter, we address restoration of tropical and subtropical grasslands. While grassland restoration in the temperate zone is well established in practice and research, the same is not true for the tropics and subtropics, where this is a much more recent field. Thus, the body of literature on restoration of these systems is rather scarce: from many regions, no studies are available. On the other hand, we often can draw from experiences made in ecosystems under different climatic conditions that share ecological features, such as the tallgrass prairie, where conservation and restoration have been important issues for a long time. Additionally, many parallels exist to restoration of tropical savannas, at least when it concerns the herbaceous layer of savannas (see Chapter 21).

Distribution of tropical and subtropical grasslands

Where do we find tropical and subtropical grasslands? No consistent classification of grasslands worldwide is at hand that can easily be used for our purpose. The term 'tropical grassy biomes' (Bond and Parr 2010; Parr *et al.* 2014; Veldman *et al.* 2015a) comprises the broad range of open (i.e. non-forest) vegetation types in tropical regions, such as grasslands, savannas, shrublands and open woodlands with up to 80 per cent tree cover. This can be easily extended to subtropical regions. Pragmatically, we here define 'grasslands' in a strict sense as those ecosystems where cover of woody plants is below 10 per cent. This means that we will, in this chapter, not consider savannas, shrublands and open woodlands, but will include grassland physiognomies situated within the savanna biome. 'Tropical and subtropical' areas occur in the region between the 38th parallels of northern and southern latitude. This thus includes, besides the tropics, the grasslands in the southern part of the US (e.g. in Texas and in Florida), southern China, all of South Africa, the southern half of Australia, and, in South America, the southernmost part of Brazil until Uruguay and the northern parts of Argentina.

Ecological characteristics of tropical and subtropical grasslands

The ecosystems that comprise the world's 'grassy biomes' share a number of ecological characteristics, despite distinct climatic conditions and a broad variety of soil conditions that influence vegetation patterns together with disturbance regimes. Recently, Joe Veldman and co-authors suggested the term 'old-growth grassland' to characterize the primary grassy biomes in the tropics and subtropics (Veldman *et al.* 2015a). The term implies that these are antique ecosystems that developed under disturbances (fire and herbivory by large animals), and/or on soil with properties that limit tree growth. The long co-evolution of plants with fire and herbivores selected for specific characteristics, and among these, the capacity to resprout from storage organs belowground or close to the soil surface after biomass removal may be the most important in a restoration context. Similarly, communities on poor nutrient soils present a high degree of root specializations (e.g. Oliveira *et al.* 2015). In consequence, a considerable part of biomass of these grasslands is belowground and the ability of such ecosystems to regenerate after disturbances is probably highly dependent on the integrity of the belowground structure. Bud banks are persistent and guarantee vegetation recovery after disturbance, in contrast to often only transient seed banks.

Species richness of these systems can be very high, as is the number of endemic species. Often, it is especially the forb component that is particularly rich in species, and many of these forbs are perennial, slow-growing and show low reproductive output and colonization ability. Nonetheless, despite these common characteristics, grasslands function in distinct ways.

On the one hand, there are systems where resource levels are rather high (productive systems). Here, diversity is often highest under intermediate frequencies or intensities of fire and grazing (by native grazers, but also by domestic animals), respectively. In the absence of disturbances, diversity tends to decrease as competitive grasses become dominant. The resilience of the system to disturbances (including to changes of grazing management) is expected to be specific to particular ecosystems, according to the evolutionary history of the grassland, allowing or not for alternative diversity states (Cingolani *et al.* 2005). A crucial issue for restoration is thus to define (and understand) the potential productivity and the evolutionary history of grazing and/or fire of the target area. Nonetheless, due to high productivity, vegetation development under restoration, if properly managed, can be expected to be rather fast.

On the other hand, the tropics and subtropics also contain grasslands that are very limited in resources and in productivity, such as the *campos rupestres* in Brazil (Oliveira *et al.* 2015) or grasslands in Southwestern Australia or in the South African Cape province. Hopper (2009) introduced the term 'old climatically buffered infertile landscapes' (OCBILs) to describe those regions where the combination of long evolutionary history (no history of glaciation) and low nutrient availability (in consequence of their situation on old land-surfaces) have led to remarkable levels of species richness and endemism. Three landscapes were originally proposed in the OCBIL concept (Southwest Australian Floristic Region, South Africa's Greater Cape, and Venezuela's Pantepui Highlands; all of them including open vegetation types), but other regions may also be considered as OCBILS, such as the *Campos rupestres* in Central Brazil (Oliveira *et al.* 2015). Here, we can expect vegetation development to be much slower, due to environmental filters that limit plant establishment and growth – and here, restoration may be especially difficult (Standish and Hobbs 2010).

In summary, the 'grassy biomes' contain distinct types of grassland: distinct in productivity, distinct in their reaction to disturbances, and, as we will discuss in more detail, also distinct in their potential to recover after degradation or land use change.

What do we want to restore?

Just as restoration of any other ecosystems, restoration of tropical and subtropical grasslands needs clear objectives and, ideally, a clearly defined reference. For ecological restoration, we expect that one of the objectives is the restoration of natural biodiversity, but additionally we may want to focus on ecosystem services such as carbon storage, erosion control or grassland productivity for meat production: restoration objectives will vary among projects, depending on characteristics of the degraded system and the specific context. At any rate, reference systems are necessary. For grasslands at sites with limiting soil conditions, these may be rather easy to define (if pre-disturbance conditions are known, or if conserved sites still exist in a close spatial context). For those grasslands whose characteristics depend to a greater extent on disturbances, it is not always easy to establish such a reference in an objective way, as we often do not know what a 'natural' disturbance regime would be like. In savannas, such as in Africa and in the Brazilian Cerrado, the fire regime is a key factor in determining the density of woody species, and thus needs to be considered in defining the reference state and then in restoration management. In the case of grazing, the situation may be even more complex. In the subtropical South Brazilian grassland region, we can expect that the introduction of cattle in the 17th century by the Spanish led to considerable changes in terms of composition and structure of the grasslands; likely, fires were more important in the past, and native (now extinct) herbivores densities were much lower than those of cattle today. It is well known for the study region that fire favours tussock grasses, while grazing favours prostrate species (Overbeck *et al.* 2007), but frequency and intensity of both disturbances matter as well. As it is impossible to know what grasslands looked like before human colonization of the region, the pre-Columbian state of grasslands is not a tangible restoration objective. In other words, uncertainties exist in the establishment of restoration objectives.

On the other hand, these uncertainties may actually be turned into something useful in the sense that they give a range of options for the definition of restoration aims of a concrete project, depending not only on the ecological but also on the socio-economic context. On private lands, the establishment of extensive grazing regimes that allows economic returns may be an important restoration strategy that can be compatible with other objectives, e.g. those related to biodiversity. The inclusion of management allows integrating socio-economic aspects and interests and motivations of landusers and other stakeholders into restoration projects, issues much debated in the restoration community, but often still not implemented sufficiently in practice. In countries like Brazil, the consideration of anthropogenic management in conservation and ecological restoration often is still considered a taboo – maybe not surprising in a country where the focus of conservation and restoration traditionally had been on forest (e.g. Overbeck *et al.* 2013).

Strategies and techniques for restoration of tropical and subtropical grasslands

Ecological restoration (i.e. re-establishment of the ecosystem's integrity in terms of species composition, community structure and ecosystem functions) can be achieved following two distinct strategies, passive and active restoration, that are used to overcome biotic and abiotic filters at restoration sites. Figure 22.1 gives an overview of principal targets and strategies mentioned in the text, separately for systems with high or low productivity, and considering contrasting levels of degradation.

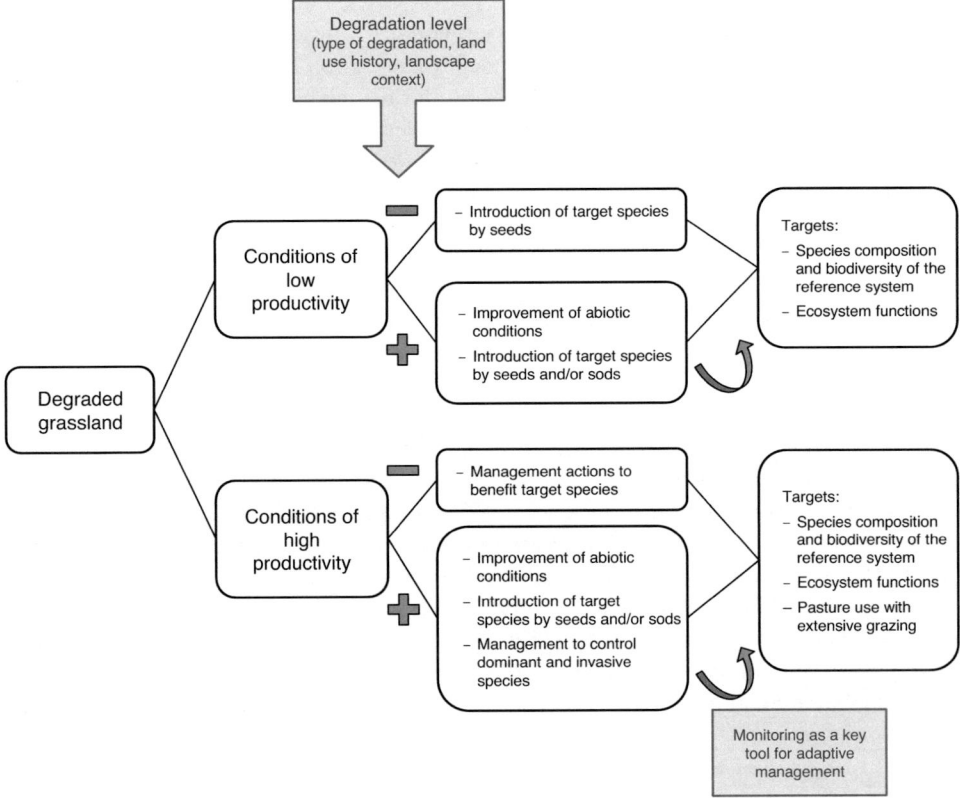

Figure 22.1   Schematic representation of the principal targets and strategies for restoration of tropical and subtropical grasslands. Selection depends on the location of the site on the gradient from productive to non-productive conditions and on the level of degradation suffered by the system

Passive restoration

Spontaneous recovery of grasslands (i.e. success of passive restoration) will vary with the potential of species to persist during the degradation processes, or with their ability to reach the sites under recovery. We know that many of the characteristic or dominant species of tropical and subtropical grasslands are poorly represented in the seed bank, which indicates a low recovery potential after degradation. Besides, the seed bank may have been depleted after long periods of other land uses. Underground storage organs likely will have been destroyed by any kind of land use change. In the coastal grasslands of south-eastern South Africa, areas that recovered spontaneously after use for pine plantations returned to grasslands, but with markedly lower species richness; species missing were exactly those typical for old growth grasslands *sensu* Veldman *et al.* (2015a); that is, resprouting forbs with low colonization ability (Zaloumis and Bond 2011). Seed limitation has been reported as a major constraint for recovery of degraded states in the *Campos rupestres*, tropical mountain grasslands on poor soil in the Brazilian Cerrado biome (Le Stradic *et al.* 2014a). Additionally, these biotic filters may be exacerbated by extremely limiting environmental conditions: high solar radiation and shallow and nutrient-poor soils may further complicate plant regeneration.

Obviously, species with efficient dispersal, such as the wind-dispersed Asteraceae, have greater ability to reach degraded sites (da Silva Menezes *et al.* 2015), while species from other groups may not, which explains the dominance of ruderal species, and often of annuals, in early stages of vegetation recovery, at least under productive conditions. In addition to the proper biology of the species, features of the surrounding landscapes are important in defining successional pathways: where exotic and invasive species are present, they may become a serious impediment to reach the reference conditions. Tognetti *et al.* (2010), in their study of old-field succession in the Argentinian Pampa, observed that exotics retained high cover values even after long periods of time and could effectively arrest succession. They also found low cover of native perennial tussock grasses, which may be a consequence of higher fitness of their exotic counterparts under the current conditions, even more so if the latter managed to colonize the sites earlier, but also due to seed limitation of these native species. However, generalizations seem difficult: Fensham *et al.* (2015) reported, for grasslands in Queensland, Australia, the same gradual substitution of annuals by perennials, despite slow development of some perennials, specifically of the grasses. However, in their study, exotics were much less successful and did not impede recovery of communities to high similarity with the reference sites.

Passive restoration apparently works better under productive conditions, where vegetation development is a lot faster than under limiting site conditions. But does 'passive restoration' or 'spontaneous recovery' of tropical and subtropical grasslands mean exclusion of all human interference, in the same way it does in forest restoration? As the maintenance of structure of diversity of these grasslands is linked to management or disturbances (e.g. Lezama *et al.* 2014), the ecological processes intrinsic to grasslands at distinct productivity levels will also need to be considered in restoration. Otherwise, trajectories will likely not lead to the reference conditions but will end in grasslands dominated by few highly competitive species or go towards systems dominated by shrubs and, ultimately, trees that resemble the original system only very little (Figure 22.2). The boundary between passive and active restoration may not always be easy to establish for some tropical and subtropical grassland systems.

Figure 22.2 Vegetation recovery without management on former grassland sites after use as pine plantations for approximately 10 years does not lead to reestablishment of grassland. Instead, both pine and native wind–dispersed shrubs become abundantly established. This area in the highland grasslands near Cambará do Sul, Rio Grande do Sul, Brazil experienced light levels of cattle grazing, but clearly more restoration interventions are necessary

Source: photo by G. Overbeck

Active restoration

Where recovery of vegetation is limited, active introduction of species is necessary. For subtropical and tropical regions, knowledge of the autoecology of grassland species still is fragmentary: we know little about germination requirements for most species, or how many seeds of a given species are necessary for successful establishment (Le Stradic *et al.* 2014a, 2015b). A variety of techniques and methods exist for species introduction in ecological restoration, and most have been successfully employed in many projects both in European grasslands (Kiehl *et al.* 2010) and in the North American prairie (Packard and Mutel 2005); these techniques need to be tested under tropical/subtropical conditions. Seeding or planting of native species allows for selection of species that seem most appropriate, e.g. considering specific sets of functional traits (see following text). However, there is a major problem: in most tropical and subtropical countries, availability of seeds for restoration is very low or inexistent, so seeds still need to be collected specifically for a restoration project.

On the other hand, the transfer of hay containing seeds of native species or of sods from other sites allows for the introduction of subsets of local communities, usually within the specific context of the restoration site. Few experiments of this kind have been conducted so far, and results are not always promising. Establishment success in experiments using hay transfer was extremely low in the study on former quarry sites in the *Campos rupestre* by Le Stradic *et al.* (2014a), probably due to a combination of low germination rates, i.e. properties of the species, and environmental conditions (Figure 22.3). Here, possibly plantings of species that are able to colonize sites by vegetative spreading or that can serve as facilitators for others may be more successful (Le Stradic *et al.* 2014b). The effectiveness of turf (or sod) transplant for restoration has been tested for metallophyte grassland communities in tropical Africa (Le Stradic *et al.* 2015a). The established communities showed similar cover and species number to reference communities, but differed from them in community structure. Success varied according to soil depth, and, as in the South African grasslands studied by Zaloumis and Bond (2011), specifically species with large underground root systems or storage organs were difficult to establish. Nonetheless, turf transplants performed much better than topsoil transplants; however, it is questionable if the technique is appropriate for larger areas (Le Stradic *et al.* 2015a). Likely, chances of success of hay transfer and turf transplant are higher under productive conditions where vegetation development is faster, but we clearly need more studies before being able to generalize about this.

Until now, we have only considered biotic filters. However, in most cases degradation affects also abiotic features of the ecosystem. After intensive use (e.g. by agriculture), soils may be more fertile than that of original grassland vegetation, which may lead to establishment of ruderal species that are not typical of previous natural communities. On the other hand, where topsoil

Figure 22.3 Grasslands under harsh environmental conditions, such as this site of the *Campos rupestres* in the Serra do Cipó, Minais Gerais, Brazil, are extremely difficult to restore, due to the combination of biotic and abiotic constraints, paired with extremely high diversity in terms of species, life forms and (micro-)habitats, and very specific plant adaptations

Source: photo by G. Overbeck

had been removed (e.g. after mining), conditions may be too extreme for establishment, in consequence of both chemical and physical factors. After severe degradation or for grasslands under extremely harsh conditions, such as the *Campos rupestres*, biotic filters may be exacerbated by extremely limiting environmental conditions: high solar radiation, shallow and nutrient-poor soils may further complicate plant regeneration. In fact, these harsh conditions and thus slow vegetation development are characteristics of these systems – and make them much harder to be restored than more productive grasslands where development is faster (but other problems may exist in them, e.g. invasive species; see below).

At any rate, the establishment of similar pre-degraded abiotic site conditions is usually a prerequisite for species establishment in degraded areas, just as in restoration of most systems. While studies still are scarce, many methods that are well established in temperate systems can be expected to be useful in tropical and subtropical regions. In particular grasslands in the subtropics show considerable affinities to prairie vegetation in the temperate zone, and for this region, long-term experience with restoration exists (e.g. Packard and Mutel 2005). One generally applicable strategy may be to first create vegetation cover with a few native species, usually grasses, and then try to introduce more species in subsequent restoration phases. Often, restorationists want to quickly reestablish the species composition of the reference area. In regions where no seeds are available commercially, this is not a sensible restoration aim. Rather, the aim can be to introduce the ecological processes that may then increase species richness, such as dispersal by animals in grazed systems. In other regions, it seems important to re-introduce the characteristic fire regimes, and important to create the desired balance between grassy and woody vegetation strata.

Invasive species as a challenge

In tropical and subtropical grasslands around the world, invasive species are a major challenge in conservation and restoration. Tropical grasses are aggressive invaders in tropical regions, while temperate forbs, shrubs and grasses are more problematic in higher latitude subtropics. Landscape features, highly determined by land use changes, and local impacts are two important drivers of local invasions by non-native species. Under the current more or less fragmented landscape, propagule pressure of invasive species can be very high, but the locally scaled invasion process is always dependent on the invasibility of the community. Degraded grasslands, where the previous plant community structure has been destroyed, may be very susceptible to invasions, principally in fragmented landscapes. Once a degraded grassland area is invaded, the restoration challenge increases.

Tropical grass invaders are commonly fire-prone species. Degraded grasslands that have been invaded by such species (e.g. *Urochloa* sp. in open Cerrado physiognomies) may maintain a positive feedback to fire (Gorgone-Barbosa *et al.* 2015): the traits of the invader promote fire, from which the species itself can recover rapidly via seeding and resprouting, constraining the recolonization of perennial native grasses or fire-sensitive forbs. At such sites, management with fire should be cautiously combined with actions to inhibit the regrowth of invasive species. Besides the reestablishment of disturbances (mostly fire and grazing), under close monitoring, changes of abiotic conditions (e.g. the soil C:N ratio) have also been considered in the control of invasive species, together with the approach of enhancing cover of the remnant native component – 'legacy species' of the original community – in grasslands under restoration (Cole *et al.* 2016).

A recent meta-analysis (Price and Pärtel 2013) found that resident species can reduce colonization and performance of functionally similar invasive forbs, but less so grass invaders, especially in artificially assembled communities. However, more specifically in a restoration context, there is evidence that demonstrates the effectiveness of adding grass species with similar traits to reduce populations of the problematic grass invader (Hulvey and Aigner 2014). Thus, knowledge of functional traits of native and exotic species in restoration projects with invasion pressure can help to define the set of species to be introduced, as proposed under the 'biotic filter' perspective (*ibid.*). This may be especially successful when the manipulation of abiotic conditions also favours native resident species that are to be enhanced by restoration.

Invasive grasses are extremely problematic, but other life forms can have equally negative impacts (Figure 22.4). Exotic shrubs, such as from the Fabaceae, may invade grasslands previously dominated by grasses and completely change community structure. Examples are *Acacia nilotica* in grasslands in Queensland, Australia (Brown and Carter 1998), or *Ulex europaeus* in many regions of the world, including the tropics and subtropics (Lowe *et al.* 2000). This kind of invasion usually is related to shifts in the disturbance regime or environmental conditions (e.g. Brown and Carter 1998) and can lead, in the case of dense stands of the invasive species, to the complete loss of ecological functions of the previous grassland ecosystem. Restoration of these systems is a big challenge.

Additionally, we should also consider that currently nonaggressive alien species might benefit from climate change, causing their expansion, and that ongoing land-use change in the tropics and subtropics may increase plant invasions: restorationists working in a scenario of global climate changes will need to be prepared to deal with more invasive species.

Figure 22.4 Gorse (*Ulex europaeus*) is a globally problematic invasive shrub. Where it invades grasslands in southern Rio Grande do Sul, Brazil (municipality of Pedras Altas), composition, structure and functioning of the original ecosystem is changed completely. For this kind of degradation, no adequate restoration strategy has been developed so far in this region

Source: photo by G. Overbeck

Opportunities – from the functional perspective to adaptive approaches

Ecological restoration today goes beyond restoring biodiversity and species composition, and includes ecosystem functioning and services. A functional perspective that explicitly considers the link between plant traits and ecosystem properties or responses to changed environmental conditions (Lavorel and Garnier 2002) can help to link these goals and to better understand the restoration process. Restoration projects are considered opportunities to test the relationship between community parameters (e.g. species and functional diversity and ecosystem functions; Díaz *et al.* 2007). Studies in tropical and subtropical grasslands are especially interesting in this regard, since they are often very species-rich and may be easily associated with many ecosystem services.

Particularly relevant plant traits for restoration aims are those related to plant fitness, establishment and competitiveness. Leaf traits (e.g. specific leaf area, dry matter content, N and P content) are frequently associated with resource acquisition and growth (Larson *et al.* 2015; Cole *et al.* 2016), and are rather easy to obtain ('soft traits') in comparison to 'hard traits' (e.g. plant physiology parameters) or even seed traits. However, traits associated with germination and emergence may be crucial for restoration outcomes: In drylands in western North America, 90 per cent of the variation in seedling survival among grasses was explained by germination and emergence rates, whereas leaf and root traits explained very little (Larson *et al.* 2015).

When considering the landscape scale, fragmentation and isolation of source and restoration areas may greatly influence grassland assembly processes, making traits related to dispersal ability the most responsive to restoration (Helsen *et al.* 2013). Even though some studies are available, the lack of understanding of the role of traits in driving recruitment and vegetation recovery processes as well as the lack of trait availability for tropical and subtropical grasslands are considerable barriers toward an increased use of these traits in restoration projects. Additionally we should consider a broadening of the view beyond the vegetation component. In later monitoring phases of restoration projects traits of other trophic levels than plants can also be used to measure the success of restoration actions – after all, successful restoration means a lot more than restoring the vegetation.

The monitoring of restoration processes based on functional traits at the community level can be done regarding mean trait values (functional composition), usually weighted by species abundances, and functional diversity indices, based on a set of traits (which should be chosen according to specific aims). Response traits can indicate constraints in site conditions, whereas effect traits can better refer to specific ecosystem functions that may or may not be recovering. Ilunga wa Ilunga *et al.* (2015), for instance, showed clear differences in functional trait expression between primary and secondary grasslands on metal-rich substrates in the Congo. Additionally, functional composition may tell us about mass effect processes (e.g. productivity), while functional diversity can be used to inform us about species complementarity and redundancy. By monitoring restoration areas under this perspective, practitioners can intervene in the ongoing recovery process, following the idea of adaptive management, if the trajectory is not moving towards the restoration targets.

But adaptive management must not be restricted to monitoring of vegetation development in terms of functional traits or functional composition. Rather, it can be employed in restoration management in general, allowing for the testing of management alternatives in restoration projects (e.g. Zedler 2005). As discussed previously, the use of management is important in the restoration of many tropical and subtropical grasslands, be it to control biomass development, to reduce performance of invasive species and enhance that of natives, or to establish the typical ecological processes that govern these systems. As we have limited experience with the effects of different treatments, the adaptive approach will increase chance of success while allowing for learning during ongoing and future projects.

In many cases of grassland restoration, at least when under productive conditions, some kind of management will persist even when the system can be considered restored. In fact, it may be hard to tell where restoration ends and 'normal' management begins. At any rate, the effects of management used for restoration or conservation purpose need to be monitored.

Conclusions

Restoration of grasslands only recently has gained a more prominent place on the conservation agenda in many tropical and subtropical countries, where restoration traditionally has focused on forests. The continuing pressure on open ecosystems that are often less protected than forests will cause further losses of natural grasslands – and likely increase restoration needs in the future, often in severely fragmented landscapes. The development of restoration techniques specific for tropical or subtropical regions – or the adaptation of approaches already applied successfully elsewhere – is urgent. We need more experimentation under field conditions: at the moment, generalizations about the effectiveness of techniques still are difficult, as too few studies have been conducted. A number of other important issues in restoration have not been tackled at all so far in tropical and subtropical grasslands. One example is the question of genetic diversity in restoration, which needs to be studied as soon as seed material becomes available. Also, the available studies – and admittedly this chapter – focus on vegetation, while other trophic levels or ecosystem services are still poorly considered – both regarding restoration techniques and monitoring of restoration success.

Despite the still rather small empirical basis, we can clearly see that restoration requirements, challenges and opportunities differ considerably among regions, the main contrast apparently being site productivity. Even though productive sites face challenges as well, e.g. by invasive species, the potential for restoration seems rather high, if the restoration process is properly managed. Nonetheless, some species groups, and among them the typical 'old growth' grassland species with their low dispersal abilities, will need special attention. In less productive environments, under harsh abiotic conditions, the available studies show that restoration is extremely difficult, due to a combination of abiotic constraints and the biology of the grassland species themselves. For these systems, we need to be prepared for restoration efforts to be lengthy and likely costly. The problem is that attempts to 'improve' environmental conditions in order to facilitate plant establishment may risk changing the system in a way that the original limiting conditions will not exist anymore.

One common problem for restoration of grasslands in the tropics and subtropics is the lack of seeds available for restoration. Considerable investments in terms of research and infrastructure will be necessary to be able to restore grasslands on a larger scale, but if this was tackled, we likely will see a big advance in restoration of tropical and subtropical grasslands.

References

Bond, W. J., and C. L. Parr. 2010. Beyond the forest edge: Ecology, diversity and conservation of the grassy biomes. *Biological Conservation* 143: 2395–2404.

Brown, J. R., and J. Carter. 1998. Spatial and temporal patterns of exotic shrub invasion in an Australian tropical grassland. *Landscape Ecology* 13: 93–102.

Cingolani, A. M., I. Noy-Meyr, and S. Díaz. 2005. Grazing effects on rangeland diversity: A synthesis of contemporary models. *Ecological Applications* 15: 757–773.

Cole, I. A., S. M. Prober, I. D. Lunt, and T. B. Koen. 2016. A plant traits approach to managing legacy species during restoration transitions in degraded temperate eucalypt woodlands. *Restoration Ecology* 24(3): 354–363.

Da Silva Menezes, L., S. C. Müller, and G. E. Overbeck. 2015. Scale-specific processes shape plant community patterns in subtropical coastal grasslands. *Austral Ecology* 41: 65–73.

Díaz, S., S. Lavorel, F. de Bello, F. Quétier, K. Grigulis, and T. M. Robson. 2007. Incorporating plant functional diversity effects in ecosystem service assessments. *Proceedings of the National Academy of Sciences* 104: 20684–20689.

Fensham, R. J., D. W. Butler, R. J. Fairfax, A. R. Quintin, and J. M. Dwyer. 2015. Passive restoration of subtropical grassland after abandonment of cultivation. *Journal of Applied Ecology* 53(1): 274–283.

Gorgone-Barbosa, E., V. Pivello, S. Bautista, T. Zupo, M. Rissi, and A. Fidelis. 2015. How can an invasive grass affect fire behavior in a tropical savanna? A community and individual plant level approach. *Biological Invasions* 17: 423–431.

Helsen, K., M. Hermy, and O. Honnay. 2013. Spatial isolation slows down directional plant functional group assembly in restored semi-natural grasslands. *Journal of Applied Ecology* 50: 404–413.

Hopper, S. 2009. OCBIL theory: Towards an integrated understanding of the evolution, ecology and conservation of biodiversity on old, climatically buffered, infertile landscapes. *Plant and Soil* 322: 49–86.

Hulvey, K. B., and P. A. Aigner. 2014. Using filter-based community assembly models to improve restoration outcomes. *Journal of Applied Ecology* 51: 997–1005.

Ilunga wa Ilunga, E., G. Mahy, J. Piqueray, M. Séleck, M. N. Shutcha, P. Meerts, and M.-P. Faucon. 2015. Plant functional traits as a promising tool for the ecological restoration of degraded tropical metal-rich habitats and revegetation of metal-rich bare soils: A case study in copper vegetation of Katanga, DRC. *Ecological Engineering* 82: 214–221.

Kiehl, K., A. Kirmer, T. W. Donath, L. Rasran, and N. Hölzel. 2010. Species introduction in restoration projects – Evaluation of different techniques for the establishment of semi-natural grasslands in Central and Northwestern Europe. *Basic and Applied Ecology* 11: 285–299.

Larson, J. E., R. L. Sheley, S. P. Hardegree, P. S. Doescher, and J. J. James. 2015. Seed and seedling traits affecting critical life stage transitions and recruitment outcomes in dryland grasses. *Journal of Applied Ecology* 52: 199–209.

Lavorel, S., and E. Garnier. 2002. Predicting changes in community composition and ecosystem function from plant traits: Revisiting the Holy Grail. *Functional Ecology* 16: 545–556.

Le Stradic, S., E. Buisson, and G. W. Fernandes. 2014a. Restoration of Neotropical grasslands degraded by quarrying using hay transfer. *Applied Vegetation Science* 17: 482–492.

Le Stradic, S., E. Buisson, D. Negreiros, P. Campagne, and G. Wilson Fernandes. 2014b. The role of native woody species in the restoration of Campos Rupestres in quarries. *Applied Vegetation Science* 17: 109–120.

Le Stradic, S., M. Séleck, J. Lebrun, S. Boisson, G. Handjila, M.-P. Faucon, T. Enk, and G. Mahy. 2015a. Comparison of translocation methods to conserve metallophyte communities in the Southeastern D.R. Congo. *Environmental Science and Pollution Research* 23(14): 13681–13692.

Le Stradic, S., F. A. O. Silveira, E. Buisson, K. Cazelles, V. Carvalho, and G. W. Fernandes. 2015b. Diversity of germination strategies and seed dormancy in herbaceous species of campo rupestre grasslands. *Austral Ecology* 40: 537–546.

Lezama, F., S. Baeza, A. Altesor, A. Cesa, E. J. Chaneton, and J. M. Paruelo. 2014. Variation of grazing-induced vegetation changes across a large-scale productivity gradient. *Journal of Vegetation Science* 25: 8–21.

Lowe, S., M. Browne, S. Boudjelas, and M. De Poorter. 2000. *100 of the World's Worst Invasive Alien Species: A Selection from the Global Invasive Species Database*. Gland: IUCN.

Oliveira, L. B., E. M. Soares, F. Jochims, T. Tiecher, A. R. Marques, B. C. Kuinchtner, D. S. Rheinheimer, and F. L. F. de Quadros. 2015. Long-term effects of phosphorus on dynamics of an overseeded natural grassland in Brazil. *Rangeland Ecology and Management* 68: 445–452.

Overbeck, G. E., S. C. Müller, A. Fidelis, J. Pfadenhauer, V. D. Pillar, C. C. Blanco, I. I. Boldrini, R. Both, and E. D. Forneck. 2007. Brazil's neglected biome: The South Brazilian Campos. *Perspectives in Plant Ecology, Evolution and Systematics* 9: 101–116.

Overbeck, G. E., J. Hermann, B. O. Andrade, I. I. Boldrini, K. Kathrin, A. Kirmer, C. Koch, J. Kollmann, S. T. Meyer, S. C. Müller, C. Nabinger, G. E. Pilger, J. P. P. Trindade, E. Vélez-Martin, E. A. Walker, D. G. Zimmermann, and V. D. Pillar. 2013. Restoration ecology in Brazil – Time to step out of the forest. *Natureza and Conservação* 11: 92–95.

Overbeck, G. E., E. Vélez-Martin, F. R. Scarano, T. M. Lewinsohn, C. R. Fonseca, S. T. Meyer, S. C. Müller, P. Ceotto, L. Dadalt, G. Durigan, G. Ganade, M. M. Gossner, D. L. Guadagnin, K. Lorenzen, C. M. Jacobi, W. W. Weisser, and V. D. Pillar. 2015. Conservation in Brazil needs to include non-forest ecosystems. *Diversity and Distributions* 21: 1455–1460.

Packard, S., and C. F. Mutel (eds). 2005. *The Tallgrass Restoration Handbook*. Washington, DC: Island Press.

Parr, C. L., C. E. R. Lehmann, W. J. Bond, W. A. Hoffmann, and A. N. Andersen. 2014. Tropical grassy biomes: Misunderstood, neglected, and under threat. *Trends in Ecology and Evolution* 29: 205–213.

Price, J. N., and M. Pärtel. 2013. Can limiting similarity increase invasion resistance? A meta-analysis of experimental studies. *Oikos* 122: 649–656.

Standish, R., and R. Hobbs. 2010. Restoration of OCBILs in south-western Australia: Response to Hopper. *Plant and Soil* 330: 15–18.

Tognetti, P. M., E. J. Chaneton, M. Omacini, H. J. Trebino, and R. J. C. León. 2010. Exotic vs. native plant dominance over 20 years of old-field succession on set-aside farmland in Argentina. *Biological Conservation* 143: 2494–2503.

Veldman, J. W., E. Buisson, G. Durigan, G. W. Fernandes, S. Le Stradic, G. Mahy, D. Negreiros, G. E. Overbeck, R. G. Veldman, N. P. Zaloumis, F. E. Putz, and W. J. Bond. 2015a. Toward an old-growth concept for grasslands, savannas, and woodlands. *Frontiers in Ecology and the Environment* 13: 154–162.

Veldman, J. W., G. E. Overbeck, D. Negreiros, G. Mahy, S. Le Stradic, G. W. Fernandes, G. Durigan, E. Buisson, F. E. Putz, and W. J. Bond. 2015b. Tyranny of trees in grassy biomes. *Science* 347: 484–485.

Zaloumis, N. P., and W. J. Bond. 2011. Grassland restoration after afforestation: No direction home? *Austral Ecology* 36: 357–366.

Zedler, J. 2005. Restoring wetland plant diversity: A comparison of existing and adaptive approaches. *Wetlands Ecology and Management* 13: 5–14.

23

TROPICAL FOREST RESTORATION

David Lamb

Introduction

Methods of restoring tropical forests are broadly similar to those used to restore forests in other biogeographical regions. That is, they involve identifying the relevant species to use and finding ways of assembling these to create diverse and sustainable new forest communities resembling the historical ecosystems once present at a particular site. But tropical regions also have several features that make this task a little different to that faced in most temperate regions. One is that tropical ecosystems often have higher levels of biodiversity, especially those in moister regions with short dry seasons. This means it can be difficult to identify all the species originally present because many are present in very low densities and are hard to find. One immediate consequence of this is it can be very difficult to make fully representative seed collections to begin the restoration process. A second difference is that temperatures are mostly warmer so that that growth and reproduction is often rapid and inter-specific competition can begin at a relatively early stage of the restoration process. Successional changes can be rapid and adaptive management is needed from a very early stage. A third factor is that deforestation is still ongoing in many tropical areas, in contrast to much of the temperate world where deforestation often occurred in the distant past. This means knowledge of the relevant biota and ecosystems is commonly incomplete and ecosystems are being destroyed before we understand the species involved, their taxonomy or how they interact or function. Each of these features has obvious implications for the ways in which restoration can be undertaken in tropical regions.

But, in addition to these bio-physical features, there are several socio-economic factors that also make the task of tropical forest restoration different to that commonly practised in most temperate regions. One is that many of the people living in tropical landscapes are poor and sometimes lack food security. A second is that many landholders are likely to be unfamiliar with the technologies needed to undertake forest restoration and may be sceptical that restoration is something that will benefit them or their families. And, thirdly, policies and institutional frameworks concerning land tenure and land use planning are often fragmentary and still evolving.

These various constraints mean that ecological restoration aimed primarily at restoring former ecosystems and conserving biodiversity may be difficult to achieve except on public land that is owned and managed by state agencies such as national parks or watershed management authorities. On the other hand, other forms of restoration such as multi-species plantings that offer some form of financial or material benefit may be attractive to a wider range of land-holders, especially those with smaller farms, because they improve household incomes, enhance agricultural sustainability and reduce risks inherent in relying on a single product sold into a single market. This means that the task for restoration practitioners working in the tropics is to develop a variety of restoration approaches that suit the various ecological situations found in different types of degraded tropical landscapes and that, at the same time, match the circumstances and aspirations of people living at these sites. In many landscapes a variety of such approaches are likely to be needed.

Initiating forest restoration

The three most common methods used to initiate tropical forest restoration are to rely on natural regeneration, to re-establish plants by directly sowing seeds or to plant seedlings. The main advantages and disadvantages of each of these are summarized in Table 23.1.

Natural regrowth

Forests can sometimes recover on degraded landscapes if the sites are simply protected from further disturbances (Janzen 2002; Chazdon 2014). Their capacity to do so largely depends on there being a residual population of old stumps or roots and a soil seed bank populated by native species. This condition can be met in many recently deforested landscapes but may be less common in agricultural areas that have been used over a longer period. This is because repeated cropping is likely to diminish this pool of residual plants. Natural regrowth is also more likely if there are still patches of natural forest nearby, together with fauna able to distribute seed from these forests across the cleared landscape. Much less colonization will take place if these remnants are small or distant. Natural regrowth can be especially effective in drier climates because the plants present as old stumps, roots or seed can lie dormant until environmental conditions become favourable (e.g. following rain); in such situations planted seedlings often fail because they are taken into the field at the wrong time or there is no follow-up rain.

Despite the financial advantages of natural regeneration there are several disadvantages. One is that it is difficult to control the species composition in the new forest. For example, it may contain only a small proportion of those originally present (at least in the early regrowth stages), the composition may be dominated by one or two species and invasive exotic species may become common. A second potential disadvantage is that natural regrowth might be seen as regenerating forest by site managers but as vacant wasteland by others. This means it may be at risk of being burned or cleared and used for other purposes unless care is taken to protect it. Finally, regrowth may be too sparse or patchy with clumps of new tree growth in some places while regrowth is completely absent in others. In such cases direct seeding or planted seedlings may be more reliable forms of forest restoration.

Planted seedlings

A second means by which reforestation can be done is by planting seedlings. Planting seedlings raised in nurseries is the best way of establishing species that produce only small amounts of seed. Seedlings also have the advantage that the preferred species can be established at particular locations, proportions and densities and that fertilization and weed control is more easily undertaken to enable restoration at highly degraded sites. Planting is obviously more expensive

Table 23.1 Some advantages and disadvantages of alternative methods of initiating forest restoration

Method	Advantages	Disadvantages
Natural regrowth	Low cost. Can treat large areas. Need little ecological knowledge of biota. Less dependent on knowledge of appropriate seasonal conditions for planting seedlings or sowing seed.	May be difficult to prevent further disturbances because it might not be obvious that a site is being restored. Rarely able to overcome legacy of past environment changes such as topsoil loss or soil compaction. Difficult to restore all flora; site may become dominated by only a few species (some of which may be exotic weeds). May result in sparse or patchy distribution of trees. Rate of regrowth may be slow.
Planting seedlings	More reliable. Makes efficient use of scarce seed resources. Can ensure preferred species are established. Can plant at required locations and densities. Weed control easier to manage.	More costly to raise seedlings in nurseries and plant in field. May be difficult to collect seed and raise seedlings of all species in species-rich forests.
Direct seeding	Relatively low cost. Seed can be dispersed from the air meaning inaccessible (e.g. steep) sites can be treated.	Not always effective. Need large quantities of seed since establishment rates often low. Hence may only be possible to sow some of the original biota because of limited seed availability. Seed of some species may need to be buried to achieve germination success. There may be only a short 'window' of time when seed should be sown. Weed control is difficult so best used at sites that have been entirely cleared before seeds are sown (e.g. by fire or ploughing).

than natural regrowth (or direct seeding – see following text) but the degree of control it provides makes it the preferred option in many situations. Not all species are easily re-established by planting. In some cases this may be because the seeds are difficult to collect (too few parent trees or infrequent flowering times). Alternatively, it may be difficult to restore some indigenous species at degraded sites because of changes in environmental conditions (e.g. topsoil loss, soil compaction). Likewise some rainforest species are difficult to establish in open-field conditions and appear to require some temporary shelter to facilitate their establishment and further research is needed to clarify which species benefit from nurse trees, how much cover is needed and how long this cover should be maintained. In many facets of restoration ecology, there is a good understanding of the principles but the sheer amount of variation within socioecological systems means it is hard to keep pace and come up with technical approaches that meet all possible combinations. For many tropical forests, small-scale variation across landscapes is often large and exacerbates this problem.

Directly sown seed

A third way of initiating restoration is to directly sow seed of the preferred species thereby avoiding the need to depend on residual plants or seed being dispersed into the site from external sources (Engel and Parrotta 2001; Doust *et al.* 2008; Schmidt 2008; Tunjai and Elliott 2012). The advantage of direct seeding is that it avoids the patchiness problems found in some natural regrowth situations and is cheaper than planting seedlings since the costly nursery stage is avoided. Conversely, the process is relatively inefficient because only a very small proportion of seeds that are sown actually produce seedlings; the remainder are either eaten by predators, fail to germinate or the seedlings die soon after germination because of drought or weed competition. This means large volumes of seed are usually required to achieve appropriate seedling densities. This, in turn, means it may be difficult to use direct sowing alone to restore diverse forest communities because it is impossible to collect sufficient seed of most species. However, direct sowing may be a useful way of initiating forest successions where further development can be carried out by colonists brought to the site by wind or seed dispersers. It may also be a complementary tool to supplement further development of successions that have already been initiated by natural regrowth or planted seedlings.

Forest restoration in practice

There are a number of silvicultural variants of each of these three basic approaches that differ in the proportions of various species used, the sequence in which these species are introduced into the new succession and in the ways in which facilitation and competition are managed. Some of the more common tropical restoration methods are summarized in Table 23.2 with more detailed accounts given by Griscom and Ashton (2011), Lamb (2011), Elliott *et al.* (2013) and Stanturf *et al.* (2014).

Methods relying primarily on natural regrowth

Natural regeneration

Natural regrowth can be an attractive option when there is evidence that coppice growth is common or that newly germinated tree seedlings are widespread (Viera and Scariot 2006). But what is the threshold condition when direct seeding or planted seedlings will be a safer option? Shono *et al.* (2007) suggest there should be 200–800 seedlings per ha that are taller than 15 cm if natural regeneration is to be relied upon. But situations differ and any guidelines need to reflect the extent of competition from weeds and the rate at which new regrowth trees are being recruited. Once regrowth is under way the main tasks are to monitor it to ensure that an appropriate successional trajectory is maintained and protect it from further disturbances. Any protection programme is likely to require some form of community involvement. In some situations landholders may wish to incorporate natural regrowth into their existing farming practices.

Enrichment with seedlings or seed

Natural regrowth areas can also be dominated by a relatively small number of species or lack species of particular conservation interest. Such compositional limitations can be overcome by enriching the regrowth with additional trees of the preferred species. This might be done by

Table 23.2 Different approaches to restoring tropical forests

Approach	Methodology
Natural regeneration	
Natural regeneration	Assess whether regrowth is sufficient and not excessively patchy; if this regrowth is adequate, protect site from further disturbances and remove exotic species.
Enrich regrowth with seeds or seedlings of preferred species	Distribute seeds or plant seedlings in localized areas with insufficient natural regeneration; create canopy gaps in forest regrowth and plant seedlings.
Plant perch trees to facilitate additional seed dispersal	Plant fast-growing trees, preferably in small groups, able to act as perches for seed-dispersing birds.
Planted seedlings	
Framework species method	Initiate successional development with small number (<30 species) of fast-growing pioneer species and poorly dispersed species.
Maximum diversity plantings	Plant as many species as possible; complement these with additional plantings as seed of further species become available.
Under-plant beneath nurse trees	Use a single tolerant nurse species to exclude weeds and facilitate the subsequent establishment of less tolerant species unable to be established in the open.
Multi-species plantings to generate incomes	Grow mixtures (mostly <20 species) of commercially attractive species.
Direct seeding	
Sow seeds from ground or from aircraft	Disperse seeds into land from which weeds have been removed (by ploughing, fire or weedicides).
Combing direct seeding and other reforestation methods	Enrich natural regrowth or plantings with seed sown beneath an existing canopy. Alternatively, enrich directly sown forest with planted seedlings.

planting these in newly created canopy gaps or strips cut through the regrowth that ensures sufficient light reaches seedlings at the ground level. This usually means enrichment is best done when regrowth is still relatively short and it is less feasible when regrowth trees are already tall because bigger gaps are then needed to increase light levels at the ground level (Adjers *et al.* 1995; Lamb 2011). Supplementary plantings might also be carried out in areas where there are too few naturally regenerating trees. Alternatively, seeds of preferred species might be distributed in these more open patches.

Establish perch trees to facilitate additional seed dispersal

Much natural regrowth depends on seeds being dispersed into the site. The rate at which this occurs can be increased by establishing trees providing perches for seed-dispersing birds. Fast-growing trees can act as focal points for bird movement across a site and hasten seedling colonization. There are no prescriptions specifying the number of trees to use or the spacing between them. However there is evidence that clumps of trees are preferable to single trees because they provide larger and more resilient habitat areas for seed–dispersing wildlife (Zahawi *et al.* 2013; Corbin *et al.* 2016).

Methods relying primarily on planted seedlings

Deliberately planting seedlings of the required species is more reliable than relying on natural regeneration but, as noted earlier, is not without problems. A number of approaches have been developed to deal with these problems.

Framework species method

The framework species method seeks to deal with the problem of restoring large numbers of species when it may be difficult to get their seed (Goosem and Tucker 2013). The approach takes advantage of the capacity of even modest plantings involving a relatively small number of species to catalyse further successional development when there are patches of residual forest nearby together with seed dispersers able to cross the intervening landscape. This method involves planting seedlings of a mixture of up to, say, 20 species, including some fast-growing and short-lived pioneer species (perhaps 30 per cent of the total). Other species to use are those found in more mature successional phases and especially species that are poorly dispersed or that flower and fruit only infrequently and so may not otherwise reach the site. By planting these at relatively high densities (e.g. 2,500 seedlings per ha) the pioneers will grow quickly and close canopy thereby excluding grasses and other weeds. The slower growing but shade-tolerant later successional-stage species may be temporarily supressed by these pioneers but, over time, are eventually able to grow up and join the upper canopy layer when the short-lived pioneer species senesce. The regenerating forest quickly becomes a target for seed-dispersing wildlife that brings in seed of additional species and thereby help accelerate successional recovery (Figures 23.1 and 23.2). Not all colonists are necessarily beneficial and the edge of new forests can sometimes be invaded by weed species because side light penetrates beneath the forest canopy. In such cases it may be useful to plant a boundary row or two of a deep-crowned species (i.e. those where leaves persist to ground level) to seal the edge and prevent such weed invasion.

Maximum diversity plantings

Enrichment of plantings via colonization may not occur in heavily and extensively cleared landscapes lacking patches of remnant forest or where there are now too few wildlife species or populations still able to carry seed across deforested areas. It can also be difficult where most species are dispersed by wind since most tree seed dispersed in this way usually only move over relatively short distances. In such sites the only alternative is to plant as many native species as possible and to continue this in successive years as seed of further species becomes available (e.g. as flowering occurs). Apart from simply using species that are readily available, priority might be given to ensuring a diversity of life forms (e.g. trees, shrubs and lianas) and functional types (e.g. shade tolerant and shade intolerant, nitrogen fixers, species providing resources for wildlife). These subsequent plantings may be in adjoining areas or take the form of enrichment plantings. In this case there is less need to emphasize planting pioneer species and the proportions of these may be around 10 per cent of the total. This enables some periodic canopy gap creation which allows seedlings of the planted trees to regenerate and grow into the canopy layer. The prospect of full restoration clearly depends on the capacity to locate and establish all the original species.

Figure 23.1   Forest restored using framework species method in northern Thailand at age 8 years. At this stage most of the tree species present would have been planted as seedlings

Source: photo by Steve Elliott

Figure 23.2   Forest restored using framework species method in northern Thailand at age 16 years after planting (same site as Figure 23.1). By this stage a number of wildlife have begun using the site and the tree species present include those planted and others that have colonized the site

Source: photo by Steve Elliott

Under-plant beneath nurse trees

Not all primary forest species can be planted in the open and some will only regenerate beneath a tree canopy. For example, many (though not all) members of the tropical Asian Dipterocarpaceae require some initial shade to become established. In such cases restoration can be achieved by first growing a monoculture of a pioneer species to act as a nurse crop and then under-planting this with the desired species. The nurse species facilitates their establishment by reducing radiation levels and eradicating weeds. Species differ in the degree of shade initially needed and how long this is required before the nurse trees can be removed (because they become competitive inhibitors rather than facilitators). A two-stage establishment system like this is clearly more complicated than one where everything is planted at the same time. Nonetheless, they can be arranged so that harvesting of the nurse trees helps fund future restoration thereby enabling the new forest area to expand. One such system is described by McNamara *et al.* (2006) involving *Acacia* as nurse trees that provide shade and also have the capacity to fix nitrogen and improve soil fertility. Of course the nurse trees may also facilitate the colonization of the site by additional species from any nearby natural forest remnants.

Multi-species plantings to generate incomes

The final method involving seedlings is one that seeks to strike a balance between the ideal of restoring an original forest ecosystem and the need to provide improvements in the livelihoods of people living in the area. In this case the primary motive for using multi-species plantings has less to do with conservation and rather more to do with diversifying the variety of goods being produced thereby reducing a household's exposure to economic risks. Of course such plantings are also more ecologically resilient than simple monocultures and are able to provide a variety of ecosystem services such as improved watershed protection and wildlife habitats. Their complexity means they are unlikely to be used by large industrial timber companies but this may not be such an impediment for small landholders. The numbers and types of species used are likely to be dictated by local market conditions. One model described by Nguyen *et al.* (2014) used around 20 species of which 13 were relatively fast-growing timber trees that could be harvested before 20 years to provide an early cash flow leaving slower growing and more valuable native species to grow into larger sawlogs (Table 23.3). These longer-lived species might be all clear-felled when they reach a marketable size or, depending on demand, individual trees in the forest may be opportunistically logged leaving the plantings to evolve into an unevenly-aged forest.

Methods based on direct seeding

Direct seeding has not been widely used for tropical forest restoration (except in some mining rehabilitation projects) although it continues to be of interest to researchers because of its potentially significant cost-effectiveness.

Seed sown from ground or aircraft

Seed can be sown by hand, modified agricultural machinery or from the air (Jonson 2010; Lamb 2014). The advantage of agricultural machinery is that this can bury seed and this usually improves seedling establishment rates. On the other hand, aircraft may allow restoration to begin at sites that are otherwise difficult to access and to cover a relatively large area within the short time that weather conditions are favourable for germination and seedling establishment.

Table 23.3 A multi-species plantation system designed for use by smallholders in the Philippines. The species differ in their ecological traits (especially shade tolerance, growth rates, wood densities, potential to fix nitrogen) and in their economic value. Selective harvesting (i.e. thinning) gradually reduces the diversity of species present but also generates a steady cashflow thereby increasing the overall attractiveness of the plantings to landholders

Product	Time of thinning (years)	Number of species	Density of trees per ha	Possible species able to generate particular products within this timeframe*
Firewood	6–10	3–5	450	*Casuarina equisetifolia, Gmelina arborea, Gymnostoma rumphianum, Leucaena leucocephala, Melia dubia, Samanea saman*
Poles	8–12	2–3	200	*Dracontomelon dao, Gmelina arborea, Gymnostoma rumphianum, Pterocarpus indicus, Terminalia macrocarpa*
Fast-growing timber	14–18	3–5	250	*Agathis philippinensis, Gmelina arborea, Parashorea plicata, Shorea contorta, Shorea palosapis, Shorea polysperma, Swietenia macrophylla, Tectona grandis*
Slow-growing timber	>20	3–10	200	*Calophyllum blancoi, Calophyllum lancifolium, Dipterocarpus grandiflorus, Dipterocarpus kerrii, Dipterocarpus kunstleri, Hopea dalingdingan, Hopea malibato, Hopea plagata, Podocarpus rumphii, Pterocarpus indicus, Shorea negrosensis, Shorea polysperma, Vitex parviflora*
Total		11–23	1100	

Note: *Species include some of those currently used in smallholder plantings in various parts of the Philippines.
Source: based on Nguyen *et al.* (2014)

Even though the diversity of species that can be established using direct seeding may be limited it can be useful as a means of initiating a tree cover thereby allowing successional development based on recolonization from natural forest remnants.

Combining direct seeding and other reforestation methods

In each of the methods outlined before there may be scope for opportunistic species enrichment using direct seeding beneath established tree canopies (Cole *et al.* 2011). Likewise, there may be opportunities to plant seedlings of species with only limited seed availability into forests recently established using direct seeding.

Management of the new forests

Forests established by planting seedlings or sowing seed on a single occasion will be even-aged and have uniform canopy layers. Successional development should mean that that these even-aged forests gradually change to become uneven-aged forests having multiple canopy layers. But this will only occur when there is sufficient light able to reach the forest floor to allow naturally regenerated seedlings of the planted trees as well as colonists from external sources to develop in the understorey. The main way of ensuring this transition and accelerating successional development is to foster the generation of canopy gaps. This will automatically occur when using the Framework Species Method but, otherwise, may have to be achieved by deliberately creating canopy gaps (which is the basis of the nurse tree method described previously). Of course the extent to which any intervention such as this is actually needed depends on the density of the canopy layer. Species-rich understories have been observed in some monocultural timber plantations where the species involved have relatively open crowns (Keenan *et al.* 1997).

Attempts to re-assemble plant communities can sometimes lead to unexpected outcomes. Some species may be lost from the succession because they are out-competed or fail to reproduce while others can become over-abundant because they are more successful in competing for resources or can produce comparatively large numbers of seedlings. Knowledge of relative growth rates and shade tolerances can be useful in predicting possible outcomes but the nature of competitive interactions in multi-species forests is always uncertain, especially at degraded sites where topsoils have been lost and the relative competitive abilities of various species may have changed. The only option for managers is to practise adaptive management based upon a programme of regular monitoring (Lindenmayer and Likens 2009). This may indicate that, at some stage, the site needs to be enriched with certain species or that the populations of some over-abundant species should be reduced. Both can be expensive options if carried out over large areas.

How successful have previous attempts at tropical forest restoration been?

Although tropical forest restoration is a relatively new practice, there is evidence from a variety of locations and ecological situations indicating that successional development can be rapid once the process has been initiated provided sites have not been too heavily degraded. Under these circumstances there is every prospect of species-rich tropical forests broadly similar to those once present being restored at previously deforested sites, at least in landscapes still retaining some residual natural forest (Keenan *et al.* 1997; Janzen 2002; Elliott *et al.* 2013; Chazdon 2014). The caveat is that while botanical restoration seems promising, there is less evidence concerning the restoration of wildlife species and the reassembly of food webs. A number of studies have found evidence of certain wildlife species recolonising newly restored forest areas especially when natural forests able to act as sources of colonists are nearby (Chazdon *et al.* 2009; Catterall *et al.* 2012). These species are likely to be critical for future successional development because many animals form mutualistic relationships with plant species. But the restoration of forest habitats does not necessarily always lead to the restoration of wildlife populations. Indeed, in some parts of the tropics, wildlife species have been hunted to near-extinction and the residual forests are 'silent' and unable to act as sources of colonists. It is not just the large predators that have been lost but herbivores as well. In such situations the newly restored forests are unlikely to acquire a significant component of their original biodiversity. The long-term consequences of this biological impoverishment for successional development are unclear.

Scaling up – forest restoration at a landscape scale

It is important that ecological restoration be undertaken on a large scale if we are to generate the conservation and functional outcomes which are the common objectives of much restoration. This is because this more extensive scale is the one at which many key ecological processes operate.

Ecological issues

Undertaking ecological restoration on a large scale poses particular problems. One is that large areas usually have a large variety of ecological niches. This means that the restoration 'target' varies and that planting the same species assemblage across the whole area is unlikely to be appropriate. But, at the same time, it is usually difficult to identify the appropriate species to use at these various sites, especially in degraded landscapes, because so little is usually known about the basic ecological preferences of most tropical flora. In practice this problem may be more apparent than real since many species are likely to be re-established through being dispersed to particular sites by wind or wildlife than by planting seedlings or sowing seed. But the extent to which this occurs depends on the landscape context and the capacity of this type of colonization to cover all sites. Some of the practical issues involved in implementing forest restoration at a landscape scale are discussed by Rodrigues *et al.* (2011) in the context of a large-scale forest restoration programme in Brazil.

A second problem is that the capacity of a new forest to be functionally effective depends on just where in a landscape it is established. For example, restoration that improves connectivity between natural forest remnants or that enlarges small, recently isolated forest remnants is likely to be more effective than scattered small random plantings in agricultural areas. Likewise, erosion will be more effectively controlled by restoration done on hill slopes or along riparian strips than by plantings on flatter lands.

Thirdly, functional effectiveness is also likely to depend on the scale of restoration at these target sites. For example, many hydrological processes will not be greatly influenced by simply restoring a few hectares of additional forest, however well this is done. Likewise, the maintenance of viable populations of many wildlife species depends on them having sufficiently large and contiguous areas of suitable habitat. Small areas of carefully restored forest may be enough to maintain populations of certain plant species but may not be sufficient to sustain viable populations of species requiring larger home ranges. Some of the ecological issues involved in designing how much and where restoration should be done are discussed further in Holl and Aide (2011) and Lamb (2014).

Socio-economic issues

Large-scale forest restoration will also have socio-economic consequences and the prospect of these may cause some landholders to resist the implementation of forest restoration. This means large-scale forest restoration should only carried out after developing a supportive policy and institutional framework (Mayers and Bass 2004; Guariguata and Brancalion 2014; Lamb 2014). Policies define the objectives and ways in which forest restoration should be carried out while institutions are the organizations that ensure these policies are implemented. A variety of each will be needed. Some policies might seek to remove any current disincentives for restoration (e.g. by granting land tenure or by removing excessive bureaucratic regulatory environments). Some might specify when ecological restoration must be done (e.g. where ecological restoration is legally needed to offset biodiversity losses occurring elsewhere, to protect watersheds or to restore land degraded by mining). Others might include measures that actively promote restoration as a new regional land use (e.g. through the provision of subsidies and incentive payments to actively encourage landholders to undertake restoration at particular areas or by encouraging the development of markets for the supply of ecosystem services generated by restoration).

The institutions will necessarily include some operating at a national scale such as national land use planning bodies or national inter-agency committees seeking to balance national reforestation or conservation goals and future food security. Others will be those dealing with more regional or local issues. These might include provincial bodies involving government agencies and local stakeholders that implement decisions made at a national level and distribute funds such as subsidies or incentive payments. They might be bodies that assist in the identification of priority areas for restoration and where compensation payments might be needed to enable restoration to be carried out. But they may also include growers clubs to share silvicultural knowledge or marketing groups to assist landholders' benefit from the sale of goods or ecosystem services. Ostrom (2010) described this network of complementary institutions as a polycentric institutional framework.

Role of land ownership in restoration of tropical forests

Ecological restoration is always difficult but the combination of high levels of biodiversity and the widespread occurrence of rural poverty make tropical ecological restoration especially so. The initial costs can be significant (especially for smallholders) and most benefits often flow to the community as a whole rather than to individual households. This means that many types of ecological restoration are likely to be more easily done on state-owned land or on parts of the estates of large landholders who are able to afford to take a long-term perspective rather than on the farms of smaller landholders. But even this can be difficult. In theory the state should be able to protect and manage natural regeneration occurring over large areas but in practice this has been hard for governments to achieve (e.g. it can be politically difficult to exclude landless squatters). Likewise despite the increasing prevalence of national reforestation programmes many governments may find it difficult to undertake complex multi-species plantings such as the framework species method over large areas because it necessarily involves investing significant financial and human resources over an extended period. In practice much large-scale reforestation still involves relatively small numbers of species and hardly qualifies as ecological restoration.

What might be the role of other, smaller landholders? Most tropical landscapes are occupied by many small landholders and their collective impact could be large if they undertook forest restoration on part of their land. But, to the extent that these households are willing to undertake any form of reforestation, most are likely to be more interested in reforestation that generates an economic benefit rather than simply restores biodiversity. Until recently, the only reforestation option available to such landholders appeared to be the same as those used by industrial plantation growers, namely planting monocultures using fast-growing tree species. Such plantations may be financially attractive for landholders able to form an out-grower relationship with an industrial timber mill but may be much less financially attractive when this is not the case. The multi-species plantings of the type shown in Table 23.3 represent another option for these landholders. These types of plantings have the potential to provide economic benefits and reduce a household's exposure to risk but also generate a wider variety of ecosystem services for the wider community (Aerts and Honnay 2011).

For those interested in restoring degraded tropical forest lands the best option may be to recognize that full ecological restoration over large areas may be difficult to achieve and that a more realistic target may be to aim for a variety of different types of reforestation using several of the approaches described in Table 23.2. This may seem a dispiriting outcome but it is important to recognize that restoration is necessarily a long-term process and that changes are under way that may favour its wider adoption. One is that migration from the country to the city is common in most countries, increasing opportunities for natural regrowth to take place on some former farmland. At the same time there is increasing interest in expanding forest cover in order to supply various ecosystem services. Some of these may be supplied by floristically simple forests but many, such as carbon storage or watershed protection, will require new forests having relatively high levels of biodiversity in order to be resilient and self-sustaining and functionally effective. In short, the need for ecological restoration is likely to increase with time. Some of the issues involved in assessing future restoration 'success' are reviewed by Le *et al.* (2012).

Conclusion

The restoration of tropical forests is difficult because of the high levels of biodiversity these contain, especially those receiving high rainfall and having limited dry seasons. It is even more difficult when this is being attempted over large areas. In practice a variety of methods are likely to be needed to initiate forest restoration in different parts of a landscape depending on the ecological situation and the economic circumstances of landholders. A number of these have been developed but these need to be fine-tuned according to the attributes of local species. Ecologists often assume that the benefits of restoration are self-evident but landowners are necessarily forced to judge the opportunity costs of restoration and compare the merits of restoration with alternative land uses. Policies and institutions are needed to ensure that both the benefits and costs of restoration are shared between individual landholders and the broader community.

Case studies

Case study 1: natural regeneration in Niger has been incorporated into local farming systems

Most of the agricultural areas of Niger are in the dry zone of sub-Saharan Africa. Colonial administrators sought to 'modernize' agricultural practices by clearing trees from farms to create broad fields. This led to increased erosion, a reduction in crop yields and a decline in rural livelihoods. But, prompted by studies done by non-governmental organizations, a change in government policy has led to the encouragement of woody regrowth on parts of farmers' lands. The regrowth has reduced erosion and diversified household incomes by producing firewood. The natural regeneration is better described as an increase in rural tree growth than as forest restoration because it is occurring within and around field crops. Nonetheless, the policy change has been responsible for tree regrowth over 5 million ha at no cost to the government and has been described as the 'regreening of the Sahel' (Sendzimir *et al.* 2011).

Case study 2: tropical forest restoration on a large scale in Thailand

An area of degraded farmland near Khao Yai National Park in central Thailand has been the centre of a large-scale restoration project covering around 800 ha. Restoration was initiated using a small number of species with readily available seed. These were introduced by planting seedlings or sowing seed. The species mix included native species as well as a few exotic species such as *Leucaena leucocephala*. The number of species planted was modest but the sites have been colonized and enriched by species brought in from the nearby park by wildlife. Some 69 species were planted but the area now contains around 232 species. This has led to a closed canopy and the gradual loss of the exotic species used to initiate the succession. The site is now being colonized by a number of large vertebrate species and has become a popular location for visits by eco-tourists (Lamb 2011).

Case study 3: policies and institutions to encourage forest restoration in Brazil

National and provincial policies in Brazil are designed to encourage forest restoration on extensively cleared agricultural lands. In some places this means restoring forests until 20 per cent cover is reached. In the state of Sao Paulo these restored forests should contain at least 80 species. A variety of institutional arrangements have been developed to achieve these policy goals. One of the most interesting is an organization known as the Atlantic Forest Restoration Pact which includes non-government conservation organizations, private companies, research institutions and state agencies. The group has around 80 projects covering 60,000 ha under way and aims to restore 15 million ha of tropical forests in the Atlantic Forest region by 2050. This body helps arrange supplies of seedlings, advises on methodologies to use depending on site conditions and acts as a knowledge bank to gather feedback and share information among members about reforestation methods and marketing as well as monitoring successional development (Brancalion *et al.* 2013; Melo *et al.* 2013; Richards *et al.* 2015).

References

Adjers, G., S. Hadengganan, J. Kuusipalo, K. Nuryanto, and L. Vesa. 1995. Enrichment planting of dipterocarps in logged-over secondary forests – effect of width, direction and maintenance method of planting line on selected Shorea species. *Forest Ecology and Management* 73: 259–270.

Aerts, R., and O. Honnay. 2011. Forest restoration, biodiversity and ecosystem functioning. *BMC Ecology* 11: 1–10.

Brancalion, P. H., R. A. Viani, M. Calmon, H. Carrascosa, and R. R. Rodrigues. 2013. How to organize a large-scale ecological restoration program? The framework developed by the Atlantic Forest Restoration Pact in Brazil. *Journal of Sustainable Forestry* 32: 728–744.

Catterall, C. P., A. N. D. Freeman, J. Kanowski, and K. Freebody. 2012. Can active restoration of tropical rainforest rescue biodiversity? A case with bird community indicators. *Biological Conservation* 146: 53–61.

Chazdon, R. 2014. *Second Growth: The Promise of Tropical Forest Regeneration in an Age of Deforestation.* University of Chicago Press, Chicago, IL.

Chazdon, R. L., C. A. Peres, D. Dent, D. Sheil, A. E. Lugo, D. Lamb, N. Stork, and S. E. Miller. 2009. The potential for species conservation in tropical secondary forests. *Conservation Biology* 23: 1406–1417.

Cole, R. J., K. D. Holl, C. L. Keene, and R. A. Zahawi. 2011. Direct seeding of late-successional trees to restore tropical montane forest. *Forest Ecology and Management* 261: 1590–1597.

Corbin, J. D., J. R. Robinson, L. M. Hafkemeyer, and S. N. Handel. 2016. A long-term evaluation of applied nucleation as a strategy to facilitate forest restoration. *Ecological Applications* 26: 104–114

Doust, S. J., P. D. Erskine, and D. Lamb. 2008. Restoring rainforest species by direct seeding: tree seedling establishment and growth performance on degraded land in the wet tropics of Australia. *Forest Ecology and Management* 256: 1178–1188.

Elliott, S., D. Blakesley, and K. Hardwick. 2013. *Restoring Tropical Forests: A Practical Guide.* Royal Botanic Gardens, Kew.

Engel, V. L., and J. A. Parrotta. 2001. An evaluation of direct seeding for reforestation of degraded lands in central Sao Paulo state, Brazil. *Forest Ecology and Management* 152: 169–181.

Goosem, S., and N. Tucker. 2013. *Repairing the Rainforest* (2nd edn). Wet Tropics Management Authority and Biotropica, Cairns, Australia.

Griscom, H. P., and M. S. Ashton. 2011. Restoration of dry tropical forests in Central America: a review of pattern and process. *Forest Ecology and Management* 261: 1564–1579.

Guariguata, M. R., and P. H. Brancalion. 2014. Current challenges and perspectives for governing forest restoration. *Forests* 5: 3022–3030.

Holl, K. D., and T. M. Aide. 2011. When and where to actively restore ecosystems? *Forest Ecology and Management* 261: 1558–1563.

Janzen, D. 2002. Tropical dry forest: Area de Conservacion Guanacaste, northwest Costa Rica. Pages 559–583 in M. Perrow and A. Davy (eds), *Handbook of Ecological Restoration.* Cambridge University Press, Cambridge.

Jonson, J. 2010. Ecological restoration of cleared agricultural land in Gondwana Link: lifting the bar at 'Peniup'. *Ecological Management & Restoration* 11: 16–26.

Keenan, R., D. Lamb, O. Woldring, T. Irvine, and R. Jensen. 1997. Restoration of plant diversity beneath tropical tree plantations in northern Australia. *Forest Ecology and Management* 99: 117–132.

Lamb, D. 2011. *Regreening the Bare Hills: Tropical Forest Restoration in the Asia-Pacific Region*. Springer, Dordrecht.

Lamb, D. 2014. *Large-Scale Forest Restoration*. Earthscan Routledge, Abingdon.

Le, H. D., C. Smith, J. Herbohn, and S. Harrison. 2012. More than just trees: assessing reforestation success in tropical developing countries. *Journal of Rural Studies* 28: 5–19.

Lindenmayer, D. B., and G. E. Likens. 2009. Adaptive monitoring: a new paradigm for long-term research and monitoring. *Trends in Ecology & Evolution* 24: 482–486.

Mayers, J., and S. Bass. 2004. *Policy that Works for Forests and People: Real Prospects for Governance and Livelihoods*. Earthscan, London.

McNamara, S., D. V. Tinh, P. D. Erskine, D. Lamb, D. Yates, and S. Brown. 2006. Rehabilitating degraded forest land in central Vietnam with mixed native species plantings. *Forest Ecology and Management* 233: 358–365.

Melo, F. P., S. R. Pinto, P. H. Brancalion, P. S. Castro, R. R. Rodrigues, J. Aronson, and M. Tabarelli. 2013. Priority setting for scaling-up tropical forest restoration projects: early lessons from the Atlantic Forest Restoration Pact. *Environmental Science & Policy* 33: 395–404.

Nguyen, H., D. Lamb, J. Herbohn, and J. Firn. 2014. Designing mixed species tree plantations for the tropics: balancing ecological attributes of species with landholder preferences in the Philippines. *PLoS ONE*; e98600. doi: 10.1371/journal.pone.0098600.

Ostrom, E. 2010. Polycentric systems for coping with collective action and global environmental change. *Global Environmental Change* 20: 550–557.

Richards, R.C., J. Rerolle, J. Aronson, P. H. Pereira, H. Gonçalves, and P. H.S. Brancalion. 2015 Governing a pioneer program on payment for watershed services: stakeholder involvement, legal frameworks and early lessons from the Atlantic forest of Brazil. *Ecosystem Services* 16: 23–32

Rodrigues, R. R., S. Gandolfi, A. G. Nave, J. Aronson, T. E. Barreto, C. Y. Vidal, and P. H. S. Brancalion. 2011. Large-scale ecological restoration of high-diversity tropical forests in SE Brazil. *Forest Ecology and Management* 261: 1605–1613.

Schmidt, L. 2008. *A Review of Direct Sowing versus Planting in Tropical Afforestation and Land Rehabilitation*. Forest and Landscape Denmark, University of Copenhagen, Horsholm.

Sendzimir, J., C. P. Reij, and P. Magnuszewski. 2011. Rebuilding resilience in the Sahel: regreening in the Maradi and Zinder regions of Niger. *Ecology and Society* 16: 1.

Shono, K., E. A. Cadaweng, and P. B. Durst. 2007. Application of assisted natural regeneration to restore degraded tropical forestlands. *Restoration Ecology* 15: 620–626.

Stanturf, J. A., B. J. Palik, and R. K. Dumroese. 2014. Contemporary forest restoration: a review emphasizing function. *Forest Ecology and Management* 331: 292–323.

Tunjai, P., and S. Elliott. 2012. Effects of seed traits on the success of direct seeding for restoring southern Thailand's lowland evergreen forest ecosystem. *New Forests* 43: 319–333.

Viera, D. L. M., and A. Scariot. 2006. Principles of natural regeneration of tropical dry forest for restoration. *Restoration Ecology* 14: 11–20.

Zahawi, R. A., K. D. Holl, R. J. Cole, and J. L. Reid. 2013. Testing applied nucleation as a strategy to facilitate tropical forest recovery. *Journal of Applied Ecology* 50: 88–96.

24

THE RESTORATION
OF CORAL REEFS

Boze Hancock, Kemit-Amon Lewis and Eric Conklin

Introduction

A threatened ecosystem

Coral reefs are areas of exceptional biodiversity, social value and are among the most economically valuable ecosystems on the planet. They provide numerous benefits to people including producing fish for food and markets, protecting coasts from storms, providing recreational benefits and they are a critical source of incomes from the tourism industry with 23 countries deriving more than 15 per cent of GDP from reef-related tourism (Burke *et al.* 2011; Rinkevich 2015). Over 275 million people live in the immediate vicinity of coral reefs and are highly dependent on the services they provide but coral reefs are not just important to coastal communities and small island states. Inland communities depend on the fish produced or the sand generated from coral reefs for the production of concrete or as the beach of a holiday destination. Despite our dependence on coral reefs, they are under threat with 75 per cent rated as threatened globally (Burke *et al.* 2011).

Causes of decline

Threats to coral reefs tend to fall into similar categories around the world, whether they are local or globally derived. Local threats can be categorized as overfishing and destructive fishing, watershed-based pollution (mainly eutrophication and increased sedimentation), marine-based pollution and biogenic threats (Hughes *et al.* 2010; Burke *et al.* 2011). Overfishing generally describes the reduction in the abundance of herbivores which exert top-down control on algal growth. Herbivores such as the parrot fish, surgeonfish, tangs, unicornfish, urchins and Trochid gastropods exert grazing pressure on macro- algae and algal turfs and reducing the abundance of grazers can allow their unchecked growth. When anthropogenic nutrients are added these act to further increase the growth of algae, adding to the likelihood of a phase shift from coral dominated system to an algal dominated one (Hughes *et al.* 2010). An increased sediment load from coastal development, poorly designed runoff or agriculture can further stress corals tending to favour the growth of algae over coral. The impact of individual chemical pollutants tends to be more difficult to determine. An example is the effect of oxybenzone, an active ingredient in many sunscreens and a breakdown product of plastics. Recent work points to this compound being toxic to the planula larvae of many coral species and suggests that the trace concentrations common in popular recreational areas could be contributing to recruitment failure (Downs *et al.* 2015). Biogenic threats include the predators such as crown of thorns starfish, drupellid snails, and coral diseases. An often overlooked but frequently important cause of decline in coral reef condition is coral mining and destructive fishing practices. Mining has commonly included taking coral blocks as building material or to burn for the production of lime to produce mortar.

The 'local' scale impacts on coral reefs which tend to lower a reef's capacity to withstand additional insults are increasingly exacerbated by global impacts from climate change: increased water temperatures, ocean acidification and increasing storm intensity. Bleaching events caused by periods of increased water temperatures have had devastating impacts on coral communities in most regions of the world. Globally 60 per cent of coral reefs are under threat from local sources and the number rises to 75 per cent when global impacts are considered (Burke *et al.* 2011).

Addressing the declines

In the face of such ubiquitous local and global threats the most appropriate management approach to conserve the remaining reefs is to maximize the resilience of the coral reef ecosystem. The resilience of an ecosystem is its capacity to absorb recurrent disturbances or shocks and adapt to change without fundamentally switching to an alternative stable state (Hughes *et al.* 2010). An entire web-based network designed to support coral reef managers while they work to build resilience in coral reefs has been developed with an abundance of information on the topic (www.ReefResilience.org). The network extends to over a thousand coral reef managers in over 70 countries. Given the scale and importance of coral reefs it is critically important that the remaining reefs are conserved to the degree possible.

It is also apparent that in general those reefs that have suffered substantial changes are not reverting to their former state over time (Horoszowski-Fridman *et al.* 2015; Rogers *et al.* 2015). Thus there is a growing need to supplement management measures with active intervention, or restoration. It needs to be stressed that the restoration approaches tend to be labour intensive and, therefore, expensive and influence relatively small areas, so the growing need for restoration does not reduce the ongoing need to manage the existing reef for maximum resilience in the face of multiple threats.

Restoration

Scope

Coral restoration is a young and rapidly developing field. As such a definition of what constitutes coral restoration is still developing. The distinction between conservation and restoration is a somewhat artificial and, therefore, arbitrary line that divides the continuum of practices between managing the stressors impacting coral biology and the active manipulation of that biology toward a goal that includes improved coral status. For this chapter we use the active manipulation of components of the reef as the measure of what constitutes restoration. This could include: the addition or manipulation of recruitment surfaces, the transplant of colonies with or without the asexual multiplication of the colony and nursery grow-out, the introduction of sexual recruits, as well as the active maintenance of algal growth and the herbivore community required to control it.

Artificial substrates

For degraded reefs that have lost much of the structural complexity of coral dominated reefs it is tempting to hypothesize that simply adding structure will restore many of the ecosystem services originally provided by the coral community (Rogers *et al.* 2015). There is unresolved debate around the function of these 'artificial reefs' and they are not considered in the current scope of ecological restoration. There are two situations where artificial structures may become restoration tools. The first is where they are designed to provide settlement substrate in situations that are deemed to be substrate limited. Early initiatives were based on the concept of artificial reefs with structures better known as 'fish-aggregating devices' deployed on non-reef platforms mainly to enhance fish catches (Chou *et al.* 2009). More recently, the practice has been extended to reef platforms. The utility of this approach is questionable at best, as recruitment failure has not been linked with substrate limitation.

The second situation is where artificial structure is placed on a coral reef to provide a particular service, such as shoreline protection, but the artificial reef design also incorporates secondary ecological objectives realized through repopulation with coral. Shoreline protection is a major and valuable service of coral reefs estimated in the billions of dollars each year (Beck and Shepard 2012). Healthy fringing reefs at sea level cause waves to break, dissipating wave energy. The friction caused by the corals of the back reef further reduces wave energy as it travels toward the shore. Ferrario *et al.* (2014) determined that coral reefs consistently dissipate 97 per cent of the incident wave energy, with an 86 per cent reduction by the reef crest alone. As coral reefs degrade they become less capable of resisting the erosive force of waves (by adding new growth) and begin to erode. The eroding reef not only becomes deeper than an actively growing one but also presents less structure and, therefore, less friction. Both the increased depth and reduced friction presented by degraded reefs allow increased wave energy to reach the shore, increasing shoreline erosion and run-up flooding. While shoreline erosion and flooding are generally addressed using traditional engineered shoreline armouring, recognizing that the root cause of the erosion is largely related to reef health has prompted trials of artificial reefs in conjunction with coral gardening methods to provide hybrid solutions that may be more cost effective and also provide greater ecological benefits. An example of this type of re-engineered reef design can be found in Grenville Bay, Grenada, where pilot submerged breakwater structures have been placed on the existing reef crest to reduce wave energy but also include internal cavities that provide habitat for many reef organisms and host transplanted coral to enhance ecological functional and structural performance (Reguero *et al.* in review). This style of combined engineering and coral restoration may become more common with continued sea level rise and coral reef degradation (TNC, unpublished data).

Coral gardening

Coral gardening is the marine analogue of silviculture and is often described as a two stage process (Horoszowski-Fridman *et al.* 2015). The first stage involves producing and growing large numbers of coral colonies by fragmenting coral tissue into small sections, growing those sections then repeating the process of asexual reproduction and grow-out in underwater nurseries. The second stage involves the out-planting of colonies from the nursery to degraded reef. Transplanting corals has a long history (see Young *et al.* 2012) but the large-scale nursery-based production of coral colonies has really developed since the turn of the century. With the global decline in condition of coral reefs the development of coral nurseries has occurred in parallel in many countries, from East Africa, the Red Sea, the Seychelles, most of the countries between

Japan and Singapore, several of the Pacific islands, Mexico, Florida and the Caribbean (*ibid.*; Horoszowski-Fridman *et al.* 2015). With this diversity of geographies there has also been a variety of species propagated. The majority of the effort has been directed toward the faster growing species that have historically been abundant, provide structure and are habitat species for fish and invertebrates. More recently the focus has expanded to a broad suite of reef forming species.

Caribbean case study

Coral restoration in the Caribbean was brought into sharp focus in 2005 when reefs throughout the region were severely impacted by a mass coral bleaching event, triggered by prolonged exposure to above normal water temperatures. The bleaching observed in 2005 caused some direct mortality and was followed by an increased incidence of disease. Miller *et al.* (2009) and others observed signs of thermal stress in over 90 per cent of corals at sites in the US Virgin Islands with some associated mortality. Subsequent disease increased the coral area loss by as much as 13 times. Deep reefs, which did not bleach as severely as shallow reefs, also had an increase in the incidence of white syndrome diseases and a 22 per cent loss of coral cover. Years later, many impacted reefs throughout the Caribbean are yet to recover.

Coral reefs of the Caribbean are further impacted by a list of chronic and some novel stressors. Sediment from coastal development, nutrient and chemical pollution degrade water quality and can eventually lead to bleaching, outbreaks of diseases and direct mortality. The overharvest of 'pot fish' (a number of fish species, many of which are herbivorous coral reef inhabitants), and the die-off of the herbivorous long-spined sea urchin (*Diadema antillarum*) have hastened the phase shift from coral to algal-dominated reefs. In addition to the removal of herbivores, destructive fishing practices and a steady increase in the frequency of vessel groundings have contributed to the degradation and/or loss of coral reef habitat. The newest threat that has been identified on Caribbean coral reefs is the introduction of the invasive Indo–Pacific lionfish. Without natural predators populations of these ambush predators have exploded, adding predation pressure to the herbivorous fish populations and further decreasing the grazing pressure.

In the Caribbean, threat abatement alone has not resulted in recovery of coral coverage and it is widely accepted that the successful conservation of Caribbean reefs will require enhancing the resilience of local reef sites while actively propagating important reef-building corals with the aim of increasing their abundance and genetic diversity.

Throughout the Caribbean, elkhorn (*Acropora palmata*) and staghorn (*A. cervicornis*) corals were historically two of the most important reef builders. These two species are the fastest growing and their natural capacity for asexual fragmentation would lead to vast thickets of both. Broad-scale decline of Caribbean acroporids has resulted in disconnected populations with low genetic diversity and recruitment failure. As a result, reef sites are now more vulnerable to population collapses. These species were the first in the region to be added to the endangered species list (Bruckner 2002). Early recognition of the importance of the acroporids in the Caribbean, along with their increasing scarcity, made this group the initial focus of restoration in the region. Since 2000 a coalition of Caribbean scientists and partners has developed coral gardening techniques for acroporid corals to a point where one coral donor or fragment of opportunity can produce 100 coral colonies in just over one calendar year. In the Caribbean the historical importance of the 2 species of *Acropora*, their increasing scarcity, and an accelerating interest in restoration, prompted a coalition led by The Nature Conservancy to publish the *Caribbean* Acropora *Restoration Guide: Best Practices for Propagation and Population*

Enhancement (Johnson *et al.* 2011). Active coral reef restoration is gaining traction in the Caribbean with over 60 *Acropora* restoration projects established in over 15 countries and US territories (Young *et al.* 2012).

Coral gardening phase I: nurseries

Clearly defining the goal(s) of the coral restoration activity (species recovery, restoration of fish habitat, shoreline protection, etc.) is the first step in developing a coral restoration programme. The restoration goals will help to define target species, will inform the site selection process, and will be used to determine the type of nursery to construct. The primary aim and benefit of the nursery to coral restoration projects is to provide optimal conditions for survival, growth and the capacity for further fragmentation by controlling physical and biological stressors (storm surge, predators, competitors, etc.). Even colonies to be transplanted without further fragmentation can gain growth and survival benefits from a period of 'conditioning' in a nursery before outplanting (dela Cruz *et al.* 2015).

Obtaining the necessary permit(s) and approval(s) is a necessary and often time consuming process to be completed before restoration activity begins. Some countries permit collection of stock material from a limited number of wild growing corals. Others only allow collection of fragments of opportunity, corals broken from the main colony that would be likely to die if left unattended. Fragments should be collected, transported to the nursery and allowed to acclimate to the new conditions prior to being further fragmented. Coral should be collected and manipulated during the cooler months as they typically stress during the summer as a result of high sea temperatures, and excessive handling and fragmenting of corals that are stressed may lead to mortality. Good record keeping is essential to allow for adaptive management and mixing of genotypes during the transplanting process. If collecting from wild growing colonies is permitted, the donor colonies should also be monitored for any impacts that may have resulted from harvesting fragments.

Selecting an appropriate site(s) is critical to the survival of corals in a nursery. It is generally recommended that multiple sites be piloted as small-scale nurseries to determine how corals respond to the site prior to expanding to larger nurseries. Multiple nursery sites will also help to spread the risk of wide-scale impacts from any one event. The depth of the coral nursery should match the depth at which corals naturally occur in the wild. Deeper nurseries are better protected from storm events, but they typically result in slower growth rates and may limit the bottom time available for monitoring and maintenance. Open sandy areas are ideal as coral predators are typically absent and it eliminates the risk of damage to corals or other sensitive benthic habitat by divers while within the nursery. The nursery should be big enough to include space for structures, potential for expansion and adequate room around each structure to comfortably accommodate working divers.

The proximity of healthy reefs and wild populations of the target corals is encouraging as healthy colonies on nearby reefs provide a good indication that nursery-grown corals will survive. Nearby reef sites could also serve as donor sites for coral fragments. Areas with high abundances of algae and predators (corallivorous worms and snails) should be avoided where possible. Predators and competitors will increase the rate at which routine maintenance must occur to prevent loss of corals. Nurseries should be located in areas where there is a constant flow or turnover of water provided it is also a practical working environment for divers. Marine managed areas (MMAs) and other low-use areas are ideal for preventing impacts to nurseries from recreational activities such as anchor damage. Signage or the provision of moorings may reduce the likelihood of accidental anchor damage. Remote sites may provide the best conditions for coral nurseries, however, they increase the cost of operation and tend to result in poorly maintained nurseries.

Regardless of type, each nursery should have a small quarantine area established nearby but down current from the nursery. Corals of questionable health can be isolated while they are assessed and treated.

Once the nursery is established and stocked the initial inspections should be frequent in order to determine an appropriate monitoring and maintenance schedule. At minimum nurseries should be monitored for growth, survivorship, the incidence and prevalence of bleaching/paling, disease, algal overgrowth, breakages, predation, temperature and pH. Measurements of total coral length or volume should be taken initially for at least a subsample of colonies, and again prior to transplanting or further propagation. Measurement methodology will differ according to the species of coral within the nursery (Johnson *et al.* 2011).

There is a wide range of materials and nursery structures that have been utilized. The design of a nursery and nursery structures should be based on the local conditions as well as the species of corals to be grown. It may help to organize the nursery with monitoring in mind. For example the lay-out could be designed to help track the location of individual genotypes as the number of colonies expands. Nursery construction can be described in several general categories: block, frame or floating (Figure 24.1). These different types are described in the following:

Figure 24.1 Staghorn coral being grown in (a, b) block, (c) frame and (d) floating nurseries

- *Block nursery.* Cinderblocks have been used as a base upon which a PVC and/or concrete pedestal or 'puck' are attached. The coral fragment is then affixed to the puck with a two-part non-toxic marine epoxy. Block structures are typically anchored to the seafloor using steel rebar. Covering the steel rebar with a PVC sleeve can slow the degradation of the steel and shorten maintenance time. When attaching pedestals and pucks to cinderblocks, spacing should allow for growth of corals and for maintenance activities.

- *Frame nursery.* Mesh frames made out of metal, plastic or PVC are attached to the seafloor, typically with anchors or re-bar stakes. Corals are then attached using epoxy or cable ties. It should be assumed that all frames will require maintenance to repair damage and control algal growth. Corals on frame structures should be spaced to prevent crowding and allow room for maintenance activities.

- *Floating nursery.* Floated lines or PVC 'trees' are anchored to the sea floor using duckbill anchors or mooring screws. Corals are then suspended from the line or tree branch and can be attached using monofilament line, cable ties, or coated wire with corals spaced to prevent crowding and damage. When constructing PVC trees, for example, one should consider adequate branch sizes and using alternate lengths of line to attach corals. Suspending the nursery structures high in the water column is an efficient use of vertical space with decreased predation, increased water flow, and decreased maintenance generally resulting in improved growth and survival compared to nurseries on the sea floor (Johnson *et al.* 2011). Consequently, floating nurseries are becoming the design of choice where there is sufficient depth and conditions allow. Floating nurseries also introduce the potential to anchor equipment in deeper water allowing the corals to be lowered prior to storm events to avoid the highest wave energy.

Coral gardening phase II: outplanting

The goal of outplanting has primarily been to establish enough colonies from a variety of different genotypes to enhance the potential for sexual reproduction and recruitment and/or increasing the species cover on recipient reefs (Figure 24.2). Water quality is a pervasive threat and is a factor to consider when selecting sites for transplanting corals. The depth of the transplant plot should match the depth at which corals naturally occur in the wild and, where possible, be similar to the depth at which the corals grew within the nursery. Hard coral reef structure provides a stable substrate for coral transplants. Loose rubble and sand that are transported by storms and swell events will damage growing colonies.

The condition of the recipient reef is important as it indicates the likelihood of survivorship of transplanted corals. Transplant sites should have adequate space to allow for growth of coral colonies after transplant. A site with moderate coral coverage will provide space for transplanted corals and the assurance that corals can survive on the reef. A healthy herbivore community (diadema, scarids, and acanthurids) will help to graze algae that can compete against corals for space while predators and competitors will increase the potential for mortality. Hot spots for bleaching, paling, and diseases should be avoided as a number of reef dwelling organisms serve as vectors for the transport of coral diseases. Marine managed areas and other low use areas are ideal for preventing impacts to transplanted corals from recreational activities such as anchor damage. Favourable sites are those where known stressors on adjacent reefs have been reduced or eliminated.

Corals to be transplanted should be healthy, actively growing and robust in colour (no signs of disease, paling, bleaching or lesions). A mixture of the genetic material from nurseries should be used at each transplant site and multiple sites used to further spread risk. Corals can be

Figure 24.2   Outplanting sites in the US Virgin Islands showing staghorn coral at outplanting with the
fresh epoxy cement appearing (a) as white dot at the base and (b) after 1.5 years, and
(c) staghorn coral at outplanting and (d) 2.5 years later

attached to the reef either directly or on its nursery mounting (cement puck, plastic pin, etc.)
using masonry nails, coated wire, cable ties or epoxy. At least a subsample of the transplanted
corals should be tagged with unique identifiers for monitoring purposes.

Monitoring and maintenance of outplant sites is important in order to assess growth and
survivorship, and maintain transplant plots. One of the most common deficiencies in coral reef
restoration is a lack of capacity for long-term monitoring, though good examples of monitoring outplant sites are beginning to enter the literature (Horoszowski–Fridman *et al.* 2015).
Assessing the condition of outplant reefs should optimally be done on a continuing basis.
Practical advice and examples of nursery management and outplanting are available in Edwards
and Gomez (2007), Edwards (2010) and Johnson *et al.* (2011).

Building resilience: other considerations for conservation

Active coral restoration is one important component that should be considered for any coral
conservation programme. However, reefs will continue to decline and restoration projects will
be more likely to fail if the human–induced stressors to the reef are not addressed. Therefore,
coral restoration should be a part of a larger effort to improve the health of coral reefs. While
individual countries should initiate programmes to reduce their carbon emissions, climate
change is a global issue that will require a global approach to truly affect change. In the short

term countries can be proactive by developing appropriate legislation and programmes. These should include programmes to improve fishing practices that reduce physical disturbance to reefs and re-establish healthy herbivore populations. The establishment and effective management of MMAs that serve as refuges for corals, fishes and other coral reef dwelling organisms is important. Similarly, efforts to improve development practices on land and improve waste-water treatment standards will reduce the amount of sediments and pollutants that stress reefs.

The success of coral reef conservation depends heavily on the buy-in and support of a wide range of stakeholders. In many instances, the conservation of coral reefs will require behavioural changes. To that end coral restoration, particularly gardening, has proved to be an effective way to engage and educate the public, creating support for parallel conservation measures and increasing the likelihood of their success. Communication plans should be inclusive of all facets of the community (youth, adults, fishers, politicians, etc.). Each member of the community can become an ambassador for coral reef restoration and conservation and support action towards improving the health of the natural resources.

Sexually produced recruits

One of the features common to many degraded reef systems is low or failed recruitment, generally measured as the appearance of small colonies (<2 cm) in diver surveys. This has been widely observed despite the continued spawning of the remaining colonies (Caribbean: Miller and Szmant 2005; Red Sea: Linden and Rinkevich 2011; Hainan Islands, China: Dr Yuyang Zhang, South China Sea Institute of Oceanology, Guangzhou, China, unpublished data). One enticing avenue for increasing recruitment success is to harvest the spawn from broadcast spawning species or the planula larvae from brooding species and manipulate the recruitment process. This approach does require developing or adapting larval rearing and transport systems for the field so that competent or settled larvae can be returned to the reef. Dr Margaret Miller of the National Oceanic and Atmospheric Administration (NOAA) has investigated the potential of collecting the gametes of *Acropora palmata* and *Orbicella sp.* during spawning events and culturing them. Larvae were then either introduced to enclosures over dead coral skeletons during settlement or set to various substrates in the lab for transfer to the reef (Miller and Szmant 2005). Miller and Szmant achieved a survivorship of between 1 and 3 per cent over the first three months and concluded that the techniques hold promise, but require additional development.

Brooding species may prove to be more straightforward for enhancing sexually generated recruits as they do not require care and maintenance during fertilization and the maturation of larvae. This approach has been aided by the demonstration that planula can be transported for several days between collection sites and rearing facilities. In the Red Sea *Stylophora pistillata* is a dominant brooding species that is important to reef function, being fast growing, robust, tolerant of high temperatures and important to a number of reef dwelling fish and invertebrates. The species has been the focus for developing restoration techniques in the Red Sea (Linden and Rinkevich 2011; Horoszowski-Fridman *et al.* 2015). Linden and Rinkevich (2011) developed a technique for setting larvae that had been trapped in the field from individual breeding colonies to a flexible preconditioned polyester film substrate. Recruits were grown in the lab for approximately 4 weeks then removed from the setting film and transferred to plastic pins that could be transported to the nursery for additional grow-out in the field. The polyester setting substrate allows for simple transfer of very small recruits (approx. 2 mm to 4 mm). By separating individual colonies for grow-out the technique avoids issues of post-settlement competition between colonies or chimerism, the fusion of different spat of the same species, which reduce growth and survival. In addition the juvenile colonies benefit from the enhanced survival and growth rates experienced in the nursery setting allowing rapid development to sexual maturity. The authors report production rates of approximately 15 colonies at 5 months of age per person hour. This can be viewed as the current cost of obtaining the increased genetic diversity delivered by sexually produced recruits as well as avoiding any negative impacts from the initial harvest of living coral material for grow–out and subsequent re-fragmentation.

Using herbivory to facilitate restoration: Hawai'i case studies

The coral reefs of the Hawaiian Islands are confronted by many of the same threats and stressors as tropical reefs found elsewhere in the world, with a growing population and urbanization leading towards increases in nutrients and other pollutants at the same time that fishing pressure in those waters is on the rise. This situation has given rise to some famous examples of 'phase shift' from a reef whose benthic community is dominated by calcifying (i.e. reef-building) corals and crustose-coralline algae to one dominated by fleshy macroalgae.

The specific combination of environmental stressors and algae responsible for reef degradation in Hawai'i varies from site to site, as do the opportunities available to address the impacts. Over the long term, the ultimate solutions combine better watershed management of land-based inputs and better fisheries management to protect the herbivores that control algal competition with corals and maintain suitable settlement substrate for new coral recruits. However, these approaches are often very costly and/or controversial and can take many years to implement. While efforts are underway to enact these more sweeping measures, alien algal species continue to expand their range in Hawai'i and decades of coral growth is being lost as invasive algae overgrow and kill colonies. In response, scientists and managers of coral reefs in Hawai'i have been developing short- and mid-term management tools and approaches centred around herbivory to control algal invasions and restore effected coral reefs while allowing time for large-scale and long-term management changes to be implemented.

The Kahekili Herbivore Fishery Management Area

Summertime blooms of both alien red algae (*Acanthophora spicifera*) and native green algae (*Cladophora sericea*) on the reefs of Kaanapali on the island of Maui were associated with the loss of almost half of the coral cover in the area from the late 1990s into the early 2000s. These blooms were thought to be due to some combination of both increased nutrient inputs into the area from a nearby wastewater treatment facility and diminished herbivore stocks that had been subjected to heavy fishing pressure. As a management response to these blooms, the Hawai'i Department of Land and Natural Resource's Division of Aquatic Resources (DAR) developed a plan to protect the herbivores of the area from fishing in the hope that herbivore stocks would then increase and be better able to contain algal blooms, enabling the corals of the area to recover.

After conducting a series of meetings with Maui residents to garner support for a 'herbivore replenishment area', in 2009 DAR established the Kahekili Herbivore Fishery Management Area (KHFMA). All take of herbivorous fish (surgeonfish, parrotfish, or rudderfish) and sea urchins is prohibited within its approximately 2-mile stretch of costal coral reef, though fishing for other reef species is allowed. This was a forward-thinking strategy, and to the best of our knowledge, the first time the take of herbivores has been prohibited specifically for coral reef restoration. Comprehensive monitoring of fish and the benthic community was also established in order to track changes through time.

With 4.5 years of data analysed to date, surgeonfish numbers have increased by almost half, while numbers of the faster-growing parrotfish have more than doubled (DAR and NOAA, *unpublished data*). The size distribution of parrotfishes is also changing, with larger size classes that were once extremely rare becoming increasingly common. It is likely that herbivore populations will continue to increase over the coming years, particularly the longer-lived surgeonfishes. With the increase in herbivores, particularly parrotfish, turf algal cover has started to be replaced by crustose coralline algae (Figure 24.3). Crustose coralline algae (CCA) are calcifying (i.e. reef building) algae and a preferred settlement substrate for new coral recruits. While coral cover itself had not increased significantly within the 4.5 years of post-closure data, that is not surprising given the slow-growing nature of corals. The declines in the cover of turf algae and the increases in CCA cover are the types of intermediate changes expected in the benthic community before seeing coral cover improve as corals do not have to compete with algal turfs and have preferred settlement areas.

The Moku o Lo'e Marine Laboratory Refuge algal removal

At the Marine Laboratory Refuge in Kāne'ohe Bay on the island of O'ahu, herbivore numbers were low and several large, persistent blooms of alien algae species (*A. spicifera*, *Gracilaria salicornia*, and members of the *Kappaphycus/Eucheuma* species complex) had been dominating reef slope habitats for several years. Alien algae became invasive on the leeward reef slopes where they formed as much as 70 per cent of the shallow-water reef slope community, directly smothering the corals that used to dominate that environment, filling in the topographic complexity

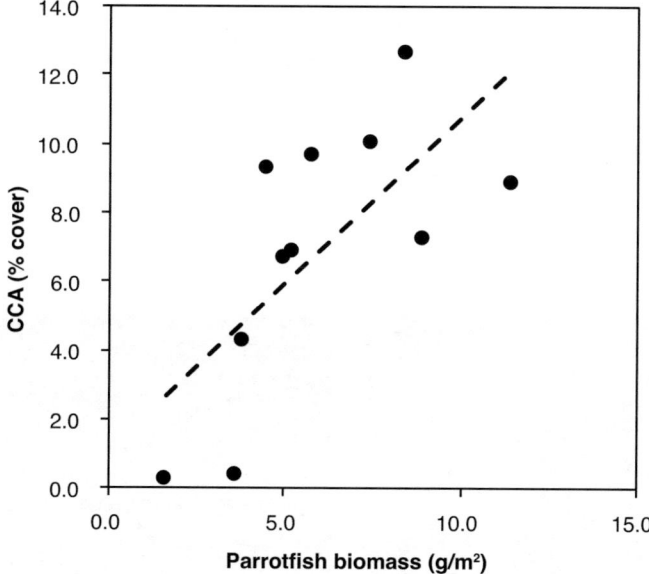

Figure 24.3 Strong positive relationship between parrotfish biomass and CCA coverage. Each point represents one survey

Source: Unpublished data courtesy of NOAA and DAR

and lowering the habitat quality for corals and fish alike (Conklin 2007; Stimson 2013). Beginning in the late 1990s, effective enforcement of the Marine Laboratory Refuge was implemented and the number and size of herbivores increased to double that seen on neighbouring reefs. However, the large quantities of invasive algae that had previously accumulated on the reef slope was beyond the ability of even this protected herbivore population to control.

In 2004, a collaboration between the State's DAR, researchers from UH, and The Nature Conservancy (TNC) developed a new tool to address this issue, the Super Sucker, a custom-built underwater vacuum that allows divers to efficiently remove large quantities of alien algae from the reef without harming the underlying reef structure or native species (Figure 24.4). In 2005 and 2006, members of this collaborative group operated the Super Sucker on the leeward slope of Moku o Lo'e, removing almost 10,000kg of accumulated alien algae, most of which was *G. salicornia*. Coverage of *G. salicornia* was decreased from >70 per cent to just over 10 per cent. Over the following months, the coverage of alien algae on the leeward reef slope of Moku o Lo'e decreased further, until by six months after the last Super Sucker removal from the area, the alien algae cover was 0.4 per cent. Over that same time period, alien algae cover remained at ~50 per cent at a nearby control site. It is now 10 years after the last Super Sucker removal from Moku o Lo'e, and the reef slope remains free of the alien algae that had dominated that habitat for several years prior to algae removal.

While there is not experimental documentation of all the factors that may have contributed to this phase shift reversal on the Moku o Lo'e reef slope, it appears likely that removing the bulk of the accumulated alien algal biomass decreased algae levels below a threshold where the protected herbivore community could successfully control the remaining algae. Monitoring of the benthic community at this site by UH and TNC has verified the persistence of this relatively algae-free state, and data suggest that coral cover is starting to increase in the absence of competition with invasive algae. In addition, much of the structural complexity that existed on these reefs that was being filled in by mats of invasive algae is now clear.

Whereas the example of the KHFMA showed that protecting herbivores is a promising management tool for coral reef restoration, the experience at Moku o Lo'e suggests that herbivore protection alone may not be enough to restore areas that are already significantly degraded. In those instances, a combination of both increasing the abundance of herbivores and reducing algal biomass to manageable levels may be necessary.

Figure 24.4 (a) Diver removing invasive alien algae from the reef using the Super Sucker. (b) The Super Sucker in Kāne'ohe Bay

Removing algae and enhancing herbivory in Kāneʻohe Bay

Both the KHFMA and Moku o Loʻe examples involve the closure of areas to at least the taking of herbivores as the means to increase herbivory. However, implementing new fisheries regulations can be controversial, and is often either not possible or will take a long time to implement. In Hawaiʻi, the need to halt the spread of invasive alien algal species required an approach that could be effectively deployed on a shorter time scale. Different invasive alien algae, *Eucheuma denticulatum* and *Kappaphycus alvarezii*, were causing phase shifts on other reefs within Kāneʻohe Bay where herbivores were not protected. A means of enhancing herbivory that wasn't reliant on fishing regulations needed to be developed.

Researchers at UH began in the early 2000s to investigate the potential of the native collector urchin, *Tripneustes gratilla*, as a biocontrol agent. These sea urchins are common throughout the Hawaiʻi Islands, have limited movement, readily eat invasive algae, and their abundance is easily manipulated (Stimson *et al.* 2007). In 2008, DAR biologists removed *Kappaphycus* and *Eucheuma* from a patch reef in Kāneʻohe Bay without enhancing herbivory in any way. Within a matter of months, the alien algae had regrown to pre-removal levels (Figure 24.5). In 2009, they again removed *Kappaphycus* and *Eucheuma* from the reef, but this time placed over 1,000 *T. gratilla* on half of the reef. The following year algal cover had again rebounded on the half of the reef without urchins while the half of the reef with urchins maintained a very low algal abundance for the duration of the experiment.

Based on the success of this and similar experimental efforts, DAR and TNC are now working to deploy this strategy of Super Sucker algal removal followed by treating the reef with *T. gratilla* across the bay. Priority sites for implementation are selected to stop the alga *Eucheuma denticulatum* from spreading, and to restore reefs with high coral cover and diversity. DAR has

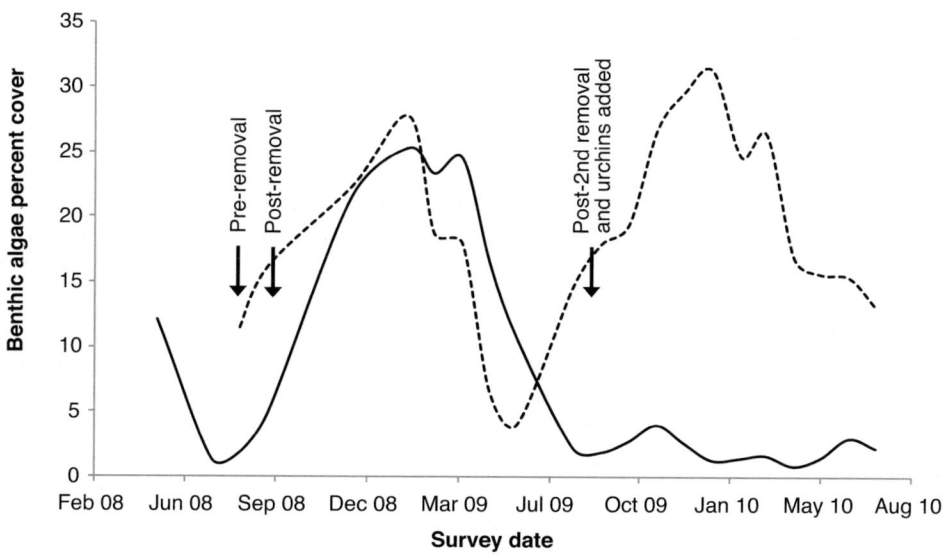

Figure 24.5 Changes in the percentage cover of alien algae over time following removal with the Super Sucker with (solid line) and without (dashed line) the addition of native urchins as a biocontrol agent to enhance herbivory

Source: Unpublished data courtesy of DAR

developed an urchin hatchery facility to provide urchins for the restoration, and TNC is partnering with Ocean Institute to develop a second hatchery to meet the restoration demand.

Researchers from UH and Hawai'i Pacific University (HPU) have assessed the impact of
increased urchin densities on coral recruitment to address fears that their intensive grazing may
be inhibiting coral recruitment. This research found that urchins selectively avoid coral recruits,
keeping substrate clear of invasive algae while allowing coral recruits to settle and thrive.
Treated reefs have gone from algal-dominated reefs where corals are being actively overgrown,
smothered, and killed by alien algae species, to reefs in which algal cover is kept in check, corals
that were directly competing with algae can recover, and abundant settlement substrate is available for new coral recruits to settle and grow.

While the long-term solutions for controlling alien algae within Kāne'ohe Bay still lie with
decreasing land-based nutrient inputs that fertilize algal growth and allowing natural populations of herbivores to recover, the combined strategy of reducing algal biomass with the Super
Sucker and enhancing herbivory with cultured urchins is proving to be a successful short- to
mid-term solution. And the experience of Moku o Lo'e suggests that the areas most heavily
degraded and overgrown with alien algae probably need algal biomass reduction by the Super
Sucker even if water quality improves and herbivore populations are restored.

Herbivory and restoration

A suite of tools and approaches have been developed in Hawai'i for using herbivory to facilitate coral reef restoration. Managing herbivores and the ecological function of herbivory is a
valuable tool for ecological restoration as in the examples of protecting herbivores at Kahekili
or using the Super Sucker to reduce algal biomass at Moku o Lo'e. In some instances, more
intensive management is required, as with the sea urchins in Kāne'ohe Bay, and this can be
viewed as a short- to mid-term solution while working to develop the longer-term strategies.
There is no one-size-fits-all solution, either in terms of the approach needed in a given area,
or that is feasible from a management perspective. However the suite of tools available seems
to be a robust set of options that can be tailored to the particulars of a given situation.

Summary

Coral restoration is a relatively recent but rapidly expanding component of the overall coral
reef conservation landscape. Threats to coral reefs are widespread, and active restoration of coral
reefs has become a necessary component of maintaining coral diversity and cover in many areas
of the world. Coral gardening is being used to augment or replace the natural recruitment of
corals where recruitment is not keeping pace with the rate of destruction or where recruitment itself is reduced or failing. Algae removal and active manipulation of herbivore abundance
is being used to reverse the process of 'phase shift' to algal dominated reefs. Manipulating the
sexual reproduction of corals holds promise in some circumstances and restoration is beginning
to be used to increase the ecological function of artificial structures such as those used to recreate the wave barrier function of coral reefs. The coral restoration techniques described before
can best be viewed as short- to medium-term measures necessary to buy the time required to
demonstrate the true value of coral reefs to society and address the stressors responsible for the
declines. The requirement for restoration in the future will be determined by our ability to
address the threats to their continued health.

Acknowledgements

The authors would like to thank Dr Ivor Williams of NOAA and Dr Russell Sparks of HI DAR for permission to use the previously unpublished data represented in Figures 24.3 and Dr Brian Neilson of HI DAR for permission to use the data in Figure 24.5. We thank D. Millton for assistance preparing those figures and Dr P. Kramer for suggested improvements to the text.

References

Beck, M. W. and Shepard, C. (2012). Coastal habitats and risk reduction. In P. Mucke (ed.), *World Risk Report 2012*, pp. 32–41. Berlin: Alliance Development Works.

Burke, L., Reytar, K., Spalding, M. and Perry A. (2011). *Reefs at Risk Revisited*. Washington, DC: World Resources Institute. Retrieved from http://pdf.wri.org/reefs_at_risk_revisited.pdf.

Bruckner, A. W. (2002). *Proceedings of the Caribbean* Acropora *Workshop: Potential Application of the US Endangered Species Act as a Conservation Strategy*. Technical Memorandum NMFS-OPR-24. Silver Spring, MD: NOAA.

Chou, L. M., Yeemin, T., Yaman, A. R. B. G., Vo, S. T., Alino, P. and Suharsono (2009). Coral reef restoration in the South China Sea. *Galaxea: Journal of Coral Reef Studies* 11: 67–74.

Conklin, E. J. (2007). The influence of preferential foraging, alien algal species, and predation risk on the interaction between herbivorous fishes and reef macroalgae. PhD dissertation, University of Hawai'i.

Dela Cruz, D. W., Rinkevich, B., Gomez, E. D. and Yap, H. T. (2015). Assessing an abridged nursery phase for slow growing corals used in coral restoration. *Ecological Engineering* 84: 408–415.

Downs, C. A., Kramarsky-Winter, E., Segal, R., Fauth, J., Knutson, S., Bronstein, O., Ciner, F. R., Jeger, R., Lichtenfeld, Y., Woodley, C. M., Pennington, P., Cadenas, K., Kushmaro, A. and Loya, Y. (2015). Toxicopathological effects of the sunscreen UV filter, Oxybenzone (Benzophenone-3), on coral planulae and cultured primary cells and its environmental contamination in Hawaii and the US Virgin Islands. *Archives of Environmental Contamination and Toxicology* 70(2): 265–288.

Edwards, A. J. (ed.) (2010). *Reef Rehabilitation Manual*. St Lucia, Australia: Coral Reef Targeted Research & Capacity Building for Management Program.

Edwards, A. J. and Gomez, E. D. (2007). *Reef Restoration Concepts and Guidelines: Making Sensible Management Choices in the Face of Uncertainty*. St Lucia, Australia: Coral Reef Targeted Research & Capacity Building for Management Program.

Ferrario, F., Beck, M. W., Storlazzi, C. D., Micheli, F., Shepard, C. C. and Airoldi, L. (2014). The effectiveness of coral reefs for coastal hazard risk reduction and adaptation. *Nature Communications* 5: 3794.

Horoszowski-Fridman, Y. B., Brethes, J., Rahmani, N. and Rinkevich, B. (2015). Marine silviculture: Incorporating ecosystem engineering properties into reef restoration acts. *Ecological Engineering* 82: 201–213.

Hughes, T. P., Graham, N. A. J., Jackson, J. B. C., Mumby, P. J. and Steneck, R. S. (2010). Rising to the challenge of sustaining coral reef resilience. *Trends in Ecology and Evolution* 25: 633–642.

Johnson, M. E., Lustic, C., Bartels, E., Baums, I. B., Gilliam, D. S., Larson, L., Lirman, D., Miller, M. W., Nedimyer, K. and Schopmeyer, S. (2011). *Caribbean* Acropora *Restoration Guide: Best Practices for Propagation and Population Enhancement*. Arlington, VA: The Nature Conservancy.

Linden, B. and Rinkevich, B. (2011). Creating stocks of young colonies from brooding coral larvae, amenable to active reef restoration. *Journal of Experimental Marine Biology and Ecology* 398: 40–46.

Miller, M. and Szmant, A. (2005). Lessons learned from experimental key-species restoration. In W. F. Pretch (ed.), *Coral Reef Restoration Handbook*, pp. 219–234. Boca Raton, FL: CRC Press.

Miller, J., Muller, E., Rogers, C., Atkinson, A., Whelan, K., Patterson, M. and Witcher, B. (2009). Coral disease following massive bleaching in 2005 causes 60% decline in coral cover on reefs in the US Virgin Islands (USVI). *Coral Reefs* 28: 925–937.

Reguero, B., Beck, M.W., Agostini, V., Kramer, P. and Hancock, B. (in review). How reef degradation increases coastal risk and a restoration solution in small island states: a case study from Grenada. *Ecological Engineering*. Ms. Ref. No.: ECOLENG-D-16-01382.

Rinkevich, B. (2015). Novel tradable instruments in the conservation of coral reefs, based on the coral gardening concept for reef restoration. *Journal of Environmental Management* 162: 199–205.

Rogers, A., Harborne, A. R., Brown, C. J., Bozec, Y. M., Castro, C., Chollett, I., Hock, K., Knowland, C.

A., Marshell, A., Ortiz, J. C., Razak, T., Roff, G., Samper-Villarreal, J., Saunders, M. I., Wolff, N. H. and Mumby, P. J. (2015). Anticipative management for coral reef ecosystem services in the 21st century. *Global Change Biology* 21: 504–514.

Stimson, J. (2013). Consumption by herbivorous fishes of macroalgae exported from coral reef flat refuges to the reef slope. *Marine Ecology Progress Series* 472: 87–99.

Stimson, J., Cunha, T. and Philippoff, J. (2007). Food preference and related behavior of the browsing sea urchin *Tripneustes gratilla* (Linnaeus) and its potential for use as a biological control agent. *Marine Biology* 151: 1761–1772.

Young, C. N., Schopmeyer, S. A. and Lirman, D. (2012). A review of reef restoration and coral propagation using the threatened genus *Acropora* in the Caribbean and western Atlantic. *Bulletin of Marine Science* 88(4): 1075–1098.

25

ECOLOGICAL RESTORATION IN AN URBAN CONTEXT

Jessica Hardesty Norris, Keith Bowers and Stephen D. Murphy

Urban centres, though they often arise on what was once a vibrant and complex landscape, are major sinks for natural resources. In a strict sense, ecological restoration seeks to re-establish the key ecological structures and functions of times past and return to a self-organizing and self-sustaining system. Where sufficient land and resources persist to allow ecological restoration projects at scale, the approaches and strategies employed in urban restoration projects largely mirror those discussed in other chapters. For example, a marsh restoration project in a coastal city may have complicating factors – altered hydrology or man-made substrates – but generally the principles discussed in Chapter 19 would apply.

It is a challenge, then, to define what belongs in a chapter about urban restoration as unique and distinct from the detailed discussions of the systems as found in Chapters 8–24. To the extent that urban restoration is simply a geographic category of ecological restoration, it may not merit separate treatment. However, a strict ecological restoration approach is impractical in most urban areas, largely because of the scale of the environmental disturbances in cities and the need to explicitly consider the human context of the post-restoration system (Gobster 2010).

Urban ecological restoration is, however, garnering attention and generating excitement, even as it challenges some of the theory of ecological restoration (Ingram 2008). Many consider it a coherent concept that is set apart by an approach to restoration that is more narrowly functional than that of non-urban ecological restoration. This chapter very briefly introduces a few of the most influential ideas in the evolving concept of urban ecological restoration and then presents a set of examples of urban restoration projects.

Defining urban ecological restoration

The discipline of urban ecology is still an emerging sub-discipline of ecology, though now it is fairly well established as study of 'the interactions of organisms, built structures, and the physical environment, where people are concentrated' (Forman 2014). This definition distinguishes urban ecology by explicitly including both humans as an organism whose interactions are studied and built structures as part of the environment. Urban ecological restoration can similarly be distinguished from non-urban restoration by its necessary inclusion of humans and buildings into project design and implementation, as well as by two inescapable characteristics of the ecology of cities, fragmentation and the degree of alteration.

The influence of humans as stakeholders and beneficiaries of the ecological functions restored by urban ecological restoration is part of what defines the practice. Public interest in parks and greenspace drove many of the earliest efforts at urban ecological restoration, though the focus was often on open, mown areas (Zmyslony and Gagnon 2000). Today, post-industrial and urban ecological restoration is often likely to focus less on lawns and parks and more on 'natural' areas, especially those that provide an outlet for people to enjoy what is left of 'natural' fragments, though there is certainly a diversity of opinions on the benefits of open naturalized green spaces (Hobden *et al.* 2004; Manuel 2003). Social justice and human benefits are inextricably entwined with urban ecological restoration (Palamar 2010), in part because economic and social thresholds alter the goals of restoration for local stakeholders.

Secondly, the built environment has a pervasive influence on urban restoration. Buildings, dams, channels, sewage infrastructure, rail and road corridors and myriad other components of the built environment are inextricably part of urban ecological restoration projects, both as opportunities for supporting ecological function and as determinants of ecological processes.

Fragmentation is perhaps the dominant ecological condition of the urban landscape, and though it is often a factor in non-urban restoration projects, the intensity of fragmentation in urban areas defines the practice of ecological restoration there. Fragmentation also creates problems because much of the available habitat is juxtaposed with urban structures, leaving abrupt transitions between habitat and urban built features. Accordingly, most of the habitat is exposed to edge effects such as drastic changes in temperature, humidity, light, nutrient and water exchanges. The pervasiveness of fragmentation in urban areas and its role as a key driver of ecological processes has a profound effect on the typical scale of urban restoration projects. In tiny fragments of open space, along streets or on rooftops, meaningful restoration takes on a different dimension. Professionals are actively engaged in urban restoration projects that are literally in people's own backyards (Rudd *et al.* 2002). Such small-scale restoration can be relevant to the larger scale if species or ecological functions can connect across these nodes, but fragmentation severely constrains the ecological potential of many sites (Naveh 2005; Quon *et al.* 1999).

The last element in this very brief list of distinguishing characters of restoration in the urban context is the level of alteration. Urban ecosystems require different foci than non-urban systems because of the degree and additive effects of the disturbance driving the need for restoration. For example, while invasive species are frequently considerations in ecological restoration projects, invasives and ornamentals are heavily involved in every single urban restoration. Urban systems may be simultaneously intensely fragmented, filled with impervious surfaces, subject to pollution, and hot, with hydrology that has been completely overhauled by dewatering, channelization and other development.

Such changes in microclimate, species composition, connectivity and hydrology in urban environments are often so drastic that it is impossible to find reference systems or any comparable species assemblages. This scrum of constraints and influences not only sets urban ecological restoration apart from non-urban, it also changes the very project of restoration. Alongside classical restoration projects directed at reestablishing the key ecological structures and functions of times past and returning areas to a self-organizing and self-sustaining system, urban restoration projects are more often directed at protecting sensitive species, enhancing habitat for a wide range of species, and integrating human history into cultural landscape restoration (Westphal *et al.* 2010). Urban ecological restoration projects are often designed to restore specific ecological services, a type of restoration also known as process-based, structural, or functional restoration (Howell *et al.* 2012). Restoration efforts might be narrowly focused on restoring specific ecological functions such as retaining and filtering water, providing habitat for insects, birds and fish, or supporting pollination.

Select examples of urban restoration projects

The examples below emphasize the restoration approach used to address one of the primary ecological functions of each project described. In selecting them we leaned towards heavily engineered projects that have less in common with ecological restoration outside the urban context. The examples are partly intended to serve as an intuition pump or discussion point about what constitutes urban ecological restoration. Rather than attempting to be representative, in selecting examples we sought to demonstrate the diversity of approaches to restoring ecological functions within our cities as well as the creativity that doing so can inspire.

Shading and heat mitigation: green façade at National Wildlife Federation headquarters, Reston, Virginia by Greenscreen

The headquarters of the National Wildlife Federation (NWF) was envisioned as a building that adhered to the conservation mission of the NWF. In compliance with current standards, the headquarters achieved an Energy Star partnership, and won a 2002 AIA/COTE Top Ten Award. The landscaping is a tapestry of native plant materials that extends to the vertical plane through a prominent greed façade made of greenscreen trellis panels.

The three-storey curtain wall is a glazed façade for offices and public areas facing south over the parking area. The panels are mounted to a structural steel frame that is four feet from the face of the building. Standard-sized panels help maximize use of materials and minimize costs. It is covered with native vines that shade and cool the southern elevation of the headquarters in the summer, and when the deciduous vines drop their leaves in the fall, the interior office space benefits from passive solar warming. The vertical landscape is anchored by native plantings that are part of the site's stormwater management system and extend across the entire front façade.

The living wall offers practical, environmental and educational value. In the design process, HOK's designers evaluated the benefits of an architectural brise-de-soleil or louvre system versus a green façade of native vines to shade the glass wall, and confirmed that a deciduous 'sunscreen' improved energy performance as efficiently as an architectural option with potentially added benefits and at a lower cost. Birds and insects use the vegetation of the wall, and two nearby biofiltration ponds that drain the parking area complement the urban habitat provided. The client, NWF, leads tours of students through the building, in compliance with their outreach mission. From inside the building, there are intimate views of nesting activity in the vines. In addition, the changing of the vines with the seasons gives the building a responsive, vital quality.

The benefits provided by the wall at NWF Headquarters are clear, but it is difficult to say how well the wall serves some of the benefits that are often touted for green walls. For example, the organization Green Roofs for Healthy Cities lists reduction of the Urban Heat Island Effect and Improved Exterior Air Quality as benefits of green walls, but these processes occur at very large scale and it is empirically hard to measure the contribution of any single strategy. There are performance data specifically for the thermal properties of green façades, which show that foliated panels reduce the surface temperatures of walls (Tilley *et al.* 2012). Ambient air temperatures were also reduced by vegetated façades due to evapotranspiration of the leaf canopy, cooling interior temperatures by approximately 4°C and wall surface temperatures by 7–14°C compared with bare walls.

Figure 25.1   Green façade at National Wildlife Federation headquarters
Source: photo by D. Hill

Bird nesting habitat: roof-nesting tern initiative by Audubon Florida, Eckerd College, and the Florida Shorebird Alliance

Coastal habitats are under intense development pressure throughout the world, and beach-nesting bird populations are accordingly in decline at a broad scale. In coastal communities such as the Tampa Bay area in Florida, which has grown by about a third in the last 25 years, breeding habitat for beach-nesting birds is limited and birds that do nest on open beaches typically experience a high degree of disturbance from human recreation. Least terns have made a remarkable adaptation to the lack of beach habitat: entire colonies nest on gravel rooftops.

But these roofs, though they offer some advantages, also have perils. Chicks can become overheated and often fall off unprotected roof edges to their deaths. Audubon Florida, with a diverse team of partners, chapter members and staff, work to support the nesting colonies. The efforts began as volunteer monitoring efforts by people who visited the nesting sites and returned fallen birds to the roofs, but soon conservationists realized that additional efforts were possible and needed.

After several breeding seasons of nest-checking, conservationists partnered with local researchers to gain a more nuanced understanding of the threats and benefits of rooftop nesting. Researchers examined the selection of roofs by terns in order to better understand what made roofs attractive to them. A multi-year study found that only distance to water was a significant predictor of whether a rooftop was used by the terns (Forys and Borboen–Abrams 2006). The team then redoubled efforts to develop modifications that would make the roofs safer for birds.

Figure 25.2   Least tern chick on its way to a check-a-boom
Source: photo by D. Kandz

Where the property owner is amenable, a rooftop where birds nest year after year can be 'chick-proofed' before nesting begins in the spring. Installing fencing around low-edge rooftops and covering drain pipes with screens helps prevent chicks from falling off the roof. In some cases, conservationists have attempted to provide extra shade to mitigate the extreme temperatures that can be reached on the rooftops.

Although rooftop nesting is a fascinating example of an adventitious ecological service provided in the built environment and a powerful communication and outreach tool, it is not a good long-term solution to the lack of nesting habitat because gravel rooftops are being phased-out or replaced over time with lightweight and more reflective roof surfacing materials. Roof-nesting terns also tend to face stressors, such as travelling longer distances to foraging areas (Forys *et al.* 2013).

Groundwater recharge: natural organic recycling machine in Hassalo on 8th, Portland, Oregon by Biohabitats for American Assets Trust

Urbanization affects a wide swath of above- and belowground urban processes and hydrological systems. One of the greatest of these is the substitution of natural for impervious surfaces that bypass the natural processes of filtration and groundwater recharge (Jacobson 2011). Hassalo on 8th is a redevelopment project in Portland's Lloyd neighbourhood, an expanse of parking lots and office buildings with little residential space. The area is served by a combined sewer system, in which rainwater is conveyed along with sewage and ultimately treated as wastewater.

Such combined sewers, when they overflow ('combined sewer overflows' or 'CSOs') can be responsible for the majority of contaminants entering urban waterways, even if there are only few overflow events each year (Phillips *et al.* 2012).

Envisioned as an EcoDistrict, Hassalo on 8th is a four-block urban redevelopment project that combines a series of sustainability measures such as renewable energy, green roofs, a bike hub and access to mass transportation. Three new residential towers create substantial new housing opportunities, and the exiting office tower was renovated to improve building performance. The two-block long pedestrian way that runs through the development has been dubbed 'water street' since the wastewater, stormwater, and rain-fed water feature systems are integrated directly into the streetscape, bringing green infrastructure directly to the public's attention. It is one of the first urban neighbourhoods to treat and recycle its wastewater on site and discharge it into the ground rather than into the sewer system. The onsite wastewater system, nicknamed NORM (Natural Organic Recycling Machine), is a decentralized treatment and reuse system designed to divert 100 per cent of the wastewater from the three new buildings away from the municipal sewer.

NORM, which was designed in collaboration with GBD Architects, Glumac, and PLACE studio, treats 170,000 litres per day to State of Oregon Class A reuse standards through a series of trickling filters and constructed wetlands. The trickling filters are integrated into the design of a plant-filled pedestrian corridor at the centre of the project, while the wetlands provide the unifying element of the landscaping. The entire treatment system is constructed into a one-block pedestrian way informally dubbed 'water street'. Treated, disinfected wastewater is reused to meet the majority of non-potable demand for toilet flushing, running the buildings' cooling systems, and landscape irrigation. Any excess, unused treated wastewater is dispersed into dry wells for crucial groundwater recharge in a largely impervious urban area. Stormwater from the Hasslo on 8th project is also treated through a variety of green infrastructure methods such as green roofs, stormwater planters and a block-long water feature. A 227,000 litre underground cistern captures roof runoff and filters and recycles it through the water feature. A series of pedestrian bridges cross the water feature to provide access to the various buildings along this block while bringing the experience of open, flowing clean water into the core of the project.

The use, reuse and discharge of water in urban developments is clearly a very specialized example of urban ecological restoration, if it is considered a restoration project at all. But such projects do restore ecological processes such as groundwater recharge that are usually absent from densely developed sites. They can also bring functional green space and the experience of water into the urban fabric.

Figure 25.3 Trickling filters and constructed tidal wetland cells

Source: photo by Biohabitats

Pollination: Urban Pollinator Project by a consortium of universities, local government and NGOs

At least 1,500 species of insects, including bumble bees, solitary bees, honeybees, hoverflies, wasps, butterflies and moths, pollinate plants in the UK. Insect pollination is important for the production of many crops, including apples, strawberries and tomatoes, and also for many wild plant species. However, pollinators are under threat because of the loss of natural habitat, intensive modern farming practices that promote crop monocultures, the extensive use of pesticides, and emerging diseases (Vanbergen *et al.* 2013). Some research has shown that flower-rich areas within the urban matrix can support substantial numbers of native pollinators (e.g. Lopezaraiza-Mikel *et al.* 2007), and the Urban Pollinators Project set out to find out which pollinators are found in UK urban areas and how urban sites can be enhanced to improve pollinator diversity and abundance.

The first phase of the research examined how pollinator communities in UK urban areas compare to those in agro-ecosystems and protected areas and compared pollinators among urban habitats (Baldock *et al.* 2015). The researchers then worked in partnership with local government and conservation NGOs to create 60 urban meadows containing flowers high in pollen and nectar in four UK towns and cities: Bristol, Edinburgh, Leeds and Reading. One of the most important characteristics of this initiative was its large spatial extent. Furthermore, its rigorous experimental design sets it apart from most pollinator projects, which are often led by landscape architecture firms who do not have the resources to conduct controlled experiments or replicate their plantings.

Figure 25.4 A perennial flower meadow in Leeds, UK, which was created as part of the Urban Pollinators Project

Source: photo by M. Goddard

The team created both annual and perennial flower meadows of 300 m² in public parks, school playing fields, university grounds and along roadsides. The meadows were established in 2012 and 2013 and varied in shape depending on the layout of the site. In most locations meadows were established and maintained by local government parks or greenspace departments, with some sites managed by local contractors. Although identical protocols were used at all sites, the flowering composition of the meadows varied markedly, not just between cities but between sites within the same city. Factors such as site topology and management history are likely to be important.

Pollinators were monitored at all sites in 2013, with insects identified to species level and the flower species each insect visited recorded. This effort was one of the largest science-driven coordinated efforts to support pollinators in an urban environment, and it will likely guide future efforts in supporting pollinator communities.

Nutrient assimilation: Baltimore Harbor floating wetlands by Biohabitats for Waterfront Partnership of Baltimore, Inc.

Along many urban waterfronts, tidal marshes have been replaced by bulkheads and piers. Now that many waterfronts are transforming from industrial shipping centres into mixed use and public open space, their degraded water quality and habitat value are attracting increasing attention. In 2009, the Waterfront Partnership of Baltimore undertook a comprehensive effort to create a swimmable, fishable harbour. One pilot project to improve water quality and habitat was floating treatment wetlands.

The wetland design took several things into consideration. The functional goals of the floating wetlands are to remove nutrients, process and metabolize nutrients, reduce eutrophication, sequester heavy metals, improve water quality, and provide habitat, functions that are common in natural wetland systems. The floating wetlands draw nutrients from the surrounding waters and assimilate them into the biofilm (Streb 2013). In addition to the technical goals, the team sought a design that would be readily replicable at a low cost. The selected design used recycled plastic soda bottles collected from the harbour for buoyancy.

Putting floating wetlands in the tourist centre of the city was conceived in part as a public and outreach activity, and accordingly, the designers recruited volunteers from the community. An interpretive sign explains the importance of wetlands in estuarine ecosystems. As outreach tools, the wetlands are considered an unmitigated success by city officials and the area of wetlands was expanded tenfold in 2012.

Quantifying their nutrient removal and other ecological function is more complicated. Floating wetlands with emergent macrophytes are known to remove copper and zinc at rates that compare favourably to surface flow wetlands built to treat stormwater (Tanner and Headley 2011). In Baltimore floating wetland microcosms were rapidly colonized by organisms that provided a monitoring opportunity. The National Aquarium of Baltimore assessed the colonizers' capacity to absorb nutrients, finding them effective at absorbing nutrients. However, the long-term fate of the nutrients thus absorbed by the wetlands remains unknown.

One of the project challenges was an indirect product of its novelty. Regulators were cautious about granting permits, with a primary concern that floating wetlands might become recognized as a substitute for natural wetlands for purposes of wetland mitigation. Opening the door to this sort of overextension of the concept was an unexpected risk of the project, though it was resolved through a diligent examination of the permitting language.

Figure 25.5   Floating wetlands in Baltimore harbour
Source: photo by Biohabitats

Fish habitat: Biohuts by Ecocean

In many urban waterways, fish and the habitats on which they depend are in decline. In natural systems, young fish spend most of their time hidden in submerged vegetation and taking refuge in marshes, shallows and on rocky shores. In urban areas, natural shoreline and shallow aquatic habitats have largely been replaced by vertical walls, docks, jetties, ports and other infrastructure to accommodate boats, commerce and other human needs. Such structures do not afford young fish adequate foraging opportunities and shelter to escape predation. Urban restorationists have made various attempts to replace escape habitat for young fish. Biohuts are one such attempt to make shoreline infrastructure more hospitable to young fish by reintroducing the essential nursery functions of shelter and food.

The Biohut, which was created to be a versatile, easily replicable habitat augmentation tool, uses a double cage system to improve the survival rate of young fish that settle in shoreline areas before they move to deeper water. The Biohuts are slim, simple, two-layered wire cages. They are inexpensive and have been replicated widely on both sides of the Atlantic and in various types of habitat. The models deployed in the Mediterranean have dimensions of about 50 × 80 × 10 cm and weigh approximately 18 kg. The inner cage of the double cage system is filled with oyster shells that attract rapid colonization by microorganisms such as algae and copepods, providing forage for juvenile fish. The outer cage is empty and provides only shelter. The mesh of the outer cage allows only fingerlings to enter, and thus provides a predator-free zone.

The Biohut can be attached to wood, concrete or steel bulkheads. Each requires three holes in the dock, two at the top and one at the bottom to provide ballast. Other attachment systems, such as magnets, exist depending on the target surface. Installation requires two people, a diver and an assistant at the surface.

In the Mediterranean Sea, more than 650 Biohuts have been deployed in 30 French marinas and at the Marseille International Port. These habitats are also used in Baltimore MD and in Morocco. The preliminary results of a controlled study showed that the biodiversity in the vicinity of Biohut was about 150 per cent that of the control sites (28 species versus 19). However, because they are designed to support only early life stage on the life of a mobile organism, the ultimate effects in the fish populations is not yet known. The first peer-reviewed publications are expected to be released in 2016.

Figure 25.6　Installed Biohut

Source: photo by Ecocean

Water retention: the WaterShed House green roof by University of Maryland students, installed at Pepco Watershed Sustainability Center

Stormwater is a pervasive driver of urban ecological restoration efforts, and stormwater 'best management practices', designed to retain and treat stormwater, are perhaps the most common form of ecological restoration in the urban context. Because of the interest and financial incentives, this area has led to some of the most rigorous evaluation of any restoration projects (Bratieres *et al.* 2008).

The diversity of approaches to retaining water in the urban landscape has also led to some of the most creative approaches to supporting ecological functions within the built context. One such example is the green roof of the 'WaterShed House', which was designed in response to the annual Solar Decathalon design competition and today stands as the centrepiece of the Pepco Watershed Sustainability Center. The WaterShed House is designed to fit into the natural water cycle, helping conserve and filter water. Its butterfly roof helps direct rainwater into constructed wetlands along the entrance to the house, which filters and stores rainwater for use in irrigation and other non-potable functions.

The building is an interesting innovation on common green roof designs. Instead of the traditional flat green roof, the WaterShed's roof cants at a 10 degree slope and sheds water more quickly that traditional green roof media. It is created from a LiveRoof Hybrid Green Roof System with LiveRoof Lite modules. Fully planted and saturated, the weight of LiveRoof Lite modules is about 15–17 pounds per square foot (6.8–7.9 kg per 0.9 m^2), which is less than Maryland's average snow load. This innovation sidesteps one of the most common drawbacks to installing a green roof, the need for special structural support for the increased weight.

Green roofs are thought to reduce urban heat islands (Getter *et al.* 2001), provide additional insulation and reduce water loss to storm sewers (Whitinghill *et al.* 2015). Green buildings also promise to intercept and bioremediate or impound airborne and precipitation-borne pollutants (Rowe 2001). These claims, as so many others in urban restoration efforts, are challenging to

Figure 25.7 Green roof on the WaterShed House at the Pepco Watershed Sustainability Center
Source: photo by M. Beal

evaluate at a local scale, though the existing evidence seems to support the idea that micro-scale approaches at the building roof level can be a worthwhile first step to restoration in densely developed areas.

References

Baldock, K. C. R., Goddard, M. A., Hicks, D. M., Kunin, W. E., Mitschunas, N., Osgathorpe, L. M., Potts, S. G., Robertson, K. M., Scott, A. V., Stone, G. N., Vaughan, I. P. and Memmott, J. (2015) Where is the UK's pollinator biodiversity? The importance of urban areas for flower-visiting insects, *Proceedings of the Royal Society B*, vol 282, article 20142849.

Bratieres, K., Fletcher, T. D., Deletic, A. and Zinger, Y. (2008) Nutrient and sediment removal by stormwater biofilters: a large-scale design optimisation study, *Water Research*, vol 42, no 14, pp. 3930–3940.

Forman, R. T. T. (2014) *Urban Ecology: Science of Cities*, Cambridge: Cambridge University Press.

Forys, E. A. and Borboen-Abrams, M. (2006) Roof-top selection by least terns in Pinellas County, Florida, *Waterbirds*, vol 29, no 4, pp. 501–506.

Forys, E. A., Poppema-Bannon, A., Krajcik, K. and Szelistowski, W. A. (2013) Roof-nesting least terns travel to forage in brackish/marine waters, *Southeastern Naturalist*, vol 12, no 1, pp. 238–242.

Getter, K. L., Rowe, D. B., Andresen, J. A. and Wichman, I. S. (2011) Seasonal heat flux properties of an extensive green roof in a Midwestern US climate, *Energy and Buildings*, vol 43, pp. 3548–3557.

Gobster, P. H. (2010) Introduction: urban ecological restoration, *Nature and Culture*, vol 5, no 3, pp. 227–230.

Hobden, D. W., Laughton, G. E. and Morgan, K. E. (2004) Green space borders – a tangible benefit? Evidence from four neighbourhoods in Surrey, British Columbia, 1980–2001, *Land Use Policy*, vol 21, no 2, pp. 129–138.

Howell, E. A., Harrington, J. A. and Glass, S. B. (2012) *Introduction to Restoration Ecology*, Washington, DC: Island Press.

Ingram, M. (2008) Urban ecological restoration, *Ecological Restoration*, vol 26, no 3, pp. 175–177.

Jacobson, C. (2011) Identification and quantification of the hydrological impacts of imperviousness in urban catchments: a review, *Journal of Environmental Management*, vol 92, no 6, pp. 1428–1448.

Lopezaraiza-Mikel, M. E., Hayes, R. B., Whalley, M. R. and Memmott, J. (2007) The impact of an alien plant on a native plant-pollinator network: an experimental approach, *Ecology Letters*, vol 10, pp. 539–550.

Manuel, P. M. (2003) Cultural perceptions of small urban wetlands: cases from the Halifax Regional Municipality, Nova Scotia, Canada, *Wetlands*, vol 23, no 4, pp. 921–940.

Naveh, Z. (2005) Epilogue: toward a transdisciplinary science of ecological and cultural landscape restoration, *Restoration Ecology*, vol 13, no 1, pp. 228–234.

Palamar, C. (2010) From the ground up: why urban ecological restoration needs environmental justice, *Nature and Culture*, vol 5, no 3, pp. 277–209.

Phillips, P. J., Chlamers., A. T., Gray, J. L., Kolpin, D. W., Foreman, W. T. and Wall., G. R. (2012) Combined sewer overflows: an environmental source of hormones and wastewater micropollutants, *Environ. Sci. Technol.*, vol 46, no 10, pp. 5336–5343.

Quon, S. P., Martin, L. R. G. and Murphy, S. D. (1999) Ecological rehabilitation: a new challenge for planners. *Plan Canada*, vol 39, no 4, pp. 18–21.

Rowe, D. B. (2011) Green roofs as a means of pollution abatement, *Environmental Pollution*, vol 159, no 8-9, pp. 2100–2110.

Rudd, H., Vala, J. and Schaefer, V. (2002) Importance of backyard habitat in a comprehensive biodiversity conservation strategy: a connectivity analysis of urban green spaces. *Restoration Ecology*, vol 10, no 2, pp. 368–375.

Streb, C. (2013) Building floating wetlands to restore urban waterfronts and community partnerships, *National Wetlands Newsletter*, vol 35, no 2.

Tanner, C. C. and Headley, T. R. (2011) Components of floating emergent macrophyte treatment wetlands influencing removal of stormwater pollutants, *Ecological Engineering*, vol 37, no 3, pp. 474–486.

Tilley, D., Price, J., Matt, S. and Marrow, B. (2012) *Vegetated Walls: Thermal and Growth Properties of Structured Green Facades*, Toronto: Green Roofs for Healthy Cities.

Vanbergen, A. J. and the Insect Pollinators Initiative (2013) Threats to an ecosystem service: pressures on pollinators, *Frontiers in Ecology and the Environment*, vol 11, no 5, pp. 251–259.

Westphal, L. M., Gobster, P. H. and Gross, M. (2010) Models for renaturing brownfield areas, in M. Hall (ed.), *Restoration and History: The Search for a Usable Environmental Past*, pp. 208–217, New York: Routledge.

Whittinghill, L. J., Rowe, D. B., Cregg, B. M. and Andresen, J. A. (2015) Comparison of stormwater runoff from sedum, native prairie, and vegetable producing green roofs, *Urban Ecosystems*, vol 18, pp. 13–29.

Zmyslony, J. and Gagnon, D. (2000) Path analysis of spatial predictors of front-yard landscape in an anthropogenic environment, *Landscape Ecology*, vol 15, no 4, pp. 357–371.

PART III

Management and policy issues

26

INTERNATIONAL LAW AND POLICY ON RESTORATION

An Cliquet

Introduction

Law and policy can be an important trigger to get restoration activities done 'on the ground'. However restoration-based laws often facilitate remediation or compensation, instead of demanding effective, feasible, and knowledge-based *in situ* restoration or rehabilitation (Aronson *et al.* 2011). When no clear guidance is given in law and policy, 'restoration' activities might not have the intended result, or might even have unwanted side-effects. More guidance on ecological restoration in laws and policies is needed. Clear principles of practice should be at the basis of restoration efforts on the ground (Suding *et al.* 2015). Restoration obligations in legal instruments can roughly be divided into three categories:

- Restoration obligations can be found in nature conservation laws, imposing obligations on states or stakeholders to conserve and/or restore the environment to a favourable conservation status – a 'command and control' approach. These obligations are often linked to the management obligations for protected areas, but are not limited to this. Examples include the *in situ* obligations in the Biodiversity Convention.[1]
- Restoration obligations also can be found in laws that oblige compensatory measures in case of infrastructure projects. Compensatory measures are understood as measures that restore, create or enhance an area of habitat or a species population in order to compensate for residual damage caused by a plan or project. Examples include mitigation banking in the US and the compensation obligations for the Natura 2000 network in the EU Habitats Directive.[2]
- Restoration can be imposed in the framework of a liability regime for environmental or ecological damage, in which the responsible party for ecological damage is required to restore the environment. Examples can be found in specific legal regimes, such as the Convention on Civil Liability for Oil Pollution Damage,[3] or the EU Environmental Liability Directive.[4]

The focus in this chapter will mostly be on the first category.

International environmental law and policy can play an important role in providing obligations and guidelines for regional and national strategies, plans and laws on restoration.

Restoration has already been included in international legal instruments and policy for several years, either explicit or implicit, the latter comprising restoration under their conservation obligations, for instance by including obligations to reach a favourable conservation status for species or habitats. Recently, attention for restoration at the international level has increased and more concrete and explicit provisions and targets on restoration have been elaborated.

This chapter will focus on some of the major international legal and policy instruments that are relevant for restoration, but it is not intended to give a complete overview. A more thorough overview of international law on restoration can be found in Telesetsky *et al.* (2017). This chapter will be limited to the major multilateral conservation conventions, including the Biodiversity Convention, the Ramsar Convention,[5] the World Heritage Convention[6] and the Bonn Convention on Migratory Species[7] (and its related agreements). Numerous other international and regional instruments are relevant for restoration. To provide examples, this chapter will contain a reflection on the possibilities and commitments for restoration of drylands in the framework of the Desertification Convention,[8] and the initiatives for restoration of forests that have recently been developed, mostly in the framework of climate change regulations.

This chapter reviews the obligations on restoration in these instruments, but also if and how restoration is defined and if there are any concrete guidelines, principles or standards developed under these instruments. Not only the convention texts are important, but even more important are the decisions taken by the Conference of the Parties (COPs). For most multilateral environmental conventions, these COP decisions are usually considered to be legally non-binding. That doesn't mean these decisions are without value. Even if the character of COP decisions might not always be legally binding, they establish at least a political commitment. Also, in a Dutch court case a Ramsar COP decision was used to interpret the meaning of a binding provision of the Ramsar Convention (Verschuuren 2008).

Restoration under the Biodiversity Convention

Restoration in the Convention text and decisions

The Biodiversity Convention includes some provisions which refer to restoration as a means to fulfil the Convention objectives. Article 8,f provides that each party, as far as possible and as appropriate shall 'rehabilitate and restore degraded ecosystems and promote the recovery of threatened species, *inter alia*, through the development and implementation of plans or other management strategies'. However, the Convention text itself does not contain any concrete targets, or further guidelines and although the Convention holds a large list of definitions in its article 2, the terms 'restore' and 'rehabilitate' are not defined.

Article 14, §2 includes the obligation for the Conference of the Parties to 'examine, on the basis of studies to be carried out, the issue of liability and redress, including restoration and compensation, for damage to biological diversity'. In the Convention text itself, no further explanation is given on what is understood by restoration or compensation, and what the difference is between the two terms. Some further explanation can be found in a report by the Secretariat:[9] primary restoration is seen as the preferred focus or approach of redress over compensatory measures. In the report some further guidance and examples from state practice on primary restoration are given.

The Biodiversity Convention is not the only legal instrument dealing with liability for environmental damages. Specific regimes have been elaborated in international environmental law, including for example liability regimes for oil pollution or nuclear energy. A common approach in these instruments is that the responsible parties for ecological damage are required to take restoration measures. Most often, the amount of compensation is determined by the restoration cost approach, meaning that the amount of compensation to be paid is determined by the cost of reasonable measures for restoring the environment back to its original state (e.g. in the Convention on Civil Liability for Oil Pollution Damage and the Convention on Civil Liability for Nuclear Damage[10]). Some of these instruments allow for compensation through the costs for the introduction of the equivalent of damaged or destroyed components into the environment (e.g. in the Convention on Civil Liability for Nuclear Damage, the Convention on Civil Liability for Damage resulting from Activities Dangerous to the Environment[11] and the Nagoya – Kuala Lumpur Supplementary Protocol on Liability and Redress to the Cartagena Protocol on Biosafety[12]).

Several CBD COP decisions elaborate on restoration. Since the enactment of the Biodiversity Convention, thirteen Conference of Parties have been organised, and several decisions have been taken that include provisions on restoration of (for example) forests, dry and sub-humid lands, inland water systems, mountain areas, marine and coastal areas (Telesetsky *et al.* 2017)[13] and on the issues of liability and restoration.[14]

Restoration in the strategic plan 2011–2020 and Aichi Targets

The most specific targets for restoration can be found in the Aichi Biodiversity Targets (Biodiversity Convention Strategic plan 2011–2020)[15] of the Biodiversity Convention. Three targets are of particular interest in this regard.

According to Target 11, by 2020,

> at least 17 per cent of terrestrial and inland water, and 10 per cent of coastal and marine areas, especially areas of particular importance for biodiversity and ecosystem services, are conserved through effectively and equitably managed, ecologically representative and well connected systems of protected areas and other effective area-based conservation measures, and integrated into the wider landscapes and seascapes.

Although Target 11 doesn't explicitly mention restoration, the words 'conservation' and 'effectively managed' should be able to be interpreted to include restoration activities, if a protected area is in an unfavourable conservation status. 'Effectively managed' is the degree to which protected area management protects biological and cultural resources and achieves the goals and objectives for which the protected area was established.[16] Effectively managed' should include planning measures to ensure ecological integrity and the protection of species, habitats and ecosystem processes.[17]

The reference to 'well-connected systems of protected areas' in Target 11 is relevant for restoration. Creating or restoring functional linkages between protected areas and their surrounding regions is essential if we are to strengthen ecological coherence and resilience for both biodiversity conservation and sustainable development. An action plan should include a plan for improving the protected area network, including one for restoring degraded protected areas and establishing ecological corridors.[18]

While one may assert that restoration of protected areas is included in Target 11, it lacks specific language about restoration and guidelines about issues such as to what level protected areas should be restored, is restoration due in all of the protected areas and in the whole of the protected area. The effectiveness of the management of the protected area depends on the goals and objectives for which the protected area has been designated. However, a lot of protected areas are 'paper' protected areas and lack clear or sufficient conservation and restoration objectives.

Because of a lack of clear definitions, it will be difficult to measure the progress on restoration targets. The Global Biodiversity Outlook (GBO) 4 Report (Secretariat of the CBD 2014), gives a mid-term assessment on the progress towards the implementation of the Aichi Targets. In the evaluation of Target 11 on protected areas, nothing specific is mentioned about restoration in protected areas. The Report gives an evaluation of the quantitative aspect (percentage of protected areas). However, for the quality of the management, progress is much harder to establish. A minority of protected areas enjoy sound management, although this appears to be improving over time given the limited information available. National actions reported to the CBD indicate that most countries have targets relating to improvement of protected area coverage, although a minority have quantitative targets and relatively few address issues of ecological representativeness, connectedness or management effectiveness (*ibid.*).

Some key actions are proposed to enhance the implementation of the Aichi Targets. Relevant actions with regards to Target 11 include the expanding of protected area networks and other effective area-based conservation measures to become more representative of the planet's ecological regions, of marine and coastal areas (including deep sea and ocean habitats), of inland waters and of areas of particular importance for biodiversity; and improving and regularly assessing management effectiveness and equitability of protected areas and other area-based conservation measures (*ibid.*). However, these proposed key actions add little to already existing legal obligations and commitments and contain no further guidance relating to restoration in or between protected areas.

There are two targets that explicitly reference restoration. Target 14 includes the obligation that 'ecosystems that provide essential services, including services related to water, and contribute to health, livelihoods and well-being, are restored and safeguarded'. Restoration is defined as 'the process of actively managing the recovery of an ecosystem that has been degraded, damaged or destroyed as a means of sustaining ecosystem resilience and conserving biodiversity'.[19] Under the proposed actions and milestones, the guide for Target 14 mentions that avoiding degradation through conservation is preferable to restoring an ecosystem after a disturbance. Given the emphasis on safeguarding in this target, the programme of work on protected areas provides relevant guidance on the types of actions which could be taken to fulfil this target. Further restoration activities, such as forest and wetland landscape restoration, are already under way in many parts of the world and increasingly will be needed to re-establish ecosystem functioning and the provision of valuable services. Consolidating policy processes and the wider application of these efforts could contribute significantly to the achievement of the objectives of the Convention and this target specifically.[20]

Target 15 is the most explicit reference to restoration:

> By 2020, ecosystem resilience and the contribution of biodiversity to carbon stocks has been enhanced, through conservation and restoration, including restoration of at least 15 per cent of degraded ecosystems, thereby contributing to climate change mitigation and adaptation and to combating desertification.

The restoration of degraded habitats represents an opportunity to both improve ecosystem resilience and to increase carbon sequestration. In 2010, two-thirds of the planet's ecosystems could be considered degraded. The global potential for forest landscape restoration alone is estimated to be in the order of 1 billion hectares, or about 25 per cent of the current global forest area. Therefore there is a large potential for the increased use of restoration.[21] Although the 15 per cent target has a strong link to climate change and carbon storage, and gives an important impetus for forest restoration, it is not limited to this, because the target refers to ecosystem resilience in general and also other ecosystems besides forests can play an important role in carbon storage.

Taking into consideration that nearly two-thirds of the planet's ecosystems could be considered degraded, the target is not limited to protected areas. We would rather argue that the target for restoration within protected areas is already included within Target 11 and that Target 15 is an additional target for restoration, and should thus predominantly be realized outside protected areas. However, there are no guidelines on how much of the 15 per cent target should take place within or outside of protected areas.

Although Target 15 sounds promising, the danger exists that in order to reach the 15 per cent target, governments will tend to take restoration measures that are already legally required by international or national laws, which will only occur within already protected areas, with no added value.

Even in Targets 14 and 15, the exact same list of shortcomings exists as for Target 11 (see earlier comment). At the COP 10 in 2010, when the Aichi Targets were decided, the COP in its Decision X/9 decided to consider the identification of ways and means to support ecosystem restoration at its 11th meeting in 2012, including the possible development of practical guidance on ecosystem restoration and related issues.[22] This means that targets on restoration have been established, but without a full understanding of what those targets entail. Target 15 may have been written in such a way that the goal cannot be reached because it cannot be measured (see Jørgenson 2013 for deeper analysis).

On the restoration Targets 14 and 15, the GBO 4 states that restoration is under way for some depleted or degraded ecosystems, especially wetlands and forests, sometimes on a very ambitious scale, as in China. Many countries, organizations and companies have pledged to restore large areas. Abandonment of farmland in some regions including Europe, North America and East Asia is enabling 'passive restoration' on a significant scale. Despite restoration and conservation efforts, there is still a net loss of forests, suggesting no overall progress on this component of the target. The combined initiatives currently under way or planned may put us on track to restore 15 per cent of degraded ecosystems by 2020, but it is hard to assess and it is not certain that this part of the target will be met (Secretariat of the CBD 2014). The evaluation of GBO 4 was confirmed and strengthened in an analysis by Tittensor *et al.* (2014), which clearly stated that the progress on Target 15 cannot be measured as there are no indicators available for extrapolation.

According to GBO 4, the key actions to achieve Target 15 on restoration include that governments need to identify opportunities and priorities for restoration, including highly degraded ecosystems, areas of particular importance for ecosystem services and ecological connectivity, and areas undergoing abandonment of agricultural or other human-dominated use. Strategies to restore at least 15 per cent or more of degraded areas should be expanded and further developed, including methods like environmental permitting procedures and market instruments such as wetland mitigation banking, payments for ecosystem services and other mechanisms. The contribution of biodiversity to carbon sequestration through state or private sponsored passive and active afforestation programmes, such as the REDD+ mechanism must be increased. Where feasible, restoration should be made an economically viable activity, by coupling income generation to restoration activities (Secretariat of the CBD 2014).

Ecosystem restoration work by the COP after the Aichi Targets

Since the setting of the targets in 2010, further work on restoration has been undertaken in the framework of the Subsidiary Body on Scientific, Technical and Technological Advice

(SBBTA)[23] and the COPs, including information documents and notes by the Executive Secretary[24] and COP decisions.

COP Decision XI/16[25] of 2012 urges Parties to make concerted efforts to achieve the Aichi Targets 14 and 15, and contribute to the achievement of all other Aichi Biodiversity Targets through ecosystem restoration by a range of activities. This includes identifying degraded ecosystems that have the potential for ecosystem restoration bearing in mind that such areas may be occupied by indigenous and local communities.[26]

The Decision does not give precise guidelines on where to take restoration measures. Furthermore, there are still uncertainties about the concept of restoration. The Decision requests the Executive Secretary to develop clear terms and definitions of ecosystem rehabilitation and restoration and clarify the desired outcomes of implementation of restoration activities[27] and to develop a tool for collating and presenting baseline information on the condition and extent of ecosystems, in order to facilitate the evaluation of Target 15 so as to assist Parties in identifying ecosystems whose restoration would contribute most significantly to achieving the Aichi Biodiversity Targets.[28]

The Executive Secretary was also requested to convene regional and subregional capacity-building and training workshops and expert meetings.[29] The Secretariat has carried out several subregional workshops and initiated several global studies.[30] At the eleventh Conference of the Parties, governments from India, South Korea and South Africa, together with the secretariats of several international conventions and NGOs agreed on the Hyderabad Call for a Concerted Effort on Ecosystem Restoration,[31] which called upon governments, governmental and other organizations and private parties 'to make concerted and coordinated long-term efforts to mobilize resources and facilitate the implementation of ecosystem restoration activities on the ground for sustaining and improving the health and well-being of humans and all other species with whom we share the planet'. Although this is a voluntary initiative, it shows increased interest for restoration at the international level. In response to the Hyderabad Call, South Korea developed the Forest Ecosystem Restoration Initiative,[32] which was welcomed by the 12th Conference of the Parties of the CBD.[33] Another initiative that has been welcomed at the COP 12 is the Caring for Coasts Initiative, as part of a global movement to restore coastal wetlands.[34]

In Decision XII/19 of 2014 on Ecosystem conservation and restoration[35] the Conference of the Parties notes with concern that not enough progress has been made towards Targets 14 and 15. Parties are invited to take into account an ecosystem approach to promote ecosystem restoration. Priority should be given to avoiding or reducing ecosystem losses, to promote ecosystem restoration activities, in particular large-scale restoration activities, noting also the cumulative benefits of small-scale restoration activities that can collectively contribute to biodiversity conservation, climate-change adaptation and mitigation, and reducing desertification. Furthermore Parties are invited to develop and strengthen monitoring of ecosystem degradation and restoration, with a view to supporting adaptive management and reporting on progress towards the Aichi Biodiversity Targets, in particular Targets 5, 14 and 15. At COP13 (December 2016) Decision XIII/5 adopted a short-term action plan on restoration.

Restoration under other global conventions

Ramsar Convention

In the Ramsar Convention on Wetlands of International Importance restoration is not mentioned explicitly. Implicitly, restoration measures can be required. According to article 3 the Parties 'shall formulate and implement their planning so as to promote the conservation of the wetlands included in the List, and as far as possible the wise use of wetlands in their territory'. These conservation measures could include restoration measures. Furthermore article 4, §2 includes a compensation obligation for the loss of a listed wetland:

> where a Contracting Party in its urgent national interest, deletes or restricts the boundaries of a wetland included in the List, it should as far as possible compensate for any loss of wetland resources, and in particular it should create additional nature reserves for waterfowl and for the protection, either in the same area or elsewhere, of an adequate portion of the original habitat.

Within the subsequent strategic plans of the Convention, wetland restoration has been recognized as one of its strategic objectives.[36] In the Strategic Plan 2009–2015 wetland restoration is mentioned as one of the strategies and key activities. Priority wetlands and wetland systems where restoration or rehabilitation would be beneficial and yield long-term environmental, social, or economic benefits should be identified. The necessary measures to recover these sites and systems have to be implemented. By 2015 all Parties have to identify priority sites for restoration; restoration projects are under way or completed by at least half the Parties.

The Strategic Plan 2016–2024 states that to achieve the mission of the Convention it is essential that vital ecosystem functions and the ecosystem services they provide to people and nature are fully recognized, maintained, restored and wisely used.[37] Under the Strategic Goals 2016–2024, several targets address restoration. According to Target 5 'The ecological character of Ramsar sites is maintained or restored, through effective planning and integrated management'. Target 12 states: 'Restoration is in progress in degraded wetlands, with priority to wetlands that are relevant for biodiversity conservation, disaster risk reduction, livelihoods and/or climate change mitigation and adaptation'. In Annex 1 of the Strategic Plan, baselines and indicators are given for the different strategic goals. Based on the national reports to COP 12, 68 per cent of Contracting Parties have identified priority sites for restoration and 70 per cent of Contracting Parties have implemented restoration or rehabilitation programmes. Indicators for this target include the percentage of Parties that have established restoration plans (or activities) for sites, and the percentage of Parties that have implemented effective restoration or rehabilitation projects.

Several recommendations[38] and resolutions[39] refer to restoration. Especially Resolution VIII.16 on principles and guidelines for wetland restoration is relevant as extensive guidelines to restoration of wetlands are given in the annex to the resolution. The importance of restoration of wetlands for human well-being and in adapting to climate change has been stressed in resolutions.[40] Specifically with regard to wetlands on the Ramsar List of wetlands of international importance, various resolutions have been taken which refer to restoration of Ramsar sites.[41] Restoration has also been explicitly mentioned in resolutions regarding specific ecosystems, such as peatlands[42] and mangroves.[43]

World Heritage Convention

The World Heritage Convention does not mention 'restoration' in the Convention text itself. Several articles in the Convention mention 'rehabilitation'. Article 5, §2 includes the obligation for State Parties to take the appropriate legal, scientific, technical, administrative and financial measures necessary for the identification, protection, conservation, presentation and rehabilitation of World Heritage. The World Heritage Committee can receive requests from State Parties for international assistance with respect to property forming part of the cultural or natural heritage situated in their territories, and included or potentially suitable for inclusion in the World Heritage List or World Heritage List in Danger. The purpose of such requests may be to secure the protection, conservation, presentation or rehabilitation of such property (article 13, §1). International assistance by the World Heritage Committee and by the World Heritage Fund can be of reasons of rehabilitation of World Heritage (articles 22 and 23).

Also the Operational Guidelines pay attention to restoration of World Heritage, in its chapter on reactive monitoring. Reactive Monitoring is the reporting by the Secretariat, other sectors of UNESCO and the Advisory Bodies to the Committee on the state of conservation of specific World Heritage properties that are under threat.[44] Based on information by the State Parties or others, the Committee has several possibilities. When the Committee considers that the property has seriously deteriorated, but not to the extent that its restoration is impossible, it may decide that the property be maintained on the List, provided that the State Party takes the necessary measures to restore the property within a reasonable period of time. The Committee may also decide that technical co-operation be provided under the World Heritage Fund for work connected with the restoration of the property, proposing to the State Party to request such assistance, if it has not already been done.[45]

Sites can also be placed on the List of World Heritage in Danger. When considering the designation of a property on the List of World Heritage in Danger, the Committee shall develop, and adopt, as far as possible, in consultation with the State Party concerned, a desired state of conservation for the removal of the property from the List of World Heritage in Danger, and a programme for corrective measures.[46] Although the Operational Guidelines do not specifically mention restoration, the 'corrective' measures can include restoration of sites. If the restoration measures prove to be successful, the site can be removed from the List of World Heritage in Danger, based on a regular review of the state of conservation of properties on the List of World Heritage in Danger.[47]

Deletion of a site from the World Heritage List can be done for the following reasons: where the property has deteriorated to the extent that it has lost those characteristics which determined its inclusion in the World Heritage List; and where the intrinsic qualities of a World Heritage site were already threatened at the time of its nomination by action of man and where the necessary corrective measures as outlined by the State Party at the time, have not been taken within the time proposed.[48] Thus, a failure to take restoration measures might lead to the removal of a site from the World Heritage List.

Convention on Migratory Species

The Convention on Migratory Species, dealing with the protection of transboundary migratory species, includes the obligation for Parties that are Range States of a migratory species listed in Appendix I 'to conserve and, where feasible and appropriate, restore those habitats of the species which are of importance in removing the species from danger of extinction'.[49] No further guidance or concrete targets are given in the Convention text. Annex II of the Convention lists migratory species which have an unfavourable conservation status and which require international agreements for their conservation and management, as well as those which have a conservation status which would significantly benefit from the international cooperation that could be achieved by an international agreement. Range States have to conclude agreements for these species.[50]

In several of these agreements obligations have been included to restore the habitats of the migratory species. The Wadden Sea Seals Agreement[51] provides that the Parties shall explore the possibility of restoring degraded habitats and of creating new ones.[52] According to the African-

Eurasian Waterbird Agreement[53] the Parties shall identify sites and habitats for migratory water-birds occurring within their territory and encourage the protection, management, rehabilitation and restoration of these sites.[54] Parties to the Agreement on the Conservation of Albatrosses and Petrels[55] are expected to conserve and, where feasible and appropriate, restore those habitats which are of importance to albatrosses and petrels.[56] The Gorilla Agreement[57] obliges Parties to identify sites and habitats for gorillas occurring within their territory and ensure the protection, management, rehabilitation and restoration of these sites.[58]

Desertification Convention

The Desertification Convention does not include an explicit reference to restoration. However, combating desertification includes activities which are part of the integrated development of land in arid, semi-arid and dry sub-humid areas for sustainable development which are aimed at rehabilitation of partly degraded land.[59] The objective of the Convention is to combat desertification and mitigate the effects of drought in countries experiencing serious drought and/or desertification, which will involve long-term integrated strategies that focus simultaneously, in affected areas, on improved productivity of land, and the rehabilitation, conservation and sustainable management of land and water resources, leading to improved living conditions, in particular at the community level.[60]

In the subsequent work of the Convention attention has explicitly been paid to restoration. In 2012 a policy brief was enacted by the Secretariat of the Convention, proposing Zero Net Land Degradation[61] as a Sustainable Development Goal for the Rio+20 UN Conference on Sustainable Development in 2012.[62]

In the Rio + 20 Outcome Document, world leaders recognized the need for urgent action to reverse land degradation. In view of this, they will strive to achieve a land–degradation neutral world in the context of sustainable development (para 206). They reaffirm their resolve in accordance with the United Nations Convention to Combat Desertification to take coordinated action nationally, regionally and internationally, to monitor, globally, land degradation and restore degraded lands in arid, semi–arid and dry sub–humid areas (para 207).[63]

The Convention has established an Intergovernmental Working Group to develop concrete options for achieving the target, aimed at preventing and reversing land degradation through good land management and restoration. The target proposed by the Open Working Group on sustainable development goals has been agreed upon at the UN General Assembly meeting in September 2015: 'By 2030, combat desertification, restore degraded land and soil, including land affected by desertification, drought and floods, and strive to achieve a land degradation-neutral world'.[64]

Restoration of forests

Additional impetus for ecological restoration of forests is coming from current concerns about global climate change. Under the framework of the Climate Change Convention (UNFCCC)[65] a new mechanism is being developed, Reducing Emissions from Deforestation and forest Degradation in developing countries (REDD). REDD is a mechanism to provide funding for developing countries for climate mitigation activities and sustainable management of forests. The initial focus of REDD was on reducing emissions from deforestation and forest degradation, but was broadened to also include the role of conservation, sustainable management of forests and enhancement of forest carbon stocks in developing countries (known as REDD+). REDD+ includes several activities, including the enhancement of forest carbon stocks, which can include forest restoration. REDD was first introduced at the Conference of the Parties of 2005, and the item came back at the subsequent COPs.[66] The Paris Agreement of 2015, decided at COP 21,[67] gives a formal legal basis to COP decisions relating to REDD+.

Despite the absence until 2015 of a formal legal agreement on REDD+, the Conference of the Parties in 2010 asked the countries to undertake REDD+ action on a voluntary basis. In order to guide this work, the UN developed the UN-REDD programme.[68] Other support is coming from the Forest Carbon Partnership Facility (FCPF)[69] at the World Bank. Also, various bilateral agreements have been concluded, as well as projects supported by NGOs and the private sector.

Beside the Climate Change Convention, other initiatives have been taken on the restoration of forests. Within the Global Partnership on Forest and Landscape Restoration, which is a partnership between governments, international organisations, communities and individuals, the Bonn Challenge was accepted in 2011.[70] The Bonn Challenge calls upon the members to restore 150 million ha of deforested and degraded lands by 2020. The Bonn Challenge is not a new global commitment but rather a practical means of realizing other existing international commitments, including the CBD Aichi Target 15, the UNFCCC REDD+ goal, and the Rio+20 land degradation neutral goal. As of this moment (March 2017) 148.38 million of 150 million hectares has been committed to.[71] However it is not clear how many of these commitments have already been realised in practice.

In 2014, an additional commitment was taken in the New York Declaration on Forests,[72] to restore an additional 200 million hectares by 2030. The Declaration was signed by various governments, companies, civil society and indigenous organisations. It is also a non-legally binding document.

Although the increased attention for forest restoration certainly has its merits, other ecosystems such as non-forested peatlands can also play a major role in carbon sequestration. Given the current rate of climate change, our attention is too narrowly limited to forests. Also, if too narrowly focused on forests and failing to acknowledge the importance of other ecosystems, a danger exists of perverse effects for example by planting trees in grassland ecosystems (Veldman *et al.* 2015). Furthermore, there are some challenges for restoration under programmes such as REDD+ that need to be addressed (Alexander *et al.* 2011).

Conclusion

Attention to ecological restoration in the international community has increased significantly in the last few years. Although restoration, often as part of conservation, was already included in several multilateral environmental agreements, albeit implicitly, a more explicit and more detailed attention to restoration became fully developed, mainly in COP decisions of the agreements. Probably the most concrete target that has been formulated regarding restoration is the Aichi Target 15, aiming at 15 per cent restoration of degraded ecosystems. It is at least the achievement of the Aichi Targets and other international recommendations and decisions on restoration that restoration has been brought to the forefront, leading in turn to various commitments and initiatives at the regional and national level.

However, in most international instruments, there is an absence of any reference to a definition of restoration and clear guidelines on restoration of degraded ecosystems. Failure to clearly define and reference a term means that attempts to monitor and evaluate implementation become difficult. Additionally, most of the instruments are considered 'soft' law and non-legally binding.

Within this context, it is not surprising that the mid-term assessment on the progress of implementation of the Aichi Targets have found it difficult to impossible to measure the progress on the realization of the 15 per cent target. Given continued degradation of ecosystems and ecosystem services, prevention is laudable but insufficient. Conservation efforts are too late and will no longer suffice; increased efforts on restoration are necessary in order to reach the international commitments. Despite existing uncertainties in restoration ecology, further delay on the implementation of restoration commitments is no longer feasible. It is doubtful that often voluntary commitments will be sufficient to timely stem the tide of ecological degradation.

A legally binding instrument on restoration is needed. A possible option in this regard is an additional protocol under the Biodiversity Convention, even though enacting binding legal instruments at the international level is often a lengthy and difficult process. A legally binding instrument should include a clear definition, binding targets, minimum quality standards to which ecosystem restoration should adhere and monitoring obligations. Restoration obligations should not be limited to restoration as compensation for infrastructure works or restoration as part of reparation obligations under liability regimes, but should be extended and become a more explicit part of nature conservation laws. Specific and practical guidelines for national governments for the implementation of restoration obligations should be provided for in the frameworks of international and regional environmental instruments. Obligations on restoration need to be included in regional or national and local legislation – and this will be resisted by many governments, hence is easier to propose than to enact. We have witnessed the steep rise of restoration ecology as an emerging field in science, even leading some to call it 'toward an era of restoration in ecology' (e.g. Suding 2011). The question is if it will also be the era of ecosystem restoration in law. We strongly believe this should be the case.

Notes

1 Convention on Biological Diversity (CBD), Rio de Janeiro, 5 June 1992, www.cbd.int.
2 Directive 92/43/EEC of 21 May 1992 on the conservation of natural habitats and of wild fauna and flora, OJ L 206, 22 July 1992.
3 International Convention on Civil Liability for Oil Pollution Damage, London, 29 November 1969, as amended by the 1992 Protocol.
4 Directive 2004/35/CE of the European Parliament and of the Council of 21 April 2004 on environmental liability with regard to the prevention and remedying of environmental damage, OJ L 143, 30 April 2004.
5 Convention on Wetlands of International Importance especially as Waterfowl Habitat, Iran, 2 February 1971, www.ramsar.org.
6 Convention Concerning the Protection of the World Cultural and Natural Heritage, Paris, 16 November 1972, http://whc.unesco.org.
7 Convention on the Conservation of Migratory Species of Wild Animals, Bonn, 23 June 1979, www.cms.int.
8 United Nations Convention to Combat Desertification in Countries Experiencing Serious Drought and/or Desertification, Particularly in Africa, Paris, 17 June 1994, www.unccd.int.
9 Liability and redress in the context of paragraph 2 of article 14 of the Convention on Biological Diversity. Synthesis report on technical information relating to damage to biological diversity and approaches to valuation and restoration of damage to biological diversity, as well as information on national/domestic measures and experiences. Note by the Executive Secretary, UNEP/CBD/COP/9/20/Add.1, 20 March 2008.
10 Convention on Civil Liability for Nuclear Damage, Vienna, 21 May 1963, amended by the 1997 Protocol.
11 Convention on Civil Liability for Damage resulting from Activities Dangerous to the Environment, Lugano, 21 June 1993.
12 Supplementary Protocol on Liability and Redress to the Cartagena Protocol on Biosafety, Nagoya – Kuala Lumpur, 15 October 2010.

13 For an overview, see SBBTA, Progress report on ecosystem restoration and related Aichi targets. Note by the Executive Secretary, UNEP/CBD/SBSTTA/17/7, 5 September 2013.

14 See for example COP Decision IX/23. Liability and redress, 2008; this decision welcomes the synthesis report prepared by the Executive Secretary on technical information relating to damage to biological diversity and approaches to valuation and restoration of damage to biological diversity, UNEP/CBD/COP/9/20/Add.1, 20 March 2008.

15 COP Decision X/2. Strategic Plan for Biodiversity 2011–2020, 2010.

16 Explanatory Guide on Target 11 of the Strategic Plan for Biodiversity, 2012, p. 16, www.cbd.int/protected/tools/default.shtml.

17 Quick guide to target 11 of the Aichi Biodiversity Targets, 2012, www.cbd.int/doc/strategic-plan/targets/T11-quick-guide-en.pdf.

18 Explanatory Guide on Target 11 of the Strategic Plan for Biodiversity, 2012, pp. 22–23.

19 Quick guide to the Aichi Biodiversity targets. Ecosystems and essential services safeguarded, 2012, https://www.cbd.int/doc/strategic-plan/targets/T14-quick-guide-en.pdf.

20 *Ibid.*

21 Quick guide to the Aichi Biodiversity targets. Ecosystems restored and resilience enhanced, 2012, www.cbd.int/doc/strategic-plan/targets/T15-quick-guide-en.pdf.

22 COP Decision X/9. The multi-year programme of work for the Conference of the Parties for the period 2011–2020 and periodicity of meetings, 2010, (a) ix.

23 SBSTTA, Ways and means to support ecosystem restoration. Note by the Executive Secretary, UNEP/CBD/SBSTTA/15/4, 5 August 2011; SBSTTA, Recommendation XV/2. Ways and means to support ecosystem restoration, UNEP/CBD/SBSTTA/REC/XV/2, 7 December 2011; SBBTA, Progress report on ecosystem restoration and related Aichi targets. Note by the Executive Secretary, UNEP/CBD/SBBTA/17/7, 5 September 2013; SBBTA, Report on issues in progress: ecosystem conservation and restoration. Note by the Executive Secretary, UNEP/CBD/SBBTA/18/14★★, 29 May 2014.

24 Ecosystem restoration. Note by the Executive Secretary, UNEP/CBD/COP/11/21, 12 August 2012; Available guidance and guidelines on ecosystem restoration. Note by the Executive Secretary UNEP/CBD/COP/11/INF/17, 15 August 2012; Available tools and technologies on ecosystem restoration. Note by the Executive Secretary, UNEP/CBD/COP/11/INF/18, 4 September 2012; Most used definitions/descriptions of key terms related to ecosystem restoration. Note by the Executive Secretary, UNEP/CBD/COP/11/INF/19, 6 September 2012; Ecosystem Conservation and restoration. Note by the Executive Secretary, UNEP/CBD/COP/12/22, 15 August 2014.

25 COP Decision XI/16, Ecosystem restoration, 2012.

26 *Ibid.*, 1(c).

27 *Ibid.*, 5(i).

28 *Ibid.*, 5(g).

29 *Ibid.*, 5(a).

30 See on reports: Ecosystem Conservation and restoration. Note by the Executive Secretary, UNEP/CBD/COP/12/22, 15 August 2014.

31 Hyderabad Call for a Concerted Effort on Ecosystem Restoration, https://www.cbd.int/doc/restoration/Hyderabad-call-restoration-en.pdf.

32 Forest ecosystem restoration initiative. Note by the Executive Secretary, UNEP/CBD/COP/12/INF/19.

33 UNEP/CBD/COP/DEC/XII/19.

34 UNEP/CBD/COP/DEC/XII/19, para 6.

35 COP Decision XII/19. Ecosystem conservation and restoration, 2014.

36 Strategic plan 1997–2002 (Resolution VI.14. The Ramsar 25th Anniversary Statement, the Strategic Plan 1997–2002, and the Bureau Work Programme 1997–1999, 1996), Operational Objective 2.6. To identify wetlands in need of restoration and rehabilitation, and to implement the necessary measures; Strategic Plan 2003–2008 (Resolution VIII.25. The Ramsar Strategic Plan 2003–2008, 2002), Operational objective 4. Restoration and rehabilitation); Strategic Plan 2009–2015 (Resolution X.1. The Ramsar Strategic Plan 2009–2015, 2008 and adjusted for the 2013–2015 triennium by Resolution XI.3, 2012), Strategy 1.8. Wetland restoration.

37 Resolution XII.2. The Ramsar Strategic Plan 2016–2024, 2015.

38 Recommendation 4.1. Wetland Restoration, 1990; Recommendation 6.15. Restoration of wetlands, 1996.

39   Resolution VII.17. Restoration as an element of national planning for wetland conservation and wise use, 1999; Resolution VIII.16. Principles and guidelines for wetland restoration, 2002.

40   Resolution VIII.3. Climate change and wetlands: impacts, adaptation, and mitigation, 2002; Resolution X.3. The Changwon Declaration on human well-being and wetlands, 2008; Resolution X.24. Climate change and wetlands, 2008, para 18.

41   For example: Resolution VIII.8. Assessing and reporting the status and trends of wetlands, and the implementation of Article 3.2 of the Convention, 2002; Resolution X.16. A Framework for processes of detecting, reporting and responding to change in wetland ecological character, 2008, flowchart.

42   Resolution VIII.17. Guidelines for Global Action on Peatlands, 2002; Resolution XII.12. Call to action to ensure and protect the water requirements of wetlands for the present and the future, 2015.

43   Resolution VIII.32. Conservation, integrated management, and sustainable use of mangrove ecosystems and their resources, 2002.

44   UNESCO, Operational Guidelines for the Implementation of the World Heritage Convention, WHC. 13/01, July 2013, para 169.

45   *Ibid.*, para 176, b.

46   *Ibid.*, para 183.

47   *Ibid.*, para 190–191.

48   *Ibid.*, para 192.

49   Convention on Migratory Species, Article III, 4, a.

50   *Ibid.*, Article IV, 1 & 3.

51   Agreement on the Conservation of Seals in the Wadden Sea, Bonn, 16 October 1990.

52   Article VII, § 4, Wadden Sea Agreement.

53   Agreement on the Conservation of African-Eurasian Migratory Waterbirds (AEWA Agreement), The Hague, 15 August 1996.

54   Article III, § 2, c, AEWA Agreement.

55   Agreement on the Conservation of Albatrosses and Petrels (ACAP Agreement), Canberra, 19 June 2001.

56   Article 3, § 1, a, ACAP Agreement.

57   Agreement on the Conservation of Gorillas and their Habitats, Paris, 25 April 2008.

58   Article III, § 2, b, Gorilla Agreement.

59   Article 1, b, ii, Desertification Convention.

60   Article 2, Desertification Convention.

61   Zero Net Land Degradation, UNCCD Secretariat policy brief, May 2012.

62   United Nations Conference on Sustainable Development (Rio+20), Rio de Janeiro, 20-22 June 2012, https://sustainabledevelopment.un.org/rio20.

63   Outcome document Rio + 20, The future we want, Rio De Janeiro, A/CONF.216/L.1, 19 June 2012, endorsed by a General Assembly Resolution 66/288, A/RES/66/288★, 11 September 2012.

64   http://www.un.org/sustainabledevelopment/biodiversity.

65   United Nations Framework Convention on Climate Change, Rio de Janeiro, 9 May 1992.

66   For an overview of the relevant COP decisions, see http://unfccc.int/files/land_use_and_climate_change/redd/application/pdf/compilation_redd_decision_booklet_v1.1.pdf.

67   Decision 1/CP.21. Adoption of the Paris Agremeent, FCCC/CP/2015/10/Add.1 (2015).

68   See www.un-redd.org.

69   See www.forestcarbonpartnership.org.

70   See www.bonnchallenge.org/content/challenge.

71   See www.bonnchallenge.org.

72   New York Declaration on Forests, New York, 23 September 2014, www.un.org/climatechange/summit/wp-content/uploads/sites/2/2014/07/New-York-Declaration-on-Forest-%E2%80%93-Action-Statement-and-Action-Plan.pdf.

References

Alexander, S., Nelson, C., Aronson, J., Lamb, D., Martinez, D., Harris, J., Higgs, E., Lewis, R. III, Finlayson, M., Erwin, K., Hobbs, R., Covington, W., Murcia, C., Kumar, R., Cliquet, A. and De Groot, R. (2011) Opportunities and challenges for ecological restoration within REDD+, *Restoration Ecology*, vol. 19/6, pp. 683–689.

Aronson, J., Brancalion, P.H.S., Durigan, G., Rodrigues R.R., Engel, V.L., Tabarelli, M., Torezan, J.M.D.,

Gandolfi, S., de Melo, A.C.G., Kageyama, P.Y., Marques, M.C.M., Nave, A.G., Martins, S.V., Gandara, F.B., Reis, A., Barbosa, L.M. and Scarano, F.R. (2011) What role should government regulation play in ecological restoration? Ongoing debate in São Paulo, Brazil, *Restoration Ecology*, vol. 19, pp. 690–695.

Jørgensen, D. (2013) Ecological restoration in the Convention on Biological Diversity targets, *Biodiversity Conservation*, vol. 22, pp. 2977–2982.

Secretariat of the CBD (2014) *Global Biodiversity Outlook 4*, Secretariat of the CBD, Montréal.

Suding, K. (2011) Towards an era of restoration in ecology: successes, failures, and opportunities ahead, *Annu. Rev. Ecol. Evol. Syst.*, pp. 465–487.

Suding, K., Higgs, E., Palmer, M., Callicott, J. B., Anderson, C. B., Baker, M., Gutrich, J. J., Hondula, K. L., LaFevor, M. C., Larson, B. M. H., Randall, A., Ruhl, J. B. and Schwartz, K. Z. S. (2015) Committing to ecological restoration: efforts around the globe need legal and policy clarification, *Science*, vol. 348, pp. 638–640.

Telesetsky, A., Cliquet A. and Akhtar-Khavari A. (2017) *Ecological Restoration in International Environmental Law*, Routledge, Abingdon.

Tittensor, D. *et al.* (2014) A mid-term analysis of progress toward international biodiversity targets, *Science*, vol. 346, no. 6206, pp. 241–244.

Veldman, J. W., Overbeck, G. E., Negreiros, D., Mahy, G., Le Stradic, S., Wilson Fernandes, G., Durigan, G., Buisson, E., Putz, F. E. and Bond, W. J. (2015) Tyranny of trees in grassy biomes, *Science*, vol. 347, pp. 484–485.

Verschuuren, J. (2008) Ramsar soft law is not soft at all: discussion of the 2007 decision by the Netherlands Crown on the Lac Ramsar site on the island of Bonaire, http://archive.ramsar.org/pdf/wurc/wurc_verschuuren_bonaire.pdf.

27

GOVERNANCE AND RESTORATION

Stephanie Mansourian

Introduction

In the context of Aichi Target 15 of the United Nations Convention on Biological Diversity (CBD) which is to restore 15 per cent of all degraded ecosystems on Earth by 2020, there is a palpable increase in global attention to ecological restoration. While the twentieth century saw strides in our technical competence on restoration, the complexity of the task is frequently compounded by governance challenges. With an increased attention to restoration comes a concomitant need to better understand related governance matters, as well as to promote those elements of governance that will support restoration.

Restoration brings both opportunities and costs to people living near the resource in question, but also to those further away. Who decides what to restore, where and at what cost has an impact on the livelihoods of people, both proximate and distant. It is in this context that this chapter discusses the issue of governance. The emphasis here will be on forests for which there is more data, but other ecosystems will also be briefly considered where relevant.

What is governance?

Definitions of governance abound, and much has been written about the topic, notably in the context of natural resources. This brief overview seeks to summarize the main literature on the topic.

Environmental – and in fact broader – governance sets the framework for effective management of natural resources. While it does not equate to management, it provides the elements that contribute to the success (or failure) of natural resource management more generally.

Governance determines who takes decisions, and how these decisions are made and applied. It includes, rights, obligations, institutions, policies, and ways in which decisions related to the environment are taken and implemented (Colfer and Pfund 2010; Hydén and Mease 2004; Lemos and Agrawal 2006). A simple yet elegant definition provided by Chhotray and Stoker (2009) is that governance 'seeks to understand the way we construct collective decision-making'. For the United Nations Environment Programme (UNEP) environmental governance comprises 'the rules, practices, policies and institutions that shape how humans interact with the environment' (UNEP undated).

Important shifts in governance of natural resources have appeared in the last few decades: while until the 1970s, governance was synonymous with the 'command and control' nature of central governments, trends indicate a shift towards decentralization and the expanding role of both civil society and the private sector in natural resource governance (Agrawal *et al.* 2008). Gunningham (2009) defines this 'new governance' in the context of the environment using the terms 'participatory', 'devolved', 'transparent', 'flexible' and 'consensus-building'. This perceptible shift away from centralized government raises numerous challenges beyond the ecosystem in question. Ownership of the resource is one critical challenge, but not the only one (e.g. Agrawal *et al.* 2008). Another important challenge is the different levels of governance that impact on a resource. For example, at the international level global demand for a given mineral and how it is regulated (or not) impacts the local forest near the mine in question. Equally at the national level, corruption impacts natural resource governance. At the local level, conflicts between neighbouring communities over forest access and use rights may also lead to negative impacts on the forest. The different roles of stakeholders and how they relate to each other and organize themselves to take decisions (and how they are enabled and empowered to do so) concerning natural resources is yet another governance challenge.

Improvements in our understanding of the role of governance in natural resource management led to the definition of 'good governance' with a number of principles proposed by several organizations and authors. These normative wish lists include: participation, transparency, responsiveness, consensus-building, effectiveness and efficiency, accountability and strategic vision (FAO and Profor 2011; see also www.iog.ca). The Global Accountability Framework also adds 'participation, evaluation and complaint mechanisms' as further key criteria for good governance (Lloyd *et al.* 2007). For Gale (2008) transparency, openness, balance, accountability, deliberation, efficiency, science and risk are key principles for good governance. Moore *et al.* (2010) showed that many of these frameworks share the following fundamental principles for good governance: accountability, transparency, participation and predictability.

Diagnostic tools to analyse the pre-existing governance context have been developed, particularly as concerns the forest sector (e.g. see Mayers *et al.* 2005). Concurrently, minimum governance standards have also been developed as a pre-condition for investment as is the case for example, with the European Union's desire to halt illegal logging, through the Forest Law Enforcement, Governance and Trade (FLEGT) process (McDermott *et al.* 2012).

Many large conservation organizations have developed frameworks for assessing environmental governance. These can be general – for example, IUCN's Commission on Environmental, Economic and Social Policy (CEESP) developed 'The Natural Resource Governance Framework' (CEESP 2016). Alternatively, they can be specifically developed for a particular resource, such as WRI's 'Governance of Forests Initiative Indicator Framework' (Davis *et al.* 2009). The UN's Food and Agriculture Organization developed a 'Framework for Assessing and Monitoring Forest Governance' (FAO and Profor 2011; see Figure 27.1). The World Bank also developed the 'Analytical Framework for Forest Governance Reforms (FFGR)' (World Bank 2009).

The emergence of processes linking restoration to carbon markets, such as REDD+ (reducing emissions from deforestation and forest degradation, conservation, sustainable management of forests, and enhancement of forest carbon stocks), will have most certainly spurred on the development of these frameworks. Governance in the forest sector has come under particular scrutiny in the last few decades. Serious concerns emerging because of REDD+ relate in particular to land rights, with the risk that indigenous or other rural communities might be dispossessed of their rights and access to forest resources in the name of climate change mitigation under the scheme (e.g. Phelps *et al.* 2010; see also Chapter 26 in this volume).

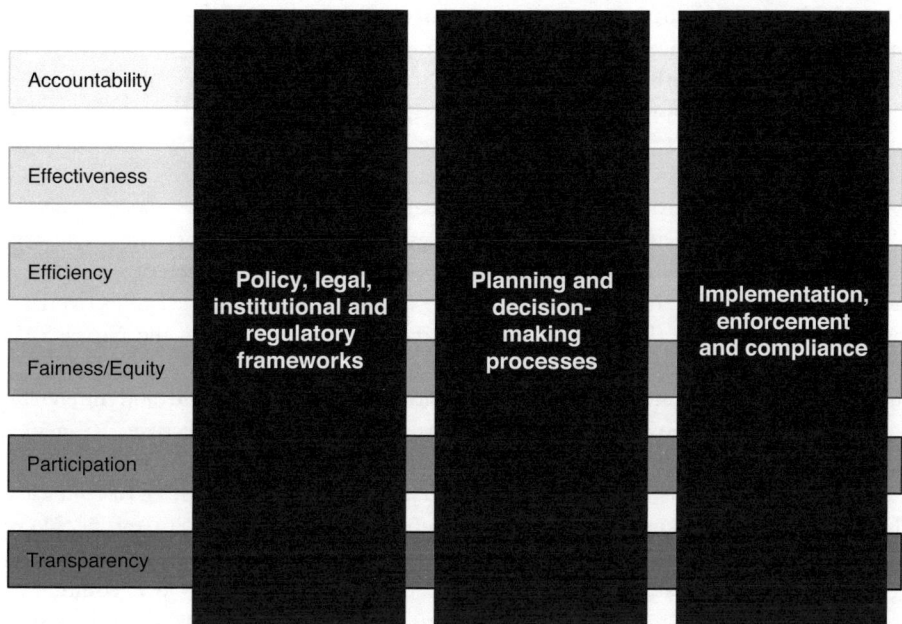

Figure 27.1   FAO's three pillars and six principles of forest governance

Source: FAO and Profor (2011), reproduced with permission

In the broadest sense, governance encompasses the socio–economic and political contexts within which conservation and natural resource management, including restoration, take place. This context needs to be understood, at times altered, and definitely contended with in any effort to protect, manage or restore a natural resource.

Evolution of governance in natural resource management

In traditional and pre-colonial societies, it is not uncommon for decisions related to natural resources to be taken – in an 'informal' fashion – by non-state actors (Agrawal and Lemos 2007). However, in western societies, the idea of environmental governance not being the exclusive remit of nation states, can be traced back to the 1970s and to the UN Stockholm Conference on the Human Environment (1972) (Chhotray and Stoker 2009). The next couple of decades, through to the 1990s, saw the increasing desire for international collaboration, standards and regulations on the environment, reflected by the numerous environmental conventions developed in this period – for example, the 1972 UNESCO Convention Concerning the Protection of the World Cultural and Natural Heritage, the 1973 Washington Convention on International Trade in Endangered Species (CITES), the 1979 Bonn Convention on the Conservation of Migratory Species of Wild Animals and the three Rio Conventions in 1992. More recently, and recognizing the importance of local level governance and the role of civil society (which can certainly in part be attributed to globalization, the Internet and the expansion of democracies around the world), models to improve the engagement of civil society and the private sector in formal governance arrangements have appeared.

One significant shift that occurred in the course of the twentieth century with respect to natural resource governance was the concept of protected areas as a way of governing an area with the specific goal (with different degrees of 'strictness') to protect nature. While the existence of some form of protection of land for its natural values dates back centuries, the international legal designation and collaborative effort towards the creation of protected areas dates back to the early days of the Convention on Biological Diversity (Mulongoy and Chape 2004). Although frequently poorly managed, the actual designation of areas (specifically, forests, wetlands or marine areas) as 'protected' has had significant implications on nature conservation but also beyond. Today (as of April 2016), a total of 217,155 protected areas have been 'officially' designated for their natural values, the single largest conscious, collective land-use decision in recent history (UNEP-WCMC and IUCN 2016; Mulongoy and Chape 2004).

However, in many cases, the decision to set aside land for protection was taken by governments in a top-down fashion, leading to negative impacts on people and often also on biodiversity. Such poor governance contexts have translated into ineffective management of protected areas and conflict with displaced or neighbouring communities as highlighted in a number of reports (West *et al.* 2006; Schmidt-Soltau and Brockington 2007). In contrast, there is also documented evidence of the successful impact both on biodiversity and people of the designation of these areas as protected (Ferraro *et al.* 2011; Andam *et al.* 2010; Ostrom and Nagendra 2007). Instances of co-management or effective engagement of local communities in protected areas' decision-making emerging towards the end of the twentieth century have highlighted the relevance and importance of a positive governance framework for successful conservation outcomes.

Recently, the recognition of differing governance realities, led to the further categorization of different forms of governance for protected areas:

(a) governance by government (at various levels);
(b) shared governance (i.e., different rightsholders and stakeholders collaborating to share governance);
(c) governance by private individuals and organizations; and
(d) governance by indigenous peoples and/or local communities (Borrini-Feyerabend *et al.* 2013).

The focus in this case is on the model of governance as determined by the lead actors involved in decision-making related to the protected area.

In both forest and fisheries management, there has been a significant shift in governance towards the use of voluntary market schemes such as eco-labelling and certification, to improve governance and ultimately to attest to the sustainability of the resource (Gulbrandsen 2004).

Lemos and Agrawal (2006) identify four important recent trends shaping the discourse on environmental governance:

1 globalization,
2 decentralized environmental governance,
3 market- and individual-focused instruments, and
4 governance across scales.

Globalization, they argue, can have both positive and negative effects on environmental governance as it may lead to over-exploitation of resources, excessive waste and pollution, but it may also lead to improved standards and global regulatory mechanisms.

Decentralized environmental governance includes greater participation and involvement of civil society in environmental governance, a process which can be traced back to the post-colonial period. For example, decentralization has led to various forms of co-management of natural resources. Decentralization may also signify strengthening of powers of local governments. The role of market and individual-focused instruments, including certification and eco-labelling, has grown in the last three or so decades and results largely from a recognition of government failure to regulate key environmental resources (McDermott *et al.* 2012; Gulbrandsen 2004).

Finally, the reduction in the importance of the exclusive role of the state has been replaced by a recognition of the diversity of levels – from international down to local – at which environmental problems need to be addressed and therefore, the level at which their governance operates. One other important trend of recent relevance with significant impact on governance is the 'triumph of the democratic ideal' (Chhotray and Stoker 2009) that has promoted the wider participation of multiple stakeholders in a growing number of countries.

Why is governance important for restoration?

Restoration practitioners have often neglected governance. Yet increasingly, it is understood as being a critical element to the success of forest restoration (e.g. the special issue of the journal *Forests* (Guariguata and Brancalion 2014) on governance and forest restoration).

Deforestation and forest degradation have frequently been attributed to poor governance (Speth and Haas 2006) as has the over-exploitation of fisheries (e.g. Allison 2001). The lack of appropriate measures to ensure the effective governance of forests or fisheries contributes to their loss but crucially it can also hinder their restoration. Understanding the root causes of environmental degradation and loss is essential for successful restoration, and frequently these may be traced back to a range of governance failures. They could be related for example, to poor regulation, lack of enforcement, corruption, unclear land rights, poor or non-existent participatory mechanisms, among others. Such governance failures can be highly complex, and exist at the local level (e.g. lack of representative institutions), at the national level (e.g. perverse incentives such as subsidies for commodity crops), or even at the international level (e.g. ineffective global regulation, as exists with fisheries).

More specifically, I highlight here six important issues that relate to the intersection between governance and restoration:

- governance as regulation;
- governance as transformation;
- governance processes;
- governance structures;
- actors in governance; and
- multiple levels of governance.

Governance as regulation

Governance can be perceived as the regulatory context within which restoration takes place. It regulates the way in which restoration can be implemented. Regulation consists notably of laws, policies, but also of soft laws and incentives. Traditionally associated with command and control forms of government, governance as regulation in the context of restoration can equally be made up of international agreements such as the CBD's Aichi Targets, or traditional forms of regulating the use of natural resources and their regeneration (e.g. Steurer 2013). While regulation was considered the exclusive domain of government, recent arrangements, including networked governance or collaborative governance, recognize the spread of responsibility among a diversity of actors (Jedd and Bixler 2015; Lockwood *et al.* 2010).

An example of governance as regulation is the Scottish forest policy which includes specific restoration objectives (see Table 27.1).

Governance as transformation

Changes in governance, notably away from centralized governments towards decentralized mechanisms, empowerment of civil society and the growing role of the private sector, create new opportunities for restoration to be embedded within these new arrangements. Changing or influencing governance arrangements or facilitating new elements of the governance architecture at different levels may be an important component to aid in the implementation and ensure the sustainability of restoration. So for example, at a global level the recent inclusion of restoration within the CBD's targets is likely to have significant impacts on restoration worldwide notably through the development of supportive legislation and increased financial flows. In contrast, continued demand (and related subsidies and incentives) for products such as palm oil, beef or biofuels, reduces the incentive for restoration around the tropical world (e.g. Mansourian *et al.* 2014).

Table 27.1 Some components of governance impacting on restoration

Administrative scale	Components of governance	Examples
Global	International commitments and targets	Aichi Target 15 to restore at least 15 per cent of degraded ecosystems by 2020 (www.cbd.int). Bonn Challenge to restore 150 million ha of degraded and deforested lands by 2020 (www. forestlandscaperestoration.org)
National	Policies	The Scottish forest strategy (2006) has explicit objectives to encourage the restoration of forest wetlands, to support woodland expansion for the restoration of degraded landscapes and to contribute to landscape-scale habitat restoration projects (Scottish Executive 2006).
National	Incentives	Fiscal incentives are provided in Paraguay to restore forests on private lands under the Forest Law (Forestry Law 422/73) (Mansourian *et al.* 2014). Financial incentives supporting fishing fleets around the globe lead to overfishing (and deter from restoring fish stocks) (Allison 2001).
Local	Tenure	In Vietnam, privatization of 'bare hills' for plantations, have in many cases marginalized the poor who were dependent on non-timber forest products (McElwee 2009).
Landscape/local	Tenure and rights	In Madagascar's Fandriana-Marolambo landscape, restoration was not allowed in an area that was set to become a protected area for fear that the community would then lay claim to the land once trees had been planted (Mansourian and Vallauri 2012).

Governance processes

Governance arrangements include processes such as dispute settlement mechanisms, negotiation platforms, mediation etc. (Bingham *et al.* 2005) which can be used to support the restoration effort. Particularly at large scales, such processes can facilitate objective-setting, implementation and monitoring, and will contribute to ensuring sustainability of the restoration effort. For example, in Madagascar's Fandriana-Marolambo landscape – a large-scale forest restoration initiative (see Mansourian and Vallauri 2012) – several local facilitators had to be hired in the course of the project in order to better negotiate implementation practicalities of the project with the local populations (Roelens *et al.* 2010).

Governance structures

Governance structures can for example be joint bodies that represent both civil society and government; or they can be private sector initiatives such as voluntary certification initiatives. Specific 'structures' of governance might also be promoted in the context of restoration, in particular, structures that bring together stakeholders from diverse levels and sectors. Thus, for example, new local-level decision-making structures might be established to define the species and sites to restore and to monitor progress. New governance structures that can be helpful for landscape-scale restoration might include groups of actors from different sectors with a stake in the landscape.

For example, the Roundtable on the Crown of the Continent (an area spanning over 7 million ha in western USA and Canada) brought together over 100 stakeholders representing indigenous communities, government and NGOs connected by a common landscape (Wyborn and Bixler 2013). Particularly successful governance structures that could provide useful lessons are the 'wildlife conservancies' in southern Africa where the government has devolved wildlife use rights to communities that are organized as an institution (the 'conservancy') with a constitution, registered members, a committee and locally-agreed boundaries (e.g. Corbett and Jones 2000).

Actors in governance

Recognizing the diminished role of the state and traditional political processes and the rise of civil society and market actors in governing resources (Pistorius and Freiberg 2014), more complex forms of governance are being defined involving multiple actors. An initial swing in governance from government to the private sector was witnessed in the 1990s with the advent of voluntary regulation in the form of standards such as those certifying sustainable fisheries or timber. While their focus is limited – certifying that the resource was sourced from a 'responsibly managed' location as defined by a number of criteria – these approaches have expanded exponentially (albeit more so in developed than developing countries, in part because of their complexity and cost). A parallel transformation occurred in many countries with the empowerment of civil society and in particular indigenous communities.

A number of studies (Charnley and Poe 2007; Agrawal and Lemos 2007) point to the positive relationship between communities' clear access and rights to resources and the quality of the resulting management of those natural resources. In a large-scale review by Pagdee *et al.* (2006) both well-defined property rights and strong community institutions were found to be vital to ensure effective community forest management. Nevertheless, as cautioned by Ostrom and Nagendra (2007) when addressing complex social-ecological systems, no one size fits all.

Today, more complex constellations of actors are being seen as energizing the movement to better govern natural resources. In the context of restoration, these diverse actors mirror the different scales of the challenge: from international donors, companies and investors, to national policy-makers and organizations, and down to local landowners, resource users, farmers, fisherfolk, indigenous communities and small businesses among other stakeholders (Figure 27.2).

Modes of organizing these diverse actors and engaging with them have been defined as networked governance which emphasizes the linkages between different stakeholders (Jedd and Bixler 2015). 'Collaborative governance' also reflects the recognition that increasingly the collaboration across multiple actors contributes to governance and that this should be both harnessed and somehow regulated (Lockwood *et al.* 2010). In polycentric governance, multiple actors translate into multiple decision-making centres (Ostrom 2010).

Multiple levels of governance

Decisions to restore an ecosystem can be taken at different levels, from the international right down to the local level (Table 27.1). For example, international targets such as the CBD Aichi Targets or the Bonn Challenge to restore 150 million ha of forests are all international decisions to restore. At the national level, governments can also set targets to restore a percentage

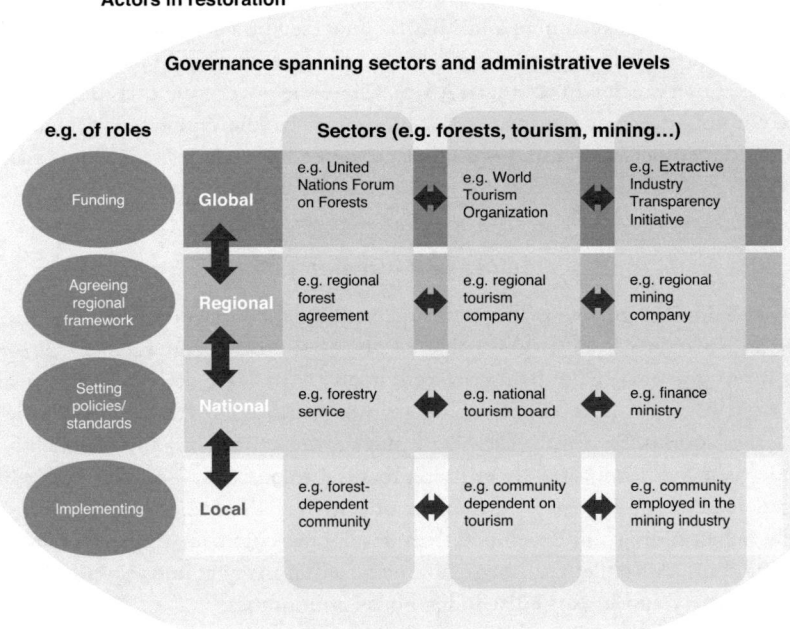

Figure 27.2   Actors in restoration. Actors include those at different levels and from different sectors (only some sectors shown here as an example). They may be involved in different ways, for example, providing funding, agreeing on frameworks, setting policies or implementing restoration actions

of their territory (e.g. in 2011, Rwanda committed to restore 2 million ha of forests) or to restore a particular area. At the site level, either a community or an NGO might also decide to restore a particular area. For example, in Tanzania the international conservation organization, WWF, has supported restoration around village forest reserves (Mansourian and Vallauri 2012).

Equally, industrial actors such as logging companies or mining companies might also decide – for a number of reasons, notably legal requirements, but also to meet their corporate social responsibility – to restore certain ecosystems or parts of their concessions. Importantly, while ultimately the technical restoration interventions will need to be local, governance interacts with these interventions in many ways. Both the imperative for restoration and the impact of restoration interface with many levels (Baker and Eckerberg 2013).

Emerging challenges

A number of key emerging challenges can be identified in the context of restoration (particularly forest restoration) that relate to governance.

Generating new values

Restoration, particularly natural regeneration, may take place without any change in designation of land. Evidence exists of traditional and historical forms of land and fisheries governance for example, that allowed for periods of set aside to enable natural regeneration or re-stocking. Yet the change in ecosystem quality resulting from restoration may generate new interests.

Light and Higgs (1996) refer to the concept of restoration generating new values (unlike protected areas that 'simply' conserve existing values). The landscape thus acquires new attributes and in some cases a new designation altogether (e.g. from degraded lands to forests or woodlands). Value creation is particularly critical as it begs the question of 'value for whom?' and triggers real or potential winners and losers. On a small scale, for instance on private land, this is not such an issue. However, on a larger scale and on land that is either public or communal or has mixed tenure, or indeed does not have clear designation, then value addition generates particularly important governance challenges. It may lead to conflicts (for example, where forest has regenerated on previously abandoned land, improving soil fertility for new farming communities) and it may also hamper the restoration effort (e.g. Buckley and Crone 2008). Mechanisms such as payments for ecosystem services (PES) might prove useful to address these challenges.

Competing land use

Increasingly, in a world suffering land scarcity and competition for resources, forest restoration takes place on land that has been, is being or could be used for other purposes. In this context, the decision to restore landscapes implies not using the land for alternative purposes.

Competing land uses are particularly acute in areas where there is either conflict over tenure rights or tenure rights are unclear. As more and more instances of 'land grabs' are being documented (e.g. Borras and Franco 2010; Barr and Sayer 2012) where land that is untitled is being diverted to other (generally market-based) uses, particularly for the food industry or biofuels, this poses a challenge for forest restoration. Poor governance favours these land grabs.

Tenure and rights

Tenure and rights are important dimensions of the governance-restoration relationship. In some countries (e.g. Madagascar) those planting trees appropriate rights to the land on which those trees are planted. This in itself turns restoration into a political tool and can also prove a strong deterrent for restoration. Equally, when restoring forests, those owning the trees may be different from those owning the right to harvest their fruit and other non-timber forest products. This is particularly true within communities where different members may hold different rights. For example, in Morocco, the state owns argan trees, even though they are on private land; men own the rights to harvest the trees, and women have rights over the products from the trees (Biermayr-Jenzano *et al.* 2014).

The fact that restoration frequently takes place on land owned by diverse stakeholders, for a diversity of reasons (e.g. legal obligation by a mining company, community desire to restore an indigenous species, government imperative to reduce landslides, external company's investment in carbon credits through planting of trees etc.) represents a significant challenge. This can be contrasted with protected areas where the legal framework is generally already in place, and equally importantly, ownership of the resource is also legally established (and generally limited to one entity) – even if, at times, it may be contested. Security of tenure also has implications for the likelihood of investing in land and land improvement, notably by planting trees or promoting natural regeneration.

The issues of tenure and rights have come increasingly to the forefront with the REDD+ debate, with the perceived risk of REDD+ leading to recentralization of resources (e.g. Phelps *et al.* 2010).

Scaling up

To date the majority of forest restoration efforts have been small in scale (reportedly under 100 ha, Melo *et al.* 2013; Menz *et al.* 2013). The Aichi Target and related global initiatives purport to change this, scaling up restoration significantly to landscape scales (or larger scales). In this context, governance takes on new importance. From their experience in Brazil, Melo *et al.* (2013) highlight a number of governance-related lessons for scaling up restoration, including a clear legal environment, economic incentives, a network of stakeholders with shared restoration interests and collectivized activism.

As increasing attention is being given to the landscape level which acts as a useful unit for reconciling social and environmental concerns, governance challenges need to be better incorporated in these initiatives. Increasingly governance within landscapes is becoming a topic of research (see Görg 2007; Van Oosten *et al.* 2014; Mansourian *et al.* 2014; Guariguata and Brancalion 2014; Kozar *et al.* 2014). Nevertheless, landscapes do not reflect a clear administrative unit (thus, they are omitted from Figure 27.2, albeit being of increasing relevance) which in itself exacerbates governance challenges.

However, scaling up also signifies extra costs. A recent study to estimate costs of complying with the Aichi Target 15 on restoration, suggests that at a cost of between US$500 and 1500 per ha, a total of somewhere between US$45–75 billion would be needed by 2020 (Pistorius and Freiberg 2014).

Looking to the future

Until recently governance has been poorly understood in the context of restoration. This is changing and it is increasingly being perceived as a critical element of success and sustainability in any restoration effort, particularly at large scales. The issues are complex however, and further tools may be needed to facilitate assessment and implementation.

This chapter has attempted to highlight the major trends currently being discussed in terms of governance of natural resources more generally and how these apply to restoration, using essentially (but not exclusively) literature on forest restoration.

Effective, lasting and large-scale restoration efforts will require additional engagement in governance by practitioners. Governance arrangements at all levels will need to be better understood and in some cases, a priority in the restoration effort may be to alter or influence a governance element that could be a stumbling block to the restoration effort and its sustainability. For example, insecure tenure or conflict over land resources may need to be clarified before engaging in forest restoration, while enforcement of existing regulations may be required for effective fisheries restoration.

As the Aichi Targets and other global commitments are propelling forest restoration (and ecosystem restoration more generally) to the forefront of environmental conservation efforts, there is likely to be an increase in financial flows for restoration. There is an opportunity to influence governance using these financial flows, notably by making them conditional on improvements in specific dimensions of governance. For example, recognizing the specific land rights of indigenous communities might be a pre-condition to funding restoration.

A number of existing instruments might be used to promote restoration and others might need to be developed. In the same way that initially an increased attention on the need to conserve areas for nature protection led to the definition of legal instruments for protecting these areas, and that the recognition of the need to identify forest products stemming from sustainably managed forests led to the development of certification tools, there may be opportunities to define new tools to not only assess but also to influence and define good governance in the context of restoration projects or programmes.

Acknowledgements

I thank P. J. Stephenson, Anne Sgard, Stuart Allison and Stephen Murphy for their constructive feedback on an earlier version of this chapter.

References

Agrawal, A. and Lemos, M. C., 2007. A greener revolution in the making? Environmental governance in the 21st century. *Environment: Science and Policy for Sustainable Development* 49(5): 36–45.

Agrawal, A., Chhatre, A. and Hardin, R., 2008. Changing governance of the world's forests. *Science* 320: 1460–1462.

Allison, E. H., 2001. Big laws, small catches: global ocean governance and the fisheries crisis. *Journal of International Development* 13: 933–950.

Andam, K. S., Ferraro, P. J., Sims, K. R., Healy, A. and Holland, M. B., 2010. Protected areas reduced poverty in Costa Rica and Thailand. *PNAS* 107(22): 9996–10001.

Baker, S. and Eckerberg, K., 2013. A policy analysis perspective on ecological restoration. *Ecology and Society* 18(2): 17.

Barr, C. M. and Sayer, J. A., 2012. The political economy of reforestation and forest restoration in Asia–Pacific: critical issues for REDD+. *Biological Conservation* 154: 9–19.

Biermayr-Jenzano, P., Kassam, S. N. and Aw-Hassan, A., 2014. *Understanding Gender and Poverty Dimensions of High Value Agricultural Commodity Chains in the Souss-Masaa-Draa Region of South-Western Morocco.* Working paper, mimeo. Amman, Jordan: ICARDA.

Bingham, L. B., Nabatchi, T. and O'Leary, R., 2005. The new governance: practices and processes for stakeholder and citizen participation in the work of government. *Public Administration Review* 65(5): 547–558.

Borras Jr, S. and Franco, J., 2010. From threat to opportunity-problems with the idea of a code of conduct for land-grabbing. *Yale Human Rights and Development Journal* 13: 507.

Borrini-Feyerabend, G., Dudley, N., Jaeger, T., Lassen, B., Pathak Broome, N., Phillips, A. and Sandwith, T., 2013. *Governance of Protected Areas: From Understanding to Action*. Best Practice Protected Area Guidelines series no. 20. Gland, Switzerland: IUCN.

Buckley, M. C. and Crone, E. E., 2008. Negative off site impacts of ecological restoration: understanding and addressing the conflict. *Conservation Biology* 22(5): 1118–1124.

CEESP, 2016. *The Natural Resource Governance Framework*. Retrieved from www.iucn.org/commissions/commission-environmental-economic-and-social-policy/our-work/knowledge-baskets/natural.

Charnley, S. and Poe, M. R., 2007. Community forestry in theory and practice: where are we now? *Annual Review of Anthropology* 36: 301–336.

Chhotray, V. and Stoker, G., 2009. *Governance Theory and Practice: A Cross-Disciplinary Approach*. Basingstoke: Palgrave Macmillan.

Colfer, C. and Pfund, J.-L. (eds), 2010. *Collaborative Governance of Tropical Landscapes*. London: Earthscan.

Corbett, A. and Jones, B. T. B., 2000. *The Legal Aspects of Governance in CBNRM in Namibia*. Windhoek: Government of Namibia.

Davis C., Daviet, F. and Nakhooda, S., 2009. *Governance of Forests Initiative Indicator Framework*. Washington, SC: WRI.

FAO and Profor, 2011. *Framework for Assessing and Monitoring Forest Governance*. Rome: FAO.

Ferraro, P. J., Merlin, M., Hanauer, M. M. and Sims, K. R. E., 2011. Conditions associated with protected area success in conservation and poverty reduction. *PNAS* 108(34): 13913–13918.

Gale, F., 2008. Tasmania's Tamar Valley pulp mill: a comparison of planning processes using a good environmental governance framework. *Australian Journal of Public Administration* 67: 261–282.

Görg, C., 2007. Landscape governance: The 'politics of scale' and the 'natural' conditions of places. *Geoforum* 38(5): 954–966.

Guariguata, M. R. and Brancalion, P. H., 2014. Current challenges and perspectives for governing forest restoration. *Forests* 5(12): 3022–3030.

Gulbrandsen, L. H., 2004. Overlapping public and private governance: can forest certification fill the gaps in the global forest regime? *Global Environmental Politics* 4(2): 75–99.

Gunningham, N., 2009. The new collaborative environmental governance: the localization of regulation. *Journal of Law and Society* 36(1): 145–166.

Hydén, G., and Mease, K., 2004. *Making Sense of Governance: Empirical Evidence from Sixteen Developing Countries*. Boulder, CO: Lynne Rienner Publishers.

Jedd, T. and Bixler, R. P., 2015. Accountability in networked governance: learning from a case of landscape scale forest conservation. *Environmental Policy and Governance* 25(3): 172–187.

Kozar, R., Buck, L. E., Barrow, E. G., Sunderland, T. C. H., Catacutan, D. E. Planicka, C., Hart, A. K. and Willemen, L., 2014. *Toward Viable Landscape Governance Systems: What Works?* Washington, DC: EcoAgriculture Partners, on behalf of the Landscapes for People, Food, and Nature Initiative.

Lemos, M. C. and Agrawal, A. 2006. Environmental governance. *Annual Review of Environmental Resources* 31: 297–325.

Light, A. and Higgs, E. S., 1996. The politics of ecological restoration. *Environmental Ethics* 18(3): 227–247.

Lloyd, R., Oatham, J. and Hammer, M., 2007. *The Global Accountability Framework*. London: One World Trust.

Lockwood, M., Davidson, J., Curtis, A., Stratford, E. and Griffith, R., 2010. Governance principles for natural resource management. *Society and Natural Resources* 23(10): 986–1001.

Mansourian, S. and Vallauri, D. 2012. *Lessons Learnt from WWF's Worldwide Field Initiatives Aiming at Restoring Forest Landscapes*. Marseille: WWF France.

Mansourian, S., Aquino, L., Erdmann, T. K. and Pereira, F., 2014. A comparison of governance challenges in forest restoration in Paraguay's privately-owned forests and Madagascar's co-managed state forests. *Forests* 5(4): 763–783

Mayers, J., Bass, S. and Macqueen, D., 2005. *The Pyramid: A Diagnostic and Planning Tool for Good Forest Governance*. London: IIED.

McDermott, C. L., van Asselt, H., Streck, C., Assembe Mvondo, S., Duchelle, A. E., Haug, C., Humphreys, D., Mulyani, M., Shekhar Silori, C. and Suzuki, R., 2012. Governance for REDD+, forest management and biodiversity: existing approaches and future options. In J. A. Parrotta, C. Wildburger and S. Mansourian (eds), *Understanding Relationships between Biodiversity, Carbon, Forests and People: The Key to Achieving REDD+ Objectives*, IUFRO World Series Volume 31, pp. 115–137. Vienna: International

Union of Forest Research Organizations.

McElwee, P., 2009. Reforesting 'bare hills' in Vietnam: social and environmental consequences of the 5 million hectare reforestation program. *Ambio: A Journal of the Human Environment* 38(6): 325–333.

Melo, F. P., Pinto, S. R., Brancalion, P. H., Castro, P. S., Rodrigues, R. R., Aronson, J. and Tabarelli, M., 2013. Priority setting for scaling-up tropical forest restoration projects: early lessons from the Atlantic Forest Restoration Pact. *Environmental Science and Policy* 33: 395–404.

Menz, M. H. M., Dixon, K. W. and Hobbs, R. J., 2013. Hurdles and opportunities for landscape-scale restoration. *Science* 339: 526–527.

Moore, P., Greiber, T. and Beig, S., 2010. *Strengthening Voices for Better Choices*. Gland: IUCN.

Mulongoy, K. J. and Chape, S. P. (eds) 2004. *Protected Areas and Biodiversity: An Overview of Key Issues*. Montreal and Cambridge: CBD Secretariat and UNEP-WCMC.

Ostrom, E., 2010. Beyond markets and states: polycentric governance of complex economic systems. *Transnational Corporations Review* 2(2). Retrieved from https://ssrn.com/abstract=2019699.

Ostrom, E. and Nagendra, H., 2007. Tenure alone is not sufficient: monitoring is essential. *Environmental Economics and Policy Studies* 8: 175–199.

Pagdee, A., Kim Y.-S. and Daugherty, P. J., 2006. What makes community forest management successful: a meta-study from community forests throughout the world. *Society and Natural Resources* 19: 33–52.

Phelps, J., Webb, E. L. and Agrawal, A., 2010. Does REDD+ threaten to recentralize forest governance? *Science* 328: 312–313.

Pistorius, T. and Freiberg, H. 2014. From target to implementation: perspectives for the international governance of forest landscape restoration. *Forests* 5: 482–497.

Roelens, J. B., Vallauri, D., Razafimahatratra, A., Rambeloarisoa, G. and Razafy, F. L., 2010. *Restauration des paysages forestiers: Cinq ans de réalisation à Fandriana-Marolambo, Madagascar*. Paris: WWF France.

Schmidt-Soltau, K. and Brockington, D., 2007. Protected areas and resettlement: what scope for voluntary relocation? *World Development* 35(12): 2182–2202.

Scottish Executive, 2006. *The Scottish Forestry Strategy*. Edinburgh: Forestry Commission Scotland.

Speth, J. G. and Haas, P. M., 2006. *Global Environmental Governance*. Washington, DC: Island Press.

Steurer, R., 2013. Disentangling governance: a synoptic view of regulation by government, business and civil society. *Policy Sciences* 46(4): 387–410.

UNEP, undated. Environmental governance. Retrieved from http://unep.org/pdf/brochures/EnvironmentalGovernance.pdf (accessed 7 April 2015).

UNEP-WCMC and IUCN, 2016. *Protected Planet Report 2016*. UNEP-WCMC and IUCN: Cambridge UK and Gland, Switzerland.

Van Oosten, C., Gunarso, P., Koesoetjahjo, I. and Wiersum, F., 2014. Governing forest landscape restoration: cases from Indonesia. *Forests* 5(6): 1143–1162.

West, P., Igoe, J. and Brockington, D., 2006. Parks and peoples: the social impact of protected areas. *Annu. Rev. Anthropol.* 35: 251–277.

World Bank, 2009. *Roots for Good Forest Outcomes: An Analytical Framework for Governance Reforms*. Washington, DC: World Bank.

Wyborn, C. and Bixler, P., 2013. Collaboration and nested environmental governance: scale dependency, scale framing, and cross-scale interactions in collaborative conservation. *Journal of Environmental Management* 123: 58–67.

28

RESTORATION, VOLUNTEERS AND THE HUMAN COMMUNITY

Stephen Packard

A thriving volunteer programme can be a major ecological (and political) asset, but it's not easy to build. Such a programme, in some situations, may make the difference between success and failure. Creating and supporting a fine programme requires special kinds of proficiency, resources, and leadership.

Especially for public lands in or near metropolitan areas – the influence, expertise, and commitment of volunteers (and the community support that they may help foster) can help determine:

- whether project components get approved in the first place;
- whether needed permits are issued;
- how well the work is funded;
- whether the press and community support the mission; and
- whether long-term follow-up is adequate.

Collaborative volunteer programmes are fundamentally different from many standard volunteer programmes and need different kinds of leadership. Many professionals are surprised by how much decision-making authority is delegated to highly experienced volunteer stewards in some of the best programmes. Leadership restoration volunteers may in time become dedicated experts who make technical decisions on a regular basis. In an advanced programme, some volunteers become colleagues and spokespeople as full partners.

A collaborative rather than 'top–down' approach is crucial. Training and empowering volunteer leaders and volunteer experts opens the door to initiative, creativity, and dedication far beyond what any organization could afford to hire, especially at this stage in the history of restoration. Some volunteers have worked for decades to be become experts in some fields. The respect and freedom granted them are crucial parts of the volunteers' motivation and 'compensation'. Some organizations with the most advanced programmes regularly entrust some expert volunteers with broader authorization and flexibility than some professional staff in some situations.

This chapter is directed to those who make three sets of decisions. The following first section speaks to those who decide whether or not, and at what phase, a project will support the creation of a collaborative volunteer programme (often called a 'volunteer community'). Subsequent sections are directed towards those who will hire and supervise the leader(s) of that programme, and those who will launch and lead the programme day to day (including the leadership volunteer themselves).

Recommendations to upper-level decision-makers

Volunteers may cost more than they're worth. Or they can be crucial to the success of the project. To determine whether a collaborative volunteer programme is worth it to you, consider these questions.

1   How important is community support for getting needed funding and permits?
2   Can your goals be achieved by the level of funding and staff available? If not, you may want to (a) choose more realistic goals, (b) find a way to increase professional resources available, or (c) plan and implement a volunteer programme that can do what otherwise couldn't be done.
3   How important is public support and media? What advantages would accrue from community spokespeople and educational 'human interest' stories about dedicated local volunteers?
4   How secure is the long-term plan? Will there be the necessary resources from year to year and decade to decade? How important might community support and volunteer commitment be to the continuing evolution of the restored natural community?

One way to approach these questions is to explore how ambitious the effort will be and what the long-term vision is. Aiming for a high diversity of plants and animals (especially species of conservation concern) may enhance community support. But such quality can be expensive.

Early planning

If you've decided on strong community involvement, involve local community leaders, conservation activists, and the prospective volunteer facilitator in the early planning. They may improve some fundamental components of the project. They may suggest major or subtle changes in language, sequencing, etc. that can have powerful leverage. Most important, they will be partners.

One key element is the 'spirit' of the 'people component' of the mission. This spirit needs attention, even in early planning stages. What you want is a lofty vision supported by a collaborative, get-things-done, agreeable work ethic. One thing you do *not* want from staff or volunteers is a 'crusading', 'fundamentalist', 'holier than thou' kind of environmentalism that, among other problems, is prone to public contention and internal bickering.

Would a pilot project help tell the story and close the deal? A positive atmosphere surrounding a much smaller project in the same area could go a long way towards selling a larger project. The smaller project could have its own little 'kick-off' and messaging (see the following text). If the pilot were to have expensive expert guidance but otherwise be run collaboratively with true volunteers, then media and community interactions could thoroughly set the stage for the larger project.

On the other hand, it's a mistake to ask for volunteers for a project that is not approved and may be subject to delays. People who drop out are not likely to drop back in later. People who become adversarial advocates may foster an unhelpful spirit of controversy around the project.

Planners of restoration may adopt a variety of approaches to sequencing. Those in charge may choose, for example, to plant ten percent of its area each year for ten years. Or it may choose to start the work in the whole area at the same time and gradually improve diversity and quality over years or decades. Volunteer strategies should be part of these decisions.

Quality restoration typically takes years to initiate and decades to mature. Some decades-old projects still consider themselves to be in early stages. Poorly planned initiatives often identify funding for a few years and assume that minor custodial care will be sufficient for the subsequent decades. In some cases you can almost hear backroom planners say, 'Let's do the parts that are easy to fund and then dump the mess on local staff or volunteers.' In certain situations the 'low maintenance' prediction may be accurate. In others, especially those with ambitious biodiversity and conservation goals, substantial inputs may be needed for the foreseeable future. Unfortunately, there are examples where the restoration makes progress for a while and then deteriorates, or a project may even be abandoned as invasives overwhelm resources, commitment, and patience.

Another problem for organization decision-makers to consider: you may not be in that position ten or twenty years from now. A collaborative volunteer programme may be expensive to initiate yet more than worth it as the investment pays increasing dividends decades after decade. However, a community of stewards will thrive only with support from future decision-makers. Be creative about adopting effective means of communicating strategy and commitments to future leaders. In the short run, take pains to truly support the volunteer facilitation staff – to protect them from other parts of the organization where current and future staff may be slow to understand the unusual nature of the partnership with the volunteers.

Recommendations to the person who hires and supervises the volunteer 'coach' or 'boss'

What kinds of people make the best leadership for collaborative volunteer projects?

The principle facilitator is typically a paid staff member, but it can be a dedicated volunteer who wants to make a major commitment. Regardless, this person has to be smart, nice, and dedicated. Those three words pack great power. A 'smart' person is preferably knowledgeable, but the people we're looking for are quick to admit what they don't know and talented at finding out where to go for various types of expertise. A 'nice' person, in this context, is fundamentally generous and has excellent social skills. You can be well intentioned but not really 'nice' in this context without those social skills. You have to understand what the other people need, what they perceive, how to help them grow and their spirits to blossom. A person too 'self-confident' or 'full of themself' often can't do it – even though knowledgeable and dedicated.

An alternative is to have the volunteers report to a dedicated conservation ecologist whose knowledge and commitment inspires the volunteers – who then themselves handle many of the interpersonal, political, and organizational challenges, through accomplished individuals with the needed skills.

In my experience, professionally trained 'volunteer coordinators' are often the worst choice for leadership. Though they may be excellent at what they do, they are likely to try to create the wrong kind of programme for this purpose. At a conference that featured a collaborative volunteer programme, a puzzled representative of a federal agency revealed his experience with this comment in preface to a question. He said, 'At first I couldn't understand how volunteers could be given so much authority and do so well by it. After a while it came to me. Volunteers can work wonders if you keep them away from the volunteer coordinators.' In his agency, he was familiar with outstanding volunteerism mostly in a few 'hardship' situations where a dedicated scientist or ranger or other member of a very small staff worked directly with the volunteers, instead of through a staff volunteer coordinator. These volunteers were motivated by and learned from the vision and passion of the expert. The professional volunteer coordinator may believe in motivation through donuts, embossed certificates, and wanting busy smart people to sit through tedious 'volunteer appreciation events'.

Perhaps even more importantly, professional volunteer coordinators may have learned to look for the wrong people and make the wrong decisions for this kind of initiative. Many programmes are designed for relatively passive (by-the-book, 'easy to work with') volunteers who are able to withstand the delays, reverses, and limitations of bureaucracies. These people may be expensive to supervise, relative to what they can accomplish. In contrast, what a collaborative programme wants is volunteer leaders who will do the supervision and who themselves may organize their own more meaningful appreciation events.

In other words, for this kind of initiative, we are looking for volunteer leaders who are motivated by the actual mission. This kind of person is impatient for real results, creative about new approaches, viscerally dedicated to the ecosystem, and has good judgment on when to seek official approval and when not to bother people. Such people move on if bureaucracy prevents them from making the kinds of contribution they're capable of making.

Effective volunteer facilitators have only casual contact with most volunteers and spend most of their time recruiting, training, and empowering volunteer leaders. A big part of the work of a good volunteer facilitator is to protect the leadership volunteers from bottlenecks and agency politics.

It's worth time and trouble to find the unusual person who can facilitate a strong programme of this kind. She or he will likely have to work hard for modest pay. But the satisfaction is huge. Sometimes the right person is a bit old for an apparently 'entry level' job but has 'special circumstances'. Perhaps he or she is emerging from an early adulthood headed in the wrong direction by pursuit of flawed idealism and now is eager to finally sink their teeth into something important for the long haul. Perhaps they've had some success at another career trajectory but realize they want more vision, optimism, and real-world results in their lives.

In the best cases, the facilitator finds volunteer leaders who become admired life-long friends. He or she recognizes people ready to commit themselves deeply and helps them do it in a way that works for them.

Assuming for the purposes of this chapter, that the volunteer facilitator has now been selected, the next section will be addressed to that person.

Recommendations to volunteer facilitator and leaders

Many projects start with a 'kick-off' event that can be utterly critical. This event is worth thorough preparation and a lot of work. The principles of the volunteer component should have been agreed to in a way that will empower the facilitator and emerging volunteer leaders to start right in to run their own part of the operation. The most important preparation work is inspired outreach that attracts many potential leadership volunteers.

During the kick-off, define the project in the minds of the public and potential volunteer leaders. In good graphics and summarizing language convey the vision, the challenge, and that this important mission truly needs dedicated, trained volunteers to succeed. Convey why the event is historic and that people who want to lend a hand can be a part of that history.

Have a kick-off event at the right time of year. Seasonality issues will differ depending on geography and ecosystem type. For example, in northern Illinois, late March and early September are good times. Fall and spring are key times for 'getting to know you' events and stewardship tasks (planting, controlling weeds, gathering seeds). They allow for three months of opportunity for potential leaders to bond with the site and their colleagues under the most welcoming conditions. In mature projects, winter becomes a major time for strategic volunteer brush control (and bonfire burning with cookouts and socializing in the snow), but new people may stop coming when wintry weather challenges them. Summer can also put people off, partly because of heat, mosquitoes, and chiggers, but also because the work needed may be more demanding and thus is less accessible to new people. Examples of summer work include applying herbicides to problem invasives and technically complex care for endangered and rare species.

It's sometimes a good idea to headline the kick-off with an inspiring speaker who believes in the project (but has the political good sense to get out of the way after speaking). It's never a good idea to burden the event with bureaucrats or officials who 'need to be included' for political reasons. Find some other way to honour them. The speeches and displays at the event should be brief enough to 'leave people wanting more' when the time comes to hike through the ecosystem and sign up for possible responsibilities.

A detailed example of such a kick-off may make this critical event easier to understand. When the Forest Preserve District of Cook County and The Nature Conservancy decided to work together on a major restoration of the Poplar Creek Prairie, the event had just two speakers: Professor Robert F. Betz and me. Professor Betz was a respected veteran of many restoration projects large and small (e.g. the Fermi National Accelerator Laboratory prairie restoration). He told the crowd of the plight of the nearly vanished eastern tallgrass prairie, regaled them with a couple of brief stories of problem solving and dramatic success, and ended with a vision. It went something like:

> You know how when you get a new car or a new appliance, at first you're proud and take such good care of it. And then one day it gets its first scratch, and bit by bit it ages and starts to go downhill? Well look around at this beautiful site and realize today is the exact opposite. This is the worst you'll ever see it. Year by year, month by month, and day by day, thanks to many of you, it will get richer, more beautiful, more important, more worthy of your pride. Thank you.

I said rather little. As the person who was to be the volunteer facilitator, I just needed to establish myself as the guy to talk to – and briefly describe next steps. I had intended to announce a walking tour where I'd have the opportunity to answer specific questions and listen to the ideas of people who seemed to be especially interested. But we had an unexpected problem. Although we had put notices in every newspaper and group newsletter we could reach and sent notes to every possibly interested name we could round up, we had expected perhaps 15 or 20 people to show up. The invitation had said, 'Come if you want to volunteer'. Eighty people showed. I quickly found substitute tour leaders to replace me and then made this announcement:

> Thanks for the impressive turnout. We had expected to do some initial planning with the group as a whole, on the hike. But, congratulations, this is too big a turnout for that. We need a smaller group to develop some plans for our larger group. How many people, in addition to volunteering, would be willing to help make the plan and lead?

An impressive twelve people raised their hands. I said, 'Great, meet me right now, under that tree over there, and we'll get down to work.' Then someone else took over and organized tour groups for the seventy. (The tour leaders would then end with the announcement that the planning group would reach out to everyone before next weekend with a volunteer schedule.)

Thus, for a few minutes our group under the tree discussed our abilities and passions. Then we stopped for initial comments. One person said, 'I could write the newsletter.' One said, 'I know how to do seed gathering, I could lead that.' Another said, 'I did controlled burns for the Forest Service, I could head that up.' Soon we had weekly organized projects of many kinds. The burn crews, of course, had professional staff as training wheels for a while, but in time, the expert volunteer crew did burns on their own. One of the most important offers came a few days later: 'I'm VP for personnel at a tech company. I could head up recruiting, training, community outreach, parties, and internal leadership ladders.' We were off and running.

Over the decades that project saw ups and downs in levels of funding, staff support, special appropriations for 'heavy lifting' components, etc. What it did not see was 'downs' in the commitment and confidence of the volunteer congregation. Leaders evolved and changed. Many new people joined the team. And the consistent motivator and reward was the ecosystem. As Dr. Betz predicted, Poplar Creek's prairie, woodland, and wetland communities got richer and healthier year after year. Rare plants proliferated by the thousands, then millions; rare birds returned to sing and nest and raise young; rare butterflies and dragonflies hovered over flowers and ponds.

The 'Poplar Creek Prairie Stewards', as the group chose to call itself, have been a consistent community that Forest Preserve staff have relied on. Over a quarter century, the project has grown in size, goals, quality, and reputation, with dozens of volunteer leaders and hundreds of dedicated individuals. It's crucial that they enjoy each other and are appreciated. But fundamentally, they do it because they're needed and because it works.

Challenges in early stages

The stories of successful volunteer communities are variations on the theme illustrated by Poplar Creek. What are the elements that contribute to success for this kind of conservation?

After the kick-off, the new community needs momentum and triumphs in order to jell as a group and begin its own evolution. While collaborating with the overall plan, the new people need to begin problem-solving. Staff can help in many ways, but a key is to admit and convey that paid people are not capable of doing the volunteer leader job. It is crucial that capable leaders be empowered as soon as possible.

A schedule of bi-weekly or weekly events (three hours long) often seems right for a start. 'Too much too quick' scares people away. Too infrequent events prevent the formation of a growing critical mass. Some good people may say, 'Well, I only have time for this once a month, or once a quarter, so let's start slow.' No good. Those people don't believe in this particular mission enough to make it part of their day-to-day lives. They can't lead it. Find people who believe in it. Don't let 'important' or 'expert' putative volunteer leaders intimidate the people who actually could lead day to day. Respect the people who show up; listen to their ideas; give them your time; help them succeed.

Harvesting seeds too rare to buy is often a good first focus when starting in fall. Even a project with millions of dollars in funding often misses two important components in their seed mixes. One is the highly local seed that may contain parts of the gene pool found nowhere else. The other is certain 'difficult to propagate' species that may be important to certain animals, to community structure, and to natural diversity generally. The volunteers can learn these species, seek them out (often in interesting places that add to the drama of the mission), harvest them, and sow them. Volunteers may be able to negotiate approval to gather seed from sites that would otherwise be off limits to contractors and agencies. They may be willing to search out some species that otherwise would be prohibitively expensive for contractors or agency staff to find. Some species that aren't commercially available can be introduced in small numbers by painstaking volunteer work (for example, growing bulbs in gardens and transplanting them into the ecosystem while dormant). In many cases, once introduced in small numbers, species left out of most restorations start to proliferate profusely according to their own ways. Professionals can (expensively) do this work too, but it may take years, and the results are often better as product of a volunteer community and very rewarding for the people who do it.

At early sessions, initially perhaps led by the volunteer facilitator, be quick about empowering people, ambitiously but reasonably. One potential leader may offer to start a Facebook page. Another may offer to help lead a special brush control or seed gathering project. Excellent. Jump on it. Help potential leaders be successful and seen as such.

Mentor leaders to contribute to plans. Don't do the plan yourself. Focus your time on the many people who seem to have potential to lead. Even a superhero facilitator often can't tell for a while who will come through. Many will have limitations that take some time to overcome. Some who seem great initially may fade in the long run, for a wide variety of reasons. You're actually fishing for someone in the tiny minority of people who have what it takes – and for whom this opportunity is just what they need.

Often the best leaders are humble and argue that others with more expertise should be given authority. Fine. Let them consider themselves to be more 'supporters' than 'leaders'. But help them be successful; show them publically the respect their work deserves; help the group sort out leader relationships that work. Often an 'introvert' and an 'extrovert' make an effective leadership team, if they trust and support each other. Focus on the mission. Volunteers quickly support others who are advancing a mission all believe in.

Although it should go without saying, never give volunteers trivial work 'to keep them busy'. That's a sure way to lose the volunteers with the most capability and integrity. If some snafu stands in the way of the most important work, and you must offer lesser work in the meantime, be honest about the process, reasons, etc. Do not be cynical and pessimistic, of course. Resolve the snafu as quickly as possible.

As volunteers develop abilities and achieve goals, *celebrate them* in the ways that will be most appreciated by the individuals and the group. Perhaps quietly; perhaps publically; perhaps both. Do the humble work that will support the leaders' successes. Later, members of their teams will do that work, but for now *you* find needed tools, or expertise, or take on some of the drudge tasks. Your most important job is to facilitate the success of the volunteer leaders.

Mistakes to avoid

A longer list of cautions would be easy to make. But the seven examples below illustrate general principles.

1   *Approving bad ideas.* What do you do when a new leader excitedly proposes something that is 'not that great' an idea? There's a temptation to say, 'Well that really isn't a priority' (discouraging?) and there's an opposing temptation to say 'Sure, let's try it' (possible learning and increased commitment; possible disappointment and disruption?). One good alternative may be to explore 'tweaks' that might transform the proposal into something good. Another possibility is to suggest comparing that approach with a more tried-and-true one – possibly even proposing to write up the results for a contribution to the literature or blogosphere.

2   *Wasting early time on schools, companies, and churches.* It's tempting to shy away from the one-to-one relationships that are key to community. Schools, companies, churches, and other

groups may be valuable components in the longer run, but early on, focus on individuals who'll be regulars. Find the project's own people.

3   *Lack of clear chain of authority.* In agencies where a collaborative volunteer programme is not well understood, any staff person may believe that they have the authority to start, stop, or overrule the work of any volunteer. Instead, it is critical that the each lead volunteer 'reports' through one person who can approve or not. If other staff people want to make changes, they should work through that chain of command.

4   *Avoid bureaucratic 'start', 'stop', and 'do the opposite'.* People in large organizations often have to put up with a lot of waiting and reversals. Although in many respects it is good practice to 'treat volunteers like unpaid staff', in this case it is not. Every effort should be made by the volunteer facilitator to protect the volunteers from this kind of thing. Staffs have to put up with it. Volunteer dedication deserves and requires a higher standard.

5   *Avoid staff resentment.* One frequent result of efforts to avoid problem 4 is that some staff may begin to resent the volunteers as 'pampered'. But one of the payments to volunteers is that they are spared as much of the idiocy of bureaucracy as possible. Some staff actually work as volunteers for other organizations which give them more freedom to think, solve problems, and see quick results. Good relations between staff and volunteers is a priority that requires creativity. Talk with people; promote understanding.

6   *The 'individualist hero' error.* In celebrating triumphs of volunteerism, beware of the temptation to present the steward as a crusader against the world. The public loves (and thus media rushes to celebrate) the lone hero who triumphs. Some element of that can be in the PR. But overall, it's misleading and provokes resentment among partners. Ecological restoration is so complex that deep expertise and teamwork are usually required. Plans deserve advice and review by the most dedicated and knowledgeable people (possibly professors, long-time volunteer experts, specialist staff, entrepreneurs). Celebrate them too. Celebrate collaboration.

7   *Undue deference to counterproductive experts.* Many professors and other 'experts' give counterproductive advice. Some respected project leader needs to mediate and point the team towards conservationist scientists with true expertise on practical questions. The project needs a core of scientific expertise that is respected. All concerned need to know that the overall decision-making process is well-founded.

Watch your language!

Because this work depends on collaboration, facilitative language is key. Expert use of volunteer language is not 'common sense'. Here's a bad example illustrating three common errors: 'Thanks for helping us. We're really glad we decided to use volunteers. You've saved us a lot of money.'

The first fault in the bad example is the verb 'use'. To speak of 'using' people is alienating. In common language, when people 'use' others, they are taking advantage of them manipulatively.

The second error often results from quick briefings. Some official has been told that the volunteers have saved the organization money. 'Bottom line' is all some officials have time for. But it's the wrong bottom line. Better: 'With our collaborative volunteer programme we have set higher goals, met higher standards, and accomplished more. We can all be proud'. Conservation volunteers are motivated by the opportunity to make things better than they otherwise would be. That's what they want to be recognized for. Often volunteer facilitators and leaders find ourselves needing to apologize for short-sighted comments of this sort. The programme volunteers and staff can deal with it. But the less miscommunication the better.

Another way to put it is that the well-intentioned decision maker has lost sight of the programme's vision. A volunteer's motivation is probably not principally to save the taxpayers money. It's to help nature, the environment, and or people's quality of life.

The third problem is the 'we' versus 'you' contrast. The language should be inclusive. When someone says, 'We want to thank you', that person is saying that those being addressed are not one of 'us'. Better language: 'We are gathered to celebrate major steps forward in this partnership of staff and volunteers. We have a lot to be proud of.'

Considerations as the programme matures

A project with strong agency and public support will over the years grow in staff, volunteers, and advisors. These people will have the experience, wisdom, and ability to notice opportunities or problems and design initiatives far better than what can be suggested generically here. But a few principles, special cautions, and success stories seem worth outlining.

Ongoing education

Occasional 'field seminars' and indoor educational events can attract experts who will inform and inspire participants. Diversified educational opportunities promote a diversified team. Let volunteers pick and choose what skills they want to master. Many people will thrive best in a few chosen areas that may be as diverse as chain-sawing, monitoring breeding birds by ear, designing seed mixes, or identifying and monitoring rare sedges or dragonflies.

Educational events should avoid academic self-promoter and contentious types. Instead, seek out dedicated conservationists who genuinely appreciate the goals and the work. Also feature volunteer and staff 'heroes'. Doing so motivates the people chosen, encourages them to step back and re-think their work from a broader perspective, and provides role models for other staff and volunteers.

Do not demand that the volunteers attend an 'educational' event featuring something they already know or could read in an email. People are negatively motivated by requirements that take away the time they have to contribute. Do offer occasional events or publications that help people feel the coherence and energy of the cause. Do provide opportunities for subsets of people to gain expertise in any of the special skills that may be helpful. For education, the best approach is to 'offer' more than 'require'. If it's really good people will come.

Image, media, and motivation

Good media (including, of course, social media) can be highly effective at supporting the stewards, the staff, the agency, and the cause.

Some organizations or regions have annual or biennial 'conservation award' events that lead to multiple local news stories featuring local personalities, local preserves, and key values and issues. Widespread recognition by neighbours, co-workers, friends, relatives, and local officials can be highly motivational to the volunteers and staff so recognized.

It takes work to write many little press releases and round up good photos for each, but the rewards are worth it. Desirable 'by-products' often include less misuse and damage to preserves, more support for the agencies that own the preserves, better reporting about related issues, more volunteers attracted to the project, and more motivation in the hearts the people honoured. Most neighbours and local officials may have little idea that important and needy ecological sites are in their midst, much less that local people care about them.

Language that has been carefully formulated is important when dealing with the press, partner agencies, and among the various parts of larger agencies. Most leadership volunteers are happy to take coaching in these situations, because they want their work to succeed. 'Good communication skills' seem to come more easily to some people, and it is those people who should be put forward when community spokespeople are needed.

Newspapers and TV – and now social media, especially – often look for controversy, of course. It is in our interest to help them find better ways to tell ecology and stewardship stories appealingly. Some of the best and most striking advice ever given to me came from a high official of a major agency at the point (just before an election) when a bitter advocacy dispute (over whether to purchase threatened land) was settled. To seal the deal, we had to go public, but controversy could hurt us. I spoke for the advocates. The official advising me said, 'Steve, when you talk with the media, don't forget: It's a wonderful world, and we're all working together.'

That recommendation has a lot of applications. Griping and cynicism come naturally to many people, even some environmentalists. Cynics are not the best 'messengers'. It's worth serious effort to establish a spirit of generosity and optimism in the volunteer community. Sometimes that leads to the need for 'inside' and 'outside' language. In large organizations, one part of the organization may intentionally make trouble for another, and it's important for good leaders to strategize and repair damage. Thus volunteer leaders and key staff may need to speak freely among themselves about conflicts while with most people using language that focused on the vision.

Although it's crucial to tell volunteers enough 'inside politics' to protect them from infighting, we don't want a volunteer crew that feels it is at loggerheads with parts of the staff. Unlike participants in some advocacy organizations, the kinds of people who thrive as restoration volunteers don't want an atmosphere of politics. These volunteers have little in common with traditional environmental protestors. Motivation by anger and fear is brief for most people. To attract the decades of commitment we seek for volunteer leaders, motivation by hope, vision, and ongoing success works best.

Both staff facilitators and leadership volunteers need to understand and be sympathetic to the challenges of upper-level decision-makers and the staff in other parts of the organization. Helping make their work as uncomplicated and un-stressful as possible is one of your major responsibilities. Much of what leadership volunteers need to know is beyond the scope of this chapter, but they definitely need to understand the challenges of the staff – and how to represent the landowners in whatever authority is delegated to them.

Almost all the time, don't go public with interpersonal conflict. It's a distraction. The real dramas of conservation and restoration focus on winning battles against degradation and creatively developing solutions.

Coaching tips

As a trainer, you can be a tough and demanding coach, with some people, sometimes. *But* as a rule, when you're seeking to empower someone, you need to be more facilitative and less directive. Listen to their ideas; listen to their view of what is needed and what would work. Perhaps they have a slightly different (or very different) route of getting there, but possibly the end result would be more-or-less the same. They perhaps understand their own approach better and would feel more committed to it – and more proud of its success. If so, then perhaps they'd have more self-confidence and would work harder to succeed. Go with it.

As mentioned earlier, some staff, trained in traditional 'volunteer coordination', focus their motivation efforts on certificates, memento gifts, 'recognition parties', etc. The leadership volunteers, who should be their focus, would much more appreciate real help that benefits the work they believe in. The leadership volunteers should be the chairs and spokespeople at events for volunteers generally. It's fine if they give recognition (even awards) to staff. But it undermines the relationship you want the volunteer leaders to have with most volunteers if staff members are the principal leaders at volunteer meetings. Partner agency staff can sometimes be helpful in these situations.

Volunteer impacts on levels of staff and funding

One of the arguments against volunteer programmes that has been used by some professionals is that decision-makers might see volunteerism as an excuse not to hire staff or to provide funding. In the case of collaborative volunteer programmes, my experience has been the opposite. Programmes and projects not supported by volunteers have tended to stagnate or lose funding and staff. Those with strong volunteer constituencies have tended to get the highest levels of staffing and funding.

Volunteer relations with funding and partner agencies

After decades of a restoration programme in which the volunteers did the lion's share of the work, the Forest Preserve District of Cook County realized that there was sufficiently widespread support to dramatically expand the programme. Some of the volunteers and staff had recommended a prairie restoration project considerably more ambitious than the agency had done before. It would require cutting large numbers of mature trees, hundreds of acres of weed control, and expensive seed planting. Volunteer bird monitoring had documented a dramatic loss of breeding grassland birds, a conservation priority. With strong volunteer support, collaborating NGOs launched major outreach programmes that engaged community groups, churches, bird clubs, and neighbours. A partner agency, the Openlands Project, contracted with professionals for the initial 'heavy-lifting' of tree-cutting, prairie seed acquisition, and invasives control. Over time the project grew from 275 to more than 900 acres and attracted more than US$5 million in outside funding.

Another ambitious project, the Orland Grassland, proceeded similarly at first. Its prairie and savanna became a central feature of the Village of Orland Park's image and culture. The new library soon sported a 26-foot mural by a National Geographic illustrator showing plants and animals (and human visitors) thriving as they would in years ahead. The local Congresswoman saw that it was good and secured US$7 million in additional restoration funding. Over the years, the Forest Preserve District hired new science and land management staff to support the project. Today the Orland Grassland Volunteers sponsor tours and festivals every spring, summer, and fall, while continuing to find and control isolated weeds, gather rare seed, and do other restoration work that can't so easily be done by staff. Trained volunteers annually monitor plant, bird, and butterfly transects. Others lead grammar and high school trips, put out newsletters and social media, serve on committees, and generally act as the Orland Grassland outreach.

With the region's strongest volunteer programmes, this Cook County (Illinois) agency embarked in 2014 on a 'blue ribbon' planning process that emerged with a board-approved plan to expand restoration contract funding from US$3.5 million to US$40 million a year, to hire 500 full-time 'intern level' young adults for the restoration programme, to expand its programme from about 7,000 acres to 55,000 acres, and a long list of similar lofty aspirations which will need to be worked out in reality over the fullness of time but which have already dramatically increased numbers and quality of staffing and increased the prescribed burn programme from about 3,000 acres per year to about 7,000 acres.

Small agencies, private lands and not-for-profits

Any preserve can have a strong volunteer programme with the right leadership. The Nature Conservancy can boast a fine example at its Nachusa Grasslands south of Rockford Illinois. It began with a few hundred acres, including just half a dozen acres of high quality prairie. This preserve in twenty-five years has initiated restoration on thousands of acres and now contains hundreds of acres that have achieved high quality. The site benefits from a level of expertise, detail-work, and funding that could not be achieved in any other way.

Although the Conservancy has on site the most advanced equipment, major components of the work are done by hand, because that way results in the best conservation. For example, despite tractors and harvesting equipment, most seed is gathered by hand, because so many hands are willing and ready, and because the seed of most species can't be successfully harvested mechanically. The goal of the preserve is full natural biodiversity, and many obscure species are needed, whether or not tractor-pulled machinery can harvest them.

The preserve is divided into units, many in the 20- to 40-acre range. For each unit, a steward makes the major week-to-week decisions and supervises the volunteer work. Such work includes much of the weed control, planting strategies, seed gathering, and broadcasting. Some stewards have been restoring their units for decades; new stewards are being recruited all the time.

Summary

Collaborative restoration volunteer programmes can be well worth the costs – but to thrive the staff needs to empower leadership volunteers as partners. An ambitious and far-sighted programme requires many people with a variety of skill-sets. Such programmes work best (and in some cases only) with understanding and support from four principal types of people: (1) upper-level decision-makers; (2) the people who will hire and supervise the volunteer facilitator; (3) dedicated volunteer facilitators; and (4) wise volunteer leaders, all of whom understand or at least respect this kind of programme. When all four work together, conservation can be successful for decades, ultimately centuries, which means for generations.

Bibliography

Bonney, R., J. L. Shirk, T. B. Phillips, A. Wiggins, H. L. Ballard, A. J. Miller-Rushing and J. K. Parrish. 2014. Next steps for citizen science. *Science* 343: 1436–1437.

Dickinson, J. L., B. Zuckerberg and D. N. Bonter. 2010. Citizen science as an ecological research tool: challenges and benefits. *Annuals of Ecology, Evolution, and Systematics* 41: 149–172.

Dickinson, J. L., J. Shirk, D. Bonter, R. Bonney, R. L. Crain, J. Martin, T. Phillips and K. Purcell. 2012. The current state of citizen science as a tool for ecological research and public engagement. *Frontiers in Ecology and the Environment* 10(6): 291–297.

Packard, S. and C. Mutel (eds) 1997. *Tallgrass Restoration Handbook: for Prairies, Savannas and Woodlands.* Washington, DC: Island Press.

29

BUILDING SOCIAL CAPACITY FOR RESTORATION SUCCESS

Elizabeth Covelli Metcalf, Alexander L. Metcalf and Jakki J. Mohr

Introduction to social dimensions of restoration

One of the greatest challenges to successful restoration is the integration and participation of key stakeholders including NGOs, citizens, private landowners, restoration practitioners, and many others. Often, restoration success is defined solely by ecological factors. However, evidence continues to mount suggesting success may ultimately be driven by social factors (Metcalf *et al.* 2015; Grimble and Wellard 1997). Ecologists, social scientists, and natural resource managers alike have called for a more integrated approach to restoration that balances social goals along with ecological goals (Palmer and Bernhardt 2006). This integrated approach is particularly essential for large-landscape restoration efforts where resources span multiple jurisdictions, including public and private land, the outcomes affect a multitude of stakeholders, and decisions are open to public scrutiny. Achieving restoration success at this scale requires a social-ecological systems (SES) approach. According to the Stockholm Resilience Centre (undated), 'there are no natural systems without people, nor social systems without nature. Social and ecological systems are truly interdependent and constantly co-evolving'. It has become clear that management agencies and restoration practitioners must be willing to actively engage stakeholders in decision-making processes and balance social goals alongside ecological goals if restoration efforts are to succeed.

Restoration, at any scale, must consider the linkages between social and ecological dimensions. In some instances, restoration efforts may involve a sole landowner, public or private (e.g. the US Forest Service, a rancher). More commonly, restoration efforts span complex landownership mosaics. Affected landowners might include private entities such as amenity owners, ranchers, or timber companies; county/city, state, or federal government lands (public lands); or NGOs and other non-profits that own land for conservation or other purposes. Even when restoration occurs exclusively on public lands, actions are subject to oversight by the public (e.g. the National Environmental Policy Act, public bonding, local ordinances). In any of these scenarios, ecological goals are realized only after navigating a complex web of social perceptions, objectives, decision-making processes, and constraints. Restoration success depends not only on re-establishing ecological integrity to the system and enhancing resilience, but also on the respectful inclusion of stakeholder input and the building of social capacity for implementation. To ensure success of a restoration project, natural resource managers need to invest in relationships and actively involve diverse interests as problems are defined, objectives established, alternatives considered, and projects implemented. This chapter describes ways to build social capacity, discusses the foundational principles of engagement, and highlights tools practitioners can use to meaningfully engage stakeholders in restoration projects.

Building social capacity for restoration success

Building social capacity has been a prominent concept in the natural resource literature, including areas of natural hazards, risk perceptions, environmental change, and climate change (Kuhlicke *et al.* 2011; Lebel *et al.* 2006). The terms 'capacity' and 'capacity building' have a broad set of definitions; however, there are core elements that can be applied to restoration contexts. According to Kuhlicke *et al.* (2011), social capacity 'is the context-related ability to decide and to behave successfully in a certain situation in order to anticipate, respond to, cope with, recover from or adapt to the negative impact of an external stressor as well as to employ necessary resources' (*ibid.*: 807). An important aspect of this definition is the ability for individuals or communities to act and respond to changes (Thomson and Pepperdine 2003). Capacity building is a form of intervention that is aimed at discovering, enhancing, and developing resources and abilities of individuals and communities (Kuhlicke *et al.* 2011; Lebel *et al.* 2006). Building social capacity means strengthening skills, knowledge, and competencies of stakeholders including practitioners, agencies, and community members that enhance restoration efforts over time. Capacity building requires management agencies and restoration practitioners to work intimately with affected stakeholders. To achieve 'buy-in' from diverse stakeholders, it is not nearly enough that people simply understand the goals of restoration efforts; they must have been included during the process as those goals were established, believed the processes to be fair, been encouraged to understand competing points of view, carefully weighed alternatives, seen their interests reflected in project goals, agreed with the overall plan, and, in some cases, been willing to allow restoration activities to occur on their own land and invest in their maintenance over time.

An essential step to building social capacity is to establish a fair and inclusive process for stakeholder engagement – to engage stakeholders early and often. Managers involved with restoration projects should consider implementing public engagement processes which go above and beyond legally required minimums to achieve meaningful stakeholder engagement. Often, restoration that occurs on US federal lands may require National Environmental Policy Act (NEPA) or similar state-level reviews. According to the US Environmental Protection Agency, using the NEPA process, 'federal agencies evaluate the environmental and related social and economic effects of their proposed actions. Agencies also provide opportunities for public review and comment on those evaluations' (Environmental Protection Agency undated). This critical piece of legislation mandates public involvement, but in most cases is manifest as public comment (Steelman 1999). While public comment is valuable at gathering input, relying solely on this method of engagement can leave the public feeling as though their opinions and values were not fully considered in the planning process (*ibid.*). For example, some may believe the alternative actions presented for comment were developed behind closed doors leaving them excluded from the 'real' decision-making process, even if initial input was gathered during a scoping phase. When alternatives are not meaningfully modified in response to public comment, stakeholders can feel confused as to the decision-making process or, worse, ignored. Despite the importance of NEPA, the base-level of public engagement required by the law has not prevented lawsuits, dissatisfaction with outcomes, decreased participation, and distrust in government (e.g. Lachapelle *et al.* 2003). Natural resource managers should consider developing strategies to embrace more comprehensive public engagement processes which are inclusive, fair, deliberative, and transparent. Engaging the public in these meaningful ways can help managers enhance stakeholder knowledge, develop better alternatives that balance multiple interests, build wide agreement, and empower critical actors. This social restoration capacity leads to better decisions, reduces uncertainties and legal challenges, and helps achieve restoration success in complex social–ecological landscapes.

One of the challenges to effectively engaging stakeholders is ensuring adequate representation of diverse interests. There is no single solution for engaging the public; rather, there are a suite of approaches which are context specific and require continual trial and error. Social capacity is built through early, ongoing, and responsive engagement strategies employed through all phases of a restoration project (Figure 29.1). These efforts are guided by foundational principles of engagement including trust, collaborative communication, inclusivity, and transparency – elements that have been observed as critical to restoration success. This list of foundational principles is not exhaustive and is meant to orient practitioners toward meaningful engagement. These principles inform three phases of engagement – initial, adaptive, and continued – each with discrete prescriptive tools and considerations for restoration management throughout a project. Because restoration can occur over a variety of temporal scales, practitioners should recognize that continued engagement strategies must interact with adaptive strategies to address shifting demands which evolve based on intermediate outcomes, monitoring, evaluation, and shifting social dynamics.

Foundations of stakeholder engagement

Creating social capacity in restoration projects requires a strong foundation built on trust, collaborative communication, transparency in the process, and inclusion of all stakeholders. These aspects of social capacity are interconnected and difficult to treat as separate entities. However, keeping these foundational aspects at the forefront of a restoration project will help

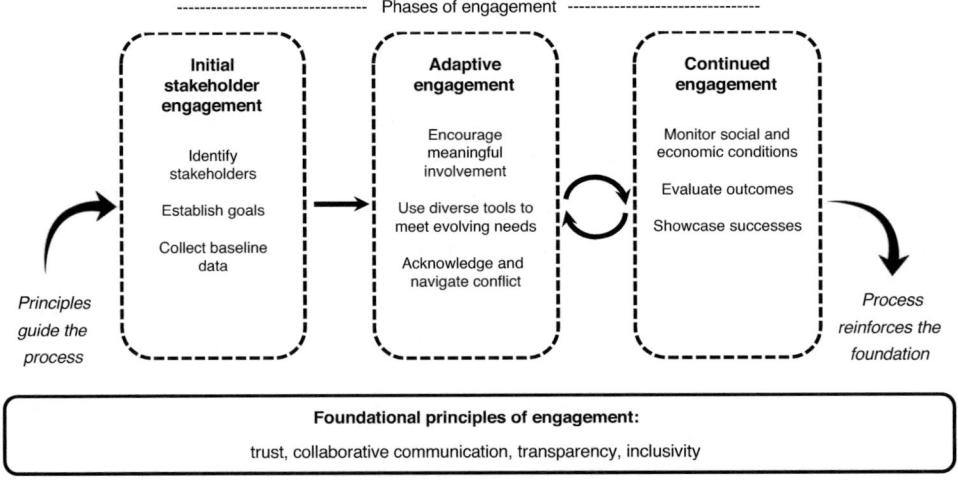

Figure 29.1 Building social capacity for restoration success through phased engagement

lead to a successful outcome. We encourage practitioners to use these foundational elements as a starting point to guide meaningful engagement with stakeholders. At any point during a restoration project it may be necessary to revisit these elements and determine if process could be adapted to reinforce these principles.

Trust

A critical component of social capacity is trust. Trust can be defined as, 'the willingness of a party to be vulnerable to the actions of another party based on the expectation that the other will perform a particular action important to the trustor, irrespective of the ability to monitor or control that party' (Mayer *et al.* 1995: 712). In the natural resource literature, trust has been shown to reduce conflict, encourage cooperation between actors, decrease costs associated with management, lessen uncertainty among stakeholders, and help decision-makers navigate collaborative efforts involving multiple stakeholders (Olsen and Shindler 2010). Agencies and practitioners working toward restoration success must actively build and sustain trusting relationships with each other and among all stakeholders. An example of the critical nature of building trust can be seen in the case study of the Clark Fork River restoration effort (Box 29.1).

Trust can be built (or eroded) for several different reasons. For example, you may believe that a person holds your best interests at heart, but you cannot trust them with a task because of past instances of incompetence. You may trust others because they have always followed through on their promises in the past, even though you don't seem to value the same things. In other circumstances, you may not trust someone at all, but believe regulatory or legal oversight will ensure their actions are acceptable.

These examples serve to differentiate specific dimensions of trust that managers should be aware of when engaging with stakeholders. Trust dimensions include: dispositional, affinitive, rational, and procedural (Stern and Coleman 2015). *Dispositional trust* is a person's natural inclination to trust or distrust an entity; some people find it easy to trust while others find trust incredibly difficult. *Affinitive trust* is a cognitive based assessment by the trustor of the trusted and is often an emotional judgement of shared values, connectedness, and benevolence; people tend to trust others who they can relate to, who are charismatic, who value similar things in life, and who respect their interests. *Rational trust* is a calculated assessment of potential outcomes from the trusted; based on past evidence, how likely is it the expected outcome will be realized? *Procedural trust* is a belief that the process, procedures, or governing rules will ensure satisfactory outcomes, despite the integrity or trustworthiness of individual actors. Restoration managers should consider these dimensions individually and take proactive steps to build layers of trust throughout restoration projects. Each domain is uniquely important to fostering productive relationships and successful restoration efforts. Trust takes time to build, but only moments to destroy. Building and maintaining trust should be on the forefront of everyone's mind during restoration efforts.

Since trust is so complex, managers may need to employ several strategies for building and maintaining trustful relationships with stakeholders. Trust often means not leading with a proposed solution and instead embracing a process by which stakeholders can collectively define the problem, identify fundamental objectives for the restoration, develop alternatives, and consider tradeoffs among predicted outcomes before agreeing on a path forward. Only after careful, collaborative consideration will stakeholders trust the process and its leaders enough to feel ownership of restoration proposals and contribute to restoration success. Without trust, it is highly unlikely actors within an SES will embrace and successfully implement any restoration plan.

Box 29.1 Understanding the social complexities of large landscape river restoration

The Clark Fork River in western Montana is the largest Superfund complex in the United States. As a result of over 100 years of contamination from copper mining in Butte, MT, the Clark Fork River needed extensive remediation to remove heavy metals from the water. Restoration followed the remediation and focused on restoring floodplain function, riparian vegetation, and fish populations on the main stem and tributaries. The Superfund complex is broken into four units and the project includes multiple phases. One of the early phases focused on the removal of the Milltown dam, including excavation of sediment and post-dam removal channel restoration (Woelfle-Erskine *et al.* 2012).

The project incorporated a diversity of stakeholders including state agencies (e.g. Montana Department of Environmental Quality), federal agencies (e.g. Environmental Protection Agency), NGOs (e.g. Trout Unlimited), community groups (e.g. Watershed Restoration Council), the Confederated Salish and Kootenai Tribes, and communities and private landowners. The landowners are a particularly interesting group since they are legally required only to allow remediation on their property, making restoration completely voluntary. This complex array of stakeholders can make public engagement a challenge.

To evaluate perceptions of the project, a series of in-depth interviews with key stakeholders was conducted in 2013. Stakeholders were asked a variety of questions related to the Superfund project including project goals, components of restoration success, information sources, the nature of public engagement, and hurdles to successful restoration.

Both trust and public engagement and their relationship to project success were salient themes that emerged from the data. Trust was particularly interesting as it was not explicitly asked about in the questions. Stakeholders identified that trust needed to be 'gained' or 'built' for project success. Trust varied across the project landscape with some communities feeling they had more trust than others. For example, the Milltown dam project was contentious at the onset. However, as the project progressed and stakeholders were included in the project planning, more trust was gained and consequently the dam removal was seen as a success. Trust was not established in other communities along the Clark Fork River. Places like Deer Lodge, MT and Opportunity, MT had concerns about restoration negatively impacting their ranching livelihoods or that their community was going to be the dumping ground for mining waste from Milltown. Many of these stakeholders did not feel they had been engaged in the process and felt they lacked trust in the organizations overseeing the remediation and restoration.

Similarly, public engagement processes were found to be successful in the Milltown dam removal with stakeholders feeling their voices were heard and that the process was exemplary. However, the same feeling did not exist with landowners who were not situated in the Milltown area. Many of these landowners have working ranches and depend on the river for their livelihoods. This group of stakeholders perceived agencies to have already made decisions for their land without soliciting public comment. This perception led to the erosion of trust in the process and ultimately may negatively impact project success.

Findings from this study highlight the importance of having trust and effective stakeholder engagement in restoration projects. The ultimate success of the Clark Fork River Superfund project is dependent on the diversity of stakeholders feeling they were part of the process.

Collaborative communication

Building trust that helps foster social capacity can be challenging; however, employing principles from collaborative communication may assist throughout the restoration project. Collaborative communication refers to the degree to which communication between parties is open, honest, and constructive (Mohr and Nevin 1990). Collaborative communication is an important correlate of trust, commitment, and satisfaction. We believe that in a restoration context, collaborative communication is essential for building social capacity.

According to Mohr, Fisher and Nevin (1996), collaborative communication is comprised of four dimensions: communication frequency, bi-directionality, degree of formality/informality, and content. Collaborative communication (as opposed to 'autonomous' communication) requires a certain amount of frequency of interaction between parties, as well as two-way flows of dialogue (not being 'talked at'), a mix of both formal (say, meetings and correspondence) and informal interactions (e.g. social interactions, encounters in the community, etc.), and content that is based on genuine information sharing (rather than coercive, threatening, or hostile).

Studies from the business world show that when parties are interdependent (i.e. achieving their respective goals requires dependence on each other and each has a certain degree of power in the relationship), collaborative communication can serve as a form of governance that is as effective as outright ownership or authoritative control. Moreover, collaborative communication engenders positive affective outcomes: Collaborative communication is central to trust, which in turn is related to cooperation, commitment, and decreased conflict (Geyskens *et al.* 1998). On the other hand, when communication is not collaborative (i.e. too infrequent, one-way, overly formal or overly informal, with threatening, hostile, or coercive content), trust can be eroded and ultimately deter development of the social capacity needed for restoration projects to be successful.

Transparency and inclusivity

To build a strong foundation of stakeholder engagement, transparency must be embraced throughout the process and all potential stakeholders must be included. The idea of 'procedural fairness' may be particularly salient in this instance. When stakeholders feel they have been included in the process and understand the dynamics of how decisions are made, they will be more satisfied with the outcome (Smith and McDonough 2001). This is particularly true in a restoration context; the more stakeholders are part of the process, the more likely they will be to support restoration efforts. This can lead to long-term success of restoration goals and help foster trust.

Inclusivity must be achieved through active recruitment of stakeholders and broad opportunities for participation. Stakeholders excluded from the process may undermine restoration plans by raising concerns at the last minute that were not properly addressed during deliberations. Others may object on procedural grounds that the decision-making process was not representative. Practitioners should actively seek to identify and invite all stakeholders, *particularly* those who may have strong or even polarizing views; involving them (and doing so early) will help ensure restoration proposals have considered all viewpoints and decision-making processes are respected by all parties. Being inclusive means not only extending invitations, but providing accessible meeting times/locations and making sure decisions reflect input from all perspectives.

Transparency is paramount to building trust and avoiding perceptions of bias. Online tools can help tremendously and should be employed liberally. All meeting times, locations, agendas, and minutes should be posted openly and in a timely manner. Open access to information demonstrates fairness and honesty while diffusing suspicions of ulterior motives or perceptions of biased processes. Information to post online might include: full meeting schedules, memberships of groups/committees, decision or voting procedures, budget and funding information, etc. Maximizing transparency demonstrates that power and decision-making authority is not hidden or removed from the stakeholder engagement process.

Initial stakeholder engagement

Knowing your stakeholders and engaging them in meaningful ways takes work and investment. There are many tools available to accomplish these tasks; each situation will require a unique combination of research and engagement processes to measure and enhance social capacity. It is beneficial to engage local leaders early and often in the restoration process – a move which signals an eagerness and willingness to include all perspectives while considering restoration goals. Below we outline several examples of how to engage stakeholders in the project. These are not exhaustive; rather, they provide a starting point for managers looking to understand their stakeholders.

Identify stakeholders and establish goals

The best way to get to know stakeholders and establish wide agreement around restoration goals is to talk to people. Initial discussions with key stakeholders can help restoration professionals and managers quickly and efficiently identify issues and concerns with proposed restoration and desired goals and outcomes. Using local leaders, key personnel at agencies, public officials, or particularly active members of the public who have unique insight into issues of interest are all relevant perspectives to consider. Gathering information from this connected and informed subset of the population will provide insight to critical issues, potential pitfalls, or areas of mutual interest among stakeholders. Two approaches that can be used are *key informant interviews* and *social network analysis*.

Key informant interviews

Key informant (KI) interviews provide a useful point of entry for in-depth community research and facilitate the gathering of data that cannot be ascertained with secondary data or quantitative survey techniques (Luloff 1990). KIs are knowledgeable about the area, existing social linkages, and can provide guidance on likely adaptive responses of local actors in restoration projects. In-depth interviews with KIs can provide practitioners with critical information about influential stakeholders, perceptions of trust, and assessments of communication strategies.

Lessons learned from KIs should not be considered gospel; checking back in with more broadly inclusive audiences can help reinforce messages heard from KIs, correct information misperceived, and add additional clarification when necessary. A good method for checking KI conclusions is facilitated group discussions, where the general public is invited to hear preliminary findings of KI interviews and asked to provide feedback and additional information, using small group discussions led by facilitators. Combining these two approaches creates a powerful tool for engaging stakeholders at multiple scales and collecting detailed input for restoration processes and planning.

Social network analysis

Social network analysis (SNA) has received recent attention in the natural resource literature as a way to identify actors and the patterns of relationships among actors. It is a participatory approach to identifying key stakeholders and networks that could 'increase the likelihood of collective action and successful natural resource management' (Prell *et al.* 2009). There are several ways to collect information to create social networks. One approach is through interviews with stakeholders or key informants. Other methods have been the use of focus groups or other group settings like public meetings. The patterns of relationships between actors are depicted visually in a web-like structure demonstrating how actors are connected, via lines drawn between actors who interact. SNA can also assess the strength of the relationships, or ties, between parties. Actors connected through strong ties tend to influence one another, share similar views, offer emotional support, communicate effectively, and are more likely to trust one another (*ibid.*). In a restoration setting, identifying social networks can help managers understand the influential stakeholders within a system who should be engaged early in a restoration effort. SNA can also identify the areas that need to be strengthened before or during a restoration project.

To start the data collection process, whether it is through SNA or through KIs, researchers must work with management agencies and practitioners to identify affected landowners and other affected NGOs, community leaders, and government officials. This list provides a starting point for data collection, and each person interviewed should be asked about their recommendations for other parties who are affected by the restoration efforts.

Collect baseline data

Baseline data is required for long-term monitoring and to provide an understanding of the social and economic landscape for restoration. Results from KIs or SNA can be considered a form of baseline data, as these provide general attitudes and perceptions about restoration projects. However, additional data may be needed. For example, if there is a large-scale restoration project occurring, managers and professionals may need to rely on economic metrics (i.e. median home/real estate values, revenue generated from tourism) to measure the success of a project. A case study of pine invasions in the southern hemisphere (Box 29.2) provides an opportunity to collect baseline data as a way to assess and monitor attitudes associated with restoration. Several large databases already exist that may provide generic baseline data. For example, census data may be particularly relevant to projects in the United States and provide the most comprehensive list of social variables (http://factfinder.census.gov) as well as the Bureau of Labor Statistics (www.bls.gov) for those interested in how employment is impacted by restoration. Several international databases also exist for baseline data including Eurostat (http://ec.europa.eu/eurostat) that provides economic data for countries in the European Union.

Box 29.2 An opportunity to engage stakeholders in an international restoration context

Successful restoration must consider the social as well as the ecological dimensions of the systems in which the project operates. Interactions between these dimensions can often complicate efforts to accomplish seemingly simple goals. Throughout the Southern Hemisphere, pine invasions of native ecosystems effect social and ecological changes which must be understand and incorporated into successful restoration plans.

In Argentina, pines (mostly from the genus *Pinus*) have spread beyond their plantation boundaries to invade native, naturally treeless landscapes. Efforts to control these invasions have been met with varied ecological results and mixed reviews by people and communities. To achieve restoration success, local managers must consider both ecological and social dimensions of introductions, invasion dynamics, various impacts of shifting species composition, and responses to control efforts.

Pine introductions to Argentina (and many other Southern Hemisphere countries) in ecosystems conducive to pine growth were driven primarily by social goals. Although pines are not native to the Southern Hemisphere, natural adaptations in their native ranges make them fast growers and resistant to fire and other threats in introduced ecosystems. Introduced pines have grown fast, their spread driven by a social desire for economic returns and jobs offered by a burgeoning timber industry (Simberloff *et al.* 2010). While introductions were primarily socially driven, pines have spread outside plantation borders and into native ecosystems because of ecological interactions with native species and the underlying biogeochemistry of the region.

As they have spread, pine invasions have had a variety of social and ecological impacts. Ecologically, these impacts have modified native systems through biodiversity loss, increasing fire intensity, and decreasing water availability through increased evapotranspiration rates compared to grass or shrubland dominated ecosystems. Social impacts have included increased jobs and economic activity, shifting landscape aesthetics, reductions in culturally valuable species (e.g. *Araucaria*), and changing recreation opportunities and experiences. While ecological impacts have been fairly well documented, the social impacts are more anecdotal due to limited human dimensions research.

From what is known, attitudes toward pine invasions have varied substantially. Some perceive impacts negatively while others welcome them. For example, although water availability decreases in pine-invaded systems, humans tend to assume that pines are associated with increased rainfall and view pines as a sign of healthy ecosystems (van Wilgen and Richardson 2012). Others see biodiversity loss as problematic, especially when culturally sensitive species are threatened. Changing aesthetics may be welcomed by foreign tourists who respond favourably to views reminiscent of North American and European vistas with pines encroaching up steep mountainsides; these responses may have a positive impact on the tourism industry. Overall, attitudes likely vary depending on awareness, sophistication of ecological understanding, and individuals' personal objectives. Labelling pine invasions as 'good' or 'bad' depends on the interactions among all these factors. For some, the economic returns provided by plantation timber production may justify any and all ecological impacts.

Successful restoration efforts must be ecologically sound and socially acceptable. Even when a situation appears horrendous from an ecological perspective, others may view the situation favourably. Efforts to restore a system to a previous ecological state may be undesirable from social perspectives. Even when there is agreement 'something' must be done, specific restoration tools may not be socially acceptable. Further, social and ecological dimensions are intricately linked, with each affecting the other as conditions change. For example, as restoration efforts are implemented and ecological health improved, social licence for those efforts may decrease as the problem is viewed as less of a threat. Efforts to control pine invasions in Argentina, which may initially share wide support, may eventually impede on profits of pine plantations thus undermining support as jobs are lost. Including stakeholders from all perspectives in the planning and implementation process can help restoration efforts successfully navigate these complex social and ecological dynamics and interactions.

Adaptive engagement

Once key stakeholders are identified and their perceptions of the restoration project are understood, agencies and professionals can use a variety of tools to collaboratively build restoration solutions and enhance social capacity. These tools should be leveraged throughout the restoration process with enough flexibility to address evolving needs. Implementing these tools at appropriate intervals can help build trust and reinforce the foundation needed for stakeholder engagement and capacity building. However, several factors may stymie stakeholder participation, the most notable being conflict. Competing stakeholder viewpoints and opinions on how restoration should be implemented can lead to conflict and derail project success. Below we outline some of the strategies for capacity building and the challenges to participation focusing heavily on conflict.

Encourage meaningful involvement using diverse tools

As restoration has become more prevalent on the landscape, public demand for participation in decisions has increased. Agencies and organizations are being forced to adapt to this change as it becomes clear decisions cannot be made in a vacuum (Smith and McDonough 2001). There are several ways to create open and meaningful communication among stakeholders. Large, public meetings provide one tool that has traditionally been used in natural resource management. To build social capacity for restoration, practitioners should try to move meetings beyond basic information dissemination or question and answer sessions. Creative meeting structures can provide opportunities for meaningful deliberation where alternative viewpoints can be heard and stakeholders can learn from one another (Petts 2007). In this way, public meetings can help foster understanding of the process and help create trust between those proposing the restoration and those who will ultimately be impacted. There are several open access references or guides that can assist managers and practitioners as they design public engagement processes. For example:

- *Bureau of Land Management (BLM) Desk Guide to Collaboration*
- *Collaborating for Resilience: A Handbook*
- *Assessing Public Involvement in an Open Government Era*
- *Collaborative Learning Guide For Ecosystem Management*
- *Collaborative Stakeholder Engagement and Appropriate Dispute Resolution.*[1]

Additionally, agencies and professionals may want to employ one-on-one approaches to engaging stakeholders. Individual voices are easily lost in large public meetings or people may not feel comfortable talking in front of crowds. One-on-one contact with the most impacted stakeholders may be needed. For example, if restoration is happening on private lands, it is essential that agency representatives meet face-to-face with landowners who may be impacted by the project. Calling landowners on the phone and committing time to meet them and visit their property can be a helpful first step toward a trusting and understanding relationship.

Lastly, keeping stakeholders and the public informed via written communication can help with capacity building. This can come in the form of newspaper articles, newsletters (electronic and print), and websites. It is important to employ all of these strategies at multiple times during the restoration project. Continuous check-ins with impacted stakeholders and updated information on websites can help build social capacity over time.

Different stakeholders may be engaged at varying levels – some more than others. Employing a variety of approaches, in concert, can provide opportunities for different audiences to be involved at their own desired level. Public involvement can span a spectrum from informational, the least involvement, all the way to empowering, the most engaged level of involvement. Lukensmeyer and Torres (2006) suggest five levels of involvement:

1   Inform: where the public is provided with information to understand the issue, what actions could be taken, and the solutions ultimately selected by decision-makers.
2   Consult: where the public is asked for feedback on issue analysis, potential alternatives, or decisions before they are finalized.
3   Engage: where decision-makers involve the public at all steps throughout the process to make sure public concerns are heard and incorporated.
4   Collaborate: where decision-makers partner with the public at all stages of the process to understand the problem, develop and weigh alternatives, and decide future action.
5   Empower: where stakeholders not only are involved in decision-making processes, but are given full control over the process and final decisions.

Higher levels of involvement require substantial investment of time and effort from stakeholders. Some stakeholders may be willing to invest in collaborative efforts, while others simply want an opportunity to see what decisions have been made or submit comments. No single level of involvement on the spectrum is ideal. In fact, using multiple levels concurrently may be the best way to satisfy stakeholders with different abilities and desires to be involved. Straus (2002) suggests using different 'concentric circles' of involvement based on stakeholder interest and abilities. For example, it may make sense to have a core team of collaborators who are engaged at all levels throughout the process as well as a set of advisory committees that are consulted at key stages throughout the process and, finally, processes for informing the broader public and soliciting feedback at critical junctures. Such tiered approaches to involvement can help meet diverse demands for inclusion and deliberation regarding restoration planning and implementation.

Acknowledge and navigate conflict

One of the most detrimental challenges to stakeholder participation and engagement is conflict. Conflict is an integral part of any decision-making process and includes numerous stakeholders with potentially differing goals, values, and expectations (Jehn *et al.* 1999). Conflict can be *relationship based* and include disagreements between participants on an interpersonal level; *task based,* where stakeholders fundamentally do not agree with the goals of a project; or *process based,* when stakeholders feel excluded from the decision-making process. Overcoming conflict can be difficult and may ultimately impact the success of the restoration project. Although there are several ways to work through and overcome conflict, we encourage readers to lean on the foundations of stakeholder engagement and consider trust, collaborative communication, transparency, and inclusion as ways to avoid or diffuse conflict. In particularly heightened situations, using an objective mediator (someone with no vested interests in the outcome) can help de-escalate conflict. An outside mediator is not always necessary, but lessons from the world of facilitation can be extremely helpful to meeting organizers navigating even minor conflict (e.g. Moore 2013).

Continued engagement

To build social capacity over the lifetime of a project requires monitoring, evaluation, and ways to demonstrate and communicate success. Many restoration efforts occur over a long time frame which may be frustrating to stakeholders (Metcalf *et al.* 2015). However, open communication with stakeholders about restoration project progress and success can help ease some of this frustration. As projects mature, continued engagement ensures stakeholders are celebrating their achievements, but also continuing to adapt to meet shifting needs.

Socio-economic monitoring and evaluation

Restoration success depends on adequate monitoring of social impacts as well as ecological metrics. Social monitoring can be qualitative or quantitative; restoration managers and professionals may want to engage local communities in regular stakeholder discussions to gather periodic feedback on projects and their impacts. Engaging a second round of KIs may help. Other strategies may include additional public meetings or one-on-one contact to gather information about progress on restoration goals. Population surveys of nearby residents and/or project partners can provide robust measures of social impact, track changes over time, reveal important relationships among social variables, and help managers invest in specific measures aimed at enhancing restoration capacity.

Showcase successes

Showcasing success can be instrumental in building social capacity. Visiting project field sites with stakeholders can provide a sense of accomplishment and connect stakeholders to the restoration project and each other. However, successes do not have to be communicated solely through field trips. Other strategies may include utilizing local media to share stories of success; asking independent film makers to create stories around project themes; or creating interpretive signs along frequently travelled corridors. Other ways to showcase success include working with local artists to create artistic renditions of the restored landscape to display publically, presenting at local chapters of civic organizations (Rotary, Chamber of Commerce, etc.), and engage local schools to learn from the restoration project. While these approaches may be outside of managers' and practitioners' typical duties, they will pay dividends for building restoration capacity.

Conclusion

Collaborating with stakeholders in a restoration context takes time and patience. This is particularly evident for large landscape restoration efforts involving complex social ecological systems that have a variety of stakeholders with competing interests in the project. By understanding how the natural system is connected with the social system, restoration managers can begin to develop appropriate engagement and outreach strategies to build social capacity and restoration success. This chapter outlined key elements to begin that process by highlighting 'Foundations of Engagement'. We hope this overview has provided managers and practitioners with a heightened appreciation for the interrelationships among trust, collaborative communication, inclusivity, and transparency and their importance for building social capacity.

Stakeholder engagement should begin early in a project and continue throughout the project lifespan. There are several tools that managers and practitioners can use to enhance stakeholder engagement. A critical first step is involving stakeholders as initial goals are established for the restoration project. Additionally, collecting baseline social data is important for measuring success of a project and demonstrating meaningful outcomes. Throughout the project, trust can be built with continued information sharing and inclusion of interested stakeholders. Lastly, monitoring the impacts of a restoration project and showcasing successes can help build lasting connections between the public and restored sites or landscapes.

This chapter outlined the strategies for building social capacity in restoration projects that will ultimately impact project success. Although it does not provide an exhaustive list of ways to build capacity, it does provide foundational elements and tools to help restoration managers and professionals understand how long-term stakeholder engagement can impact the success of restoration projects.

Note

1   These guides are available online at the following URLs at the time of writing: *BLM Desk Guide to Collaboration*, www.blm.gov/style/medialib/blm/wo/Law_Enforcement/nlcs/partnerships.Par. 36368.File.dat/Collaboration_Desk_Guide.pdf; *Collaborating for Resilience*, www.adelphi.de/sites/ default/files/mediathek/bilder/en/publications/application/pdf/collaborating_for_resilience_hand- book.pdf; *Assessing Public Involvement in an Open Government Era*, www.businessofgovernment.org/ sites/default/files/Assessing%20Public%20Participation%20in%20an%20Open%20Government%20 Era.pdf; *Collaborative Learning Guide For Ecosystem Management*, www.wellsreserve.org/sup/ downloads/collaborative_learning_guide.pdf; *Collaborative Stakeholder Engagement and Appropriate Dispute Resolution*, www.blm.gov/style/medialib/blm/wo/Planning_and_Renewable_Resources/ adr_conflict_prevention.Par.69360.File.dat/BLM%20Field%20Guide%20-%20Collaboration%20 and%20ADR%20Field%20Guide%20-%202009-11-12.pdf.

References

Environmental Protection Agency (undated) *What is the National Environmental Policy Act?* Retrieved from www.epa.gov/nepa/what-national-environmental-policy-act (accessed 17 January 2017).

Geyskens, I., Steenkamp, J. B. E. and Kumar, N. (1998) Generalizations about trust in marketing channel relationships using meta-analysis. *International Journal of Research in Marketing*, vol 15, no 3, pp. 223–248.

Grimble, R. and Wellard, K. (1997) Stakeholder methodologies in natural resource management: a review of principles, contexts, experiences and opportunities. *Agricultural Systems*, vol 55, no 2, pp. 173–193.

Jehn, K. A., Northcraft, G. B. and Neale, M. A. (1999) Why differences make a difference: a field study of diversity, conflict and performance in workgroups. *Administrative Science Quarterly*, vol 44, no 4, pp. 741–763.

Kuhlicke, C., Steinführer, A., Begg, C., Bianchizza, C., Bründl, M., Buchecker, M., De Marchi, B., Di Masso Tarditti, M., Höppner, C., Komac, B., Lemkow, L., Luther, J., McCarthy, S., Pellizzoni, L., Renn, O., Scolobig, A., Supramaniam, M., Tapsell, S., Wachinger, G., Walker, G., Whittle, R., Zorn, M. and Faulkner, H. (2011) Perspectives on social capacity building for natural hazards: outlining an emerging field of research and practice in Europe. *Environmental Science and Policy*, vol 14, no 7, pp. 804–814.

Lachapelle, P. R., McCool, S. F., and Patterson, M. E. (2003) Barriers to effective natural resource planning in a 'messy' world. *Society and Natural Resources*, vol 16, no 6, pp. 473–490.

Lebel, L., Anderies, J. M., Campbell, B., Folke, C., Hatfield-Dodds, S., Hughes, T. P. and Wilson, J. (2006) Governance and the capacity to manage resilience in regional social-ecological systems. *Ecology and Society*, vol 11, no 1, article 19.

Lukensmeyer, C. J. and Torres, L. H. (2006) Today's leadership challenge: engaging citizens. *Public Manager*, vol 35, no 3, pp. 26–31.

Luloff, A. E. (1999) The doing of rural community development research. *Rural Society*, vol 9, no 1, pp. 313–328.

Mayer, R. C., Davis, J. H. and Schoorman, F. D. (1995) An integrative model of organizational trust. *Academy of Management Review*, vol 20, no 3, pp. 709–734.

Metcalf, E. C., Mohr, J. J., Yung, L., Metcalf, P. and Craig, D. (2015) The role of trust in restoration success:

public engagement, and temporal and spatial scale in a complex social-ecological system. *Restoration Ecology*, vol 23, no 3, pp. 315–324.

Mohr, J. and Nevin, J. R. (1990) Communication strategies in marketing channels: a theoretical perspective. *The Journal of Marketing*, pp. 36–51.

Mohr, J. J., Fisher, R. J. and Nevin, J. R. (1996) Collaborative communication in interfirm relationships: moderating effects of integration and control. *The Journal of Marketing*, vol 60, no 3, pp. 103–115.

Moore, L. 2013. *Common ground on hostile turf: stories from an environmental mediator*. Washington, DC: Island Press.

Olsen, C. S. and Shindler, B. A. (2010) Trust, acceptance, and citizen–agency interactions after large fires: influences on planning processes. *International Journal of Wildland Fire*, vol 19, no 1, pp. 137–147.

Palmer, M. A. and Bernhardt, E. S. (2006) Hydroecology and river restoration: ripe for research and synthesis. *Water Resources Research*, vol 42, no 3, pp. 1–4.

Petts, J. (2007) Learning about learning: lessons from public engagement and deliberation on urban river restoration. *The Geographical Journal*, vol 173, no 4, pp. 300–311.

Prell, C., Hubacek, K. and Reed, M. (2009) Stakeholder analysis and social network analysis in natural resource management. *Society and Natural Resources*, vol 22, no 6, pp. 501–518.

Simberloff, D., Nuñez, M. A., Ledgard, N. J., Pauchard, A., Richardson, D. M., Sarasola, M., Van Wilgen, B. W., Zalba, S. M., Zenni, R. D., Bustamante, R. and Peña, E. (2010) Spread and impact of introduced conifers in South America: lessons from other southern hemisphere regions. *Austral Ecology*, vol 35, no 5, pp. 489–504.

Smith, P. D. and McDonough, M. H. (2001) Beyond public participation: fairness in natural resource decision making. *Society and Natural Resources*, vol 14, no 3, pp. 239–249.

Steelman, T. A. (1999) The public comment process: what do citizens contribute to national forest management? *Journal of Forestry*, vol 97, no 1, pp. 22–26.

Stern, M. J. and Coleman, K. J. (2015) The multidimensionality of trust: applications in collaborative natural resource management. *Society and Natural Resources*, vol 28, no 2, pp. 117–132.

Stockholm Resilience Centre (undated) What is resilience? Retrieved from www.stockholmresilience.org/research/research-news/2015-02-19-what-is-resilience.html (accessed 16 June 2016).

Straus, D. (2002) *How to make collaboration work: powerful ways to build consensus, solve problems, and make decisions*. San Francisco, CA: Berrett-Koehler Publishers.

Thomson, D. and Pepperdine, S. (2003) *Assessing community capacity for riparian restoration*. Canberra: Land and Water Australia.

Van Wilgen, B. W. and Richardson, D. M. (2012) Three centuries of managing introduced conifers in South Africa: benefits, impacts, changing perceptions and conflict resolution. *Journal of Environmental Management*, vol 106, pp. 56–68.

Woelfle-Erskine, C., Wilcox, A. C. and Moore, J. N. (2012) Combining historical and process perspectives to infer ranges of geomorphic variability and inform river restoration in a wandering gravel bed river. *Earth Surface Processes and Landforms*, vol 37, no 12, pp. 1302–1312.

30

ECOLOGICAL RESTORATION

A growing part of the green economy

Keith Bowers and Jessica Hardesty Norris

This chapter focuses on the businesses that are engaged in designing and implementing restoration projects. As discussed in the preceding chapters, public-sector land managers, private landowners, and conservation NGOs all play important roles in the field of restoration. There is substantial overlap among these actors, and the differences in their roles are increasingly blurred by project teams that include all of the above. Nevertheless, there are several aspects of the field unique to businesses. This chapter focuses on the emergence and development of the restoration industry, gives a brief overview of the management of restoration firms, and reflects on the role of innovation and emerging trends in restoration.

Over 30 years ago, just as the Clean Water Act regulations came online, the first author (Keith Bowers) started one of the first businesses to specialize in ecological restoration, the company that became Biohabitats. Biohabitats' experience is largely based in the US, though our engagement with the international restoration community has been steadily increasing in recent years, in part because of the first author's role in SER and the IUCN thematic group on ecosystem restoration. In this chapter, we have attempted to include examples from both outside the US and outside our own experience.

The industry of ecological restoration

The modern concept of ecological restoration is the legacy of Aldo Leopold's efforts to bring back the land around his house near the Wisconsin River, his efforts to encourage restoration at the University of Wisconsin Arboretum that led to the creation of Curtis Prairie, and the school of thought and work that emerged from his leadership there – though some restoration practices reach back even farther to the work of landscape architects and ecologists such as Jens Jensen and Frederic Clements (Jordan and Lubick 2011). However, it was not until more recently in the evolution of the field that a strong role for private business and an opportunity for entrepreneurs emerged.

Some restoration tasks, such as grading and excavation, have been performed by businesses since the earliest days of restoration, but businesses with specific expertise in restoration did not flourish until the market began to grow. Strong restoration markets, whether characterized by particular geographies or specific industries, are created where clients are highly motivated by a combination of regulations, financial interests, values. Restoration business opportunities increase where there is an organized oversight of regulatory compliance and where there is a stable source of financing for restoration activities. The demand for restoration services also increases with project scale, because businesses are particularly well-suited to projects that need to staff up and bring expertise to bear at short notice. For example, restoring the Everglades in southern Florida and large-scale prescribed burning has opened up business opportunities for firms that can quickly mobilize the substantial quantities of people and resources required by such large projects. Landscape-scale restoration often benefits from the capacity and flexibility afforded by businesses because few other sectors can host sustained planning, design and construction services in-house.

Industry drivers

The most important policies and regulations that drive the restoration economy are covered in Chapters 26 and 27 of this volume. Our view of the field from the vantage point of Biohabitats suggests that four major conditions have driven the development of the US industry since the early 1980s, which are echoed in international markets. Foremost, as discussed in Chapter 28, was the twofold regulatory impetus of US federal government's no-net-loss policy for streams and wetlands, which was created under the Clean Water Act (CWA), and the protections that were established under the Endangered Species Act (ESA). The mandate to restore land to other uses after mining activities (as originally established in the Under Surface Mining Control and Reclamation Act of 1977) has also created restoration opportunities (e.g. Zipper *et al.* 2011).

Secondly, since the late 1970s there has been an increase in people sharing a strong land ethic, which has created a demand for ecological restoration as preservation of a cultural and environmental legacy. For example, in the Midwest, the notion of prairie restoration has not hinged on CWA prescriptions or even preservation of endangered species, but rather on local leadership and a pervasive land ethic that has created a growing demand for restoration (Friederici 2006).

Another driver is the recognition that humans rely on natural resources, the services they offer, and the functions they protect (Gómez-Baggethun *et al.* 2010). Chapter 40 outlines the benefits provided by natural systems, benefits which are increasingly understood to include cultural benefits (Daniel *et al.* 2012). For example, in a recent Biohabitats wetland restoration project in Delaware City, the client was an economic development non-profit. They were trying to attract ecotourists and recognized that wetland restoration combined with invasive species management would enhance their efforts.

Examples of the emerging market for restoration that protects natural resources are not limited to the US. On the international stage, non-profit organizations such as Working for Water use ecosystem restoration to remove water-intensive invasive species and employ other techniques to reduce the demand for water and the stress on sources of drinking water in South Africa. If such efforts continue to be adopted at broader scales or by governments, we anticipate the emergence of a stronger international market for specialized restoration businesses.

A fourth driver in the development of the ecological restoration industry has been the incorporation of ecological restoration as a strategy in conservation. Over the last century, there has been a deep-rooted change in the philosophy of conservation and our thinking about how to protect natural resources, even since the 'natural regulation' philosophy of the 1960s (Chase 1987). From the hands-off management of the first parks in the 1800s, the field of conservation has evolved to recognize that intensive management and restoration is often necessary to support ecological functions. Land managers have sought and developed greater expertise in restoration because the need for it is inescapable (Benayas *et al.* 2009). This philosophical shift is being reinforced on a global scale. For example, the Conference of the Parties in the Convention on Biological Diversity, which was established in the early 1990s, formally urged its parties to employ ecosystem restoration to reach biodiversity targets only twenty years later in 2010's Decision XI/16.

Each of these drivers has gained influence and created additional market opportunities since the emergence of the field of ecological restoration. Our technical and cultural approach to restoration has also become more complex in recent decades, spurred by greater environmental awareness and the threat of climate change.

Growth of the restoration industry

It is challenging to summarize the scope and scale of the restoration economy because the term restoration is inconsistently applied and restoration activities are spread across a variety of sectors and not grouped according to census data (US Census Bureau 2013). Nevertheless, there have been local efforts to document the growth of the sector in certain geographies or to track the effects of policy. For example, Moseley and Nielsen-Pincus (2010) found in their analysis of the Oregon Plan for Salmon and Watersheds that about 230 jobs are created per year from Oregon plan investments. Spending on several major restoration initiatives are summarized in required reports to Congress; for example, between 2010 and 2014, the US Environmental Protection Agency received approximately US$1.657 billion in Great Lakes Restoration Initiative funds (EPA 2014).

The first robust quantitative estimate of the dollars invested and jobs provided by the restoration industry across the US was published in 2015 (BenDor *et al.* 2015). According to a national survey of restoration businesses, the annual sales in restoration are approximately US$9.5 billion, and the sector employs ~126,000 people (*ibid.*). There was no prior attempt to provide a national estimate of the size of the field, though the US Census Bureau's reporting on the revenue from remediation services offers a proxy statistic of the growth of the field (Wagmer and Shropshire 2009). Its 250 per cent growth from 1998 to 2013 indexes the national growth in revenue from the full suite of restoration services. This agrees with our personal experience, though we have only anecdotal information. The growth of Biohabitats in staff and annual revenue is a response to the development of the market. We are made aware of more restoration opportunities every year, and it is easy to witness the proliferation of restoration companies. When Biohabitats emerged in 1982, there were few opportunities but also little competition, because few firms specialized in any aspect of restoration. Today that situation has undergone a radical change and the market is vibrant.

Sample firm histories

This section further describes the restoration market as its emergence was experienced from the vantage point of prominent ecological restoration consultants from the mid-Atlantic region, Midwest and west coast of the US. Please refer to their websites for a comprehensive view of their histories and current services.

Biohabitats: Baltimore, MD

In the mid- to late 1980s, two separate but related events provided a burgeoning market for ecological restoration services in the mid-Atlantic region of the US. On a national level, the

US Army Corps of Engineers began more robust enforcement of non-tidal wetland regulations under the CWA, requiring compensatory mitigation for filling wetlands and waters of the US. On a local level, the state of Maryland passed the Chesapeake Bay Critical Area Protection Act, which required the preparation of Critical Area Plans for all jurisdictions adjacent to the Bay. These plans, among other things, called for the conservation and restoration of wetlands, riparian and wildlife habitat.

Biohabitats' first projects included preparing over two dozen Chesapeake Bay Critical Area Plans and a contract to restore 11 acres of tidal marsh over the newly constructed I-95 Fort McHenry tunnel in Baltimore's inner harbour. This project propelled Biohabitats into the emerging field of wetland mitigation and restoration design.

Wetland restoration as mitigation for impacts to tidal and non-tidal wetlands was a mainstay for Biohabitats throughout the 1990s and into the early 2000s. During this time, though, other types of ecological restoration projects gained market share. Due in large part to the CWA, the restoration of stream and river systems followed closely behind wetland mitigation. Restoring the geomorphic and biological characteristics of streams and associated riparian areas that had been impacted by agriculture, urban development and flood control efforts gained momentum throughout the east coast, mid-west and northwest parts of the US. Both private land development entities and public government agencies were in need of firms that could assess, design and implement river restoration projects. Biohabitats responded to this need by hiring experts in fluvial geomorphology, hydrologic and hydraulic engineering and fisheries biology.

As wetland mitigation and stream restoration gained a strong foothold in the restoration market, other types of ecological restoration projects began to emerge. Biohabitats began seeing more and more projects related to invasive species management, forest restoration, grassland restoration and coastal restoration. These markets developed due to federal, state and local regulations along with a desire by both private land owners and public agencies to enhance the value of their land. The increasing complexity of stormwater management regulations also prompted many jurisdictions to undertake watershed management planning, often prescribing wetland, stream and riparian restoration as a principle component to improving water quality and managing water quantity.

Recently, Biohabitats has found more opportunities in urban restoration. With the advent of the Leadership in Energy and Environmental Design (LEED) programme, sustainable development began to permeate much of the land development industry. Biohabitats has found a niche in the sustainability market by offering a combination of ecological restoration, ecological engineering and green infrastructure services in the built environment.

Pizzo and Associates: Leland, IL

Pizzo and Associates was created in 1988 specifically to focus on ecological restoration, because Jack Pizzo recognized the need for specialized expertise in the Chicago area. The firm's growth and development reflect both national and local trends in restoration markets.

Over the last three decades, Pizzo and Associates' focus and client base has changed. To their early core market of private clients working to comply with federal regulations, they have added more clients reacting to local ordinances. As happened elsewhere in the 1980s, federal regulations created an initial market for restoration. In the early years, Pizzo and Associates had a broad client base in developers who hired firms to plan new construction for compliance with the CWA and ESA. In those days, there was little formal guidance on what constituted mitigation and restoration, and non-technical rules of thumb began to proliferate, often lingering until regulators around Chicago began to take a more nuanced approach to issues such as stormwater management. Some of the less technically informed techniques (e.g. using non-native seed mixes; creating sterile, ecologically depauperate stormwater ponds) were only recently extirpated by local regulations that encourage more sensitive approaches. Pizzo sees this boom in sophisticated local regulation as one of the most important developments in their restoration market. It has happened largely since 2000 and caused an uptick in the number of municipal clients.

The local geography of their firm has also been important. Cook County, which includes Chicago, has a system of Forest Preserves that is a network of over 27,000 hectares of forest, prairie, and wetlands. The 1990s saw substantial controversy in the restoration community and the management of the Forest Preserve District. After a series of unpopular management projects in the early 90s, which included controlled burns, vocal citizen opposition propelled a moratorium on fire management and a generally conservative stance toward any kind of active management. This attitude slowly began to change in 1997, when most of the moratorium was lifted. Pizzo and Associates, along with many firms, suffered in the economic downturn of 2007, but they were bolstered by the renewed interest and broader understanding of management on the part of the Forest Preserve District, which began to take a much more active approach to controlling invasive species and implementing fire management. Pizzo and Associates weathered the economic crisis with some substantial restoration work in the forest preserves, and their portfolio has today returned to a robust mix of client types.

H. T. Harvey and Associates: San Francisco, CA

The formation of H. T. Harvey and Associates anticipated the emergence of a restoration market by several years. Their roots go back to a 1970 contract with the Sierra Club, which hired H. Thomas Harvey and a few other ecology professors to assess recreational impacts in the High Sierra. H. T. Harvey retains a strong focus on science and has a full complement of researchers on staff, but today they also have a strong ecological restoration practice.

With the passage of a series of environmental laws in the early 1970s, the California market grew considerably, as did the firm. Their California location played a strong role in their development because the state is environmentally progressive and passed an aggressive series of laws that matched or exceeded federal regulations. H. T. Harvey and Associates were involved in some of these earliest efforts in impact assessment, mitigation planning, and conservation, markets related to restoration that grew steadily for decades.

The firm's restoration practice had its roots in San Francisco Bay. As the great salt extraction operations that once dominated shoreline habitat lost their economic feasibility, Californians recognized the opportunity for tidal marsh restoration, and Harvey quickly developed expertise that led to five decades of salt marsh restoration projects.

One of the emerging market trends in recent years that has affected H. T. Harvey has been the downturn in private clients who are designing mitigation for their development sites. Mitigation banks, which specialize in selling mitigation credits to developers to help them comply with US Army Corps of Engineers permits, have proliferated in California and across the country, and many clients are no longer seeking help from restoration designers. While this trend is obviously of concern to companies specializing in restoration design, it is also of concern because mitigation may take place in watersheds or systems far removed from the original impacts (BenDor *et al.* 2009).

US markets for ecological restoration

Focal systems

The few available data suggest that restoration of aquatic systems dominates the US restoration market. Twenty-six percent of the restoration professionals in BenDor *et al.*'s (2015) survey reported that they were engaged in wetland, stream, or marine restoration compared to just 12 per cent in terrestrial systems. The other large categories, contaminant clean-up (13%) and mitigation banking (12%), were not specific to a type of site and may have included hydrological systems – mitigation banking is usually focused on wetland restoration (Bronner *et al.* 2013). A quick survey of the biggest ecosystem restoration sites in terms of timescale and dollars invested reveals that most are aquatic systems: Chesapeake Bay, Coastal Louisiana, Columbia River, Great Lakes, San Francisco Bay Delta, South Florida Everglades, and Upper Mississippi River (Vigmostad *et al.* 2005). Not one of these large-scale restoration projects is solely terrestrial, though large-scale restoration efforts, such as the coordinated grassland restoration in the Great Plains by the partners of the Northern Plains Conservation Network, certainly exist.

Stream, wetland and coastal restoration

Restoration of marine and freshwater aquatic systems has characterized the largest and most comprehensive restoration projects in the US since the mid-1970s. The market is driven by the CWA and by the benefits that humans derive from water for consumption, industry, irrigation, recreation and aesthetics. Flooding, storm surges and other natural disasters exacerbated or created by poor water management also create more political pressure and result in greater allocations of funding than the threats addressed by terrestrial restoration, such as invasive species.

Terrestrial restoration

The terrestrial restoration markets in the US and across the globe encompass forest, prairie grassland and dryland opportunities. In the US, the terrestrial share of the overall market is smaller than that of aquatic systems (BenDor *et al.* 2015), but there are ecological restoration firms that are able to specialize entirely on terrestrial systems. For example, prescribed burning has given rise to specialist businesses and become more fully professionalized in recent years, with the first Fire Ecology and Management degree offered in 2007 by the University of Idaho (Kobziar *et al.* 2009). Regulations that govern restoration vary by state and province, and as they increase, the attendant business markets will likely grow overall.

The US division of the restoration market between aquatic and terrestrial projects does not represent a consistent international trend. For example, terrestrial restoration in Australia that attends the mining industry occupies a large proportion of the market there. In Peru, a current USAID-driven large-scale reforestation initiative employs private forestry businesses, but there is no corresponding effort to restore aquatic systems.

Geography

In the US, restoration firms tend to cluster near coastal population centres, though they often have regional, national or international capacity. Restoration businesses often rely on a steady diet of smaller local projects, and they therefore usually locate near cities. That, paired with the aquatic focus of many of the regulations that drive the restoration market, mean that firms are more often based on coasts, near large water bodies such as the Great Lakes or in large aquatic systems such as the Ohio River. Restoration tends to happen where opportunities are bolstered by popular support, and markets tend to be stronger where people are paying attention to the ecological state of their surroundings, whether because of their economic dependence on the system or because of broader values. Accordingly, many firms have been launched near economies that depend heavily on water bodies such as the San Francisco and Chesapeake Bays.

Internationally, the US and Australia are hot spots of restoration in the private market. Europe's active restoration community counts many fewer private firms among its practitioners, perhaps because of deep-seated cultural differences about businesses. Academic and government institutions more often lead Europe's pioneering or experimental restoration projects.

Running a restoration business

The diversity of restoration businesses

'Restoration businesses' are far from a homogeneous group. Businesses might specialize in providing native plant material, designing restoration initiatives, implementing restoration projects, or managing landscapes over the long term. Some firms encompass two or more of these services. The entrepreneurs who participate in the market include local native plant nurseries, mid-sized restoration consultants, and multinational consulting firms that house a dozen disciplines of pertinent technical staff. Because there is no single model of a restoration business, there is no simple way to describe operating an ecological restoration business.

Nevertheless, all of the businesses that work in restoration require some degree of specialized competency. For example, restoration construction firms retain or acquire a degree of background in ecology so that they can make minor design modifications without compromising the ecological integrity of the design. Nurseries that specialize in native plant material for restoration must retain staffs that are knowledgeable in the composition and ecology of native plant communities.

Each type of restoration business has a unique business model, so the other common themes are reduced to those that are common to all business. It is a list summarized in the first few chapters of any Introduction to Business textbook, but it also generally defines a mindset and culture that is clearly distinct from academic or non-profit actors in our sector. Regardless of specialty, the leaders of restoration businesses face the basic task of determining the type of restoration business they wish to pursue, matching available resources of skills with market needs, creating solid business practices, managing financial performance, and managing risk.

Types of clients

Clients who seek the services of firms specializing in restoration can be private, public, or non-profit (Clewell and Aronson 2011). In California and other environmentally progressive regions, there tend to be more business opportunities with private and NGO clients whose motivations reach beyond regulatory compliance. Depending on the political and economic climate, the market balance between private clients and clients in federal, state, municipal governments can vary widely. In the aftermath of the 2008 economic crisis, for example, Biohabitats saw a reduction in its business with private clients, which took over 6 years to recover. Many restoration firms target certain types of clients, for example, by actively seeking land developers with mitigation needs or by focusing on large-scale reforestation and replanting efforts for federal land managers.

Business development and competition

Though specialized restoration businesses might get most of their work through word of mouth or being invited onto project teams, those who provide more comprehensive services are in a constant process of making contact with clients and seeking new work. In addition to responding to formal solicitations from prospective clients, firms attend trade conferences and professional meetings where they meet potential clients as they engage in technical exchange.

It is important to understand that, in the course of a given year, restoration companies are likely to be teammates as often as they are competitors. This creates a certain amount of collegiality among firms. Even as two firms are competing head-to-head in an interview for one county, they might be collaborating on a contract for a neighbouring municipality. Most restoration projects are conducted by teams of firms that lend specific areas of expertise. A restoration firm, acting as a prime contractor, may assemble a team of specialty subcontractors to undertake a complex restoration project. For example, a large engineering firm looking at restoring a river for salmonid habitat in the Pacific Northwest might retain a local fisheries biologist as a subcontractor to offer advice on the placement of larger woody debris. A firm specializing in removing invasive species might bring in a public outreach specialist to run community workshops during the design phase. Skill sets may even be duplicative. Biohabitats sometimes invites subcontractors into proposal efforts for large on-call contracts, which might have several separate tasks, simply to add depth and redundancy for some of the technical roles.

The client can also play a role in creating teams. For example, on a recent restoration proposal, Biohabitats was invited to team with a large transportation firm that was seeking a contract as prime contractor. Although that firm lost the contract, the client asked the selected firm to bring Biohabitats on to help with certain aspects of the site hydrology. This kind of unpredictable fluidity in teaming dynamics creates a more congenial, collaborative dynamic among firms than one might expect among competitors.

One concrete example of teaming realized by Biohabitats was the Lizard Hill Mine restoration (Figure 30.1). Like many firms specializing in restoration, Biohabitats has a diversity of technical expertise on staff. For the reclamation of Lizard Hill Sand Mine in the coastal plain of Maryland, the Maryland State Highway Administration contracted Biohabitats to restore an abandoned sand mine to a mosaic of non-tidal wetlands. Biohabitats, in turn, subcontracted survey and geotechnical evaluations and a wetland specialist with experience in sand seepage systems. Biohabitats' role shifted from restoration designer to construction management once the design was complete. By retaining our services through the construction process, MD State Highway Administration was able to rely on us for the unforeseen design adjustments that ensured the success of the project.

Contractual frameworks

As discussed before, it is important to keep in mind the diversity of business types in ecological restoration. A restoration business might have a specific and discreet role such as growing and planting a suite of native species, or might be responsible for all of the facets of a project from baseline assessments to long-term management. For restoration design and construction firms, one important distinction is between the traditional and design-build approach.

When private firms are hired to lead a restoration design, especially by public sector clients, they usually follow a formal approach much like that of architects. In typical restoration projects using the traditional procurement method (Figure 30.1), the design firm may produce alternative concept plans through dialogue with the client, one of which is selected for

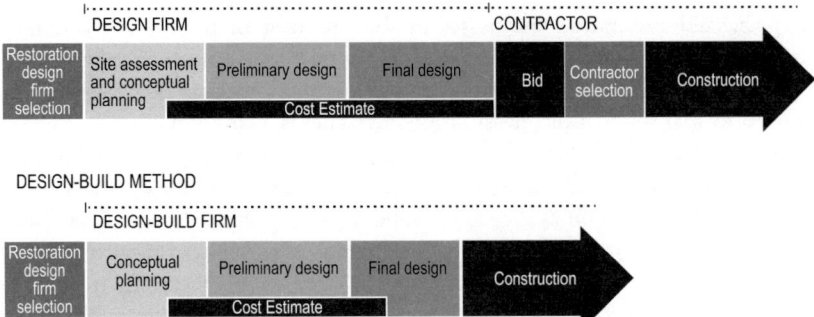

(a) TRADITIONAL PROCUREMENT METHOD

DESIGN FIRM | CONTRACTOR

| Restoration design firm selection | Site assessment and conceptual planning | Preliminary design | Final design | Bid | Contractor selection | Construction |

Cost Estimate

DESIGN-BUILD METHOD

DESIGN-BUILD FIRM

| Restoration design firm selection | Conceptual planning | Preliminary design | Final design | Construction |

Cost Estimate

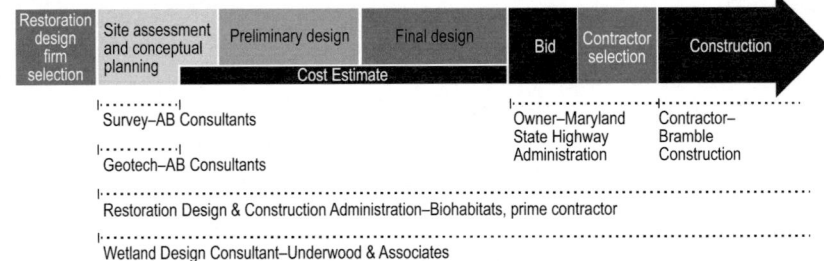

(b) TEAMING EXAMPLE
Lizard Hill Sand Mine Reclamation

| Restoration design firm selection | Site assessment and conceptual planning | Preliminary design | Final design | Bid | Contractor selection | Construction |

Cost Estimate

Survey–AB Consultants

Geotech–AB Consultants

Owner–Maryland State Highway Administration

Contractor–Bramble Construction

Restoration Design & Construction Administration–Biohabitats, prime contractor

Wetland Design Consultant–Underwood & Associates

Initial conditions

Immediately after restoration

Figure 30.1 (a) Here is a diagram showing the difference between the traditional procurement method vs a design–build procurement method. (b) This shows the team building process as conducted at the Lizard Mine restoration site

preliminary design. Permitting and cost-estimates proceed alongside design until the final recommendations or full construction drawings are complete with specifications for the entire project. Contractors then submit competing bids on the project, and typically the contractor with the lowest bid is awarded the contract. The design firm may then have some involvement or supervisory role during construction if they are contracted to provide construction administration services.

In contrast, some clients prefer to use a design-build method (Figure 30.1), a procurement mechanism that can save time and money. Rather than designing and building the project under two separate contracts, the client hires a single firm who takes responsibility for both design and construction. In the design-build model, the final design typically takes less time, there is no bid and selection phase, and the design-build firm has greater flexibility to respond quickly to challenges as they arise during construction. Many firms prefer a design-build approach if it is possible, because restoration is a complex process that benefits from flexibility and adaptation throughout its course. On the other hand, clients may have reservations about design-build methods because they establish the full contract price during the beginning stages of the project and could create a temptation to economize on construction costs to the detriment of the project. Procurement regulations sometimes use this as a reason to prohibit design-build methods.

Career planning and management

In this section, Keith Bowers writes from personal experience as a manager and someone who has hired hundreds of employees in this field, partially in an attempt to distil his best advice for early-career professionals interested in restoration.

Design and consulting companies that specialize in ecological restoration typically work in fluid and multidisciplinary teams that shift between projects. Teaming dynamics among firms are described previously, and they are echoed within firms as various specialties are brought to bear throughout the course of a project. Ecological restoration firms that also include implementation and management services can also work this way, though each company's organizational structure and processes are different.

Entry-level positions

In ecological restoration, there are always opportunities for people who have exceptional technical proficiency combined with a passion to make the world a better place. The typical design and consulting disciplines that we hire for are terrestrial and aquatic ecology, restoration ecology, landscape ecology, conservation biology, geology and geomorphology, soil science, civil, water and ecological engineering, landscape architecture, environmental planners, and people with technical skills in computer-aided design and drafting, Geographic Information Systems, construction management and surveying. Because our work is consistently interdisciplinary, we put a particular value on people who can demonstrate some flexibility and problem-solving abilities that cross outside of their main discipline. The type of training that we most often see in applicants but cannot place is a basic ecology or biology degree that is not grounded in applied ecology or conservation biology. Many of our 'entry level' positions are filled by people who have some professional experience that is related or can support restoration work, even if they have not worked directly in the field of ecological restoration before. For example, environmental scientists with experience as consultants doing wetland delineations and other site assessments might be interested in switching to a specialized firm that would provide them the opportunity to focus on ecological restoration. In practice, of course, hiring also depends heavily on intangible qualities, such as a sincere passion to support the mission of our company.

Team members

After our team members have years of experience, they begin to take on more opportunities to help the firm acquire business. Ultimately, some of the most valuable team members take on multiple roles throughout the company; working as a technical team member on one project, leading other projects as a project manager, and participating in business development activities to help secure new work. There are many career paths in the ecological restoration field, with some leaning toward technical mastery in specific subject areas and others who take on more managerial or leadership roles.

Restoration companies that focus on implementation and management are often structured with managerial and support staff that serve a sales people, project managers, estimators, shop stewards and purchasing and inventory managers. They are accompanied by the project crews that perform the day-to-day installation and management work in the field. What typically separates the ecological restoration implementation and management company from that of a conventional landscape contractor, besides the type of work, is the professional and technical expertise on staff, such as backgrounds in terrestrial and aquatic ecology, wildlife biology, geomorphology, fire ecology and forestry.

Research and innovation in the private sector

Because the audience for this book will be largely academic, we wanted to include a specific treatment of research in the private sector. Ecological restoration includes a wide swath of activities by professionals and craftspeople with varying degrees of training in science. By the very nature of this work, there is undoubtedly much informal research that happens on a day-to-day basis. Increasingly, ecological restoration firms are recognizing the need to formalize this research, often in conjunction with academic and research institutions, as well as non-government agencies, in an effort to advance both the profession and practice of restoration. For example, Biohabitats and H. T. Harvey have created specific technical practices in the research of restoration ecology and the development of ecological restoration practices. As the profession of ecological restoration grows, it will be critical for companies to team with private and public partners to experiment and test new restoration techniques, learn from successes and failures and pass this information on to the next generation of companies.

Evaluating restoration

Although well-planned monitoring and adaptive management are accepted as best practices, unfortunately most restoration projects suffer from a lack of robust monitoring, and many are not structured to facilitate an adaptive management framework. While many US government agencies require monitoring as a condition for certain projects that are designed to meet regulatory requirements, the vast majority of projects that take place out of a regulatory framework go unmonitored. Even some within the regulatory framework are characterized by monitoring that lacks consistency, robustness and follow-through. While many restoration design, implementation and management companies want to do monitoring, a lack of funding may impede their efforts. For these reasons, most projects are not monitored, or the task falls to a heterogeneous band of data collectors that range from university laboratories with an investment in restoration research to citizen scientists to locally invested NGOs such as Riverkeepers to summer interns.

Box 30.1 The definition of 'science'

Among practitioners of restoration, the term 'science' is sometimes used where it might be more appropriate to use the term 'data'. Rather than a truly scientific approach that includes hypothesis testing, 'science' and 'science-driven' gets used to describe the simple monitoring efforts and data collection that should be included in every restoration project.

Developing new techniques

In some ways, the business model is very well-suited to fostering innovation in restoration. Businesses with active restoration practices may implement hundreds of projects a year, which provide opportunities to develop new techniques. Furthermore, the core values of many private firms can be aligned with innovation to support opportunistic tinkering.

Private practitioners advance the field significantly through trial and error and stepwise modifications of their basic approaches, even if their approach is infrequently through formal hypothesis testing. Often as not, small improvements in techniques and models happen so fast that no one invests the time for methodical evaluations.

There are hundreds of unprepossessing examples of such innovations, such as fixtures to modify large bore drill seeders for the smaller seeds of native grasses and shrubs. Other less tangible 'inventions' are also the products of businesses that encourage innovation. Data models that streamline site selection or facilitate the process of selecting species and creating a schedule for planting are also examples of small but important innovations. These sorts of tools provide a valuable service and should be widely disseminated through the discipline to increase efficiency. Biohabitats and others try to use outlets such as the Creative Commons to share the best methods in restoration.

Dissemination and reporting

In this era of technologies that become outdated within months, progress in restoration techniques is comparatively slow, and practitioners have a critical role in improving and innovating. Improvements to restoration techniques are slow to disseminate for several reasons. First, it may take years to see the limitations and opportunities to improve a given restoration design. Once a solution is developed, it may be too late to incorporate it into projects for which design has already started, even those for which completion is still years away. In addition, many clients are conservative in an attempt to be good stewards of taxpayer dollars, and they may prefer designs that are widespread and familiar.

Advances in the practice of ecological restoration come in waves, with many firms quick to pick-up on new restoration approaches and technologies. However, opportunities to put these approaches and practices into place is often encumbered by outdated procurement processes, regulatory frameworks, funding and aversion to perceived and real risks. The spread of innovative restoration techniques often follows a typical cycle of technology adoption. Firms that are innovators and early adopters tend to be less risk adverse, using a nimble, adaptive management approach to their designs. Larger companies practising ecological restoration often times have

451

access to more funding then their smaller counterparts, though they are typically more risk adverse.

Reporting on such information varies widely across the sector. Private firms are not typically compensated for the time that would be required to go through peer review, so unlike academics, whose institutions support their research and publication, few practitioners contribute to academic journals. Instead, practitioners often produce articles for publications in trade magazines, white papers or grey literature that reports monitoring data and innovations in the field. Case studies in trade publications are often more widely read by practitioners than peer-reviewed journals. There have been various efforts to change this status quo by journals such as *Ecological Restoration*, which requires a short summary of the practical application of the findings reported in every full-length submission. It is difficult to measure the effect of such efforts, but we do know that there is still a lack of communication between private and academic restoration ecologists.

Research collaborations

The disconnect between research and practitioners is itself not sustainable. We need practitioners informing experimental designs; we need researchers working with us to evaluate projects; and we need to pool our resources to find the best solutions out there. Some of these challenges can be addressed by purposeful collaboration between private companies and like-minded universities and research institutions. Others would benefit from third party support, such as grantors that explicitly require collaborations that bridge not only disciplines but also sectors.

Looking ahead

The business of ecological restoration has a bright and promising future. As the world recognizes the connection between healthy robust ecological processes and a healthy robust economy, demand for ecological restoration will continue to grow, and businesses are poised to meet the demand. Ecological restoration is not a passive movement, it is an active endeavour that requires myriad of talents, skill sets, and avocations, supporting livelihoods and enriching cultures. The business of ecological restoration will continue to play a primary role in the green economy.

Ecological restoration alone will not be able to stem the tide of new and growing challenges such as climate change, water scarcity and the loss of habitat. What is needed is a combination of approaches that are nimble, adaptive and efficient. Combining approaches from the disciplines of restoration ecology and conservation biology with ecological engineering, biomimicry and ecological economics, for example, will lead to solutions that may not have been possible from any one of these disciplines alone. Businesses have the ability to respond quickly to market needs, assembling talent and skill sets that can seamlessly work through a cross-disciplinary approach, all the while contributing to the growth and well-being of the economy.

As a result, the future will see the practice of ecological restoration become more widespread across the landscape and more integrated into other land development and design initiatives. From remediation of brownfields to designing buildings and sites that regenerate ecological processes, to urban infrastructure that includes pollinator pathways, riparian corridors and rooftop biodiversity, the business of ecological restoration will continue to evolve and grow. And like the many different ecosystems that contribute to the complexity and diversity of life, so will the many different types of businesses that embrace and carry out restoration.

References

Benayas, J. M. R., Newton, A. C., Diaz, A. and Bullock, J. M. (2009) Enhancement of biodiversity and ecosystem services by ecological restoration: a meta-analysis. *Science* 325(5944): 1121–1124.

BenDor, T., Sholtes, J. and Doyle M. W. (2009) Landscape characteristics of a stream and wetland mitigation banking program. *Ecological Applications* 19: 2078–2092.

BenDor, T., Lester, T. W., Livengood, A., Davis, A. and Yonavjak, L. (2015) Estimating the size and impact of the ecological restoration economy. *PLoS ONE* 10(6): e0128339.

Bronner, C. E., Bartlett, A. M., Whiteway, S. L., Lambert, D. C., Bennett, S. J. and Rabideau, A. J. (2013) An assessment of US stream compensatory mitigation policy: necessary changes to protect ecosystem functions and services. *JAWRA Journal of the American Water Resources Association* 49(10): 449–462.

Chase, A. (1987). *Playing God in Yellowstone*. Orlando, FL: Harcourt Brace and Company.

Clewell, A. and Aronson, J. (2011) *Ecological Restoration: Principles, Values and Structure of an Emerging Profession*. Washington, DC: Island Press.

Daniel, T., Muhar, A., Arnberger, A., Aznar, O., Boyd, J., Chan, K., Costanza, R., Elmqvist, T., Flint, C., Gobster, P., Grêt-Regamey, A., Lave, R., Muhar, S., Penker, M., Ribe, R., Schauppenlehner, T., Sikor, T., Soloviy, I., Spierenburg, M., Taczanowska, K., Tam, J. and von der Dunkj, A. (2012) Contributions of cultural services to the ecosystem services agenda. *Proceedings of the National Academy of Sciences USA* 109(23): 8812–8819.

EPA (2014) *US Environmental Protection Agency Great Lakes Restoration Initiative Report to Congress and the President*. Washington, DC: Environmental Protection Agency.

Friederici, P. (2006) *Nature's Restoration: People and Places on the Front Lines of Conservation*. Washington, DC: Island Press.

Gómez-Baggethun, E., Rudolf de Groot, R., Lomas, P. and Montes, C. (2010) The history of ecosystem services in economic theory and practice: from early notions to markets and payment schemes. *Ecological Economics* 69(6): 1209–1218.

Jordan, W. and Lubick, G. (2011) *Making Nature Whole*. Washington, DC: Island Press.

Kobziar, L. N., Rocca, M. E., Dicus, C. A., Hoffman, C., Sugihara, N., Thode, A. E. and Morgan, P. (2009) Challenges to educating the next generation of wildland fire professionals in the United States. *Journal of Forestry* 107(7): 339–345.

Moseley, C. and Nielsen-Pincus, M. (2009) *Economic Impact and Job Creation from Forest and Watershed Restoration: A Preliminary Assessment*. Eugene, OR: Institute for a Sustainable Environment.

US Census Bureau (2013) Service annual survey historical data. Retrieved from www.census.gov/services/sas/historic_data.html (accessed 24 June 2015).

Vigmostad, K., Mays, N., Hance, A. and Cangelosi, A. (2005) *Large-Scale Ecosystem Restoration: Lessons for Existing and Emerging Initiatives*. Washington, DC: Northeast Midwest Institute.

Wagmer, B. and Shropshire, R. (2009) *An Estimation of the Economic Impacts of Restoration in Montana*. Helena, MT: Montana Research and Analysis Bureau.

Zipper, C. E., Burger, J. A., Skousen, J. G., Angel, P. N., Barton, C. D., Davis, V. and Franklin, J. A. (2011) Restoring forests and associated ecosystem services on Appalachian coal surface mines. *Environmental Management* 47(5): 751–756.

31

RESTORATION AND
MARKET-BASED INSTRUMENTS

Alex Baumber

Market-based instruments (MBIs) have become increasingly prevalent in the environmental policy sphere in recent decades and their application to ecological restoration reflects this global trend. MBIs can take a variety of forms, from simple grants through to complex offsetting and trading schemes. When implemented carefully, they can allow providers of ecological restoration services to capture a greater share of the economic benefits produced by their projects, as well as attracting new sources of investment into ecological restoration. However, MBIs also bring with them the risk that the diverse range of ecosystem services and functions provided by restoration activities may be commodified, over-simplified or traded-off against environmental degradation at other locations. This chapter explores the various classes of MBIs, the extent to which they have been used to promote ecological restoration and the advantages and disadvantages they offer.

MBIs cover a broad range of policy instruments including grants, subsidies, taxes, charges, penalties, certification programmes and tradable permit schemes. Their unifying feature, according to the OECD (2007), is that they seek to address market failures relating to 'environmental externalities'. This term refers to the environmental costs or benefits related to an action that are not experienced directly by those undertaking the action and are not captured in traditional markets for goods and services. In the case of ecological restoration, the externalities in question primarily consist of the environmental benefits from restoration projects that may be felt well beyond the immediate restoration site, such as biodiversity conservation, soil protection, carbon sequestration and the provision of clean drinking water. These benefits are rarely captured in traditional markets for goods and services and, as a result, do not flow back to those undertaking restoration activities in the form of economic returns. MBIs have the potential to address this market failure and provide incentives to undertake further restoration work.

MBIs are often viewed as a more efficient alternative to 'command and control' measures. For example, a tax on greenhouse gas emissions or the creation of a market to trade emission permits may be promoted as a more efficient alternative to an inflexible emissions cap being placed on every enterprise. A key economic principle behind such arguments is that enterprises are better placed than the government to determine the most cost-effective way to reduce their impact. Indeed, where markets for tradable permits have been created, an enterprise may determine that the most cost-effective way to reduce their impact is to continue emitting while paying someone else to stop (or to provide an offset by sequestering carbon in trees or soil). However, when looking at MBIs from the perspective of ecological restoration, the argument that they are a more efficient alternative to command-and-control measures has limited relevance. This is because governments rarely compel landholders to undertake restoration activities through regulation (except in limited cases such as mine-site restoration). Instead, the use of MBIs for ecological restoration tends to focus more on factors such as enhancing the cost-effectiveness of the limited pools of funds available for restoration work, the creation of economic incentives to undertake environmentally-beneficial activities and the potential for restoration activities to offset environmental damage elsewhere.

This chapter will progress from the simplest forms of MBIs, such as government grants, to the more complex market-based arrangements that can be used to direct payments to the providers of restoration services. Most of the measures discussed in this chapter fall under the category of PES – payments for ecosystem (or environmental) services (Wunder 2005; OECD 2010). However, this chapter also considers options that do not strictly qualify as PES, such as penalties for failing to restore degraded landscapes and incentives to design production systems that combine restoration with commercial harvest.

Simple MBIs – grants, penalties and taxation approaches

Government grants for restoration projects are common in many countries. These offer a simple way to incentivize restoration activities and compensate those undertaking them for the public benefits (i.e. positive externalities) they provide. Grants may be provided by municipal, state or provincial governments, along with schemes operating at a national level, such as the National Landcare Programme in Australia or the various restoration programmes run by the Environmental Protection Agency (EPA) and the Fish and Wildlife Service (FWS) in the United States. Financial incentives to undertake restoration can also be provided through the taxation system, such as the tax concessions permitted for the creation of Voluntary Conservation Easements in the USA.

Grants programmes may also be run by non-government organizations (NGOs), with this option being very common in the USA, where a wide range of foundations offer grants aimed at local areas or specific habitat types. In developing countries, international NGOs such as World Wildlife Fund (WWF) provide an important source of funding for restoration projects, along with inter-governmental agencies such as the Global Environmental Facility (GEF) operated by the United Nations Development Programme (UNDP).

Generally speaking, grants programmes are aimed at voluntary restoration activities and aim to cover some or all of the costs involved. They vary in terms of whether they are intended to assist only with the direct costs of restoration activities or to also cover the opportunity costs of taking land out of agricultural production or, in some cases, to generate a profit for the landholder. Depending on the programme, there may be specific rules about what kinds of costs can be covered using programme funds (e.g. materials and labour), and which cannot (e.g. administrative and opportunity costs).

Apart from grants and tax breaks, a range of other simple incentives can be used to encourage landholders to protect or restore ecosystems, including access to credit and increased security of tenure. In Brazil, access to agricultural credit and insurance has been used to provide an incentive for Amazon landholders to comply with forest protection laws (Butler 2011). Enhanced security of land tenure can also act as an incentive in situations where tenure is insecure, with an example being the Sumberjaya pilot programme aimed at watershed protection in Indonesia (OECD 2010). These types of non-monetary incentives can often have a significant impact on landholder decision-making and complement the monetary incentives offered under grants programmes or tax breaks.

An alternative to using grants, tax breaks or other incentives to encourage voluntary restoration is the use of involuntary negative incentives such as taxes, fines or other financial penalties to compel certain stakeholders to restore ecosystems. This option is suitable only where restoration is a regulatory requirement or a condition that has been placed on a development approval. Restoration bonds used in the mining sector are a notable example of this approach, with the bond acting as both an incentive for the mining company to restore land to an acceptable condition as well as a source of funds to correct any damage if the company fails to comply.

Enhancing cost-effectiveness – reverse auctions

Grant applications are commonly assessed by a panel or committee who must decide which of the many applications they receive represent the best value for money. Considerable research has focused on ways to enhance the cost-effectiveness of public investment in restoration activities (e.g. Pannell 2008; Crossman and Bryan 2009) and MBIs have the potential to assist with this goal. A particular challenge for grants programmes is 'information asymmetry', whereby landholders bidding for grants have a better understanding of the true costs of restoration activities than the government agencies assessing them, which can lead to bids being inflated above the minimum level that the landholder would be willing to accept (OECD 2010).

Auction approaches offer a means of overcoming the risks posed by information asymmetry and improving the cost-effectiveness of grants for ecological restoration. One prominent example of a grants scheme that employs an auction approach is the BushTender programme in the Australian state of Victoria. This process is more accurately described as an inverse or reverse auction. Unlike a traditional auction, where multiple bidders compete to purchase a single item, the BushTender process involves multiple providers of restoration services competing for a fixed pool of government funds (DEPI 2014). Landholders offer to protect and restore areas of remnant native vegetation, with these competing bids being rated in terms of their likely biodiversity benefit relative to the cost (i.e. the amount of funding requested by the bidder). Bids with the highest cost-effectiveness ratings receive the limited funds available.

The metric used in the BushTender scheme to express the predicted benefit from each bid is known as the Biodiversity Benefit Index (BBI), which has a maximum score of 100 per cent that takes into account the proposed management practices and the regional conservation significance of the site. The predicted gain in BBI is multiplied by the area of the proposed site to provide a predicted gain in terms of 'habitat hectares' (Figure 31.1). For example, a 100 hectare site that is managed in such a way as to improve its BBI from 50 to 70 per cent (i.e. a gain of 20 per cent) would result in an overall gain of 20 habitat hectares. Proposals are ranked for cost-effectiveness based on the funds requested and their predicted gain. This then links to a vegetation quality assessment method which is able to monitor the actual change in 'habitat hectares' over time by comparing the site to a benchmark based on a mature, long-undisturbed site of the same vegetation type, taking into account factors such as landscape context and the presence of large trees, understorey plants and logs (DSE 2004).

Cost-effectiveness is promoted under a reverse auction approach due to the competitive nature of the bidding process and the fact that the bidders do not know what level of cost-effectiveness will be required to win funding. In theory, this should reduce the likelihood that bidders will ask for more than the minimum level they are willing to accept, for fear of missing out to a more competitive bidder. The OECD (2010) analysed a number of case studies

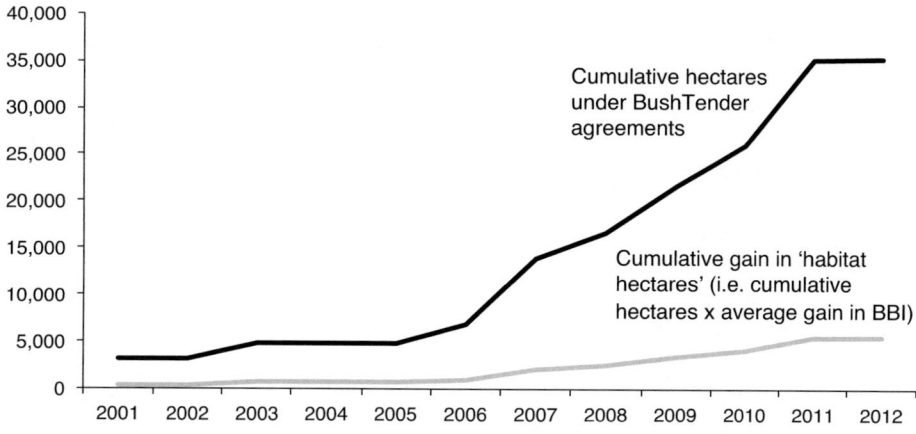

Figure 31.1 Cumulative hectares under BushTender agreements and gain in habitat hectares
2001–2012

Source: DEPI (2014)

where reverse auctions have been used to distribute environmental grants, including the Tasmanian Forest Conservation Fund in Australia, the Conestoga watershed protection scheme in the USA and the Sumberjaya watershed pilot in Indonesia. They found a strong case for reverse auctions enhancing the cost-effectiveness of grants programmes, including a seven-fold increase in phosphorous reduction per dollar spent in the Conestoga example and a 52 per cent cost-effectiveness gain in the case of the Tasmanian Forest Conservation Fund compared to allocating grants on a 'first-come first-served' basis. The Sumberjaya pilot programme in Sumatra, Indonesia, is notable for the fact that it was NGO-funded (World Agroforestry Centre) and that it involved active revegetation rather than simply the protection of remnant vegetation.

A reverse auction approach is also central to one of the most prominent conservation programmes aimed at agricultural landscapes – the Conservation Reserve Programme (CRP) in the USA. The CRP is aimed at taking highly erodible and environmentally sensitive crop-land out of production and contributes to ecological restoration through reduced soil disturbance, reduced chemical use and re-establishment of grasses and trees. A reverse auction is used to select the vast majority of participating CRP land under the general sign-up process, which weighs up competing bids using an Environmental Benefits Index (EBI).

A recent review by the US Department of Agriculture found that the use of auctions to distribute conservation payments can be more effective in terms of reducing costs and maximizing environmental benefits than other mechanisms, such as offering a single fixed price to landholders (Hellerstein *et al.* 2015). However, they also suggested reforms to some elements of the CRP, particularly around the use of bid caps, which are designed to prevent landholders making excessive profits. Easing restrictions on how grant money may be used (e.g. for direct costs vs profits) has the potential to make a scheme more attractive to entrepreneurial land-holders who can provide cost-effective restoration for a profit, but who would not apply if the scheme only covers a portion of direct costs.

The CRP provides a notable example of a grants scheme that is explicitly designed to cover the opportunity costs of taking land out of agricultural production. This reflects the fact that a key goal behind the development of the CRP was to support farmer incomes by simultaneously providing an alternative income source and reducing the US farm surpluses that were putting downward pressure on crop prices. The European Union's Common Agricultural Policy (CAP) is another example of a scheme developed to protect farmer incomes through subsidies and the 'setting-aside' of farmland. Historically, the CAP has not had the same focus on environmental objectives that the CRP has had in the US, but recent reforms have made 'restoring, preserving and enhancing biodiversity' a specified aim of the CAP. This includes the exploration of new market-based approaches such as a pilot programme to link landholder payments directly to measurable improvements in habitat quality and biodiversity (European Commission 2014).

A controversial aspect of auction approaches is the need to weigh up competing bids on a common scale, such as the Environmental Benefits Index (EBI) of the CRP or the Biodiversity Benefit Index (BBI) of the BushTender programme. Assessing all bids on a common scale requires that diverse outcomes relating to biodiversity, soils and water must be weighed against one another, potentially disadvantaging projects with unique outcomes that cannot easily be compared to other projects. One solution to this problem under the CRP is to allow a relatively small number of sites with unique characteristics to join through a non-competitive continuous sign-up process, which is aimed at protecting land with the greatest conservation value, regardless of whether such sites would rank highest in terms of cost-effectiveness.

Other challenges around auction approaches include the risk that the predicted benefits will never be realized, the risk that offering payments will deter voluntary action that would have taken place without any payment (known as 'crowding-out') and the risk that payments will be made for projects that would have happened anyway (Hellerstein *et al.* 2015). This latter problem may be referred to as a failure to ensure the 'additionality' of conservation projects receiving funding and can reduce the cost-effectiveness of an auction scheme. Ensuring that predicted benefits are realized can require expensive monitoring and verification processes, as well as mechanisms for rescinding payments in cases of non-compliance.

Tradable permit and offset schemes

Grants for ecological restoration, whether they involve an auction approach or not, fall under the broad category of payments for environmental (or ecosystem) services (PES). According to Wunder (2005), the criteria for PES are that the arrangement is voluntary, involves at least one 'seller' and one 'buyer', and is conditional on the delivery of a well-defined environmental service (or land use activity likely to secure that service). The examples discussed so far mostly follow the model of a single buyer (generally a government agency) paying a range of sellers for the public benefits that result from their ecosystem management. However, PES schemes can also be designed in such a way as to allow multiple buyers to compete for ecosystem services and for the benefits to be privately rather than publicly owned.

In theory, PES approaches that involve multiple buyers and multiple sellers should result in more efficient allocation of resources by enhancing competition. However, from the perspective of those planning ecological restoration activities, they also offer another key benefit – an alternative funding source that sidesteps the traditional reliance on grants from government agencies or environmental NGOs. Harnessing the capacity of businesses and wealthy individuals to pay for the services they derive from managed ecosystems offers the potential to greatly expand the pool of funding available for restoration activities.

PES schemes can operate according to either the 'beneficiary pays' or the 'polluter pays' principle. The most common beneficiaries involved in making payments under beneficiary pays schemes are government agencies (on behalf of the public), but some PES schemes have been successful at encouraging other beneficiaries to pay for ecosystem services as well. Costa Rica in particular has become well known internationally for its PES model, which has succeeded in directing voluntary payments from private companies (mostly hydroelectric plants) to landholders managing land for watershed protection, biodiversity conservation, carbon sequestration and landscape beauty (Porras *et al.* 2013). The demand in this case stems from a desire by corporations to be seen as socially responsible. A system of certificates for ecosystem services enables efficient over-the-counter transactions rather than having to rely on costly and time-consuming one-on-one negotiations between companies and landholders. While the main impetus behind Costa Rica's embrace of PES was a desire to slow deforestation rates (resulting in 860,000 ha of forest being protected between 1997 and 2012), the programme has also resulted in the active reforestation of 60,000 ha and the natural regeneration of another 10,000 ha (*ibid.*).

When it comes to polluter pays schemes, the most notable options with implications for ecological restoration are tradable offset schemes involving carbon and biodiversity. Carbon trading schemes generally operate by requiring emitters of greenhouse gases to hold permits covering their emissions, with additional permits or credits able to be purchased from landholders who sequester carbon through restoration activities. Biodiversity offsets involve developers being permitted to clear or degrade ecosystems provided that they restore a commensurate ecosystem elsewhere. The key pre-requisites for promoting restoration activities through a tradable offset scheme are:

1   demand for credits to offset environmentally damaging activities, which may be created by a regulatory requirement or voluntary decisions by businesses to offset their impacts;
2   a system to verify that restoration projects are able to provide the required ecosystem services (e.g. carbon sequestration or habitat value) and award credits accordingly; and
3   a market mechanism to allow trading to take place between those providing the environmental services and those wishing to undertake damaging activities.

Australia provides an example of a country that has experimented with a variety of MBIs involving carbon and restoration over recent years, as shown in Box 31.1. Ironically, rather than following a progression from simpler to more complex schemes over time, the trend in Australia has been the opposite due to political considerations.

In addition to national schemes, such as the example discussed in Box 31.1, the United Nations Framework Convention on Climate Change (UNFCCC) also provides for international carbon trading under the Clean Development Mechanism (CDM) and Joint Implementation (JI) provisions of the Kyoto Protocol. The CDM offers the potential for investment money to flow from developed to developing countries for reforestation and afforestation projects. However, out of more than 1600 CDM projects created by 2010, only four were for reforestation or afforestation, with Thomas *et al.* (2010) arguing for CDM reforms to provide greater flexibility, simpler methodological and documentation procedures and a switch in focus from adjudicating to facilitating CDM reforestation projects.

In the case of biodiversity offsets, the demand stems from developers wishing to undertake environmentally-damaging activities that would not ordinarily be permitted under biodiversity protection legislation. While there may be a loss of biodiversity at the development site, the use of offsets is designed to ensure that there is 'no net loss' overall. An example is the BioBanking

Box 31.1 The evolution (or regression) of marked-based instruments for carbon in Australia

In the lead-up to the 2007 federal election, a bipartisan political consensus emerged that Australia should employ a market-based cap-and-trade approach to reducing greenhouse gas emissions in line with its commitments under the Kyoto Protocol. This scheme, which came to be known as the Carbon Pollution Reduction Scheme (CPRS), would have placed emissions caps on large emitters but allowed trading between them, such that those with excess emission permits could sell them to those wishing to increase emissions. Alternatively, emitters could offset their emissions by purchasing offsets from reforestation projects.

The CPRS was progressed by the newly-elected Labor Government through a 2008 White Paper and 2009 negotiations with the opposition Liberal/National coalition, almost making it through parliament before the coalition switched leaders to the anti-CPRS Tony Abbott. Further progress was delayed until after the 2010 election, when a new parliamentary balance allowed Labor to negotiate a revised carbon pricing model. Unlike the CPRS, which involved placing caps on emitters but letting the price of permits 'float' according to demand, the revised model placed no caps on emitters but instead required them to pay a fixed price for permits to the Government (i.e. a 'carbon tax'). This price was set initially at AU\$23 per tonne of CO_2-equivalent and was set to rise to AU\$25.40/tCO_2-e within three years before transitioning to a floating price (Commonwealth of Australia 2011). However, this transition to a floating price was never realized, as the scheme was scrapped after the Abbott-led coalition won the 2013 election on a platform of 'scrapping the carbon tax'.

In terms of market-based instruments, the change from the CPRS to the carbon tax represented a simplification from a multiple buyer/multiple seller model to one in which there were multiple buyers of permits but only one seller (the Government) and the price was fixed. However, despite the lack of a competitive market for emissions permits, a competitive market for offsets was created to complement the carbon tax. Under this arrangement, multiple providers of offsets were able to sell to multiple emitters wishing to reduce their carbon tax liability for whatever price the two parties agreed on (with the carbon tax acting as the effective maximum price for offsets). Reforestation and revegetation projects were able to earn offset credits under the Carbon Farming Initiative (CFI), which recognized the sequestration value of eligible activities following approved methodologies, such as permanent environmental plantings, human-induced regeneration and farm forestry.

The abolition of the carbon tax in 2014 represented a retreat from placing either emissions caps or taxes on emitters. However, it did not result in the total abandonment of market-based approaches, as the newly-created Emissions Reduction Fund (ERF) employed a reverse auction approach to distribute Government funds to providers of emission reductions or sequestration. Importantly (from a restoration perspective), the ERF incorporates the key elements of the CFI, allowing reforestation and regeneration projects to be eligible for ERF payments. Indeed, in the initial ERF auction in April 2015, sequestration projects represented around 60 per cent of the 47 million tonnes of abatement purchased by the Australian Government (Clean Energy Regulator 2015).

Both sides of politics in Australia have argued that their preferred model is the most efficient option for reducing greenhouse emissions at the lowest cost. While the transition from cap-and-trade to carbon tax to reverse auction may not be what most advocates of market-based instruments would anticipate or recommend, a commitment to some form of market-based approach has been an enduring element of Australia's climate change policy in the period 2007–2015.

scheme in place in the Australian state of New South Wales (NSW). After NSW strengthened its regulations on native vegetation clearing in 2005, the BioBanking scheme was introduced to enable developers to clear or degrade vegetation for particular projects – provided that biodiversity outcomes were enhanced elsewhere. Plant regeneration is one of the activities that can be used to generate biodiversity credits under the scheme, along with controlling grazing, retaining fallen timber, managing fire and controlling pests and weeds (Department of Environment and Climate Change 2008). Other jurisdictions that have implemented biodiversity offset schemes include the USA, which pioneered the 'no net loss' concept for wetlands in the 1970s, and Brazil, which allows landholders to use offsets to meet their requirements for retaining forested habitat (Doswald *et al.* 2012).

The advantages and disadvantages of tradable offsets for carbon or biodiversity depend on the perspective from which they are viewed. For businesses facing restrictions on carbon pollution or land clearing, they offer a cheaper and more flexible approach than having to comply with hard regulatory limits on their activities. The broader economy may also benefit from the lower cost of compliance and this may in turn make it politically more feasible to tighten caps in future years. An example of this is the European Union's emission trading scheme, in which the lower-than-anticipated costs of abatement in Phase I made it easier to convince member states to tighten their emission caps in subsequent phases.

From the perspective of those planning restoration projects, tradable offsets represent a potential new source of funding, but it is one that can bring with it a number of challenges. One key challenge is ensuring equivalence between the damaging activity and the restorative one. It is easier to make a case for equivalence in relation to carbon trading, as the Earth's atmosphere is an interconnected global commons and the locations at which CO_2 is added or removed is not particularly important. However, this is not the case for biodiversity outcomes, which are very much dependent on the location at which habitat restoration occurs. Furthermore, the complex and imprecise nature of biodiversity science can make it challenging for offset schemes to appropriately value biodiversity outcomes (Burgin 2008).

Biodiversity offsetting schemes may attempt to ensure equivalence through a complex set of rules. For example, under the NSW BioBanking scheme, developers wishing to destroy habitat receive BioBanking statements that detail not only the number of credits that must be surrendered to offset the habitat destruction, but also the type of credit required (ecosystem or species credits) and the vegetation types in which those credits can be generated. Offset ratios also vary between projects, with the clearing of certain habitat types requiring the protection or restoration of an area several times larger. However, despite these measures, Gibbons and Lindenmayer (2007) suggest that biodiversity offsets are likely to be successful in achieving 'no net loss' only in circumstances where clearing is restricted to relatively simple vegetation types, and where time lags between destruction and regeneration of habitat do not represent a significant risk.

While carbon offsets face lesser concerns around equivalence than biodiversity offsets, they can present difficult choices for restoration providers in terms of balancing the goal of carbon sequestration (which is valued in the carbon offset market) and other goals relating to biodiversity or other benefits (which may be desired by the project planners but have no market value). Focusing only on how carbon sequestration can be maximized may lead to monocultures of single-species, single-age plantations that comply with Kyoto rules but offer little in the way of habitat value. Furthermore, the issues of 'additionality' and 'crowding out', which were discussed in relation to auctions, also represent a key challenge for tradable offset schemes.

The latest frontier in the establishment of offset markets for environmental services is land degradation. Under the framework of the United Nations Convention to Combat

Desertification and the Sustainable Development Goals to be introduced in 2015, targets have emerged around 'zero net land degradation' or 'land degradation neutrality'. This has clear similarities with the 'no net loss' provisions that underpin biodiversity offset schemes in countries such as the USA and Australia. While the development of a scheme for international land degradation offsets presents a potential opportunity to direct funding from activities that degrade soil fertility to those that restore them, it faces many of the challenges faced by other PES schemes involving tradable offsets. These include ensuring the reliability of trades, defining clear quantifiable units of measure, ensuring equivalence across a wide range of land types and the risk of time lags or delayed benefits (Tal 2015).

One advantage for new schemes around land degradation or other issues is access to the considerable body of literature that has built up over the past decade providing policy advice on the implementation of PES schemes, including guides published by the Centre for International Forestry Research (Wunder 2005; Fripp 2014) and the Department for Environment, Food and Rural Affairs in the UK (DEFRA 2013).

Combining restoration and commercial production

The final group of MBIs to be explored in this chapter do not involve the creation of new markets in environmental services but rather focus on existing markets and seek to promote production systems that jointly deliver commercial products and environmental services. A number of different terms may be used to describe this basic concept in different contexts, including multifunctionality (OECD 2001) and conservation through sustainable use (Baumber *et al.* 2012).

One mechanism for giving preference to products that provide associated environmental benefits is certification against an industry sustainability standard. Figure 31.2 shows the logos of three prominent sustainability certification schemes operating in different industry sectors. These are the Forest Stewardship Council (FSC), which is prominent in the forestry sector, the Rainforest Alliance, which utilizes the standards of the Sustainable Agriculture Network (SAN), and the Roundtable on Sustainable Biomaterials (RSB), which has developed a set of standards for use in the biofuel sector.

The three standards used as examples here differ somewhat in the emphasis they place on different issues. The FSC standards have a strong focus on preventing over-harvesting of forests and forest clearing for plantation establishment. The RSB standards include provisions on life-cycle greenhouse gas emissions and food security, both of which have been prominent issues in the biofuel sector. The SAN standards have the strongest focus on social factors such as workers' rights, reflecting the position of the Rainforest Alliance as a key advocate for 'fair trade'.

Just as the FSC, RSB and SAN standards differ in their approach to environmental and social protections, they also differ in the degree to which they promote ecological restoration activities. For the most part, all three standards follow a benchmark of 'maintain or enhance' when it comes to environmental values, but there are some notable provisions that require land managers to undertake active restoration. This is particularly the case for the SAN standards, which require plantations or farms to:

- 'establish and maintain vegetation barriers between the crop and areas of human activity' (SAN 2010: 20);
- 'dedicate at least 30% of the farm area for conservation or recovery of the area's typical ecosystems' (*ibid.*); and

Figure 31.2   Trademarks that may be displayed on products certified by the Rainforest Alliance, Forest Stewardship Council and the Roundtable on Sustainable Biomaterials

Source: reproduced by kind permission of the Rainforest Alliance, Forest Stewardship Council and the Roundtable on Sustainable Biomaterials

- 'use and expand its use of vegetative ground cover to reduce erosion and improve soil fertility' (*ibid.*: 42).

The previously mentioned provisions represent an attempt to harness consumer demand for fairly-traded and environmentally-friendly agricultural products to promote ecological restoration outcomes. However, they are largely based on the notion that restoration areas are separate to production areas. In contrast, examples can also be found of production systems in which production and conservation outcomes are more closely integrated, such as:

- The cork oak forests of the western Mediterranean, which have been shaped by human management over a variety of spatial and temporal scales to create diverse mosaic habitats that support endangered species such as the Iberian lynx, the Iberian imperial eagle and the Barbary deer (WWF 2006; Urbieta and Marañón 2008).
- The damar agroforests of Sumatra, which are planted ecosystems based around the damar tree (*Shorea javanica*), and provide not only resin for the production of incense, varnish, paint and cosmetics, but also offer a range of environmental services and act as a buffer for the World Heritage-listed Bukit Barisan Selatan National Park (Kusters *et al.* 2008).
- Short-rotation cropping of poplar and willow in Europe for bioenergy, which has been shown to increase soil organic matter, improve water quality and enhance biodiversity (Simpson *et al.* 2009), as well as filtering wastewater (Schroeder 2012) and removing metals such as cadmium and zinc from contaminated soils (Laureysens *et al.* 2005).
- Mallee eucalypts that have been being trialled as short-rotation crops in Western Australia to produce bioenergy, eucalyptus oil and other products, while helping to mitigate dryland salinity (Stucley *et al.* 2012).

In cases where commercial production and the provision of environmental services are strongly connected, market-based initiatives that promote the commercial product may also help to promote the associated environmental services. The two bioenergy-based examples discussed previously are of particular interest due to the global proliferation of market-based incentives for the production of liquid biofuels and other forms of bioenergy. Table 31.1 lists a number of different policy instruments that have been used to promote bioenergy, along with examples of where they have been used and ways in which they could be modified to incorporate a preference for feedstock production systems that offer associated environmental benefits.

The kinds of land use options discussed earlier require compromises between environmental and economic objectives and raise the question of whether they should be characterised as 'ecological restoration'. To some, ecological restoration should be aimed at restoring 'naturalness' and be designed to 'compensate for human influence on an ecological system in order to return the system to its historic condition' (Jordan 1994: 32). To others, the very idea of naturalness is subjective and problematic. Lindenmayer *et al.* (2008: 82) argue that human perspectives will inevitably differ on what constitutes appropriate vegetation structure and condition and that, in landscapes long influenced by humans, 'naturalness may not even be an appropriate characteristic to consider'. Similarly, Australia's 2006 State of the Environment Report emphasizes that successful restoration may require that 'absolute concepts of naturalness be abandoned in favour of management for specific objectives' (Beeton *et al.* 2006: 44).

Establishing plantations for a combination of commercial production and ecosystem enhancement may not fit within everyone's vision of ecological restoration. However, it is important to recognize that all forms of restoration require the prioritization of certain ecosystem attributes over others, either explicitly or implicitly. Restoration goals may revolve around

Table 31.1 Bioenergy support measures with the potential to promote environmental services

Policy option	Example	Potential modifications to promote environmental services
Tax breaks for biofuel producers	Brazil's biodiesel support scheme, which offers larger tax breaks for 'social fuel' that comes from small family farmers	While Brazil's scheme seeks to deliver a social benefit, a similar model could be used to preference production systems with restoration outcomes
Mandates requiring the use or supply of biofuels	EU Renewable Energy Directive, which provides greater support to fuels from non-food cellulosic crops through a system of 'double-counting'	A similar model of multiple-counting could be used to preference production systems with restoration outcomes
Mandates requiring the supply of renewable electricity (including bioenergy)	UK Renewables Obligation, which includes 'banding' that provides higher levels of support for certain options (e.g. energy crops)	Similar to multiple-counting under biofuel mandates, the level of support for biomass crops for electricity could be based on the environmental services provided
Feed-in tariffs requiring electricity companies to pay a fixed price for bioelectricity	German feed-in tariffs, which have incorporated a bonus for biomass from land managed for landscape preservation	Higher feed-in tariffs could be applied to biomass crops with restoration outcomes

the enhancement of one particular ecosystem attribute or function, such as erosion control, salinity mitigation or habitat provision, or they may involve the enhancement of multiple ecosystem attributes simultaneously. Any policy measures that are aimed at delivering on-ground environmental outcomes as a co-product of a commercial production system need to give careful consideration to which ecosystem functions should be prioritized over others and how to ensure these outcomes are not compromised by commercial pressures to maximize production.

Conclusion

As the range of schemes offering payments for ecosystem services and other market-based instruments continues to expand, more and more ecological restoration activities are likely to be established or modified in accordance with the incentives offered by these schemes. This presents an opportunity to increase the cost-effectiveness of restoration spending and to increase the funding available for restoration projects, but it also brings with it risks that certain projects will be compromised, simplified, under-valued or traded off against environmental destruction elsewhere. These opportunities and risks are likely to multiply as schemes progress in complexity from simple grants to single-buyer markets to markets involving multiple buyers and sellers, such as tradable offset markets around carbon, biodiversity or land degradation.

Markets for restoration services may be able to internalize some of the environmental externalities that currently go unvalued in traditional markets, but it is unlikely they will ever be able to value all of the outcomes that restoration can offer, at least not to the satisfaction of all stakeholders. Controversy around MBIs is largely unavoidable and stems from an inherent conflict between the diverse and often unique outcomes that restoration projects provide and the market requirement that outcomes be substitutable. Unlike commodities like wheat or oil, the outcomes of ecological restoration projects are context-specific and cannot be loaded onto ships and traded across the globe. Every restoration project produces a unique combination of outcomes for biodiversity, soils, water and climate that operate across a variety of scales and will be valued differently by different stakeholders.

A key challenge that will always remain for MBIs is striking a balance between having sufficient substitutability to keep a market functioning while recognizing the inherent differences between restoration projects in different contexts. However, debating how this balance should be struck and how MBIs could be improved need not stand in the way of providers of restoration services capitalizing on the opportunities that MBIs can provide. In many cases, it may not matter much to those undertaking restoration projects whether the scheme that has been set up is the most efficient one possible or whether the outcomes at one site are perfectly substitutable for those at another. Instead, what is likely to matter more is whether the scheme has created additional incentives for restoration and made additional sources of funding available. As shown in this chapter, many MBIs around the world have shown the capacity to do this – even if further work could be done to better align the incentives they provide, reduce barriers to participation and reduce the risk of perverse outcomes.

References

Baumber, A., Merson, J., Diesendorf, M. and Ampt, P. (2012) Revegetation, bioenergy and sustainable use in the New South Wales central west, in J. Merson, R. Cooney and P. Brown (eds), *Conservation in a Crowded World: Case Studies from the Asia-Pacific*, pp. 186–204, NewSouth Publishing, Sydney.

Beeton, R. B., Buckley, K. I., Jones, G. J., Morgan, D., Reichelt, R. E. and Trewin, D. (2006) *Australia: State of the Environment 2006*, Department of the Environment and Heritage, Canberra.

Burgin, S. (2008) BioBanking: an environmental scientist's view of the role of biodiversity banking offsets in conservation, *Biodiversity and Conservation*, 17, 807–816.

Butler, R. (2011) *Could Palm Oil Help Save the Amazon?* Retrieved from http://news.mongabay.com/2011/0614-amazon_palm_oil.html (accessed 14 April 2015).

Clean Energy Regulator (2015) Auction results: April 2015, retrieved from www.cleanenergyregulator.gov.au/ERF/Published-information/auction-results/auction-results-april-2015 (accessed 22 June 2015).

Commonwealth of Australia (2011) *Securing a Clean Energy Future: The Australian Government's Climate Change Plan*, Department of Climate Change and Energy Efficiency, Canberra.

Crossman, N. D. and Bryan, B. A. (2009) Identifying cost-effective hotspots for restoring natural capital and enhancing landscape multifunctionality, *Ecological Economics*, 68, 654–668.

DEFRA (2013) *Payments for Ecosystem Services: A Best Practice Guide*, Department for Environment, Food and Rural Affairs, London.

Department of Environment and Climate Change (2008) *BioBanking Assessment Methodology*, Department of Environment and Climate Change, NSW Government, Sydney.

DEPI (2014) *BushTender*, retrieved from www.depi.vic.gov.au/environment-and-wildlife/environmental-action/innovative-market-approaches/bushtender (accessed 21 April 2015).

Doswald, N., Barcellos Harris, M., Jones, M., Pilla, E. and Mulder, I. (2012) *Biodiversity Offsets: Voluntary and Compliance Regimes – A Review of Existing Schemes, Initiatives and Guidance for Financial Institutions*, UNEP-WCMC, Cambridge.

DSE (2004) *Native Vegetation: Sustaining a Living Landscape, Vegetation Quality Assessment Manual – Guidelines for Applying the Habitat Hectares Scoring Method Version 1.3*, State of Victoria, Department of Sustainability and Environment, East Melbourne.

European Commission (2014) *Call for Proposals: Pilot on-Farm Projects to Test Result-Based Remuneration Schemes for the Enhancement of Biodiversity*, European Commission, Brussels.

Fripp, E. (2014) *Payments for Ecosystem Services (PES): A Practical Guide to Assessing the Feasibility of PES Projects*, CIFOR, Bogor, Indonesia.

Gibbons, P. and Lindenmayer, D. B. (2007) Offsets for land clearing: no net loss or the tail wagging the dog? *Ecological Management and Restoration*, 8, 26–31.

Hellerstein, D., Higgins, N. and Roberts, M. (2015) *Options for Improving Conservation Programmes: Insights From Auction Theory and Economic Experiments*, Economic Research Service, United States Department of Agriculture, Washington, DC.

Jordan, W. R. I. (1994) Sunflower forest, in A. D. J. Baldwin, J. De Luce and C. Pletsch (eds), *Beyond Preservation: Restoring and Inventing Landscapes*, pp. 17–34, University of Minnesota Press, Minneapolis, MN.

Kusters, K., Ruiz Perez, M., De Foresta, H., Dietz, T., Ros-Tonen, M. A. F., Belcher, B., Manalu, P., Nawir, A. A. and Wollenberg, E. (2008) Will agroforests vanish? The case of damar agroforests in Indonesia, *Human Ecology* 36, 357–370.

Laureysens, I., De Temmerman, L., Hastir, T., Van Gysel, M. and Ceulemans, R. (2005) Clonal variation in heavy metal accumulation and biomass production in a poplar coppice culture, II: vertical distribution and phytoextraction potential, *Environmental Pollution*, 133, 541–551.

Lindenmayer, D., Hobbs, R. J., Montague-Drake, R., Alexandra, J., Bennett, A., Burgman, M., Cale, P., Calhoun, A., Cramer, V., Cullen, P., Driscoll, D., Fahrig, L., Fischer, J., Franklin, J., Haila, Y., Hunter, M., Gibbons, P., Lake, S., Luck, G., MacGregor, C., McIntyre, S., Nally, R. M., Manning, A., Miller, J., Mooney, H., Noss, R., Possingham, H., Saunders, D., Schmiegelow, F., Scott, M., Simberloff, D., Sisk, T., Tabor, G., Walker, B., Wiens, J., Woinarski, J. and Zavaleta, E. (2008) A checklist for ecological management of landscapes for conservation, *Ecology Letters*, 11, 78–91.

OECD (2001) *Multifunctionality: Towards an Analytical Framework*, Organisation for Economic Co-operation and Development, Paris.

OECD (2007) *Business and the Environment Policy Incentives and Corporate Responses*, Organisation for Economic Co-operation and Development, Paris.

OECD (2010) *Paying for Biodiversity: Enhancing the Cost-Effectiveness of Payments for Ecosystem Services*, Organisation for Economic Co-operation and Development, Paris.

Pannell, D. J. (2008) Public benefits, private benefits, and policy mechanism choice for land-use change for environmental benefits, *Land Economics*, 84, 225–240.

Porras, I., Barton, D. N., Miranda, M. and Chacón-Cascante, A. (2013) *Learning from 20 Years of Payments for Ecosystem Services in Costa Rica*, International Institute for Environment and Development, London.

SAN (2010) *Sustainable Agriculture Standard, Version 4*, Sustainable Agriculture Network.

Schroeder, W. (2012) Capacity of poplar and willow clones to withstand high levels of wastewater application, presented at 24th Session of the International Poplar Commission, Dehradun, India, retrieved from www.fao.org/forestry/80796/en.

Simpson, J. A., Picchi, G., Gordon, A. M., Thevathasan, N. V., Stanturf, J. and Nicholas, I. (2009) *Short Rotation Crops for Bioenergy Systems: Environmental Benefits Associated with Short-Rotation Woody Crops*, IEA Bioenergy Task 30, International Energy Agency, Paris.

Stucley, C., Schuck, S., Sims, R., Bland, J., Marino, B., Borowitzka, M., Abadi, A., Bartle, J., Giles, R. and Thomas, Q. (2012) *Bioenergy in Australia: Status and Opportunities*, Bioenergy Australia, Killara NSW.

Tal, A. (2015) The implications of environmental trading mechanisms on a future Zero Net Land Degradation protocol, *Journal of Arid Environments*, 112, Part A, 25–32.

Thomas, S., Dargusch, P., Harrison, S. and Herbohn, J. (2010) Why are there so few afforestation and reforestation Clean Development Mechanism projects?, *Land Use Policy*, 27, 880–887.

Urbieta, I. R. and Marañón, M. A. Z. T. (2008) Human and non-human determinants of forest composition in southern Spain: evidence of shifts towards cork oak dominance as a result of management over the past century, *Journal of Biogeography*, 35, 1688–1700.

Wunder, S. (2005) *Payments for Environmental Services: Some Nuts and Bolts*, Center for International Forestry Research (CIFOR), Bogor, Indonesia.

WWF (2006) *Cork Screwed? Environmental and Economic Impacts of the Cork Stoppers Market*, WWF/MEDPO, Rome, Italy.

32

PROFIT MOTIVATIONS AND ECOLOGICAL RESTORATION

Opportunities in bioenergy and conservation biomass

Carol L. Williams

Purpose

Practitioners, researchers and policy-makers are undoubtedly aware of costs associated with ecological restoration and the challenges of limited budgets. They may not understand how for-profit entities in the private sector can be strong allies for achieving ecological restoration particularly at very large scales. This chapter provides a portrait of profit motivation as a positive influence in ecological restoration within managed and semi-natural ecosystems, based on practitioners' experience in North America. The chapter begins with a rationale for why and how private economic gain can motivate and support ecological restoration. Case studies are provided as illustrative examples and a brief discussion of commonalities and distinctions among them is provided to highlight influences and constraints on profit motivations in ecological restoration.

Introduction: the conservation–economy nexus

Imagine a broad, level plain in the mid-latitude, rain-fed interior of North America, where in early spring prairie soils have been tilled. Lying bare and exposed they await the annual planting of corn (*Zea mays*, L.) and soybean (*Glycine max*, L.). Imagine innumerable, precisely parallel rows of tilled earth extending to a point so distant they appear to converge on a far-off horizon. The rows are unpunctuated by hedge or fence, or stream tumbling overland. As far as the eye can see there are no trees, except one large shade oak spared from the saw. There are no livestock and no pasture upon which they might graze. At field entrances grass has been sown to fortify the soil against passage of heavy equipment such as 32-row planters, and 600-horsepower, air-conditioned, satellite-equipped combine harvesters. Although it is early spring the air is silent; there is no bird song, no hum of bees. There are no people. Everywhere you look you see a vastly simplified landscape that is perhaps the most productive agricultural system in history.

This vignette is a typical scene in central Iowa at the heart of the US Corn Belt (Figure 32.1). It has been made possible through heavy reliance on external inputs of mineral fertilizers,

Figure 32.1   Row-crop agriculture in central Iowa, USA, has greatly simplified landscapes compared to pre-settlement conditions

Source: photo by C. L. Williams

synthetic pesticides, genetically modified organisms and fossil fuels but such reliance impinges upon long-term sustainability. Although these systems have achieved productivity unimaginable among previous human generations, conventional row crop agriculture has had unintended consequences affecting the environment, life and livelihood with estimated costs in the billions of dollars annually (Pimentel *et al.* 1995; Rabotyagov *et al.* 2014). Moreover, preservation and management of native ecosystems, biodiversity and wildlife species in the context of such agricultural landscapes has come to be planned within a framework of 'land sparing' in which lands are set aside from agricultural production and agronomic intensity is increased on previously converted lands. There is a growing realization, however, that these approaches to ecological and environmental resource protection are insufficient (Matson and Vitousek 2006; Perfecto and Vandermeer 2010).

Agricultural productivity can be maintained and ecological and environmental resources can be protected in landscapes like the US Corn Belt when agroecosystems are diversified (Liebman *et al.* 2013 and references therein). Diversification of crops and cropping systems, reintegration of livestock with the land, and configuration of landscape elements in intentional designs for production of multiple benefits can help restore ecosystem services and wildlife habitats while supporting agricultural productivity (Figure 32.2; Jordan and Warner 2010; Liebman *et al.* 2013; Lovell and Johnston 2009). That is, restoration and maintenance of ecological resources in highly industrialized landscapes can be achieved while maintaining production systems by widening the range of goods and services produced so that private gain and public

Figure 32.2   Crop diversity, pasturing livestock, windrows and vegetated buffers help improve
multifunctionality of agricultural landscapes

Source: photo by Todd Stradford

benefit are both supported (Figure 32.2; Jordan and Warner 2010). Agricultural producers are motivated to produce ecological and environmental goods and services when there is no perception of undue difficulty in changing practices, when there is strong evidence that new practices will be as successful as current practices and when there is perceived economic benefit to them (Ma *et al.* 2012; Nassauer *et al.* 2011; Swinton *et al.* 2015).

In the US, policy to drive transitions toward greater landscape multifunctionality (Jordan and Warner 2010) is lacking. Many federal programmes for conservation in agriculture have recently been defunded or have been funded at reduced levels due to budget austerity (Stubbs 2014). Conservation agencies have been similarly affected by budget austerity and are likely to continue to face scarcity for the foreseeable future (Burke 2013). Hence, public policy, public programmes and government agencies alone are unlikely to drive the needed change. We know from history that farmers respond quickly to market signals – they rapidly adopt new crops and cropping systems in an effort to grow what is most profitable. Many Corn Belt farmers are college educated in agronomy or other agricultural science; yet, incongruous to environmental impacts of conventional agriculture, they often self-identify as 'stewards of the land' (Comito *et al.* 2013). Therefore, the greatest promise for restoring ecosystem services at large scales (i.e. landscapes) may be novel approaches to conservation that (1) preserve the production function of working lands, are not unduly burdensome and are economically advantageous for farmers, (2) appeal to farmers' stewardship attitudes, and (3) verifiably preserve or restore ecological resources and services at desired scales. That is, innovation at the *conservation–economy nexus* is needed for large-scale solutions to maintain soils, water, wildlife, ecosystems and the services they provide.

The conservation–economy nexus is where private gain and public benefit coincide. Public benefit may or may not have a monetary value or the value may be immeasurably high. Examples include scenic vistas, wildlife habitats, safe drinking water and clean air. Private gain always has a monetary value although exchange of currency may not occur. For example, a farmer may agree to harvest vegetation in a public conservation area in exchange for the harvested material. The public benefits from maintenance of wildlife habitat. The farmer benefits from use of the harvested material which has a known monetary value based on current market prices. The conservation–economy nexus is the opportunity to develop win-win relationships. Of course, much depends on careful planning and management of these relationships and their outcomes.

Bioenergy and conservation biomass as drivers of ecological restoration

Bioeconomies may help catalyse transition to greater multifunctionality of managed ecosystems (Atwell *et al.* 2011). *Bioeconomy* is used here to indicate economic activity that relies on renewable resources from agriculture or other ecosystems that goes beyond primary production via value-added processing and marketing. That is, bioeconomies produce more than commodities; they produce higher-value products and potentially ecosystem services. Expansion of grassland commerce is often suggested as a bioeconomic opportunity. Grass crops are of little or no food value to humans and can be converted to biopower and advanced biofuels as well as renewable materials and chemicals that have comparatively high value. Low-input diverse grasslands hold promise as dedicated crops for bioenergy while delivering conservation benefits although there are many challenges to scale-up for commercialization and long-term sustainability (Tilman *et al.* 2006; Werling *et al.* 2014). Commodity markets for dedicated perennial grass crops are lacking which requires that supply chains be created for each grass-based bioenergy facility (Tallaksen 2011). Transitions toward greater agricultural multifunctionality via bioenergy may therefore evolve at regional and local scales.

Periodic harvest of biomass from conservation lands has been considered for bioenergy uses (Adler *et al.* 2009; Fargione *et al.* 2009). Conservation lands, whether privately or publically owned, require management for maintaining cover types and meeting various conservation goals. For example, grasslands may be harvested to prevent encroachment by woody vegetation. In other cases, invasive non-native species may be harvested as part of efforts to construct or restore a more desired ecosystem. Biomass harvested as a habitat management action has the potential to be used in a variety of conversion technologies much like dedicated perennial grass crops (Lee *et al.* 2013). Land managers may be able to offset habitat management costs by exchanging harvested materials for services or they may have access to local or regional markets to offer biomass for sale.

Production of perennial grasslands as dedicated bioenergy crops and economic use of biomass collected from periodic harvest of conservation lands are two examples of *conservation biomass*. Conservation biomass is the production of biomass as a crop intentionally for ecosystem services benefits (in addition to profit), or the harvest of non-dedicated biomass for the purposes of restoring ecosystems or ecosystem services and for which utilization of the resulting biomass is intended to return an economic gain. For biomass to be conservation biomass its production or harvest must result in improved ecosystem function or services compared to baseline (or fulfil a conservation goal) and the biomass must be put to an economically beneficial use. Conservation biomass illustrates how the value of biomass can support restoration and maintenance of ecological resources.

Case studies: Missouri, Wisconsin, Manitoba

The cases described in this section were selected as illustrations of conservation biomass and the deliberate exploitation of economic gain to achieve ecological restoration. The cases demonstrate varied contexts and some commonalities among disparate examples. Conservation biomass examples are generally too few to permit a more theoretically robust discussion here; hence generalizability is yet to be realized. The author is affiliated with the first two case studies; the third is described from the literature.

Native grassland biomass in anaerobic digestion for renewable natural gas: Missouri, USA

North-central Missouri, near the geographical centre of the US, is bordered by the state of Iowa to the north, and two of the world's largest river basins on the west and east, the Missouri and Mississippi, respectively. Prairie and upland forest were the main pre-settlement vegetation. The region is home to one of the highest densities of concentrated animal feeding operations (CAFOs) for hog production in the country. The number and size of hog CAFOs has resulted in a variety of ecological and environmental problems most notably conversion of agricultural lands to large manure lagoons for manure storage, manure spills from open lagoons, expanded corn production to enable more field application of manure, nutrient run-off from fields sprayed with manure and subsequent impacts to upland and aquatic ecosystems.

Murphy Brown Missouri (MBM), the hog production subsidiary of Smithfield Foods, Inc., operates 63 sow farms and nine grow/finish farms with an excess of two million hogs across five of the region's counties – an area of approximately 6,734 km². Local farmers operate additional grow/finish farms under contract. In addition to CAFOs, each farm has one or more manure lagoons 2–3 ha in size each (i.e. hold 20–45 million gallons of wastewater). The operations have been subject to multi-million dollar regulatory fines for manure spills and other wastewater violations that fouled waterways. In 1999, the operations were subject to the largest environmental settlement of any hog producing operation in the country.

In 2013, Roeslein Alternative Energy (RAE), based in St. Louis, Missouri, partnered with MBM to develop a multi-phased anaerobic digestion project to generate 50 million diesel gallon equivalents of biogas annually while establishing native grasslands on 40,500 ha of agriculturally marginal soils currently planted in row crops or *Festuca* spp. (including MBM and other private lands). RAE's purpose (in addition to profit) is to restore ecosystem services (e.g. wildlife habitat, soil erosion prevention, water quality improvement and climate change mitigation). The first phase of the project, nearly completed, involves covering 88 manure lagoons (Figure 32.3) and establishing high-diversity prairie on an initial 202 ha of MBM land. Covering the lagoons transforms them into passive anaerobic digesters which hastens decomposition of organic materials and prevents entry of precipitation, thus increasing their storage capacity and reducing the amount of manure spray acres needed to dispose of manure. The second phase, begun in 2014, involves flaring of methane captured from the covered lagoons. This earns RAE environmental credits for preventing greenhouse gas emissions from the lagoons. The next phase, in development, is construction of above-ground anaerobic bioreactors that will receive grey water from the lagoons (not manure solids or slurry) to which will be added prairie biomass harvested from the 40,500 ha of restored marginal lands. This process results in additional biogas which improves commercial viability of the project while restoring ecosystem services to working lands within the region. A diverse mixture of native grass and forb species is being designed to maximize biomethane yield while also providing high quality

Figure 32.3 Northern Missouri anaerobic digestion project, phase one: covering of livestock manure lagoons to capture methane emissions and prevent entry of precipitation

Source: photo by Roeslein Alternative Energy

habitat for native pollinator species. Certified Missouri-sourced ecotype seed are being used exclusively for the project.

Monitoring of ecosystems services restoration has been proposed by a consortium of professionals from across the Midwest. This consortium is seeking competitive grants for funding of demonstration/outreach and monitoring activities. Monitoring is necessary to verify the type and amount of ecosystem services being delivered by restored acres compared to baseline (row crops or cool-season grass monoculture). This verification is essential for RAE's next step: negotiating additional environmental credits for ecosystem services gained through production of the dedicated prairie biomass (e.g. carbon sequestration). RAE has already sold nearly one third of the biogas that will be produced. Buyers include companies regulated under statutory renewable fuel standards. RAE continues to seek opportunities for expanding its model for reducing the environmental and ecological impacts of existing CAFOs while contributing to national renewable fuel goals. (For more information on this project, see Bilek 2014; Fletcher 2015.)

Economic benefits of waterfowl habitat harvest for conservation goals: Wisconsin, USA

The Leopold Wetland Management District of the US Fish and Wildlife Service (FWS), headquartered in Portage, Wisconsin, manages over 5,260 ha of waterfowl production areas (WPAs) in 17 Wisconsin counties. WPAs include uplands (i.e. grasslands) managed as waterfowl nesting habitats. Fire is the preferred habitat management tool but the District is unable to apply fire at a desired scale. In 2012, the District considered haying as a habitat management tool to supplement fire schedules. However, the District does not possess the equipment necessary for windrowing, aggregating and removing biomass from WPAs. The FWS has Service-wide initiatives to reduce its carbon footprint. In response, the District considered use of biomass harvested from WPAs to heat its buildings. They determined their space heating needs would require an amount of biomass harvested from approximately 2 ha of WPA annually – far less than would be harvested as part of a rotational harvest regime for habitat management. The key questions for the District, then, were who would do the harvest, what would be done with the many tonnes of surplus biomass annually and could the District offset cost of WPA harvest through the value of the biomass?

The District's first objective was to determine whether haying meets waterfowl breeding habitat management goals. In 2013 the District partnered with University of Wisconsin (UW) experts in grassland ecology and wildlife ecology to conduct the necessary research. The District and UW personnel established a Memorandum of Understanding to formalize their collaborative partnership at the institutional level. Soon afterward UW researchers designed a landscape experiment involving 12 WPAs, four in each of three landscape types: row crop dominant, row crops and grasslands in equal proportion, and equal mix of row crops, grasslands and wetlands. Two WPAs in each landscape type would be harvested annually while two would receive status quo management (e.g. fire if scheduled and if conditions permit), so that six WPAs were harvested annually and six were not harvested. Only portions of each WPA would be harvested to mimic the patch-burn patterns used by the District. The same patches were to be harvested each year. All 12 WPAs were to be monitored for waterfowl productivity, plant community dynamics, grassland bird productivity, and beneficial insect dynamics.

UW researchers did not have access to equipment necessary to harvest WPAs. Therefore, additional partners were required. During the experimental design phase of project planning, District and UW personnel recruited a dairy farmer and several custom harvesters to complete annual harvest of six WPAs on a pilot basis in which materials were exchanged for services. WPAs are not like farm fields – they have very uneven surfaces, many contain ant colonies that form large mounds, and all WPAs contain hazards such as old fences, abandoned pipes, shrubby areas and wetlands that must be mowed around. Despite these challenges, harvesters believed there would be sufficient gain to make participation worthwhile. Harvesters were instructed on protocols for harvest dates and stubble height to protect nesting waterfowl and grassland birds. Average harvested area was 21 ha. The dairy farmer harvested two WPAs annually during the three-year project period (Figure 32.4). Five custom operators each harvested one of the remaining four WPAs for one or more years.

Habitat management objectives included prevention of woody vegetation encroachment and maintaining grassland plant diversity. According to District personnel these objectives were generally accomplished. Based on results of the experiment, the District has indicated that harvest by private partners is an important option for them for maintaining WPAs as grassland habitats. They indicate no perceived administrative burden that would preclude future haying of WPAs by private partners to achieve habitat management goals. Unfortunately, pelletizing of biomass for space heating proved too labour intensive for the District's small workforce to be feasible. Harvest partners unanimously report favourable experiences in harvesting the WPAs and would consider doing so again in the future. Some of the harvest partners used the biomass themselves while some brokered it to others. All of the biomass harvested from WPAs was used as animal bedding, mostly for dairy livestock. (For more information on this project, see Charland *et al.* 2014.)

Cattail harvest for water quality, bioproducts and bioenergy: Manitoba, Canada

Lake Winnipeg is the tenth largest freshwater lake in the world and is deeply interconnected with Manitoba's economy, recreation and culture. It has become one of the most eutrophic

Figure 32.4 Waterfowl Production Area harvest, Wisconsin, USA
Source: photo by C. L. Williams

large lakes in the world due to excessive amounts of nutrients from agriculture, urban waste-water and other sources within the surrounding landscape (Grosshans *et al.* 2013). The Netley–Libau Marsh (NLM) is a 250km² freshwater coastal wetland at the mouth of the Red River and the south end of Lake Winnipeg. It consists of shallow lakes, channels and wetland areas through which the Red River flows on its way to Lake Winnipeg. While providing only 11 per cent of annual water flow into Lake Winnipeg, the Red River Basin contributes 30 per cent of the annual loads of nitrogen and over 60 per cent of annual phosphorus loads to the lake (*ibid.*). Unfortunately, NLM's ability to provide nutrient filtering and storage has been severely compromised by drainage, dredging, and other landscape changes over the past century (*ibid.*). Since the 1970s, Lake Winnipeg water levels have been managed by Manitoba Hydro for hydroelectric production which has also contributed to changes in NLM's ecological functioning (*ibid.*). Restoration of NLM and other wetlands of Lake Winnipeg are a high priority for addressing water quality (*ibid.*).

Cattail (*Typha* spp.) is a fast-growing, competitive, emergent aquatic plant. Because of these characteristics cattails have become over-abundant in NLM. Cattails can absorb phosphorus and other agricultural nutrients from litter and sediment layers in wetlands, including those found in NLM (Grosshans *et al.* 2013). Phosphorus and other nutrients in cattail biomass will return to sediment and waterways when the plants die and decompose. Harvesting cattails will prevent this release thus permanently removing nutrients from aquatic environments and reducing loads to aquatic systems (Grosshans *et al.* 2013, 2015). Studies in NLM indicate that harvesting cattails can remove 20–60 kg of phosphorus and up to 160 kg of nitrogen per hectare (Grosshans *et al.* 2015). This amount of nutrient uptake has convinced researchers that if harvest is applied as a widespread management tool in NLM there could be measureable water quality improvements with benefits for Lake Winnipeg (*ibid.*). The economic viability of nutrient management via cattail harvest, however, depends on developing markets for cattail biomass (*ibid.*).

Cattail biomass has many qualities for which it has been recommended to catalyze local and regional bioeconomies in Manitoba (*ibid.*). Biochar and bioenergy feedstocks are being promoted as potential major economic uses of cattail biomass (*ibid.*). Studies by the International Institute of Sustainable Development (IISD) indicate that cattails are suitable for biofuel and biomaterial products because of very good heating values, and densification and fiber properties (*ibid.*). If used as fuel, harvested cattail biomass could displace coal for energy production and produce valuable carbon and other greenhouse gas offsets (Grosshans *et al.* 2013). Water quality trading schemes are being considered for the Lake Winnipeg Basin and cattail harvest could be a mechanism for earning environmental credits (Voora *et al.* 2010).

In 2005, IISD began work on the concept of harvesting cattails in NLM with the goal of capturing phosphorus and reducing algal blooms in Lake Winnipeg. Since then, IISD has worked with partners throughout Manitoba to develop new bioenergy and bioproduct value chains (Grosshans *et al.* 2015). In 2012, IISD collaborated with the Prairie Agricultural Machinery Institute and the La Salle Redboine Conservation District in Manitoba to evaluate the use of commercially available agricultural equipment to demonstrate that cattails can be harvested economically on a large scale. A pilot-scale commercial harvest of cattail was conducted at Pelly's Lake, Manitoba, using conventional agricultural equipment—a watershed-scale proof-of-concept. Approximately 300 tonnes of cattail biomass was successfully harvested capturing a total of 230 kg of phosphorus. The NLM project demonstrates that cattail harvesting can deliver a host of benefits to Lake Winnipeg and the Manitoban economy (*ibid.*). IISD finds the underlying principles they developed and deployed in the project can be replicated in many other locations with co-benefits for water quality, wetland health and community economic development.

Synthesis: commonalities and distinctions

The cases discussed earlier illustrate how economic gain can be an essential ingredient in accomplishing ecological restoration at large scales. While the essential nature of economic gain may not be a limiting factor in all ecological restoration work, cost certainly is. Tapping into the profit motivation of people and organizations who are in a position to assist restoration efforts may help offset restoration costs, or even help catalyse bioeconomic development where broader ecological restoration and resource conservation goals are envisioned or intended. The case studies presented in this chapter are but a small handful, but they represent a nascent conservation biomass industry that understands the profit motivation and who are ready to act upon it.

Scope, focus and stimulus of ecological restoration

The ecological restoration goals and objectives among the three case studies differ according to scope, focus and stimulus (Table 32.1). In its pilot project to prove feasibility of commercial scale cattail harvest, IISD's partners harvested just 89 ha. However, the restoration goal involves a 250 km² wetland with ambition for applying the nutrient capture concept over a much broader area of ditches and channels across the Lake Winnipeg basin. In Wisconsin, the FWS's

Table 32.1 Summary of case studies in conservation biomass

Project	Grassland feedstock for biogas	Waterfowl habitat harvest for habitat management	Cattail harvest for nutrient uptake and water quality
Date	2013–present	2013–2015	2005–present
Location	Missouri, USA	Wisconsin, USA	Manitoba, Canada
Biome(s)	Prairie, row crop agriculture	Constructed grassland	Freshwater coastal wetland
Size			
Experiment/Pilot	202 ha	378 ha	n/a
Commercial	405 km²	52.6 km²	250 km²
Goal			
Objective(s)	Convert cool-season grasslands on marginal soils to diverse, warm-season grasslands	Remove shrubby vegetation, increase grasslands diversity, maintain waterfowl nesting habitats	Remove biomass for nutrient removal for water quality improvement and marsh rehabilitation
Profit motive			
Who	Farmers and/or grasslands services company	Farmers; in materials–for-services exchange	Farmers, custom harvesters; biomass processors
What	Price/tonne of biomass; long-term land lease	Value of biomass for livestock bedding	Harvest contracts; price/tonne of biomass; price/unit of value-added bioproduct or bioenergy
Stimulus	Sale of biogas; environmental credits	Statutory priority	Water pollution regulation
Public benefit(s)	Methane capture, soil erosion prevention, water quality improvement, carbon sequestration, odour abatement	Waterfowl production	Water quality improvement, aquatic ecosystem improvement, economic stimulus
Collaborative effort (Y/N)	Y	Y	Y
Partners required (Y/N)	Y	Y	Y

goal is maintenance of 5,260 ha of WPAs across an approximately 2,620 km² area. Their experiment involved harvest of 126 ha across six experimental units within a 10,310 ha study area. In Missouri, RAE's bioenergy project, when mature, will require dedicated grassland crops from 40,468 ha of land with restoration of ecosystem services as an outcome on all of these lands. Their initial efforts have involved conversion of approximately 202 ha of marginal lands from cool season grass monocultures to high-diversity warm-season grass crops.

The Wisconsin and Manitoba case studies possess ecological restoration goals that are specific and driven in whole or in part by government authority and regulatory mechanisms (Table 32.1). In the Missouri case, the ecological restoration goals are multiple and diffuse, and are driven by a business case with broad vision for agricultural transformation. In contrast, the Portage District of the FWS is obligated by priority-making authority based in federal law to maintain WPAs for waterfowl nesting habitat. Other priorities cannot supersede or interfere with the mandated priority for the WPAs. Administrators of the WPAs are held accountable for meeting the mandated priority and therefore their stimulus is statutory and direct.

Phosphorus and nitrogen are regulated nutrients in the Lake Winnipeg basin although NLM is not a source of these regulated nutrients. IISD and its partners are not regulated entities but NLM's restoration is seen as a critical piece in an integrated mosaic of actions necessary to achieve the restoration goal within a complex ecological-regulatory framework involving point and nonpoint sources of pollution (Grosshans *et al.* 2015). There is no regulatory penalty for IISD or its partners if they do not meet the strategic ecological restoration objective of nutrient removal from NLM. Therefore, IISD and its partners have an indirect but regulatory-related stimulus for their ecological restoration activities.

RAE is not a regulated entity in Missouri but it is leading a bioenergy project with a key partner who is a regulated entity for agricultural nutrients and water quality in Missouri. RAE's business case is symbiotic with MBM's need for manure management improvements and avoidance of regulatory offenses, but RAE is not obligated in a regulatory framework. Payments for ecosystem services are one part of RAE's overall strategy for commercial viability but these payments are anticipated to be a small percentage of their overall cash flow. RAE's restoration goals are driven in-part by the personal vision of its owner who is dedicated to prairie restoration and the existence of a more benevolent agriculture in the Midwest. The motivation for RAE's restoration goal therefore, is indirect and non-regulatory.

Profit motivations

There are distinct profit motivations in all three case studies (Table 32.1) but they differ in at least three ways: whose profit motivation is tapped to complete the restoration objective, whether the use of conservation biomass was low- or high-value, and whether there was (or will be) a supply chain involved. In the Wisconsin case, the FWS did not have a profit motivation as they could not complete the restoration objective (i.e. harvest). Farmers and custom harvesters of the study area were motivated to complete harvest activities voluntarily for the FWS because they perceived value in the harvested biomass. All of the harvesters in the WPA experiment used (or sold) the biomass for a low-value use (e.g. livestock bedding). In instances where the harvested biomass was sold the supply chain was very short – involving the buyer and the harvester who delivered the biomass to the buyer. None of the harvest partners in the WPA experiment attempted processing of the biomass into a higher-value product. Demand for the harvested biomass was proven.

Similar to the Wisconsin case, RAE does not have the capability of implementing their restoration objective (i.e. convert cool-season grass monocultures and row crops on marginal lands to diverse warm-season grasslands). RAE does not have the profit motivation for completing the restoration action. Instead producers of dedicated grassland biomass have (or are anticipated to have) the perception of gain in growing a dedicated biomass crop on their marginal lands under long-term contracts with RAE or a third-party biomass services company. In the Missouri case there is no supply chain unless a third-party supplier becomes involved. There is no processing of the biomass for value addition. RAE's need for grassland biomass is increasing as their multi-phased project develops. That is, the demand is proven.

In Manitoba, the IISD does not have a profit motivation because they cannot implement the restoration objective themselves (i.e. harvest cattails). For the broader restoration goal of NLM restoration and Lake Winnipeg water quality improvement, there is a vision for basin-wide bioeconomic development and supply chain creation (Grosshans *et al.* 2015). According to this vision the supply chain begins with cattail harvesters who are motivated by payment for harvest activities (i.e. low-value proposition). Cattail harvesters will need to be paid, presumably by companies who will process and market the biomass as higher-value products such as building materials and solid fuel (i.e. high-value proposition). At present, demand for cattail biomass is low or non-existent (i.e. unproven). Much work is yet to be done to generate demand for cattail biomass in the Lake Winnipeg basin. The Manitoba model includes low- and high-value profit motivations in a longer supply chain model as a means for sustaining ecological restoration of NLM and other wetlands and marginal lands within the Lake Winnipeg watershed.

Multiple benefits, public benefit, and adaptiveness

In each of the case studies, the leaders and many project partners were keenly aware of complexity in their situation, that there were multiple and often inter-related factors leading to the need for ecological restoration, and those solutions would be complex as well. All three case studies demonstrate willingness to design solutions that solved multiple challenges with multiple benefits to multiple parties (*sensu* Aronson and Alexander 2013). In each case there was awareness of the opportunity to deliver public benefits as well as private gain. In each case public gain was an intentional component of the integrated solution. In the Wisconsin case, public benefit was mandated. In the Missouri case, public benefits from grassland restoration are an entirely voluntary priority by RAE. That is, providing public benefit is not essential to fulfilling RAE's organizational purpose (i.e. profit). In Manitoba, public benefits derived from cattail harvest are also a voluntary undertaking for IISD, in the sense that they are not mandated by statute or regulatory authority to provide public good. IISD is a non-profit organization where their mission is to 'promote human development and environmental sustainability through innovative research, communication and partnerships'. High priority of public benefit fulfils IISD's organizational purpose. In all three cases the principles were willing to invest considerable amounts of time, personnel and other resources in exploration, information gathering, work-shopping, and adaptive decision-making. In each case study, the leaders were committed to learning and applying knowledge gained to future decision-making.

Collaboration and need for private partners

Economic and socio-political conditions may be forcing 'out of the box' thinking on behalf of public land managers and others who are obligated to ecological restoration by statutory or regulatory means. Collaborative approaches to ecological restoration, like those illustrated here, have gained popularity during budget constraints and shifting political priorities (Thompson *et al.* 2001). Ecological services companies and entrepreneurial spirits of all stripes are joining collaborative efforts to solve vexing multi-faceted challenges in ecological restoration. There is growing awareness of collaboration as a *necessary* ingredient in ecological restoration efforts aimed at multifunctionality of large landscapes such as the cases presented here (Scarlett 2013).

In each of the cases, private partners were required to complete the restoration goal and objectives (Table 32.1). For RAE, a for-profit corporation, acquiring and maintaining strategic partnerships in operations and financing, and subcontracting of tasks are basic competencies.

For the FWS, the involvement of private partners is not routine for their main operations though FWS personnel do have experience in interactions with private contractors or other non-government personnel while conducting official business for the FWS. Partnering with private companies or individuals by FWS must follow bureaucratic policies and procedures. IISD is a global organization experienced in coalition-building and management, and leadership in projects and networks of public, non-profit and for-profit organizations.

Conclusion

Set-aside approaches to ecological conservation in agricultural systems and industrialized landscapes are insufficient for protecting ecological and environmental resources. To maintain agricultural productivity and protect ecological resources there needs to be a transition toward greater multifunctionality of agricultural systems and the landscapes that contain them. Public policy and agency programmes alone are unlikely to catalyse the needed transition. Exploration of, and innovation at the conservation–economy nexus is needed for large-scale, market-driven solutions to maintain soils, water, wildlife, ecosystems and the services they provide. Bioenergy and conservation biomass are opportunities for achieving ecological restoration while enabling economic development. Much depends, however, on careful planning and management of restoration activities involving profit motivations.

The case studies in this chapter illustrate how economic gain is an essential ingredient in accomplishing the work of large-scale ecological restoration. They hint at a nascent conservation biomass industry that understands the nature and necessity of profit motivation and who are poised to act upon it. Ecological restoration planners and managers should be prepared to reach out to this industry and seek mutually beneficial solutions to complex restoration challenges.

The case studies represent a diversity of stimuli for ecological restoration and a spectrum of project size and potential impact. The profit motivations are quite different among the case studies as well. In each case the restoration work depends upon economic gain for partners intimately involved in restoration efforts and potentially those in supply chains far removed from restoration activities and project locations. The cases demonstrate that public benefit, even when not mandated by regulation, is a powerful component in conservation biomass. The cases also demonstrate how collaboration is an essential means for accomplishing restoration goals and objectives.

Success for each of the three cases required farmers or other agribusinesses as partners. In their experiments and pilot studies the three cases were able to recruit farmers and agribusiness (or quasi-agribusinesses) to participate. This demonstrates a willingness of farmers and others in the agricultural sector to participate in new ways of doing things. Their participation in these case studies may well signal a willingness to make wider contributions in the transition to greater multifunctionality of agricultural landscapes in particular.

The rationale for conservation biomass made in this chapter and the case studies are but a starting point. Socio-economic and socio-ecological research in the future should examine the factors that foster success in transitioning to greater multifunctionality of agriculture and large landscapes, test hypotheses, and help build a general theory of conservation biomass based on empirical observation. These advancements could be useful to policymakers, land managers, and restoration planners by increasing knowledge of design of conservation biomass-based projects. Such insight could help restoration planners and managers in their strategic thinking.

Acknowledgements

The author gratefully acknowledges the suggestions for manuscript improvement made by Paul Charland. Much gratitude is given to David Williams for loving support in the preparation of this chapter.

Disclaimer: the findings and conclusions in this chapter are those of the author solely and do not necessarily represent the views of the International Institute of Sustainable Development, Roselein Alternative Energy, LLC, or the US Fish and Wildlife Service.

References

Adler, P., M. Sanderson, P. J. Weimer, and K. P. Vogel (2009) 'Plant species composition and biofuel yields of conservation grasslands', *Ecological Applications* vol 19 pp. 2202–2209.

Aronson, J., and S. Alexander (2013) 'Ecosystem restoration is now a global priority: Time to roll up our sleeves', *Restoration Ecology* vol 21 pp. 293–296.

Atwell, R., L. A. Schulte, and L. M. Westphal (2011) 'Tweak, adapt, or transform: Policy scenarios in response to emerging bioenergy markets in the US Corn Belt', *Ecology and Society* vol 16 article 10, retrieved from www.ecologyandsociety.org/vol16/iss1/art10 (accessed 15 June 2015).

Bilek, A. (2015) 'A visionary model for biogas projects', The Great Plains Institute, Minneapolis, MN, retrieved from www.betterenergy.org/blog/visionary-model-biogas-projects (accessed 14 June 2015).

Burke, M. (2013) 'Why we need more – and not less – conservation funding in the federal budget', *Forbes*, retrieved from www.forbes.com/sites/monteburke/2013/11/22/why-we-need-more-and-not-less-conservation-funding-in-the-federal-budget (accessed 9 June 2015).

Charland, P., J. Lutes, S. Otto, D. Duncan, C. Gratton, R. Jackson, T. Meehan, G. Radloff, C. Ribic, K. Shinners, C. Williams, and D. Williams (2014) 'Harvesting vegetation for habitat management and bioenergy: A landscape-level experiment to understand ecological and social responses', retrieved from www.wgbn.wisc.edu/sites/default/files/WPAProject_Brief_final_AUG_2012_0.pdf (accessed 14 June 2015).

Comito, J., J. Wolseth, and L. Morton (2013) 'Stewards, businessman, and heroes?: Role conflict and contradiction among row-crop farmers in an age of environmental uncertainty', *Human Organization* vol 72 pp. 283–292.

Fargione, J. E., T. R. Cooper, D. J. Flaspohler, J. Hill, C. Lehman, T. McCoy, S. McLeod, E. J. Nelson, K. S. Oberhauser, and D. Tilman (2009) 'Bioenergy and wildlife: Threats and opportunities for grassland conservation', *BioScience* vol 59 pp. 767–777.

Fletcher, K. (2015) 'Grass to gas', *Biomass Magazine*, retrieved from biomassmagazine.com/articles/11468/grass-to-gas (accessed 14 June 2015).

Grosshans, R. E., P. Gass, R. Dohan, D. Roy, H. D. Venema, and M. McCandless (2013) 'Cattail harvesting for carbon offsets and nutrient capture: A "lake-friendly" greenhouse gas project', International Institute for Sustainable Development, Winnipeg, Manitoba, Canada, retrieved from www.iisd.org/pdf/2013/cattail_harvesting_carbon_offsets.pdf (accessed 15 June 2015).

Grosshans, R. E., L. Greiger, J. Ackerman, S. Gauthier, K. Swystun, P. Gass, and D. Roy (2015) 'Cattail biomass in a watershed-based bioeconomy: Commercial-scale harvesting and processing for nutrient capture, biocarbon and high-value bioproducts', International Institute for Sustainable Development, Winnipeg, Manitoba, Canada, retrieved from www.iisd.org/publications/cattail-biomass-watershed-based-bioeconomy-commercial-scale-harvesting-processing-nutrient-capture-biocarbon-high-value-bioproducts (accessed 9 June 2015).

Jordan, N., and K. D. Warner (2010) 'Enhancing the multifunctionality of US agriculture', *BioScience* vol 60 pp. 60–66.

Lee, D. K., E. Aberle, C. Chen, J. Egenolf, G. Kakani, R. L. Kallenbach, and J. C. Castro (2013) 'Nitrogen and harvest management of Conservation Reserve Programme (CRP) grassland for sustainable biomass feedstock production', *GCB Bioenergy* vol 5 pp. 6–15.

Liebman, M., M. J. Helmers, L. A. Schulte, and C. A. Chase (2013) 'Using biodiversity to link agricultural productivity with environmental quality: Results from three field experiments in Iowa', *Renewable Agriculture and Food Systems* vol 28 pp. 115–128.

Lovell, S. T., and D. M. Johnston (2009) 'Designing landscapes for performance based on emerging principles in landscape ecology', *Ecology and Society* vol 14 p. 44, retrieved from www.ecologyandsociety.org/ vol14/iss1/art44 (accessed 15 June 2015).

Ma, S., S. M. Swinton, F. Lupi, and C. B. Jolejole-Foreman (2012) 'Farmers' willingness to participate in payment-for-environmental-services programmes', *Journal of Agricultural Economics* vol 63 pp. 604–626.

Matson, P. A., and P. M. Vitousek (2006) 'Agricultural intensification: Will land spared from farming be land spared for nature?', *Conservation Biology* vol 20 pp. 709–710.

Nassauer, J. I., J. A. Dowdell, Z. Wang, D. McKahn, B. Chilcott, C. L. Kling, and S. Secchi (2011) 'Iowa farmers' responses to transformative scenarios for Corn Belt agriculture', *Journal of Soil and Water Conservation* vol 66 pp. 18A–24A.

Perfecto, I., and J. Vandermeer (2010) 'The agroecological matrix as alternative to the land-sparing/agricultural intensification model', *Proceedings of the United States of America National Academy of Sciences* vol 107 pp. 5786–5791.

Pimentel, D., C. Harvey, P. Resosudarmo, K. Sinclair, D. Kurz, M. McNair, S. Crist, L. Shpritz, L. Fitton, R. Saffouri, and R. Blair (1995) 'Environmental and economic costs of soil erosion and conservation benefits', *Science* vol 267 pp. 1117–1123.

Rabotyagov, S. S., C. L. Kling, P. W. Gassman, N. N. Rabalais, and R. E. Turner (2014) 'The economics of dead zones: Causes, impacts, policy challenges, and a model of the Gulf of Mexico hypoxic zone', *Review of Environmental Economic and Policy* vol 8 pp. 58–80.

Scarlett, L. (2013) 'Collaborative adaptive management: Challenges and opportunities', *Ecology and Society* vol 18 article 26, retrieved from www.ecologyandsociety.org/vol18/iss3/art26 (accessed 15 June 2015).

Stubbs, M. (2014) 'Conservation provisions in the 2014 Farm Bill (P.L. 113-79)', Report 7-5700, Congressional Research Service, Washington, DC, retrieved from nationalaglawcenter.org/wp-content/uploads/assets/crs/R43504.pdf (accessed 4 June 2015).

Swinton S. M., N. Rector, G. P. Robertson, C. B. Jolejole-Foreman, and F. Lupi (2015) Farmer decisions about adopting environmentally beneficial practices, pp. 340–359 in S. K. Hamilton, J. E. Doll and G. P. Robertson, eds, *The Ecology of Agricultural Ecosystems: Long-Term Research on the Path to Sustainability*. Oxford University Press, New York, New York, USA.

Tallaksen, J. (2011) 'Chapter 6: Guidelines for developing a sustainable biomass supply chain', in *Gasification: A comprehensive demonstration of a community-scale biomass energy system*. West Central Research and Outdoor Center, University of Minnesota-Morris, Morris, Minnesota.

Thompson, J. N., O. J. Reichman, P. J. Morin, G. A. Polis, M. E. Power, R. W. Sterner, C. A. Gough, R. Holt, D. U. Hooper, F. Keesing, C. R. Lovell, B. T. Milne, M. C. Molles, D. W. Roberts, and S. Y. Strauss (2001) 'Frontiers of ecology', *BioScience* vol 51 pp. 15–24.

Tilman, D., J. Hill, and C. Lehman (2006) 'Carbon-negative biofuels from low-input high-diversity grassland biomass', *Science* vol 314 pp. 1598–1600.

Voora, V., M. McCandless, D. Roy, H. D. Venema, B. Osborne, and R. Grosshans (2010) 'Water quality trading in the Lake Winnipeg Basin: A multi-level trading system architecture', International Institute of Sustainable Development, Winnipeg, Manitoba, Canada, retrieved from www.iisd.org/pdf/2010/water_quality_trading_lake_wpg_basin.pdf (accessed 15 June 2015).

Werling, B. P., T. L. Dickson, R. Issacs, H. Gaines, C. Gratton, K. L. Gross, H. Liere, C. M. Malmstrom, T. D. Meehan, L. Ruan, B. A. Robertson, G. P. Robertson, T. M. Schmidt, A. C. Schrotenboer, T. K. Teal, J. K. Wilson, and D. A. Landis (2014) 'Perennial grasslands enhance biodiversity and multiple ecosystem service in bioenergy landscapes', *Proceedings of the National Academy of Sciences of the United States of America* vol 111 pp. 1652–1657.

PART IV

Ecological restoration for the future

PART 7

Geographical Information for the Future

33

ECOLOGICAL RESTORATION AND ENVIRONMENTAL CHANGE

Stuart K. Allison

Introduction

One of the greatest challenges facing the practice of ecological restoration in the twenty-first century is the rapid pace and global scale of current and projected environmental change. There are many sources for these changes to the environment – conversion of ecosystems to other types and uses, habitat fragmentation, declines in species populations, extinctions of species, the human assisted movement of species from their original ecosystem to new ecosystems on a global scale, pollution, and global climate change. Almost all of these changes are either directly or indirectly related to continuing growth and movement of the human population – a population which is projected to continue growing until at least the middle of the twenty-first century.

Restorationists, like many groups of people, have been aware of these changes for many years but only recently have begun to fully grasp the rapidity and scale of environmental changes. The *SER Primer on Ecological Restoration* (published in its current form in 2004) lays out a clear definition of ecological restoration and a set of 9 characteristics of restored ecosystems. The authors of the primer acknowledged that ecological restoration requires a long-term commitment of space, people and resources because the length of time necessary to achieve a successful restoration is difficult to predict and in some cases restored sites will require perpetual care. But climate change, invasive species and continued human activity were only mentioned in passing as factors that may require continuous monitoring and maintenance work at restored sites. Just a couple of years later, Harris *et al.* (2006) wrote that the developing consensus about the scale and rapidity of projected climate change will force all restorationists to reconsider their goals for ecological restoration projects and in some cases whether restoration is even feasible. They noted that many restorations may take 100 years or more to reach maturity – especially when restoring forests and ecosystems with complex food webs – and that conditions that may arise within the next 100 years are likely to be very different from conditions that exist today as we initiate many restorations. In the intervening years, there has been considerable discussion within the restoration community about how to respond to rapid environmental change, but consensus about how to adjust our goals and methods has been slow to develop – if consensus has developed at all.

It is difficult to pick an exact beginning date for the field of ecological restoration because many related activities have occurred around the world for centuries, but the field began to coalesce with the practice of prairie restoration and mine reclamation in the 1930s (Hall 2005; Allison 2012). From that point on, the practice of ecological restoration had a strong element of historicity as restorationists frequently used historical reference ecosystems as a guide when attempting to return an ecosystem to an undegraded state (Higgs *et al.* 2013). Frequently the use of the past as a guide assumed a rather static environment in which changes were gradual and limited to a somewhat predictable and narrow range of historic variability (Millar 2014; Stanturf *et al.* 2014). However, we have come to understand that past environments changed constantly, sometimes abruptly – average temperature has been reported to change as much as 4°C in a 1 to 3 year period during the past 20,000 years (Millar 2014). Our measures of historic variability are largely based on conditions during the Little Ice Age (1400–1920 CE), a period of time with a relatively narrow range of cool global temperatures that does not match current or predicted future climates (*ibid.*). In that context, we must ask whether references based on past, historical records and ecosystems will allow us to develop realistic and reliable goals for restoration today and in the future.

Frequently restorations based on historic references have focused on re-establishing the pre-disturbance species composition of degraded ecosystems. But the species composition of current ecosystems has arisen fairly recently in ecological time and individual species appear to respond independently to environmental change (Millar 2014). Given what we know about rates of environmental change and the dynamic nature of ecosystems, we must ask whether restoring the species composition of ecosystems is always a realistic goal. Should we be more focused on restoring the structure of an ecosystem – so that a degraded grassland is replaced by a restored grassland, a degraded evergreen forest by a restored evergreen forest, etc.? Or should we focus on restoring ecosystem functions such as primary production, nutrient cycling, the hydrological cycle, resilience to disturbance, etc.? A critical question for all restorationists thus becomes what should we restore – composition, structure, function, all three or some combination of them – and at what scale should we restore it? The answers to that question will be dependent upon the ecological setting and on the social climate surrounding the restoration (Society for Ecological Restoration 2004; Hallett *et al.* 2013).

The emergence of novel ecosystems

A key feature of restorationists' understanding of the scale and rapidity of environmental change has been the recognition in the last 10 to 15 years of the formation of so-called novel or no analogue ecosystems (Hobbs *et al.* 2006). There are many factors which can lead to the development of novel ecosystems, but the main ones are as follows. Humans intentionally and unintentionally move species all around the world so that almost every ecosystem receives a constant input of non-native species. A particularly well documented case is the flora of New Zealand which has 2065 native species but which now has at least 24,774 non-native species growing there, over 2200 of which are considered to be naturalized and growing in the wild with no human care (Norton 2009). According to some experts, San Francisco Bay has the highest amount of non-native species of any aquatic habitat in the world (Molnar *et al.* 2008). The effects of non-native species are apparent in all ecosystems, on all continents. The other key factor governing the formation of novel ecosystems is climate change (Hobbs *et al.* 2006). But other factors that also play a role include land use change and pollutants – which can be directly toxic or things like airborne nitrous oxides which may cause excess nutrient input. These factors can both eliminate native species and alter ecosystem composition, structure and function so that the previously existing ecosystem is forced into a new state (Hobbs *et al.* 2009).

In some ways, the presence of ecosystems that appear to be novel compared to existing ecosystems is nothing new. Because climate change is a ubiquitous feature of the earth's environmental history, we don't have to look very far in the past to see species assemblages in ecosystems – especially in temperate and boreal regions – that are unlike any we observe today (Fox 2007). Just looking back a few hundred years in the fossil record is enough to find ecosystems that have no analogue to currently existing ecosystems. Because the current rate of environmental change is greater than we can detect in the geological past and is likely to continue as long as current human activities continue, we can expect that environmental change will occur well into the future (MEA 2005). It is likely that for many, perhaps most, locations on earth, future environmental conditions will be so different from today that we will consider them to have no analogue to current conditions at that site (Fox 2007).

How can or should restorationists respond to the observed and projected changes in ecosystems and the formation of novel ecosystems? One response is to think and plan at larger scales than just what happens at a particular restoration site. The global landscape – and although I use the term landscape, please understand that I also mean to include aquatic environments – is increasingly a mosaic of ecosystems that occur along a continuum from nearly intact ecosystems largely undegraded by the activities of modern industrial age humans, through varying levels of degradation to completely destroyed such as brownfields and post-mining sites (Hobbs *et al.* 2014). Thus it is vital to plan restorations so that they are situated within the context of the entire landscape, not just the boundaries of an individual restoration site. Restorations must be planned so that at the landscape level we can improve the situation so that we are better able to preserve biodiversity, the potential for evolution and ecological processes. Individual projects are seldom large enough to provide the kind of scale necessary to ensure the health and integrity of species populations and ecosystem functions (*ibid.*).

Because we anticipate that the environment will continue to change in the future, many restorationists are increasingly turning to the methods and philosophy of adaptive management (Hansen *et al.* 2010). Adaptive management is based on the idea of studying the system as it is being managed or restored, and adjusting management as the project is on-going (Holling 1978). It requires constant monitoring so that the restoration plan, techniques and strategies can be modified to fit changing conditions on the ground – conditions that may change both because the restoration is maturing and due to changes occurring in the environment on a larger landscape and even global scale. Adaptive management is an especially challenging, although important, method because it requires us to essentially build a bicycle while we try to ride it (Hansen *et al.* 2010).

A potential problem with use of adaptive management is that initial plans are frequently based on the historic range of variability (HRV) for a particular ecosystem. HRV comes from an examination of historic data to determine which conditions existed during the past for a particular ecosystem or site. However there are reasons to be sceptical about the value of HRV when planning management and restoration strategies for the future. HRV is frequently based on data gathered during a period of relatively stable global environment, the Little Ice Age from 1400 to about 1920, which had conditions cooler than exist today and cooler than are likely to exist in the foreseeable future (Millar 2014). Extending HRV to include longer periods of time – say up to 1000 or 2000 years – introduces such large amounts of variability that the management guidelines become almost meaningless (*ibid.*). Using different time periods – say 1000 to 1400 – where the quality of the data is less reliable is problematic because it may introduce more uncertainty to our models (*ibid.*). Thus any models for environmental conditions used to plan adaptive management schemes will have to be carefully calibrated with both expectations based on past history and what will be necessary and possible for maintaining restored ecosystems in the future. Our knowledge and understanding of past environmental conditions will have to serve as a way to connect the past with the rapidly approaching future (Higgs *et al.* 2013).

One of our largest concerns with the rate of environmental change is that many species will have limited ability to migrate in order to live in environments that have the best suite of conditions for their survival and reproduction. Even though it is now clear that past climate changes could be very large over short periods of time, most species living within the past 2 million years were able to migrate to new locations which better supported them (Millar 2014). However, the current landscape is highly fragmented with many areas completely dominated by human created ecosystems – urban, agricultural, or in other ways highly modified and managed – so that species cannot simply move from one location to another – they may find their migration path blocked by modified ecosystems that are uninhabitable by them (MEA 2005). While migration to new locations in the current world is difficult for almost all species, we are especially concerned about species with poor dispersal abilities, such as plants which are literally rooted to one spot for their adult lives, other sessile species and small sized species. At the current rate of change, plants would have to migrate about 5 km per year towards the poles to keep pace with changing climate zones as the earth warms. This is about 10 times faster than the current average rate of plant migration (Stanturf *et al.* 2014).

Given the problems with limited migration ability, slow migration times and hostile intermediate environments that many species will face now and in the future, we may have to engage in assisted migration – also known as managed relocation (Richardson *et al.* 2009; Minteer and Collins 2010). With assisted migration we actively move species from their current location to a new site where we think they will prosper in the future. There are many concerns about the use of assisted migration not the least of which are questions about how well we can predict the best locations for species in the future, the hazards of introducing species to new areas and thus spreading non-natives into new ecosystems, and the potential for hybridization between native and migrated species diluting the gene pool and adaptations of both species. Because many scientists and restorationists feel an urgent need to act now to preserve species, it should not be surprising that assisted migration is already under way with some experimental movements of trees from Florida to North Carolina and also movements of trees further north within the province of British Columbia (Richardson *et al.* 2009; Minteer and Collins 2010). We should be very cautious about the use of assisted migration because we aren't certain how well such introductions will work and the effects they will have on the destination. But I'm also fairly certain that assisted migration will become an important tool for restorationists. It is not something we need to embrace immediately because if there is one thing we know about human ecology, it is that we are very good at transporting species to new places when we want to do so.

Yet another challenge is that the process of restoring highly modified, novel ecosystems may have unintended consequences. Because non-native species, especially species considered to be invasive because of their rapid growth and high competitive abilities, are thought to be an especially large problem in novel ecosystems, we often focus on removing invasive, non-native species from degraded ecosystems (Hobbs *et al.* 2006). However, the removal of invasive species may sometimes result in additional changes to the ecosystem that pushes it even further away from the desired predisturbance state. For example, removal of highly invasive cordgrass *Spartina anglica* from coastal salt marshes in Tasmania resulted in greatly elevated rates of erosion at the site where *Spartina* was removed. The erosion increased turbidity of the estuary and likely had other effects on marine life (Sheehan and Ellison 2015). Because many novel ecosystems are likely to have altered structure and changed rates of ecological functions along with different species composition, it may be difficult to predict what conditions will arise during the restoration process. It behoves us to be cautious in our restoration efforts and to be ready to modify our plans as we monitor the restoration.

The discussion of novel ecosystems has been controversial within restoration circles (Hobbs *et al.* 2013; Murcia *et al.* 2014). I think the controversies have developed because of two things. The original discussions of novel ecosystems pointed out that some degraded ecosystems have become so different that they are indeed novel or no analogue when compared to currently existing or historic ecosystems. Further the authors concluded that in some cases it may be impossible to fully restore novel ecosystems to historic conditions either because it is ecologically impossible or because full restoration would be too expensive in terms of resources and human effort to be feasible (Hobbs *et al.* 2006; Hobbs *et al.* 2009). This conclusion led some people to feel that recognition of novel ecosystems and the challenges they present is a kind of surrender to degradation and that in fact, we can restore any ecosystem if we have the time and resources necessary (Clewell and Aronson 2013; Woodworth 2013). Second there is a fear that recognition of novel ecosystems is a form of environmental engineering in which we are choosing which ecosystems we think best serve our needs, and thus talking about novel ecosystems perpetuates a human-centric view of how to treat the environment (Clewell and Aronson 2013). While we may not like them, there is no sense in denying the existence of novel ecosystems, the fact that they are an inevitable result of widespread environmental change, and that their presence forces us to reconsider what it is possible to accomplish via ecological restoration.

The only constant is change

As previously noted, the global environment constantly changes. What is new today, in the post-industrial revolution world, is that in many ecosystems the main driver of environmental change is now human activity (Sanderson *et al.* 2002; MEA 2005). Human activity may drive environmental change either directly or indirectly, but in either case human needs for food, freshwater, fuel and building materials have drastically increased post-1950 as the human population has grown, and satisfying those needs has consequences for the entire planet (Sanderson *et al.* 2002; MEA 2005). Humans directly and indirectly use over 40 per cent of global net primary productivity. We also use 60 per cent of total freshwater runoff and consume over 35 per cent of the total productivity of the ocean's continental shelves (Sanderson *et al.* 2002). Sixty per cent of all ecosystem services measured are being either degraded or used unsustainably (MEA 2005). The human population is projected to continue growing until at least 2050, which means that human needs will continue to grow, resulting in ever more consumption and domination of the earth's resources and ecosystem services (*ibid.*). The human footprint is huge and growing. It is increasingly difficult to find any places on the earth that are not dominated by human activity (Sanderson *et al.* 2002).

As we have begun to quantify and understand the scale of human domination of the earth's ecosystems, it has become increasingly clear that this domination is resulting in such extensive change that ecosystems are likely to experience abrupt, non-linear shifts in their structure and function (MEA 2005; Millar 2014). The potential for abrupt shifts in ecosystem structure and function leads many to worry that we may be approaching a time − or indeed are already at the time − when ecosystems are poised at tipping points which, if crossed, will result in drastic changes in ecosystems − such that the ecosystem is now so different from before that it has become a novel ecosystem (Hobbs *et al.* 2006, 2009). The problem with tipping points is that we usually can't recognize them until they have been crossed, so that our ability to discuss and describe tipping points is always in retrospect. A classic case in point is the collapse of the

Atlantic cod fishery off the coast of Newfoundland in the 1990s. We couldn't predict that the collapse was iminent and thus did not have a way to prevent it and could only react once the fishery had collapsed (MEA 2005).

Unfortunately, unless there are drastic changes to human activities, the trends of declines and losses of ecosystem structure, function and services are likely to continue throughout the twenty-first century (*ibid.*). The complications of trying to establish global standards for atmospheric carbon emissions have led to practices that sometimes run counter to the goals of limiting carbon emissions (Galatowitsch 2009). Given those complications, will we be able to make significant changes in our behaviour and activities that will limit or eliminate the coming environmental change? There are reasons to be pessimistic about our ability to make large enough changes quickly enough to alter current environmental trajectories. The 2014 State of the Climate Report (Blunden and Arndt 2015) warns that the current trend of continuing warming of the oceans cannot be stopped even if we are able to achieve our desired goals for reduced carbon emissions. The effects of ocean warming will influence the earth for hundreds of years due to rising sea level and warmer sea surface temperature increasing the risk of severe storms and extreme weather patterns – extreme relative to conditions in the recent past (*ibid.*). This most recent State of the Climate Report paints a gloomy picture and indicates that environmental change in many ways has become a runaway train. Given the pace, scale and inevitability of environmental change, what can restorationists do to help restore damaged ecosystems and maintain desirable ecosystem structures, composition and functions?

Responses to environmental change

Our responses to environmental change must be grounded in understanding two basic facts about our limitations. The first is to acknowledge the degree of uncertainty about the actions we may take in order to conserve and restore ecosystems. When we set out to do something, we begin with things we know and we also know there are things we don't know. But even more worrisome are the unknown unknowns – the things we don't know we don't know (Allison 2012). Inevitably it is the unknown unknowns which arise and bite us just when we think we have solved a problem. Around 1900 when the US Forest Service implemented its policy of absolute fire prevention and suppression the policy made perfect sense. Forest fires destroyed trees and forests and thus had to be prevented in order to preserve forests and stands of trees. However after about 80 years of fire suppression, forest ecologists recognized that fire was an essential part of forest ecology and that fire prevention created many unintended negative consequences. We have to make our decisions about how to restore and manage ecosystems with humility, realizing that what makes sense to us today may turn out to be incorrect once we have learned about the unknown unknowns.

Second, we have to remember that the pace of changes wrought by technologically advanced humans is far more rapid than the pace of changes arising from wild evolutionary and ecological forces. In a matter of a few days, a bulldozer can strip away plants and soil that took centuries to develop on a site. It is unrealistic of us to expect that we can fully restore a damaged site in a short period. We have to have the patience to initiate a plan, carry it out and allow time for ecology and evolution to complete the process we have started – a process that may take centuries. I think it is typical of post-industrial humans to be impatient and to hope for instant results – we must remember the environment works at a different pace and is indifferent to our desires.

Acknowledging our limitations almost forces us to adopt the methods of adaptive management in which we learn and adjust our methods and goals as we proceed with our plans and restorations. Adaptive management occurs within both a scientific and social framework in which scientists and managers learn and adapt as the plan unfolds and in which stakeholders also help inform changes in goals and plans as they respond to the results of management and restoration activity. The use of adaptive management will require us to:

- 'learn to live with change and uncertainty';
- 'combine different types and sources of knowledge for learning how to manage effectively';
- 'create opportunities for self-organization so that both ecological and social systems are more resilient in the future'; and
- 'nurture resilience so that we are able to promote renewal, learning, growth and new forms of management' (Folke *et al.* 2005 as summarized in Allison 2012).

The practice of ecological restoration involves an obvious contradiction – we actively manipulate the environment in order to return an ecosystem to a condition we hope will be less degraded and free to develop along an evolutionary trajectory which is less constrained by human activity. Restoration activity is almost always expensive and labour intensive. One response to this contradiction is to practise passive restoration as a choice (Zahawi *et al.* 2014). Passive restoration is perhaps more accurately thought of as natural regeneration – we do nothing more than mark off an area as a restoration site and allow succession to occur. It is sometimes thought of as being less expensive and more natural due to the lack of active human manipulation beyond something like erecting fencing to set off the area. Passive restoration is thought to be more common on a global scale than active restoration and may be a key part of maintaining biodiversity in the twenty-first century (*ibid.*). However passive restoration is not free – there may be substantial costs for fencing, monitoring and protection so that the site is not used for some other purpose. It is a slow process and because site conditions may be different than they were when the previous ecosystem developed, the outcome is uncertain and passive restoration may not result in the ecosystem having the composition, structure and function that we desire (*ibid.*).

Acknowledging the built-in contradiction with ecological restoration has also led some to move in the opposite direction to passive restoration and advocate for an even more active, intentional form of restoration which they term intervention ecology (Hobbs *et al.* 2011). Their starting point is that the rapid pace of environmental change makes the traditional goal of ecological restoration as a return to past historic conditions less and less tenable. Humans already intervene in the environment in many ways – some positive – we hope ecological restoration is a positive intervention – and some negative. Intervention ecology suggests that we actively embrace trying to develop a positive set of interventions that will maintain biodiversity and key ecosystem properties such as structure and function. Intervention ecology is not focused on recreating the past, but more in making certain the future environment will continue to support species and functions that are critical to ecological health and integrity – key goals of the *SER Primer on Ecological Restoration* (Society for Ecological Restoration 2004). Given our recognition of many unknown unknowns in our interactions with the environment and how the environment operates even without our interventions, the key question becomes, do we know enough to intervene wisely? Will we have the skills necessary to plan and carry out interventions that accomplish our goals of maintaining biodiversity and ecosystem structure and function? Given that intervention ecology is such a new idea, the jury is still out on answering those questions. I suspect that we should continue to approach restoration with a great amount of humility and suggest we should seek to further refine our understanding and ability in ecological restoration before engaging in a more aggressive approach to managing the environment.

Our recognition of the limitations and contradictions inherent in ecological restoration force us to accept that even in the face of extremely rapid and large-scale environmental change, any practice that refers to itself as restoration must maintain a connection to the pre-disturbance ecosystem we are attempting to restore. We cannot rely on past historical ecosystems as the primary framework for ecological restoration. Instead, historical knowledge will be most useful to us as a guide to the possibilities of what may be achieved via restoration (Higgs *et al.* 2013). We may think of historicity as a virtue that helps to add value to restoration as we struggle to maintain a connection to the past in our restorations done in the future (*ibid.*).

One way of maintaining this connection is to strive for our restorations to produce authentic ecosystems. 'An authentic ecosystem is described as: *a self-regulating ecosystem with the expected level of biodiversity and expected complexity of ecological interactions, given historic, geographic and climatic factors*' (Dudley 2011: 138; emphasis original). Restored ecosystems are unlikely to be − and it would be unrealistic to attempt given current trajectories of climate change − replicas of historic ecosystems, but if they are authentic, if they fit the proper ecological context and allow evolutionary processes to operate as we would expect based on history, climate and geography we will have achieved successful restorations that expand our definition of what good restoration can be (Figure 33.1). This expansion of our understanding of good restoration will be vital in the twenty-first century. We need to restore with an eye to the future because restorations are slow to mature and we want them to be viable throughout the lifetime of the restoration (Allison 2002, 2012). Our restorations must be flexible, adaptable and future oriented because restoration must produce ecosystems that function well regardless of what the future brings (Choi 2007; Hanberry *et al.* 2015).

As we move forward there are several steps that we must accomplish to help us produce good, authentic restorations in the face of uncertainty and constant environmental change (Allison 2012):

1   We need to demonstrate the benefits of restorations that have already been done so that the general public better understands the value of ecological restoration. As we plan restoration work for the future we must be certain that each plan provides an explicit statement of goals, benefits and establishes monitoring protocols so we can determine whether the restoration is progressing as desired.

2   Our restoration plans must be flexible and we must be ready to change them as necessary as we learn more from our monitoring programmes and as the environment changes around the restoration site. We will need to adopt the methods of adaptive management to help ensure we understand how our ecosystems are functioning and how to respond as the ecosystem matures (or not) over time.

3   We will have to combine restoration with conservation so that we both restore as many damaged ecosystems as possible over as large an area as possible while also preserving as many wild and semi-wild ecosystems as possible.

4   We will almost certainly have to adopt assisted migration as a part of our restoration plans, especially as ecological restoration becomes more forward looking. But we will have to be cautious in its application to ensure that our attempts to maintain species in proper climatic zones do not cause unforeseen problems.

5   We should do as much as we can to promote ecological restoration in urban and other highly domesticated areas in order to promote biodiversity, ecosystem function and services as broadly as possible. We must do so in order to have restoration reach more people,

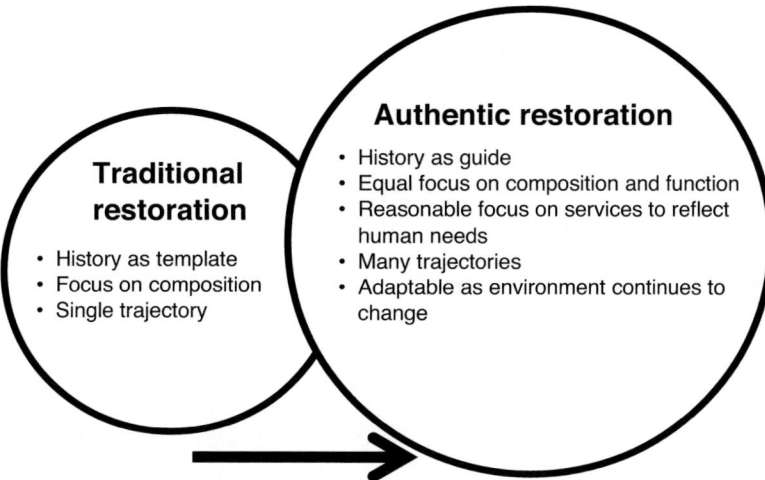

Changes in restoration models over time

Figure 33.1   Traditional models for ecological restoration have been based on using history as a template for restoration. There has been a strong emphasis on species composition and following a singular pathway to return the ecosystem to its past condition. Our recognition of the rapid pace and global extent of environmental change has led to developing a new model of ecological restoration which I am calling authentic restoration. Authentic restoration is a more flexible approach to ecological restoration which uses past history, along with climate and geography, as a guide to what is possible in the restoration. There is a greater emphasis on function and services so that the restored ecosystem will satisfy both ecological and human cultural needs. Authentic restoration has clearly evolved from traditional restoration because both are based on the need to repair damaged and destroyed ecosystems

Source: redrawn and modified from Fig. 1 in Higgs *et al.* (2013)

especially people who may not otherwise have a chance to encounter anything approaching a functioning ecosystem. We must also do so because in order to achieve our goals we will have to restore every place it is possible to restore.

6   We will have to become effective spokespeople for the value and benefits of ecological restoration. We must engage as many people as possible in this important work and we will need lots of stakeholder engagement in order to achieve restoration on a large enough scale to have significant benefits. I once heard a talk by Wes Jackson in which he said it is important to preach to the choir. We need to speak to fellow restorationists in order to boost their morale and keep them engaged – being effective spokespeople among ourselves is just as important as reaching out to new audiences.

The size of the human footprint is so huge that whether we want to be or not, we are now the major ecological force in many, if not most, ecosystems (Sanderson *et al.* 2002). Everything we do or do not has consequences for the global ecosystem. Years ago, one of my professors told our class, 'You can't not choose. Failure to make a choice is a choice.' In today's rapidly changing world, we can't simply sit back and watch the changes. That is in fact a choice in favour of an increasingly altered, depauperate world (Allison 2007). It is clear that we readily alter the world, but as Daniel Botkin suggested, we have to decide whether our alterations will make the world better or worse than it is now (Botkin 1990). Intentional, active, forward thinking restoration to produce authentic ecosystems is one of the best choices we can make to help leave the world a better place than it is today.

References

Allison, S. K. 2002. When is a restoration successful? Results from a 45-year-old tallgrass prairie restoration. *Ecological Restoration* 20: 10–17.

Allison, S. K. 2007. You can't not choose: embracing the role of choice in ecological restoration. *Restoration Ecology* 15: 601–605.

Allison, S. K. 2012. *Ecological Restoration and Environmental Change: Renewing Damaged Ecosystems.* Abingdon: Earthscan from Routledge.

Blunden, J. and D. S. Arndt (eds). 2015. State of the climate in 2014. *Bulletin of the American Meteorological Society* 96: S1–S267.

Botkin, D. B. 1990. *Discordant Harmonies: A New Ecology for the 21st Century.* New York: Oxford University Press.

Choi, Y. D. 2007. Restoration ecology to the future: a call for a new paradigm. *Restoration Ecology* 15: 351–353.

Clewell, A. and J. Aronson. 2013. The SER Primer and climate change. *Ecological Management and Restoration* 14: 182–186.

Dudley, N. 2011. *Authenticity in Nature: Making Choices About Naturalness in Ecosystems.* Abingdon: Earthscan from Routledge.

Folke, C., T. Hahn, P. Olsson and J. Norberg. 2005. Adaptive governance of social-ecological systems. *Annual Review of Environment and Resources* 30: 441–473.

Fox, D. 2007. Ecology: back to the no-analog future? *Science* 316: 823–825.

Galatowitsch, S. M. 2009. Carbon offsets as ecological restorations. *Restoration Ecology* 17: 563–570.

Hall, M. 2005. *Earth Repair: A Translatic History of Environmental Restoration.* Charlottesville, VA: University of Virginia Press.

Hallett, L. M., S. Diver, M. V. Eitzel, J. J. Olson, B. S. Ramage, H. Sardinas, Z. Statman-Weil and K. N. Suding. 2013. Do we practice what we preach? Goal setting for ecological restoration. *Restoration Ecology* 21: 312–319.

Hanberry, B. B., R. F. Noss, H. D. Safford, S. K. Allison and D. C. Dey. 2015. Restoration is preparation for the future. *Journal of Forestry* 113: 425–429.

Hansen, L., J. Hoffman C. Drews, and E. Mielbrecht. 2010. Designing climate-smart conservation: guidance and case studies. *Conservation Biology* 24: 63–69.

Harris, J. A., R. J. Hobbs, E. Higgs and J. Aronson. 2006. Ecological restoration and global climate change. *Restoration Ecology* 14: 170–176.

Higgs, E., D. A. Falk, A. Guerrini, M. Hall, J. Harris, R. J. Hobbs, S. T. Jackson, J. M. Rhemtulla and W. Throop. 2013. The changing role of history in restoration ecology. *Frontiers in Ecology and the Environment* 12: 499–506.

Hobbs, R. J., S. Arico, J. Aronson, J. S. Baron, P. Bridgewater, V. A. Cramer P. R. Epstein, J. J. Ewel, C. A. Klink, A. E. Lugo, D. Norton, D. Ojima, D. M. Richardson, E. W. Sanderson, F. Valladares, M. Vilà, R. Zamora and M. Zobel. 2006. Novel ecosystems: theoretical and management aspects of the new ecological world order. *Global Ecology and Biogeography* 15: 1–7.

Hobbs, R. J., E. Higgs and J. A. Harris. 2009. Novel ecosystems: implications for conservation and restoration. *Trends in Ecology and Evolution* 24: 599–605.

Hobbs, R. J., L. M. Hallett, P. R. Ehrlich and H. A. Mooney. 2011. Intervention ecology: applying ecological science in the twenty-first century. *Bioscience* 61: 442–450.

Hobbs, R. J., E. Higgs and C. M. Hall (eds). 2013. *Novel Ecosystems: Intervening in the New Ecological World Order.* Hoboken, NJ: Wiley-Blackwell Publishers.

Hobbs, R. J., E. Higgs, C. M. Hall, P. Bridgewater, F. S. Chapin III, E. C. Ellis, J. J. Ewel, L. M. Hallett, J. A. Harris, K. B. Hulvey, S. T. Jackson, P. L. Kennedy, C. Kueffer, L. Lach, T. C. Lantz, A. E. Lugo, J. Mascaro, S. D. Murphy, C. R. Nelson, M. P. Perring, D. M. Richardson, T. R. Seastedt, R. J. Standish, B. M. Starzomski, K. N. Suding, P. M. Tognetti, L. Yacob and L. Yung. 2014. Managing the whole landscape: historical, hybrid and novel ecosystems. *Frontiers in Ecology and the Environment* 12: 557–564.

Holling, C. S. 1978. *Adaptive Environmental Assessment and Management.* London: John Wiley & Sons.

MEA. 2005. *Ecosystems and Human Well-Being: Synthesis.* Washington, DC: Island Press.

Millar, C. I. 2014. Historic variability: informing restoration strategies, not prescribing targets. *Journal of Sustainable Forestry* 33: S28–S42.

Minteer, B. A. and J. P. Collins. 2010. Move it or lose it? The ecological ethics of relocating species under climate change. *Ecological Applications* 20: 1801–1804.

Molnar, J. L., R. L. Gamboa, C. Revenga and M. D. Spalding. 2008. Assessing the global threat of invasive species to marine biodiversity. *Frontiers in Ecology and the Environment* 6: 485–492.

Murcia, C., J. Aronson, G. H. Kattan, D. Moreno-Mateos, K. Dixon and D. Simberloff. 2014. A critique of the 'novel ecosystem' concept. *Trends in Ecology and Evolution* 29: 548–553.

Norton, D. A. 2009. Species invasions and the limits to restoration: learning from the New Zealand experience. *Science* 325: 569–571.

Richardson, D. M., J. J. Hellmann, J. S. McLachlan, D. F. Sax, M. W. Schwartz, P. Gonzalez, E. J. Brennan, A. Camacho, T. L. Root, O. E. Sala, S. H. Schneider, D. M. Ashe, J. R. Clark, R. Early, J. R. Etterson, E. D. Fielder, J. L. Gill, B. A. Minteer, S. Polasky, H. D. Safford, A. R. Thompson and M. Vellend. 2009. Multidimensional evaluation of managed relocation. *Proceedings of the National Academy of Sciences* 106: 9721–9724.

Sanderson, E. W., M. Jaiteh, M. A. Levy, K. H. Redford, A. V. Wannebo and G. Woolmer. 2002. The human footprint and the last of the wild. *Bioscience* 52: 891–904.

Sheehan, M. R. and J. C. Ellison. 2015. Tidal marsh erosion and accretion trends following invasive species removal, Tamar Estuary, Tasmania. *Estuarine, Coastal and Shelf Science* 164: 46–55.

Society for Ecological Restoration. 2004. *The SER Primer on Ecological Restoration.* Tucson, AZ: Society for Ecological Restoration.

Stanturf, J. A., B. J. Palik, M. I. Williams, R. K. Dumroese and P. Madsen. 2014. Forest restoration paradigms. *Journal of Sustainable Forestry* 33: S161–S194.

Woodworth, P. 2013. *Our Once and Future Planet: Restoring the World in the Climate Change Century.* Chicago, IL: University of Chicago Press.

Zahawi, R. A., J. L. Reid and K. D. Holl. 2014. Hidden costs of passive restoration. *Restoration Ecology* 22: 284–287.

34

INVASIVE SPECIES AND ECOLOGICAL RESTORATION

Joan Dudney, Lauren M. Hallett, Erica N. Spotswood and Katharine Suding

Exotic species invasions have dramatically increased over the past century (Pyšek and Richardson 2010), and are considered one of the greatest threats to endangered and vulnerable species worldwide (Levine *et al.* 2003). Although conservationists and restoration ecologists typically prioritize invasive species management, several issues preclude success. In many settings, it is impractical if not impossible to restore systems to pre-disturbed states, especially with increasing rates of directional changes, such as climate change, development and eutrophication (Dukes 2011; Seastedt *et al.* 2008). There is also growing evidence that exotic species can serve important functional roles within ecosystems (Davis *et al.* 2011). Given these constraints, it is becoming more difficult to develop practical invasive species restoration goals.

A number of conceptual frameworks for invasive species control have been proposed, including integrated pest management that emphasizes the use of multiple control methods, adaptive management for broader conservation planning and approaches that focus on managing multiple species and multiple landscapes simultaneously. A growing theme is to integrate monetary value with other ecological and social information (Larson *et al.* 2011) by viewing invasive species management within an ecosystem services or biodiversity offsetting context. In this chapter, we focus on invasive species management within a restoration setting, highlighting important factors to consider at the site level (Figure 34.1). Throughout the three main stages of management – assessment, prioritization, and control – we review key factors and present case studies to illustrate practical implementations of each stage. We conclude with a discussion of invasive species management moving forward.

Assessment

Before deciding which species to control, it is important to develop broader objectives for invasive species management at the restoration site. Removal projects can be more time-intensive, difficult and costly than expected (Larson *et al.* 2011), leading to unanticipated failures. Understanding the social and ecological constraints on the system allows managers to develop pragmatic and successful plans. A careful analysis of known invasive species control methods, public opinions, project duration, local partnerships and budget limitations can help managers identify the social constraints. Gathering data about the site's ecology, such as the current or historic cause of habitat degradation, the number and distribution of invasive species at the site

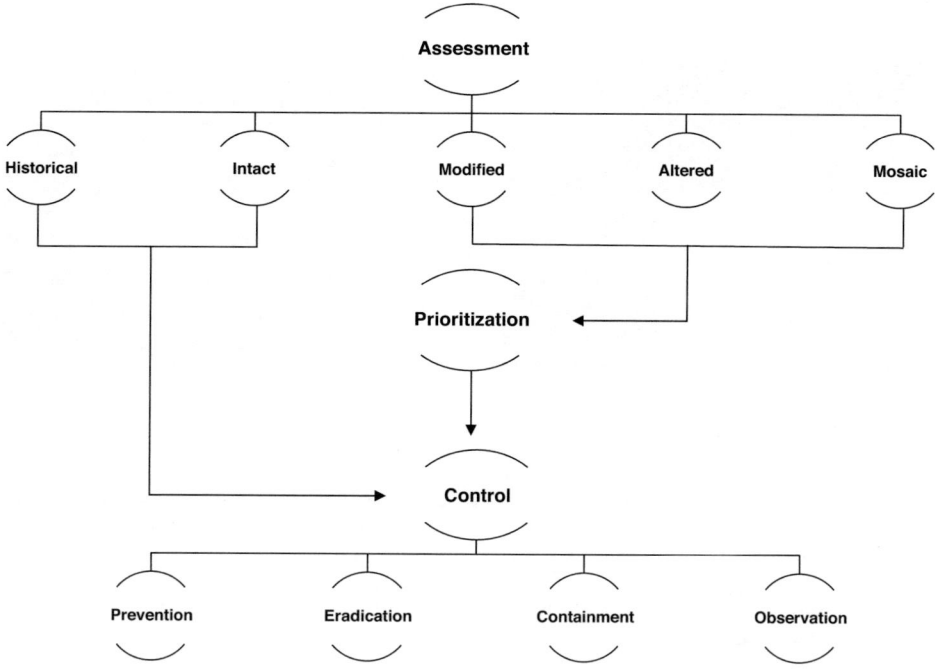

Figure 34.1 Invasive species management framework. Identifying social and ecological constraints and using this information to classify sites during the assessment stage helps managers navigate species prioritization and control options. By identifying the major factors that affect management outcomes during the three main stages, achievable goals may be developed and control plans successfully implemented

and the interaction of the restoration site within the surrounding landscapes, can also strengthen site assessment. As managers begin to compile this information, it may be useful to classify the site along a continuum of degradation (after Hobbs *et al.* 2009). Although often applied more broadly to restoration sites, site classifications can specifically focus on invasive species management (Figure 34.2). Here we suggest five classifications (historical, intact, modified, altered and mosaic) that can facilitate pragmatic goal setting.

Historical site

Historical sites are devoid of exotic or invasive species and the structure and function of the ecosystem reflect prior conditions. For example, fire frequencies match historical records, trophic levels are unaltered and/or human impacts, such as road construction or nitrogen deposition, are very limited. Keeping historical sites free of exotic species is the primary management objective and conducting weed risk analysis and large-scale monitoring plans for early detection can be effective control strategies. Few places on earth remain free of human influence, but those that retain their historical heritage may persist with proactive management. Such ecosystems are typically remote and/or high in elevation. The Atacama Desert in Chile, for instance, is still considered relatively undisturbed and exotic species free, though scientists predict that increased CO_2 will change the abiotic environment and encourage exotic plant

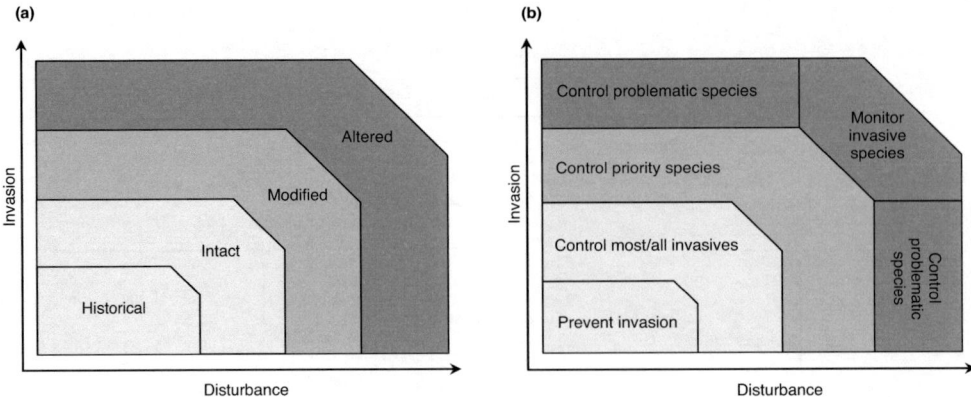

Figure 34.2 Site classifications and their corresponding management goals. (a) Placing sites into categories that reflect the extent of degradation and invasive species helps managers develop pragmatic goals. (b) Control objectives should shift as the disturbance and number and distribution of invasive species increases. For example, if the site is relatively intact and much of the historical biota is still present, then more traditional restoration objectives of invasive species removal are feasible. If the site is continuously disturbed and highly invaded, perhaps the only realistic goal is to monitor problematic species

Source: modified from Hobbs *et al.* (2009)

invasions in deserts (Smith *et al.* 2000). Because invasive species are spreading rapidly world-wide, emphasizing management of more pristine areas may be an important goal for conservation and restoration in the future.

Intact site

Within an intact site, the vast majority of the biota is native and the structure and function of the system remains unaltered. Natural disturbance regimes may shift, such as fire frequencies and nitrogen deposition, but these anthropogenic changes do not severely disrupt the biota or ecosystem processes. Traditional restoration goals of rehabilitating historic biodiversity are more feasible, as the number of different invaders on the landscape is limited compared to more degraded sites. It may even be realistic to curtail the spread of all invasive species to some degree. Many redwood forests in northern California are examples of intact ecosystems. Though fire regimes have been altered and human traffic is on the rise, only a few exotic species threaten the landscape, such as New Zealand mudsnails, *Potamopyrgus antipodarum*, and barred owls, *Strix varia*. Alpine zones are another example of intact systems and have dramatically reduced rates of species invasion compared to lower elevations (Pauchard and Alaback 2004). However, propagule pressure, continued human disturbance and climate change will likely facilitate species invasions in the future. Preventing exotic species from spreading into higher elevations can be more cost-effective compared to post-invasion control. Wilderness-wide management plans may still be impractical, however, and carefully targeting areas where invasive species are likely to spread is often more feasible.

Modified site

A modified site is one that contains a mixture of invasive and native species, though retains much of its original structure and function. These sites are typically influenced by human disturbance, such as urbanization and globalization, and restoration is generally more costly or even impossible. For example, the Great Lakes of North America have been invaded by over 180 exotic species within the last two centuries, some of which endanger underwater flora communities as well as commercial fisheries (Vander Zanden and Olden 2008; Barbiero *et al.* 2009). Despite the strong presence of invasive species, much of the ecosystem functioning in the Great Lakes region remains intact, suggesting that timely and strategic management could increase native species composition. Feasible management goals can include the removal of problematic organisms and the acceptance of ubiquitous or naturalized species that provide important services. Problematic species comprise those that disrupt ecosystem functioning, cause further degradation or alter important characteristics, such as vegetation structure or aesthetic qualities. Species that facilitate invasion of other exotics are also key management targets in modified systems. For example, nitrogen enrichment following the invasion of nitrogen fixing plants has been well-documented (Scherer-Lorenzen *et al.* 2008). Understanding which species have higher impacts on ecosystem health is an important strategy to improve restoration outcomes in sites with abundant invasive species.

Altered site

An altered site is one that is dominated by exotic and invasive species. The degree to which ecosystem structure and function resembles historic conditions is probably low, though likely varies depending on the severity and type of disturbance. Restoration is often very costly or infeasible and novel approaches for management are advised. The goals of invasive species management within an altered site should be highly strategic and pragmatic, focusing on species impacts and their interactions with the broader landscape. Targeting problematic invasive species that reduce ecosystem functioning, such as water retention or carbon cycling, may be a feasible goal even in highly disturbed sites. For example, the giant reed, *Arundo donax*, threatens riparian areas in California by increasing flammability in historically fire retardant areas (Coffman *et al.* 2010). Reducing fire risk is important for the overall functioning of the ecosystem, a goal that is less about origin and more about the species' impacts. In addition, monitoring problematic species or preventing new exotic species' introductions in highly trafficked and disturbed sites could be an important goal when considering the site's effects on surrounding landscapes and opportunities to prevent further degradation.

Mosaic site

In many cases, the ecology of restoration sites is highly variable and the invasive species distributions are fragmented across the landscape. Sites may contain habitats across multiple states of degradation from historical to altered. It is useful to consider such sites as a mosaic of different states that each requires goal setting reflective of this variability. For example, invasive fruit bearing species in Hawaii are often spread by frugivorous birds (Simberloff and Holle 1999) and to protect surrounding areas from invasion, it may be more cost-effective to control a highly altered area in order to protect less modified ecosystems. Characterizing the degree of degradation and invasive species threats across the landscape can help managers prioritize species and make informed decisions about control strategies.

Prioritization

Creating priority lists for invasive species is a useful management strategy, especially for modified, altered, or mosaic sites where invasive species are often well established. Various effective classification trees have been developed to assist managers and policy makers with species prioritization (Randall *et al.* 2008; Skurka Darin *et al.* 2011; Verbrugge *et al.* 2012). Lists are often created using multiple criteria, such as impact, potential for spread, and feasibility of control. Although such prioritization is useful for regional management strategies, the high variability among restoration sites, as well as their corresponding social and ecological constraints, often necessitates site-specific decision-making. To extend the discussion of prioritization, we do not present an exhaustive list of important factors, but instead highlight three emerging factors that warrant careful consideration when deciding which species to manage at a restoration site.

Non-target ecosystem impacts

Management success is often contingent on understanding the potential for non-target ecosystem impacts following invasive species removal (D'Antonio and Meyerson 2002). For example, has the invader been incorporated into trophic levels or do native species rely on the exotic species for habitat? The endangered California clapper rail is an archetype of non-target impacts, as its population dramatically declined following removal of invasive *Spartina* in the San Francisco Bay Area. When managers realized that the bird was using invasive *Spartina* as habitat, they responded by slowing plant biomass removal until native habitat was restored (Lampert *et al.* 2014). The interactions between invasive species and the local ecology at restoration sites can pose difficult choices for managers. Though easy solutions may be elusive, it is still important to consider the negative and positive impacts of invasive species when developing prioritization lists.

The act of invasive species control can also have non-target effects on native species diversity and ecosystem function. For example, much of the Galapagos's humid highlands have been invaded by the exotic ground cover, *Tradescantia flumiesis*. When managers used chemical control to contain the population, the compounds also reduced native species cover, thus providing a window of opportunity for other invasive species to spread (Gardener *et al.* 2013). Disturbance due to mechanical control can also facilitate invasion by other exotic species. Although topsoil removal has been touted as an effective tool to reduce annual grass propagule pressure, it can also reduce the native seed bank and alter the microbial community (Buisson *et al.* 2006). Thus, assessing and weighing additional ecosystem effects of removal strategies is an important step when deciding how to manage invasive species at a restoration site.

Recovery constraints

Invasive species may also alter ecosystem attributes and functions (e.g., geomorphology, hydrology, microbial communities, and disturbance regimes) that impede native species recovery (Corbin and D'Antonio 2011). These changes may persist on the landscape following removal and favour invasive over native species. For example, coastal dunes dominated by the invasive *Acacia longifolia* have greater litter accumulation, available nitrogen and microbial biomass. These legacies intensify with time following invasion (Marchante *et al.* 2008), creating a positive feedback that favours *A. longifolia* over other native species. Invasive species establishment may also directly inhibit native species. Garlic mustard, *Alliaria petiolata*, a widespread non-native species in North American forests, produces a long-lasting compound that impedes symbiotic associations between mychorrizae and native plants (Perry *et al.* 2005). In addition, invasive species can impede management goals for ecosystem function and native species recovery, particularly when they alter disturbance regimes. Invasive annual grasses in the Western United States, for instance, have increased fire frequency which threatens public safety and prevents native species recovery (Brooks *et al.* 2004). Understanding the recovery constraints affecting native species' re-establishment following invasive species' removal will likely improve decision-making and help managers identify which species are more feasible to control.

Re-invasion risk

Invasive species removal efforts may be rendered ineffective if managers cannot prevent subsequent re-invasion. Re-invasion is particularly likely if sites are located near a source population, highlighting that delineating species distributions within the surrounding landscape is integral for management success (Holl *et al.* 2003). Depending on jurisdiction and cost, managers may opt to scale-up invasive control efforts in invaded landscapes or prioritize removal in sites far from source populations. Life history of the invasive species may also affect prioritization decisions. For example, re-invasion may be unavoidable for sites located near populations of highly fecund or widely dispersing invasive species (Larios *et al.* 2013).

The risk of re-invasion may also be high for sites affected by land-use or global change that favours invasive species over native species. Sites that experience ongoing local disturbance may be at particular risk of re-invasion by invasive species with 'weedy' traits (fast growing, high fecundity) (Larios and Suding 2013). In addition, global change can alter interactions between native and exotic species (Dukes 2011). Invasive species often benefit from increased resource availability, and consequently may be more likely to re-invade areas affected by high nitrogen deposition or increased water availability due to climate change. Similarly, environmental change can also reduce the efficacy of initial species removal. Canadian thistle, *Cirsium arvense*, recovers from herbicide more quickly in areas with elevated CO_2, probably due to increased below-ground investment (Ziska *et al.* 2004). Distribution modelling can be useful for predicting areas where invasive-native species interactions may change, and site-level experimental tests can help managers predict re-invasion risk before implementing large-scale restorations.

Control

In this section, we highlight some of the most important factors to consider when deciding how to control invasive species (Figure 34.3). The target population's distribution plays an integral role in management outcomes and should be delineated before choosing among the different approaches. Control plans are best implemented with adaptive systems that enable managers to incorporate new information and respond to unforeseen site constraints. Monitoring is an excellent method for identifying success and failure that can be used to improve strategies and inform future projects (Zavaleta *et al.* 2001). In addition, active restoration that rehabilitates ecosystem structure and function may do more to control an invasive species population than removal alone. In general, the best control strategies are those that incorporate pertinent information gathered during the assessment and prioritization stages.

Four strategies – prevention, eradication, containment and observation – are commonly used by managers (Figure 34.3). The literature describing these approaches is extensive, and the definitions are highly variable and overlapping (Pyšek and Richardson 2010). Here we use control broadly to encompass the ideas of prevention, eradication, containment and observation, and in the following sections we describe their respective strategies within a restoration context.

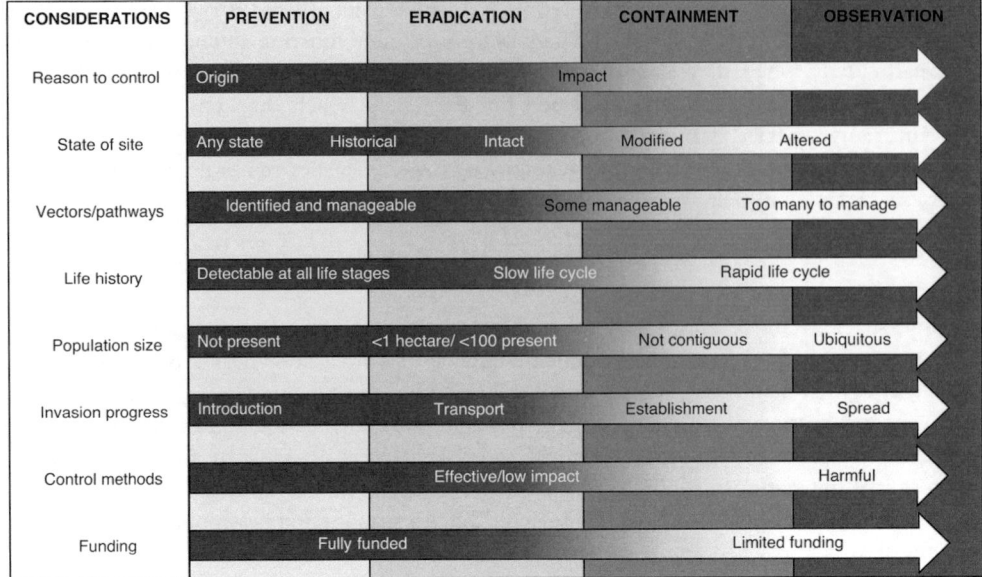

Figure 34.3 Factors to consider for invasive species control. The important considerations (listed on the left) shift with reference to four common control strategies: prevention, eradication, containment and observation. For example, eradication is more successful if populations are small, organisms are large with few offspring, vectors and pathways can be monitored. Should the project and species not match all the considerations for a control technique, move right to the next strategy. If most factors align with eradication except life history, for instance, it may still be impossible to eradicate the species and containment strategies are probably better

Source: adapted from Shackelford *et al.* (2013)

Prevention

Prevention is focused on precluding known or incipient invasive species from spreading into un-invaded habitats. Preventative strategies can be the most cost-effective, especially considering the financial consequences of invasive species establishment (Pimentel *et al.* 2005). Because it is difficult to predict invasiveness of exotic species, removing all arriving non-native species where identified can help curtail the establishment of future invasions (i.e. guilty until proven otherwise). The two common methods for detecting invasive species are surveys and remote sensing, though significant limitations, such as small or cryptic offspring, dense vegetation structure, dormancy, herbivory and observer error, prevent more widespread applications (Emry *et al.* 2011). To enhance the efficacy of early detection and mapping techniques, rigorous, frequent surveys in conjunction with citizen science programmes is recommended (Jordan *et al.* 2012). The California Invasive Plant Council, for example, implemented the Don't Plant a Pest programme in 2003 that distributed brochures throughout the state highlighting important invasive species to avoid planting in gardens and landscaping projects (see www.cal-ipc.org/landscaping/dpp).

A relatively new technique, pathway and vector management, can also bolster prevention strategies (Pyšek and Richardson 2010). By identifying the possible pathways (i.e., ports, roads, and nurseries) and vectors (i.e., hikers, cattle, big machinery) leading to invasion, managers can improve monitoring protocols in order to remove incipient invasions (Hulme 2009) or contain expanding populations. Pathway and vector management strategies, as well as other control efforts, are best implemented with clear plans for monitoring, both to ascertain long-term success rates and to inform future management practices (Zavaleta *et al.* 2001).

Eradication

The goal of eradication is to eliminate all individuals and/or the seed bank of an invasive species in a target area. Eradication decisions should be based on the invasive species' population size and distribution, and individuals that are farthest from population epicentres should be targeted first because they are likely less established and more feasible to eradicate (Rejmánek and Pitcairn 2002). Typically, successful eradications are associated with small, isolated populations less than one hectare in size. Species that are highly visible with lower fecundity and dormancy rates are also possible to eradicate. If one or more of these criteria are not met, such species are better managed by a containment strategy (Ramsey *et al.* 2009).

The literature is contradictory in its support of eradication efforts. In many cases, once a species is established it is almost impossible to remove (Vander Zanden and Olden 2008). Certain situations, however, may warrant eradication efforts. For instance, various island eradication efforts throughout the world have succeeded. New Zealand land managers effectively restored native bird diversity to small islands by removing all invasive predators (Veitch and Clout 2002). In addition, eradications of small plant populations have also been successful, especially when detected early (Rejmánek and Pitcairn 2002). Eradication is often most efficacious when followed by prevention strategies, as pathways and vectors may continue to facilitate species spread (Veitch and Clout 2002).

Containment

The goal of containment is to preclude or slow the spatial expansion of established invasive species populations. Containment is useful for species that do not meet all criteria for eradication (i.e., widespread distributions, prohibitively high removal costs or populations not conducive for monitoring surveys). Effective containment strategies ideally delineate the invasive species' distribution and important habitats threatened by the invasion. Thus managers can identify and remove smaller satellite populations that are spreading into ecologically valuable areas. Because restorations are often concentrated on degraded landscapes, containing problematic populations may be the most useful strategy, as eradications are extremely costly and often ineffective when populations are well established.

Invasive grasses greatly reduce biodiversity in serpentine grasslands, and various containment strategies are used to decrease invasive grass densities, such as grazing, fire, and mowing (Weiss 1999). Graminoid eradications in these systems are probably impossible because removal techniques often miss small populations, abutting landscapes continually provide propagules and human influence, such as increased nitrogen deposition, is difficult to control. Though containment strategies do not completely remove the invasive species population, management can still effectively restore native biodiversity to degraded landscapes when used strategically.

Observation

Observation approaches are useful when an invasive species is too widely distributed, difficult to remove or prohibitively costly to control. In such cases, it may be more realistic to monitor the population, watching that it does not spread further and affect nearby areas. For example, *Cronartium ribicola* is a lethal invasive pathogen affecting white pines throughout the United States. Although historic containment strategies failed, more recent developments, such as planting resistant trees or removing infected limbs, may prove effective at local levels. Until widespread application is feasible, however, observation provides essential information for future control techniques and scientific discoveries that bolster management strategies.

Future directions

Invasive species management can benefit from harnessing emerging technology to improve detection and control techniques. Broader advances in technology are already being adapted, and this trend is likely to continue into the future. For example, the internet is enabling broader collaborations among diverse sets of stakeholders, and big data analytics is driving analyses of larger genetic, remote sensing and citizen science datasets. Social media and mobile apps enable more citizens to become involved in invasive species detection and research. In addition, spatial data coupled with computing power facilitates spatially explicit simulations, allowing dynamic optimization of scarce dollars. In this section, we highlight four areas where recent technological advancements have improved or will likely improve invasive species management strategies.

Remote sensing

Remote sensing has emerged as a promising technology that can aid in mapping infested areas and monitoring spread. Currently, remote sensing has been used to create distribution maps to target the management of early invasion and model future invasion risk (Bradley 2014). The most common approach is to use differences in spectral signatures (typically with hyperspectral data) to differentiate invasive plants from surrounding vegetation. Forest canopy and single storey ecosystem (e.g., grasslands) invaders are more easily detected, while technological advancements are still needed to distinguish understorey species. Other obstacles to widespread implementation include economic expense and poor resolution quality for small patch identification, especially if individuals are small or early detection is the primary goal. The future is likely to see improvements in satellite technology and reductions in the cost of aerial images, which could expand the utility and efficacy of remote sensing.

Citizen science

Available technology has facilitated greater public participation in ecological research. By increasing the number of citizen science programmes, managers engage larger audiences that potentially improve early detection and control of invasive species (Dickinson *et al.* 2012). Widespread adoption of mobile devices has increased the accuracy of species location information, and mobile apps are currently experimenting with a broad range of methods to engage the public. Mobile apps include platforms for learning about the natural world and can assist users with species identification. Leafsnap (www.leafsnap.com), for example, uses image recognition of leaf photos to help identify trees in the eastern United States, while at the same time gathering occurrence and location data for species of interest. Browser based visualization tools and social media can subsequently facilitate the dissemination of such user data.

In some cases, citizen science projects are less expensive than traditional scientific research

(Gardiner *et al.* 2012). While technology offers enormous promise for enabling citizen science, constraints continue to arise. Perhaps the biggest challenge is ensuring data quality. Volunteers, for example, can make more mistakes in identifying organisms than professionals, which can lead to inflated species richness estimates. Although future advances in technology are needed to increase accuracy and decrease costs, citizen science programmes show great promise for invasive species management. Many programmes already exist that facilitate more widespread implementation of early detection efforts. Citizen science in conjunction with big data analytics will continue to advance our abilities to control invasive species.

Genetic tools

Contemporary genetic tools have been increasingly useful in invasive species research and benefit from advancements in computational and statistical approaches. Understanding how historical invasions progressed can improve predictive capacity for new invasions, as well as consequent impacts on host communities. Environmental DNA using barcoding has been successfully applied to aquatic systems (Dejean *et al.* 2012), where water samples can help detect incipient fish and amphibian invasions. In terrestrial systems, genetic tools have quantified dispersal pathways (Medley *et al.* 2015) and identified loci contributing to species adaptation during invasion (Vandepitte *et al.* 2014). In addition, metagenomics approaches, including Roche/454 pyrosequencing, can create whole community interaction networks more easily (Pompanon *et al.* 2012). These networks (i.e., host–parasite, predator–prey and food webs) enable scientists to characterize and measure invasive species impacts at the community level. Network patterns may also help uncover the conditions that helped communities resist invasion, improving predictions of when and how invasions occur.

Advancements in computational power, as well as new statistical approaches for analysing genetic data, have paralleled the rapid development of genetic tools. Improvements relevant for invasion ecology include approximate Bayesian computation, which allows invasion routes to be inferred from molecular and historical data. Similarly, discriminant analysis of principal components is a powerful approach to assign individuals to populations, broadening the application of population genetics to better define invasion pathways. Coupling genetic and computational tools with GIS based landscape analysis is also improving reconstruction of invasive species dispersal histories. The ability to evaluate how landscape features and environmental parameters affect dispersal is key to predicting how invasive species will spread in the future and provides important information for management strategies.

Collaborations

Invasive species management projects increasingly involve broader collaborations across agencies and international boundaries. Digital and online resources, for instance, play a greater role in all stages of weed management, from detection to strategic planning and control. Online databases that aggregate information on the identity, impact and location of invasive species provide useful management information. These databases function at the state, national and international levels, including the global invasive species database (www.issg.org/database/welcome), the National Invasive Species Information Center in the United States (www.invasivespeciesinfo.gov/index.shtml), and the Delivering Alien Invasive Species for Europe (DAISIE) database (www.europe-aliens.org).

Collaborative efforts are likely to extend beyond databases and use online platforms to integrate information from multiple stakeholders. The California Invasive Plant Council is attempting this approach through an online tool, CalWeedMapper, which consolidates material from various stakeholders about invasive species distributions, impacts and management activities in California. The website provides additional resources as well, such as upcoming conferences and training programmes, policy information, relevant scientific research and priority areas for research and management. This integrative approach recognizes that engagement with a broader community of partners is necessary for effective management.

Conclusion

Worldwide, it is hard to escape issues of invasive species when embarking on ecological restoration. In this chapter, we described various strategies to address the challenges that arise with invasive species management in restoration, highlighting three main stages – assessment, prioritization and control (Figure 34.1). At each stage, we encourage the consideration of how climate and human modifications may affect what restoration goals can be set and accomplished, including exotic species impacts on local species interactions and ecosystem functioning. Future management efforts will likely see an emphasis on more integrative conceptual frameworks and complex optimization tools to guide decision-making. Increased globalization and climate change are expected to accelerate the rates of invasion, challenging the ability of managers to keep pace with new invasions as they occur. Funding limitations will also likely persist, as well as the high costs of invasive species control. Thus, a key to future management efficacy in ecological restoration will be to build upon and document records of success in assessing, prioritizing and controlling invasive species.

References

Barbiero, R. P., Balcer, M., Rockwell, D. C. and Tuchman, M. L. (2009) Recent shifts in the crustacean zooplankton community of Lake Huron, *Canadian Journal of Fisheries and Aquatic Sciences*, 66(5), pp. 816–828.

Bradley, B. A. (2014) Remote detection of invasive plants: a review of spectral, textural and phenological approaches, *Biological Invasions*, 16(7), pp. 1411–1425.

Brooks, M. L., D'Antonio, C. M., Richardson, D. M., Grace, J. B., Keeley, J. E., DiTomaso, J. M., Hobbs, R. J., Pellant, M. and Pyke, D. (2004) Effects of invasive alien plants on fire regimes, *BioScience*, 54(7), pp. 677–688.

Buisson, E., Holl, K. D., Anderson, S., Corcket, E., Hayes, G. F., Torre, F., Peteers, A. and Dutoit, T. (2006) Effect of seed source, topsoil removal, and plant neighbor removal on restoring California coastal prairies, *Restoration Ecology*, 14(4), pp. 569–577.

Coffman, G. C., Ambrose, R. F. and Rundel, P. W. (2010) Wildfire promotes dominance of invasive giant reed (*Arundo donax*) in riparian ecosystems, *Biological Invasions*, 12(8), pp. 2723–2734.

Corbin, J. D. and D'Antonio, C. M. (2011) Gone but not forgotten? Invasive plants' legacies on community and ecosystem properties, *Invasive Plant Science and Management*, 5(1), pp. 117–124.

D'Antonio, C. and Meyerson, L. A. (2002) Exotic plant species as problems and solutions in ecological restoration: a synthesis, *Restoration Ecology*, 10(4), pp. 703–713.

Davis, M. A., Chew, M. K., Hobbs, R. J., Lugo, A. E., Ewel, J. J., Vermeij, G. J., Brown, J. H., Rosenzweig, M. L., Gardener, M. R., Carroll, S. P., Thompson, K., Pickett, S. T. A., Stromberg, J. C., Tredici, P. D., Suding, K. N., Ehrenfeld, J. G., Philip Grime, J., Mascaro, J. and Briggs, J. C. (2011) Don't judge species on their origins, *Nature*, 474(7350), pp. 153–154.

Dejean, T., Valentini, A., Miquel, C., Taberlet, P., Bellemain, E. and Miaud, C. (2012) Improved detection of an alien invasive species through environmental DNA barcoding: the example of the American bullfrog *Lithobates catesbeianus*, *Journal of Applied Ecology*, 49(4), pp. 953–959.

Dickinson, J. L., Shirk, J., Bonter, D., Bonney, R., Crain, R. L., Martin, J., Phillips, T. and Purcell, K. (2012) The current state of citizen science as a tool for ecological research and public engagement, *Frontiers in Ecology and the Environment*, 10(6), pp. 291–297.

Dukes, J. S. (2011) Responses of invasive species to a changing climate and atmosphere, in D. M. Richardson (ed.), *Fifty Years of Invasion Ecology: The Legacy of Charles Elton*, pp. 345–357. Chichester: Wiley-Blackwell.

Emry, D. J., Alexander, H. M. and Tourtellot, M. K. (2011) Modelling the local spread of invasive plants: importance of including spatial distribution and detectability in management plans, *Journal of Applied Ecology*, 48(6), pp. 1391–1400.

Gardener, M. R., Trueman, M., Buddenhagen, C., Heleno, R., Jäger, H., Atkinson, R. and Tye, A. (2013) A pragmatic approach to the management of plant invasions in Galapagos, in L. C. Foxcroft, P. Pyšek, D. M. Richardson, and P. Genovesi (eds), *Plant Invasions in Protected Areas*, pp. 349–374. Dordrecht: Springer Netherlands.

Gardiner, M. M., Allee, L. L., Brown, P. M., Losey, J. E., Roy, H. E. and Smyth, R. R. (2012) Lessons from lady beetles: accuracy of monitoring data from US and UK citizen-science programs, *Frontiers in Ecology and the Environment*, 10(9), pp. 471–476.

Hobbs, R. J., Higgs, E. and Harris, J. A. (2009) Novel ecosystems: implications for conservation and restoration, *Trends in Ecology and Evolution*, 24(11), pp. 599–605.

Holl, K. D., Crone, E. E. and Schultz, C. B. (2003) Landscape restoration: moving from generalities to methodologies, *BioScience*, 53(5), pp. 491–502.

Hulme, P. E. (2009) Trade, transport and trouble: managing invasive species pathways in an era of globalization, *Journal of Applied Ecology*, 46(1), pp. 10–18.

Jordan, R. C., Brooks, W. R., Howe, D. V. and Ehrenfeld, J. G. (2012) Evaluating the performance of volunteers in mapping invasive plants in public conservation lands, *Environmental Management*, 49(2), pp. 425–434.

Lampert, A., Hastings, A., Grosholz, E. D., Jardine, S. L. and Sanchirico, J. N. (2014) Optimal approaches for balancing invasive species eradication and endangered species management, *Science*, 344(6187), pp. 1028–1031.

Larios, L. and Suding, K. N. (2013) Restoration within protected areas: when and how to intervene to manage plant invasions?, in L. C. Foxcroft, P. Pyšek, D. M. Richardson, and P. Genovesi (eds), *Plant Invasions in Protected Areas*, pp. 599–618. Dordrecht: Springer Netherlands.

Larios, L., Aicher, R. J. and Suding, K. N. (2013) Effect of propagule pressure on recovery of a California grassland after an extreme disturbance, *Journal of Vegetation Science*, 24(6), pp. 1043–1052.

Larson, D. L., Phillips-Mao, L., Quiram, G., Sharpe, L., Stark, R., Sugita, S. and Weiler, A. (2011) A framework for sustainable invasive species management: environmental, social, and economic objectives, *Journal of Environmental Management*, 92(1), pp. 14–22.

Levine, J. M., Vila, M., D'Antonio, C. M., Dukes, J. S., Grigulis, K. and Lavorel, S. (2003) Mechanisms underlying the impacts of exotic plant invasions, *Proceedings of the Royal Society B: Biological Sciences*, 270(1517), pp. 775–781.

Marchante, E., Kjøller, A., Struwe, S. and Freitas, H. (2008) Short- and long-term impacts of *Acacia longifolia* invasion on the belowground processes of a Mediterranean coastal dune ecosystem, *Applied Soil Ecology*, 40(2), pp. 210–217.

Medley, K. A., Jenkins, D. G. and Hoffman, E. A. (2015) Human-aided and natural dispersal drive gene flow across the range of an invasive mosquito, *Molecular Ecology*, 24(2), pp. 284–295.

Pauchard, A. and Alaback, P. B. (2004) Influence of elevation, land use, and landscape context on patterns of alien plant invasions along roadsides in protected areas of south-central Chile, *Conservation Biology*, 18(1), pp. 238–248.

Perry, L. G., Johnson, C., Alford, E. R., Vivanco, J. M. and Paschke, M. W. (2005) Screening of grassland plants for restoration after spotted knapweed invasion, *Restoration Ecology*, 13(4), pp. 725–735.

Pimentel, D., Zuniga, R. and Morrison, D. (2005) Update on the environmental and economic costs associated with alien-invasive species in the United States, *Ecological Economics*, 52(3), pp. 273–288.

Pompanon, F., Deagle, B. E., Symondson, W. O., Brown, D. S., Jarman, S. N. and Taberlet, P. (2012) Who is eating what: diet assessment using next generation sequencing, *Molecular Ecology*, 21(8), pp. 1931–1950.

Pyšek, P. and Richardson, D. M. (2010) Invasive species, environmental change and management, and health, *Annual Review of Environment and Resources*, 35(1), pp. 25–55.

Ramsey, D. S. L., Parkes, J. and Morrison, S. A. (2009) Quantifying eradication success: the removal of feral pigs from Santa Cruz Island, California, *Conservation Biology*, 23(2), pp. 449–459.

Randall, J. M., Morse, L. E., Benton, N., Hiebert, R., Lu, S. and Killeffer, T. (2008) The invasive species assessment protocol: a tool for creating regional and national lists of invasive nonnative plants that negatively impact biodiversity, *Invasive Plant Science and Management*, 1(1), pp. 36–49.

Rejmánek, M. and Pitcairn, M. J. (2002) When is eradication of exotic pest plants a realistic goal, in C. R. Veitch and M. N. Clout (eds), *Turning the Tide: The Eradication of Invasive Species: Proceedings of the International Conference on Eradication of Island Invasives*, pp. 249–253. Gland: IUCN.

Scherer-Lorenzen, M., Venterink, H. O. and Buschmann, H. (2008) Nitrogen enrichment and plant invasions: the importance of nitrogen-fixing plants and anthropogenic eutrophication, in D. W. Nentwig (ed.), *Biological Invasions*, pp. 163–180. Berlin: Springer.

Seastedt, T. R., Hobbs, R. J. and Suding, K. N. (2008) Management of novel ecosystems: are novel approaches required?, *Frontiers in Ecology and the Environment*, 6(10), pp. 547–553.

Shackelford, N., Hobbs, R. J., Heller, N. E., Hallett, L. M. and Seastedt, T. R. (2013) Finding a middle-ground: The native/non-native debate, *Biological Conservation*, 158, pp. 55–62.

Simberloff, D. and Holle, B. V. (1999) Positive interactions of nonindigenous species: invasional meltdown?, *Biological Invasions*, 1(1), pp. 21–32.

Skurka Darin, G. M., Schoenig, S., Barney, J. N., Panetta, F. D. and DiTomaso, J. M. (2011) WHIPPET: A novel tool for prioritizing invasive plant populations for regional eradication, *Journal of Environmental Management*, 92(1), pp. 131–139.

Smith, S. D., Huxman, T. E., Zitzer, S. F., Charlet, T. N., Housman, D. C., Coleman, J. S., Fenstermaker, L. K., Seemann, J. R. and Nowak, R. S. (2000) Elevated CO_2 increases productivity and invasive species success in an arid ecosystem, *Nature*, 408(6808), pp. 79–82.

Vandepitte, K., Meyer, T., Helsen, K., Acker, K., Roldán-Ruiz, I., Mergeay, J. and Honnay, O. (2014) Rapid genetic adaptation precedes the spread of an exotic plant species, *Molecular Ecology*, 23(9), pp. 2157–2164.

Vander Zanden, M. J. and Olden, J. D. (2008) A management framework for preventing the secondary spread of aquatic invasive species, *Canadian Journal of Fisheries and Aquatic Sciences*, 65(7), pp. 1512–1522.

Veitch, C. R. and Clout, M. N. (eds) (2002) *Turning the Tide: The Eradication of Invasive Species: Proceedings of the International Conference on Eradication of Island Invasives*. Gland: IUCN.

Verbrugge, L. N., van der Velde, G., Hendriks, A. J., Verreycken, H. and Leuven, R. (2012) Risk classifications of aquatic non-native species: application of contemporary European assessment protocols in different biogeographical settings, *Aquatic Invasions*, 7(1), pp. 49–58.

Weiss, S. B. (1999) Cars, cows, and checkerspot butterflies: nitrogen deposition and management of nutrient-poor grasslands for a threatened species, *Conservation Biology*, 13(6), pp. 1476–1486.

Zavaleta, E. S., Hobbs, R. J. and Mooney, H. A. (2001) Viewing invasive species removal in a whole-ecosystem context, *Trends in Ecology and Evolution*, 16(8), pp. 454–459.

Ziska, L. H., Faulkner, S. and Lydon, J. (2004) Changes in biomass and root: shoot ratio of field-grown Canada thistle (*Cirsium arvense*), a noxious, invasive weed, with elevated CO_2: implications for control with glyphosate, *Weed Science*, 52(4), pp. 584–588.

35

RESTORATION AND RESILIENCE

Elizabeth Trevenen, Rachel Standish, Charles Price and Richard Hobbs

What is resilience and why is it relevant to restoration?

Resilience is a term used in a variety of contexts, from human health and psychology through to ecology and conservation biology. Resilience was introduced to the ecological literature as a 'measure of the persistence of systems and their ability to absorb change and disturbance and still maintain the same relationships between populations or state variables' (Holling 1973). By this definition, resilience has the potential to inform ecosystem management and restoration because it can potentially help to predict ecosystem recovery to a discrete disturbance event. The concept of resilience has been widely adopted by ecologists, environmental managers and policy-makers. Maintaining or restoring ecosystems that are resilient to human-induced global change has become one of the primary goals of modern-day intervention and stewardship. However, the concept has, over the years, become increasingly vague and has often been misused, rather than being a truly meaningful concept driving research or informing ecosystem management (Brand and Jax 2007; Myers-Smith *et al.* 2012).

Despite conceptual vagueness, its intuitive appeal is evident in its wide-spread adoption by both policy-makers and environmental managers. For example, it has been included in *The Society for Ecological Restoration (SER) International Primer on Ecological Restoration* (Society of Ecological Restoration 2004) as a key attribute restoration practitioners should aim to restore or maintain. A recent review of the Primer emphasized the relevance of resilience-based goals for modern-day restoration practice (Shackelford *et al.* 2013). In this context, resilience is seen as an ecosystem property important for maintaining desired ecosystem states (Gunderson *et al.* 2010; Walker and Salt 2012). However, more clarity around the concept is needed in order to meaningfully apply it in a management setting (Beisner 2012). In particular, confusion about how to define, measure and predict resilience (Myers-Smith *et al.* 2012; Beisner 2012) has largely prevented the application of the concept to the practice of restoration and ecosystem management more generally (Standish *et al.* 2014).

A major difficulty in the application of resilience concepts to management has been a lack of guidance on which concepts might be useful and how these might be measured. Two of the more popular concepts, ecological resilience and engineering resilience, are defined differently. Ecological resilience acknowledges the potential for thresholds between alternative stable states, and self-reinforcing feedback mechanisms (Holling 1973) whereas engineering resilience focuses on the time taken for an ecosystem to recover to a steady-state after disturbance (Pimm 1984). Each definition is relevant to the management of ecosystems subject to periodic, discrete disturbances. Conceptually, ecological resilience could help managers to decide if ecosystems are likely to recover from a disturbance and engineering resilience could help define a time scale for recovery. Recent ecological literature has paid increasing attention to resilience and allied concepts, particularly resistance: the degree to which a state variable is changed following disturbance (*ibid.*) and stability: the ability of a system to return to an equilibrium state after a temporary disturbance; the more rapidly it returns and the less it fluctuates, the more stable it would be (Holling 1973). Thus there are promising signs of coalescence of potentially useful ideas and approaches that may allow the translation of the concepts into practical measures for ecosystem management.

In a restoration context, the most pressing questions concerning resilience are these – how much disturbance can an ecosystem absorb before switching to another state? Where is the threshold marking the switch between ecosystem states? And will ecosystems recover from disturbance without management intervention? If not, then is the scale of intervention required beyond that which managers can hope to achieve? Resilience, resistance, stability and recovery are all potentially important concepts for answering these questions. If assessed quantitatively, resilience should help managers decide whether or not intervention will be required to push an ecosystem back towards the pre-disturbance state, and to guide restoration efforts. In this chapter we explore resilience and allied concepts for their potential to be implemented in practical settings to achieve restoration goals.

Resilience in theory and practice

Implicit in the definition of resilience is the acknowledgement of the possibility of alternative stable states and the existence of ecological thresholds that can prevent the recovery of the often desirable pre-disturbance state. This conceptual framework has considerable appeal in restoration, in theory, and sometimes in practice. The latter is due in large part to the intuitive appeal of the concept of thresholds for people working to restore degraded ecosystem states (Suding and Hobbs 2009). Whether measured, or implied but not measured, thresholds provide some guidance on where and when to intervene in degraded ecosystem states to restore desirable ecosystem states.

Alternative stable states are different end points of community assembly, each having unique species compositions, which can develop despite similar environmental conditions and regional species pools (Lewontin 1969). Thresholds help to identify truly alternative stable states; thresholds occur where small changes in environmental conditions result in large changes to community composition (Suding and Hobbs 2009). In the context of resilience, thresholds offer a means to quantify how much disturbance an ecosystem can absorb before switching to another state (Figure 35.1). Thus, the presence of thresholds can help to determine whether intervention might be necessary to prevent the undesirable transition, or to promote the return of the pre-disturbance state (Harris and Hobbs 2001). Considerable attention has been given to how thresholds might be measured and predicted with varying degrees of success (Scheffer *et al.* 2012; Qian 2014). Most progress has been made in aquatic systems where changes in a single variable can indicate a shift between easily recognizable alternative ecosystem states (Carpenter *et al.* 2011; Hansen *et al.* 2013). Thresholds in terrestrial systems have proven harder to quantify, although significant progress has been made in some systems. For example, a long-term study of rangelands on the Jornada Experimental Range in New Mexico has enabled the identification of early-warning signals of impending thresholds to grazing disturbance that

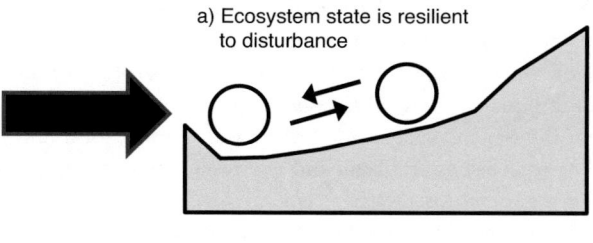

a) Ecosystem state is resilient
to disturbance

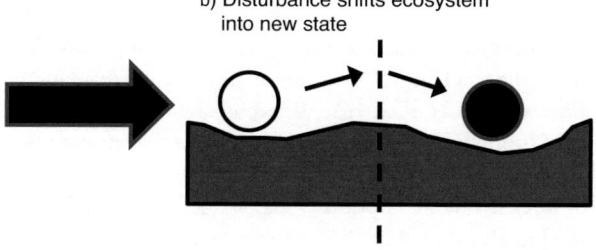

b) Disturbance shifts ecosystem
into new state

Figure 35.1 A simple ball-and-cup conceptual diagram of Holling's resilience to a disturbance (the black arrow). An ecosystem state is resilient to the disturbance if it returns to the pre-disturbance state (white-filled ball) after disturbance (a) In contrast, if the ability of the ecosystem state to absorb disturbance is exceeded, then it will shift into an alternative stable state (black-filled ball) after disturbance (b) The alternative stable states will have different relationships between populations or state variables (Holling 1973), which could manifest as differences in ecosystem structure, function or feedbacks (Suding and Hobbs 2009). The threshold between alternative stable states is indicated by the dashed line

managers can use to avert undesirable transitions to degraded states (Bestelmeyer *et al.* 2013). In another example, detailed research on the links between fuel load, fire intensity and vegetation states in juniper woodland suggests that managers can force transition to desirable grassland states by managing fuel loads according to thresholds associated with encroachment of juniper (Twidwell *et al.* 2013). While these recent examples demonstrate the potential for thresholds to be quantified for the purposes of management and restoration, the requirement to 'push' ecosystems past the threshold in order to find it, in addition to the requirement for extensive and detailed data, tends to render the exercise unrealistic in most cases. Locating thresholds to disturbance may also be more complicated in highly diverse ecosystems, and those subject to multiple, interacting disturbances. Consequently it is more common for thresholds to be implied rather than measured (e.g. Trueman *et al.* 2014).

In reality, management and restoration of complex ecosystems often relies on imperfect ecological knowledge of their responses to multiple, sometimes novel, disturbances. Locating ecological thresholds to disturbances characteristic of the current human-dominated era is challenging at best. At the same time, there can be an urgent requirement to act to prevent potentially irreversible and undesirable ecosystem changes. A pressing example of this reality can be found in coral reef ecosystems. The switch between algal-dominated and coral-dominated ecosystem states is well documented and there is evidence to suggest the switch is due to a combination of factors, including overfishing, pollution, coral bleaching, coral disease and predation by the crown-of-thorns starfish, which vary depending on where in the world the reef occurs (Bellwood *et al.* 2004). Perhaps owing to the sheer number of potentially interacting factors contributing to the loss of resilience and the shift to the undesirable algal-dominated state, in addition to the cryptic nature of the change leading up to the shift, researchers have suggested focusing on the management of key functional groups rather than attempting to identify early-warning signals of critical thresholds (*ibid.*). A recent test of this approach using long-term data from the Seychelles provided empirical support for a link between cross-scale redundancy of herbivorous fish assemblages and reef resilience to bleaching induced by a short-term increase in sea temperature (Nash *et al.* 2015). Consequently, these data suggest that preventing a switch to an algal-dominated state can be achieved by ensuring that no size classes of reef herbivores are disproportionately depleted by fishing (*ibid.*).

There is general agreement that it is more efficient for ecologists and land managers to prevent switches to undesirable ecosystem states because these undesirable states can be virtually immune to restoration efforts (e.g. Hobbs *et al.* 2013). Indeed, highly degraded states may themselves be very resilient to change, hence requiring large management inputs to return to a more desirable state (Standish *et al.* 2014). Where changes are not reversible, either because of the intensity of the underlying ecological changes or because these ecological changes render restorative action impractical or unsupported by the available resources, then it is probably necessary to consider the wider options for management (Hobbs *et al.* 2013). While the setting and achieving of goals other than the recovery of pre-disturbance state is largely outside the scope of this chapter it is relevant to mention the ongoing debate about whether or not ecological thresholds are truly irreversible (Murcia *et al.* 2013). It is a weighty issue to resolve because of the significant implications to restoration practice – the extent to which ecosystem changes are reversible largely determines the success or otherwise of management interventions. Thus, the outcome of this debate has broader implications regarding the future relevancy of the pre-disturbance state as the primary desired goal of restoration efforts. It is an open question whether the concept of resilience is applicable to the management of any re-defined desirable states. If the answer is yes, then resilience management could conceivably follow similar principles but look quite different if the new desirable state was controlled by a different set of variables to the old desirable (pre-disturbance) state.

In all, attempts to measure ecological resilience by identifying alternative stable states and locating thresholds has proved challenging. This challenge is due in part to the lack of a unified or agreed upon approach to locating thresholds; instead methodological approaches tend to be system specific and so not transferable to other systems. Added to this challenge is the requirement for detailed and intensive collection of data to locate thresholds that suits some systems and disturbance regimes more than others. Therefore, we argue that traditional approaches aimed at determining thresholds should be accompanied by theoretical approaches (*sensu stricto* Marquet *et al.* 2014) to quantify resilience because such a unified and simple approach is likely to gain traction in restoration practice and management. There are two significant advantages of theoretical approaches, one is their predictive capacity regarding the behaviour of resilient ecosystems, and the second is the simplicity of the data required to test these predictions although the latter advantage is largely absent for some theoretical approaches such as viability theory (Martin *et al.* 2011) and panarchy theory (Allen *et al.* 2014). Other promising theoretical avenues for measuring resilience include scaling approaches (Kerkhoff and Enquist 2007), the discontinuity hypothesis (Nash *et al.* 2014) which generate diagnostic measures of resilience based on community or ecosystem structure and function. Despite the aforementioned advantages of theoretical approaches for measuring resilience, these are yet to be adopted in management, possibly because of perceived complexity or lack of data. It is also the case that some of the more recent approaches are still in conceptual development and we discuss the potential for their application to management later in the chapter.

Underpinning all theoretical approaches aimed at measuring resilience should be a clear conceptualization of resilience along with a comprehensive understanding of the ecological factors and conditions that result in its expression. Conceptualizations of ecological resilience (Holling 1973) and engineering resilience (Pimm 1984) have helped guide research aimed at understanding the ecological factors and conditions that impart resilience, however, the terminology has not been consistent with much research referring to ecological resilience as 'ecosystem stability'. Nonetheless progress has been made and likely mechanisms of resilience will continue to emerge (e.g. Shackelford *et al.* 2017). In the next section we will discuss the ecological properties that are proposed to contribute to resilience. We follow this discussion with an evaluation of how these properties can be used to guide theoretical efforts towards quantifying resilience.

Approaches to resilience management

Managing for resilience requires a clear understanding of the attributes that confer resilience. So far, it is generally accepted that what makes an ecosystem resilient is complexity and adaptability (DesJardins *et al.* 2015). However, knowledge of the specific ecological factors and environmental conditions underpinning these attributes is limited. Numerous case studies have identified different kinds of biodiversity that contribute to resilience, and specifically for switches between ecosystem states to result from biodiversity loss. However, it is recognized that ecosystem resilience is not a product of high biodiversity alone (indeed, some low-diversity systems appear to be highly resilient), and that other ecological factors contribute to resilience. Connectivity is another attribute likely integral to resilience (*ibid.*; Shackelford *et al.* 2017). Similar to the story for complexity and adaptability, the mechanism by which connectivity fosters resilience is largely unknown, with some authors suggesting that a complex adaptive systems approach would offer critical insights (Filotas *et al.* 2014). In this section we will discuss current ideas on the attributes that appear to confer resilience, and how we might intervene to modify ecological factors and environmental conditions so as to maintain these attributes.

Resilience and diversity

There is mounting evidence from aquatic, marine and terrestrial studies that biodiversity loss impairs and destabilizes ecosystem functioning, ultimately leading to reduced resilience (Cardinale *et al.* 2012). This evidence is further supported by the modelled effects of diversity on temporal stability which are consistently positive (Lehman and Tilman 2000). While a positive relationship between diversity and resilience is widely accepted, it is acknowledged that the relationship is complicated, and identifying which mechanisms are underlying the relationship has proved challenging, with many explanations emerging (Tilman *et al.* 2014).

Initially it was hypothesized that high species diversity allowed for species to compensate for other species lost or negatively impacted by disturbance, thereby stabilizing ecosystems in the face of disturbance (MacArthur 1955). It was later clarified that functional redundancy, or the existence of multiple species within an ecosystem that fulfil the same functional role, was instead the primary mechanism by which ecosystems maintained functioning and thus were resilient in the face of disturbance (Díaz and Cabido 2001; Sundstrom *et al.* 2012). Recently another form of diversity integral to resilience has been identified: response diversity, which is diversity in the way species respond to environmental fluctuations, also referred to as the 'insurance hypothesis' (Griffin *et al.* 2009). Response diversity has been shown to partly explain the stabilizing effect of species diversity on ecosystem functioning in several aquatic communities

(Leary and Petchey 2009; Thibaut *et al.* 2012). In particular, diversity in species response times to disturbance has been noted as an important contributing factor determining effects on stability (Loreau and Mazancourt 2013). Overall, there is evidence to support the idea of functional redundancy and response diversity contributing to resilience because these attributes help to buffer the impact of discrete disturbance events, and potentially longer-term environmental changes, on ecosystem dynamics.

Other mechanisms by which diversity is thought to increase resilience include the diversity-invasion theory and over-yielding. The diversity-invasion theory suggests that highly diverse systems are more difficult to invade, because invaders only have access to unconsumed resources, and the levels of unconsumed resources decline as diversity increases. This phenomenon has been demonstrated in consumer-resource interaction models and diversity-interaction models based on empirical data (Thébault and Loreau 2003). For studies on diversity and invasion, functional diversity of the resident community is thought to be a better indicator of biotic resistance than species diversity, with niche pre-emption (priority effect) and niche partitioning (diversity effect) being suggested as the mechanisms underlying this relationship (Byun *et al.* 2013). Over-yielding on the other hand, is a phenomenon where species in biodiverse communities are seen to yield more compared with community members grown in monoculture or less-diverse species mixes. The greater performance of species grown in diverse communities has been seen to have a stabilizing effect on ecosystem functioning and net primary production (Hector *et al.* 2010).

What then does this mean for restoration and management? Clearly, the implication is that restoration should aim to restore diverse communities wherever possible, with the requisite levels of diversity determined through reference to historical or nearby undisturbed communities. Obviously a restoration project in the Brazilian Atlantic forest will require higher levels of diversity than a project in the heathlands of Scotland. Considerations affecting the degree of diversity to be targeted include the practicalities of establishing the required array of species, which include availability of seed or seedlings and their relative ease of germination and establishment (Perring *et al.* 2015). Important questions in this regard focus on the likely minimum level of diversity required to confer resilience to the restored system: unfortunately, this question has rarely been asked in a practical setting. The likelihood of positive or negative interactions among established species and the potential for successional development are also important considerations (e.g. Cramer *et al.* 2008).

Connectivity and modularity

The complex adaptive systems framework posits that the resilience of a complex system can be directly influenced by the level of connectedness among the entities within the system (Lansing 2003). Connectivity can relate to biotic entities (e.g. species networks) and abiotic entities (e.g. water flow through landscapes). Connectivity may confer more or less resilience depending on the context. Highly connected systems may have low resilience because all entities will be sensitive to changes happening in any part of the system. Research into the role of keystone species on trophic structure have demonstrated this phenomenon, with the changes in the abundance of a few well-connected species resulting in large downstream effects and in some cases the collapse of the community (Mills *et al.* 1993). In such cases, where connectivity is high, higher diversity will not necessarily prevent a switch to an alternative ecosystem state when entities are lost or negatively affected by a disturbance. On the other hand, higher levels of connectivity may allow the movement of species through the landscape, thus allowing recolonization of areas where species go locally extinct. At the same time, connected landscapes can also allow increased spread of disease, invasive species and disturbances such as fire, thus potentially reducing overall resilience.

Modularity is relevant to this discussion too. Modular structures, which are structures that contain aggregates of well-connected entities, and which are weakly connected to neighbouring aggregates, have been put forward as being resilient in terms of their structural organization according to the complex adaptive systems framework (Newman 2003). Too much modularity however, can play against resilience, for example if a system's capacity to adjust to change depends on the coordination and regulation of the entities present in the modules (Fischer and Lindenmayer 2007). A mix of high local connectivity and weak connectivity at broader scales may be the most effective strategy in designing resilient landscapes. Again, however, there are few empirical studies that provide concrete guidance as to how this can be realistically achieved in restoration projects. The current trend towards large-scale restoration projects (Perring *et al.* 2015) offers the potential to test different approaches.

Future directions

Managers will need to quantify resilience so progress can be evaluated in order to successfully manage a degraded or restored system for resilience. In the previous section, properties and conditions that result in resilience were discussed. These included multiple measures of diversity, connectivity and modularity. While these attributes might be used to indicate resilience, their quantification may prove difficult in the same way that quantifying thresholds can be difficult (i.e. data intensive). This impracticability highlights the need to identify diagnostic measures of resilience. This need, to some extent, has been the holy grail of resilience research, and what most theoretical efforts have set out to accomplish. The field of research is currently very active. Here, we discuss some recent ideas in ecology that may inform the development of effective measures of resilience.

One promising approach is the use of macroscopic scaling relationships for their potential to reflect community or ecosystem structure and functioning (Kerkhoff and Enquist 2007). Scaling relationships are typically larger scale emergent properties of systems that result from the interaction of individuals, species and functional types and environmental constraints. Examples include species–area, species-individual, and size-abundance relationships. Exogenous or endogenous disturbance events result in quantifiable changes in scaling relationships (i.e. a shift in the slope or intercept of a power law). For example, the species–area relationship, which describes how species richness scales with area, emerges from interactions between the species present, the number of individuals per species, and the resources available to support populations of species. Changes in any one of these properties (i.e. a change in resource availability due to intensive grazing or drought) may result in changes to spatial structure as reflected in a new species–area relationship (i.e. one differing in slope or intercept from that observed pre-disturbance; De Bello *et al.* 2007). Scaling relationships could allow for differences between systems (e.g. restored and references systems) and within systems (structural reorganization) to be quantified. Thus, scaling relationships could be a useful tool to guide restoration practices and resilience management.

Despite their exceptional complexity and variety, communities can exhibit striking regularities in the way they self-organize (Brown 1984). This phenomenon often takes the form of scaling relationships, where regularities between systems can be illustrated by a convergence in, for example, the values of the scaling exponent. The presence of these regularities suggests that there may be general, underlying principles governing the structural organization of ecological systems, shaping them in a self-similar manner (West *et al.* 2009). Some researchers have suggested that the state to which ecological systems converge can be considered an attractor and point to evidence suggesting that these attractors represent equilibrium states (Kerkhoff and Enquist 2007). Consistent with this notion, functions that describe the scaling relationship of disturbed and successional systems are reported to diverge from pre-disturbance scaling relationships, and in time attract back towards them, reflecting system resilience.

The Kerkhoff and Enquist paper published in 2007 shows how scaling relationships could be used to help understand and measure resilience. They compared the power law scaling relationship of mean tree size and total stem density for plants within disturbed and undisturbed plots of lowland forests in Nicaragua (Boucher *et al.* 2000) and *Quercus*-dominated forests in Costa Rica (Kappelle 1995; Figure 35.2). The scaling relationships of disturbed systems diverged from those of pre-disturbed sites, however over time conditions within disturbed sites recovered, seemingly following a trajectory back towards the structural patterns measured in the pre-disturbed sites. Additionally, they compared this trajectory to the self-thinning line also referred to as the energetic equivalence rule (EER) which depicts the regular decrease in plant population density as its members increase in size. The EER represents a saturated plant community at equilibrium, where total resource use equals rates of supply and where recruitment and growth are continuously offset by mortality (Kerkhoff and Enquist 2006). Kerkhoff

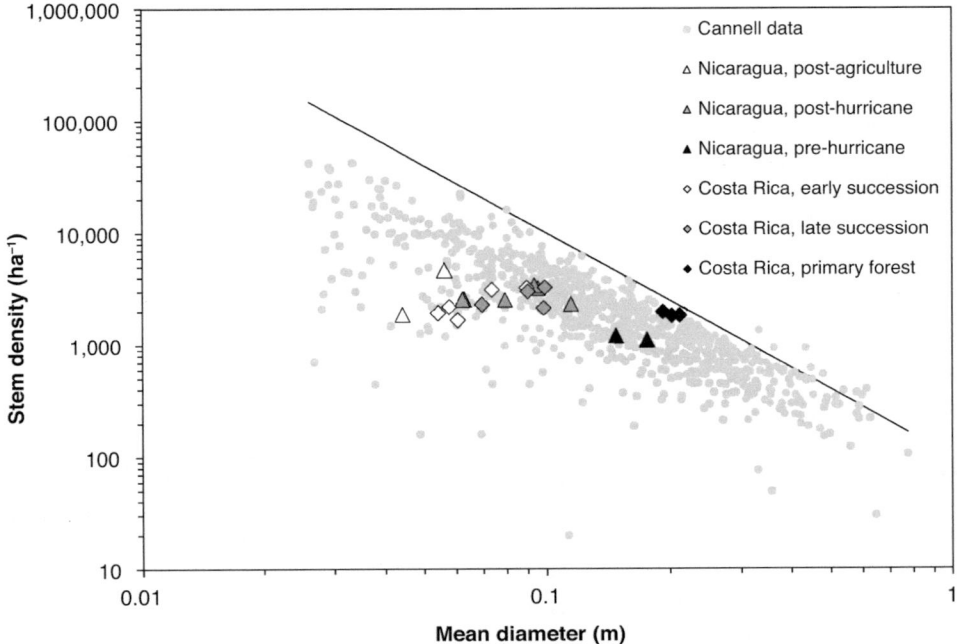

Figure 35.2 A cross-site comparison of stem density as a function of mean stem diameter, showing data from stands in Nicaragua (triangles; Boucher *et al.* 2000) and Costa Rica (diamonds; Kappelle 1995). Shading represents different degrees of disturbance (Nicaragua) or stages of succession (Costa Rica). Grey circles are reference data (Cannell 1982), and the line is the energetic equivalence rule (EER) prediction of −2, plotted here for reference. Note that more disturbed sites (white and grey symbols) tend to lie farther from the centroid of the data, for which the EER line appears to represent a reasonable constraint

Source: reproduced from Kerkhoff and Enquist (2007)

and Enquist suggested that the EER appeared to be acting as an attractor, supporting the idea that the scaling relationship represented a stable equilibrium, and moreover, that divergence and return to these attractors could be used to quantify resilience (Kerkhoff and Enquist 2007; Anfodillo *et al.* 2013; Rosenfeld and Tonn 2014).

It has been argued that models other than those based on power law functions can describe scaling relationships with a higher degree of accuracy and therefore may be more appropriate for measuring changes in ecosystem structuring and functioning. Specifically, the maximum entropy theory of ecology (MaxEnt), which uses information theory to generate predictions about ecosystem energetics, structure and functioning, is reported to outperform power law scaling at predicting and describing species–area relationships across various taxa, habitat types and spatial scales (Harte and Newman 2014; Wilber *et al.* 2015). The information required to develop some MaxEnt predictions is generated from a few, simple, commonly measured variables: the number of species, the number of individuals, and the area of the system. Additional predictions can be derived from less commonly measured variables such as the metabolic rate summed over individuals (derived from proxies, e.g. organism size), or the number of trophic links in a network (Harte 2011; Williams 2010). A valuable feature of maximum entropy theory is that it predicts a universal scale collapse of all species–area curves onto one universal curve (Harte *et al.* 2009). This collapse occurs when the log-log slopes of the species–area relationships of different communities are plotted as a function of the ratio of the average total abundance at that scale to average species richness at that scale. The scale collapse can accurately predict the species–area distributions within numerous systems ranging from aquatic, marine to terrestrial (Harte 2011). The systems whose MaxEnt species–area relationships deviate from the universal curve may be systems undergoing some level of structural reorganization. For example, deviations have been reported for Hawaiian arthropod communities in stages of succession (*ibid.*) and moth communities undergoing rapid change after a disturbance (Kempton and Taylor 1974). Whether these deviations can reliably predict resilience to disturbance is an open question that requires further empirical testing. This testing is necessary before MaxEnt can be further developed as a guiding tool for resilience management.

Conclusion

Resilience has obvious importance for the management and restoration of ecological systems. Conceptualizations of ecological resilience and engineering resilience have primarily guided efforts into understanding resilience, however much still remains unknown. Questions regarding ecological thresholds, how much disturbance a system can absorb before switching to another state and recovery of a system after disturbance are still largely unanswered. Determining alternative stable states and thresholds has proved challenging, and although there are examples where the identification of thresholds and agents responsible for switches between ecosystem states has been successful, much of this understanding has come from observing systems moving into alternative stable states, which is not a practical approach for systems with unknown thresholds to disturbance. Because the quantification of resilience is desirable for management, empirical approaches aimed at determining thresholds could be complemented by theoretical approaches that have a predicative capacity. To construct a framework for quantifying resilience, a clear conceptualization of resilience is needed, and a comprehensive understanding of the attributes that contribute to resilience. Research into identifying and understanding such attributes, though still in its infancy, has so far determined diversity to be a key contributor to resilience, and therefore restoration to restore biological diversity appears to be a logical first step towards restoring resilience.

Research has also identified the positive effects of modularity on resilience, and although there are few directives on how to intervene to maintain or create a modular structure, large-scale restoration projects offer unique test-beds for developing ideas. Heterogeneity in landscapes is known to contribute to the coexistence and persistence of similar species (Tilman 1994) and in doing so, heterogeneity may foster functional redundancy and response diversity within a system (DesJardins *et al.* 2015). Additionally, heterogeneity has been associated with increased modularity (Mumby and Hastings 2008; Galstyan and Cohen 2007). From this, restoration projects could strive to create heterogeneous landscapes in an attempt to promote resilience. A key aspect of this is aiming to have a variety of management approaches, that is, not doing the same thing everywhere.

While the presence of diversity, heterogeneity, modularity and connectivity within a system may indicate resilience, quantification of the capacity of these attributes to buffer a system against a disturbance is challenging and impractical. Ecological proxies or signatures of resilience that are easily measured remain hard to find. However, research into the use of scaling relationships and the universal scale collapse as potential attractors of resilience, appears promising, though in need of empirical verification. A key feature of these approaches is that they allow for ecological systems to be compared against each other, regardless of their spatial scale, the disturbance regime, the taxa that have been censused or the size of the resident organisms. Whether this can be made applicable to particular restoration and management situations remains to be determined.

In summary, the ideas and concepts around resilience have an intuitive appeal and obvious relevance to restoration, and have caught the imagination of policy makers and managers alike. The science behind these ideas remains mostly conceptual, but there are ongoing efforts to hone the concepts and find measurable and manageable ecosystem characteristics that relate to resilience. The development of these approaches is an exciting prospect for people working at the nexus of ecological theory and restoration practice and motivated to halt the ongoing decline in global biodiversity.

References

Allen, C. R., Angeler, D. G., Garmestani, A. S., Gunderson, L. H. and Holling, C. (2014). Panarchy: theory and application. *Ecosystems*, *17*(4), 578–589.

Anfodillo, T., Carrer, M., Simini, F., Popa, I., Banavar, J. R. and Maritan, A. (2013). An allometry-based approach for understanding forest structure, predicting tree-size distribution and assessing the degree of disturbance. *Proceedings of the Royal Society of London B: Biological Sciences*, *280*(1751), 2012–2375.

Beisner, B. (2012). Alternative stable states. *Nature Education Knowledge*, *3*(10), 33.

Bellwood, D. R., Hughes, T. P., Folke, C. and Nyström, M. (2004). Confronting the coral reef crisis. *Nature*, *429*(6994), 827–833.

Bestelmeyer, B. T., Duniway, M. C., James, D. K., Burkett, L. M. and Havstad, K. M. (2013). A test of critical thresholds and their indicators in a desertification prone ecosystem: more resilience than we thought. *Ecology Letters*, *16*(3), 339–345.

Boucher, D. H., Vandermeer, J. H., de la Cerda, I. G., Mallona, M. A., Perfecto, I. and Zamora, N. (2000). Post-agriculture versus post-hurricane succession in southeastern Nicaraguan rain forest. *Plant Ecology*, *156*(2), 131–137.

Brand, F. S. and Jax, K. (2007). Focusing the meaning(s) of resilience: resilience as a descriptive concept and a boundary object. *Ecology and Society*, *12*(1), 23.

Brown, J. H. (1984). On the relationship between abundance and distribution of species. *American Naturalist*, *124(2)*, 255–279.

Byun, C., Blois, S. and Brisson, J. (2013). Plant functional group identity and diversity determine biotic resistance to invasion by an exotic grass. *Journal of Ecology*, *101*(1), 128–139.

Cannell, M. G. (1982). *World Forest Biomass and Primary Production Data*. Academic Press, London.

Cardinale, B. J., Duffy, J. E., Gonzalez, A., Hooper, D. U., Perrings, C., Venail, P., Narwani, A., Mace, G. M., Tilman, D. and Wardle, D. A. (2012). Biodiversity loss and its impact on humanity. *Nature, 486*(7401), 59–67.

Carpenter, S. R., Cole, J. J., Pace, M. L., Batt, R., Brock, W., Cline, T., Coloso, J., Hodgson, J. R., Kitchell, J. F. and Seekell, D. A. (2011). Early warnings of regime shifts: a whole-ecosystem experiment. *Science, 332*(6003), 1079–1082.

Cramer, V. A., Hobbs, R. J. and Standish, R. J. (2008). What's new about old fields? Land abandonment and ecosystem assembly. *Trends in Ecology and Evolution, 23*(2), 104–112.

De Bello, F., Lepš, J., Sebastià, M.-T. and Pärtel, M. (2007). Grazing effects on the species–area relationship: variation along a climatic gradient in NE Spain. *Journal of Vegetation Science, 18*(1), 25–34.

DesJardins, E., Barker, G., Lindo, Z., Dieleman, C. and Dussault, A. C. (2015). Promoting resilience. *The Quarterly Review of Biology, 90*(2), 147–165.

Diaz, S. and Cabido, M. (2001). Vive la difference: plant functional diversity matters to ecosystem processes. *Trends in Ecology and Evolution, 16*(11), 646–655.

Filotas, E., Parrott, L., Burton, P. J., Chazdon, R. L., Coates, K. D., Coll, L., Haeussler, S., Martin, K., Nocentini, S. and Puettmann, K. J. (2014). Viewing forests through the lens of complex systems science. *Ecosphere, 5*(1), article 1.

Fischer, J. and Lindenmayer, D. B. (2007). Landscape modification and habitat fragmentation: a synthesis. *Global Ecology and Biogeography, 16*(3), 265–280.

Galstyan, A. and Cohen, P. (2007). Cascading dynamics in modular networks. *Physical Review. E, 75*(3 Pt 2), 036109–036109.

Griffin, J. N., O'Gorman, E. J., Emmerson, M. C., Jenkins, S. R., Klein, A.-M., Loreau, M. and Symstad, A. (2009). Biodiversity and the stability of ecosystem functioning. *Biodiversity, Ecosystem Functioning and Human Wellbeing: An Ecological and Economic Perspective*, 78–93.

Gunderson, L. H., Allen, C. R. and Holling, C. S. (eds) (2010). *Foundations of Ecological Resilience.* Washington, DC: Island Press.

Hansen, G. J., Ives, A. R., Vander Zanden, M. J. and Carpenter, S. R. (2013). Are rapid transitions between invasive and native species caused by alternative stable states, and does it matter? *Ecology, 94*(10), 2207–2219.

Harris, J. and Hobbs, R. J. (2001). Clinical practice for ecosystem health: the role of ecological restoration. *Ecosystem Health, 7*(4), 195–202.

Harte, J. (2011). *Maximum Entropy and Ecology: A Theory of Abundance, Distribution, and Energetics.* Oxford University Press, Oxford.

Harte, J. and Newman, E. A. (2014). Maximum information entropy: a foundation for ecological theory. *Trends in Ecology and Evolution, 29*(7), 384–389.

Harte, J., Smith, A. B. and Storch, D. (2009). Biodiversity scales from plots to biomes with a universal species–area curve. *Ecology Letters, 12*(8), 789–797.

Hector, A., Hautier, Y., Saner, P., Wacker, L., Bagchi, R., Joshi, J., Scherer-Lorenzen, M., Spehn, E., Bazeley-White, E. and Weilenmann, M. (2010). General stabilizing effects of plant diversity on grassland productivity through population asynchrony and overyielding. *Ecology, 91*(8), 2213-2220.

Hobbs, R. J., Higgs, E. S. and Hall, C. (2013). *Novel Ecosystems: Intervening in the New Ecological World Order.* Wiley-Blackwell, Chichester.

Holling, C. S. (1973). Resilience and stability of ecological systems. *Annual Review of Ecology and Systematics, 4*, 1–23.

Kappelle, M. (1995). *Ecology of Mature and Recovering Talamancan Montane* Quercus *Forests, Costa Rica.* University of Amsterdam, Amsterdam.

Kempton, R. and Taylor, L. (1974). Log-series and log-normal parameters as diversity discriminants for the Lepidoptera. *The Journal of Animal Ecology, 43*(2), 381–399.

Kerkhoff, A. J. and Enquist, B. J. (2006). Ecosystem allometry: the scaling of nutrient stocks and primary productivity across plant communities. *Ecology Letters, 9*(4), 419–427.

Kerkhoff, A. J. and Enquist, B. J. (2007). The implications of scaling approaches for understanding resilience and reorganization in ecosystems. *Bioscience, 57*(6), 489–499.

Lansing, J. S. (2003). Complex adaptive systems. *Annual Review of Anthropology, 32*, 183–204.

Leary, D. J. and Petchey, O. L. (2009). Testing a biological mechanism of the insurance hypothesis in experimental aquatic communities. *Journal of Animal Ecology, 78*(6), 1143–1151.

Lehman, C. L. and Tilman, D. (2000). Biodiversity, stability, and productivity in competitive communities. *The American Naturalist, 156*(5), 534–552.

Lewontin, R. C. (1969). The meaning of stability. *Brookhaven Symposia in Biology, 22*, 13–23.

Loreau, M. and Mazancourt, C. (2013). Biodiversity and ecosystem stability: a synthesis of underlying mechanisms. *Ecology Letters*, *16*(s1), 106–115.

MacArthur, R. (1955). Fluctuations of animal populations and a measure of community stability. *Ecology*, *36*(3), 533–536.

Marquet, P. A., Allen, A. P., Brown, J. H., Dunne, J., Enquist, B. J. *et al.* (2014). On theory in ecology. *BioScience*, *64*(8), 701–710.

Martin, S., Deffuant, G. and Calabrese, J. M. (2011). Defining resilience mathematically: from attractors to viability. In G. Deffuant and N. Gilbert (eds), *Viability and Resilience of Complex Systems*, pp. 15–36. Kluwer Academic Publishers, Boston, MA.

Mills, L. S., Soulé, M. E. and Doak, D. F. (1993). The keystone-species concept in ecology and conservation. *Bioscience*, *43*(4), 219–224.

Mumby, P. J. and Hastings, A. (2008). The impact of ecosystem connectivity on coral reef resilience. *Journal of Applied Ecology*, *45*(3), 854–862.

Murcia, C., Aronson, J., Kattan, G. H., Moreno-Mateos, D., Dixon, K. and Simberloff, D. (2013). A critique of the 'novel ecosystem' concept. *Trends in Ecology and Evolution*, *29*(10), 548–553.

Myers-Smith, I. H., Trefry, S. A. and Swarbrick, V. J. (2012). Resilience: easy to use but hard to define. *Ideas in Ecology and Evolution*, *5*, 44–53.

Nash, K. L., Allen, C. R., Angeler, D. G., Barichievy, C., Eason, T., Garmestani, A. S., Graham, N. A., Granholm, D., Knutson, M. and Nelson, R. J. (2014). Discontinuities, cross-scale patterns and the organization of ecosystems. *Ecology*, *95*(3), 654–667.

Nash, K. L., Graham, N. A., Jennings, S., Wilson, S. K. and Bellwood, D. R. (2015). Herbivore cross scale redundancy supports response diversity and promotes coral reef resilience. *Journal of Applied Ecology*, *53*, 646–655.

Newman, M. E. (2003). The structure and function of complex networks. *SIAM Review*, *45*(2), 167–256.

Perring, M. P., Standish, R. J., Price, J. N., Craig, M. D., Erickson, T. E., Ruthrof, K. X., Whiteley, A. S., Valentine, L. E. and Hobbs, R. J. (2015). Advances in restoration ecology: rising to the challenges of the coming decades. *Ecosphere*, *6*(8), article 131.

Pimm, S. L. (1984). The complexity and stability of ecosystems. *Nature*, *307*(5949), 321–326.

Qian, S. S. (2014). Ecological threshold and environmental management: a note on statistical methods for detecting thresholds. *Ecological Indicators*, *38*, 192–197.

Rosenfeld, J. S. and Tonn, W. (2014). Modelling the effects of habitat on self-thinning, energy equivalence, and optimal habitat structure for juvenile trout. *Canadian Journal of Fisheries and Aquatic Sciences*, *71*(9), 1395–1406.

Scheffer, M., Carpenter, S. R., Lenton, T. M., Bascompte, J., Brock, W., Dakos, V., Van De Koppel, J., Van De Leemput, I. A., Levin, S. A. and Van Nes, E. H. (2012). Anticipating critical transitions. *Science*, *338*(6105), 344–348.

Shackelford, N., Hobbs, R. J., Burgar, J. M., Erickson, T. E., Fontaine, J. B., Laliberté, E., Ramalho, C. E., Perring, M. P. and Standish, R. J. (2013). Primed for change: developing ecological restoration for the 21st century. *Restoration Ecology*, *21*(3), 297–304.

Shackelford, N., Starzomski, B. Banning, N., Battaglia, L. L., Becker, A., Bellingham, P., Bestelmeyer, B., Catford, J., Dwyer, J., Dynesius, M., Gilmour, J., Hallett, L. M., Hobbs, R. J., Price, J. Sasaki, T., Tanner, E. V. J., Standish, R. J. (2017). Isolation predicts compositional change after discrete disturbances in a global meta-study. *Ecography*. DOI: 10.1111/ecog.02383

Society of Ecological Restoration. (2004). *The SER International Primer on Ecological Restoration*. Society for Ecological Restoration International, Tucson, AZ.

Standish, R. J., Hobbs, R. J., Mayfield, M. M., Bestelmeyer, B. T., Suding, K. N., Battaglia, L. L., Eviner, V., Hawkes, C. V., Temperton, V. M. and Cramer, V. A. (2014). Resilience in ecology: abstraction, distraction, or where the action is? *Biological Conservation*, *177*, 43–51.

Suding, K. N. and Hobbs, R. J. (2009). Threshold models in restoration and conservation: a developing framework. *Trends in Ecology and Evolution*, *24*(5), 271–279.

Sundstrom, S. M., Allen, C. R. and Barichievy, C. (2012). Species, functional groups, and thresholds in ecological resilience. *Conservation Biology*, *26*(2), 305–314.

Thébault, E. and Loreau, M. (2003). Food-web constraints on biodiversity–ecosystem functioning relationships. *Proceedings of the National Academy of Sciences*, *100*(25), 14949–14954.

Thibaut, L. M., Connolly, S. R. and Sweatman, H. P. (2012). Diversity and stability of herbivorous fishes on coral reefs. *Ecology*, *93*(4), 891–901.

Tilman, D. (1994). Competition and biodiversity in spatially structured habitats. *Ecology*, *75*(1), 2–16.

Tilman, D., Isbell, F., and Cowles, J. M. (2014). Biodiversity and ecosystem functioning. *Annual Review of Ecology, Evolution and Systematics, 45(1)*, 471.

Trueman, M., Standish, R. J. and Hobbs, R. J. (2014). Identifying management options for modified vegetation: Application of the novel ecosystems framework to a case study in the Galapagos Islands. *Biological Conservation, 172*, 37–48.

Twidwell, D., Fuhlendorf, S. D., Taylor, C. A. and Rogers, W. E. (2013). Refining thresholds in coupled fire–vegetation models to improve management of encroaching woody plants in grasslands. *Journal of Applied Ecology, 50*(3), 603–613.

Walker, B. and Salt, D. (2012). *Resilience Practice: Building Capacity to Absorb Disturbance and Maintain Function.* Island Press, Washington, DC.

West, G. B., Enquist, B. J. and Brown, J. H. (2009). A general quantitative theory of forest structure and dynamics. *Proceedings of the National Academy of Sciences, 106*(17), 7040–7045.

Wilber, M. Q., Kitzes, J. and Harte, J. (2015). Scale collapse and the emergence of the power law species–area relationship. *Global Ecology and Biogeography, 24*(8), 883–895.

Williams, R. J. (2010). Simple MaxEnt models explain food web degree distributions. *Theoretical Ecology, 3*(1), 45–52.

36

ECOLOGICAL RESTORATION AND ECOSYSTEM SERVICES

Robin L. Chazdon and José M. Rey Benayas

Ecosystems are the life support system for all of earth's creatures. As originally defined by Daily (1997), *ecosystem services* are 'the conditions and processes through which natural ecosystems, and the species that make them up, sustain and fulfill human life'. Collectively speaking, ecosystem services (ES) are the benefits that humans receive from nature. These include both goods (products) and services (processes) that generate livelihoods and support economic activity; provide food, water and shelter; cleanse the air; regulate climate; mitigate extreme weather events; and provide spiritual and recreational value to our lives. The list of goods and services that ecosystems provide is virtually endless; what is or is not an ecosystem service is determined by the context of human needs. For example, one large deer can provide all the meat needed by a family for an entire year. But deer also mediate tick-borne human disease and consume backyard vegetable gardens, ornamental plantings, and natural vegetation. One person's ES can be another person's 'disservice'.

As ecosystems become degraded, transformed, or simplified, the quantity, quality, and value of the goods and services they produce often declines. The Millennium Ecosystem Assessment (MEA 2005) documented deterioration of 60 per cent of ecosystem services globally. Costanza *et al.* (2014) compared changes in the estimated dollar value of the earth's ecosystem services in 1997 and in 2011. They concluded that global land-use changes over these 15 years resulted in an estimated loss of US$4.3 to US$20.2 billion per year.

In this chapter, we first address the question: 'Can ecological restoration reverse losses in ecosystem services due to degradation or conversion to simplified agricultural production systems?' We present five case studies to illustrate how changes in different ES are measured to assess outcomes of active and passive restoration in a range of ecosystems. We then present results of meta-analyses that examine the recovery of ES and biodiversity during ecological restoration. We conclude with a discussion about the scope and promise of ecosystem service research in the broad socio-ecological context of ecosystem and landscape restoration.

What are ecosystem services?

The *SER Primer on Ecological Restoration* (Society for Ecological Restoration 2004) defines ecological restoration as 'the process of assisting the recovery of an ecosystem that has been degraded, damaged, or destroyed'. We focus here on the recovery of ES in ecosystems that have been transformed or destroyed, and where active restoration interventions or natural regeneration processes (assisted or unassisted) are taking place. Assessing the recovery of ES requires a baseline for comparison. In many cases, restored areas are compared to a 'reference' ecosystem that has not been subjected to degradation or destruction and is a proxy for the intact historical ecosystem. In other cases, reference ecosystems no longer exist, and recovery is defined operationally as a desired state or condition that represents significant improvement over the state prior to restoration.

The MEA (2005) brought the concept of ES to the world stage and into the policy arena. Ecosystem services were divided into four general categories (Figure 36.1):

1   provisioning services such as food, fresh water, fuel, fibre and genetic resources;
2   regulating services such as climate, water and disease regulation as well as pollination and seed dispersal;
3   cultural services such as educational, aesthetic, spiritual and cultural heritage values as well as recreation and tourism; and
4   supporting services such as soil formation, nutrient cycling and primary production.

Although biodiversity is not classified as an ES *per se*, many assessments consider both attributes together, as they are often closely associated (Mace *et al.* 2012). The biotic composition of ecosystems forms the basis of the ecosystem functions that supply ES. But the extent to which particular ES are linked to biodiversity remains an open question, particularly in the context of ecological restoration (Bullock *et al.* 2011). Moreover, the wide range of ecological processes that form the basis for ES often leads to trade-offs as well as synergies (Bennett *et al.* 2009;

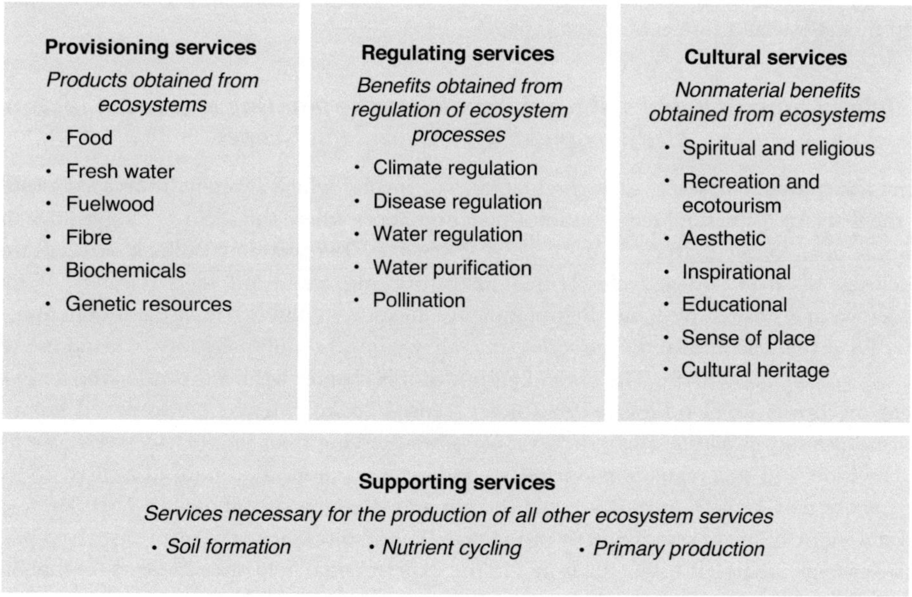

Provisioning services

Products obtained from ecosystems

- Food
- Fresh water
- Fuelwood
- Fibre
- Biochemicals
- Genetic resources

Regulating services

Benefits obtained from regulation of ecosystem processes

- Climate regulation
- Disease regulation
- Water regulation
- Water purification
- Pollination

Cultural services

Nonmaterial benefits obtained from ecosystems

- Spiritual and religious
- Recreation and ecotourism
- Aesthetic
- Inspirational
- Educational
- Sense of place
- Cultural heritage

Supporting services

Services necessary for the production of all other ecosystem services

- Soil formation   • Nutrient cycling   • Primary production

Figure 36.1   Types of services provided by ecosystems. Supporting services are the basis for production of all other ecosystem services

Source: based on MEA (2005)

Nelson *et al.* 2009; Rey Benayas and Bullock 2012). Trade-offs can lead to conflicts among stakeholders who assign different values to different ES (Howe *et al.* 2014). In many cases, increases in provisioning services correlates with decreases in regulating and cultural services (Raudsepp-Hearne *et al.* 2010).

During the restoration process, different types of ES recover at different rates. For example the recovery of soil carbon storage may lag behind the recovery of aboveground carbon storage (Farley *et al.* 2004; Lü *et al.* 2014). Those services that recover more quickly are most useful as indicators of early restoration success, but may not be useful indicators of long-term restoration success. Thus, in assessing the success of ecological restoration interventions in achieving recovery of ES, it is important to consider services in different categories that recover over different time scales.

The definition and assessment of ES formalizes a utilitarian framework for management and valuation of natural processes (Palmer *et al.* 2014a). As restoration ecology grew from its origins in conservation and emphasis on restoration of the historical composition of species, a shift towards emphasis on ES has strengthened the linkage with natural resource management interests and institutions (Bullock *et al.* 2011). Recovery of ES, rather than reconstruction of prior species composition, therefore provides the criteria for success for many restoration interventions today, particularly for restoration efforts focused at the landscape scale.

Case studies

Here we review five case studies that demonstrate how ecological restoration in different types of ecosystems and using different approaches can impact recovery of a wide range of ES. Our case studies focus on Mediterranean agricultural landscapes; agricultural production landscapes in Spain; degraded croplands and barren lands in China; stream watersheds in the Chesapeake Bay watershed, USA; and carbon storage in soils and vegetation during tropical forest regeneration in wet lowland forests of Costa Rica.

Habitat provision and soil regulation following planting of wooded islets in Mediterranean agricultural landscapes

Many agricultural landscapes lack parental trees or shrubs, which severely limits seed availability, the first step for natural regeneration in circumstances where the socio-ecological dynamics promote abandonment (Rey Benayas *et al.* 2008; Rey Benayas and Bullock 2012; Ramos-Palacios *et al.* 2014). In actively farmed fields, planting woodland islets (clusters of trees), hedgerows and isolated trees has the potential to enhance wildlife, agricultural production, and other ES at the field and landscape scales since they compete minimally for farmland use (Rey Benayas and Bullock 2015). The second author of this chapter has been conducting an experiment on former cropland located in Toledo (central Spain), where he introduced holm oak (*Quercus ilex* susbsp. *ballota*, a late successional, slow-growing tree) in 1993 to investigate habitat provision and soil regulation during secondary succession. The indicators used to assess these ecosystem services were oak establishment and measures of soil quality variables.

Fifty-eight holm oak individuals more than 1 year old that grew from dispersed acorns outside the planted islets have established in the experimental field after 21 years (Figure 36.2), resulting in a density of 65.1 oaks per ha with an average establishment rate of 3.3 oaks ha^{-1} yr^{-1}. Height, basal diameter and crown projected area of the established oaks averaged 49.2 ± 64.3 cm, 20.1 ± 31.3 mm, and 0.44 ± 1.09 m^2, respectively (Rey Benayas *et al.* 2015). However, the woodland islets remain with the original planted surface of 100 m^2. Initial oak

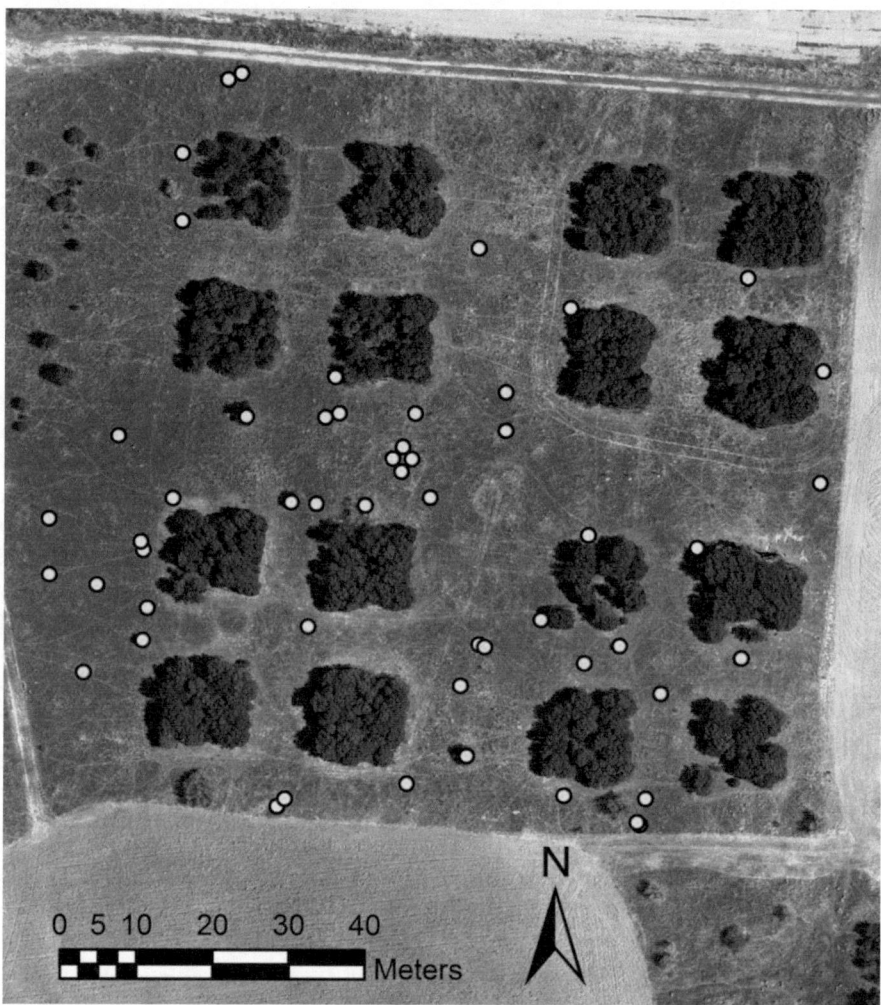

Figure 36.2   Position of the 58 naturally established oaks more than 1 year old in 'La Higueruela' (Toledo, Spain) experimental field where sixteen 100 m² woodland islets were planted in 1993

Source: Rey Benayas *et al.* (2015)

regeneration triggered by small islets planted in Mediterranean abandoned farmland is slowed by high acorn predation, seedling herbivory, stressful microclimatic conditions and competition from the established islets. Regardless, these islets are viable tools for regeneration of Mediterranean oak woodland.

Cuesta *et al.* (2012) experimentally tested the response of soil chemical properties to secondary succession (old field) and to *Q. ilex* plantation when this study cropland was abandoned 13 years ago. Carbon and NH_4^+–N concentrations and availability of mineral N were higher in the planted woodland islets than in the old field. However, soil pH, total N, P, K and NO_3^-–N concentrations, mineralization rates, and available PO_4^{3-}–P were similar. Previous management practices to establish the woodland islets and current environmental conditions affected soil chemistry. Irrigation increased K and P concentrations and NH_4^+–N availability, herb biomass enhanced C concentration, and composition of the herbaceous community increased N availability if it was dominated by legumes but reduced it if it was dominated by grasses and forbs.

As oak woodland regeneration is slow, this woodland restoration approach allows for the coexistence of open farmland species and woodland species. Mediterranean pastures are highly diverse (Cayuela *et al.* 2008). A recent survey of the herbaceous plant communities around the islets found 73 species, 10 of which are perennial.

The Fields for Life project – a practitioner's perspective

The International Foundation for Ecosystem Restoration (FIRE, www.fundacionfire.org) aims at translating academic knowledge to ecosystem restoration in the real world. It provides leadership in implementing restoration actions in farmland habitat and farmland stewardship in Spain by means of its Fields for Life project, which targets reconciliation of agricultural production and wildlife enhancement. Since 2008, this project has revegetated more than 9 km of hedgerows and more than 3 ha of woodland islets of different sizes with roughly 15,000 seedlings of 32 native species, introduced 11 artificial ponds and several hundred artificial nests for insectivorous birds and 121 for birds of prey, and has completed 12 signed stewardship agreements with land owners, among other achievements, including the participation of hundreds of volunteers in such actions. To date, the project covers ca. 4000 ha. One hypothesis of this project is that strategic restoration of agroecosystems aimed at increasing populations of insectivorous birds can reduce the incidence of agricultural arthropod pests and thereby provide effective biological control.

The effects of the above mentioned restoration actions on the diversity and abundance of insectivorous birds as well as on the quantity and quality of crop harvests were analysed. Nest box occupancy with hatched eggs or recruit production was, on average, as high as 26 per cent during the first year and 53 per cent during the fourth year. The predation rate of sentinel caterpillars was up to 33 per cent higher near active nest boxes than in the corresponding distant areas without nest boxes. Estimated arthropod consumption by birds was as high as 1009.06 kg per year in the approximately 200 ha vineyards of Abadía Retuerta (Valladolid). Economic analysis suggests that using insectivorous birds as part of the trademarked 'Fields for Life' approach developed by FIRE can reduce the cost of combatting agricultural pests while increasing revenues from wine and oil production. In addition to pest regulation, this approach can provide other ecological benefits. For example, a planted hedgerow in Campo de Montiel was found to contain 107 herb species. Additionally, a meta-analysis of 13 studies based on 51 indicators of harvest quantity and quality showed that these measures were 40 per cent more positive in the presence of insectivorous birds than in their absence. To ensure the success of this approach, field operations are conducted in conjunction with education and training for agricultural workers, as well as environmental education programmes involving volunteering, working side-by-side with agricultural workers, technical seminars, presentations in schools and student projects.

China's Grain for Green programme: mixed results on the impact of ecosystem services

Following devastating flooding of China's major river basins that killed over 3000 people, caused over $12 billion in damages, and resulted in the loss of 5 million ha of crops in 1987–

1988, China embarked on the world's largest restoration project, with the goal of mitigating soil erosion, reducing flooding and alleviating rural poverty. Also known as the Conversion of Cropland to Forest and Grassland programme or the Sloping Land Conversion programme, the Grain for Green (GFG) programme has grown to involve more than 32 million rural house-holds in the world's largest Payments for Ecosystem Services programme (Figure 36.3). The government decreed that all lands with a slope of over 25 degrees be converted to forest, terraced orchards, or grassland. By the end of 2012, restoration interventions were implemented on 27.2 million ha; 56 per cent was reforestation/afforestation on barren land, 33 per cent was reforestation/afforestation on cropland, 10 per cent was natural forest regeneration, and 1 per cent was conversion from cropland to grassland (Shi and Han 2014).

Across the entire GFG region, conversion of croplands to grasslands and tree plantations significantly increased aboveground carbon storage and soil organic carbon after 16–20 yr (Persson *et al.* 2013; Song *et al.* 2014). Although natural forests occur within the broad region of the GFG programme (Figure 36.3), many regions are too arid to naturally support tree growth. Within the Loess Plateau region, the highest gains in net primary production (NPP) were in areas restored to shrubland. Across all vegetation zones, vegetation cover increased, annual NPP increased, and soil carbon storage increased from 2000 to 2008. During this period, the entire region shifted from being a net carbon source to a net carbon sink (Feng *et al.* 2013).

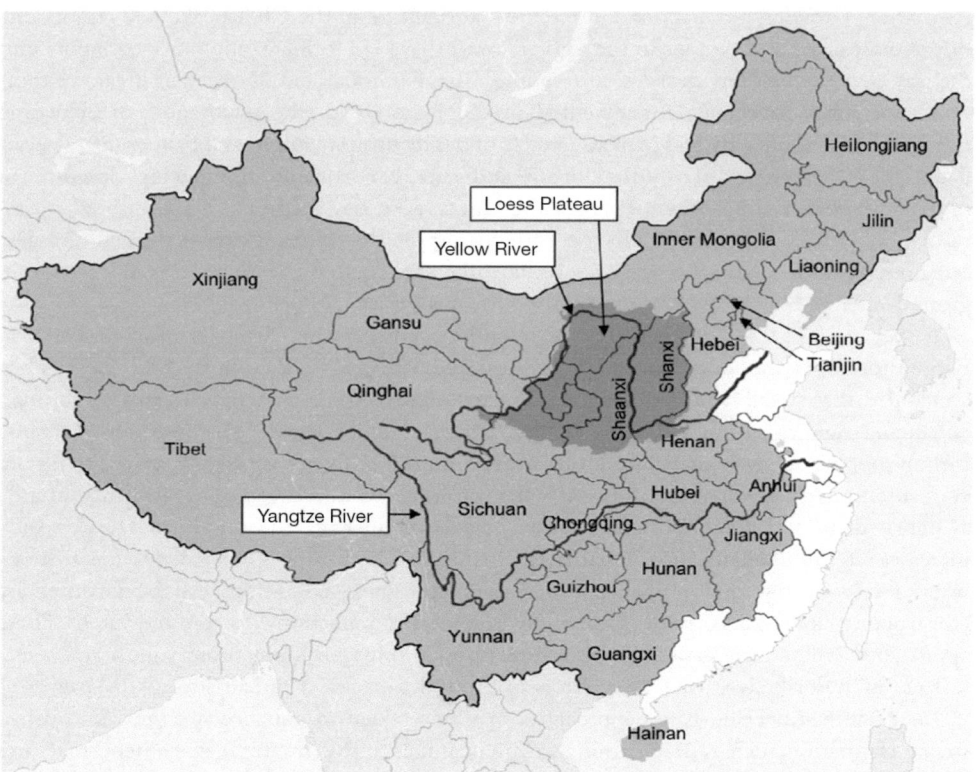

Figure 36.3   The coverage of the Grain for Green programme in China, showing the sensitive areas of the Loess Plateau in darker grey around the Yellow River

Source: Persson *et al.* (2013)

Afforestation did not uniformly have a positive effect on the provision of ES, however. Restoration efforts successfully reduced soil erosion and runoff in only half of the study areas within the Loess Plateau (Lü *et al.* 2012). As plantations of the nitrogen-fixing tree *Robinia pseudoacacia* (black locust) grew older, aboveground carbon storage and soil nitrogen increased, while soil water availability decreased (Lü *et al.* 2014). High rates of mortality of planted trees and lack of protective herbaceous ground cover in afforested areas on thick loess soils exacerbated – rather than improved – land degradation (Cao *et al.* 2011). In arid and semi-arid conditions within the Loess Plateau, afforestation caused long-term declines of soil moisture and reductions in water yield, damaging local river ecosystems and causing further soil erosion (Jiang *et al.* 2015). One lesson to be learned from the large-scale GFG programme is that recovery of multiple ES is most effective when restoration interventions are suited to the ecological constraints of local soil and rainfall conditions, and when native species and growth forms are planted.

Recovery of nitrogen processing in urban and suburban streams in the Chesapeake Bay region, USA

Altered hydrology and biogeochemistry of urban and suburban streams impair the capacity to process and remove excess nitrogen from stream water (Palmer *et al.* 2014b). Nitrogen retention is an essential service of stream ecosystems, and is therefore a major goal of stream restoration. Growing populations and intensive agriculture in the Chesapeake Bay region, one of the most urbanized regions in the United States, have led to increasing nitrogen inputs into the Bay over the past few decades. In response, a large number and diversity of stream restoration approaches have been implemented in the region with the major goal of increasing nitrogen removal capacity and increasing rates of denitrification. Filoso and Palmer (2011) evaluated the effectiveness of restored urban and suburban streams in reducing downstream nitrogen exports in the Coastal Plain of the Chesapeake Bay Region. Of six restored stream reaches, only two were clearly effective in reducing the downstream export of total nitrogen. Stream restoration approaches were not sufficiently effective in preventing nitrogen inputs from groundwater and bank seepage into restored streams (*ibid.*).

Mayer *et al.* (2013) compared nitrate retention and sediment denitrification potential in two restored streams and two degraded restored streams in Baltimore, Maryland. They concluded that stream restoration strategies can improve nitrogen removal capacity through enhancing denitrification by heterotrophic bacteria under anaerobic conditions. Organic carbon supply and levels of dissolved organic carbon (DOC) were negatively related to nitrate concentrations. Restoring stream features that promote accumulation of organic matter, such as debris dams and plant roots in riparian vegetation increase the supply of DOC, which increases rates of denitrification. Further, denitrification in streams is enhanced when groundwater levels are low and residence times are high. These conditions can be favoured by constructing low banks with greater hydrological connections to groundwater. These approaches, when complemented with reductions in nitrogen inputs from runoff and sewer leakage in watersheds, can restore nitrogen removal capacity in urban streams (Mayer *et al.* 2013). New engineering-based approaches may be needed to enhance specific ES in urban stream restoration, such as transforming stream channels to stormwater management structures designed to reduce peak flows and to enhance hydraulic retention of water flows (Palmer *et al.* 2014a).

Carbon accumulation in soils and trees during natural regeneration in wet lowland tropical forests of Costa Rica

During spontaneous regrowth of tropical forests under favourable conditions, forest structure, biomass, and species composition change dramatically, creating conditions favourable to the colonization and establishment of plant and animal species characteristic of old-growth forests (Chazdon 2014). In these cases, active restoration interventions are not needed, and multiple ES can recover through the natural regeneration process. Biomass growth is a particularly important indicator of carbon sequestration (a supporting ecosystem service), as 45–50 per cent of tree biomass is composed of carbon. Carbon is also stored in other components of the ecosystem, such as coarse and fine litter, soil organic layer, dead plant material, and deeper layers of the soil profile. Thus, regenerating tropical forests are important not only for biodiversity conservation, but also for mitigating carbon emissions through high rates of carbon sequestration and their substantial contribution to total forest cover (Pan *et al.* 2011; Chazdon *et al.* 2016).

Recovery of aboveground biomass during natural regeneration in different climate zones and forest types can take from 55 to 189 years (Chazdon 2014). Across four successional stages in wet lowland forests of the Osa Peninsula, Costa Rica, total carbon storage increases with forest age, with the overall contribution of soil carbon (0–30 cm depth) declining with age (Figure 36.4). After 30–50 years of natural regeneration, soils and trees stored 73 per cent of the carbon stored in old-growth forests of the region.

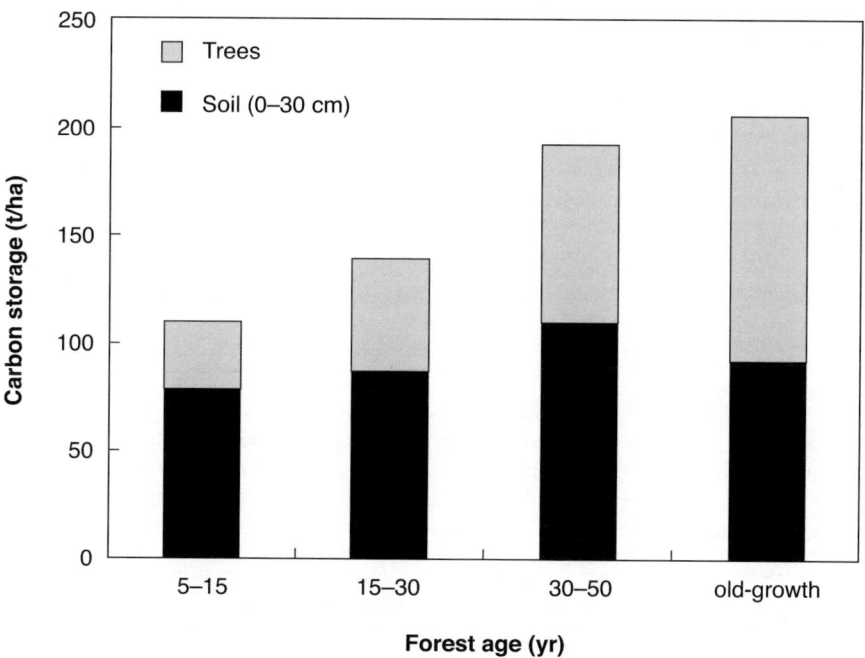

Figure 36.4 Average carbon storage in soil and tree vegetation (more than 5 cm diameter at breast height) in regenerating forests and old-growth forests in the Osa Peninsula, Costa Rica. Each stage had 4–6 replicate plots of 0.5 ha. Soil carbon was measured to a depth of 30 cm

Sources: data are from Aguilar-Arias *et al.* (2012) and E. Ortiz-Malavassi and R. Chazdon (unpublished data)

In the humid forests of the Caribbean lowlands, naturally regenerating forests 4–20 years old show mean annual increments of 8.9 and 5.3 Mg ha^{-1} yr^{-1} for aboveground biomass and carbon, respectively, with the highest rates observed in forests less than 10 years old. The rate of increase of soil carbon was 1.09 Mg ha^{-1} yr^{-1}, and the amount of carbon in the soil was greater than the amount in the aboveground vegetation (Fonseca *et al.* 2011). In the same region, Rozendaal and Chazdon (2015) found that standing tree biomass increased dramatically over time and reached 71 per cent of old-growth values within 25–41 years. Annual rates of tree biomass change declined over time, however, and were similar between 25–41 years old second-growth forests and old-growth forests. Rapid biomass and carbon accumulation in trees cannot be sustained beyond the first 20–25 years of forest regeneration succession due to declining rates of tree growth and increasing rates of mortality of fast-growing pioneer trees that dominate early successional forests.

Synthesis: enhancement of ecosystem services following ecological restoration

Meta-analysis is a statistical tool that is increasingly used in ecology to quantitatively compare results of large numbers of individual studies and to explore general trends in the variation of target response variables such as those related to delivery of ecosystem services (ES). For each response variable from individual studies, one determines a common statistical measure and those measures are compared. Here we present an overview of conclusions of three global meta-analyses to assess ecological restoration effectiveness in a wide range of ecosystem types. Typically, this effectiveness is quantified as recovery *progress* (i.e. the levels of biodiversity and ES attained in restored ecosystems and in paired degraded ecosystems), and/or recovery *completeness* (i.e. the levels of biodiversity and ES in restored ecosystems compared to those in reference ecosystems).

In a meta-analysis of a variety of ecosystem types, Rey Benayas *et al.* (2009) found that ecological restoration increased levels of supporting, regulating and provisioning ES by 25 per cent, but their values remained 20 per cent lower in restored than in reference ecosystems (Table 36.1). Similarly, restored wetlands showed an average of 36 per cent higher levels of ES than degraded wetlands, but their levels were 13 per cent lower than in natural wetlands (Table 36.1; Meli *et al.* 2014). Consistently, Moreno-Mateos *et al.* (2012) found that biogeochemical functioning, which was driven primarily by the storage of carbon in wetland soils, remained on

Table 36.1 Recovery progress (left, grey columns) and recovery completeness (right, white columns), in percentage, of services provided by restored ecosystems as compared to degraded and reference ecosystems, respectively, for a variety of ecosystem types according to published global meta-analyses

Ecosystem type	All ES types		Supporting ES		Regulating ES		Provisioning ES		Source
All	25	−20	28	−18	20	−44	−24	23	Rey Benayas *et al.* (2009)
Wetlands	43	−13	40	−16	47	−22	80	−7	Meli *et al.* (2014)
Agroecosystems	83		42		120				Barral *et al.* (2015)

average 23 per cent lower than in reference wetlands. Restoration increased levels of ES by an average of 83 per cent relative to levels in the pre-restoration agroecosystem, and restored agroecosystems showed levels of ES similar to those of reference ecosystems (Table 36.1; Barral *et al.* 2015). However, these meta-analyses indicated that enhanced delivery of ES differed among ES types. Recovery progress across ecosystem types was highest for supporting ES and lowest for provisioning services, whereas the opposite trend held for wetland and agroecosystem studies.

The meta-analyses of Rey Benayas *et al.* (2009), Meli *et al.* (2014) and Barral *et al.* (2015) tested the specific hypothesis that a change in biodiversity is positively associated with altered provision of ecosystem services by correlating biodiversity and ecosystem service response ratios across the individual studies. Results showed that biodiversity and ecosystem service response ratios were positively correlated for both restored vs. degraded and restored vs. reference comparisons in all ecosystems and in wetlands, whereas the data gathered for agroecosystems only allowed a correlation for the first comparison, which was also positive. The relationship was stronger in the degraded versus restored comparison. This difference in the observed relationships may be linked to an asymptotic relationship between biodiversity and ecosystem function, whereby increasing biodiversity from low values (i.e. those in degraded ecosystems) has relatively strong impacts on individual ecosystem functions, but the relationship plateaus at relatively high biodiversity values (as in reference ecosystems). These results suggest that when combining studies undertaken at a range of spatial and temporal scales, biodiversity is positively related to the ecological functions that underpin the provision of ecosystem services (Rey Benayas *et al.* 2009; Cardinale 2010), and indicate a win–win restoration outcome (Rey Benayas *et al.* 2009).

Meta-analyses have also investigated the effect of context factors on outcomes of ecological restoration. For a wide range of ecosystems, ES recovery progress was higher and recovery completeness was lower in terrestrial than in aquatic ecosystems and in tropical than in temperate ecosystems (Rey Benayas *et al.* 2009). Jones and Schmitz (2009) provided evidence that most ecosystems globally can, given sufficient human will and effort, recover from very major perturbations on timescales of decades to half-centuries (the average recovery time was at most 42 years for forest ecosystems and typically on the order of 10 years). They also found that recovery from anthropogenic disturbances such as agriculture and mining was generally slower than recovery from natural disturbances (hurricanes/cyclones). The extent to which restoration increased ES in degraded wetlands depended primarily on the main agent of degradation, restoration actions, experimental design, and ecosystem type. In contrast, the choice of specific restoration actions alone explained most differences between restored and natural wetlands (Meli *et al.* 2014).

Based on a meta-analysis of wetland ecosystem restoration, Moreno-Mateos *et al.* (2012) concluded that recovery was very slow or post-disturbance systems moved towards alternative states that differ from reference wetland conditions. Further, large wetland areas (>100 ha) and wetlands restored in warm (temperate and tropical) climates recovered more rapidly than smaller wetlands and wetlands restored in cold climates. For agroecosystems, restoration outcomes depended on restoration strategy (it was higher under land separation than under land sharing schemes) but not on restoration approach (active vs. passive) or restoration age, whereas climate type could not be tested (Barral *et al.* 2015).

The current momentum for forest restoration has triggered a number of global assessments looking at the recovery of services provided by these ecosystems. Piotto (2008) compared tree growth in monocultures and mixed plantations and found that mixing tree species generally increases plantation growth rate. Further, he suggested that mixed tree plantations can play an important role in satisfying economic needs by shortening rotations yet adding other ecological benefits. Conversely, Bonner *et al.* (2013) found a general tendency for aboveground biomass accumulation in the tropics to be marginally higher in plantations (approx. 11 Mg ha^{-1} yr^{-1}) than in secondary forests (approx. 8 Mg ha^{-1} yr^{-1}).

Global meta-analyses highlight the importance of comprehensive, multi-factorial assessment to determine the ecological status of degraded, restored and natural ecosystems and thereby evaluate the effectiveness of ecological restorations. Ecological restoration is generally effective to enhance biodiversity and ES provision, but this effectiveness is also highly context dependent. Future research on ecosystem restoration should seek to identify which restoration actions work best for specific habitats.

Restoring ecosystems and ecosystem services: socioecological challenges and complexities

Ecosystem services are a human construct. Ecosystems, however, are fundamentally natural systems, although they may be highly modified by human activities and may contain non-native species. Restored ecosystems, like agricultural systems, are socioecological systems that blend natural processes with human agency. Ecosystem restoration is also a human construct. Ecological restoration aims to meet the needs of the local community or of a broader set of stakeholders. It is often undertaken with the explicit goal of increasing the provision of multiple ES in areas where both ecosystem functioning and biological diversity have been greatly diminished by intensive or unsustainable land use, often causing life-threatening problems for human populations. It is possible to restore some ES without restoring native biodiversity, as in the case of commercial tree plantations (Brockerhoff *et al.* 2013). But another goal of ecological restoration is to restore components of native biodiversity by recreating habitats and conditions that are suitable for occupation by a diversity of organisms that lived there in the past and could not otherwise live there now or in the future (Wiens and Hobbs 2015). Restoration can be a means to achieve conservation or a means to rehabilitate a dysfunctional ecosystem. A major challenge is to develop restoration interventions that can achieve both goals simultaneously.

Which approach is taken depends on the role that biodiversity plays in the restoration of ES. Ultimately, the functioning of organisms forms the mechanistic basis for all ES, but these functions are constrained by the biophysical context and by the choice of restoration interventions, such as the diversity, abundance, and functional characteristics of species that are planted, assisted, or that regenerate spontaneously. It is therefore reasonable to expect a strong relationship between biodiversity (which encompasses taxonomic, functional, and genetic aspects of diversity) and ES during the process of ecological restoration, as we showed earlier in this chapter. As particular ES are often based on a composite of several ecosystem functions, it is likely that biodiversity effects on multiple ES will be stronger than those that influence a single ES (Meli *et al.* 2014). The supply of ES can also be influenced by path dependence and legacy effects, however, making it difficult to predict trajectories, particularly for multiple ES (Bennett *et al.* 2009). These issues highlight the importance of considering species selection in restoration interventions. Use of multiple native species, as in the case of high-diversity tropical forest restoration plantings (Rodrigues *et al.* 2011), can increase the likelihood of a stable and long-term supply of multiple ES. Yet, as Meli *et al.* (2014) point out, knowledge regarding the relationship between biodiversity and ES is patchy and incomplete.

An assessment of key biotic and abiotic attributes that are linked to 11 ES found that 24 of 28 attributes were important for the provision of one or more ES (Harrison *et al.* 2014). Five attributes showed particularly strong importance: species abundance, species richness, species size/weight, community/habitat area, and community/habitat structure. A number of species-level traits were also found to contribute to certain ES; species abundance was important for pest control, pollination, and recreation whereas species richness was important for timber production and freshwater fishing. Although most effects of biodiversity attributes were positive, some negative impacts were also observed, particularly in the case of abundant invasive species with rapid growth rates and extensive root systems. Increased community/habitat area, vegetation structure, and aboveground biomass generally reduced the provision of freshwater, owing to the high rates of water consumption (Harrison *et al.* 2014). Moreover, since invasive species often become dominant in early successional ecosystems, they can have a strong positive or negative impact on particular ES during restoration. So the answer to the question, 'does biodiversity matter to ES?' is: it depends, but it usually does (Wiens and Hobbs 2015).

The financial and opportunity costs of different approaches to restoring ES strongly influence the scale and type of interventions. Restoration interventions targeted within relatively small areas can be highly effective in restoring multiple ES at local scales, but due to high costs these approaches cannot be up-scaled to larger areas such as entire watersheds. As shown in the case study of nutrient retention in stream water, local interventions are not always sufficient to reduce downstream nitrogen exports. Larger-scale watershed approaches are needed to reduce excessive nitrogen inputs to stream ecosystems (Filoso and Palmer 2011). Similarly, diverse tree planting schemes are effective for restoring forests along riparian zones, but cost over US$3,000 per ha. Up-scaling this approach to restoring the entire Atlantic Forest Biome is estimated to cost between US$49 and US$77 billion (Calmon *et al.* 2009; Rodrigues *et al.* 2011).

Restoration of ecosystem services also depends on who owns the land and the opportunity costs of modifying existing land uses. Landholders are almost never willing to pay any of the cost of restoring ES, and many countries have developed programmes to pay landholders to protect or restore ES on their land (Wunder 2007). Key issues for large-scale ecological restoration are financial support and education to promote landholder and public awareness and training (Rey Benayas and Bullock 2012; De Snoo *et al.* 2013). To reward the total or social value, tax deductions for landholders who restore land and donations to non-profit organizations that run restoration projects, payment for environmental services, and direct financing measures related to restoration activities should be implemented widely (Rey Benayas and Bullock 2015).

For example, restoration interventions to enhance ES on agricultural land need to be financially beneficial as well as socially acceptable to farmers and landowners. Farmers are often reluctant to implement restoration actions that can increase ES on their farms. First, farmers do not understand or foresee benefits for agricultural production and, simultaneously, they perceive risks for crops. They may believe, for instance, that a planted hedgerow will favour the spread of crop pests rather than provide habitat for natural enemies of such pests or pollinators. They may not understand how hedgerows can function as windbreak that reduce soil erosion, desiccation, and crop damage. The second major reason has to do with the aesthetics of crop fields or pastures. In the view of many farmers and ranchers, fields should be 'clean' (i.e. with nothing other than the cultivated plants), and most often farmers that have 'unkempt' crop fields or ranchers with 'dirty' pastures are criticized in their local communities. And third, individual farmers generally react to the private use-value of biodiversity and ecosystem services assigned in the marketplace and thus typically ignore the 'external' benefits of conservation that accrue to wider society (Jackson *et al.* 2007). To overcome this reluctance, we recommend efforts to educate and show farmers that strategic revegetation and other restoration actions benefit wildlife and wildlife-based ecosystem services that can enhance crop production (Rey Benayas and Bullock 2015).

Finally, the different values placed on ES by different stakeholders pose challenges to farmers, land managers, and government agencies responsible for restoring ecosystems. Monetized ES, such as timber, fish, and crops, are more likely to be prioritized over non-monetized ES. Policies are therefore needed to provide incentives for restoration interventions, long-term monitoring, and adaptive management to yield a balanced portfolio of ES for the benefit of multiple stakeholders and future generations.

References

Aguilar-Arias, H., Ortiz-Malavassi, E., Vílchez-Alvarado, B. and Chazdon, R. L. (2012) Biomasa sobre el suelo y carbono orgánico en el suelo en cuatro estadios de sucesión de bosques en la Península de Osa, Costa Rica. *Revista Forestal Mesoamericana Kurú*, 9, 22–31.

Barral, P., Rey Benayas, J. M., Meli, P. and Maceira, N. (2015) Quantifying the impacts of ecological restoration on biodiversity and ecosystem services in agroecosystems: a global meta-analysis. *Agriculture, Ecosystems and the Environment*, 202, 223–231.

Bennett, E. M., Peterson, G. D. and Gordon, L. J. (2009) Understanding relationships among multiple ecosystem services. *Ecology Letters*, 12, 1394–1404.

Bonner, M. T. L., Schmidt, S. and Shoo, L. P. (2013) A meta-analytical global comparison of aboveground biomass accumulation between tropical secondary forests and monoculture plantations. *Forest Ecology and Management*, 291, 73–86.

Brockerhoff, E. G., Jactel, H., Parrotta, J. A. and Ferraz, S. F. (2013) Role of eucalypt and other planted forests in biodiversity conservation and the provision of biodiversity-related ecosystem services. *Forest Ecology and Management*, 301, 43–50.

Bullock, J. M., Aronson, J., Newton, A. C., Pywell, R. F. and Rey Benayas, J. M. (2011) Restoration of ecosystem services and biodiversity: conflicts and opportunities. *Trends in Ecology and Evolution*, 26, 541–549.

Calmon, M., Lino, C. F., Nave, A. G., Pinto, L. P. and Rodrigues, R. R. (2009) Pacto pela restauracão da Mata Atlântica: um movimento pela restauracão da floresta, in M. A. Fujihara, R. Cavalcanti, A. Guimarães and R. Garlipp (eds), *O valor das florestas*. Terra das Artes Editora, São Paulo, pp. 331–333.

Cao, S., Sun, G., Zhang, Z., Chen, L., Feng, Q., Fu, B., McNulty, S., Shankman, D., Tang, J. and Wang, Y. (2011) Greening China naturally. *Ambio*, 40, 828–831.

Cardinale, B. (2010) Impacts of biodiversity loss. *Science*, 336, 552–553.

Cayuela, L., Rey Benayas, J. M., Maestre, F. and Escudero, A. (2008) Early environments drive diversity and floristic composition in Mediterranean old fields: insights from a long-term experiment. *Acta Oecologica*, 34, 311–321.

Chazdon, R. L. (2014) *Second Growth: The Promise of Tropical Forest Regeneration in an Age of Deforestation*. University of Chicago Press, Chicago, IL.

Chazdon, R. L., Broadbent, E. N., Rozendaal, D. M. A., Bongers, F., Zambrano, A. M. A., Aide, T. M., Balvanera, P., Becknell, J. M., Boukili, V., Brancalion, P. H. S., Craven, D., De Almeida-Cortez, J. S., Cabral, GaL., De Jong, B., Denslow, J. S., Dent, D. H., Dewalt, S. J., Dupuy, J. M., Durán, S. M., Espírito-Santo, M. M., Fandino, M. C., César, R. G., Hall, J. S., Hernández-Stefanoni, J. L., Jakovac, C. C., Junqueira, A. B., Kennard, D., Letcher, S. G., Lohbeck, M., Martínez-Ramos, M., Massoca, P., Meave, J. A., Mesquita, R., Mora, F., Muñoz, R., Muscarella, R., Nunes, Y. R. F., Ochoa-Gaona, S., Orihuela-Belmonte, E., Peña-Claros, M., Pérez-García, E. A., Piotto, D., Powers, J. S., Rodríguez-Velazquez, J., Romero-Pérez, I. E., Ruíz, J., Saldarriaga, J. G., Sanchez-Azofeifa, A., Schwartz, N. B., Steininger, M. K., Swenson, N. G., Uriarte, M., Van Breugel, M., Van Der Wal, H., Veloso, M. D. M., Vester, H., Vieira, I. C. G., Vizcarra Bentos, T., Williamson, G. B., and Poorter, L. (2016) Carbon sequestration potential of second-growth forest regeneration in the Latin American tropics. *Science Advances* 2:e1501639

Costanza, R., de Groot, R., Sutton, P., van der Ploeg, S., Anderson, S. J., Kubiszewski, I., Farber, S. and Turner, R. K. (2014) Changes in the global value of ecosystem services. *Global Environmental Change*, 26, 152–158.

Cuesta, B., Rey Benayas, J. M., Gallardo, A., Villar-Salvador, P. and González-Espinosa, M. (2012) Soil chemical properties in abandoned Mediterranean cropland after succession and oak reforestation. *Acta Oecologica*, 38, 48–55

Daily, G. (1997) *Nature's Services: Societal Dependence on Natural Ecosystems*. Island Press, Washington, D.C.

De Snoo, G. R., Herzon, I., Staats, H., Burton, R. J. F., Schindler, S., van Dijk, J., Lokhorst, A. M., Bullock, J. M., Lobley, M., Wrbka, T., Schwarz, G. and Musters, C. J. M. (2013) Toward effective nature conservation on farmland: making farmers matter. *Conservation Letters*, 6, 66–72.

Farley, K. A., Kelly, E. F. and Hofstede, R. G. (2004) Soil organic carbon and water retention after conversion of grasslands to pine plantations in the Ecuadorian Andes. *Ecosystems*, 7, 729–739.

Feng, X., Fu, B., Lü, N., Zeng, Y. and Wu, B. (2013) How ecological restoration alters ecosystem services: An analysis of carbon sequestration in China's Loess Plateau. *Scientific Reports*, 3, 2846.

Filoso, S. and Palmer, M. A. (2011) Assessing stream restoration effectiveness at reducing nitrogen export to downstream waters. *Ecological Applications*, 21, 1989–2006.

Fonseca, W., Benayas, J. M. R. and Alice, F. E. (2011) Carbon accumulation in the biomass and soil of different aged secondary forests in the humid tropics of Costa Rica. *Forest Ecology and Management*, 262, 1400–1408.

Harrison, P., Berry, P., Simpson, G., Haslett, J., Blicharska, M., Bucur, M., Dunford, R., Egoh, B., Garcia-Llorente, M. and Geamănă, N. (2014) Linkages between biodiversity attributes and ecosystem services: a systematic review. *Ecosystem Services*, 9, 191–203.

Howe, C., Suich, H., Vira, B. and Mace, G. M. (2014) Creating win-wins from trade-offs? Ecosystem services for human well-being: a meta-analysis of ecosystem service trade-offs and synergies in the real world. *Global Environmental Change*, 28, 263–275.

Jackson, L. E., Pascual, U. and Hodgkin, T. (2007) Utilizing and conserving agrobiodiversity in agricultural landscapes. *Agriculture, Ecosystems and the Environment*, 121, 196–210.

Jiang, W., Yang, S., Yang, X. and Gu, N. (2015) Negative impacts of afforestation and economic forestry on the Chinese Loess Plateau and proposed solutions. *Quaternary International*, 399, doi:10.1016/j.quaint.2015.04.011.

Jones, H. P. and Schmitz, O. J. (2009) Rapid recovery of damaged ecosystems. *Plos ONE*, 4, e5653.

Lü, N., Fu, B., Jin, T. and Chang, R. (2014) Trade-off analyses of multiple ecosystem services by plantations along a precipitation gradient across Loess Plateau landscapes. *Landscape Ecology*, 29, 1697–1708.

Lü, Y., Fu, B., Feng, X., Zeng, Y., Liu, Y., Chang, R., Sun, G. and Wu, B. (2012) A policy-driven large scale ecological restoration: quantifying ecosystem services changes in the Loess Plateau of China. *PLoS ONE*, 7, e31782.

Mace, G. M., Norris, K. and Fitter, A. H. (2012) Biodiversity and ecosystem services: a multilayered relationship. *Trends in Ecology and Evolution*, 27, 19–26.

Mayer, P. M., Schechter, S. P., Kaushal, S. S. and Groffman, P. M. (2013) Effects of stream restoration on nitrogen removal and transformation in urban watersheds: lessons from Minebank Run, Baltimore, Maryland. *Watershed Science Bulletin* 4(1).

MEA (2005) *Ecosystems and Human Well-being*. World Resources Institute, Washington, DC.

Meli, P., Rey Benayas, J. M., Balvanera, P. and Martínez-Ramos, M. (2014) Restoration enhances wetland biodiversity and ecosystem service supply, but results are context-dependent. *PLoS ONE*, 9, e93507.

Moreno-Mateos, D., Power, M. E., Comin, F. A. and Yockteng, R. (2012) Structural and functional loss in restored wetland ecosystems. *PLoS Biology*, 10(1), e1001247.

Nelson, E., Mendoza, G., Regetz, J., Polasky, S., Tallis, H., Cameron, D. R., Chan, K. M. A., Daily, G. C., Goldstein, J. and Kareiva, P. M. (2009) Modeling multiple ecosystem services, biodiversity conservation, commodity production, and tradeoffs at landscape scales. *Frontiers in Ecology and the Environment*, 7, 4–11.

Palmer, M. A., Filoso, S. and Fanelli, R. M. (2014a) From ecosystems to ecosystem services: stream restoration as ecological engineering. *Ecological Engineering*, 65, 62–70.

Palmer, M. A., Hondula, K. L. and Koch, B. J. (2014b) Ecological restoration of streams and rivers: Shifting strategies and shifting goals. *Annual Review of Ecology, Evolution, and Systematics*, 45, 247–269.

Pan, Y., Birdsey, R. A., Fang, J., Houghton, R., Kauppi, P. E., Kurz, W. A., Phillips, O. L., Shvidenko, A., Lewis, S. L. and Canadell, J. G. (2011) A large and persistent carbon sink in the world's forests. *Science*, 333, 988–993.

Persson, M., Moberg, J., Ostwald, M. and Xu, J. (2013) The Chinese Grain for Green Programme: Assessing the carbon sequestered via land reform. *Journal of Environmental Management*, 126, 142–146.

Piotto, D. (2008) A meta-analysis comparing tree growth in monocultures and mixed plantations. *Forest Ecology and Management*, 255, 781–786.

Ramos-Palacios, C. R., Badano, E. I., Flores, J., Flores-Cano, J. A. and Flores-Flores, J. L. (2014) Distribution patterns of acorns after primary dispersion in a fragmented oak forest and their consequences on predators and dispersers. *European Journal of Forest Research*, 133, 391–404.

Raudsepp-Hearne, C., Peterson, G. D. and Bennett, E. M. (2010) Ecosystem service bundles for analyzing tradeoffs in diverse landscapes. *Proceedings of the National Academy of Sciences USA*, 107, 5242–5247.

Rey Benayas, J. M. and Bullock, J. M. (2012) Restoration of biodiversity and ecosystem services on agricultural land. *Ecosystems*, 15, 883–889.

Rey Benayas, J. M. and Bullock, J. M. (2015) Vegetation restoration and other actions to enhance wildlife in European agricultural landscapes, in H. M. Pereira and L. M. Navarro (eds) *Rewilding European Landscapes*. Springer, Berlin, pp. 127–142.

Rey Benayas, J. M., Bullock, J. M. and Newton, A. C. (2008) Creating woodland islets to reconcile ecological restoration, conservation, and agricultural land use. *Frontiers in Ecology and the Environment*, 6, 329–336

Rey Benayas, J. M., Newton, A. C., Díaz, A. and Bullock, J. M. (2009) Enhancement of Biodiversity and Ecosystem Services by Ecological Restoration: a meta-analysis. *Science*, 325, 1121–1124.

Rey Benayas, J. M., Martínez-Baroja, L., Pérez-Camacho, L., Villar-Salvador, P. and Holl, K. D. (2015) Predation and aridity slow down the spread of 21 year old planted woodland islets in restored Mediterranean farmland. *New Forests*, 46, 841–853.

Rodrigues, R. R., Gandolfi, S., Nave, A. G., Aronson, J., Barreto, T. E., Vidal, C. Y. and Brancalion, P. H. S. (2011) Large-scale ecological restoration of high-diversity tropical forests in SE Brazil. *Forest Ecology and Management*, 261, 1605–1613.

Rozendaal, D. A. and Chazdon, R. L. (2015) Demographic drivers of tree biomass change during secondary succession in northeastern Costa Rica. *Ecological Applications*, 25, 506–516.

Shi, S., and Han, P. (2014) Estimating the soil carbon sequestration potential of China's Grain for Green Project. *Global Biogeochemical Cycles*, 28, 1279–1294.

Society for Ecological Restoration (2004) *The SER Primer on Ecological Restoration*. Society for Ecological Restoration, Tucson, AZ.

Song, X., Peng, C., Zhou, G., Jiang, H. and Wang, W. (2014) Chinese Grain for Green Program led to highly increased soil organic carbon levels: a meta-analysis. *Scientific Reports*, 4, 4460.

Wiens, J. A. and Hobbs, R. J. (2015) Integrating conservation and restoration in a changing world. *BioScience*, 65, 302–312.

Wunder, S. (2007) The efficiency of payments for environmental services in tropical conservation. *Conservation Biology*, 21, 48–58.

37

THE ECONOMICS OF RESTORATION AND THE RESTORATION OF ECONOMICS

James Blignaut

Introduction

It is a good principle to manage well that which is at your disposal. Whether that is a house, a road, an office complex, a river or a pasture, if it is broken, it has to be fixed. Be that a light bulb, an office chair, a dining room table, a fence or the riparian zone. In reality, however, there is a far greater intent to replace broken or malfunctioning light bulbs, office chairs and dining room tables than there is to repair a riparian zone. It is unfortunate that there is a much stronger inclination to repair and restore manufactured capital than there is to restore natural capital. All is not lost though.

The subject of restoration, and specifically the economics of restoration, has received much and increasing attention over the past two decades. After considering the growth in the study of the economics of restoration, in this chapter I reflect on four reasons why it is imperative to consider the economics of restoration even more. Thereafter a brief look into what is required to restore economics is provided. This is essential since, after all, if economics as practised is not restored, then the effort to restore natural capital will be insufficient for the challenges posed by the prevailing economics of destruction and degradation.

The economics of restoration: setting the stage

The economics of restoration is a brand new field of investigation. The term is broadly used to imply the application of economic tools and principles, specifically pertaining to environmental resource and ecological economics, to the subject matter and discipline of restoration ecology, including the restoration of natural capital. The restoration of natural capital is defined as any activity that integrates investment in and replenishment of natural capital stocks to improve the flows of ecosystem goods and services, while enhancing all aspects of human well-being (Blignaut *et al.* 2014a). Paul Hawken (Hawken 1993, 2010; Hawken *et al.* 1999) developed the terms 'restorative economics' and its opposite 'destructive economics'. Whether this is the origin of the notion of the economics of restoration can be debated, but he highlighted the underlying philosophy and the urgent need for putting together ecological restoration and economics.

Blignaut *et al.* (2014b) analysed the history of and made a few propositions as to the development of the economics of restoration going forward. Blignaut *et al.* (*ibid.*) pointed out that both the Millennium Ecosystem Assessment (MA 2005) and The Economics of Ecosystems and Biodiversity project (TEEB 2010) have done much to advance both the thinking and application of the economics of restoration. Using *Scopus* to search for and analyse papers with respect to the economics of restoration, they found that the number of papers increased from only 10 relevant papers in 1996 to 90 in 2011 and 68 in 2012. Furthermore, the number of citations increased from 3 (in 1996) to 1,549 (in 2012). The aggregate number of citations increased to 6,970 achieving an *h*-score of 37. While it was expected that the number of citations had to be low initially (since *Scopus* began detailed analysis of tracking paper citations only in 1996), the more than 500-fold increase over a 16-year-period is a strong indication of the increasing interest in papers that cover this topic.

Papers that can be classified in the general fields of economics, econometrics and finance only had 39 papers over this period. On the other hand, large number of papers were classified in the environmental sciences category (442), followed by those in agriculture and biological sciences (232), and social sciences (102). Given the infancy of the subject, and the historical link of restoration ecology and conservation biology, this trend could have been expected. The dearth of economics, econometric and finance papers are, however, a serious concern – a topic we turn to later.

They also found that:

1   while progress has been made conceptually, there are too few examples and advances made over the past two decades in the areas of valuation and financing;
2   the monitoring and evaluation of the economic effects of restoration has been weakly monitored and documented;
3   it is uncommon for decision-makers, planners and restoration practitioners to consider the long-term economic or socio-economic benefits and impacts of restoration; and
4   those engaged in ecological restoration (decision-makers, planners, practitioners) and economists often do not share a common vocabulary, or research method or approach (Blignaut *et al.* 2014b).

Why restore? An economic reflection

It is difficult to contemplate why this question should even be considered at all. It should be a no-brainer. When one is using something, especially something that belongs to somebody else, it is a good principle to fix or clean it before handing it back. The air is not our own, likewise water and land resources. People might have private rights to access and use them, but that does not absolve anybody from the fact that those are common property resources that cannot and should not be returned in a polluted and degraded form to the next user. That is just bad manners. Yet, unfortunately, this is what happens daily at a global scale. Using today's resources the way we do is akin to borrowing it from either a neighbour or tomorrow's generation. And given the way we're returning these resources to them, the current generation is pretty bad mannered.

From an ecological point of view the urgent need for restoration at a global scale to replenish the dwindling stocks of natural capital has been emphasized before and there is no need to repeat it here (see MA 2005; TEEB 2010, 2011). Over and above the moral/ethical and ecological reasons why restoration is imperative, there are at least four economic reasons. They will subsequently be discussed.

Reason 1: economic development

One of the main objectives of economic development is the preservation and enhancement or generation of income. That is the essence of what the GDP (gross domestic product) of a country is. It is the income that is generated within a country during a specific period of time. The higher the GDP of a country, the more income has been generated over a fixed time period. A rise in income, however, does not imply an improvement in welfare as it can be, and indeed is, often associated with degradation, pollution, and several other social ills. Why is it that the GDP, arguably the most important variable in economics, could have evolved in such a way as to be so misleading when it comes to measuring progress, and actually being counter-productive to measure economic development?

The answer lies in what is defined as capital and its relationship to depreciation or depletion. To calculate depreciation, depletion, or degradation, a distinction must be made between capital and income as depreciation is only possible in the case of capital goods and expenditures. To calculate capital, true or sustainable income has to be calculated. The definition used most frequently is the one of Sir John Hicks. Hicks defines income as follows:

> The purpose of income calculations in practical affairs is to give people an indication of the amount which they can consume without impoverishing themselves. Following out of this idea, it would seem that we ought to define a man's income as the maximum value which he can consume during a week, and still expect to be as well off at the end of the week as he was at the beginning. Thus, when a person saves, he plans to be better off in the future; *when he lives beyond his income he plans to be worse off*. Remembering that the practical purpose of income is to serve as a guide for prudent conduct, I think it is fairly clear that this is what the central meaning must be.
>
> *(Hicks 1946: 172; emphasis added)*

El Serafy captures the essence of these words in the following statement:

> The fundamental principle that is flouted by applying conventional national income accounting to depletable resources is the separation that must be maintained between income and capital. This principle tells us that if you liquidate your assets and use the proceeds for consumption, you are living beyond your means, and in doing so you are undermining your ability to create future income.
>
> *(El Serafy 1989: 11)*

The implication of the above statement is that if the current income stream from a resource is lower than the net present value of that resource, the country is worse off. True income may thus be thought of as the maximum a recipient can consume in a given period *without reducing possible future consumption*. This concept encompasses not only current earnings, but also changes in the asset position as follows: capital gains are a source of income; capital losses reduce income (Lutz and El Serafy 1988). As specified in the Hicksian definition of true or sustainable income, savings contribute to more future income. As the *hotelling rule* states, it is probable that if a *country leaves alone its natural resources (in other words saves it) it would appreciate at the market discount rate because of its increasing scarcity* (El Serafy 1989). The opposite of this is that if an individual lives beyond his income, he will be worse off in future (Hicks's definition of income and capital). When the reverse of the hotelling rule is implemented, namely that when natural capital is utilized today, the country forsakes future appreciation and possible income as well as the use of the resource. Therefore, if the current income stream from a resource is lower than the present net value of the resource, the country is worse off.

When dealing with goods and services that are clearly marketed, the above-mentioned principle can be applied relatively easily as all the necessary information is in monetary terms. With regard to the natural environment, the required data is not available in monetary terms and must be transformed from the available physical data into monetary terms by means of certain valuation techniques. This is important since Hicks only includes man-made (fixed) capital in his original definition. However, environmental resource and ecological economics expand this to include natural capital as well, as Solow, the 1987 Nobel Prize winner in economics, indicates:

> When it comes to measuring the economy's contribution to the well-being of the country's inhabitants, however, the conventional measures are incomplete. The most obvious omission is the depreciation of assets. If two economies produce the same real GDP but one of them does so wastefully by wearing out half of its stock of plant and equipment while the other does so thriftily and holds depreciation to 10 per cent of its stock of capital, it is pretty obvious which one is doing a better job for its citizens. Of course the national income accounts have always recognized this point, and they construct net aggregates, like net national product, to give an appropriate answer. Depreciation of fixed capital may be badly measured, and the error effects net product, but the effort is made.
>
> *(Solow 1992: 6)*

The same principle should hold for stocks of non-renewable resources and for the environmental assets like clean air and water.

The main source of the problem of the conventional national accounting practices (used in calculating the GDP) is that since it excludes environmental (and social) capital from its ambit, it does not take into account the consequences of both market and government failure appropriately and adequately. By not taking the loss by means of a depreciation allowance into account in the Systems of National Accounts, the conventional national accounting procedures allocate a disproportionate share of current income flows to the present generation. According to the Hicksian definition of income, the amount consumed now without becoming worse off in the future is clearly being infringed upon as the value of the asset base declines (Repetto *et al.* 1989: 20). The main principle underlying this approach is to *keep capital intact* (which complements the Hicksian view of capital and income). In other words, future generations should receive a capital base, man-made as well as natural capital, with a qualitative and quantitative level being at least comparable with the present situation.

Reason 2: sustainable development

As described in Blignaut *et al.* (2014a), the progress with respect to sustainable development is hindered by conflicting definitions and core concepts with respect to its meaning. This debate does not assist policy-makers or the researchers at all, and definitely not land managers and other practitioners. While it is easy to get bogged down in details, such can imply that one forgets the purpose of sustainable development, as expressed by Herman Daly 'it is both morally and economically wrong to treat the world as a business in liquidation. Instead we need to seek lasting cures for our seemingly relentless "addiction to growth"' (Daly 1991: 248). Therefore, it is necessary to find pathways to development that would benefit both people and the rest of our global ecosystem. Blignaut *et al.* (2014a: 55) propose three pathways to sustainability, namely:

- *Sustainability through appropriate technological change*, whereby, among other things, the resource and energy intensity of an economy is significantly reduced. This is also called the dematerialization of economic development (van den Bergh and Janssen 2005).
- *Sustainability through behavioural change*, whereby society's preferences and value systems revise what is considered to be wealth, the reason or rationale for living, and the way in which we live (Daly 1991; Costanza 1991; Hawken 1993).
- *Sustainability through the restoration of natural capital (RNC)*, where RNC is defined as any activity that integrates investment in and replenishment of natural capital stocks to improve the flows of ecosystem goods and services, while enhancing all aspects of human wellbeing (Aronson *et al.* 2007). The major components of RNC are: (i) ecological restoration of impaired natural ecosystems, *sensu* SER (2004); (ii) ecological improvements and refinements of resource exploitation and production systems, and (iii) development and refinement of educational and awareness-raising programs addressing the broad topics of natural capital, biodiversity, and ecosystem services.

The first two pathways mentioned (i.e. technology and behaviour) aim to reduce general and per capita demand for, and pollution and waste of, resources, as well as erosion of natural, cultural, and human capital. The third pathway (i.e. the restoration of natural capital, combined with ongoing land use and resource management adjustments after restoration has begun) aims to *increase* the stocks of natural capital. Restoration that leads to the improvement of the capital base will lead to an increase in the flow of resources and ecosystem services. This can only be possible if natural capital is maintained (see also reason 1 earlier in the chapter). While the first two pathways focus on reducing the demand side for resources (or ecosystem goods and services), restoration is arguably the only option that has the potential for increasing the supply of ecosystem goods and services.

Reason 3: restoration pays

Money, while an invention of the human mind as a medium of exchange to aid and facilitate trade, is ruling the economy of the world. It can, justifiably, be asked why a human invention to assist and to make things easier has become an objective to be served; no longer being the means to an end, but an end itself. While a relevant question, it is not one focused on here.

De Groot *et al.* (2013), using metadata on the benefits of the restoration of ecosystem goods and services of 225 case studies (De Groot *et al.* 2012) and data of 91 studies on the cost of restoration (Neßhöver *et al.* 2011), conducted a benefit–cost analysis of restoration. Their study is based on the premise that it is necessary, if not essential, to internalize the positive values of the non-marketed but quantifiable values associated with restoration. These include amenity values, carbon sequestration and climate amelioration, soil fertility and enhanced grazing capability, etc. These values are often bundled, that is they occur simultaneously, yet are linked to an individual restoration management action and/or restoration event.

The cost associated with restoration should therefore not merely be seen as an expenditure item, but as an investment that enhances the capital base from which flows are generated over a prolonged period of time. It is tragic, however, that current accounting practices consider the extraction of resources, such as timber, and the clean-up cost of pollution, as income generating items, yet the benefits of restoration are not accounted for. In conventional accounting terms therefore, restoration is an expenditure with no financial benefit whereas degrading activities are expenditures with financial returns.

Based on Neßhöver *et al.* (2011) and De Groot *et al.* (2012, 2013), Blignaut *et al.* (2014a) calculated a range of net present values and benefit–cost ratios for various ecosystems. The results are replicated here in Table 37.1.

Table 37.1 Net present values (NPV) in 2007 US$/ha and benefit–cost ratios (BCR) of 24 different benefit and cost scenarios for restoration projects in 9 ecosystems

Scenario	Discount rate (%)	Term (years)	Coral		Coastal systems		Coastal wetlands	
			NPV: $/ha	BCR	NPV: $/ha	BCR	NPV: $/ha	BCR
100% of typical cost; 60% of TEV benefits								
	−2%	20	748,279	1.4	−192,893	0.3	1,470,620	106.4
	2%	20	−218,285	0.9	−213,514	0.2	900,750	68.2
	5%	20	−631,271	0.7	−219,780	0.1	645,918	50.6
	−2%	30	2,878,569	2.5	−126,056	0.5	2,634,990	189.9
	2%	30	549,732	1.3	−189,418	0.2	1,320,531	99.5
	5%	30	−261,526	0.9	−208,180	0.1	848,012	66.1
130% of typical costs; 60% of TEV benefits								
	−2%	20	157,463	1.1	−317,564	0.1	1,463,988	81.7
	2%	20	−785,932	0.7	−318,714	0.0	895,153	52.4
	5%	20	−1,182,699	0.5	−315,380	0.0	640,832	38.9
	−2%	30	2,287,752	1.9	−283,042	0.2	2,626,639	145.8
	2%	30	−17,915	1.0	−306,268	0.1	1,314,315	76.4
	5%	30	−812,955	0.7	−309,388	0.0	842,628	50.8
100% of typical costs; 100% of TEV benefits								
	−2%	20	2,560,057	2.3	−44,440	0.8	2,465,772	177.8
	2%	20	897,630	1.5	−122,078	0.5	1,513,686	113.9
	5%	20	173,279	1.1	−153,857	0.4	1,087,831	84.6
	−2%	30	6,110,540	4.1	138,764	1.5	4,410,209	317.2
	2%	30	2,177,657	2.2	−56,030	0.8	2,214,699	166.3
	5%	30	789,520	1.4	−122,060	0.5	1,425,318	110.5
130% of typical cost; 100 % of TEV benefits								
	−2%	20	1,969,241	1.8	−169,111	0.5	2,459,140	136.6
	2%	20	329,983	1.1	−227,279	0.3	1,508,090	87.6
	5%	20	−378,150	0.8	−249,457	0.2	1,082,746	65.0
	−2%	30	5,519,723	3.2	−18,222	0.9	4,401,858	243.7
	2%	30	1,610,010	1.7	−172,880	0.5	2,208,482	127.8
	5%	30	238,091	1.1	−223,268	0.3	1,419,934	84.9
		min	−1,182,699	0.5	−318,714	0.0	640,832	38.9
		max	6,110,540	4.1	138,764	1.5	4,410,209	317.2
		average	1,029,533	1.5	−184,171	0.4	1,741,756	114.3

Source: Blignaut *et al.* (2014a), adapted from De Groot *et al.* (2013), reproduced with permission

As noted by Blignaut *et al.* (2014a) coral reefs and coastal systems (excluding coastal wetlands which include mangroves) have the highest values with respect to natural capital benefits. They do have the worst benefit–cost ratios though. This is due to the high costs associated with restoration. Woodland/shrub land and grassland/rangelands have, in stark contrast to the marine and coastal resources, relatively low 'asset' values, but are highly attractive to invest in due to their relatively low restoration costs.

Inland wetlands		*Lakes/rivers*		*Tropical forest*		*Temperate forests*		*Woodland/ shrubland*		*Grasslands/ rangelands*	
NPV: $/ha	*BCR*	*NPV: $/ha*	*BCR*	*NPV: $/ha*	*BCR*	*NPV: $/ha*	*BCR*	*NPV: $/ha*	*BCR*	*NPV: $/ha*	*BCR*
107,012	3.6	19,522	4.2	30,475	7.6	17,324	7.4	10,014	10.8	21,064	44.7
50,120	2.2	9,703	2.6	17,020	4.8	9,647	4.7	5,783	6.9	12,792	28.7
25,194	1.6	5,388	1.9	11,058	3.6	6,247	3.5	3,902	5.1	9,097	21.2
227,670	6.4	40,228	7.5	58,441	13.6	33,285	13.3	18,762	19.4	37,998	79.9
93,620	3.3	17,169	3.9	27,102	7.1	15,402	6.9	8,936	10.1	18,897	41.8
46,137	2.2	8,982	2.6	15,912	4.7	9,017	4.6	5,421	6.7	12,036	27.8
79,785	2.5	15,521	2.9	27,457	5.6	15,560	5.4	9,350	8.0	20,750	34.2
28,613	1.5	6,543	1.9	14,636	3.5	8,254	3.4	5,258	5.1	12,544	21.9
6,406	1.1	2,627	1.4	8,975	2.6	5,030	2.5	3,444	3.8	8,880	16.2
190,134	4.5	34,712	5.3	54,279	10.0	30,854	9.8	17,846	14.5	37,566	61.0
68,397	2.3	13,462	2.8	24,306	5.2	13,768	5.1	8,321	7.5	18,607	31.9
25,559	1.5	5,958	1.8	13,631	3.4	7,684	3.3	4,919	5.0	11,799	21.2
238,857	6.7	41,427	7.7	57,499	13.4	32,792	13.1	18,167	18.8	35,803	75.4
131,327	4.3	23,196	4.9	33,665	8.6	19,174	8.4	10,804	12.0	21,870	48.3
83,742	3.1	15,116	3.6	23,059	6.3	13,116	6.2	7,523	8.9	15,642	35.8
462,864	12.1	79,305	13.9	106,648	24.0	60,878	23.5	33,305	33.6	64,291	134.5
212,086	6.3	36,851	7.2	51,384	12.5	29,300	12.3	16,262	17.6	32,140	70.5
122,622	4.1	21,690	4.8	31,589	8.3	17,990	8.1	10,150	11.7	20,586	46.8
211,630	4.9	37,426	5.7	54,481	10.0	31,028	9.8	17,503	14.2	35,489	57.7
109,820	3.1	20,035	3.6	31,280	6.4	17,781	6.3	10,279	9.1	21,622	37.0
64,954	2.3	12,355	2.7	20,976	4.7	11,899	4.6	7,064	6.7	15,426	27.4
425,328	8.8	73,789	10.2	102,487	18.0	58,447	17.6	32,389	25.4	63,859	103.0
186,863	4.6	33,145	5.3	48,588	9.4	27,666	9.2	15,646	13.3	31,850	54.0
102,045	3.0	18,666	3.5	29,308	6.2	16,657	6.1	9,648	8.8	20,349	35.8
6,406	1.1	2,627	1.4	8,975	2.6	5,030	2.5	3,444	3.8	8,880	16.2
462,864	12.1	79,305	13.9	106,648	24.0	60,878	23.5	33,305	33.6	64,291	134.5
137,533	4.0	24,701	4.7	37,261	8.3	21,200	8.1	12,112	11.8	25,040	48.2

Reason 4: logic

When all else fails, there is always logic, such as in the simple yet profound words of Herman Daly:

> More and more, the *complementary factor in short supply* is remaining *natural capital*, not manmade capital as it used to be. For example, populations of fish, not fishing boats, limit fish catch worldwide. *Economic logic says to invest in the limiting factor. That logic has not changed, but the identity of the limiting factor has.*
>
> (Daly in Blignaut et al. 2014a)

When something is in short supply curtailing development, be it adequate and clean water resources, a healthy and productive climate, or fertile soils, then it requires investment. That investment is restoration.

Up-scaling the restoration effort requires the simultaneous culmination of a range of aspects as highlighted in Figure 37.1. The up-scaling of restoration implies going from plot-scale, and a research oriented mind-set of conducting restoration, to landscape-scale restoration activities covering hundreds if not thousands of hectares (measured on left-hand side vertical axis of Figure 37.1). Simultaneously the objective should be to transform a degraded area/landscape to a functioning and resilient one. A functioning, resilient, restored landscape could take on many forms depending on the circumstances and the desire as well as requirement by the people both affected by the current degradation and those who will benefit from the restoration effort. Irrespective of the type and/or form of the restoration activity and the end-point thereof, it should be a resilient system that can support people, plants, and animals sustainably over a long period of time. This resilient-gradient is indicated on the horizontal axis of Figure 37.1. The right-hand side vertical axis of Figure 37.1 provides a measure of the need to enhance the economic welfare of the local economy as indicated by its state, whether being

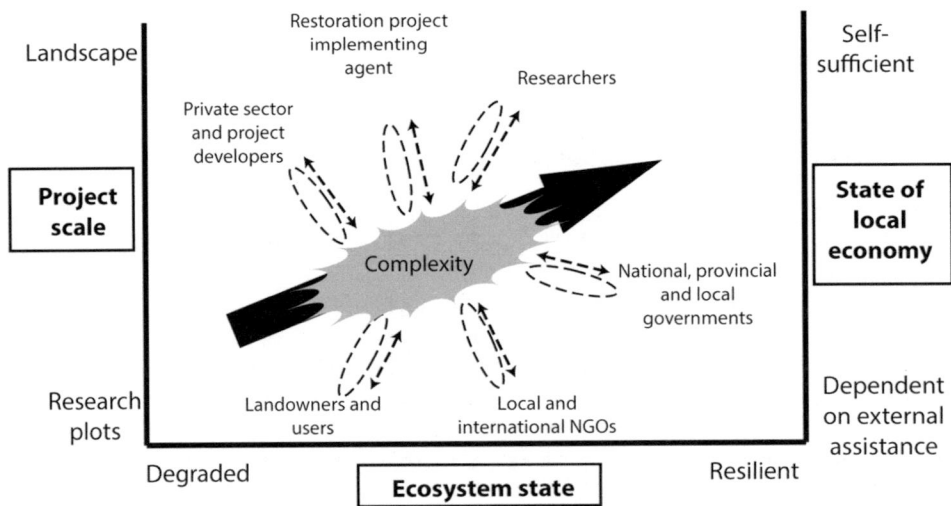

Figure 37.1 The integrative yet complex process required to up-scale restoration

dependent on outside sources for survival or being self-sufficient. Up-scaling restoration often includes the need to consider all three of these aspects in such a manner that the restoration effort benefits planet and people, and that it is affordable, following the arrow in the middle of Figure 37.1. This is a complex process involving a range of stakeholders that, in most cases, have to be consulted on an ongoing and iterative basis. The stakeholders include, but are not limited to, the private and public sectors, the restoration agency, NGOs and researchers, as well as landowners and users.

A case study

To illustrate the practical realities as well as plausible solutions, we consider the case of a restoration project outside Vanwyksdorp in the Western Cape of South Africa. The objective of this project is to restore degraded landscapes using mainly Spekboom (*Portulacaria afra*) cuttings, generating income through the sequestration of carbon and the marketing and selling of carbon credits. A 300ha pilot restoration project commenced in 2015 on five properties within a 20km radius from Vanwyksdorp, implemented under the umbrella of the Gouritz Cluster Biosphere Reserve (GCBR) (www.gouritz.com) with WESSA (www.wessa.org.za) as managing agent, and funded primarily by the European Union, but with co-funding from Working for Lands (www.environment.gov.za/projectsprogrammes/wfl). This pilot restoration project provides a learning experience with respect to the restoration of Spekboom in the Western Cape with the objective to up-scale the initiative.

The restoration in the area has the potential to impact at least 5,000–8,000 ha of degraded land in the proximity of Vanwyksdorp, but with the prospect of more than 200,000 ha in the province. The carbon sequestration potential is between 2 and 3.2 tC/ha/yr, or between 7 and 10 tCO_2/ha/yr, with the lower the estimate, the more realistic given the aridity of the area. This implies a total annual sequestration capacity of between 35,000 tCO_2/yr (7 tCO_2/ha × 5,000 ha) and 56,000 tCO_2/yr (7 tCO_2/ha × 8,000 ha). At a unit value of US\$10/tCO_2 this equates to a total annual value of between US\$350,000 and US\$560,000 per year for no less than 20, but more likely 50, years.

Developing a restoration-based carbon offset and mitigation project, however, is filled with various caveats. These caveats and how they have been addressed are discussed below.

1   *The pooling of the carbon.* It is rarely possible to develop a project of sufficient scale on one single property and/or involving one single landowner to enjoy sufficient economics of scale to warrant a project. This implies that, while land ownership is not affected, the carbon sequestrated from the property has to be pooled to generate the scale required necessitating the introduction of a 'pooling agency or facility'. To achieve such pooling requires a large degree of trust and goodwill to start with, but also a sound contractual agreement between the landowner and the pooling agency. This is because the pooling agency will be the owners of the carbon credits generated on somebody else's land. It has been decided that the GCBR will provide the pooling facility.

2   *Pairing public costs with private benefits.* Under such scenarios that restoration is conducted using public funds (i.e. tax or donor money), for both the public and private good (i.e. the sequestration of carbon, the improvement of the land's carrying capacity, improved water infiltration, reduced erosion, etc.), the private landowner can hardly claim the financial return on the carbon credit as well. When state or other public (e.g. donor) funds are being used to restore the land, care should be taken not to incentivize bad behaviour and promote further degradation in the hope of future restoration funds becoming available

from external sources. The landowner benefits from the restoration activity, for which he/she is not paying directly in cash; admittedly there might be other in-kind payments by the landowner in the process though. Restoration has the potential to enable the farmer to continue farming whereas the alternative is the possible abandonment of active farming. To overcome this conundrum, money generated from the sale of the carbon credits is reinvested in further restoration activities. The pooled carbon funds mentioned previously could therefore become a revolving fund whereby the sale of carbon credits leads to the expansion of restoration activities, increasing the reach of the restoration work.

3 *The marketing of the carbon credits.* There is a range of carbon platforms. Each of the carbon platforms or standards has a different set of validation criteria. The challenge is to select an appropriate standard for a project. Often the barriers to entry to some of these standards are so onerous that they discourage project developers to proceed. In some cases the cost of validation is so high that it cannot be absorbed by the project. I suggest using a local and 'terrestrial carbon friendly' platform.

4 *Additionality and permanence.* Irrespective of the carbon standard selected, the issue of additionality and permanence will arise. Additionality implies that the carbon sequestrated should be additional to a baseline assessment. It is therefore important that a carbon baseline assessment be conducted in advance, using appropriate and accepted carbon assessment methods. Not only should the carbon be additional to the baseline, but also should it be durable/permanent. Terrestrial restoration activities are subject to land use management practices, fires, droughts, pests and diseases – all of which impact the rate of sequestration. While some of these impacts are controllable, others are not. The one controllable item is land use management practice. That lies within the control of the landowner. The challenge is that the landowner has to commit to a certain land use practice in advance, and keep to that commitment, over a long period of time. Not only should the current landowner do so, but also the next landowner in the event that the property is sold and/or changes tenure for another reason. This is dealt with below.

5 *Land use management regime.* Selecting the most appropriate land use management regime to coincide with the restoration activity has been the subject of much debate. It is often argued that once land is subject to restoration, for example through Spekboom cuttings, the land should be fenced-off and domesticated animals (e.g. cattle, sheep and goats) excluded. Such exclusion could be construed as a form of land abandonment and is rarely socio-economically the most appropriate land use. This approach also places restoration/conservation directly against any form of active farming, income generation and food security options. While it might work in a selection of special cases, it is largely unpalatable and very elitist. Moreover, it perpetuates the conflict between active farming and conservation. By allowing the farmer to farm the land, but according to an agreed land use management plan that endorses a set of pre-defined land management actions, the farmer becomes an integral part of the restoration process. Such integrated land use management and farming practices have become known under the umbrella term conservation farming, or conservation agriculture (CA). CA with the express purpose to advance the carrying capacity of the land while improving field conditions, biodiversity, water infiltration, erosion and sediment control, and various other ecosystem goods and services, has the capacity to allow a farmer to increase the carbon stock on his farm. This can occur while actively farming the land to reduce the risk of food insecurity and the outward migration of people from towns. The development of an appropriate land use management plan for a specific farm and region that incorporates these intricacies in an integrated manner is therefore important.

6   *Scheme regulation.* One approach to secure permanence (as discussed in point 4) is to assign a specific land use management plan to a particular property and/or part thereof and to zone that under a specific municipal scheme regulation designated towards the particular type of CA as described in point 5. Zoning under such scheme regulation is not directly tied to the title deed of the property, but it does tie the designated area to the agreed land use management plan.

7   *Poverty.* While seeking to make agricultural towns work again, the need to focus on poverty reduction and income generation has to be at the forefront. This is a national strategic objective and has to be incorporated from the outset and not as an adjunct add-on.

The restoration of economics

Problems cannot be solved with the same mind-set that created them.

(attributed to Albert Einstein)

The meaning of the above dictum is very simple and straightforward: if a specific economic theory, policy, and development trajectory keeps on breaking things, such as the natural environment leading to unsustainable levels of resource use and degradation, then the problems associated with the degradation and resource use will not be resolved using the same principles and philosophy that led to the degradation and unsustainable levels of resource use.

What is therefore required is the restoration of economics. In all reality, the restoration of natural capital and the application of restoration economics only make true and long-lasting sense when economics are restored. If not, the current destructive character and abusive nature of the prevailing economic practice will persist.

Not all is doom and gloom though. There are some indications of initial changes. It has been pointed out that the concept of restoration is akin to the notion of investing in capital. That concept is known in economics and financial business analysis and is one that entrepreneurs are well familiar with. Hence, the concept can play an important bridging role in the restoration of economics. Furthermore, some advances have been made, albeit peripheral when considering the larger economic context, with respect to the development of projects related to the payment for ecosystem goods and services (PES). Large obstacles exist to a practical and economic application of PES including the establishment of property rights, execution of contracts, methods for monitoring performance, and the creation of markets (Blignaut *et al.* 2014b).

These advances, while commendable, in all fairness have been piecemeal and largely inadequate to address the current and rising challenges of unsustainable living. What is required is a fundamental change of heart in accordance with the Einstein dictum. Here we discuss one such possible fundamental change.

In distinguishing between ethics and economics, Wogaman (1986) states that it is important to distinguish between intrinsic and instrumental values. An intrinsic value is something that is good in itself and requires no further justification. Examples would be friendship, justice, music, etc. An instrumental value, on the other hand, is something that is good because it contributes to the fulfilment of an intrinsic value – it is an instrument in the realisation of an intrinsic value. Examples of instrumental values would be political power, owning a hi-fi, etc.

Economics in most cases has to do with instrumental values (i.e. commodities). These commodities are supposed to serve an intrinsic value, a greater good. Modern economics, however, has developed in such a way through the philosophy and application of the theory of the maximisation of individual utility that economics no longer concerns itself with instrumental values. Consumption has become a goal in and of itself, and hence the intrinsic value, pursued through self-interest, the instrumental value. The argument of consumption as highest good has been developed in the work done by Mill, Bentham, and Pareto in the construction of what has become known as welfare economics. Since consumption acts as intrinsic value, it gains control over the person who consumes and it directs and manipulates the thoughts and thought processes of such a person. Consumption has become the greater good, and those serving consumption the slaves. The proponents of neoclassical economics would deny this contention since they argue that neoclassical economics is value-free and morally neutral (Friedman 1953). Should one literally apply Friedman's definition of economics as a science, it would imply that an economist is a hedonist who can calculate but not think. Some would argue that it is unethical to restrict the market by introducing moral guidelines, since the outcome of the market process would be moral by definition (Von Hayek 1993). The moral solution to all economically related aspects would be to extend the boundaries of private property rights to be all-encompassing (Coase 1960). This view therefore supports the commodification of all public goods, which is a prerequisite for consumerism.

Commodification is in stark contrast with the views of Kant and Habermas regarding the presence of a context-relevant common moral denominator, which finds its expression in equality or human rights (Kant 1956; Habermas 1993), since consumerism is biased towards inequality and marginalisation of the weak in favor of the progress and self-interest of the strong. Based on Kantian principles, the common moral denominator (ethics) should provide guidance to all living beings on how to operate collectively to enable them to achieve their respective purposes. Ethics should therefore provide the moral framework within which all the relationships mentioned previously could be exercised in harmony, or with equality in rights. The primary intrinsic value that would achieve this harmony between the various cosmic objects is justice. Justice would be achieved through proper and appropriate management; hence management becomes the instrumental value.

Much has been written with regard to justice. Arguably, one of the most notable is the work of Rawls (1973), who proposes two principles:

1 Each person is to have an equal right to the most extensive basic liberty with a similar liberty for others.
2 Social and economic inequalities are to be arranged so that they are both (a) reasonably expected to be to everyone's advantage, and (b) attached to positions and offices open to all.

Based on these two principles, the application of social justice acts as a guide to how and where we should manage our dwindling natural capital stock and provides a rationale for the large-scale investment in ecological infrastructure through restoration.

Conclusion

Over the last two decades the practice of the economics of restoration has increased several-fold. While this is very encouraging, it is yet insufficient. More such collaboration among ecologists, hydrologists, economists, practitioners, landowners and users, and decision-makers is required to up-scale the effort of restoration from small plots and research-scale levels to a landscape-scale level as needed. The economic rationale for such up-scaling and the investment in restoration is sound. It is based on both theory and logic. It is hard to deny restoration's contribution to sustainable economic development as well as the fact that, when considering the value of the bundle of positive externalities it contributes to, it renders positive returns on investment.

What is required, however, to take the economics of restoration to a new level, is not improved technocratic means and ways to become more effective and efficient, even though that is necessary. What is essential is the restoration of economics. Economics' current fixation on consumption and self-interest has to make way for an approach which internalizes the need for management based on justice.

Economic analysis should be done as if people and nature matter, and conservation practised as if people matter, in an integrative manner. Then restoration will not be the 'end-of-pipe' afterthought fix, that will never be able to keep up to the prevailing trend of degradation and destruction, but an instrument integral to the upholding of an intact and well-managed system.

References

Aronson, J., Milton, S. J. and Blignaut, J. N. (2007) 'Definitions and rationale', in J. Aronson, S. J. Milton and J. N. Blignaut (eds) *Restoring Natural Capital: Science, Business and Practice*, Island Press, Washington, DC, 3–8.

Blignaut, J., Aronson, J. and De Groot, D. (2014a) 'Restoration of natural capital: a key strategy on the path to the quest for sustainability', *Ecological Engineering* vol 65, pp. 54–61.

Blignaut, J., Aronson, J. and De Wit, M. (2014b) 'The economics of restoration: looking back and leaping forward', *Annals of the New York Academy of Sciences* vol 1322, no 1, pp. 35–47.

Coase, R. (1960) 'The problem of social cost', *Journal of Law and Economics* October, vol 3, pp. 1–44.

Costanza, R. (ed.) (1991) *Ecological Economics: The Science and Management of Sustainability*, Columbia University Press, New York.

Daly, H. E. (1991) *Steady-State Economics*, 2nd edn, Island Press, Washington DC.

De Groot, R., Brander, L., Van der Ploeg, S., Costanza, R., Bernard, F., Braat, L., Christie, M., Crossman, N., Ghermandi, A., Hein, L., Hussain, S., Kumar, P., McVittie, A., Portela, R., Rodriguez, L.C., Ten Brink, P. and Van Beukering, P. (2012) 'Global estimates of the value of ecosystems and their services in monetary units', *Eco Serv* vol 1, pp. 50–61.

De Groot, D., Blignaut, J. N., Van der Ploeg, S., Aronson, J., Farley, J. and Elmqvist, T. (2013) 'Investing in ecosystem restoration pays: evidence from the field', *Conservation Biology* vol 27, no 6, pp. 1286–1293, DOI: 10.1111/cobi.12158.

El Serafy, S. (1989) 'The proper calculation of income from depletable natural resources', in Y. J. Ahmed, S. El Serafy and E. Lutz (eds) *Environmental Accounting for Sustainable Development*, World Bank, Washington, DC, 10–18.

Friedman, M. (1953) *Essays in Positive Economics*, University of Chicago Press, Chicago, IL.

Habermas, J. (1993) *Moral Consciousness and Communication Action*, MIT Press, Cambridge, MA.

Hawken, P. (1993) *The Ecology of Commerce: A Declaration of Sustainability*, HarperCollins, New York.

Hawken, P. (2010) *The Ecology of Commerce: A Declaration of Sustainability* (rev. edn), HarperCollins, New York.

Hawken, P., Lovins, A. and Lovins, L. H. (1999) *Natural Capitalism: The Next Industrial Revolution*, Earthscan, London.

Hicks, J. (1946) *Value and Capital*, Oxford University Press, Oxford.

Kant, I. (1956) *Critique of the Practical Reason* (trans. L. W. Beck), Bobbs-Merrill, New York.

Lutz, E. and El Serafy, S. E. (1988) *Environmental and Resource Accounting: An Overview*, Environment Department Working Paper no 6, World Bank, Washington, DC.

MA (2005) *Ecosystems and Human Well-Being: Biodiversity Synthesis*, Island Press, Washington, DC.

Neßhöver, C., Aronson, J. and Blignaut, J. N. (2011) 'Investing in Ecological Infrastructure' in P. Ten Brink (ed.) *The Economics of Ecosystems and Biodiversity in National and International Policy Making*, Earthscan, London, 401–448.

Rawls, J. (1973) *A Theory of Justice*, Oxford University Press, Oxford.

Repetto, R., Magrath, W., Wells, M., Beer, C. and Rossini, F. (1989) *Wasting Assets: Natural Resources in the National Income Accounts*, World Resource Institute, Washington DC.

SER (2004) *The SER International Primer on Ecological Restoration*, Society for Ecological Restoration International, Tucson, AZ.

Solow, R. (1992) 'An almost practical step towards sustainability', invited lecture on 8 October 1992 on the occasion of the 40th anniversary of the Resources for the Future, Resources for the Future, Washington DC.

TEEB (2010) *The Economics of Ecosystems and Biodiversity: Ecological and Economic Foundations*, Earthscan, London.

TEEB (2011) *The Economics of Ecosystems and Biodiversity: National and International Policy Making*, Earthscan, London.

Van den Bergh, J. and Janssen, M. A. (eds). (2005) *Economics of Industrial Ecology: Materials, Structural Change, and Spatial Scales*, MIT Press, Cambridge, MA.

Von Hayek, F. A. (1993) 'Social of distributive justice' in A. Ryan (ed.) *Justice*, Oxford University Press, Oxford, 117–158.

Wogaman, J. P. (1986) *Economics and Ethics*, Fortress Press, Philadelphia, PA.

38

BETTER TOGETHER

The importance of collaboration between researchers and practitioners

Robert Cabin

Ecological restoration is not rocket science – it's much harder! In contrast to the relative simplicity of manipulating the inanimate physical forces involved with building rockets, effective restoration inevitably involves wrestling with the far greater complexity of living species and ecosystems, thorny socio-economic-political issues, and vexing philosophical conundrums.

Consequently, restoration can be viewed as a 'social-ecological systems problem' (Ostrom *et al.* 2007; Watkins *et al.* 2015). That is, the underlying causes of ecological problems involve a complex, inseparable mixture of both biological and social components. Understanding and addressing such challenges thus requires an interdisciplinary approach that fosters robust collective analysis and decision-making processes (Davies *et al.* 2013; Keough and Blahna 2006; Redpath *et al.* 2013). Unfortunately, despite the growing awareness of the necessity of this approach to restoration and environmental problems in general, getting all the relevant stakeholders to participate and collaborate remains far easier said than done (Edwards and Gibeau 2013; Lasker and Weiss 2003; Reed 2008).

In this chapter I begin by reviewing the importance of and benefits from collaborations among restoration researchers, practitioners, and other stakeholders. I then discuss the costs and difficulties associated with these collaborations. Next, I highlight some strategies for overcoming these challenges and bridging the infamous research–practice gap. I conclude by suggesting that success in this field will increasingly require more inclusive collaborations that honour and integrate both the artistic and scientific dimensions of ecological restoration.

Importance of and benefits from collaboration

Increasing relevance

One of the highlights of the Ecological Society of America's 2012 annual meeting was a session entitled 'Growing Pains: Taking Ecology into the 21st Century'. As reported by Hampton *et al.* (2013), 'A standing-room-only crowd of approximately 500 scientists filled the room; the conference Twitter feed was dominated by reports from the session and questions from the long line of interested attendees, which spilled out into the hallway.' These authors proposed that the session's general theme could be boiled down to 'the need to dramatically reduce the insularity of our culture … Anachronistic incentive structures and practices that do not sufficiently engage other disciplines and sectors of society are hurting our capacity to maintain relevance in a fast-moving, multicultural, and highly connected society.' They argued that to address these challenges, ecologists must have more broadly-based collaborative partnerships, employ more jargon-free and inclusive communication, and incorporate more ethnic and racial diversity into their profession (they noted that one of the speakers at that session asked the audience to observe the nearly all-white sea of faces in the room).

Restoration ecology is more or less in the same boat. The combination of escalating biodiversity losses, ecological degradation and resource management conflicts have led to a rising call for restorationists to similarly reduce their insularity by reaching out to and working with a more diverse assemblage of individuals and institutions. This is partly due to a growing recognition that despite their good intentions, conservation-oriented programmes dominated by intellectually, ethnically, and/or racially homogeneous 'experts' tend to reflect the relatively narrow world views of their creators and alienate indigenous peoples, minorities and other historically underrepresented and disadvantaged groups (Wohlforth 2010).

Conversely, a wealth of scholarship and practical experience has repeatedly demonstrated that combining the knowledge, skills, values and resources of a broad array of people is essential for comprehending and addressing complex social-ecological problems such as the restoration of degraded ecosystems (Keough and Blahna 2006; Lasker and Weiss 2003; Scarlett 2013). This more inclusive, collaborative approach can also help build stronger coalitions, facilitate negotiations between competing interest groups, create a culture of collective learning and decision making, empower people to articulate and solve their own resource management problems and inspire the kind of deeper local 'buy-in' necessary to sustain restoration projects over time (Baker *et al.* 2014; Keough and Blahna 2006; Lasker and Weiss 2003). Indeed, an analysis of over 100 'ecosystem management' projects found that the 'dedication of participants' was the best predictor of their ultimate success (Yaffee 2002).

Synergies of working together

Much like the relationship between applied medical clinicians and more academic biomedical researchers, restoration researchers, practitioners and other stakeholders can work together in a mutually beneficial manner to design, implement and assess their projects and programmes. Developing and strengthening this kind of supportive, interactive relationship has in fact been a central goal of the Society for Ecological Restoration (SER) ever since its inaugural meeting in 1988 (Cabin *et al.* 2010).

As detailed in Cabin (2013a), the on-going efforts of Arthur Medeiros to restore a biologically and culturally valuable but heavily degraded forest in Hawaii provide an instructive example of the benefits of this kind of collaborative approach to restoration. 'I remember the day when a co-worker and I were taking a break from some hard physical work and admiring the beautiful ocean', Medeiros explained when I asked him about the origins of his restoration programme:

> He pointed down to the ten-acre exclosure we were trying to restore and said, 'Wow … what if this works?' I thought about that for a minute, then said, 'What if it works and no one cares?' So I started recruiting volunteers. No one came at first, then they started coming.

Working closely with the local landowners, Medeiros's work steadily expanded to include now vibrant and diverse volunteer and outreach programmes, phenomenally successful on-the-ground projects, productive interdisciplinary research and close alliances with restoration-oriented regional 'Watershed Partnerships'. 'Sure, he is passionate about the environment and Hawaii's native species', an elderly woman concluded after reflecting on Medeiros's many accomplishments. 'But he also cares deeply about the local people, the Hawaiian culture, and especially the Hawaiian kids. Not very many people who do what he does care about those things too.'

This case study illustrates some of the important synergies that can develop when researchers, practitioners and other stakeholders collaborate. First, in this and many other cases involving heavily degraded and remote field sites, it would be economically, politically and logistically difficult if not impossible for anyone working in isolation to accomplish much of anything. However, though it may proceed slowly at first, the process of people and organizations coming together can generate positive feedback loops of increasing media exposure, site visits and volunteerism, stakeholder engagement, interdisciplinary research opportunities and economic and political support.

Second, because restoration researchers, practitioners and other stakeholders tend to have some basic common needs (e.g. labour, project site security and access, and supplies and equipment), they can generally satisfy such needs far more efficiently by pooling their resources and working together. The distinctions between seemingly separate objectives such as 'research', 'restoration' and 'outreach' can even become arbitrary and irrelevant. For instance, after I spent a day with a local college class weeding and collecting data within a degraded forest that a community-based coalition was trying to restore, I realized we had seamlessly accomplished all three of these objectives (Cabin 2011).

Third, there are many less utilitarian yet perhaps equally if not more valuable benefits from working together on restoration projects. These include the gratification of collectively being able to 'make a difference', the intellectual and spiritual enrichment of interacting with a diverse community and the camaraderie of working with others who are similarly passionate about restoration despite the typically low pay and difficult work. As Medeiros put it when I mentioned the profound effect his work has had on so many different people: 'but they also have a huge effect on me. To have forty people come out, most of whom I've never seen before, and be so enthralled with what I have to say and what we're doing up there – it's an amazing thing for a scientist to experience.'

Challenges of collaborations

If working together is so great, why isn't everyone doing it, and doing it well? The short answer is that collaboration is a classic 'wicked' problem – something that is virtually impossible to unambiguously resolve because different stakeholders view both the problem and its solution differently. As is the case for many environmental conflicts in general, the people associated with any given restoration project may have very different perspectives on what the 'real' issues are and how their collective efforts should be allocated to address them. The dynamic social-ecological nature of restoration projects can force collaborators to wrestle with the often maddening interplay of ecological and interpersonal challenges as well. Unfortunately, as lamented by Edwards and Gibeau (2013), 'Even though countless books and papers have been written on collaboration, negotiation, consensus building, and dialogue, people really do not seem to know how to talk to one another.'

Different philosophies and priorities

One of the most illuminating things I ever did as a restoration researcher and practitioner in Hawaii was interview a broad spectrum of people working in the islands' greater environmental community. Part of my motivation for doing this was to better understand why there was so much conflict and tension within this community in general, and between researchers and practitioners in particular. I quickly discovered that even seemingly like-minded people were often operating under fundamentally different restoration paradigms. For example, here are excerpts (in Cabin 2011) from four people's answers to this question: 'How would you describe your overall restoration philosophy and strategy?'

- 'It's only logical that we try to save the best, most pristine places first.'
- 'Don't give up on the "basket cases" because they're saveable … don't go for the triage model.'
- 'We have to first develop a deep understanding of a place's community structure, then patiently design a plan that takes into account fine-scale processes like moisture, substrate, nutrient and disturbance gradients, locally adapted gene complexes, biogeography, and evolution.'
- 'The Hippocratic oath of ecological restoration? I don't buy it one bit – of course we're going to do harm! We've got to move fast and take big chances – we've got no choice.'

Needless to say, getting these four people to work harmoniously together on any conceivable restoration project could be a bit challenging! Such challenges can become exponentially more difficult as the number and diversity of individuals and organizations attempting to collaborate increase. In these broader coalitions, conflicts arising from opposing values, special interests and overarching world views are often exacerbated by incompatible personalities and clashing egos (Davies *et al.* 2013). Thus much like other such 'political initiatives', restoration collaborations necessarily involve the distribution of scarce resources, negotiations and trade-offs among different constituencies, power relationships and competing theoretical objectives and practical goals (Baker *et al.* 2014).

Time and money are two of the scarcest and most precious resources for many of those involved with restoration and other natural resource management projects (Cabin *et al.* 2010; Dickens and Suding 2013). Struggling to find the necessary time and money to get one's own work done while simultaneously contributing to a larger partnership that is itself trying to do too much with too few resources (e.g. academic research, on-the-ground resource management, technical and lay audience talks and reports, fundraising and education and outreach) can also generate significant conflict and tension.

The necessary process of debating and reaching consensus on even seemingly simple questions can also be laborious as well as excruciatingly circuitous and divisive: Is the most important goal of the partnership to perform cutting-edge science? Successfully restore a particular species or ecosystem? Inspire the general public and/or particular stakeholders to change their values and actions? While it may be expedient to claim 'all of the above' around the meeting table, in practice, due to limited resources and intrinsic conflicts between these different objectives, it is often not possible or even advisable to pursue them all simultaneously (Cabin 2007a).

Similar dynamics may unfold when partnerships must decide how to implement a particular management objective such as controlling a non-native invasive species: Should we use chemicals, manual labour or machinery? Which chemicals, hand tools or machines? Should we do a pilot study first? Who will do the physical work and collect and interpret the data? I will never forget the day (detailed in Cabin 2011) when, near the end of yet another tortuous meeting debating such questions, a normally patient and soft-spoken elderly local citizen finally pushed his chair back and stood up. 'If you people would just take some of the energy and hot air you expend around this table and put it to work out in the forest,' he bellowed in disgust, 'we could have cleared and planted the whole damn thing by now!'

Thus while communication and democratic participatory processes are essential for effective collaborations, they can result in 'consultation fatigue' and 'talking shops that create ambiguities and delay decisive action' (Reed 2008). Mistrust and polarization may develop, followed by participant disengagement and turnover, poor on-the-ground work and rising selfishness as busy and stressed participants increasingly focus on their own rather than the group's priorities (Davies *et al.* 2013; Keough and Blahna 2006). Factions may also develop as stakeholders jockey for position, and the empowerment of previously marginalized groups can threaten the existing power structure, which may respond by attempting to reassert and enforce its privileges and control (Reed 2008).

Researchers and practitioners as stakeholders

Researchers and practitioners are two fundamentally important 'stakeholders' associated with most restoration projects and partnerships. As a whole, researchers are a diverse group comprised of individuals who can have very different personal and professional interests, values and responsibilities. There can even be substantial diversity among researchers working within the same discipline. For example, an agricultural extension agent testing the efficacy of an herbicide on a shoe-string budget and an Ivy League biology professor investigating an abstract theory on a multi-million dollar grant may both be classified as 'research scientists'.

Nevertheless, researchers generally have some commonalities that set them apart from and potentially lead to conflicts with other groups and individuals. Perhaps most fundamentally, researchers tend to focus on 'know-why' knowledge that seeks to understand basic underlying principles behind observable phenomena (Reed 2008). Thus they value and pursue objectivity, replication, quantification and generalizable practices and theories. The more academically-oriented researchers are, the more pressure they tend to have to obtain funding from government and private institutions and publish their results in journals that value novelty and elegant intellectual accomplishments rather than applicable solutions to specific, real-world problems (Tewksbury and Wagner 2014).

Practitioners are another diverse group that also share some unifying characteristics that distinguish them from researchers and other groups. Practitioners may be natural resource managers and civil servants working for government agencies, consultants, lobbyists or employees of NGOs and other special interests. Yet it generally falls to this amorphous group to 'buy land, put up fences, set fires, put out fires, lobby politicians, negotiate with farmers, spray invasive weeds, poison rats and guard against poachers' (Anonymous 2007).

Consequently, practitioners in general, and applied practitioners in particular (i.e. those directly involved with the physical, on-the-ground work) tend to value 'know-how' knowledge that is informal, context dependent and generated over long periods of time by direct observation and practice (Reed 2008). Rather than investigate more academic questions such as 'how do potatoes grow?' they are more likely to ask 'how do we grow potatoes?' (Gonzalo-Turpin *et al.* 2008). In contrast to the reductionism and sweeping theories favoured by researchers, practitioners tend to focus on far more idiosyncratic, site-specific problems and solutions. Thus, to an applied practitioner, the researcher's 'generalized model of how to restore *an* oak savanna cannot tell us how to restore *this* oak savanna' (Simpson 2009).

The research–practice gap

Not surprisingly, given their often distinct perspectives, methodologies and responsibilities, creating and maintaining mutually beneficial relationships between researchers and practitioners can be challenging. Much like other stakeholder clashes, seemingly inconsequential preliminary conflicts between these two groups can catalyse a negative spiral of increasing distrust and disengagement.

The larger tendency of researchers and practitioners to discount, ignore and even compete with each other has been given different labels (e.g. 'the great divide', 'knowing but not doing' and the 'research-implementation gap') and examined from many different disciplines and perspectives (e.g. Arlettaz *et al.* 2010; Cabin 2011; Hulme 2011; Knight *et al.* 2008). For example, Gonzalo-Turpin *et al.* (2008) noted that 'Although an academic mode of knowledge production is essential in research for a better understanding of biological systems, it often fails to produce frameworks and methodologies having practical relevance that can be used in conservation and restoration programmes'; and a *Nature* editorial writer (Anonymous 2007) tersely concluded that 'Conservation biologists write and publish papers, which the practitioners seldom read. The practitioners, in turn, rarely document their actions or collate their data in forms useful to conservation biologists.'

Mirroring similar findings from ecology (Hulme 2011; Suding 2011), a 2009 SER survey (Cabin *et al.* 2010) revealed that only 26 per cent of the respondents believed that scientist/practitioner relationships were 'generally mutually beneficial and supportive of each other'. It also found that the 'science–practice gap' was the second and third most frequently cited category of factors limiting the science and practice of restoration, respectively ('insufficient funding' was first in both cases).

Researchers in general, and those from the natural sciences in particular, tend to have a different perspective than practitioners on the underlying factors behind the disconnect between their two groups. As summarized by Suding (2011), 'From the science side, the criticism that restoration ecology is largely ad-hoc and site specific, lacking a conceptual framework, has haunted the discipline almost since its beginning'. More academically-oriented researchers have criticized practitioners for failing to sufficiently appreciate the value and relevance of their research and more effectively incorporate it into their practices (Ntshotsho *et al.* 2015). As one research scientist (in Cabin 2013a) put it:

> Resource managers often don't realize how much of their knowledge and tools come from science. It's easy for them to take the fruits of science for granted, and say, 'Well, we already know all that' or 'That's just obvious, commonsense stuff', when the fact is that until scientists did the research, we didn't actually know or understand it at all! … Good science can help them prioritize and focus their management activities and reveal some of nature's underlying complexities.

Some further allege that because practitioners do not understand the rigorous requirements of 'real research', they do not understand why researchers can't tell them what to do or perform the kinds of more practically relevant, site-specific quick and dirty studies they frequently request (Dickens and Suding 2013). Still others have argued that there isn't a significant science-practice gap and believe that restoration science already is sufficiently practically relevant and effective at solving real-world problems (Giardina *et al.* 2007).

For their part, many applied practitioners view their work as more art than science, and question if and to what extent formal research is relevant to and necessary for their work in the first place (Dickens and Suding 2013; Suding 2011). In a parallel manner, some allege that because researchers generally do not understand the logistical difficulties and social-ecological complexities of on-the-ground work, they do not understand why it is so difficult for practitioners to more frequently and effectively incorporate 'real research' into their practices (Ntshotsho *et al.* 2015). As Arlettaz *et al.* (2010) sympathetically observed, the management recommendations in much of the conservation-oriented science literature 'are often vague and not pragmatic, and not surprisingly, practitioners are confronted with "solutions" where social and economic contexts are not properly appraised, cost effectiveness of management options is not evaluated, and management prescriptions are not quantitative or spatially explicit'.

Finally, some practitioners complain that researchers tend to take far more than they give, and that they are unable or unwilling to address their more applied needs. For example, here is how one highly skilled and passionate field technician viewed this relationship (in Cabin 2013a):

> We're always getting flooded with research applications from scientists who want to go do their research in some remote, pristine place. They always throw in some boilerplate stuff about how valuable their work might be to us managers, and we always laugh when we read those sections because they're total horseshit! ... So we bust up the forest [for the research projects], create all these new avenues for the weeds to come in, lose all that time we could have spent actually doing conservation, and what do we ever get back for our efforts in return?

And here is how a senior resource manager responded (in Cabin 2013a) when I asked him why he and his colleagues had stopped attending the 'bridging the science-practice gap' type regional meetings ostensibly designed to address this problem:

> Most of these meetings are dominated by researchers anyway ... a lot of it is cool, fascinating stuff, but it doesn't tell us anything about how to take care of the lands we are struggling to manage ... When we complained, the organizers tried to help by sending out surveys asking us to list the kinds of questions and specific problems we'd like the scientists to help us with. We'd fill them out and send them back, but nobody ever touched them, probably because they didn't involve any cutting-edge science, so guess what? We all just stopped going to those meetings.

Overcoming collaboration challenges

No silver bullet

Researchers and academics understandably strive to simplify and tame complex phenomena with eloquent theories and twelve-step programmes. Thus there is an abundance of 'how to collaborate' literature replete with generalized conceptual models, flow charts and bulleted lists. Yet the continuing struggle to make collaborations work in complex social-ecological arenas such as restoration ecology has also led many authors to argue that no single discipline, paradigm or cookie-cutter approach can possibly address the diversity of challenges associated with accomplishing this task in the real world.

For example, after discussing the 'wicked' nature of natural research management collaborations, Scarlett (2013) approvingly quotes Allen and Gould's (1986) conclusion that such problems 'cannot be solved by any multistep planning process designed to "collect more data, build bigger models, and crunch more numbers", as such thinking reflects "a naïve hope that science can eliminate politics"'. And after reviewing the dangers of 'relying on abstract cure-all proposals for solving complex problems related to achieving sustainable social–ecological systems', Ostrom *et al.* (2007) maintain that the empirical track record of this kind of 'panacea' approach to complex human/environmental problems is one of repeated failures.

Nevertheless, numerous authors have carefully analysed the attributes of the most successful social-ecological collaborations and projects and/or thoughtfully reflected on their own hard-won victories and painful failures. Interestingly, rather than arguing that their particular approaches or experiences are necessarily relevant and applicable to all situations, such authors tend to share their insights in a more collaborative spirit.

I would further humbly suggest that because those of us working in the academic/researcher portion of the restoration community typically have the most education and freedom, as well as get the grants, teach the students and read and write the papers and books such as this one, it is incumbent upon us to take the lead in improving our collaborations and bridging the research–practice gap. Consequently, most of the below suggestions and strategies for accomplishing these goals are aimed at and designed for academics and researchers.

Relationships, communication and inclusive engagement

Many authors (Beever *et al.* 2014; Davies *et al.* 2013; Dickens and Suding 2013; Redpath *et al.* 2013; Reed 2008) have identified stakeholder mistrust and unwillingness to engage and negotiate as the primary overarching challenges that must be addressed before effective collaborations can begin. These authors also note that building good personal relationships among the participants is often the key to establishing trust and creating a culture of participation, respect, empowerment and collective learning.

One of the reasons why effective collaborations can be so difficult, time-consuming and elusive is that building good personal relationships, especially among diverse participants with different values and motivations, is itself an often challenging, laborious and non-straightforward objective to accomplish. However, two relatively simple but effective techniques for getting the relationship ball rolling are regular face-to-face, small group or one-on-one meetings, and workshops, work days and field trips that enable stakeholders to work and learn together (Dickens and Suding 2013).

Open and honest communication among the participants is another essential ingredient for effective partnerships. Indeed, the first operational lesson Edwards and Gibeau (2013) share from their many years of working with people who are often hesitant to engage with others whom they perceive as different is simply to get people to talk to one another. In their experience, the only good predictor of whether or not a feasible solution will eventually emerge from the collaboration is how well people with opposing views begin a discussion of the problem-solving process.

Good communication requires good relationships and helps build them. One of the best ways to foster both is to begin by listening. For example, the previously mentioned interviews I conducted with the greater environmental community in Hawaii also proved to be an invaluable tool for helping me put myself in other people's shoes, better understand and appreciate their perspectives and establish mutual trust and respect. Once my interviewees realized that I was genuinely interested in what they had to say and was there to listen and learn rather than judge, virtually all of them were more than willing to candidly share their perspectives, passions and frustrations (most had never been asked to discuss any of these topics before).

As good relationships and communication begin to develop among particular individuals and subgroups, the next challenge often involves discussing and defining the larger partnership's overall mission, objectives, structure and operating procedures. Employing independent, mutually respected professional facilitators is often necessary to ensure that these objectives are accomplished in a fair, inclusive and transparent manner (Edwards and Gibeau 2013; Lasker and Weiss 2003; Reed 2008). These facilitators can also help steer partnerships towards designing and implementing projects that are socially acceptable, and economically and ecologically sustainable (Beever *et al.* 2014; Keough and Blahna 2006).

Bridging the research–practice gap

After reviewing the prevalence and importance of the research–practice, or knowing–doing, gap in restoration ecology and other related applied sciences such as conservation biology and ecosystem management, Knight *et al.* (2008) suggest the first step towards addressing this problem is to simply acknowledge that it is real. These and other authors (Arlettaz *et al.* 2010; Suding 2011; Tewksbury and Wagner 2014) further suggest that one effective way researchers can begin bridging this gap (and increasing the impact and rigor of their own research) is to directly contribute to real world projects.

A relatively easy way for researchers to get more involved is to at least occasionally help with the actual on-the-ground work. In addition to providing generally much-needed and greatly appreciated labour, physically working together can be an excellent way to build good relationships among the various restoration 'factions'. This is partly because it gives people a chance to demonstrate their commitment to the project while interacting more informally, without literally and figuratively having to sit in their customary chairs around the meeting table (Cabin 2011). It can also provide a valuable opportunity to let applied practitioners assume a leadership role and show off their skills, and for researchers to appreciate just how difficult it can be to meaningfully apply their theories and recommendations in the real world. Fortunately, there appears to be growing momentum to better reward researchers for working with practitioners in this more direct and applied manner (Arlettaz *et al.* 2010; Knight *et al.* 2008; Tewksbury and Wagner 2014).

Restoration researchers and practitioners appear equally frustrated by the public's general ignorance of and lack of support for their work. Consequently, some have concluded that collectively developing more broadly based political support for restoration is one of the most important things we can do to advance our science and practice and help bridge the gap between these two disciplines. Because applied practitioners and the general public typically do not read their peer-reviewed publications and attend their technical talks, researchers can help build this political support by writing relevant non-technical articles and giving more popular lay-audience talks. Using these venues to help the public understand why managers often have to implement politically and ethically unpopular acts such as killing non-native invasive species can also generate enormous good will from the larger practitioner community.

Humility and pluralism

Kozar (2010) makes the useful distinction between 'cooperative work', which involves accomplishing a task by dividing it among the participants, and 'collaborative work', which entails coordinating the efforts of the participants so that they solve the problem together. She illustrates this difference by comparing the benefits of a potluck dinner versus cooking a meal together. In the former situation, the experience will undoubtedly be richer than it would have been if everyone had just consumed their respective dishes by themselves. However, by cooking together directly, in addition to enjoying a good dinner, the participants will also learn new practical skills and form deeper inter-personal relationships.

Researchers may believe or implicitly assume that knowledge is something that is generated and disseminated by experts such as themselves. Yet much scholarship and often painful experience suggest that this is not in fact the case (Gonzalo-Turpin *et al.* 2008). Consequently, much like the difference between a potluck dinner and cooking together, several authors have argued for a shift from 'trickledown, transfer, and translate models of knowledge dissemination' (Knight *et al.* 2008) to a more truly collaborative approach in which researchers and practitioners work closely together to generate and apply knowledge that meets the needs and utilizes the skill sets of both groups (Gonzalo-Turpin *et al.* 2008).

Researchers may also passionately believe in the overarching superiority of their particular disciplines. For instance, many well-intentioned scientists argue that resource management conflicts in general should be resolved using the best available scientific data, and that restoration projects should be 'science-driven'. While this may sound good, in reality this approach can be problematic and divisive. This is because since science is a methodology, rather than a religion or philosophy, it cannot by itself tell us what to do or how to resolve our conflicts. For example, the decision to 'promote ecological restoration as a means of sustaining the diversity of life' (Society for Ecological Restoration 2004) in the first place stems from a set of personal values that lie outside the scope of science (Davis and Slobodkin 2004).

Counterintuitively, extensive research has also repeatedly shown that 'scientizing' conflict tends to lead to both greater intellectual uncertainty and political polarization (Kahan 2012; Sarewitz 2004). Thus even when disagreements seem like they could and should be addressed by existing or additional research, this research tends to reveal ever-more complexities that serve to inflame rather than resolve the differences between the opposing stakeholders and perspectives (Redpath *et al.* 2013; Sarewitz 2006). As Sarewitz bluntly concluded (in Cabin 2013b):

> If you find the science that supports what you want to do, then you think it can dictate what everyone else should do, and you don't have to be honest about the complex sources of your own motives. …the idea that scientists can act as experts dictating particular policy choices divorced from their personal worldviews and values is bullshit.

Consequently, another fundamental step that academics in general and scientists in particular can take to help bridge the research–practice gap and foster better relationships with practitioners and other stakeholders is to acknowledge the limitations and inherent subjectivities of their own disciplines, and to strive to better understand and appreciate other perspectives, values and approaches. Perhaps paradoxically, this more humble and pluralistic approach tends to result in more rather than less respect and support for researchers and their research, as well as more successful restoration collaborations and projects (Davies *et al.* 2013; Redpath *et al.* 2013; Watkins *et al.* 2015).

Conclusion

Creating and maintaining truly collaborative restoration partnerships is hard work – perhaps as hard or even harder than the restoration research and management projects themselves. There do not appear to be any foolproof how-to-collaborate theories or shortcuts, or any guarantees that any given collaboration will necessarily be successful. However, the growing difficulties associated with working in isolation, combined with the rising incentives for working with diverse stakeholder groups, may increasingly make inclusive collaborations more of a necessity than a choice.

Science and art

Whether restoration is more science than art or art than science has been the subject of some-times contentious debate (reviewed by Suding 2011; also see the exchange between Cabin 2007a, Giardina *et al.* 2007 and Cabin 2007b). But perhaps a more enlightened view is that effective restoration incorporates *both* the sciences and the arts. While this may seem like a post-modern, politically-correct idea, its roots can actually be traced back to the pioneering prairie restoration work of Aldo Leopold and John Curtis in the 1930s, which helped demon-strate and promote the idea of using restoration as an effective tool for rehabilitating and preserving nature and what we now call biodiversity (Wiens and Hobbs 2015).

Although Leopold was a dedicated and highly accomplished scientist, he also believed that there was a gap between the complexity of the 'land organism' and the ability of science by itself to comprehend this complexity and guide our ethical responsibility to 'doctor sick land'. Leopold thus urged both the scientific and practitioner communities of his day to follow his lead by ignoring the 'senseless barrier between science and art' and directly incorporating their personal experiences, intuitions and aesthetics into their work (Cabin 2011).

This kind of more holistic and inclusive approach is becoming more explicitly apparent in today's restoration literature and most successful collaborations. For instance, Schaefer and Tillmanns (2015) describe how 'listening to ecosystems' enables them to assess 'natural systems at a meta level that reveals synergisms and intangibles difficult to articulate and/or analyse scientifically, but are nevertheless critical to ecological restoration'. However, although they show how the art of listening can reveal the kinds of social-ecological complexities and local and traditional ecological knowledge that 'may evade the net of scientific biophysical invento-ries', they also point out that this approach is intended to complement rather than replace more conventional ecological research.

The on-going mountains to sea restoration programme at the Limahuli Botanical Garden in Hawaii is a similarly innovative and instructive example of a restoration approach that incor-porates both the sciences and the arts. For example, one of their first projects was to restore their seven-hundred-year-old stone walls that had been constructed to irrigate a series of taro patches (the potato-like stems of this plant were a staple food of the native Hawaiians). As Limahuli's Director at that time explained (in Cabin 2013a), when they first began restoring this ancient rockwork:

> We worked out an arrangement in which the Native Hawaiian stonemasons and the archeologists would have an equal voice, because as artisans, the stonemasons could see things in the alignments of the stones that the archeologists couldn't. Similarly, as the restoration programme progressed, I realized that we needed to bring the practi-tioners into the visioning and planning processes …

Perhaps these examples of blending art and science to listen to the land and restore those beau-tiful stone walls can serve as larger metaphors and inspiration for our ongoing collaborative efforts to restore and reconnect our remaining natural ecosystems and human cultures. They can also serve to remind us that *how* ecological restoration is done may in the end be at least as important as the end products themselves.

Summary

- Because ecological challenges involve both biological and social components, restoration can be viewed as a social-ecological problem. Understanding and addressing such problems requires an interdisciplinary and collaborative approach.
- Restoration researchers, practitioners and other stakeholders can and increasingly must work together in a mutually beneficial manner. However, how to effectively collaborate is a 'wicked' problem because different stakeholders view both the problem and its solution differently.
- Restoration collaborations involve the distribution of scarce resources and conflicts arising from opposing values, special interests and personalities. Developing mutually beneficial relationships between researchers and practitioners is a particularly difficult yet important challenge to overcome.
- Effective collaborations typically include good personal relationships, open and honest communication, trust and inclusive participation. Researchers can build more productive relationships with practitioners by contributing to on-the-ground projects, acknowledging the limitations of their own disciplines and respecting alternative perspectives and approaches.

Acknowledgements

I thank Jessica Norris and Kristen Ross for their helpful discussion and feedback on a draft version of this manuscript.

References

Allen, G. M. and Gould, E. M. (1986) 'Complexity, wickedness, and public forests', *Journal of Forestry*, vol 84, no 4, pp. 20–23.

Anonymous (2007) 'The great divide', *Nature*, vol 450, no 7167, pp. 135–136.

Arlettaz, R., Schaub, M., Fournier, J., Reichlin, T. S., Sierro, A., Watson, J. E. and Braunisch, V. (2010) 'From publications to public actions: when conservation biologists bridge the gap between research and implementation', *Bioscience*, vol 60, no 10, pp. 835–842.

Baker, S., Eckerberg, K. and Zachrisson, A. (2014) 'Political science and ecological restoration', *Environmental Policy*, vol 23, no 3, pp. 509–524.

Beever, E. A., Mattsson, B. J., Germino, M. J., van der Burg, M. P., Bradford, J. B. and Brunson, M. W. (2014) 'Successes and challenges from formation to implementation of eleven broad-extent conservation programmes', *Conservation Biology*, vol 28, no 2, pp. 302–314.

Cabin, R. J. (2007a) 'Science-driven restoration: a square grid on a round earth?' *Restoration Ecology*, vol 15, no 1, pp. 1–7.

Cabin, R. J. (2007b) 'Science and restoration under a big, demon haunted tent: reply to Giardina *et al.* (2007)', *Restoration Ecology*, vol 15, no 3, pp. 377–381.

Cabin, R. J. (2011) '*Intelligent Tinkering: Bridging the Gap Between Science and Practice*, Washington, DC, Island Press.

Cabin, R. J. (2013a) *Restoring Paradise: Rethinking and Rebuilding Nature in Hawaii*, Honolulu, University of Hawaii Press.

Cabin, R. J. (2013b) 'Science friction', *Earth Island Journal*, vol 28, pp. 49–53.

Cabin, R. J., Clewell, A., Ingram, M., McDonald, T. and Temperton, V. (2010) 'Bridging restoration science and practice: results and analysis of a survey from the 2009 society for ecological restoration international meeting', *Restoration Ecology*, vol 18, no 6, pp. 783–788.

Davies, A. L., Bryce, R. and Redpath, S. M. (2013) 'Use of multicriteria decision analysis to address conservation conflicts', *Conservation Biology*, vol 27, no 5, pp. 936–944.

Davis, M. A. and Slobodkin, L. B. (2004) 'The science and values of restoration ecology', *Restoration Ecology*, vol 12, no 1, pp. 1–3.

Dickens, S. J. M. and Suding, K. N. (2013) 'Spanning the science-practice divide: why restoration scientists need to be more involved with practice', *Ecological Restoration*, vol 31, no 2, pp. 134–140.

Edwards, F. N. and Gibeau, M. L. (2013) 'Engaging people in meaningful problem solving', *Conservation Biology*, vol 27, no 2, pp. 239–241.

Giardina, C. P., Litton, C. M., Thaxton, J. M., Cordell, S., Hadway, L. J. and Sandquist, D. R. (2007) 'Science driven restoration: a candle in a demon haunted world – response to Cabin (2007)', *Restoration Ecology*, vol 15, no 2, pp. 171–176.

Gonzalo-Turpin, H., Couix, N. and Hazard, L. (2008) 'Rethinking partnerships with the aim of producing knowledge with practical relevance: a case study in the field of ecological restoration', *Ecology and Society*, vol 13, no 2, retrieved from www.ecologyandsociety.org/vol13/iss2/art53.

Hampton, S. E., Strasser, C. A. and Tewksbury, J. J. (2013) 'Growing pains for ecology in the twenty-first century', *Bioscience,* vol 63, no 2, pp. 69–71.

Hulme, P. E. (2011) 'Practitioner's perspectives: introducing a different voice in applied ecology', *Journal of Applied Ecology*, vol 48, no 1, pp. 1–2.

Kahan, D. (2012) 'Why we are poles apart on climate change', *Nature,* vol 488, no 7411, p. 255.

Keough, H. L. and Blahna, D. J. (2006) 'Achieving integrative, collaborative ecosystem management', *Conservation Biology*, vol 20, no 5, pp. 1373–1382.

Knight, A. T., Cowlin, R. M., Rouget, M., Balmford, A., Lombard, A. T. and Campbell, B. M. (2008) 'Knowing but not doing: selecting priority conservation areas and the research-implementation gap', *Conservation Biology*, vol 22, no 3, pp. 610–617.

Kozar, O. (2010) 'Towards better group work: seeing the difference between cooperation and collaboration', *English Teacher Forum*, vol 48, no 2, pp. 16–23.

Lasker, R. D. and Weiss, E. S. (2003) 'Broadening participation in community problem solving: a multidisciplinary model to support collaborative practice and research' *Journal of Urban Health*, vol 80, no 1, pp. 14–47.

Ntshotsho, P., Prozesky, H. E., Esler, K. J. and Reyers, B. (2015) 'What drives the use of scientific evidence in decision making? The case of the South African Working for Water programme', *Biological Conservation*, vol 184, pp. 136–144.

Ostrom, E., Janssen, M. A. and Anderies, J. M. (2007) 'Going beyond panaceas', *Proceedings of the National Academy of Sciences*, vol 104, no 39, pp. 15176–15178.

Redpath, S. M., Young, J., Evely, A., Adams, W. M., Sutherland, W. J., Whitehouse, A, Amar, A., Lambert, R. A., Linnell, J. D. C., Watt, A. and Gutierrez, R. J. (2013) 'Understanding and managing conservation conflicts', *Trends in Ecology and Evolution*, vol 28, no 2, pp. 100–109.

Reed, M. S. (2008) 'Stakeholder participation for environmental management: A literature review', *Biological Conservation*, vol 141, no 10, pp. 2417–2431.

Sarewitz, D. (2006) 'Liberating science from politics', *American Scientist* vol 94, no 2, pp. 104–106.

Sarewitz, D. (2004) 'How science makes environmental controversies worse', *Environmental Science and Policy*, vol 7, no 5, pp. 385–403.

Scarlett, L. (2013) 'Collaborative adaptive management: challenges and opportunities', *Ecology and Society,* vol 18, no 3, http://dx.doi. org/10.5751/ES-05762-180326.

Schaefer, V. and Tillmanns, A. (2015) 'Listening to ecosystems: ecological restoration and the uniqueness of a place', *Ecological Restoration*, vol 33, no 1, pp. 3–9.

Simpson, T. (2009) 'A science of land individuals', *Ecological Restoration*, vol 27, no 2, pp. 115–121.

Society for Ecological Restoration Science (2004) *The SER Primer on Ecological Restoration*, Society for Ecological Restoration, Tucson, AZ.

Suding, K. N. (2011) 'Towards an era of restoration in ecology: successes and failures along the science-practice divide', *Annual Review of Ecology, Evolution, and Systematics*, vol 42, pp. 465–487.

Tewksbury, J. A. and Wagner, G. (2014) 'The role of civil society in recalibrating conservation science incentives', *Conservation Biology,* vol 28, no 5, pp. 1437–1439.

Watkins, C., Westphal, L. M., Gobster, P. H., Vining, J., Wali, A. and Tudor, M. (2015) 'Shared principles of restoration practice in the Chicago Wilderness region', *Human Ecology Review*, vol 21, no 1, pp. 155–177.

Wiens, J. A. and Hobbs, R. J. (2015) 'Integrating conservation and restoration in a changing world', *Bioscience*, vol 65, no 3, pp. 302–312.

Wohlforth, C. (2010) 'Conservation and eugenics: the environmental movement's dirty secret', *Orion*, retrieved from http://orionmagazine.org/article/conservation-and-eugenics.

Yaffee, S. L. (2002) 'Experiences in ecosystem management: ecosystem management in policy and practice', pp. 89–94 in *Ecosystem Management: Adaptive, Community-Based Conservation*, eds G. K. Meffe, L. A. Nielsen, R. L. Knight and D. A. Schenborn, Washington, DC, Island Press.

39

FEWER THAN 140 CHARACTERS

Restorationists' use of social media

Liam Heneghan and Oisín Heneghan

It's not so important to tweet as it is to tweet well.

(Fox et al. *2015)*

Introduction

By using social media, communities of users share information, opinions, personal messages, videos, photographs, and various other forms of content. Social media tools can thus serve to connect users and to facilitate reflection on diverse issues and events. Much of the scholarship on social media in recent years has been on the significance of social media on our personal affairs. However, there is a growing realization that these new forms of communication can be influential in connecting and reflecting on matters of public and political concern, including ecological restoration.

The emergence of social media, or Web 2.0, to use the alternative term coined by Darcy DiNucci in 1999 to designate his prophetic view of the interactive future of the internet, has been tied to the availability of those technologies that provide platforms to facilitate connections between users (DiNucci 1999). Initially, this involved computers connected in local networks, and subsequently computers linked via interconnected networks that constitute the internet. The widespread availability of hand-held devices, along with faster internet connection speeds has diversified both the range (measured by the diversity of tools) and the reach (measured by the number of users and the frequency of use) of specialized social media services. For example, photo and video sharing sites (Instagram, Flickr, YouTube, and so on) benefit from the enhanced quality of cameras available relatively economically to a growing number of potential users.

Social networking, one type of social media, is a prevalent element in the personal lives of many people. Facebook, the popular social networking site, reports 1.01 billion daily active users for September 2015 (http://newsroom.fb.com/company-info). Twitter, a microblogging site, though smaller in comparison to Facebook, nonetheless hosts 646 million registered users in 2014 of which 289 million are active (www.statisticbrain.com/twitter-statistics). Thus Facebook and Twitter rank among the most visited sites on the internet (Facebook is number 2 after Google, and Twitter is number 10, www.alexa.com/topsites.)

In commentary about social media there is, understandably, considerable attention given to the role of social media in our immediate personal lives. Scholars have, for example, addressed questions about how we develop and 'perform' our identities in the virtual world. Other recent papers report on such issues as: Does social media enhance or erode self-esteem (Wilcox and Stephen 2013)? Is excessive use of social media correlated with certain character traits, for example narcissistic tendencies (Seidman 2013)? Or, will reliance on virtual encounters erode people's ability to communicate face-to-face (Baek *et al.* 2013)? On a brighter note, some researchers ask if social media gives citizens leverage over authoritarian leaders (Gleason 2013).

It is difficult to evaluate the overall conclusions from this emerging body of work. There is certainly quite a pronounced generational divide between population segments who are heavily engaged with social media (so-called 'digital natives' who were born since the creation of Web 2.0) and those were not (Bowe and Wohn 2015). To illustrate this, consider the results of a survey of faculty at the Johns Hopkins Bloomberg School of Public Health in Baltimore: 53.6 per cent had used YouTube, 46.4 per cent had used Facebook, 30.4 per cent had read blogs, and 6.6 per cent had used Twitter in the past month. Time since most recent degree completion was best predictor of social media use (Keller *et al.* 2014). Thus, unsurprisingly, younger people are more connected on social media. It seems plausible that there is a relationship between the generation to which a researcher belongs and their expressions of concern about the negative impacts of newly emerging trends in technology. Indeed, Richard Louv's popular book *The Last Child in the Forest*, exemplifies this intergenerational concern (Louv 2005). Louv's volume promoted concern about 'nature deficit disorder', a suite of social and health concerns supposedly associated with children's lack of contact with nature, and exacerbated by their inclination towards those technologies that host newer media. One can reasonably speculate that restorationists, a community that often values conservation activities because it gets them out of doors and actively engaged in hands-on conservation, might be sceptical of the values of the virtual worlds of social media.

Undoubtedly, answers to questions about the impact of social media on private individuals are useful for guiding people's use of these media, or even, in some cases, can help in developing preventive therapies when hyper-use is disabling (White 2014). However, there is now also a burgeoning literature to help guide institutions, such as commercial enterprises, not-for-profits agencies, advocacy groups, professional societies, and so forth, in their management of new media (Nah and Saxton 2012). How is institutional identity integrated across multiple platforms and combined with more traditional mechanisms to disseminate information and promote community? How can advocacy groups maintain an authoritative public profile? And how can emerging social practices, ecological restoration, for example, benefit from promotion of social media?

Systematic research on the use of social networks to create communities of interest around a particular topic and to stimulate action related to that focal topic is relatively recent. However, it is already clear that some endeavours, for instance, outreach by advocacy groups and authoritative opinion leaders on issues of public health, have benefited from the use of social media (Gholami-Kordkheili *et al.* 2013; Grajales *et al.* 2014). Environmental action around issues such as climate change and the conservation of charismatic species have also seemed to benefit from dissemination of information and discussion on social media. For example, advocacy organizations, such as 350.org which galvanizes action to lower atmospheric carbon dioxide to 350 ppm, have a strong and successful social media presence.

In what follows we present preliminary data that evaluates how ecological restoration, a conservation strategy that has a large number of professional practitioners, and an even larger number of volunteer advocates, is represented on social media. Since restoration practices have provoked public controversy, we examine how some controversial practices – invasive species eradication, for example – are discussed in social fora. Furthermore, we provide a content analysis of the uses to which social media is employed by restorationists with a view to reflecting on commonalities with and differences from other environmental themes, such as the conservation plight of charismatic species. We attempt to identify 'gaps' in the use of social media by restorationists, and make suggestions for improving the means by which this practice employs emerging social media tools.

Social media defined

There is no universally agreed upon definition of social media. Perhaps this is because all media is by its very nature social, and thus the term 'social media' is on first inspection redundant. Nor is there a universally accepted classification of these new media, one that would categorize social media into discrete non-overlapping types. Though these taxonomic issues may not be hugely consequential it is certainly the case that more generic definitions will include a greater range of media forms and tools, whereas others can be quite restrictive.

One helpful way of separating social media from more traditional forms is by distinguishing between them in terms of the potential relationships they foster between content producers and consumers. In traditional media a small group of producers – journalists, television producers and so on – connect with potentially many consumers. In contrast, by using contemporary social media, many producers may connect with many consumers. Indeed, in some cases the distinction between production and consumption is confused, as in, for example, the case of Wikipedia, the online encyclopaedia. Wikipedia entries can emerge from the collaborative input of several enthusiasts, who are, presumably, all themselves users of Wikipedia. The senior author of this chapter, for example, is an occasional 'Wikipedian', as editors of Wikipedia are called, and regularly makes small changes to articles as he browses them, when they fall into his areas of expertise.

By blurring the distinction between the productive and consumptive functions of media, social media tools can undoubtedly democratize mass communication. Social media allows for a cadre of non-professional content producers to connect with potential consumers in a cost-effective manner. This is illustrated on a significant scale by the role of new media in emerging democracies, and in the revolutionary movements associated with the Arab Spring of 2010 and subsequent years (Wolfsfeld *et al.* 2013). In a lower political register, it is undoubtedly the case that in the field of ecological restoration, expertise is diffusely distributed across a range of communities. Scientific researchers, volunteer practitioners, professional restorationists, and landowners all have legitimate and sometimes competing claims about restoration practice. Social media can potentially democratize participation in discussion across these several constituencies.

Though 'many to many' connections are a hallmark of some social media, it remains the case that there is considerable variety in the types of relationship existing between the size of producer and consumer populations even in newer social media forms. Text messaging, for example, considered by some to be a social medium, permits a small producer population (e.g. me) to connect to a small consumer population (e.g. you). On the other hand, YouTube, a video-sharing website, is an example of a social media where large numbers of producers compete for the attention of a large consumer population, even if, as is often the case, the producer of individual submissions is solitary and the number of viewers is limited.

Rather than defining social media in terms of patterns of production and consumption, which is suggestive rather that delineative, Andreas Kaplan and Michael Haenlein of the ESCP European Business School provide a definition based upon the technology on which social media exchanges occurs. They define social media as 'a group of Internet-based applications that build on the ideological and technological foundations of Web 2.0, and that allow the creation and exchange of user-generated content' (Kaplan and Haenlein 2010). Indeed, it may be that the more typical way of addressing the challenge of defining social media is to describe these media as a product of the platforms that enable them. Thus, in a similar vein, White (2014) for example, defines social media as 'various online technology tools that enable people to communicate easily via the internet to share information and resources'. The sorts of communications supported include text, pictures, audio files, and so on. Finally, Wikipedia defines social media as those 'online tools and platforms' that allow users to connect with each other.

Mayfield classified social media into five basic types: blogs, social networks, content communities, wikis, and podcasts. It's possible, of course, to greatly expand upon this list of types by including such categories as 'virtual worlds' and 'online marketing tools', and so on (Mayfield 2008). However, we will restrict the discussion to Mayfield's basic categories, as these categories can absorb most social media types. Moreover, they represent those forms that are best suited to social communication about ecological restoration and among restorationists.

Blogs are among the earliest forms of Web-based communications. A blog – short for Weblog – is an online journal where typically entries are presented in reverse chronological order. Since their inception blogs have evolved to be important authoritative alternative sources of information. Like all media the quality can be mixed. At their worst they can feature unfiltered narcissistic ramblings; at their best they can be important in breaking news that would not necessarily find an outlet in mainstream media. Examples of blogs that contain relevant discussions on restoration include http://eco-restore.net, https://prairieecologist.com and http://restecology.blogspot.com. A blog hosted by the Center for Humans on Nature, on urban ecology, has featured numerous essays on restoration in cities: www.humansandnature.org/blog.

*Social network*s are a subcomponent of social media where websites and applications facilitate communications between those sharing similar interests. boyd and Ellison provide a comprehensive definition: social networks, they write, are 'web-based services that allow individuals to (1) construct a public or semi-public profile within a bounded system, (2) articulate a list of other users with whom they share a connection, and (3) view and traverse their list of connections and those made by others within the system' (boyd and Ellison 2007). These networks are arguably the fastest growing segment of social media, and Facebook, as we have noted, is now the dominant social network. The importance of social networks for conservationist advocates is on the rise, and helped by the availability of pages on Facebook for organizations and ongoing projects. For example, the North Shore Restoration Group in Chicago, a highly influential volunteer organization, has a page that advertises workdays and shares images and video of restoration work: www.facebook.com/North-Branch-Restoration-Project-280524605316037. Restoration organizations can have a social network component to their membership pages. Chicago Wilderness, a biodiversity conservation organization with over 250 institutional members, and which advocates restoration as the most effective land management practice in degraded urban habitats, includes a social media component to their membership portal: www.chicagowilderness.org/?page=MemberPortal. LinkedIn, launched in 2003, is a professional networking site. Finally, the Society for Ecological Restoration (SER), the flagship organization promoting restoration, has an active page on LinkedIn with over 4000 members at the time of writing. This is more than the reported number of members of the society, which stands at 2400 (www.ser.org/membership). Thus, through the use of social media the society reaches beyond its membership.

Content communities are those where a community of users organize, share and comment on certain types of content, for example images (Flickr, Instagram) and videos (YouTube and Vimeo). There are plenty of independent restorationists actively posting on such sites. Content can be organized and 'tagged' (i.e. assigned a metadata keyword for easy retrieval) using social bookmarking sites. (Examples of such sites include Pinterest, Reddit, Delicious and Digg.)

Wikis are websites that anyone can edit. Generally, an accessible record is maintained on the site of all changes to content. Thus the editing can be shared rapidly and effectively. The term 'wiki' comes from the Hawaiian word for 'quick'. The first wiki was invented by Ward Cunningham in 1995. As yet there is no definitive wiki for restoration, though considering the contentious nature of some terms associated with restoration, including the definition of 'restoration' itself, a wiki project could be useful.

Finally, *podcasts* are programmes, usually in the form of an audio-recording, made available for download online. Several of the more compelling environmental podcasts are listed here: www.treehugger.com/culture/the-green-room-itunes-section-for-environmental-podcasts.html.

Case studies of social network analysis

For the purposes of evaluating how social media networks have been used to connect communities of interest and to inform action especially around issues requiring some scientific guidance, we present a summary of a variety of studies. In selecting this literature we intended to be representative rather than exhaustive. Our purpose in presenting them is to illustrate the ways in which social media might promote discourse on ecological restoration. Although we recognize that there is at present little literature directly related to social media use by restorationists, nonetheless, what literature exists is instructive and will inform our following conclusions.

Though there is an interesting literature to inspect, it is clear that since social media is a relatively novel phenomenon, unsurprisingly scholarship on the subject is in its infancy. Thus, the literature on the topic is not overwhelmingly large. However because of the immense volume of communication through social media, such media will provide a treasure trove of data to social scientists in the future. It is likely that data mining of social media will be a growth area for scholarship in the coming years.

In evaluating this literature it is useful to distinguish between scholarship *of* social media and scholarship *using* social media. The former examines the ways in which social media communication occurs, and examines popular themes, and this literature can be helpful in crafting guidelines for best practices. The latter uses communicative content on social media as fodder for study, especially by sampling observations on social media as a means of getting access to social phenomenon or objects and events in the non-virtual world. Products can include analysis of the distribution of phenomena, or even the evaluation of hypotheses concerning social behaviour. An example of the former would be where a researcher establishes which hashtags (labels that designate named topics of interest) are most effective in disseminating information on climate change; an example of the latter would be using reportage on social media of phenological events to monitor climate change. We will note the distinction in the following examples.

Medical disciplines

The use of social media in the medical sciences in general, and in public health in particular, has garnered scholarly attention. Such studies can be helpful in creating an agenda for examining communication surrounding issues of ecological health and in creating a toolbox for effectively extending the impact of social media use by restorationists.

According to Nascimento and colleagues, social media can amplify the means by which we can 'share our suffering' (Nascimento *et al.* 2014). In their paper these researchers reported on a content analysis of 21,741 'migraine tweets'. The gender distribution of those who commented in real time on their migraines was 73.47 per cent female, 17.40 per cent males and 0.01 per cent transgendered. Two of the more common descriptors of these headaches were 'worst' and the 'F-word'.

To describe these sorts of interrogations on the distribution of information on social media, Eysenbach coined the useful term 'infodemiology' (Eysenbach 2009). Information gleaned in such a way can be used to inform public health practice. A related term, also coined by Eysenbach, 'infoveillance', is defined as the use of infodemiology data for surveillance purposes: a means by which a researcher (or other party) can listen in on important social trends. In the case of the migraine infodemiological study mentioned earlier the researchers were able to capture real-time commentary on suffering which avoided the memory bias that can afflict more traditional studies. Furthermore, their analysis illustrated the 'highly heterogeneous and colloquial' language used to describe migraines, including some quite colourful profanity, that the authors suggest should be taken into account in pain questionnaires.

Another significant example of an infodemiological study is from Chew and Eysenbach who monitored the use of the terms 'H1N1' and 'swine flu' over time. They found that the 2009 H1N1-related tweets served primarily to disseminate information from credible sources in addition to sharing experience of suffering (Chew and Eysenbach 2010).

The infodemiological studies discussed earlier are examples of doing social media research *of* social media content. The researchers' aim is to use social media communications as a way of getting data on how information is communicated. Such use of social media by medical practitioners when undertaken with a view to developing best practices for communication is also instructive.

Additional studies have examined practices surrounding communication between medical professionals, the use of social media in training of practitioners, and in disseminating public health information (Spallek *et al.* 2015). The use of social media in the training of medical professionals may prove to be important. For example Cheston *et al.* (2013) conducted a meta-analysis of papers investigating how the use of social media tools affected satisfaction, knowledge attitude, and skills of physicians and physicians-in-training. Social media was especially helpful in promoting learner engagement, feedback, and collaboration and professional development. Technical issues, variable participation and security concerns were the main impediments to use.

There is literature emerging that examines how social media can be used to promote professional engagement. Matta *et al.* (2014), for example, charted a 'dramatic' increase in social media use by urologists. Twitter is their preferred social network platform. Urologists and other specialists use social media as a significant 'back channel' at conferences to increase the impact of research conferences (Ross *et al.* 2011). Ferguson and colleagues (2014) examined the use of Twitter to disseminate information during the 61st Annual Scientific Meeting at the Cardiac Society of Australia and New Zealand. They found that tweets amounted to a total of 1.4 million potential impressions on Twitter users.

Finally, research on the way in which public health is communicated on social media reveals important patterns. Harris *et al.* (2014) used NodeXL to analyse 4779 Twitter followers from 59 local health departments. They coded the follows for type, location, health focus and industry. Followers were organizations. Having a public information officer on staff served a larger population.

Laranjo *et al.* (2015) found a positive effect of SNS interventions on health behaviour-related outcomes, but there was considerable variation. They provide a meta-analysis of studies that specifically target changes in health behaviour where a SNS intervention was used. Facebook was the most frequent SNS, then health-specific SNS and, finally, Twitter.

Conservation disciplines

Environmental advocates, including some restorationists, have taken to social media with gusto. To illustrate, on a single day, 16 December 2015, there were 17,816 tweets (including retweets) on climate change (Figure 39.1). However, the literature inspecting the role of social media in promoting environmental thought and practice is rather meagre. However, what literature exists illustrates that naturalists, conservationists and environmental activists are using social media creatively and to good effect.

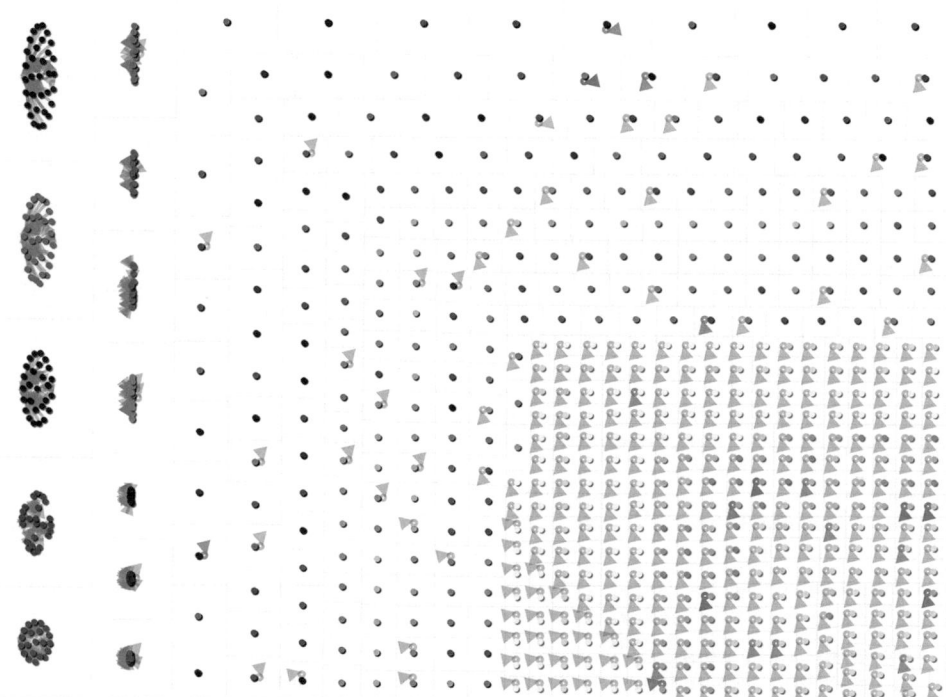

Figure 39.1 This network graph represents all tweets that include the term 'climate change' on 16 December. In this graph the most networked tweets are grouped to the left and top of the graph, with the most influential in the top left corner

Notes: The tweets and their connections are represented through points and edges, which are the lines that connect them. Because of the sheer volume of tweets, it may be hard to tell the specific interactions, but the top left groups are the more connected and influential tweets while the lower right section represents the individuals that are tweeting with little influence.

Total 'climate change' tweets (including RT, M, etc.): 17,816. Most influential tweets include:

@BernieSanders: Fifth #GOPDebate is over. Like the first, not one word about income inequality, climate change, or racial justice. The Rs are out of touch.

@CNN: Climate change is slowing Earth's rotation ever so slightly, scientists say https://t.co/PpncgRWZHQ https://t.co/356J2HU2uA

@HillaryClinton: We can't sit idly by while Republicans shame and blame women, demonize immigrants, and say climate change isn't real.

Not all environmental problems and species are created equally though, and some themes and creatures have much greater traction on social media than others. The polar bear (*Ursus maritimus*) is the most tweeted about species (Roberge 2014). Roberge concluded that the profile of lesser-known listed species needs to be raised.

There is an appetite for authoritative information about environmental issues in social media. Papworth and colleagues showed however that articles on climate change and mammals were more likely to be features in online conservation news content and delivery and are more likely to be shared and liked on Facebook (Papworth *et al.* 2015).

In addition to scholarship which illustrates which themes get the most traction on social media, and that can inform the manner in which advocates should exploit the potential of communicating via social media, other research explores how social media can be leveraged to generate new knowledge. These are examples of scholarship *on* social media in the terminology developed earlier in this chapter.

For example, Barve explored the use of photo-vouchered biodiversity occurrence data in the form of records associated with photos on social networking sites (SNS): Flickr, Facebook and Picasaweb (Barve 2014). Authoritative Digital Accessible Knowledge (DAK) about biodiversity can be important to conservation. Barve used occurrence data on the snowy owl (*Bubo scandiacus*) and monarch butterfly (*Danaus plexippus*) as proofs-of-concept and concluded that SNS could provide a new source of biodiversity data. Similarly, Cavalli *et al.* (2014) performed an online survey through Facebook which allowed them to locate 36 new burrowing owl (*Athene cunicularia*) sites. A final example: Chin (2014) demonstrated the use of social networks to solicit input from the Australian recreational SCUBA community to collect information on the porcupine ray *Urogymnus asperrimus*, a rare species. As a result an additional 29 new occurrence records were validated – doubling the number of records for this species.

No one cries over restoration on Twitter:
a content analysis of restoration tweets

In Wim Wenders's 1987 movie *Der Himmel über Berlin* (released in the US as *Wings of Desire*), angels hover around the city listening into Berliners' lonely thoughts and providing succour where they can. Twitter, comparably, can make angels of us all. This is especially true in the domain of surveillance – Twitter permits greater access to people's ruminations than at any other time in the world's history. The degree to which Twitter provides comfort, the second function of Wenders's angels, remains to be seen.

In order to hover and listen in on the conversation about ecological restoration we collected the tweets containing a set of key words associated with ecological restoration. We attempted to select a range of key terms associated with ecological restoration by rereading the *SER Primer on Ecological Restoration* (Society for Ecological Restoration 2004). We found that many potential key words were shared in common with other forms of land management practice. However, the term 'ecological restoration' is itself distinctive. Although other forms of land management and conservation practice allude to the problem of alien invasive species, concern about the removal of such species is a hallmark of restoration practice, and thus provided a second term. For the purposes of our inquiry we truncated this term to 'invasive species' in order to maximize the potential hits. Finally, since we were interested in how social media can provide a space for reflection on controversial ideas, we analysed the occurrence of the term 'novel ecosystem'. Novel ecosystems are those that have species compositions and relative abundances that have not occurred previously within a given biome. The notion is controversial since to manage systems as 'novel ecosystems' means paying less attention to the history of the system than is typically the case in restoration projects (Hobbs *et al.* 2009).

In addition to assessing the use of 'ecological restoration', 'invasive species', and 'novel ecosystems', we selected a range of terms associated with adjacent conservation practices and concerns. Interest in the fate of charismatic species, especially 'polar bear', is prevalent on social media, as we showed earlier in this chapter. The term 'polar bear' thus provide an out-group for comparison with those terms associated with restoration practice.

Between December 2015 and January 2016 we searched for our key terms on Twitter. We used the program NodeXL, an add-on for MS Excel, to manage our search and to analyse the social network connections in the results. NodeXL facilitates interactive network visualization and allows for the visual exploration of networks in addition to performing network analysis (Hansen *et al.* 2010). We prepared the network graphs in our figures as follows: There is an edge for each 'replies-to' relationship in a tweet, an edge for each 'mentions' relationship in a tweet, and a self-loop edge for each tweet that is not a 'replies-to' or 'mentions'. The graph is directed meaning that the arrows connect the edges, and show the direction associated with them. The graph's vertices were grouped by cluster using the Clauset-Newman-Moore cluster algorithm. Finally each graph was laid out using the Harel-Koren Fast Multiscale layout algorithm.

Our primary analytical approach for extracting meaning from downloaded information was a content analysis performed on the individual tweets. Each of the authors read all the tweets collected, and evaluated them based upon a set of general coding criteria.

In our analysis below we will holistically assess the content of the tweets, discussing the outlier group 'polar bears' initially, and then remarking on the tweets on restoration.

Polar bears

In the more than 10,000 tweets that we examined over the course of a month on the topic of 'polar bears', seventeen of them made a reference to crying or weeping. For example, @tayloretc tweeted: 'I started crying about polar bears this morning'. @hashbrownhalsey wrote '*starts crying over polar bears in 6th period'. Thus, polar bears come to mind and many people weep. In a similarly tender vein @wrenwhite_ wondered 'why do they gotta live such hard lives and die all the time'.

At times a fit of weep was induced by the cuteness of these predatory animals: @heyoitskaymo 'spent the last 40 minutes crying while looking at pictures of baby polar bears'. More often than not it was the conservation plight of the polar bears that provoked concern. @sofiagetler tweeted 'i also started crying in chapel because i thought about polar bears going extinct'. @gillianmoll wrote 'a guy in my cultures class talked about his friends being able to poach polar bears and they send him teeth and I started crying in class'.

Those who conjectured on the cause of these conservation concerns more often than not alluded to climate change. @lilianadiaz187 wrote 'just remembered there are like … polar bears dying bc the ice caps are melting now I'm crying again help me'. Other examples include: @deedzzzzzz 'when u dont wanna do hw so u look up what the affects [*sic*] of this wild warm weather are on the polar bears and u cry bc population decline 20 per cent'.

By virtue of the consistency of tweeted weeps – it translated into about three a week – it may be that weeping over polar bears could serve as a useful (if informal) metric, by which public concern over climate change could be evaluated.

Other themes emerge from the inspection of polar bear tweets. These can be grouped into the following categories:

- *Expressions of interest and support.* An example of a tweeter expressing general interest in polar bears includes @IvoryandBeau, who tweeted 'The older I get I feel like I am like my dad. Sitting at home watching National Geographic documentaries on polar bears'. The brief exclamation, 'I love polar bears', typically with an accompanying picture, is frequently posted.
- *Remarks on the science.* One tweeter wrote with an accompanying video, 'In this clip the population for polar bears is actually increasing by quite a healthy amounts'. Or quite emphatically @Lauriejo2 tweeted 'Global warming is NOT KILLING POLAR BEARS. Geo-engineering probable cause. Wake up!' These remarks were made in response to an article on the implications of a warmer globe for polar bears.

The most influential tweet in our December 2015 sample was the multiple retweeting of this affecting statement from @EarthfulStore, 'In 50 years our grandchildren will question if polar bears ever existed'.

Restoration keywords: invasive species, ecological restoration and novel ecosystems

Invasive species have undoubtedly captured the public imagination. 'Are Earthlings an "Invasive Species" Soon to be Exterminated?', @WakingTimes provocatively tweeted. On a less galactic scale, even Pablo Escobar, the deceased Columbian drug lord, gets a mention; or at least his pet hippopotami do. 'After Pablo Escobar's death, the 4 hippos he kept as pets have begun to breed and become an invasive species in Colum…' (@FactsInYourFace).

Viewing the content of the 2000 tweets (over two collection periods) that we evaluated in our analysis on invasive species, we assess the feed to represent a fairly mature engagement with this topic. That is, what gets disseminated and commented upon on Twitter is broadly reflective of what the community of invasion biologists generally finds significant about invasion. The discourse also reflects themes of interest to practitioner circles. In addition to the grand provocations about humans as invasive species and the factum concerning a narco-celebrity's pets going feral, this content ranges over several broad and appropriate categories. Users tweet about invasive species' definition (e.g. '@ISF_Breizh: To understand the effects of invasive species, we first have to understand what an invasive species is'), provide a roster of significant pests (e.g. 'RT @SailingSimple: In St. Bart's we participated in a Lionfish Hunt. This invasive species is destroying the reef…'), reference the implications of invasion for ecosystems (e.g. '@rfldn Answer: False! Norway maples are an invasive species and actually do more harm than good to the ecosystem. #ReForest10 #ldnont'), advertise a range of solutions (e.g. '@vtagweeds Invasive Species Banquet: If You Can't Beat 'Em, Eat 'Em'), and critical evaluations statements of policy (e.g. '@dick_shaw interesting move re #invsp Regulation with Commission rejecting it').

In contrast to tweeting about invasive species, tweeting explicitly about ecological restoration is rudimentary. Like during a coughing fit in the concert hall where a few heads may turn, and there may be a sympathetic nod or even a sharp rebuke, eyes quickly return to the front, back to the main business of the orchestra engaged in music-making upon the stage. And so it is with restoration; the primary engagement with restoration remains in the field, or in discussions at meetings, workshops, or perhaps even around a fire at a workday. Certainly no one weeps over restoration on Twitter as they do, so frequently, for polar bears.

The few hundred tweets that we reviewed on restoration can be categorized as follows, each with one or two representative tweets:

- *Advertising projects or institutional mission.* @dancingstarf1, referencing the work of the Dancing Star Foundation: 'The Foundation is involved in active ecological restoration efforts and intensive environmental field research'. @openlands, with a link to the project page: 'Major Ecological Restoration Project Begins at the Openlands Lakeshore Preserve'.
- *Promoting scientific and technical publications.* @jkinn88, with a link to a recent paper in the journal *Forest Ecology and Management*: 'Prioritizing boreal forest restoration based on disturbance regime'.
- *Sharing inspirational quotes and pictures.* @jacksiviter: 'Ecological restoration is a work of hope. #countryside #wildlife #rewilding'. Also multiple retweets of a tweet from @LynxUKTrust: 'Enigmatic #Lynx carving @GalGael for ecological restoration of Scotland Species reintroduction @LynxUKTrust'.
- *Commenting on certain practices or signature projects.* @mcewanlab: 'Castanea is such a fascinating case study in Ecological Restoration'. @stormsmart, commenting on a news article about 'beach tourism': 'Beach renourishment should not be mistaken as ecological restoration'.

Tweets about 'novel ecosystems' are, as might be expected, even sparser. On the dates we examined the feed, tweets and retweets about Marcus Collier's (University College, Dublin, @marcus_collier) engaging TEDxUCD talk on novel ecosystems dominated the feed. In addition, some tweeters linked to clarifying definitions of novel ecosystems ('A #NovelEcosystem is one that crossed an ecological brink that makes historical restoration no longer feasible…'). Communicating the relevance of novel ecosystems for urban conservation sites where traditional forms of restoration may not be feasible was also a theme in the tweets we reviewed.

Conclusion of content analysis

It is clear to us from this preliminary inspection of tweets that there is robust engagement with environmental topics using this social medium. However, some environmental topics receive considerably more attention than others. For example, there were more than 17,000 tweets about climate change in a few hours on 16 December 2015. Although we did not perform an explicit content analysis on these, nonetheless, some general remarks on them may be useful in contextualizing our comments on restoration tweeting. The more influential tweets on climate included ones from two candidates in the 2016 presidential race, Bernie Sanders and Hillary Clinton, both of whom alluded to climate change alongside other pressing social problems (income inequality, racial justice, women's issues, and the perception that immigrants are demonized; see Figure 39.1). By contrast, the number of tweets on any terms associated with ecological restoration is meagre.

An inspection of the network diagram for climate change reveals some significant patterns against which we will compare networks of tweets associated with ecological restoration. Each of our figures groups the most influential tweets to the top and left of the diagram; the tweet in the top left-hand corner is the most influential of all. In Figure 39.1 on climate change, the tweet by Bernie Sanders is not only the most influential of all, being retweeted by a large number of his 'followers' and in many cases this is followed by a flurry of additional retweets. In this way Sanders's tweet propagates through the network. This is illustrated in the figure by directed arrows that connect boxes across the grid. The network graph associated with tweets on 'polar bears' shows a more even pattern of influence across the tweets (Figure 39.2). The network is not as dominated by the top couple of tweets.

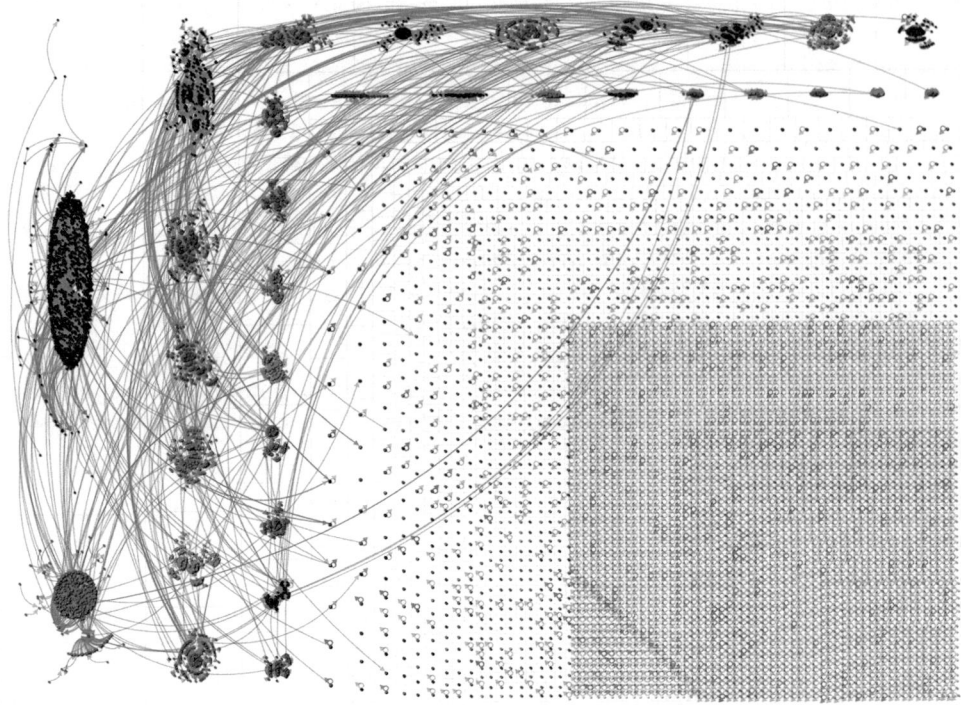

Figure 39.2   This network graph represents all tweets that include the term 'polar bear' between dates
14 to 15 December 2015

Notes: The tweets and their connections are represented through points and edges, which are the lines that connect
them. In this graph the most networked tweets are grouped to the left and top of the graph, with the most
influential in the top left corner.

Total 'polar bear' tweets (including RT, M, etc.): 1166. Most influential tweets include:

@EarthfulStore: In 50 years our grandchildren will question if polar bears ever existed.

@bedazzlingpics: Polar bears approach the camera to say hello – Photograph by Steven Kazlowski
https://t.co/eNzGzFAWhN

@photosandbacon: What do: Russian Soldiers, A Tank, and Polar Bears have in common? c.1952
https://t.co/HdDx5dHEdy

Although the number of tweets associated with 'invasive species' was small compared with
those that mentioned climate change, the behaviour of the network shows some similar
patterns to that of climate change (Figure 39.3, showing tweets from 6 to 9 December 2015).
Tweets regarding two published articles (one from *Science News* and one in *The Economist*)
dominated. By contrast, tweeting on 'novel ecosystems' and 'ecological restoration' was significantly less, and the number of retweets and mentions was small (Figures 39.4 and 39.5).

Conclusions: extending the use of social media in restoration

Ecological restorationists can use social media to connect with each other and to reflect on best
practices. However, our analysis suggests that ecological restorationists are not using social

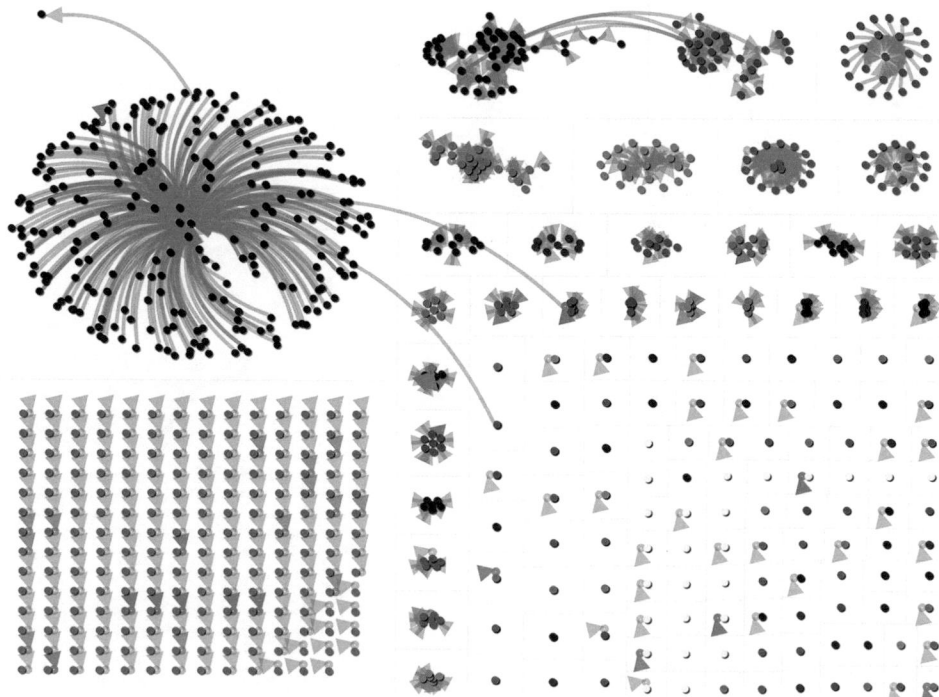

Figure 39.3 This network graph represents all tweets that include the term 'invasive species' between 6 and 9 December 2015

Notes: In this graph the most networked tweets are grouped to the left and top of the graph, with the most influential in the top left corner.

Total 'invasive species' tweets (including RT, M, etc.): 1192. Most influential tweets include:

@ScienceNews: Gene drives could wipe out malaria and take down invasive species: https://t.co/iXWWGqYfna https://t.co/0wpbMbXtbK

@TheEconomist: The EU has declared war on invasive species. This is wrong-headed https://t.co/l6UbRKIjIY https://t.co/4myt1aHVux

@RandallJBonner: DECK THE HALLS WITH PARASITES & INVASIVE SPECIES https://t.co/ZV1ReCO3NI @nwoutdoors @Outdoor_Hub @col_outdoors @OutdoorsNWMag @OUTDChannel

media to its fullest potential. We make the following suggestions for researchers *using* social media, for restoration ecologists, and for restoration practitioners.

Research 'using' social media

Examples of useful work using social media include investigating the spread of novel invasive species, and confirming the distribution of species using photo-vouchered specimens on a variety of content media. In this way researchers can access parts of the world remote from where they typically do their research. Researchers can, in addition, disseminate polls, and mine social media for attitudes about particular issues.

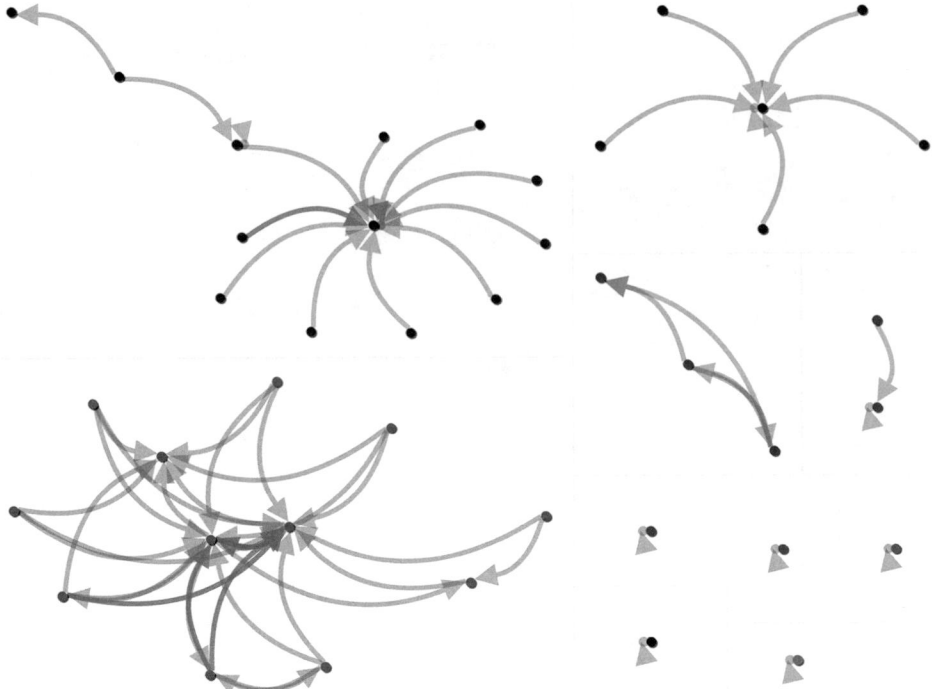

Figure 39.4   This network graph represents all tweets that include the term 'novel ecosystems' between 1 and 8 December 2015

Notes: In this graph the most networked tweets are grouped to the left and top of the graph, with the most influential in the top left corner.

Total 'novel ecosystems' tweets (including RT, M, etc.): 74. Most influential tweets include:

@TNatureOfCities: Urban spaces are newer, novel ecosystem types which implies thinking about and planning for them as hybrid systems http://buff.ly/1NLlp8L

@TEDxUCD: @marcus_collier TY and TY for a great talk on novel ecosystems, we too are looking forward to 2016!

@michaelaplein: R. Hobbs: novel ecosystems seen as 'slippery slopes' – but being pragmatic is being realistic about our possibilities #CEED2015

For restoration ecologists

Restoration ecologists might use social media in the way that their counterparts in other professions do. Veletsianos (2013) provides a framework for describing online social network practice: Sharing information about professional practice, sharing information about classroom use, requesting assistance, engaging in social commentary, engaging in digital identity and impression management, connecting with others, highlighting their engagement in other networks (see also Veletsianos 2012).

It will also be useful for restoration scientists to use social media during conference, as a 'back channel' to communicate with one another, and to increase the impact of their work on a more general public (Ross *et al.* 2011). The perceived importance of assessing the impact of work on public policy debates has led for a call to include 'tweetations' alongside more traditional citations (Eysenbach 2011; Bornmann 2014).

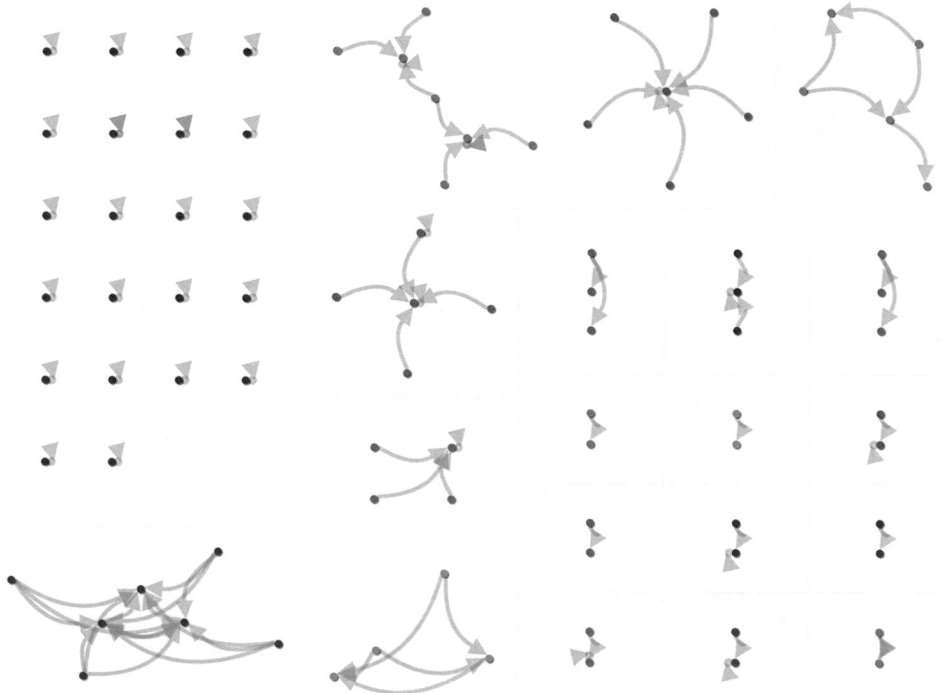

Figure 39.5   This network graph represents all tweets that include the term 'ecological restoration' between 1 and 9 December 2015

Notes: In this graph the most networked tweets are grouped to the left and top of the graph, with the most influential in the top left corner.

Total 'ecological restoration' tweets (including RT, M, etc.): 98. Most influential tweets include:

@EadhaAspen: Enigmatic #Lynx carving @GalGael for ecological restoration of Scotland Species reintroduction @LynxUKTrust

@markinflowers: 'We've inflated the derivatives & devalued the source of life'. John D. Liu speaks about ecological restoration @PlaceToBrief

@andersenetjarn: This us our new handbook on ecological #restoration on quarries. Building #NaturalCapital!

For practitioners

Practitioners often complain that researchers are out of touch, and researchers, in their turn, complain that practitioners do not always make scientifically informed decisions. Social media can help bridge this divide by providing a way of connecting researchers and practitioners (Gibbons *et al.* 2008).

At the most pragmatic levels social media can assist in advertising projects and recruiting like-minded people to engage on restoration projects. It can also facilitate discussion of methods and results. Social media can also be used to help novice practitioners become familiar with species identification and vocabulary of taxonomically relevant features (Hallman and Robinson 2015). Similarly social media can help foster consensus on definitions and terminology associated with certain practices.

References

Baek, Y. M., Y. Bae and H. Jang (2013). 'Social and parasocial relationships on social network sites and their differential relationships with users' psychological well-being'. *Cyberpsychology, Behavior, and Social Networking* 16(7): 512–517.

Barve, V. (2014). 'Discovering and developing primary biodiversity data from social networking sites: a novel approach'. *Ecological Informatics* 24: 194–199.

Bornmann, L. (2014). 'Do altmetrics point to the broader impact of research? An overview of benefits and disadvantages of altmetrics'. *Journal of Informetrics* 8(4): 895–903.

Bowe, B. J. and D. Y. Wohn (2015). 'Are there generational differences? Social media use and perceived shared reality'. In *Proceedings of the 2015 International Conference on Social Media and Society*: 1–5. Toronto: ACM.

boyd, d. m. and N. B. Ellison (2007). 'Social network sites: definition, history, and scholarship'. *Journal of Computer Mediated Communication* (13): 210–230.

Cavalli, M., A. V. Baladron, J. P. Isacch, M. S. Bo and G. Martinez (2014). 'Social networks and ornithology studies: an innovative method for rapidly accessing data on conspicuous bird species'. *Biodiversity and Conservation* 23(8): 2127–2134.

Cheston, C. C., T. E. Flickinger and M. S. Chisolm (2013). 'Social media use in medical education: a systematic review'. *Academic Medicine* 88(6): 893–901.

Chew, C. and G. Eysenbach (2010). 'Pandemics in the age of Twitter: content analysis of tweets during the 2009 H1N1 outbreak'. *PLoS ONE* 5(11): e14118.

Chin, A. (2014). '"Hunting porcupines": citizen scientists contribute new knowledge about rare coral reef species'. *Pacific Conservation Biology* 20(1): 48–53.

DiNucci, D. (1999). 'Fragmented future'. *Print* 32: 221–222.

Eysenbach, G. (2009). 'Infodemiology and infoveillance: framework for an emerging set of public health informatics methods to analyze search, communication and publication behavior on the Internet'. *Journal of Medical Internet Research* 11(1): e11.

Eysenbach, G. (2011). 'Can tweets predict citations? Metrics of social impact based on Twitter and correlation with traditional metrics of scientific impact'. *Journal of Medical Internet Research* 13(4): e123.

Ferguson, C., S. C. Inglis, P. J. Newton, P. J. S. Cripps, P. S. Macdonald and P. M. Davidson (2014). 'Social media: a tool to spread information: a case study analysis of Twitter conversation at the Cardiac Society of Australia a New Zealand 61st Annual Scientific Meeting 2013'. *Collegian* 21(2): 89–93.

Fox, C. S., M. A. Bonaca, J. J. Ryan, J. M. Massaro, K. Barry and J. Loscalzo (2015). 'A randomized trial of social media from Circulation'. *Circulation* 131(1): 28–33.

Gholami-Kordkheili, F., V. Wild and D. Strech (2013). 'The impact of social media on medical professionalism: a systematic qualitative review of challenges and opportunities'. *Journal of Medical Internet Research* 15(8): e184.

Gibbons, P., C. Zammit, K. Youngentob, H. P. Possingham, D. B. Lindenmayer, S. Bekessy, M. Burgman, M. Colyvan, M. Considine, A. Felton, R. J. Hobbs, K. Hurley, C. McAlpine, M. A. McCarthy, J. Moore, D. Robinson, D. Salt and B. Wintle (2008). 'Some practical suggestions for improving engagement between researchers and policy-makers in natural resource management'. *Ecological Management and Restoration* 9(3): 182–186.

Gleason, B. (2013). '#Occupy Wall Street: exploring informal learning about a social movement on Twitter'. *American Behavioral Scientist* 57(7): 966–982.

Grajales, F. J., III, S. Sheps, K. Ho, H. Novak-Lauscher and G. Eysenbach (2014). 'Social media: a review and tutorial of applications in medicine and health care'. *Journal of Medical Internet Research* 16(2): e13.

Hallman, T. A. and W. D. Robinson (2015). 'Teaching bird identification & vocabulary with Twitter'. *American Biology Teacher* 77(6): 458–461.

Hansen, D., B. Shneiderman and M. A. Smith (2010). *Analyzing Social Media Networks with NodeXL: Insights from a Connected World*. Burlington, MA: Morgan Kaufmann.

Harris, J. K., B. Choucair, R. C. Maier, N. Jolani and J. M. Bernhardt (2014). 'Are public health organizations tweeting to the choir? Understanding local health department Twitter followership'. *Journal of Medical Internet Research* 16(2): e31.

Hobbs, R. J., E. Higgs and J. A. Harris (2009). 'Novel ecosystems: implications for conservation and restoration'. *Trends in Ecology and Evolution* 24(11): 599–605.

Kaplan, A. M. and M. Haenlein (2010). 'Users of the world, unite! The challenges and opportunities of social media'. *Business Horizons* 53(1): 59–68.

Keller, B., A. Labrique, K. M. Jain, A. Pekosz and O. Levine (2014). 'Mind the gap: social media engagement by public health researchers'. *Journal of Medical Internet Research* 16(1): e8.

Laranjo, L., A. Arguel, A. L. Neves, A. M. Gallagher, R. Kaplan, N. Mortimer, G. A. Mendes and A. Y. S. Lau (2015). 'The influence of social networking sites on health behavior change: a systematic review and meta-analysis'. *Journal of the American Medical Informatics Association* 22(1): 243–256.

Louv, R. (2005). *Last Child in the Woods: Saving Our Children From Nature-Deficit Disorder*. Chapel Hill, NC: Algonquin Books.

Matta, R., C. Doiron and M. J. Leveridge (2014). 'The dramatic increase in social media in urology'. *Journal of Urology* 192(2): 494–498.

Mayfield, A. (2008). *What is Social Media?* London: iCrossing. Retrieved from www.icrossing.com/uk/ideas/fileadmin/uploads/ebooks/what_is_social_media_icrossing_ebook.pdf.

Nah, S. and G. D. Saxton (2012). 'Modeling the adoption and use of social media by nonprofit organizations'. *New Media and Society* article 1461444812452411.

Nascimento, T. D., M. F. DosSantos, T. Danciu, M. DeBoer, H. van Holsbeeck, S. R. Lucas, C. Aiello, L. Khatib, M. A. Bender, J.-K. Zubieta, A. F. DaSilva and U. M. G. Class (2014). 'Real-time sharing and expression of migraine headache suffering on Twitter: a cross-sectional infodemiology study'. *Journal of Medical Internet Research* 16(4): 205–215.

Papworth, S. K., T. P. L. Nghiem, D. Chimalakonda, M. R. C. Posa, L. S. Wijedasa, D. Bickford and L. R. Carrasco (2015). 'Quantifying the role of online news in linking conservation research to Facebook and Twitter'. *Conservation Biology* 29(3): 825–833.

Roberge, J.-M. (2014). 'Using data from online social networks in conservation science: which species engage people the most on Twitter?' *Biodiversity and Conservation* 23(3): 715–726.

Ross, C., M. Terras, C. Warwick and A. Welsh (2011). 'Enabled backchannel: conference Twitter use by digital humanists'. *Journal of Documentation* 67(2): 214–237.

Seidman, G. (2013). 'Self-presentation and belonging on Facebook: how personality influences social media use and motivations'. *Personality and Individual Differences* 54(3): 402–407.

Society for Ecological Restoration (2004) *The SER Primer on Ecological Restoration*. Tucson, AZ: Society for Ecological Restoration.

Spallek, H., S. P. Turner, E. Donate-Bartfield, D. Chambers, M. McAndrew, P. Zarkowski and N. Karimbux (2015). 'Social media in the dental school environment, part a: benefits, challenges, and recommendations for use'. *Journal of Dental Education* 79(10): 1140–1152.

Veletsianos, G. (2012). 'Higher education scholars' participation and practices on Twitter'. *Journal of Computer Assisted Learning* 28(4): 336–349.

Veletsianos, G. (2013). 'Open practices and identity: evidence from researchers and educators' social media participation'. *British Journal of Educational Technology* 44(4): 639–651.

White, T. R. (2014). Digital social media detox (DSMD): responding to a culture of interconnectivity. In Information Resources Management Association (ed.), *Digital Arts and Entertainment: Concepts, Methodologies, Tools, and Applications*: 1619–1635. Hershey, PA: IGI Global.

Wilcox, K. and A. T. Stephen (2013). 'Are close friends the enemy? Online social networks, self-esteem, and self-control'. *Journal of Consumer Research* 40(1): 90–103.

Wolfsfeld, G., E. Segev and T. Sheafer (2013). 'Social media and the Arab spring politics comes first'. *The International Journal of Press/Politics* 18(2): 115–137.

INDEX